Contents

Preface to second edition

The first edition of the **Collins Advanced Science** series represented a breakthrough in comprehensive support and appeal to A-level pupils and teachers that received wide critical acclaim. With the advent of AS and A2 level specifications, the second edition has been fully revised and updated to build on the strengths of the first edition.

The text has been augmented to increase the breadth of examinations board coverage. Each of the major examination boards' specifications are now represented, including the International Baccalaureate. With that in mind, examination questions from across the boards have been included for student practice at the ends of chapters.

For the teacher

Each chapter in the second edition of **Collins Advanced Science: Biology** includes an opening page that identifies the AS and A2 nature of each section in the chapter, relating to the majority of examination specifications. The section headings can be checked for the detail against individual specifications.

To support the textbook, a comprehensive **Teacher Support CD-ROM** is available. This CD-ROM includes answers to the examination questions contained within the text, illustrations for use with OHPs and white boards, further questions for in-class and homework use, together with suggestions for practical exercises.

Collins Advanced Science: Biology and Human Biology Teacher Support CD-ROM
ISBN 000 713939 4

For the student

Maintaining the book's strong support for self study, the design of the second edition updates its visual appeal. The content is written in an accessible style and there are plenty of interesting real-life applications of the subject, frequent self-test questions and 'Remember This' boxes to aid understanding. The Assignments at the end of each chapter give practice and portfolio evidence for Key Skills requirements.

Dedication

For the Boyle family: Lynda, Charlotte, James and Alex. And anyone who loves the subject. *Mike Boyle*
To my dear mother, Anne Senior, for her unfailing love, support and encouragement. With much love. *Kathryn Senior*

Examination questions
The publisher thanks examination boards for their permission to reproduce examination questions. Questions are acknowledged as follows:

AQA	The Assessment and Qualifications Alliance
Edexcel	Edexcel Foundation
IBO	International Baccalaureate Organization
OCR	Oxford Cambridge and RSA Examinations
WJEC	Welsh Joint Education Committee

Biology

MIKE BOYLE
KATHRYN SENIOR

Published by HarperCollins*Publishers* Limited

77-85 Fulham Palace Road
Hammersmith
London
W6 8JB

Browse the complete Collins Catalogue at:
www.CollinsEducation.com

First published 2002
10 9 8 7 6 5 4 3

ISBN 000 713600 5

British Library Cataloguing in Publication Data:
A catalogue record for this book is available from the
British Library.

Publishing manager: Martin Davies
Publishing coordinator: Pat Winter
Project managers: Ros Woodward and Mitch Fitton
Proofreader: Linda Antoniw
Indexer: Julie Rimington
Picture researcher: Caroline Thompson
Series designer: Caroline Grimshaw
Book designer: Ken Vail Graphic Design
Artists: Illustrated Arts, Peter Harper

Printed and bound by Printing Express Ltd., Hong Kong

You might also like to visit:
www.harpercollins.co.uk
The book lover's website

01

What's in a cell?

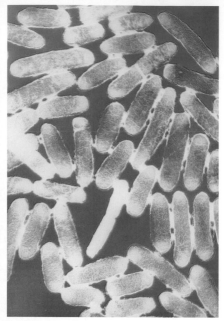

In 1996, meat infected with a bacterium called *Escherichia coli* O157 (often abbreviated to *E. coli* O157) was sold and served in Wishaw in Scotland with devastating results. In the food poisoning outbreak that followed, 501 people became ill and 21 of them died. Most of the dead were pensioners who had eaten the contaminated meat at a church lunch.

This tragic incident illustrates the delicate balance that exists between humans and the bacteria that live in and around us. Other strains of *E. coli* live quite happily in our guts, doing a useful job, but *E. coli* O157 is more unpredictable. Some people become infected and just get a bad bout of diarrhoea, but an unlucky 5% or so, can develop severe or even fatal illness after ingesting as few as 10 bacterial cells. This strain of *E. coli* produces toxins that cause cells in the lining of the intestine and in the kidney to self-destruct. The damage is particularly bad in the very old and the very young, which means that pensioners and children are particularly at risk.

The first case of *E. coli* O157 causing food poisoning wasn't reported until 1980, so where did it come from? No-one can really be sure, but experts think that the genes that code for its deadly toxins were transferred to *E. coli* from other bacteria, on short loops of DNA called plasmids. These are swapped from one cell to another during conjugation, a sort of prokaryotic sex encounter. But, whatever the mechanism, there is some hopeful news. A class of antibiotic is being developed that may be able to kill all strains of *E. coli* O157. It is not yet ready to test in people, but if all trials are completed successfully, we could have an effective treatment for *E. coli* O157 by about 2010.

Escherichia coli O157

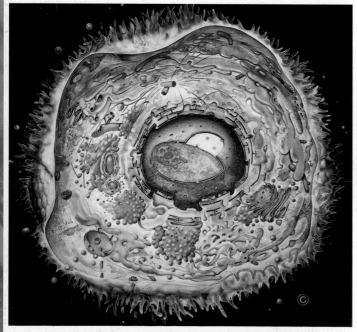

Fig 1.1 This painting illustrates something of the complexity of an animal cell. It was commissioned by the Science Museum on behalf of the Biochemical Society to provide the centrepiece of a permanent exhibition opened in 1987

1 THE COMPLEXITY OF CELLS

When you first looked down a microscope at human or animal cells, and saw something like Fig 1.9, you may have been disappointed. But appearances are deceptive. These little bags with a dot in the middle are extremely complex, and infinitely more sophisticated than you could imagine. Fig 1.1 attempts to show some of the detail that is impossible to show with a light microscope.

As complex as cells are, in your body right now cells are dying and being replaced at a staggering rate. A human body contains about 50 million million cells. Every day we make new skin cells to replace those that wear away, new red blood cells to replace those that die and new cells to replenish the lining of our digestive system. In this chapter we explore the basic types of cells, how they work and the techniques that have revealed their structure and function.

2 THE CELL THEORY

Following the invention and refinement of the microscope came the **cell theory**, a general acceptance that all living things are made of cells. Modern cell theory has three central ideas:

- The cell is the smallest independent unit of life.
- The cell is the basic living unit of all organisms: all organisms are made up of one or more cells.
- Cells arise from other cells by cell division. They cannot arise spontaneously.

As you study biology in more detail, you will discover that there are exceptions to every rule. In the case of cell theory, that exception is the **virus**. Viruses do not have a cellular structure or organisation, and whether they are actually living organisms is a subject of debate.

3 USING CELL STRUCTURE TO CLASSIFY ORGANISMS

Organisms can be classified on the basis of the internal organisation of their individual cells. With the exception of viruses, all organisms are either **prokaryotic** or **eukaryotic**. Prokaryotic cells are relatively simple, they have no separate nucleus and show little organisation. Bacteria are prokaryotes. In contrast, eukaryotic cells are larger and show much more internal organisation. Animals, plants, fungi and protoctists are all eukaryotes. The main differences between prokaryotes and eukaryotes are shown in Table 1.1.

Table 1.1 Differences between prokaryotes and eukaryotes		
	Prokaryotes	**Eukaryotes**
Organisms	bacteria	plants, animals, protoctists, fungi
Diameter of cells	0.1–10 µm	10–100 µm
Site of genetic material	DNA in cytoplasm	DNA inside distinct nucleus
Organisation of genetic material	DNA is circular; no histone proteins; DNA does not condense at cell division	DNA is linear; attached to histone proteins; condenses into visible chromosomes before cell division
Internal structure	few organelles	many organelles with complex membrane systems
Cell walls	always present	present in plants and fungi and some protoctists; never in animals
Flagella	have simple flagella	have modified cilia that consist of microtubules in a distinctive '9 + 2' arrangement (see Fig 1.29)

4 PROKARYOTES – THE FIRST ORGANISMS?

A few billion years ago, the first living organisms to evolve on Earth were probably prokaryotes. The term literally means 'before the nucleus' because the genetic material (DNA) of these organisms is not enclosed by a membrane, and therefore they do not have a true nucleus.

It is tempting to think of prokaryotes as inferior to eukaryotes, but in some ways they have achieved greater success. They have been on Earth more than twice as long as eukaryotes, they are present in greater numbers (there are more bacteria living on your skin than there are people on Earth) and they occupy an enormous number of different habitats. Some bacteria, for example, are able to live in volcanic springs at temperatures as high as 90 °C.

REMEMBER THIS

No matter how far you take the study of biology, all you need to know about microscopic distances is these two simple facts:

1 mm = 1000 µm
1 µm = 1000 nm
1 millimetre = 1000 micrometres
1 micrometre = 1000 nanometres

Cells and the molecules they contain are very small. When we study them, we must think about measurements that are minute beyond our imagination. It is easy, for example, to develop a mental block when confronted by the statement 'The nanometre is 10^{-9} m.'

The two units commonly used to describe microscopic objects are the **micrometre** (μm) (commonly called the micron) and the **nanometre** (nm). Starting with a familiar unit, the millimetre (mm), one thousandth of one millimetre is known as a micrometre (μm). The micrometre is used to describe cells and organelles. An average animal cell is 30 to 50 μm across; the nucleus has a diameter of about 10 μm. Plant cells can reach 150 μm or more in length.

When describing small cellular components and molecules, the useful unit is the nanometre (nm). A nanometre is one thousandth of a micrometre. As a rough guide, the light microscope reveals structures that can be measured in micrometres, but you need an electron microscope to see objects measured in nanometres.

Fig 1.2 Units of measurement; how they are used and how they relate to each other

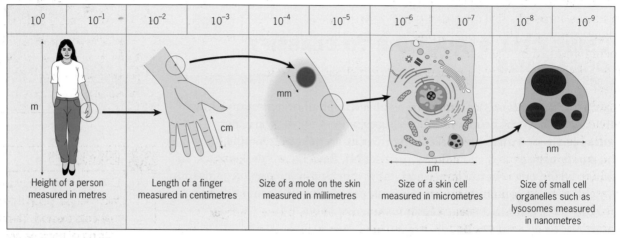

| 10^0 | 10^{-1} | 10^{-2} | 10^{-3} | 10^{-4} | 10^{-5} | 10^{-6} | 10^{-7} | 10^{-8} | 10^{-9} |

Height of a person measured in metres

Length of a finger measured in centimetres

Size of a mole on the skin measured in millimetres

Size of a skin cell measured in micrometres

Size of small cell organelles such as lysosomes measured in nanometres

---

A **a)** Estimate the diameter (in micrometres) of a full stop on this page.

b) How many nanometres are there in 1 millimetre?

BACTERIA

Most bacteria are spherical or rod-shaped cells, several micrometres long (see Figs 1.3 and 1.4) . Their rigid protective **cell wall** is made of **peptidoglycan**, a substance unique to bacteria. Beneath the cell wall is the **cell surface membrane**, which is similar in structure to the membrane of eukaryotic cells. This completely encloses the contents of the cell. Some

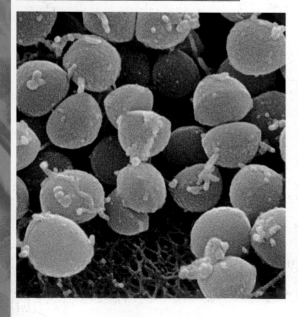

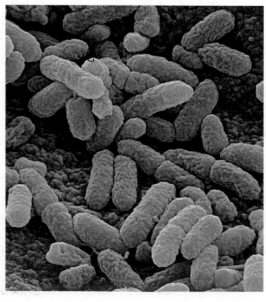

(left) **Fig 1.3** Scanning electron micrograph showing *Staphylococcus aureus* cells. These spherical cells are called cocci (singular coccus), and are one of the two main types of bacterial cell (the others are bacilli (singular baclllus); see Fig 1.4). *S. aureus* is commonly found on human skin where it usually does no harm. However, it can cause serious infection if it enters the body through a cut or wound.

(right) **Fig 1.4** Scanning electron micrograph showing *Salmonella enteritidis* cells. These sausage-shaped cells are called bacilli. This species of *Salmonella* can cause severe food poisoning.

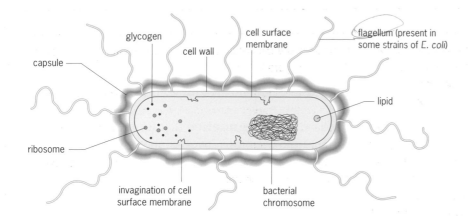

types of bacteria, such as *Neisseria meningitidis*, which can cause meningitis, also have a **capsule**. This is a sticky coat outside the cell wall that prevents the bacterium from drying out, or being digested by host intestinal enzymes or being attacked by the host's immune system.

Fig 1.5 shows the internal structure of a rod-shaped bacterium, *Escherichia coli*.

Inside the cell surface membrane, the bacterial cell is a single cytoplasmic compartment that contains DNA, RNA, proteins and small molecules. Bacteria have a circular piece of DNA in a region of the cytoplasm known as the **nucleoid**. There are smaller rings of DNA, known as **plasmids**, elsewhere in the cytoplasm. Plasmids are of great interest to biologists because they often contain genes that code for antibiotic resistance, and can be used to carry genes between cells in genetic engineering (see Chapter 30).

Bacteria feed by **extracellular digestion**. They release enzymes into the surrounding medium and absorb the soluble products. **Glycogen granules** and **lipid droplets** often occur in the bacterial cytoplasm and provide a limited store of these polymers. Bacteria can synthesise a wide variety of enzymes, and some species are able to digest unlikely substances such as oil and plastic. Proteins are synthesised on **ribosomes** (page 13), and cell respiration occurs on **mesosomes**, inner extensions of the cell surface membrane.

Some bacteria are **motile**: they can swim. They have thin fibres called **flagella** (singular flagellum) that are corkscrew-shaped and that rotate, propelling the bacteria in different directions. Other fibres, called **pili**, enable the bacteria to perform a primitive form of sex, as shown in Fig 1.6. During **conjugation**, the pili become joined and form a channel to allow plasmids to be swapped from one cell to another.

The role of bacteria in the environment is discussed in Chapter 36.

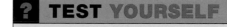

SEE QUESTIONS 1 AND 2

? TEST YOURSELF

B Some bacteria can digest, or break down, oil and plastic. Suggest how this could be useful to people.

✓ REMEMBER THIS

Until the invention of the microscope, people had no idea that cells existed. They knew nothing of bacteria and had no idea about how infectious diseases spread.

Fig 1.6 Conjugation in two bacterial cells

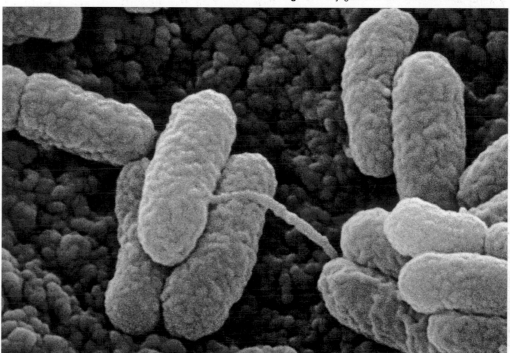

5 EUKARYOTES: ANIMAL CELLS

All species of animals, plants, **fungi** and **protoctists** are made of eukaryotic cells. The term **eukaryote** means 'true nucleus', because the DNA of eukaryotic cells is confined to a definite area inside the cell enclosed by a **nuclear envelope**.

Eukaryotic cells also have other **organelles** that form compartments. By being in a compartment the chemicals involved in a particular process, such as respiration or photosynthesis, are kept separate from the rest of the cytoplasm. This allows the chemical reactions of the process to take place quickly and efficiently. This high degree of internal organisation is one of the reasons why eukaryotic cells are larger than prokaryotic cells. The fluid that occupies the space between organelles is the **cytosol**, a solution containing a complex mixture of enzymes, the products of digestion (amino acids, sugars, etc) and waste materials.

INSIDE AN ANIMAL CELL

If you look at an animal cell under a **light microscope**, the right staining and illumination techniques, a good quality microscope and a bit of luck will allow you to see the **nucleus, nucleolus**, the **chromosomes** in a dividing cell, and even the **Golgi complex**, **mitochondria** and food storage particles. But to be sure of seeing the inside of a cell in more intricate detail, you really need an **electron microscope**.

The **electron micrograph** in Fig 1.7 shows the internal structure of an animal cell. You can see far more of the cell's components – the **endoplasmic reticulum**, the internal features of the mitochondria, the cell surface membrane, **lysosomes, ribosomes** and **cytoskeleton** are all now visible.

In this chapter, we look at the detailed structure of each individual organelle and find out how each type contributes to the function of the cell as a whole.

Table 1.2 summarises the functions of some of the major organelles and structures in the eukaryotic cell.

<table>
<tr><td colspan="4">**REMEMBER THIS**
1 mm = 1000 µm
1 µm = 1000 nm</td></tr>
</table>

Table 1.2 Summary of the functions of major eukaryotic cell organelles and structures

Organelle	Occurrence	Size	Function
Nucleus	usually one per cell	10 µm	site of the nuclear material – the DNA
Nucleolus	inside nucleus	1–2 µm	manufacture of ribosomes
Mitochondrion	numerous in cytoplasm; up to 1000 per cell	1–10 µm	aerobic respiration
Rough endoplasmic reticulum	continuous throughout cytoplasm	extensive membrane network	isolation and transport of newly synthesised proteins
Smooth endoplasmic reticulum	usually small patches in cytoplasm	variable	synthesis of some lipids and steroids
Ribosome	free in cytoplasm or attached to rough ER	20 nm	site of protein synthesis
Golgi body	free in cytoplasm	variable	modification and synthesis of chemicals
Lysosome	free in cytoplasm	100 nm	digestion of unwanted material
Chloroplast	cytoplasm of some plant cells, eg mesophyll	4–10 µm	site of photosynthesis
Vacuole	usually large, single fluid filled space in plant; smaller and more numerous in animals	up to 90 per cent of volume of whole plant cell	storage of salts, sugars and pigments; creates turgor pressure by interaction with cell wall
Cell surface membrane	encloses the cytoplasm of all cells	7–10 µm	exchange and transport of materials into and out of the cell
Cell wall	surrounds all plant cells	thickness varies	provides rigidity and strength

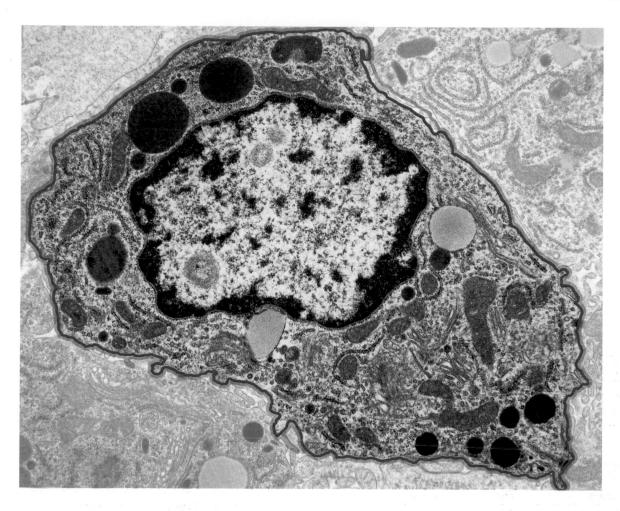

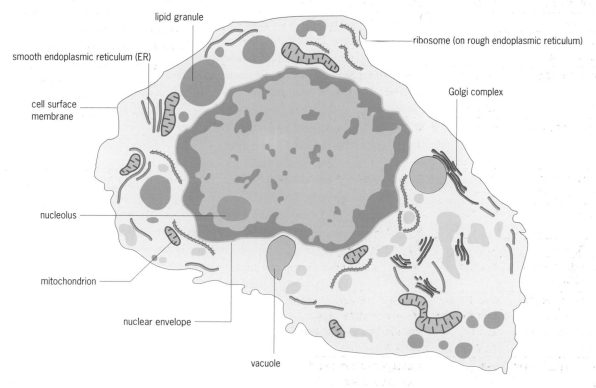

lipid granule

smooth endoplasmic reticulum (ER)

ribosome (on rough endoplasmic reticulum)

cell surface membrane

Golgi complex

nucleolus

mitochondrion

nuclear envelope

vacuole

Fig 1.7 A micrograph of a human cell with an interpretive diagram. It shows that the cytoplasm of animal cells contains a complex system of membrane-bound organelles

THE LIGHT MICROSCOPE

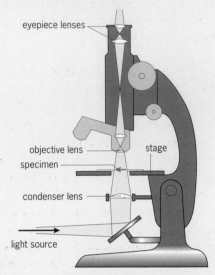

- eyepiece lenses
- objective lens
- stage
- specimen
- condenser lens
- light source

Fig 1.8 The standard compound microscope

Fig 1.8 shows the structure of a **light** or **optical microscope**. This instrument is also known as a **compound microscope** because two lenses, the **eyepiece lens** and the **objective lens**, are combined to produce a much greater **magnification** than is possible with a single lens. The total magnification is calculated by multiplying the magnification of the two lenses together. For example, if the eyepiece lens has a magnification of ×10 and the objective lens is ×50, the total magnification is ×500.

The light microscope has powers of magnification of up to ×1500, good enough to see cells, larger organelles and individual bacteria, but not powerful enough to reveal smaller structures such as cell surface membranes, viruses or individual molecules. Table 1.3 shows organelles in animal and plant cells that are visible with a light microscope. Fig 1.9 shows a light micrograph of an animal cell.

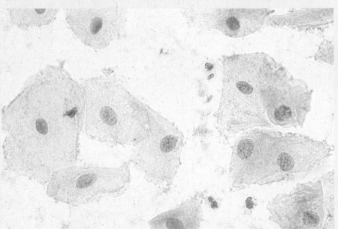

Fig 1.9 The light microscope can reveal the basic features of animal cells; this is what cheek cells look like

Table 1.3 Structures normally visible in animal and plant cells with the light microscope

Organelle	Function	Animal cells	Plant cells
Nucleus	control of cell activities	✓	✓
Vacuole	storage and support	✗	✓
Chloroplasts	photosynthesis	✗	✓
Cell wall	support	✗	✓

An important feature of a microscope is its **resolving power**, which should not be confused with the magnification. Two objects close together may appear as one single image when viewed under the light microscope. Increasing the magnification does not allow you to **resolve** the two objects into separate images; the objects just appear to be a larger single image. Resolving power is as important as magnification when investigating structural details.

The limitation of the light microscope is due to the nature of light itself. The wavelength of light determines the maximum effective magnification and the resolving power. The wavelength of visible light is around 500–650 nm and the resolving power – the resolution – of the light microscope is 200 nm (0.2 μm), so two objects separated by less than 200 nm appear as one object.

THE ELECTRON MICROSCOPE (EM)

Fig 1.10 shows the essential features of an electron microscope (EM). Invented in the 1930s, the present-day version of the EM can magnify up to 500 000 times and has a resolution of 1 nm. In other words, the EM can resolve two objects that are only 1 nm apart. As many biological molecules are larger than 1 nm, the EM can be used to study the arrangement of individual molecules that make up structures in a cell.

The development of the EM has had a huge impact on biology. A magnification of half a million means that an object the size of a full stop becomes over 200 m in diameter the length of two football pitches. Organelles that are only blurred images when viewed with a light microscope can now be studied in great detail. Many new cell structures have been discovered using electron microscopy.

While the light microscope uses lenses to focus a beam of light, the electron microscope uses electromagnets to focus a beam of electrons. The wavelength of the electrons is much smaller than the wavelength of light, so the resolving power is much greater.

The main disadvantage of the EM is that the electron beam must travel in a vacuum because, being so small,

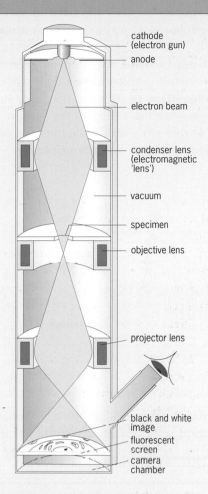

Fig 1.10 The basic features of the transmission electron microscope. The beam of electrons is focused by electromagnets before passing through the specimen, and the image appears on a fluorescent screen

electrons are scattered when they hit air molecules. Specimens for the EM must therefore be prepared (killed, dehydrated and fixed) so that they retain their structure inside a vacuum. Such harsh preparation methods can damage cells, and cause **artefacts** – features that do not exist in the living cell – to appear. For example, microsomes, tiny vesicles surrounded by ribosomes, were seen when animal cells were examined using the electron microscope. At first, cell biologists thought that these were organelles they had not noticed before, but they later

realised that microsomes were fragments of endoplasmic reticulum (see page 11) produced by the fixing process. The main differences between electron microscopy and light microscopy are shown in Table 1.4.

SCANNING AND TRANSMISSION ELECTRON MICROSCOPES

There are two main types of electron microscope – the **transmission electron microscope** and the **scanning electron microscope**.

In a transmission electron microscope (TEM), a beam of electrons is transmitted through the specimen. The specimen must be thin and it is stained using electron-dense substances such as heavy metal salts. These substances deflect electrons in the beam and the pattern that the remaining electrons produce as they pass through the specimen is converted into an image. The electron micrograph in Fig 1.7 was produced using the TEM.

We can use scanning electron microscopes (SEMs) to study relatively large three-dimensional objects. Thin sections are not necessary because the SEM records the electrons that bounce off the surface of the object rather than passing through it. Fig 1.11 shows an example of the detailed three-dimensional image that an SEM produces. Although the SEM does not have the resolving power of the TEM, it is more versatile and can be used to observe many kinds of intact structures. The person operating the SEM can move the specimen about, to look at the surface of the structure from a variety of angles – a bit like taking aerial views of the countryside.

Table 1.4 Comparing light and electron microscopes		
	Light microscope	**Electron microscope**
Illumination	light	electrons
Focused by	lenses	magnets
Maximum magnification	×1500	×500 000
Resolving power	200 nm (0.2 μm)	1 nm
Specimens	living or dead	dead
Preparation of specimens	often simple	more complex
Cost of equipment	relatively cheap	very expensive
Images in colour	yes	no (colour is added by computer)

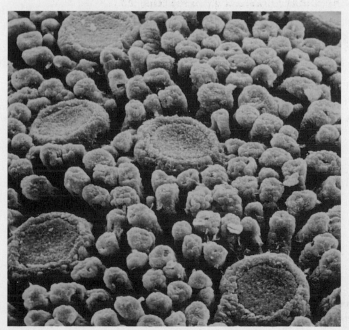

Fig 1.11 Scanning electron micrograph showing the surface of the human tongue

? TEST YOURSELF

C If a light microscope had an eyepiece lens of ×25 and an objective lens of ×40, what would the total magnification be?

SEE QUESTION 3

THE CELL SURFACE MEMBRANE

The **cell surface membrane** is the boundary between the cell and its environment. It has little mechanical strength but plays a vital role in controlling which materials pass in and out of the cell.

Although basically a double layer of **phospholipid molecules**, arranged tail to tail, the cell surface membrane is a complex structure, studded with proteins. These can be embedded in the membrane or they can penetrate the **bilayer** forming **pores** (holes or channels – see Fig 1.12 and the section on the cell surface membrane in Chapter 4) through which molecules can pass. Chapter 4 covers the structure and function of the cell surface membrane in detail.

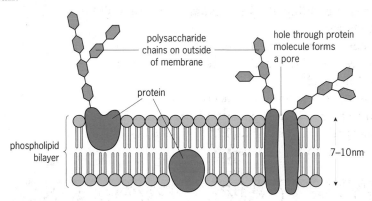

Fig 1.12 The cell surface membrane is a complex organelle but its structure can be represented by this simplified diagram

THE NUCLEUS

The **nucleus** is the largest and most prominent organelle in the cell. Almost all eukaryote cells have a nucleus – red blood cells in mammals and phloem cells in plants are an exception. Every nucleus is surrounded by a **nuclear envelope**. As Fig 1.13 shows, this consists of two membranes that are separated by a gap of 20 to 40 nm.

The nucleus is usually spherical and about 10 μm in diameter. It contains the cell's DNA, which carries information that allows the cell to divide and carry out all its cellular processes When the cell is not actively dividing (as in Fig 1.13), the DNA is spread throughout the nucleus as chromatin. Close examination of the chromatin reveals two different levels of density. Dark-staining chromatin, consisting of tightly packed DNA, is known as **heterochromatin**; the lighter, more loosely packed material is called

Fig 1.13 As the electron micrograph and the diagram show, the nucleus is bounded by the nuclear envelope, a double membrane which contains many pores, each one about 100 nm in diameter. The nuclear pores represent about 15 per cent of the total surface area of the nuclear envelope, indicating the heavy traffic of materials in and out of the nucleus. The nucleus contains the cell's genetic material that exists as chromatin, loosely packed DNA attached to proteins called histones. The nucleolus is clearly visible

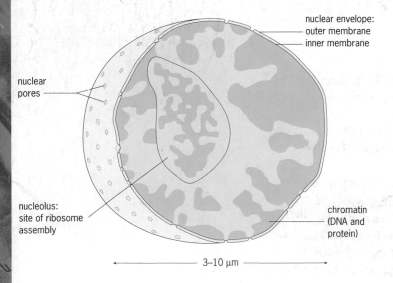

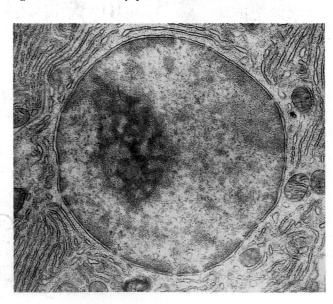

euchromatin. Euchromatin contains the DNA that is being actively read to produce proteins; in heterochromatin, the DNA is packed together, and is not being read.

Individual segments of DNA called genes contain the information necessary to make individual proteins, including the enzymes that control most of the cell's activities. In fact, a central concept in biology that is true for all cells, prokaryotes and eukaryotes, is that:

> **Many genes code for making enzymes that, in turn, control the activities of the cell.**

When a cell is dividing, its chromosomes become visible, as Fig 1.14 shows.

You can study enzymes in Chapter 5 and we look at protein synthesis in Chapter 27. We cover cell division in detail in Chapter 25.

Nuclei also have one or more nucleoli (Fig 1.15). These dark-staining, spherical structures are ribosome-producing centres: they synthesise ribosomal RNA and package it with ribosomal proteins to make ribosomes.

Fig 1.14 Before a cell divides, its DNA condenses into visible chromosomes. For further information about cell division, see Chapter 25

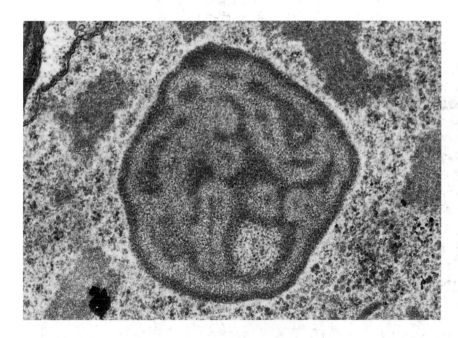

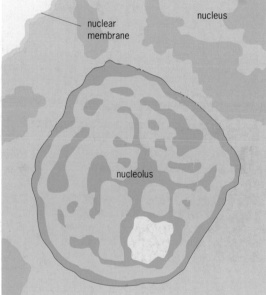

nuclear membrane

nucleus

nucleolus

Fig 1.15 The nucleolus has a highly organised structure; its DNA codes for the RNA and proteins that are used to make ribosomes

ENDOPLASMIC RETICULUM (ER)

The nuclear envelope joins with the membrane of the **endoplasmic reticulum** (ER), a system of complex tunnels that are spread throughout the cell.

On much of the outside surface of the ER in a eukaryotic cell are the sites of attachment for ribosomes. This gives it a grainy appearance (Fig 1.16) and its name, **rough ER**.

The main function of rough ER is to keep together and transport the proteins made on the ribosomes. Instead of simply diffusing away into the cytoplasm, newly made proteins are threaded through pores in the membrane and accumulate in the space called the **ER lumen** (Fig 1.17). Here, they are free to fold into their normal three-dimensional shape. Not surprisingly, a mature cell that makes and secretes large amounts of protein – such as one

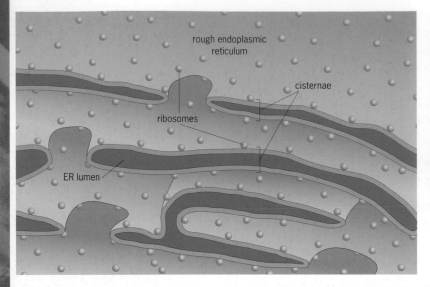
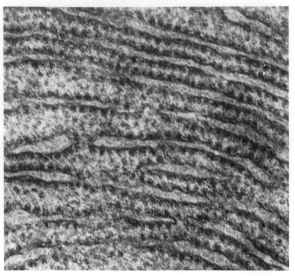

Fig 1.16 Electron micrograph (right) and interpretive diagram (left) showing rough ER. The ER is a large sheet of membrane that is folded over on itself many times, forming stacked layers called cisternae. The space inside the cisternae, the ER lumen, forms an extensive transport system throughout the cytoplasm

that makes digestive enzymes – has rough ER that occupies as much as 90 per cent of the total volume of the cytoplasm. The rough ER is also a storage unit for enzymes and other proteins.

Small vesicles containing newly synthesised proteins pinch off from the ends of the rough ER and either fuse with the Golgi complex or pass directly to the cell surface membrane.

ER with no ribosomes attached is known as **smooth ER** (Fig 1.18). Smooth ER tends to occur in small areas that are not continuous with the nuclear membrane. Smooth ER is not involved in protein synthesis but is the site of steroid (lipid hormone) production. It also contains enzymes that **detoxify**, or make harmless, a wide variety of organic molecules, and it acts as a storage site for calcium in skeletal muscle cells.

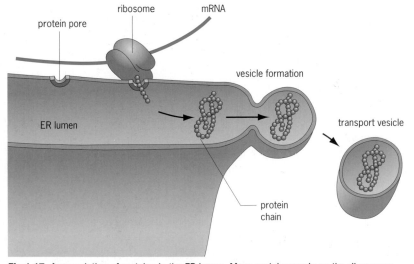

Fig 1.17 Accumulation of proteins in the ER lumen. Many proteins made on the ribosomes are threaded through pores in the ER membrane

RIBOSOMES

Ribosomes are small, dense organelles, about 20 nm in diameter, present in great numbers in the cell. Most are attached to the surface of rough ER but they can occur free in the cytoplasm, as in Fig 1.19. This artist's impression of protein synthesis shows the ribosome's distinctive shape. Ribosomes are made from a combination of ribosomal RNA and protein (65 per cent RNA : 35 per cent protein).

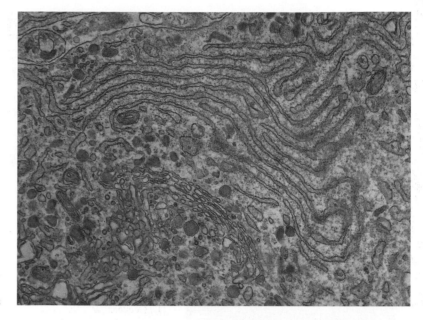

Fig 1.18 Smooth ER has no attached ribosomes. It is usually not as abundant as rough ER, although it is common in the cells of the liver, gut and some glands

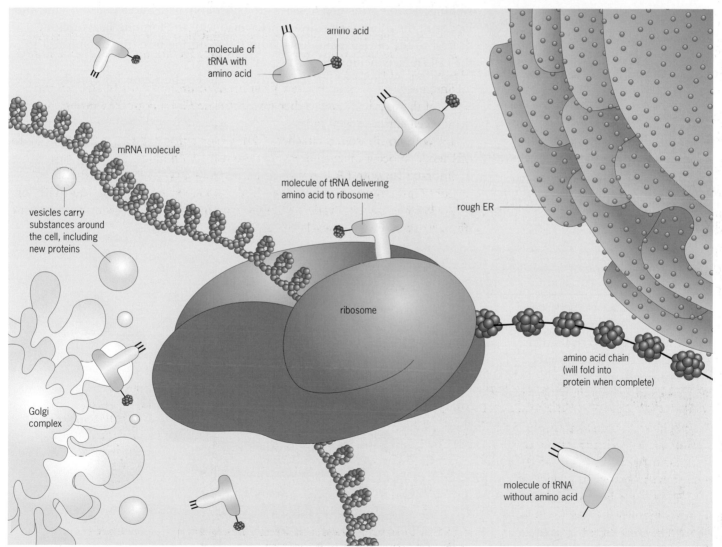

amino acid

molecule of tRNA with amino acid

mRNA molecule

molecule of tRNA delivering amino acid to ribosome

rough ER

vesicles carry substances around the cell, including new proteins

ribosome

amino acid chain (will fold into protein when complete)

Golgi complex

molecule of tRNA without amino acid

Fig 1.19 An artist's impression of protein synthesis. A ribosome can be thought of as a giant enzyme on which a protein is assembled. Transfer RNA molecules (tRNA) bring specific amino acids to the ribosome. Each one is added to the growing amino acid chain according to the code on the messenger RNA molecule

Ribosomes are involved in protein synthesis. They assemble amino acids in the right order to produce new proteins. The ribosome uses the code on messenger RNA (mRNA) to put amino acids together in chains to form specific proteins. This is another central concept in biology that you will learn more about later on:

> **A gene is a piece of DNA that codes for a particular protein. A copy of the gene in the form of messenger RNA passes out of the nucleus and travels to the ribosome where it controls protein synthesis.**

Protein synthesis is covered in detail in Chapter 27.

Generally, proteins that are to be used inside the cell are made on free ribosomes, while those that are to be secreted out of the cell are made on ribosomes that are bound to ER membranes.

THE GOLGI COMPLEX

The **Golgi complex**, also called the **Golgi apparatus**, is a tightly-packed group of flattened cavities or vesicles (Fig 1.20). The whole organelle is a shifting, flexible structure; vesicles are constantly being added at one side and lost from the other. Generally, vesicles fuse with the forming face (the one nearest to the nucleus) and leave from the maturing face (the one nearest to the cell surface membrane).

Fig 1.21 summarises the relationship between the rough ER and the Golgi complex.

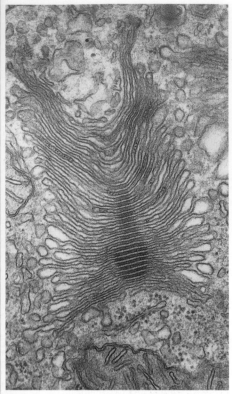

Fig 1.20 The Golgi complex with a very large number of vesicles. The forming face is at the foot and secretory vesicles are at the top

✓ REMEMBER THIS

A vesicle is a small spherical organelle, bounded by a single membrane, which is used to store and transport material around the cell.

Fig 1.21 The various functions of the Golgi complex are summarised in this diagram. Vesicles from the ER or from outside the cell bring material to the Golgi complex. After processing, material passes out of the cell or enters lysosomes

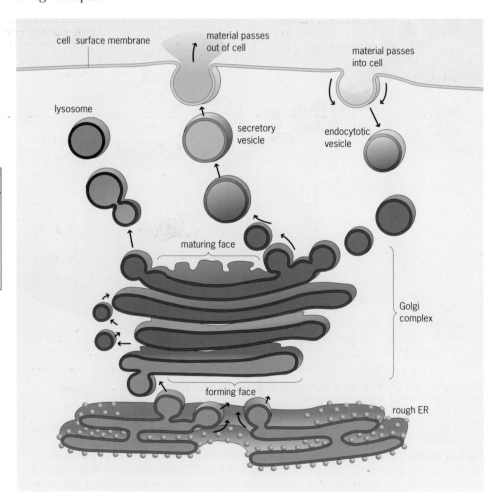

The Golgi complex appears to be involved with the synthesis and modification of proteins, lipids and carbohydrates. Studies have shown that proteins made on the ribosomes attached to ER are packaged into vesicles by the ER. Some of the vesicles join with the Golgi complex, and the proteins they contain are modified before they are secreted out of the cell.

CELL FRACTIONATION

In order to study the function of a particular organelle it is often helpful to isolate it from the rest of the cell. This can be done by cell fractionation, as outlined in Fig 1.22.

✔	**REMEMBER THIS**
	1 mm = 1000 μm
	1 μm = 1000 nm

1 Liver tissue is placed in a fluid that is an **ice-cold isotonic buffer**. When the cell is broken up, membranes are ruptured and many compounds that do not normally mix are brought together. The low temperature minimises unwanted reactions, including self-digestion by digestive enzymes. The isotonic solution prevents organelles gaining or losing water by osmosis, and the buffer prevents any changes in pH.

2 Tissue cut into small pieces   3 Tissue put into blender/homogeniser to break up whole cell   4 Mixture filtered to remove debris   5 Filtrate spun in centrifuge

Principle: The organelles will separate out in a particular order, according to their density and shape. After a time, the sediment containing a particular type of organelle can be separated from the supernatant (the liquid that contains the remaining organelles). The exact times vary from tissue to tissue.

	Organelle		Centrifuge setting (g)	Time (minutes)
First to separate out	Nuclei		800–1000	5–10
	Mitochondria		10 000–20 000	15–20
	Lysosomes			
	Rough ER		50 000–80 000	30–50
	Plasma membranes		80 000–100 000	60
	Smooth ER			
Last to separate out	Free ribosomes		150 000–300 000	> 60

Fig 1.22 Cell fractionation: how to isolate particular types of organelle from liver tissue

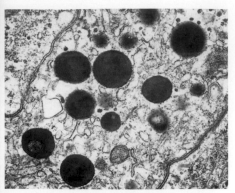

Fig 1.23 Lysosomes are simply bags of digestive enzymes. They can be distinguished from other vesicles in the cell only by using a stain specific for the chemicals inside the lysosomes

? TEST YOURSELF

F What would happen to an organism if it lost control of its lysosomes?

G If we could control lysosome activity in specific tissues, how could this be used in the treatment of cancer?

LYSOSOMES

Lysosomes are small **vesicles** 0.2–0.5 µm in diameter (Fig 1.23) that contain a mixture of digestive enzymes called **lytic enzymes**. It is important that the membrane of lysosomes remains intact because if the enzymes leak out they could digest vital molecules in the cell.

Why do cells need such potentially lethal structures? They have several uses:

- To supply the enzymes which destroy old or surplus organelles.
- To digest material taken into the cell. After a white cell has engulfed a bacterium, for example, lysosomes discharge enzymes into the vacuole (Fig 1.24) and digest the organism. This process is called **phagocytosis**.
- To destroy whole cells and tissues. Parts of tissues and organs often need to be removed after they have performed their function. The muscle of the uterus is reduced after giving birth and milk-producing tissue is destroyed after weaning. Bone is also constantly made and reabsorbed throughout life.

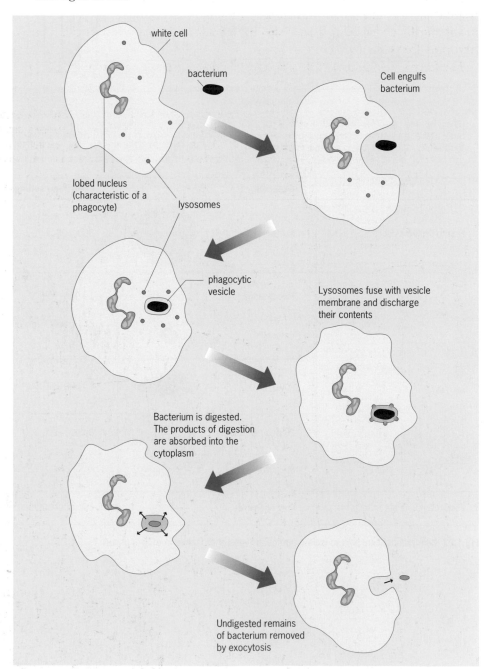

Fig 1.24 One of the functions of lysosomes is to digest material taken into the cell by the process of phagocytosis (see also Chapter 4). Here, a white cell, known as a phagocyte, is ingesting a bacterium. Lysosomes discharge their enzymes into the temporary vacuole and the bacterium is digested

Lysosomes are similar in structure to many other vesicles in the cell and are thought to be made in the same way: inactive digestive enzymes from the rough ER pass through the Golgi complex where they are activated and packaged into vesicles.

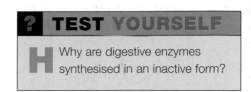

MITOCHONDRIA

Mitochondria are relatively large, individual organelles that occur in large numbers in most cells. They are usually spherical or elongated (sausage-shaped) and are 0.5–1.5 µm wide and 3–10 µm long. Their function is to make **ATP** via the process of **aerobic respiration**. ATP is a molecule that diffuses around the cell and provides instant chemical energy to the processes that require it (see Chapter 8).

As Fig 1.25 shows, mitochondria have a double membrane; the outer membrane is smooth while the inner one is folded. This arrangement gives a large internal surface area on which the complex reactions of aerobic respiration can take place. Mitochondria are particularly abundant in metabolically active cells, tissues such as muscle and tissues involved in active transport (see Chapter 4).

The function of mitochondria is covered in more detail in Chapter 8.

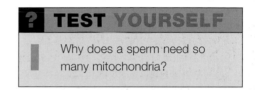

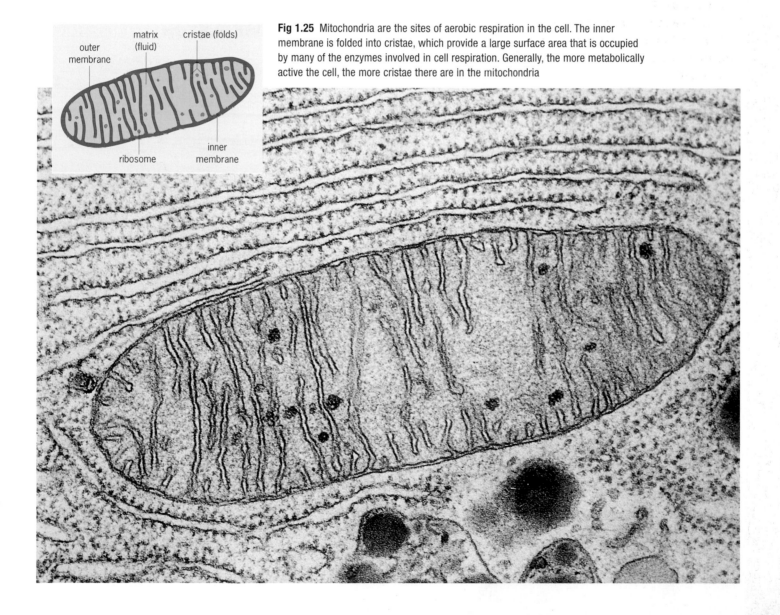

Fig 1.25 Mitochondria are the sites of aerobic respiration in the cell. The inner membrane is folded into cristae, which provide a large surface area that is occupied by many of the enzymes involved in cell respiration. Generally, the more metabolically active the cell, the more cristae there are in the mitochondria

outer membrane

matrix (fluid)

cristae (folds)

ribosome

inner membrane

SCIENCE FEATURE Autoradiography

Electron microscopes are useful tools for investigating cell structure, but they don't reveal much about the processes going on inside a living cell. After World War II, work on the atomic bomb produced chemicals called **radioactive isotopes** as a by-product. These are very useful for studying metabolism because they behave in the same way as non-radioactive atoms and their movements through the cell can be traced (Fig 1.26).

After cells are exposed to a radioactive isotope such as **carbon-14,** they are washed to remove excess label and then fixed, embedded, sectioned and mounted on microscope slides. The slides are then taken to a dark room and coated with photographic emulsion that has been heated to make it melt. After coating, the slides are cooled and kept in the dark for various time periods from a few days to several weeks. During this time, radioactive decay causes a chemical change to occur in the film of photographic emulsion. When this is developed using standard photographic techniques and the slide is examined under the light microscope, silver grains can be seen in the areas exposed by the radioactivity. The pattern of grains shows the structures of the cell that have incorporated the radioactive label.

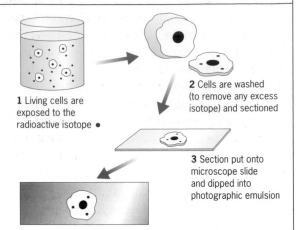

1 Living cells are exposed to the radioactive isotope ●

2 Cells are washed (to remove any excess isotope) and sectioned

3 Section put onto microscope slide and dipped into photographic emulsion

4 Slides are incubated in the dark and the isotope reacts with the light-sensitive chemicals. The emulsion is then developed like a normal photograph and, when examined under the microscope, the position of the isotope in the cell can be seen

Fig 1.26 The basic method of autoradiography

CENTRIOLES AND THE CYTOSKELETON

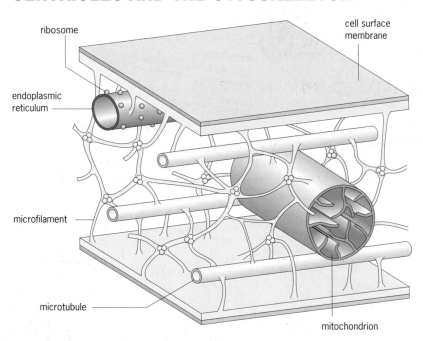

ribosome

cell surface membrane

endoplasmic reticulum

microfilament

microtubule

mitochondrion

Fig 1.27 Inside the cell is a network of tubules which forms the cytoskeleton

Between the organelles of all eukaryotic cells is a complex network of fine threads called **filaments**. These vary in size and chemical composition. As Fig 1.27 shows, they help to form a supportive scaffolding known as the **cytoskeleton** ('cell skeleton'). This:

● supports the whole cell,
● maintains cell shape, as in red blood cells,
● organises and moves organelles,
● forms the spindle during cell division,
● moves the whole cell.

This last function indicates that the cytoskeleton is not a rigid structure; it can be assembled or dismantled in seconds – as happens in a moving white blood cell.

Centrioles are short bundles of filaments, set at right angles to each other (Fig 1.28). They are found in a clear area of cytoplasm known as the **centrosome**. Centrioles occur in all animal cells but are absent from the cells of plants. For a long time their function was thought to be just the formation of the **spindle** – the cradle of threads that guide chromosomes during cell division. In addition to spindle formation, the centrioles act as the centre of formation for the whole cytoskeleton and they have therefore become known as **microtubule organising centres**.

Microtubules make up more complex external cell structures called **cilia**. These tiny hair-like projections are used for cell movement. The tail of a sperm is, for example, a modified cilium. In human cells, cilia contain a bundle of microtubules in the '9 + 2' arrangement shown in Fig 1.29.

? TEST YOURSELF

J How many of the following cells would fit into a line 1 cm long?
a) Animal cells of 20 μm
b) Plant cells of 50 μm
c) Bacterial cells of 2 μm

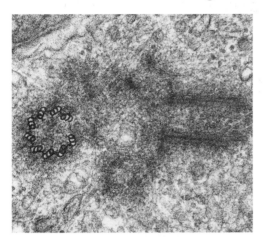

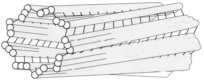

Fig 1.28 The centrioles are two bundles of fibres that form the cytoskeleton of the cell and also give rise to the spindle during cell division

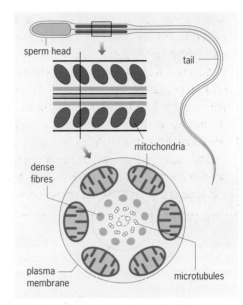

Fig 1.29 A cross-section of the tail of a human sperm showing the '9 + 2' arrangement of microtubules in its centre

6 EUKARYOTES: PLANT CELLS

Typical leaf cells from a plant, seen with the light microscope, are shown in Fig 1.30. These are **palisade mesophyll cells**.

Plant cells have several features that are not found in animal cells (see Table 1.5). A plant cell is surrounded by a cell wall made of cellulose.

Table 1.5 Comparing plant cells and animal cells

Feature Cell:	Animal	Plant
Nucleus	✓	✓
Plasma membrane	✓	✓
Mitochondria	✓	✓
Rough ER	✓	✓
Smooth ER	✓	✓
Golgi bodies	✓	✓
Lysosomes	✓	✓
Cell wall	✗	✓
Plastids, eg chloroplasts	✗	✓
Vacuoles	✗	✓
Centrioles*	✓	✓*

*Centrioles are not present in the cells of higher plants

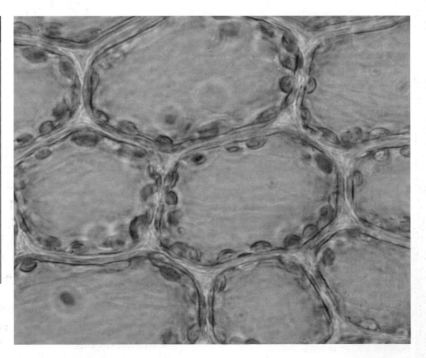

Fig 1.30 The light microscope shows some of the basic features of plant cells

A large proportion of the inside of the cell is taken up with a fluid-filled
compartment known as the **vacuole**. Together, the wall and vacuole maintain
the shape of the whole cell.

As you can see from Fig 1.31, plant cells have specialised organelles, the
chloroplasts, which enable them to make their own food by photosynthesis.
Chloroplasts bring together all the chemicals involved in photosynthesis, and
provide sites for the chlorophyll molecules so that they can absorb the
maximum amount of light.

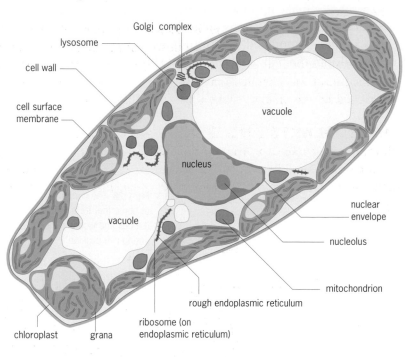

Fig 1.31 Transmission electron micrograph (above) and
explanatory diagram (right) showing a palisade mesophyll cell
from a soya bean leaf

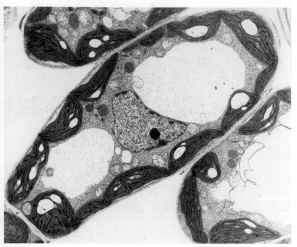

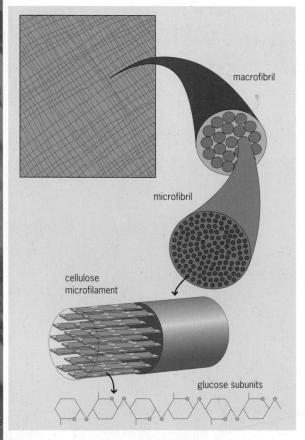

Fig 1.32 The cellulose cell wall is a composite
material. Fibrils containing thousands of cellulose
molecules are cemented together by a 'glue', a
complex mixture of compounds

THE CELLULOSE CELL WALL

Unlike animal cells, plant cells are enclosed by a **cell wall** that
consists of many **cellulose fibres** cemented together by a
mixture of other organic substances.

Cellulose is a **polysaccharide** (a sugar polymer) which consists
of long straight chains of glucose molecules bonded to adjacent
molecules by hydrogen bonds (see Chapter 3). In the cell wall,
around 2000 parallel cellulose molecules are packed together to
form **microfibrils**. These in turn are bundled together to form
fibrils, as you can see in Fig 1.32. Like fibreglass, the cell wall
has great strength because of the many strong fibres and the 'glue'
that holds them together.

The main functions of the cell wall are:
- to provide rigidity and strength to the cell. Cell walls need to
  be strong enough to resist expansion to allow the cell to
  become turgid; see Chapter 4.
- to force the cell to grow in a certain way. A particular
  arrangement of fibrils causes the cell to assume a particular
  shape – for example a long, thin tube.

VACUOLES

Plant cells also have a **vacuole**, a large, fluid-filled cavity. In mature cells the vacuole can occupy over 80 per cent of the cell volume, and is filled with a fluid called **cell sap** (Fig 1.33). The sap consists of a complex mixture of sugars, salts, pigments and waste products in water.

Vacuoles have several functions:

- They absorb water by **osmosis** and therefore swell, pushing the cytoplasm against the cell wall. In this state, a plant cell is said to be turgid.
- They store food substances such as sugars and mineral salts.
- They store pigments that give colour to plant structures such as petals.
- They can accumulate waste products and by-products of metabolism. Sometimes these chemicals may be toxic or have an unpleasant taste and are used by the plant to make it less palatable to herbivores.

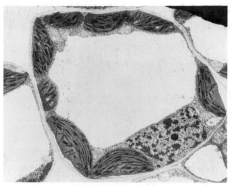

Fig 1.33 In a mature plant cell, the vacuole becomes so large that the cytoplasm is reduced to a thin layer around the cell

SEE QUESTIONS 6 AND 7

CHLOROPLASTS AND OTHER PLASTIDS

Chloroplasts are one of a group of plant cell organelles known as **plastids**. They are surrounded by a double membrane and contain an elaborate internal membrane system that houses the chemicals of photosynthesis (Fig 1.34).

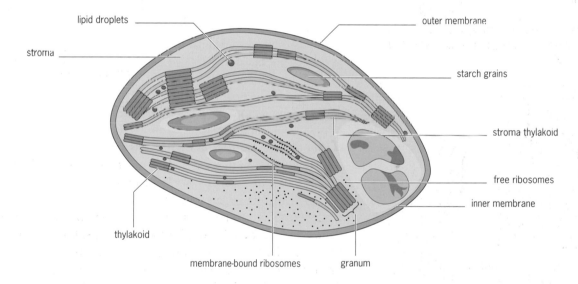

lipid droplets — outer membrane

stroma

starch grains

stroma thylakoid

free ribosomes

inner membrane

thylakoid

membrane-bound ribosomes — granum

Fig 1.34 Chloroplasts are organelles in which all the chemicals associated with photosynthesis are brought together. Chlorophyll molecules are housed on membranes to maximise light absorption. In a flowering plant, most chloroplasts are found in the palisade cells of the leaves

The structure and function of chloroplasts is covered in detail in Chapter 9.

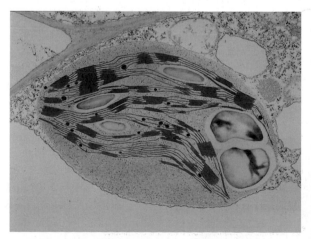

▶ EXTENSION The origin of mitochondria and chloroplasts: the endosymbiont theory

Scientists continue to speculate whether mitochondria and chloroplasts originated from inside eukaryotic cells. Many researchers think that they were originally free-living prokaryotes, similar to bacteria and blue-green bacteria, which, at some point, became incorporated into larger cells.

Let's look at the evidence. Mitochondria and chloroplasts have several similarities to prokaryotes:

● They are similar in size to prokaryotes.
● They have their own DNA, as prokaryotes do, suggesting an ability to reproduce independently.
● The DNA of both organelles is circular, like that of bacteria.
● Both organelles have their own ribosomes, smaller than those in eukaryotic cells but identical to those in prokaryotes.
● All reproduce by binary fission.
● The biochemistry of mitochondria is closer to that of aerobic bacteria than that of the eukaryotic cell; chloroplasts and cyanobacteria are even more similar.

This theory was first suggested by a team led by Lyn Margulis at Boston University, USA. Assuming the theory is true, how could these organisms have become incorporated into a larger cell without being destroyed by lysosomes?

Imagine a heterotrophic, anaerobic organism – similar to an amoeba in structure and movement – that feeds by phagocytosis and respires without oxygen, though oxygen is present in its environment. If such a cell were to ingest aerobic bacteria that it did not digest, the bacteria could respire aerobically, and in doing so could provide the host organism with far more ATP than the host could produce on its own. The host cell in turn would give the bacteria material to respire – food.

Similarly chloroplasts, which may originally have been free-living blue-green bacteria, could also have become incorporated into larger cells. If these were ingested – but not digested – they would continue to photosynthesise within the host and so provide it with organic food and oxygen – an obvious advantage. The host, in turn, could provide the blue-green bacterium with carbon dioxide and nitrogen.

Associations in which an autotroph lives inside a heterotroph, or an aerobe lives inside an anaerobe, are both very convenient arrangements. One organism produces what the other needs. This is the key to the success of several other associations, for example, lichens (algae and fungi) and corals (algae inside animals). Both host and 'passenger' gain a survival advantage: the host gets food and the smaller organism is protected from predators.

The idea of incorporation by ingestion could also explain why both organelles have a double membrane. The inner membrane would represent the original prokaryotic membrane while the outer one could represent the food vacuole into which the organism was taken.

✓ REMEMBER THIS

An **autotroph** is an organism that can make its own food: a green plant is an autotroph that make its own food by photosynthesis. A **heterotroph** is an organism that must obtain ready-made food: a cow is a heterotroph that obtains its food from grass.

SUMMARY

■ All living organisms (except viruses) are composed of either **prokaryotic** or **eukaryotic** cells.
■ **Bacteria** are prokaryotes; **plants**, **animals**, **protoctists** and **fungi** are eukaryotes.
■ Prokaryotic cells are relatively small, with very little internal organization and few internal membranes.
■ Eukaryotic cells are relatively large, with a high degree of internal organisation.
■ Eukaryotic cells have a true, membrane-bound **nucleus** and many **organelles** – membrane-bound compartments in which particular chemical processes take place.
■ The nucleus, **ribosomes**, **rough endoplasmic reticulum (ER)**, **vesicles** and **Golgi complex** in eukaryotic cells all take part in different stages in the making and packaging of chemical products. Some chemicals are used within the cell. Others – such as hormones or digestive enzymes – are released.

■ **Mitochondria** are the site of production of most of the eukaryotic cell's **ATP**, which provides the energy for other cellular processes.
■ Eukaryotic cells possess a **cytoskeleton** – a network of fine **tubules** that supports and shapes the cell.
■ Plant cells contain structures not found in animal cells: **chloroplasts**, a **cell wall**. The **vacuole** in most plant cells is much larger than the vacuoles found in an animal cells.
■ The cell can be studied using a variety of experimental techniques: the **electron microscope** reveals the structure of the cell, **cell fractionation** isolates individual organelles and, with the use of radioactive isotopes in **autoradiography**, the passage of materials through the cell can be followed.

? EXAM QUESTIONS

1

a) Explain concisely the main functions of the following structures in bacterial cells.
- **(i)** cell wall
- **(ii)** cell membrane
- **(iii)** DNA
- **(iv)** plasmids
- **(v)** ribosomes

b) Complete the table below to show the main differences between the structure of prokaryotic and eukaryotic cells.

prokaryotic	eukaryotic

OCR Biology Cells and Structures March 2001 Q2

2 The size of a cell can be measured using a microscope fitted with an eyepiece graticule. The eyepiece graticule has to be calibrated using a micrometer slide on the stage of the microscope. The figure below shows an eyepiece graticule against a micrometer slide, with a one millimetre line with subdivisions. A ×10 objective lens and a ×10 eyepiece were used on the microscope.

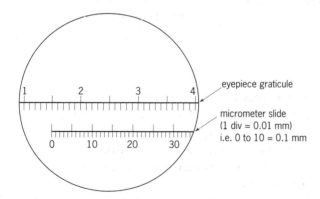

a) Calibrate the eyepiece graticule, stating the width, **in μm**, of the smallest division indicated on the graticule. Show your working.

The figure below shows the same eyepiece graticule, using the same ×10 objective lens and the same ×10 eyepiece, with some cells in focus.

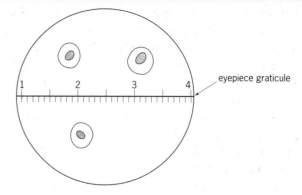

b) (i) Measure the largest diameter of each of the three cells shown. Calculate the mean diameter of the three cells, in μm. Show your working.
(ii) What would the **mean** diameter of these cells be if they were observed using the ×40 of this microscope?

OCR Biology Cells and Structures June 2000 Q5

3 The optical (light) microscope and the transmission electron microscope have both been of significant importance as tools in biological science. Complete the following paragraphs using appropriate words or phrases.

Tissues for use with both the light and the electron microscope are first ... . This prevents the cells breaking down and prepares them for further treatment. The tissues are then dehydrated using ... . Materials for use with the light microscope are embedded in wax but in the case of the electron microscope they are supported using ... . Thin sections of material are produced using a microtome fitted with a sharp metal blade for the light microscope but a blade made of ... , which gives much thinner sections, is used for the electron microscope.

The embedding material for the light microscope is removed and the sections can then be treated with stains. The presence of lignin can be demonstrated using the stain ... and the stain ... shows the presence of cellulose. Stains cannot be used on sections for use with the electron microscope but the sections are immersed in solutions of ... . There is a differential absorption of these which gives contrast in the resulting image.

The energy source in the light microscope is ... and in the transmission electron microscope is ... .

Quite often students consider that the electron microscope is a more important research tool because of magnification but in reality it is the ... of the electron microscope which is important.

OCR Biology Cells and Structures March 2001 Q1

? EXAM QUESTIONS

4

a) Complete the diagram of a mitochondrion and **label four** structures **clearly**.

b) **(i)** What process takes place in the mitochondrion?
(ii) Name a tissue where you would expect to find large numbers of mitochondria.
(iii) Explain why mitochondria are particularly important **in this tissue**.

WJEC Biology Module BI1 January 2001 Q2

5 The pie charts below show the relative amounts of types of membrane found in two types of cell.

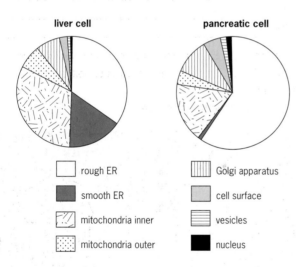

liver cell   pancreatic cell

☐ rough ER   ▥ Golgi apparatus
■ smooth ER   ▦ cell surface
▨ mitochondria inner   ▤ vesicles
▨ mitochondria outer   ■ nucleus

a) State which type of membrane is present in the greatest amount in the liver cell.

b) **(i)** Compare the relative amounts of mitochondrial membranes in the two types of cell.
(ii) Suggest **one** reason for the difference between the cell types in amount of mitochondrial membranes.

c) Using only the data in the pie charts, predict which cell type secretes more protein. Give **two** reasons for your answer.

d) Predict which type of membrane would be present in the largest relative amount in a prokaryote cell.

IBO Biology Paper 3 Standard level November 1999 QC1

6 State the name of the cell organelle, or structure, to which each of the following statements refers.

a) Fully permeable layer composed of a polysaccharide.
b) Series of membranes in the cytoplasm through which materials can be transported and in which proteins accumulate.
c) Site of ribosomal RNA synthesis.
d) Contains large quantities of chromatin.
e) Responsible for the translation of mRNA.
f) Partially permeable layer about 7.5 nm thick.
g) Structures responsible for producing the spindle apparatus.
h) Contains membrane stacks called grana.
i) Site of aerobic respiration.
j) Structure containing cell sap.
k) Proteins with a quaternary structure are assembled here.
l) Structures made of the protein tubulin.

OCR Biology Cells and Structures November 2000 Q2

7 Carefully study the drawings in the figure below made from observations of electron micrographs.

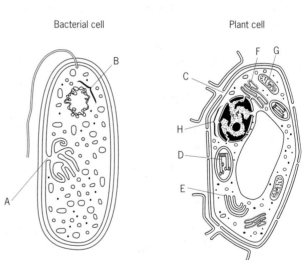

Bacterial cell   Plant cell

a) Cells are divided into two major groups according to their structures. To which group do the above cells belong?

b) Name the structures indicated by the letters **A**, **B** and **C** on the diagrams above.

c) State **one** difference between the structures labelled **B** and **H**.

d) State the **main** function of the following structures: **D**, **E**, **F** and **G**.

WJEC Biology Module BI1 June 2001 Q3

KEY SKILLS **ASSIGNMENT**

GETTING A SENSE OF PROPORTION

In this assignment we shall look at two skills that are commonly tested in examinations; scaling and matching structure to function.

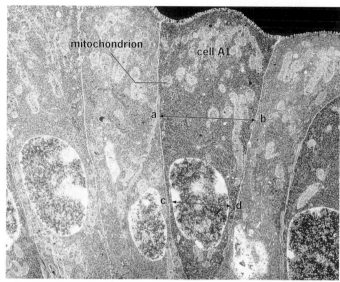

Fig 1.A1 Electron micrograph showing epithelial cells from the lining of a mammalian small intestine (×1800). Cell A1: width, line a to b; diameter of the nucleus, line c to d

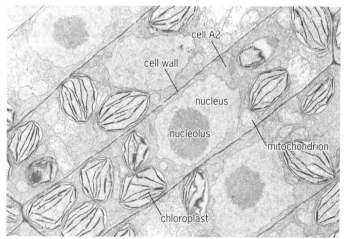

Fig 1.A2 Electron micrograph showing a palisade cell from a flowering plant (×1720)

MICROSCOPY AND SCALING

There are three aspects to scaling; the **actual size**, the **observed size** (that you can measure with a ruler) and the **magnification**. If you know two of these values, you can always work out the third one.

The formula is:

$$\text{Magnification} = \frac{\text{Observed size}}{\text{Actual size}}$$

This can be simplified even further to ⟨O / M A⟩ . In exam questions, cover up the value you need to find and the calculation you need should be clear.

1 Use the scales given (and a ruler) to work out the actual size of the cells in μm.

2 Copy and complete the table below.

Organelle	Size on paper	Actual size in the cell
Cell A1:		
Nucleus		
Mitochondrion		
Cell A2:		
Nucleus		
Nucleolus		
Mitochondrion		
Chloroplast		

3 Why do you think that the endoplasmic reticulum and the cell's surface membrane have been left out of the above table?

Fig 1.A3 shows some red blood cells.

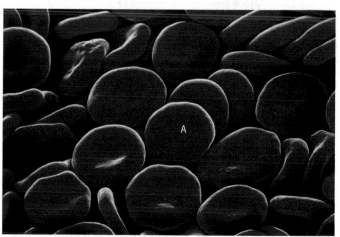

Fig 1.A3 Scanning electron micrograph of human red blood cells (×2150)

4 The magnification of Fig 1.A3 is ×2150. Calculate the diameter of the red cell labelled A. Give your answer in micrometres.

5 What evidence can you see from Fig 1.A3 that it was taken with a scanning electron microscope?

KEY SKILLS **ASSIGNMENT**

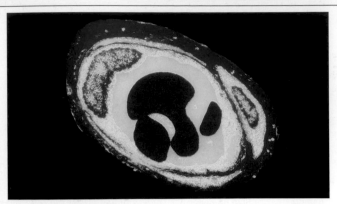

Fig 1.A4 Thin section of capillary showing different shaped red cells that have been cut across different planes (×1925)

6 Fig 1.A4 shows a blood capillary in the lungs. In order to prepare the slide for examination with the electron microscope, thin sections had to be cut. Explain how this results in the different shapes of red cells seen in the micrograph.

RELATING STRUCTURE TO FUNCTION

As we shall see in Chapter 2, cells inside an organism are specialised and perform particular functions. The structure of cells usually reflects their specialisation.

7 The **intestinal epithelial cell** (Fig 1.A1) allows rapid absorption of digested food into the bloodstream. The two mechanisms for the absorption of food molecules are diffusion and active transport. (You can look at digestion in more detail in Chapter 11.) Suggest reasons why this cell has large numbers of microvilli and mitochondria.

8 The **palisade cell** is on the upper surface of a leaf, the best location for photosynthesis. Remember that photosynthesis requires light, carbon dioxide and water and produces glucose and oxygen. Suggest how the following features help the palisade cell to perform its function:
a) the large numbers of chloroplasts,
b) the observation that chloroplasts circulate within the cell,
c) the deep, cylindrical shape of the cell.

You will gain a more complete picture of plant function when you study the structure of the whole leaf in Chapters 11 and 31.

9 **Red cells** are unusual because they lack a nucleus.
a) What is the main function of red cells?
b) Suggest an advantage of having no nucleus.
c) The cell would have a greater volume if it were spherical. Suggest an advantage of the biconvex shape.

02

Tissues and organs

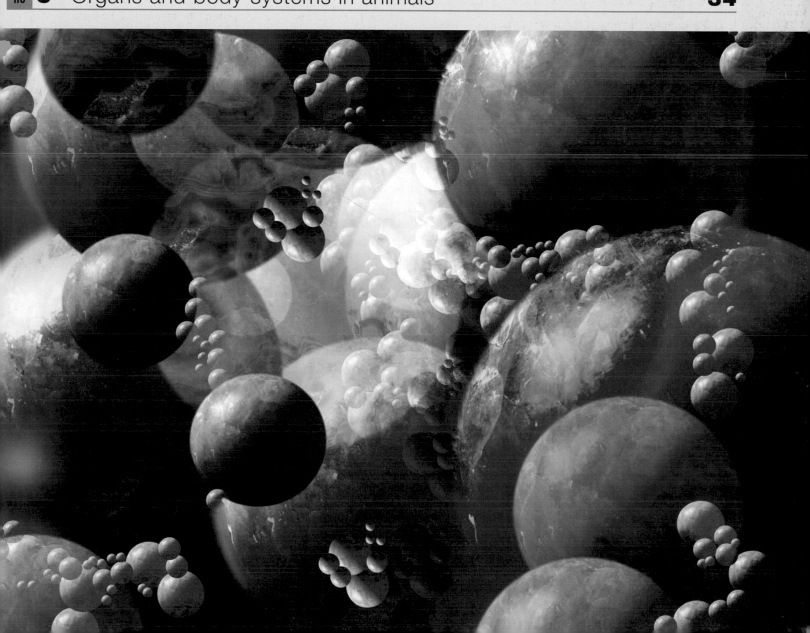

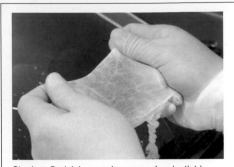

Stephen Badylak, a senior research scientist in Purdue's Department of Biomedical Engineering, holds a sheet of scaffolding material harvested from the urinary bladder of a pig. The material holds promise as a scaffold for the repair or replacement of many different body parts, such as the trachea and oesophagus in human patients.

As we learn more about how tissues interact to form organs and organ systems, we come across new opportunities to create artificial tissues to use in medicine. Artificial heart valves and tubes to keep blood vessels open have been used for quite a while, but more recently a team of bioengineers in the USA have gone one better. They have developed a way of treating the lining from the small intestine, stomach, bladder and liver of pigs and other animals to make an organic mesh that can be used to repair internal organs such as vocal cords and damaged ligaments. This mesh is also proving useful for treating skin wounds and chronic sores.

The tissues are removed and sterilised, then processed into sheets, tubes and other shapes which are grafted into areas that need to be replaced or repaired. Because the material is obtained from living tissue, it not only contains collagen, the fibrous protein that has great strength and flexibility, but also growth factors, proteins that send messages to cells to tell them where and how to grow and differentiate.

Once implanted into the body, the growth factors send instructions to the surrounding cells. The resulting new tissue growth is essentially the same shape and performs the same functions as the damaged tissues.

Fig 2.1 This photograph shows the huge difference in size between an ovum (a human egg cell) and the sperm that will fertilise it. The cytoplasm is packed with food reserves that allow the cell to divide several times before it implants in the uterus. The DNA in a fertilised egg carries all the information necessary to make a complete human baby in just 9 months

1 THE COMPLEXITY OF LIVING ORGANISMS

Most human cells are microscopic. The largest cell in the human body is the egg cell, or **ovum**. This is just visible with the naked eye. It needs to be larger than other cells because it needs space to store food reserves (Fig 2.1).

Humans, like all large organisms, are **multicellular**. Our body systems consist of complex organs and tissues (Fig 2.2) that interact and co-operate to keep all the cells of the body in the best possible environment. Different groups of cells carry out different tasks and functions: they have **differentiated** and become **specialised**. Different types of cell develop to carry out specific functions that contribute to the success of the organism as a whole. This is known as **division of labour**.

2 TISSUES

A **tissue** is a collection of similar specialised cells that work together to achieve a particular function. Different tissues combine to form more complex structures called **organs**. An organ usually has a specific function in the organism, for example to detect light, to absorb food or to produce a hormone.

Studying the detailed structure of individual tissues is a lot more interesting if you look at how the structure of a tissue relates to its function in the body of the organism. Getting familiar with normal tissues is particularly important in medicine; recognising what is normal and abnormal in a particular type of tissue can mean the difference between life and death.

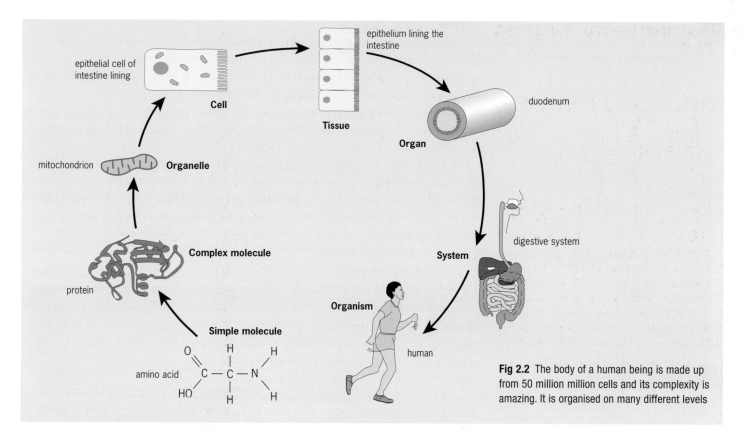

Fig 2.2 The body of a human being is made up from 50 million million cells and its complexity is amazing. It is organised on many different levels

HUMAN TISSUES

In humans, as in all mammals, each organ of the body is made up from a combination of these four basic tissue types:

- **Epithelial tissue** forms thin sheets that line and cover body structures. Epithelial tissue lines the intestines, for example.
- **Connective tissue** is tough and fibrous and forms structures that hold the body together. Ligaments and tendons are mainly connective tissue.
- **Muscular tissue** can contract to produce movement.
- **Nervous tissue** has the ability to conduct impulses, allowing communication between different parts of the body.

EPITHELIAL TISSUE

Epithelial tissues form continuous sheets that line or cover most structures and cavities in the body. Different types are classified according to the size and shape of their cells, and the number of cell layers they contain. For example, simple **squamous epithelium** contains only one layer of cells and is therefore thin and ideally suited to be a membrane where substances are exchanged, such as at the surface of alveoli. **Stratified squamous epithelium** is several cells thick and can be replaced continuously, making it perfect as a protective covering to the skin.

Epithelial tissues have very little mechanical strength; they are supported and attached to underlying connective tissue by a **basement membrane**, a continuous sheet consisting of **collagen** and other proteins.

Epithelial tissues form barriers that keep different body systems separate. They also have many other functions. For instance, the epithelial cells that line the mammalian respiratory tract are **ciliated**: they form a 'carpet' of tiny hair-like processes which trap inhaled dust and sweep it away from the lungs and out towards the throat where it is swallowed. Other epithelial tissues, such as the lining of the mammalian intestine, secrete **mucus**.

> ✔ **REMEMBER THIS**
>
> A **tissue** is a collection of similar specialised cells.
>
> **Histology** is the study of cells, tissues and organs, using microscope techniques.
>
> **Mitosis** is a type of cell division (see Chapter 25).

> **?** **TEST YOURSELF**
>
> **A** Smoking has been found to affect the lining of the respiratory system. What type of tissue is found here?

CONNECTIVE TISSUE

As their name suggests, **connective tissues** usually connect structures together. Connective tissue cells are not packed closely together but are separated by a non-cellular **matrix** (a mesh) which they secrete. (Notice the contrast to epithelial tissue where the cells are tightly packed.) The nature of the matrix accounts for the property of the tissue. In bone, for example, the matrix is a hard mineral strengthened by fibres of collagen, a tough protein (see Chapter 3).

The main types of connective tissue found in the human body are listed in Table 2.1. Cartilage, tendons, ligaments, blood and bone are all connective tissues. Each one is adapted to perform a particular function. All except blood have a structural role in the body. Blood is described as a fluid connective tissue because plasma, a fluid, separates and surrounds the cells. Because connective tissues are mainly structural, their metabolic demands (their need for food and oxygen) are relatively small.

Table 2.1 The main types of mammalian connective tissue

Connective tissue	Function
Bone	Forms the skeleton. Protects, supports the main organs of the body. Anchors the muscles
Tendon	Attaches muscles to bones
Ligament	Attaches bones to bones and provides support at joints
Cartilage	Smoothes surfaces at joints. Prevents collapse of trachea and bronchi
Adipose tissue	Stores fat and provides insulation
Blood	Transports substances around the body (see Chapter 13)
Areolar tissue	Protects organs, blood vessels and nerves. Gives strength to epithelial tissue. General 'packing tissue' arrangement

CARTILAGE

Cartilage is a tough, smooth and flexible connective tissue. There are three types of cartilage:

- **Hyaline cartilage.** This is a compressible and elastic tissue found in the nose and trachea, and at joints where it covers the ends of bones to provide smooth surfaces.
- **White fibrous cartilage.** This is found between the vertebrae and acts as a shock absorber.
- **Yellow elastic cartilage.** This is a very elastic tissue found in the ears, pharynx (throat) and epiglottis (the flap that closes over the airway during swallowing).

MUSCULAR TISSUE

Muscular tissues are contractile: they contain protein fibres, actin and myosin, which produce movement when they slide over each other (see Chapter 21).

As Fig 2.4 shows, there are three types of muscle: skeletal muscle, smooth muscle and cardiac muscle. The properties of the three types are summarised in Table 2.2.

The muscles of most vertebrates make up a large proportion of their body mass. As you might expect, active muscle tissue has a very high metabolic rate, and many mitochondria are packed between the muscle fibres.

> ⚠ **SCIENCE FEATURE** Corneal transplants with a difference

Mesh made from the lining of animal intestines is being used to repair damaged ligaments and other tissues (see opener box on page 28) and artificial skin is widely used to treat burns and slow-to-heal skin wounds. It may also soon be commonplace for people to have their damaged cornea replaced without using donated human corneas.

The technique (Fig 2.3), which has taken over 10 years to perfect, involves taking a few corneal stem cells (cells that grow to become corneal cells) from the patient's healthy eye, or from a related donor, in a painless procedure that lasts about 5 minutes. These cells are then grown in laboratory dishes under special conditions until they produce a fragile film, just one cell thick. This delicate sheet is then transferred onto a mesh made from an amniotic membrane that has been sterilised. The amniotic membranes are donated by mothers after the birth of their babies. On the mesh, the corneal cells grow into a fatter layer, between 5 and 10 cells thick. The end product is elastic like the amniotic membrane but has the biological properties of corneal tissue. The membrane is then placed over the patient's eye, and held in place with delicate stitches.

In a small trial carried out in 2000, 10 out of 14 patients who had damaged corneas, and whose traditional corneal transplants had failed, gained better eyesight with this new type of transplant. One man of 78, whose cornea became severely scarred after he suffered a rare reaction to medicated eye drops, could not drive or see a computer screen before his operation in 1998. Now, with the extra help of a contact lens, his vision is near normal in one eye and he can now drive, read and use his computer again.

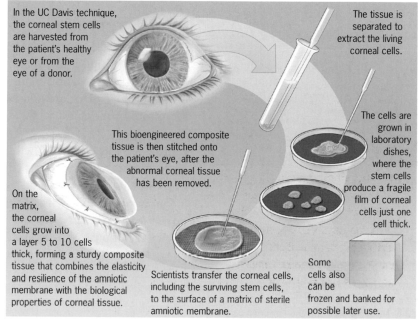

In the UC Davis technique, the corneal stem cells are harvested from the patient's healthy eye or from the eye of a donor.

The tissue is separated to extract the living corneal cells.

The cells are grown in laboratory dishes, where the stem cells produce a fragile film of corneal cells just one cell thick.

This bioengineered composite tissue is then stitched onto the patient's eye, after the abnormal corneal tissue has been removed.

On the matrix, the corneal cells grow into a layer 5 to 10 cells thick, forming a sturdy composite tissue that combines the elasticity and resilience of the amniotic membrane with the biological properties of corneal tissue.

Scientists transfer the corneal cells, including the surviving stem cells, to the surface of a matrix of sterile amniotic membrane.

Some cells also can be frozen and banked for possible later use.

Fig 2.3 The technique of corneal transplantation using stem cells grown on a membrane from a placenta

Table 2.2 Comparing the three types of mammalian muscle			
Muscle:	**Skeletal**	**Smooth**	**Cardiac**
Other names	striped, striated, voluntary	unstriped, unstriated, involuntary, visceral	heart
Site in the body	attached to skeleton	tubular organs; gut, reproductive system, glands, bronchioles	heart
Function	movement and maintenance of posture	usually controls movement of substances along tubes	heartbeat
Control	voluntary	involuntary (not under conscious control)	myogenic (self-generating)
Speed of contraction	rapid	slow, usually known as peristalsis (rhythmic squeezing movements) (see Chapter 11)	rapid
Speed of fatigue	rapid	slow	slow

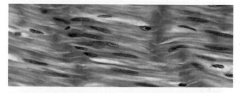

Fig 2.4 Micrographs of muscle: top, skeletal; middle, smooth; bottom, cardiac. See Table 2.2 for descriptions

NERVOUS TISSUE

Nervous tissue contains neurones, specialised elongated cells that transmit electrical impulses from one end to the other. Nervous tissue controls and coordinates the activities of the body. Changes in its internal and external environment act as stimuli that are detected by receptor cells in sense organs. From here, impulses travel along sensory neurones to the central nervous system (CNS). The CNS consists of the brain and spinal cord, both also made up mainly of neurones. The brain processes the information it receives and decides which response to make, often taking into account previous experience (memory). Impulses are sent out along motor neurones to effector organs: the muscles and glands. Control and co-ordination in the nervous system is covered in Chapters 19 and 20.

PLANT TISSUES

Like animals, plants are made up of different tissues that, together, form a whole functioning organism. Of course, there are basic differences between plants and animals, and plant tissues reflect these differences. Tissues of a multicellular plant are adapted for the following functions:

- photosynthesis
- support
- protection
- transport
- storage
- reproduction

How a plant works, growth and co-ordination in plants and plant reproduction are covered in Chapters 31, 32 and 33, respectively. The process of photosynthesis is covered in Chapter 9.

Plants contain **simple tissues** (tissues made of one type of cell) and **compound tissues** (tissues which have more than one type of cell). Simple tissues include **parenchyma**, **collenchyma** and **sclerenchyma**, compound tissues include **xylem** and **phloem**.

The basic structure of a flowering plant is shown in Fig 2.5, and the tissues of a plant are summarised in Table 2.3. You may find it useful to refer back to this table when you study the physiology of plants (in Chapters 9, 31, 32 and 33).

Fig 2.5 The basic anatomy of a flowering pant and the position of some of its major cell types

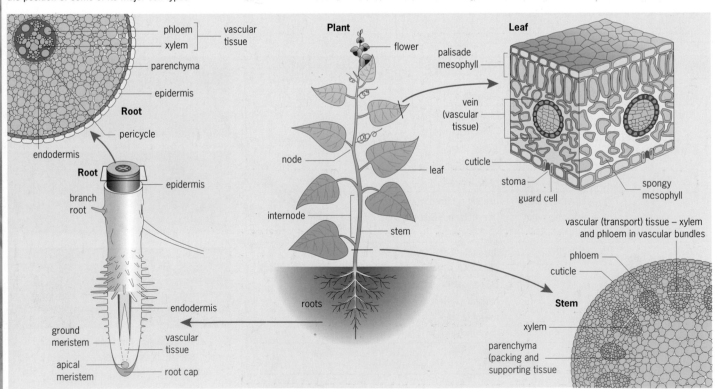

Table 2.3 Summary of the main tissues of a flowering plant

Tissue	Diagram	Distribution	Functions
Parenchyma		cortex, pith, medullary rays, vascular bundles	basic packing tissue, support and storage
Modified parenchyma:		covers all outer surfaces of plant:	
1 Epidermis		stem, leaves, roots	flattened cells protect from infection and drying out
2 Mesophyll	palisade / spongy	main part of leaf	palisade mesophyll is adapted for photosynthesis, spongy mesophyll is loosely packed to create air spaces for gas exchange
3 Endodermis	casparian strip	surrounds vascular tissue, so forms the innermost layer of cortex	selective barrier which controls movement of water and ions in and out of xylem
4 Pericycle		in roots between vascular tissue and endodermis	growth of lateral roots and secondary growth
Collenchyma		outer cortex, angle of stems and midribs	provides support
Sclerenchyma:			
1 Fibres		outer region of cortex, pericycle of stems (around vascular bundle)	support
2 Sclereids		cortex, pith, phloem, shells and stones of seeds	support/protection
Xylem*		vascular system	transport of water and dissolved ions
Phloem:			
1 Sieve tubes	companion cell / sieve tube	vascular bundles	translocation of food
2 Companion cells		vascular bundles	control of sieve tubes

*Note:
Xylem tissue contains sclerenchyma, parenchyma, tracheids and vessels.

3 ORGANS AND BODY SYSTEMS IN ANIMALS

An **organ** is a collection of tissues that work together to perform a particular function. For example, think about that most delicate and sophisticated of organs, the eye.

- The iris contains muscular tissue, as do the suspensory ligaments that control the lens.
- The retina is made from nervous tissue that detects light and sends information to the brain.
- Epithelial tissues make up blood vessels and other structures such as the capsule around the lens, and the conjunctiva that covers the front of the eye.
- The whole organ is held together and protected by connective tissue. The sclera, in particular, forms a tough, protective outer coating.

In a whole organism, organs form **systems** that carry out life processes such as digestion, excretion, reproduction. The combined systems of the body also need to create a stable internal environment to keep cells bathed in fluid that has the optimum temperature and composition.

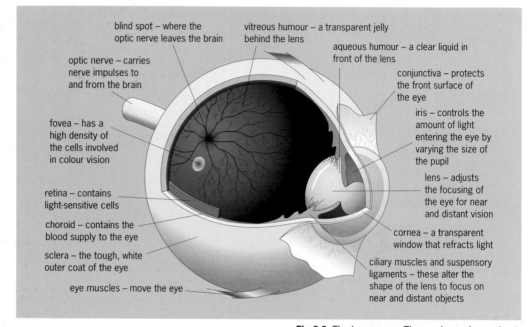

blind spot – where the optic nerve leaves the brain

vitreous humour – a transparent jelly behind the lens

aqueous humour – a clear liquid in front of the lens

optic nerve – carries nerve impulses to and from the brain

conjunctiva – protects the front surface of the eye

iris – controls the amount of light entering the eye by varying the size of the pupil

fovea – has a high density of the cells involved in colour vision

lens – adjusts the focusing of the eye for near and distant vision

retina – contains light-sensitive cells

choroid – contains the blood supply to the eye

cornea – a transparent window that refracts light

sclera – the tough, white outer coat of the eye

ciliary muscles and suspensory ligaments – these alter the shape of the lens to focus on near and distant objects

eye muscles – move the eye

Fig 2.6 The human eye. The eye is a sphere, about 2.5 cm across. Its outer wall has three layers: an inner layer, the retina; a middle layer, the choroid; and a tough outer layer, the sclera

Here is a short summary of the main systems in the human body together with their primary functions:

- The **digestive system** extracts from food the simple molecules that cells need.
- The **excretory system** removes the cell's waste products from the body.
- The **respiratory system** provides oxygen and removes carbon dioxide.
- The **circulatory system** transports all necessary substances to the cells, and removes waste products.
- The **muscles**, **skeleton** and **nervous system** combine to ensure that the organism obtains food and avoids danger.
- The **reproductive system** enables us to produce offspring and to pass on our genes.

? TEST YOURSELF

D Which body system in the list on this page does not contribute directly to the maintenance of internal conditions in the body?

SUMMARY

When you have studied this chapter, you should know and understand the following:

- The cells that make up the human body are specialised for a variety of different functions. This is called **division of labour**.
- Specialised cells form **tissues**, tissues form **organs** and organs form **organ systems**.

- A tissue is a collection of similar specialised cells.
- The tissues of the human body can be classified into four broad groups: **epithelial**, **connective**, **muscular** and **nervous**.
- The whole organism consists of all the systems working together to produce an individual that is able to control its internal conditions.

KEY SKILLS **ASSIGNMENT**

BURNS AND TISSUE GRAFTS

The severity of a burn can be measured on a three-point scale:

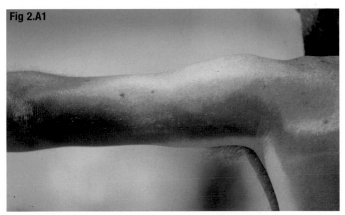

Fig 2.A1

First degree burns are minor or 'partial thickness' burns (part-way through the skin), often caused by a mild scald or over-exposure to sunlight. The main symptom is redness of the skin. Though painful, they usually heal quickly and need minimal medical treatment.

Second degree burns are partial thickness burns caused by severe scalding, contact with flames or very severe sunburn. The dermis of the skin is damaged as well as the epidermis, and redness is accompanied by blistering. This type of burn is extremely painful and usually needs medical attention.

Third degree burns are severe, full thickness burns. Though they may not be painful, third degree burns completely destroy the skin and often some of the underlying tissue such as fat, muscle and bone.

1 Why might a patient with third degree burns often be in less pain than one with less severe burns?

Estimating the damage

The extent of a burn is expressed as a percentage of the total body area the burn covers. As a rough guide, medical staff use the rule of nines, shown in Fig 2.A2. One of the main problems that occurs once a large area of skin has been damaged, is that the body loses its ability to regulate its temperature and the person's energy requirements rise to that of a marathon runner. They need huge amounts of energy to try and deal with the fluid loss, the need to keep their body temperature normal, and to fight infection.

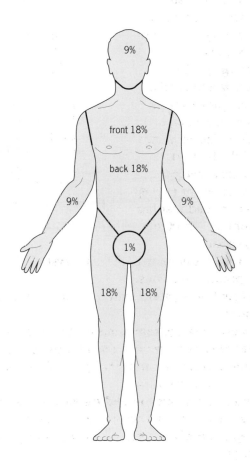

Fig 2.A2 The extent of a burn can be estimated by using the 'rule of nines'

2 **a)** A patient has burns to the whole of his back and all of both legs. What percentage of his body surface is this?
b) From your knowledge of the functions of skin, list some of the problems that will result from the loss of a large area. Refer to Chapter 15 if you need a reminder about skin function.
c) Before modern treatments were available, patients with large burns who survived the initial shock often died weeks later from starvation. Can you explain why?

KEY SKILLS **ASSIGNMENT**

Treatment

For many years, the main treatment available for burns, where the damage was not too extensive, was a **skin graft**. Skin from another region of the body was transferred onto the damaged area where it would eventually grow and provide an acceptable, if often unsightly, working replacement. Burns covering up to 40 per cent of the skin surface were usually treated by grafting techniques.

Fig 2.A3 New culture techniques allow large areas of skin to grow from a few healthy cells. Here, the cultured skin is sliced to stretch over three times its area

3 Why is it desirable to use a patient's own skin rather than a graft from another person?

During the 1990s, it became possible to culture skin outside the body (Fig 2.A3), and this is still being used and further developed. It is thought to be the most effective treatment for severe burns in the long term. Initially, however, the damage has to be covered, and this used to be done using either pig skin or skin from dead people (cadaver skin). This would be rejected within a few days, but kept the patient going until the cultured skin could be produced in large enough sheets. More recently, however, the development of artificial skin has revolutionised the treatment of burns and the death rates from the most serious burns have been reduced dramatically.

4 What advantages do you think artificial skin provides over animal or cadaver skin, or cultured skin?

Artificial skin is now produced commercially and is available in burn centres throughout the world. When a patient arrives with large area, third degree burns, surgeons have learned that the best treatment is to excise any damaged skin still left. This is a horrific operation in which the patient is 'peeled' with an instrument that does look quite like a potato peeler. The healthy underlying tissue is then covered as quickly as possible with artificial skin. This stops loss of fluid from the huge wound and also prevents microorganisms getting in to cause infection. The artificial skin also acts as a matrix for the patient's own skin to regenerate into and patients now suffer much less scarring than previously. The patient is kept sedated for long periods, so that they can recover with as little psychological trauma as possible.

5 Use the internet to research the survival rates from burns that cover between 70% and 90% of the body. How have the survival rates improved over the past 20 years or so?

The future

Researchers continue to make advances in wound healing and infection control. Newer skin substitutes, including those made from cultured cells of infant foreskins (a previously useless product of circumcision), are already or soon to be on the market. A new, aerosol spray-on suspension of the patient's own skin cells that takes only a few days to grow has also undergone trials in Australia. This method can also reduce scarring and is particularly useful for areas that are often hard to graft, such as the soles, underarms, and the palms of the hands. In Australia, more than 1200 patients have been treated using this method, but in Britain only a handful of patients have been treated so far.

03

The chemicals of life

03 The chemicals of life

Was there ever life on Mars? In 2003, instruments on a Mars lander will look for minute traces of molecular fossils as some scientists believe that similar conditions to those on Earth may have once produced organic molecules. Finding amino acids on Mars could finally solve the mystery of whether life of any sort did once exist on Mars and it could also help us understand life's origins on our own planet

All new organisms come from other organisms – they don't just arise spontaneously. So how did life first come about? Could it have developed from non-living chemicals?

Some experiments, carried out in the US in the 1950s by Stanley Miller, showed that this is a definite possibility. In his laboratory, Miller recreated the conditions thought to have been present on Earth 3500 million years ago. Then, the atmosphere contained no oxygen. Instead, it was probably a mixture of methane, ammonia, carbon dioxide, hydrogen sulphide and hydrogen. There were frequent electrical storms, and no ozone layer to protect the surface of the Earth from the Sun's ultraviolet rays.

When these conditions were recreated in the laboratory, the chemical reactions in the mixture produced several biological molecules: sugars, amino acids and bases. The findings were important because these molecules are the building blocks of nucleic acids and proteins – the chemicals on which all life is based. Since these groundbreaking experiments, further work over the past half century has suggested that one of the earliest self-replicating 'life-forms' may have been RNA. An 'RNA world' may have persisted for a long time, until conditions became milder and DNA was able to take over. However, we still have many jigsaw pieces to find before we work out how even these complex molecules made the giant step to become life.

1 STUDYING THE CHEMICALS OF LIFE

The study of the chemicals and reactions that take place inside living organisms is a science in its own right – **biochemistry**. Biochemists aim to understand how living organisms work by studying how the molecules within their cells interact. In this chapter we look at **carbohydrates**, **lipids**, **proteins** and **nucleic acids**, and describe the characteristics that make them the chemicals of life.

SOME CENTRAL THEMES IN BIOCHEMISTRY

The chemical systems that make up living things seem incredibly complicated, but there are some simple underlying patterns:

- Living organisms contain a huge range of **macromolecules**, but these large molecules are built from a small number of simple molecules.
- These simple building blocks are very similar in all organisms, suggesting that all life had a common origin.
- The characteristics of an organism are determined by the information contained in its DNA.
- The DNA contains information that the cell can use to make proteins. Many proteins are **enzymes**, which control the physical and chemical activities of an organism.
- The chemical activities that go on inside an organism can be given the general term **metabolism**.
- Metabolic reactions can be divided into two general categories: **anabolic** and **catabolic**. Anabolic reactions build up large molecules from smaller ones, while catabolic reactions do the reverse, breaking down larger molecules.

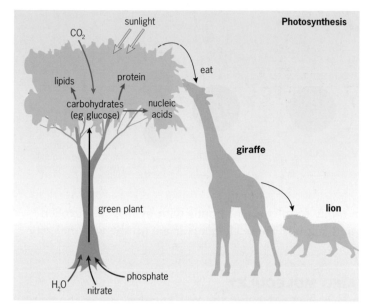

Fig 3.1 Organic molecules are normally made in plants via the process of photosynthesis. These molecules are then available to the rest of the food chain

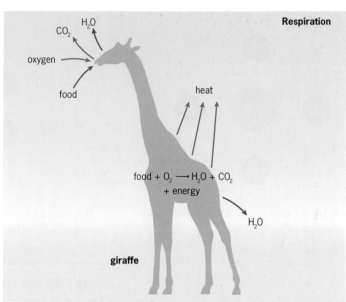

Fig 3.2 All organisms, including plants, need energy. They obtain energy by breaking down organic molecules (food). This process is respiration

- Anabolic reactions usually involve **condensation reactions** in which building-block molecules are joined together and a water molecule is released.
- Catabolic reactions, such as those that occur during digestion, usually involve hydrolysis reactions in which larger molecules are split as they react with water.
- In photosynthesis (Fig 3.1), plants use the energy from sunlight to build up organic molecules such as sugars from simple ones such as carbon dioxide and water (see Chapter 9).
- All organisms need a supply of energy, which they obtain via respiration (Fig 3.2). In respiration, organic molecules are oxidised into simpler molecules, usually carbon dioxide and water. The resulting energy is used to fuel the many energy-requiring processes within the organism.

LIFE ON EARTH IS CARBON-BASED

Why is life on Earth based on carbon? It is because this element can bond with itself repeatedly, to produce an infinite variety of molecules. Living things are composed mainly of organic molecules: proteins, carbohydrates, lipids and other molecules that have 'skeletons' made from carbon.

THE CARBON ATOM

Carbon has an atomic number of 6: it has a nucleus containing six protons and six neutrons, orbited by six electrons, two in an inner shell, the other four in an outer shell (Fig 3.3). Carbon acquires a full, and therefore stable, outer shell of eight electrons by sharing four more. So, carbon forms four covalent bonds – bonds with electrons shared between the atoms. We say it has a **valency** of 4.

Carbon can form covalent bonds with other carbon atoms to form stable chains or rings. C—C bonds are the most common; compounds that contain single carbon bonds are said to be **saturated**. C=C bonds are also frequent and C≡C are possible. Compounds with double or triple bonds are said to be **unsaturated**. This means that they are not saturated with hydrogen (more hydrogen atoms could be added at their multiple bonds). This is particularly relevant with respect to the fatty acids, as we see later in this chapter.

> ✓ **REMEMBER THIS**
>
> Water is essential to humans, as it is to all living organisms. Life evolved in water, most biochemical reactions take place in solution and all organisms that live on land – like ourselves – can do so because they have their own aquatic environment inside them.
> All of the living cells in the human body contain water and are surrounded by it; in fact, most of the human body is made of water. Even the most complex organ – the human brain – is about 85 per cent water.
> You can find more information about this important substance in the Extension box in Chapter 4.

> ✓ **REMEMBER THIS**
>
> Atoms have electrons in shells. Two electrons make the first shell full; eight electrons fill the second. Atoms with a full outer shell are stable. Atoms with a less than full outer shell are reactive and combine with others in order to fill the outer shell.

(a) Methane

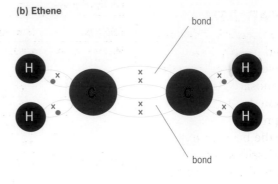

bond

• electron from H
× electron from C outer shell

Notice how the carbon shares an electron with each of four hydrogen atoms to achieve a full outer shell of eight electrons

(b) Ethene

bond

The two carbon atoms in this compound achieve a full outer shell of eight electrons by sharing electrons with hydrogen atoms and by forming a double bond

bond

Fig 3.3 Dot and cross diagrams showing two simple carbon compounds

Carbon atoms commonly form covalent bonds with hydrogen, nitrogen, oxygen, phosphorus and sulphur. These elements make up the vast majority of organic molecules.

WHAT IS AN ORGANIC MOLECULE?

The word *organic* is much used today, but the 'supermarket' use of the word to mean 'no added chemicals' is not a scientific one. Originally, the word organic meant 'of living origin', back in a time when scientists thought that the chemicals that made living things had a divine origin and could not be made by people. When scientists started to make organic molecules in the laboratory (urea was first) this idea was abandoned.

For our purposes, we can say that organic molecules are ones with carbon backbones – at least one carbon bonded to another one. Some organic molecules are enormous, with relative molecular masses in the millions. Table 3.1 shows some common examples of simple organic and inorganic compounds.

In this chapter, we look at four types of large organic molecules commonly found in living things: carbohydrates, lipids, proteins and nucleic acids.

Table 3.1 Some simple organic and inorganic compounds

Inorganic	Organic
Water (H_2O)	Ethane (C_2H_6)
Nitrogen gas (N_2)	Ethanol (C_2H_5OH)
Ammonia (NH_3)	Ethanoic acid (CH_3COOH)
Carbon dioxide (CO_2)	Glucose ($C_6H_{12}O_6$)
Hydrogen sulphide (H_2S)	Amino acid ($CHRNH_2COOH$)
Nitrate ion (NO_3^-)	

> **REMEMBER THIS**
>
> A **monomer** is a single molecule.
> A **dimer** is formed when two of the same molecules link together.
> A **polymer** is formed when many of the same molecules link together.

2 CARBOHYDRATES

Carbohydrates are organic molecules that contain three elements: carbon, hydrogen and oxygen. Carbohydrates include **sugars** and **starches** (Fig 3.4 and Table 3.2). There are three basic types of carbohydrate molecule, named according to their structure and size:

● **Monosaccharides** are single sugars.
● **Disaccharides** are double sugars (made from two monosaccharides).
● **Polysaccharides** are multiple sugars (polymers of many monosaccharides).

Carbohydrates are the first molecules made in photosynthesis. Lipids, proteins and nucleic acids are formed from carbohydrates.

Fig 3.4 The term carbohydrate covers a range of chemicals which include sugars, starches and cellulose. This picture shows food containing complex carbohydrates – these foods are staples of the human diet

MONOSACCHARIDES AND DISACCHARIDES

Monosaccharides and disaccharides are classed as sugars and usually have names ending in -**ose**, such as **sucrose** and **lactose**. In monosaccharides, the three elements carbon, hydrogen and oxygen are always present in the same ratio and they have the basic formula $(CH_2O)_n$. Monosaccharides are classified according to the number of carbon atoms they have: 3, 5 and 6 are the most usual (Table 3.3). In glucose, for example, n is 6, so its formula is $(CH_2O)_6$ or $C_6H_{12}O_6$.

Table 3.2 Some common carbohydrates

CHO type	Compound	Sub-units	Occurrence in living things
Monosaccharide	glucose		widespread
	fructose		sweet fruits
	galactose		milk
Disaccharide	maltose	$2 \times$ glucose	germinating seeds
	sucrose	glucose + fructose	fruit
	lactose	glucose + galactose	milk
Polysaccharide	starch	glucose	plants (storage)
	glycogen	glucose	animals (storage)
	cellulose	glucose	plant cell walls

Table 3.3 Common monosaccharides

Number of carbon atoms	Name	Common examples
3	triose	glyceraldehyde (see Chapter 35)
5	pentose	ribose, deoxyribose (in RNA and DNA)
6	hexose	glucose, fructose, galactose

GLUCOSE

It is useful to begin a study of carbohydrates with **glucose**: it is the main source of energy for many organisms – including humans – and most of the common polysaccharides are glucose polymers. Glucose is a hexose (6-carbon) sugar that has the formula $C_6H_{12}O_6$. All other hexose sugars, such as **fructose** and **galactose**, have the same formula (Fig 3.5).

SEE QUESTION 1

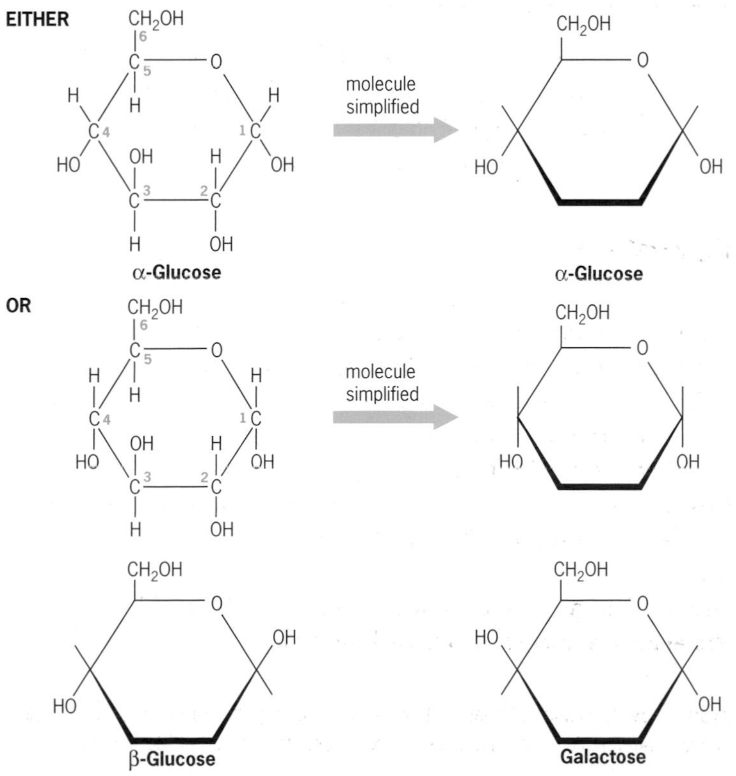

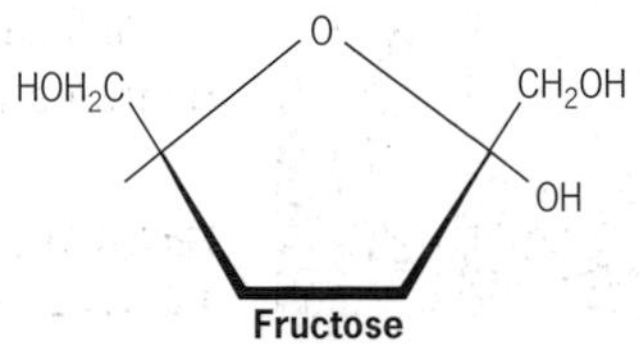

Fig 3.5 Variations on a common theme: common hexose sugars. A molecule of a glucose is shown (left) with all the atoms in place. This can be simplified to allow us to focus on the functional parts of the molecule – the OH groups on carbons 1 and 4. These groups react most commonly with other sugars. Where H on its own is joined to a C, the H can be omitted in the simpler form. The other three monosaccharides shown share the formula $C_6H_{12}O_6$ but, as you can see, they all have a slightly different structure. They are said to be **isomers** of each other. You might think these differences are too small to affect the molecule, but they do greatly affect properties such as taste (sweetness), digestibility and the nature of the polymers formed when the monomers join together

BIOCHEMICAL TESTS

Foods are usually mixtures of carbohydrates, lipids and proteins; one way to find out what type of molecules a food contains is to do biochemical tests. Some of the common ones you will use at A level are shown in Table 3.4 – these are basically the food tests you learned at GCSE level.

Table 3.4 Some simple biochemical tests

Nutrient	Reagent used	How test is carried out	Positive result
Reducing sugar	Benedict's solution	Add Benedict's solution to sample in a test-tube. Heat in a water-bath	orange-red precipitate
Non-reducing sugar	hydrochloric acid, and Benedict's solution	Once the reducing sugar test has proved negative, boil with dilute acid. Add sodium hydrogencarbonate to neutralise. Carry out reducing sugar test, as described above	orange-red precipitate
Starch	iodine solution	Add a few drops of iodine solution to the sample	blue-black staining
Lipid	ethanol	Shake the sample with ethanol in a test-tube. Allow to settle. Pour clear liquid into water in another test-tube	cloudy-white emulsion
Protein	Biuret solution	Add biuret solution to sample in a test-tube. Warm very gently	lilac/mauve colour

CHROMATOGRAPHY

Chromatography is the general name given to a number of techniques that can separate and identify the components of a mixture. Paper chromatography is a technique that separates the components of a mixture according to their solubility in a particular solvent. It is useful when only tiny amounts of each substance is present and may not be detectable by one of the biochemical tests. It can also distinguish between substances that give the same result in biochemical tests. For example, the Benedict's test gives a positive result for both glucose and fructose because both are reducing sugars but chromatography can separate them very easily.

DOING CHROMATOGRAPHY

Fig 3.6 shows a basic **chromatography tank**. The paper has a pencil line at the bottom – the samples are dotted onto the paper along this line at regular intervals. Each sample is allowed to dry before adding the next, to avoid spreading and merging of the substances under test. When all the samples have been put on the paper, the bottom end is lowered into the solvent at the bottom of the tank. The lid is put on to prevent evaporation of the solvent, and the apparatus is then left for a few hours to allow the solvent to soak up the paper.

Different substances dissolve in a particular solvent to different degrees – a substance that dissolves very well will move further up the paper, perhaps to near the top. A substance that hardly dissolves at all remains near the original spot placed on the paper. If you are separating out pigments, say, for example, from a leaf extract, it is easy to see where the different substances are on the paper. However, if the test substance is colourless, you just end up with a piece of white paper – this must be treated with a stain to reveal the spots. For amino acid or protein mixtures, the paper is sprayed with **ninhydrin** which shows up the spots. This is highly toxic, so spraying is always carried out in a fume cupboard.

Fig 3.6 Chromatography in a beaker showing separation of pigments

IDENTIFYING YOUR SUBSTANCE

Having run the chromatogram, substances can be identified by working out their R_f value.

$$R_f = \frac{\text{Distance moved by spot}}{\text{Distance moved by solvent front}}$$

R_f values vary between 0 (didn't move at all) and 1 (went as far as the solvent front). A particular compound will always have the same R_f value in

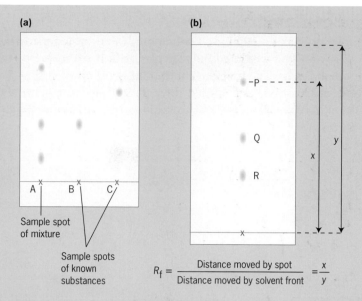

(a)

A   B   C

Sample spot
of mixture

Sample spots
of known
substances

(b)

P

Q

y

x

R

$$R_f = \frac{\text{Distance moved by spot}}{\text{Distance moved by solvent front}} = \frac{x}{y}$$

Fig 3.7 Chromatograms

a particular solvent. So, for instance, every time you run lysine in the solvent propanone, it will give you exactly the same R_f value. Fig 3.7 shows how an R_f value is calculated.

CHROMATOGRAPHY: A WORKED EXAMPLE

Paper chromatography was used to find out which amino acids were present in polypeptide X (a polypeptide is a small section of a protein). First, enzymes were used to digest the polypeptide into its amino acids. Chromatography was then carried out using a specially coated plastic sheet (this has exactly the same role as blotting paper, but gives clearer results). After running with solvent Y, the dried chromatogram was sprayed with ninhydrin.

● EXAMPLE

Q Fig 3.8 shows the chromatogram produced by polypeptide X; Table 3.5 gives R_f values of some amino acids. Which amino acids are present in polypeptide X?

A The R_f of an amino acid is

$$\frac{\text{Distance moved by the spot}}{\text{Distance moved by the solvent front}}$$

Spot A has moved x mm, the solvent front has moved y mm. Therefore the R_f value is: $x/y = 48/50 = 0.96$. This amino acid is therefore probably proline.

Spot B has moved x mm, the solvent front has moved y mm. Therefore the R_f value is: $x/y = 45/50 = 0.90$. This amino acid is therefore probably leucine.

Spot C has moved x mm, the solvent front has moved y mm. Therefore the R_f value is: $x/y = 33/50 = 0.66$. This amino acid is therefore probably tyrosine.

Spot D has moved x mm, the solvent front has moved y mm. Therefore the R_f value is: $x/y = 25/50 = 0.50$. This amino acid is therefore probably glycine.

Spot E has moved x mm, the solvent front has moved y mm. Therefore the R_f value is: $x/y = 19/50 = 0.38$. This amino acid is therefore probably glutamic acid.

? TEST YOURSELF

A Fig 3.7a shows that mixture A contains substance B. How many substances does mixture A contain? Does the mixture contain substance C? Give the reason for your answer.

B In Fig 3.7b, the R_f value of substance P is 40/50, which is 0.80. Calculate the R_f values of substances Q and R. Which substance is least soluble in the solvent used?

✔ REMEMBER THIS

If you work out an R_f value to be more than 1, you've probably got the formula upside down.

Table 3.5 R_f values of some amino acids

Amino acid	R_f value
Alanine	0.70
Arginine	0.72
Glutamic acid	0.38
Glycine	0.50
Tyrosine	0.66
Leucine	0.91
Proline	0.95

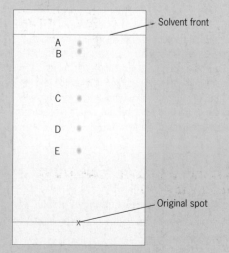

Solvent front

A
B

C

D

E

Original spot

Fig 3.8 Amino acid chromatography

TWO-WAY CHROMATOGRAPHY

Sometimes, two substances will have the same R_f values in a particular solvent, so they won't be separated out. The problem is overcome by using two-way chromatograms. The first chromatogram is run in the usual way but then the paper or plastic sheet is turned through 90° and run again with a different solvent. As Fig 3.9 shows, this reveals extra spots not distinguishable in the first chromatogram.

Fig 3.9 Two-way chromatography

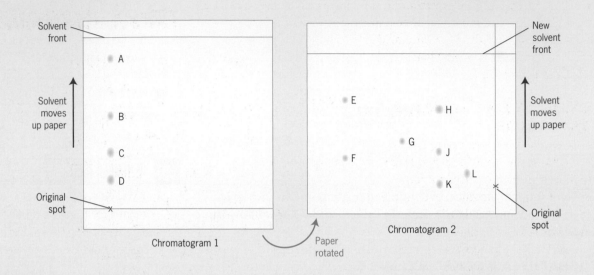

Fig 3.10 The Benedict's test can be used to estimate the amount of reducing sugars present in foods such as fruit juice. Samples with no reducing sugars remain blue, as left; those with a low concentration produce a green suspension, those with more produce yellow and orange suspensions, and juices very rich in reducing sugars produce an orange precipitate, as shown right

EXAMPLE

Q Fig 3.9 shows a one-way and two-way chromatogram run with a mixture of sugars. How many sugars were present in the mixture?

A The original chromatogram suggests only 4, but the two-way chromatogram reveals that there were actually 7. Three of the sugars must have very similar solubility in the first solvent used.

Q Which spots in chromatogram 1 contained more than one substance?

A Spots A and C. Spot A contains E and F. Spot C contains K, J and H.

Q Which substance was insoluble in the second solvent?

A Substance K because it has not moved towards the second solvent front.

? TEST YOURSELF

C In the Benedict's test, why does sucrose give a negative result before hydrolysis but a positive result after hydrolysis?

WHAT IS A 'REDUCING SUGAR'?

The term 'reducing sugar' reflects the fact that some sugars can reduce other chemicals. This basically means that the sugar can donate electrons to other substances. For a fuller explanation of reduction–oxidation reactions (redox reactions), see Chapter 35.

A standard test for a reducing sugar involves boiling the sample with Benedict's solution, a blue solution that contains copper sulphate. If a reducing sugar is present, the Cu(II) ions in copper sulphate are reduced to Cu(I) ions, resulting in an orange-red precipitate (Fig 3.10). Glucose, fructose, galactose, maltose and lactose are all reducing sugars, but sucrose is not. However, after sucrose is boiled with dilute acid to hydrolyse (split) it into its monosaccharides, it does produce a positive result.

MONOSACCHARIDES LINK BY MEANS OF GLYCOSIDIC BONDS

When two monosaccharides join together in a condensation reaction, the bond between them, a **glycosidic** link, centres around a shared oxygen atom (Fig 3.11). Two α-glucose molecules join together to make one molecule of **maltose**. Sucrose, the familiar sugar we buy in bags, consists of one molecule of glucose and one of fructose. Lactose, the main sugar in milk, is a disaccharide that contains glucose and galactose. The structures of sucrose and lactose are shown in Fig 3.12.

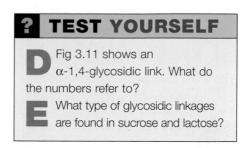

TEST YOURSELF

D Fig 3.11 shows an α-1,4-glycosidic link. What do the numbers refer to?

E What type of glycosidic linkages are found in sucrose and lactose?

SEE QUESTION 2.

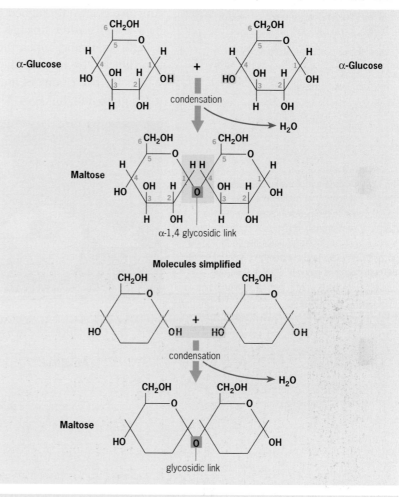

Fig 3.11 Two glucose molecules join to form maltose. Like many anabolic (building-up) reactions, this is a condensation reaction which involves the production of water

Fig 3.12 The structure of sucrose and lactose. Sucrose is made by the condensation of glucose and fructose. Lactose is made by the condensation of glucose and galactose

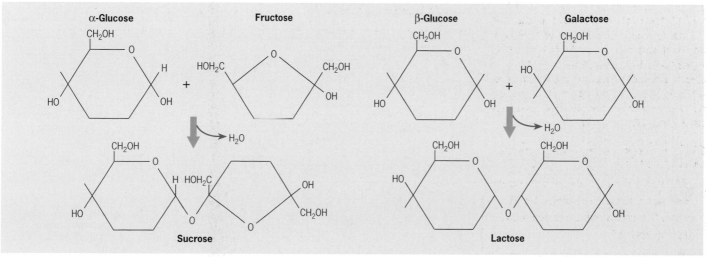

POLYSACCHARIDES

STARCH

Starch is the most abundant storage chemical in plants (Fig 3.13) and it is the single largest provider of energy for most of the world's population.

Starch has the three properties that are necessary for a storage compound. It is:

- compact
- insoluble
- readily accessible when needed.

Starch is a mixture of two compounds, **amylose** and **amlyopectin**. Amylose is an unbranched polymer in which glucose monomers are joined by α-1,4-glycosidic linkages. These bonds bring the monomers together at a slight angle and, when they are repeated many times, a spiral molecule is produced. In amylose there are six glucose residues in a turn of the spiral.

The glucose chains of amylopectin have α-1,4-glycosidic linkages and α-1,6-glycosidic linkages. This allows branching (Fig 3.14).

Fig 3.13 In many plants, starch is the main storage carbohydrate. Foods such as rice, pasta and potatoes all have a high starch content. When treated with an iodine/potassium iodide solution, a blue-black starch–iodide complex is formed. This reaction has been used here to demonstrate the presence of starch in a potato

Fig 3.14 Amylose and amylopectin, the two different polymers that make up starch. When a plant needs to break down starch to provide glucose for respiration, it removes the terminal (end) units of amylose and amylopectin to release glucose. Since the branched amylopectin molecule has more terminal glucose units that can be removed simultaneously, it can be broken down more quickly than amylose

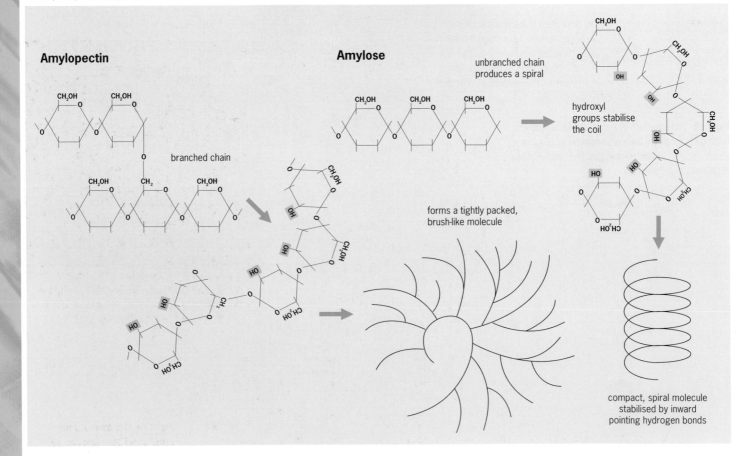

Amylopectin

branched chain

Amylose

unbranched chain produces a spiral

hydroxyl groups stabilise the coil

forms a tightly packed, brush-like molecule

compact, spiral molecule stabilised by inward pointing hydrogen bonds

GLYCOGEN

In animals, including humans, **glycogen** is the main storage carbohydrate. Its structure is similar to amylopectin, but it is even more frequently branched. In humans, glycogen is stored in large amounts in the liver and the muscles. During prolonged exercise, when the immediate supply of glucose is used up, the body restores its supplies by breaking down glycogen. If an average person goes without food, his or her glycogen stores last for about a day, but prolonged exercise such as marathon running can use all of the body's glycogen in less than two hours. When glycogen runs out, the body turns to using its lipid stores. This is why eating less while taking more exercise is the quickest way to lose weight.

One of the major changes associated with improving fitness is an increase in the amount of the enzyme glycogen synthetase in the muscles. This allows glycogen to be built up faster after it has been used (Chapter 18).

? TEST YOURSELF

H Why is it an advantage to an animal to have a storage chemical – glycogen – that is even more branched than amylopectin?

CELLULOSE

Cellulose is a structural polysaccharide: it gives strength and rigidity to plant cell walls. Individual cellulose molecules are long unbranched chains containing many β-1,4-glycosidic linkages (Fig 3.15). The molecules are straight and lie side by side, forming hydrogen bonds along their entire length. This results in strong bundles of chains called **microfibrils**.

Cellulose is probably the most abundant structural chemical on Earth, but few animals can digest it because they do not make the necessary enzyme, **cellulase**. Herbivorous animals, whose diet contains large amounts of cellulose, can deal with it because they have cellulase-producing microorganisms in their digestive system. Humans cannot digest cellulose, but we make good use of it in other ways – its strength is used in products such as paper, cotton, Lycra and nail varnish.

? TEST YOURSELF

I Why do cellulose-based materials such as cotton take a long time to rot?

CHITIN

The exoskeletons of arthropods such as insects and spiders are lightweight, strong and waterproof (Fig 3.16). These properties are provided by chitin, a polysaccharide that contains many glucosamine units. A glucosamine is formed when an amino group (NH_2) is added to a glucose molecule.

Fig 3.16 The chitin exoskeleton of an insect provides strength and flexibility without being too heavy. This may be one of the reasons why insects were the first organisms to develop flight

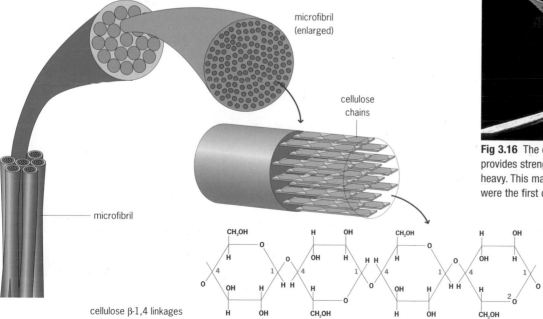

microfibril (enlarged)

cellulose chains

microfibril

cellulose β-1,4 linkages

Fig 3.15 Cellulose and the structure of the plant cell wall

3 LIPIDS

Lipids are a varied group of compounds that include the familiar **fats** and **oils**. As they are non-polar molecules, most lipids are insoluble in water but soluble in non-polar solvents such as alcohol and ether. Important exceptions are phospholipids, which have polar heads. The **emulsion test** for lipids, shown in Fig 3.17, is based on the solubility of lipids in ethanol.

Lipids contain the elements carbon, hydrogen, oxygen and sometimes phosphorus and nitrogen. They are intermediate-sized molecules that do not achieve the giant sizes of the polysaccharides, proteins and nucleic acids.

Fig 3.17 The emulsion test for lipids. The food to be tested is broken up into very small pieces, mixed with pure ethanol and shaken vigorously. Any lipid present dissolves in the alcohol. This top layer is poured off and mixed with water. If lipid is present, the mixture turns white as an emulsion, a suspension of fine lipid droplets forms. If no lipid is present, the mixture remains clear

LIPID STRUCTURE AND FUNCTION

The **triglycerides**, which act mainly as energy stores in animals and plants, are a large important group of lipids.

Triglycerides consist of one molecule of **glycerol** and three **fatty acids**. Fig 3.18 shows how the four components join by condensation reactions to form an E-shaped molecule.

SEE QUESTIONS 3 TO 5.

Fig 3.18 A triglyceride is formed from a condensation reaction between one molecule of glycerol and three fatty acids. The bonds formed are called ester bonds. The bottom right diagram shows the final shape of the triglyceride: glycerol is combined with three fatty acids to form an E-shaped molecule

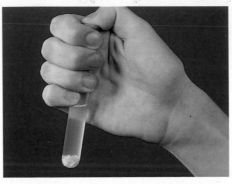
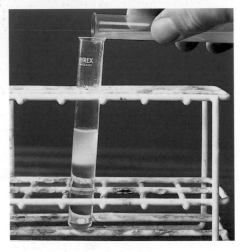

The glycerol molecule is common to all triglycerides and so the properties of different triglycerides depend on the nature of the fatty acids. Fatty acids vary in the length of their chain and in the degree of **saturation** they show.

Table 3.6 shows the chain length of some common fatty acids. Chains of about 14 to 16 carbon atoms are the most usual, but they range from 4 to over 28. A **saturated fatty acid** has the maximum amount of hydrogen and therefore has no double bonds. **Monounsaturated fatty acids** possess one C=C bond and **polyunsaturates** contain more than one.

✔	REMEMBER THIS
	Fatty acids are organic acids. They contain a carboxyl (—COOH) group.

Table 3.6 Some common fatty acids

Fatty acid	No. of carbon atoms	No. of double bonds	Abundant in	Melting point/°C
Palmitic acid	16	0	palm oil	63.1
Stearic acid	18	0	cocoa	69.6
Lauric acid	12	0	coconut, palm oil	44.2
Oleic acid	18	1	olive, rapeseed	13.4
Linoleic acid	18	2	sunflower, maize	−5.0

? TEST YOURSELF

J Look at Table 3.6. Which fatty acids are unsaturated?

K Of the fatty acids listed in Table 3.6, which are liquid at room temperature (20 °C)? What do they have in common?

! SCIENCE FEATURE Fat cells have hidden depths

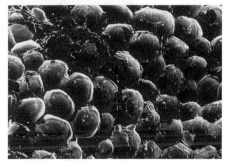

Fig 3.19 Fat cells

Fig 3.20 Leptin-deficient rat

People used to think that fat cells were just little blobs of fat surrounded by a membrane – which is certainly what they look like under the scanning electron microscope (Fig 3.19). However, we are now beginning to realise that fat cells are an important tissue in their own right, and that they help to regulate our metabolism.

Fat cells produce and secrete a wide range of proteins that act on other cells. Particularly interesting are the hormones such as leptin that act on centres in the brain that tell us to stop eating when we have had enough. (But how often do we listen?) The importance of leptin was demonstrated in the 1980s when rats that don't make any leptin were first bred. These animals look normal at birth, but they gain weight rapidly because they don't know when to stop eating. When they are a few months old, they become grossly fat, resembling spheres of fur with a head, four feet and a tail (Fig 3.20).

Other hormones produced by fat cells that also help control appetite and feeding behaviour have since been found. The interactions between them are so complex that we don't yet understand how they fit together. Even so, researchers are confidant that progress will be made and that we may have that elusive weight control pill in the next decade or so.

Fig 3.21 Lipids are vital chemicals in all living organisms. In humans, triglycerides occur mainly in adipose (fat storage) tissue, which forms under the skin and around internal organs. As well as storing energy, adipose tissue insulates against the cold, protects organs against physical damage, and contributes to a person's overall shape. The female body shape is, to a large extent, determined by adipose tissue distribution. This painting by Reubens shows the fuller female figure that was fashionable in previous centuries

Fig 3.22 There has always been considerable debate about the ideal body; it often appears to be dictated by fashion and culture, rather than common sense. The body shape of this model might by considered 'ideal' by many but it is a shape that few girls can achieve. It is also debatable whether such a low level of body fat is actually healthy

---

Some animals survive unfavourable seasons when food is scarce, by hibernating (Fig 3.23). Mammals such as dormice and bats show true hibernation: their metabolic rate is much lower than normal and their core body temperature, breathing and heartbeat rates drop significantly. Core body temperatures of around –5 °C have been recorded in hibernating bats!

Fig 3.23 Before hibernating, dormice eat large amounts of high-energy foods to build up fat stores. During hibernation they obtain the energy they need to survive the long winter from the triglyceride stored in their adipose tissue

PHYSICAL AND CHEMICAL CHARACTERISTICS OF DIFFERENT TRIGLYCERIDES

Triglycerides, which contain longer chain fatty acids and saturated fatty acids, generally form 'hard' fats such as lard and suet that are solid at room temperature. Animal tissues contain a higher proportion of saturated fats than plants. Plant lipids contain shorter chain, unsaturated or polyunsaturated fatty acids and so are 'light' oils that are liquid at room temperature. Unsaturation leads to a lowering of the melting point because double bonds produce kinks in the carbon chain. This increases the distance between molecules by stopping chains from lying parallel and so reducing weak intermolecular forces – look ahead to Table 3.8. This makes the lipid more 'fluid'.

TRIGLYCERIDES AS ENERGY STORES

Many animals store energy in the form of triglycerides: gram for gram, they yield more than twice as much energy as proteins or carbohydrates. Triglycerides are **highly reduced** compounds; they contain many C–H bonds which can yield energy during respiration. If you look at the formula for a lipid, such as $C_{14}H_{26}O_2$ you can see that there is a far greater ratio of hydrogen to oxygen than in a carbohydrate such as sucrose, which has the formula $C_{12}H_{22}O_{11}$.

PHOSPHOLIPIDS

Phospholipids have a similar structure to triglycerides but one of the fatty acids is replaced by polar **phosphoric acid** (Fig 3.24). This gives the molecule a polar head and a non-polar tail. When placed in water, phospholipids arrange themselves with their hydrophobic ('water-hating') tails pointing inwards and their hydrophilic ('water-loving') heads pointing outwards. This is vitally important because it results in double layers called bilayers. Phospholipid bilayers form the basis of all biological membranes (see Chapter 4).

Fig 3.24 The structure of a phospholipid. The phosphoric acid gives the molecule a polar head and a non-polar tail. Cell surface membranes – and other membranes in the cell – form automatically due to the behaviour of phospholipids. In water, the hydrophilic polar heads face outwards while the hydrophobic tails point inwards

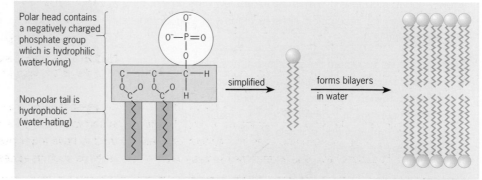

CHOLESTEROL

Many people associate **cholesterol** with heart disease, but this lipid is a perfectly normal constituent of every cell in our body (Fig 3.25 and see Chapter 4). As well as eating food that contains cholesterol, we can also synthesise cholesterol in the liver. The more there is in the diet the less the liver needs to make: vegans, who eat no animal products, are easily able to make all the cholesterol they need.

STEROID HORMONES

Steroid hormones have a similar structure to the cholesterol from which they are made. They include testosterone, progesterone and the oestrogens. The role of these hormones is discussed in Chapter 22.

WAXES

Waxes are lipids that are often used to waterproof surfaces, so preventing water loss. The cuticle of a leaf and the protective covering on an insect's body are both waxes. Waxes consist of a very long chain fatty acid joined to an alcohol molecule (not glycerol as in triglycerides). They have no nutritional value because they cannot be digested by lipases (lipid-digesting enzymes).

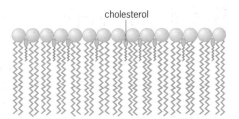

cholesterol

Fig 3.25 Cholesterol molecules are found between phospholipid molecules in membranes. Cell surface membranes are particularly rich in cholesterol, while organelle membranes contain less cholesterol

---

! SCIENCE FEATURE Steroids in athletics

Anabolic steroids are synthetic compounds that are similar in structure to **testosterone**, the male sex hormone. They increase anabolic (building-up) reactions, which enhance muscle size.

Athletes, notably weightlifters, began injecting testosterone in the 1950s to gain extra strength.

When steroids were banned, athletes looked for other natural substances that would be undetectable in blood tests. **Human growth hormone** (HGH, or **somatotropin**) and also **erythropoietin** are other illegal but potentially performance-enhancing substances. HGH increases muscle size in adults and erythropoietin increases the oxygen-carrying capacity of the blood. See the Assignment in Chapter 13.

HOW ANABOLIC STEROIDS WORK

Steroid hormones are lipid soluble and so can pass through the cell surface membrane. Studies have shown that anabolic steroids pass into the nucleus, where they force the cell to 'read' the genes that code for the muscle proteins actin and myosin. As well as increasing muscle size, athletes can train for longer, and the greater aggression that they tend to develop may also give them an added competitive edge.

The use of anabolic steroids is unfair and dangerous. To achieve a muscle-enhancing effect, athletes need to inject 10 to 40 times the normal dose, and this can have irreversible side effects. Symptoms of steroid abuse include acne, impotence, sterility, diabetes, heart disease and even liver cancer. Studies of male steroid abusers have shown that their testes may

Fig 3.26 Athletes such as Ben Johnson have risked the shame of being banned from their sport by taking steroids to enhance their performance

decrease in size by as much as 40 per cent. Their sperm count is correspondingly lowered. Female athletes who use steroids tend to have fewer periods and can develop masculine features such as excessive body hair and a deeper voice.

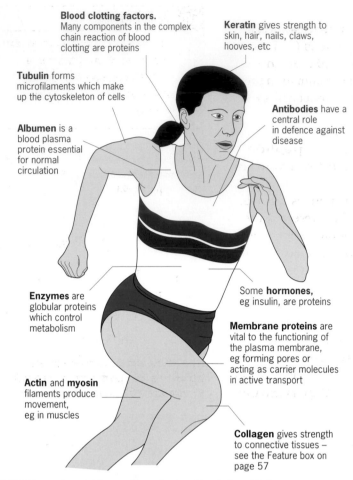

Blood clotting factors.
Many components in the complex chain reaction of blood clotting are proteins

Keratin gives strength to skin, hair, nails, claws, hooves, etc

Tubulin forms microfilaments which make up the cytoskeleton of cells

Antibodies have a central role in defence against disease

Albumen is a blood plasma protein essential for normal circulation

Enzymes are globular proteins which control metabolism

Some **hormones,** eg insulin, are proteins

Membrane proteins are vital to the functioning of the plasma membrane, eg forming pores or acting as carrier molecules in active transport

Actin and **myosin** filaments produce movement, eg in muscles

Collagen gives strength to connective tissues – see the Feature box on page 57

Fig 3.27 After water, proteins are the major constituent of our bodies

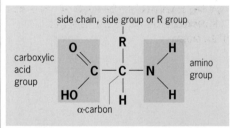

Fig 3.28 The basic structure of an amino acid

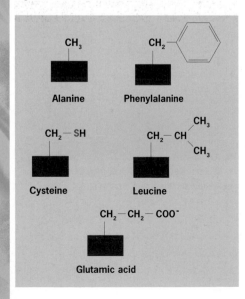

Fig 3.29 Five of the 20 amino acids found in the human body. The R groups are highlighted in red, sulphur is shown in blue. A further 200 or more amino acid structures are possible, but these need to be synthesised (made artificially). They are never found in living systems

4 PROTEINS

Proteins play a central role in the structure and metabolism of all living organisms. Protein molecules have a huge variety of shapes and sizes (the structure of carbohydrates and lipids is relatively limited by comparison). This versatility of shape is the key to the role of proteins in the cell.

THE IMPORTANCE OF PROTEINS TO LIVING ORGANISMS

We can appreciate the extent to which organisms use proteins by looking at their distribution in the human body (Fig 3.27). Consider some of the jobs that proteins need to do:

● Each cellular metabolic reaction must be catalysed by a different **enzyme**.

● Each substance that passes across a cell membrane requires a different **carrier molecule**.

● A different **antibody** is needed to combat the chemicals produced by the many (and also constantly changing) disease-causing organisms such as bacteria and viruses.

Enzymes, carrier molecules and antibodies are all proteins that are 'tailor-made' to fulfil all these requirements.

THE STRUCTURE OF PROTEINS

Proteins are large and complex molecules (Table 3.7). If a water molecule (relative molecular mass = 18) was the size of a brick, proteins would be whole buildings.

In addition to the elements carbon, hydrogen and oxygen, proteins always contain nitrogen and sometimes sulphur. The building blocks of proteins are called **amino acids**. Fig 3.28 shows the basic structure of an amino acid. As the name suggests, all amino acids contain an amino group ($-NH_2$) and a carboxylic acid group ($-COOH$). Both of these groups are attached to a central carbon atom, known as the α-carbon. The 'backbone' of an amino acid is made up from the three atoms C–C–N.

The R group varies between different amino acids. Fig 3.29 shows five of the different R groups in the 20 amino acids that make up all of the proteins in the human body.

Table 3.7 The relative molecular masses of several proteins	
Protein	**Relative molecular mass**
Insulin	5 700
Haemoglobin	64 500
Myoglobin	16 900
Hexokinase	102 000
Glycogen phosphorylase	370 000
Glutamine synthetase	592 000

AMINO ACIDS IN SOLUTION

Amino acids are readily soluble in water. When in solution they can resist a change in pH by mopping up or releasing both hydrogen ions (H^+) and hydroxyl ions (OH^-), acting as buffers, that is, regulators of pH. In whole proteins, the buffering effect is largely due to the R groups. Maintaining a constant pH is an important aspect of **homeostasis** (see Chapters 14 to 18).

PEPTIDES, POLYPEPTIDES OR PROTEINS?

It is easy to be confused about the precise meaning of the terms **peptide**, **polypeptide** and **protein**; there isn't a universally accepted rule to follow. This is how we use them in this book.

When amino acids are assembled into a short chain a peptide is formed. Longer chains are known as polypeptides. The term protein is reserved for the finished, functional molecule. Some proteins consist of one polypeptide, others consist of two or more than two. Haemoglobin, for example, contains four polypeptides.

HOW AMINO ACIDS FORM PROTEINS

Amino acids join together in long chains to form proteins by means of **peptide bonds**. Fig 3.30 shows how two amino acids join to form a dipeptide. This is an example of a **condensation reaction**.

All proteins are complex molecules and biochemists look at their structure at four different levels: **primary**, **secondary**, **tertiary** and **quaternary**.

PRIMARY STRUCTURE

The primary structure of a protein refers to the sequence, or order, of amino acids in that protein (Fig 3.31).

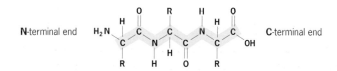

A simple primary structure of a tiny protein could be shown as:

alanine–histidine–phenylalanine–glutamine–cysteine

Real proteins usually consist of a lot more than five amino acids. The hormone insulin, for example, a relatively small protein, has 51 amino acids.

The code for the primary structure of any protein is contained in the gene or genes that code for that protein. This code determines the precise order in which amino acids are assembled. This order then dictates the way they will twist and turn to produce the precise three-dimensional shape that allows the protein to carry out its specific function in the body. The first level of three-dimensional twisting is described as the secondary structure of the protein.

SECONDARY STRUCTURE

When combinations of amino acids join together in a chain they tend to fold into particular shapes and patterns (such as spirals) in some places. These shapes form because the amino acids twist around to achieve the most stable arrangement of hydrogen bonds. The secondary structure of the protein refers to the patterns contained within the amino acid chain. Such patterns exist in different places in different proteins, producing an almost infinite variety of molecular shapes.

REMEMBER THIS

No matter how many amino acids are in the chain, if there is a —COOH group on one end, there will be an $—NH_2$ group on the other.

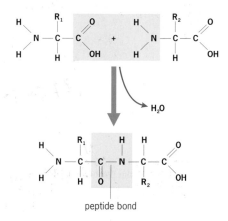

Fig 3.30 Two amino acid molecules react together in a condensation reaction (a reaction that releases water) to produce a dipeptide

Fig 3.31 When amino acids join, there is a repeating sequence of C—C—N—C—C—N— etc. This backbone runs throughout the length of the protein

SEE QUESTION 6

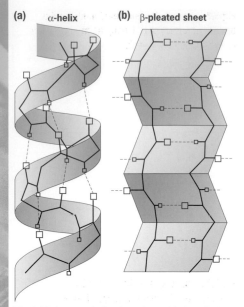

Fig 3.32 In an α-helix, the polypeptide is coiled into a helix that is held in position by hydrogen bonds between different groups in the backbone. In a β-pleated sheet, the amino acid chains lie side by side, forming a sheet that is held together by hydrogen bonds between adjacent parts of the polypeptide

✔ **REMEMBER THIS**

A **residue** is the name given to the remains of a monomer that has become incorporated into a polymer.

The main types of secondary structure in proteins are:

● The α-**helix**, a spiral, is the most common type of secondary structure (Fig 3.32a). Amino acid **residues** in the spiral twist on their axis, each residue forming a hydrogen bond with another residue four units along. These hydrogen bonds stabilise the α-helix.

● The β-**pleated sheet**, a flat structure that consists of two or more amino acid chains running parallel to each other, linked by hydrogen bonds (Fig 3.32b).

The secondary structure of a protein depends on its amino acid sequence: some amino acids (or combinations of amino acids) tend to produce α-helices, others usually make β-sheets. A few amino acids tend to produce a sharp bend in the chain, a vital function that allows the chain to fold back on itself. Fig 3.33 shows the secondary structures within the enzyme lysozyme.

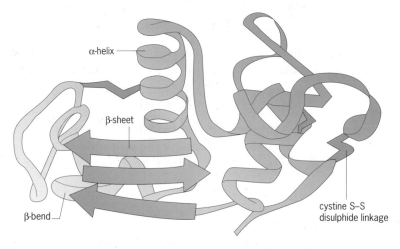

Fig 3.33 The secondary structures in the enzyme lysozyme: the α-helices and β-pleated sheets are clearly visible. The α-helices are shown as spiral ribbons, the β-pleated sheets as broad, flat arrows. This enzyme is present in tears and sweat, where it catalyses the breakdown of some bacterial cell walls

TERTIARY STRUCTURE

The tertiary structure of a protein is its overall three-dimensional shape and is produced as a result of the following:

● The sequence of amino acids that produces α-helices, β-sheets and bends at particular places along the chain.

● The hydrophobic nature of many amino acid side chains. Globular proteins are surrounded by water and so the hydrophobic side chains tend to point inwards.

Tertiary structure is maintained by attractive forces that arise when the amino acid chain folds (Table 3.8) and by **disulphide bridges**, covalent bonds that form when two **sulphur-containing cysteine residues** react. Disulphide bridges occur most often in structural proteins, where they contribute to strength.

Table 3.8 The different types of bond that maintain the secondary, tertiary and quaternary structure of a protein		
Type of bond	**Formed between**	**Relative strength**
Hydrogen bonds	H and an electronegative atom, usually O or N	weak, very common
Ionic bonds	oppositely charged ions	weak
van der Waals forces	non-specific, between nearby atoms	weak
Disulphide bridges	two SH-containing cysteine residues	strong, contribute to strength of fibrous proteins

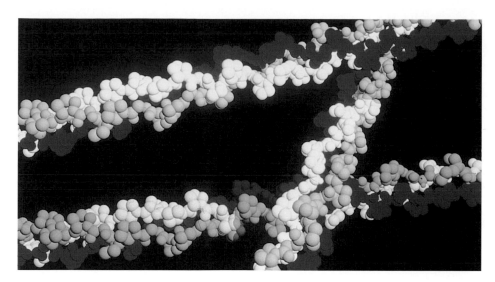

Fig 3.34 A computer-generated model of the protein molecules of collagen, the most abundant protein in the human body

We cannot over-emphasise the importance of tertiary structure for protein function. Functional proteins, such as enzymes and antibodies, must have an exact shape – and sometimes the ability to change shape – to fulfil their role in the organism. Many structural proteins depend on their tertiary structure for strength. The large number of disulphide bridges in keratin, for example, makes body structures such as hair and nails very tough.

If a protein consists of only one polypeptide, the tertiary structure is the final shape of the molecule. If, however, the protein has more than one, it has a further, quaternary level of structure.

QUATERNARY STRUCTURE

Many proteins consist of more than one polypeptide chain and sometimes also have non-protein **prosthetic** groups, often vital to the function of the protein. The quaternary structure refers to the three-dimensional structure, or **conformation**, produced when all the sub-units combine to give the final, active molecule (Fig 3.34).

FIBROUS AND GLOBULAR PROTEINS

The final three-dimensional structure of proteins results in two main classes of protein – fibrous and globular. Fibrous proteins contain polypeptides that bind together to form long fibres or sheets. They are physically tough and are insoluble in water (Fig 3.35).

Globular proteins are usually individual molecules with complex tertiary and quaternary structures. They are spherical, or globular in shape, hence the name. Most are soluble in water and they tend to have a biochemical rather than a structural function (Fig 3.36).

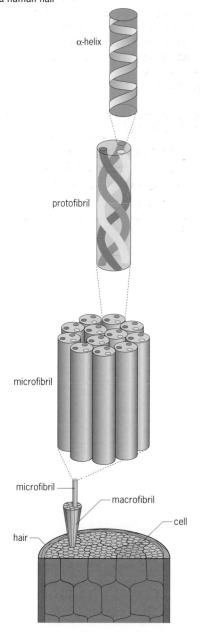

Fig 3.35 This diagram shows how the fibrous protein keratin is arranged inside a human hair

α-helix

protofibril

microfibril

microfibril

macrofibril

cell

hair

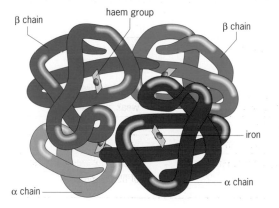

β chain

haem group

β chain

iron

α chain

α chain

Fig 3.36 Haemoglobin is an example of a globular protein. This complex molecule picks up oxygen from the lungs and releases it to respiring tissues. Each molecule consists of four polypeptide chains held together by disulphide bonds. There is a prosthetic group, the haem group, at the centre of the each polypeptide chain. The function of haemoglobin is discussed more fully in Chapter 12

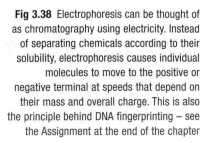

food sample Biuret solution sample goes lilac
(CuSO₄ + NaOH) if protein is present

Fig 3.37 The Biuret test is a simple laboratory test that detects the presence of peptide bonds

HOW STABLE ARE PROTEINS?

As the final shape of globular proteins is maintained by relatively weak molecular interactions such as hydrogen bonds, proteins are very sensitive to temperature increases and other changes in their environment. As the temperature goes up beyond 40 °C, molecular vibration increases and bonds that are holding together the tertiary or quaternary structure break, changing the shape of the molecule. This is known as **denaturation**. Different proteins are denatured at different temperatures. Some begin to be denatured after about 40–45 °C or even below, but many are not totally denatured until 60 °C or even higher. It is an oversimplification to say 'organisms die at temperatures over 44 °C because their proteins become denatured'. In practice organisms die because of a metabolic imbalance, caused when enzymes work at different rates – see Chapter 5.

Proteins can also be denatured by adverse chemical conditions. Chemicals that affect weak bonds alter the overall structure; even a slight change in protein shape can mean loss of function. Some proteins are particularly sensitive to changes in pH.

ANALYSING PROTEINS

We can find out if a sample contains protein using the **Biuret test** (Fig 3.37). More complex biochemical techniques can be used to tell us more about proteins in living systems. We can:

- find out which proteins are present in a mixture, such as plasma,
- work out the exact three-dimensional shape of a protein using techniques such as X-ray diffraction and nuclear magnetic resonance,
- find out which amino acids are present in a particular protein,
- analyse the exact amino acid sequence of a protein,
- use computers to predict the three-dimensional shape of a protein, using only its amino acid sequence.

IDENTIFYING INDIVIDUAL PROTEINS

The proteins present in a body fluid such as blood plasma can be separated from each other by electrophoresis. Each protein carries a particular overall electrical charge. Electrophoresis uses this fact to separate individual proteins, as Fig 3.38 explains.

WORKING OUT THE THREE-DIMENSIONAL SHAPE OF A PROTEIN

X-ray diffraction is useful here. When a beam of X-rays is fired at a crystal of purified protein, the atoms in that protein diffract (bend) the X-rays, producing a specific pattern on a photographic plate. The shape of a protein is not immediately obvious from the information produced, but a trained scientist with a computer can produce an accurate three-dimensional model of the molecule.

Fig 3.38 Electrophoresis can be thought of as chromatography using electricity. Instead of separating chemicals according to their solubility, electrophoresis causes individual molecules to move to the positive or negative terminal at speeds that depend on their mass and overall charge. This is also the principle behind DNA fingerprinting – see the Assignment at the end of the chapter

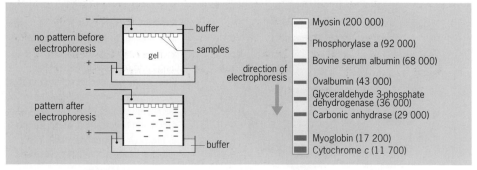

no pattern before electrophoresis

buffer
gel
samples

direction of electrophoresis

pattern after electrophoresis

buffer

Myosin (200 000)
Phosphorylase a (92 000)
Bovine serum albumin (68 000)
Ovalbumin (43 000)
Glyceraldehyde 3-phosphate dehydrogenase (36 000)
Carbonic anhydrase (29 000)
Myoglobin (17 200)
Cytochrome c (11 700)

SCIENCE FEATURE Collagen

Collagen is probably the most widespread structural protein in the human body. It is a fibrous protein that gives strength to tissues such as tendons, ligaments, bone and skin. Fig 3.39 shows the structure of collagen. A single collagen molecule contains three polypeptides – each of about 1000 amino acids – intertwined to form a triple helix. This arrangement has great strength, mainly due to the large number of hydrogen bonds that occur along the length of the polypeptides.

Collagen is secreted in an unassembled form because the instant formation of large fibres would damage the cell that made it. Complete collagen is produced as enzymes act on the individual polypeptides, causing them to twist together to form very long fibres (sometimes several millimetres long). These have the tensile strength of steel and are used to strengthen bone in much the same way as metal rods reinforce concrete. Brittle bone disease (Fig 3.40), is a genetic disorder which causes a fault in the bonding between collagen and the mineral component of bone.

The primary structure of collagen is very regular. It consists of a repeating sequence of glycine and two other amino acids, often proline and hydroxyproline. These amino acids do not cause the chain to gain the normal α-helix or β-sheet structure. Instead, they form long separate chains that allow the collagen triple helix to form. This is a good example of how the primary structure is ultimately responsible for the shape and properties of the whole protein.

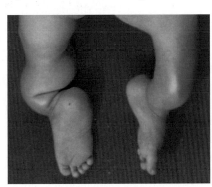

Fig 3.39 The structure of collagen

Fig 3.40 The legs of a small child suffering from 'brittle bone' disease – an unusually severe case. Frequent fractures of the leg result in the deformities

IDENTIFYING WHICH AMINO ACIDS ARE PRESENT

Protease (protein-degrading) enzymes break up the protein and then the constituent amino acids are identified using chromatography or electrophoresis. These techniques do not reveal the sequence of amino acids.

DETERMINING THE AMINO ACID SEQUENCE

The protein is digested into manageable chain lengths of amino acids and each length is analysed in turn. The polypeptide is treated with a chemical that binds to the N-terminal (N-end) amino acid but not to any of the others. The 'tagged' terminal amino acid is removed from the rest of the chain and identified by chromatography. Frederick Sanger, who won a Nobel Prize in the 1950s, was the first scientist to sequence a complete protein (insulin). Today, the process is fully automated and has become a standard technique.

COMPUTER MODELLING

As biochemists accumulate knowledge about the structure of many individual proteins, they produce computer software to help work out the shape of other complete proteins using their amino acid sequences. This is incredibly useful in genetics. Sometimes a new gene is sequenced and no one knows its function. The DNA's sequence can be given to a computer which then works out the sequence of amino acids that would be produced if the gene were expressed in a cell. It can then 'build' the final protein and compare it to other proteins with known functions.

SEE QUESTIONS 7 AND 8

5 NUCLEIC ACIDS

Nucleic acids are so called because they are slightly acidic molecules, and it because was thought originally that they occurred only in the nucleus. The two types of nucleic acid, **DNA** and **RNA**, both contain carbon, hydrogen, oxygen, nitrogen and phosphorus.

THE STRUCTURE OF NUCLEIC ACIDS

Nucleotides are the building blocks of nucleic acids. A nucleotide consists of three units (Fig 3.41):
- a **sugar** (ribose or deoxyribose),
- a **phosphate** group,
- a nitrogen-containing **base**.

As the names imply, **deoxyribo**nucleic acid has nucleotides in which the sugar is **deoxyribose**, while **ribo**nucleic acid contains the sugar **ribose**.

DEOXYRIBONUCLEIC ACID (DNA)

Deoxyribonucleic acid (DNA) is the macromolecule that carries the **genetic code**, the information for making the cell's proteins. Most of the DNA in a eukaryotic cell is in the nucleus (Fig 3.42).

The nucleotides in DNA can contain any one of four nitrogenous (nitrogen-containing) bases: **adenine**, **guanine**, **cytosine** or **thymine**. When DNA **replicates** (copies itself), it makes new strands by adding nucleotides. These are available as free molecules in the cytoplasm. Generally, cells can synthesise their own nucleotides.

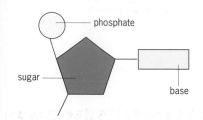

Fig 3.41 The basic structure of a nucleotide

SEE QUESTION 9

? TEST YOURSELF

L If there are 46 chromosomes in a human cell, each with an average of 5 cm uncoiled length, estimate the length of DNA contained within each cell.

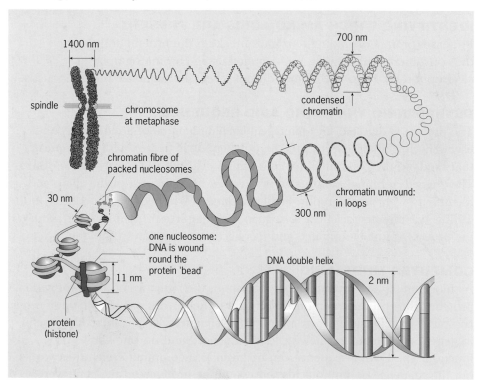

Fig 3.42 Studies of chromosomes have shown that each one is a giant, highly coiled DNA molecule. Its shape is achieved by supercoiling – coils within coils, condensing a huge amount of DNA into a tiny length. Each chromosome can contain 300 000 000 nucleotide units and, unravelled, the DNA within it would be between 5 and 10 centimetres long!

HOW DNA CARRIES THE GENETIC CODE

DNA has two remarkable characteristics:
- It is a store of genetic information.
- It can copy itself exactly, time after time.

Looking at the structure of DNA helps us to understand how it does this.

HOW THE BASES PAIR

Adenine and guanine belong to a group of chemicals called **purines** while thymine and cytosine are **pyrimidines**.

Because of the shape of the two types of molecule, each purine always bonds with only one pyrimidine. So, in DNA, adenine always bonds with thymine, and cytosine with guanine. In RNA, cytosine bonds with guanine and adenine bonds with uracil:

DNA: A=T   RNA: A=U
    G≡C   G≡C

The base pairs are held together by hydrogen bonds. There are two H-bonds between A and T (or U) and three between C and G.

Fig 3.43 shows how the base pairs within DNA fit together to form a double-stranded helix. The sides are formed by alternating sugar–phosphate units, while the base pairs form the cross-bridges, like the rungs of a ladder. Each base pairing causes a twist in the helix and there is a complete 360° turn every 10 base pairs.

There is more about how DNA stores information to act as the genetic code in Chapter 25.

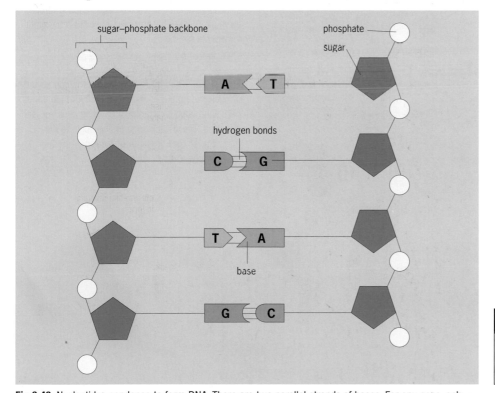

Fig 3.43 Nucleotides condense to form DNA. There are two parallel strands of bases. For any gene, only one strand of the DNA – known as the sense strand – is used to make proteins. The other side serves to stabilise the molecule. The sense strand can alternate between one gene and the next; it is not always on the same side of the molecule

---

✔ REMEMBER THIS

Hint: To remember which bases are pyrimidines and which are purines, concentrate on the letter Y: Thymine and cytosine are pyrimidines.
Adenine and guanine are purines.

SEE QUESTIONS 10 AND 11

? TEST YOURSELF

M If one half of a DNA strand had the base sequence ATCGTTACC, what sequence would the other strand have?

? TEST YOURSELF

N List the four major differences between RNA and DNA.

RIBONUCLEIC ACID (RNA)

Three of the bases in RNA – adenine, guanine and cytosine – are the same as those in DNA. The fourth is different: RNA contains uracil instead of thymine.

RNA molecules are much smaller than DNA molecules. DNA can consist of over 300 000 000 nucleotides; RNA usually consists of a few hundred. RNA is also less stable. DNA molecules are the permanent store for genetic information and last for many years. In contrast, RNA molecules have a short-term function and are easily replaced. There are three forms of ribonucleic acid (RNA) in the cell:

● **Messenger RNA** (mRNA) (Fig 3.44a) can be thought of as a mobile copy of a gene. Small lengths of mRNA are assembled in the nucleus using a single gene within the DNA as a template. When a complete copy of the gene has been produced, the mRNA moves out of the nucleus to the ribosome, where the protein is synthesised according to the code taken from the DNA.

● **Transfer RNA** (Fig 3.44b) is found in the cytoplasm and is a carrier molecule, bringing amino acids to the ribosomes for assembly into a new amino acid chain, according to the order specified on the mRNA code.

● **Ribosomal RNA** makes up part of the ribosome, a small organelle that brings together all the chemicals associated with protein synthesis.

More details about nucleic acids can be found in Chapter 25.

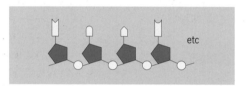

etc

Fig 3.44(a) Messenger RNA is a relatively delicate, short-lived molecule composed of a single strand of nucleotides. It carries the base sequence from the gene to the ribosome and provides a template for protein synthesis

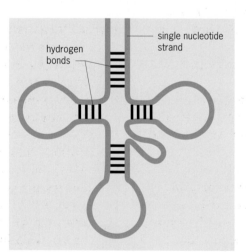

single nucleotide strand

hydrogen bonds

Fig 3.44(b) Transfer RNA is a single folded strand of nucleotides that is stabilised by hydrogen bonds, producing what is often described as a 'clover leaf' structure. The tRNA brings amino acids to the ribosome so that they can be added on to the growing amino acid chain

SUMMARY

When you have finished this chapter, you should know and understand the following:

CARBOHYDRATES

- **Carbohydrates** contain the elements carbon, hydrogen and oxygen. They are the first products made by plants in photosynthesis.
- The term **sugars** describes **monosaccharides** and **disaccharides**. Their names end in the suffix -ose.
- Monosaccharides include **glucose**, **fructose** and **galactose**. These sugars are **isomers**: they have the same formula but their atoms are arranged in different ways. Monosaccharides are linked by **glycosidic bonds**. These are formed by **condensation reactions**.
- Disaccharides, such as **maltose**, **sucrose** (cane sugar) and **lactose** (milk sugar), consist of two monosaccharides linked together.
- Most **polysaccharides** are polymers of glucose. **Starch** is used for storage in plants, **glycogen** is used for storage in animals, and **cellulose** gives strength to plant cell walls.

LIPIDS

- **Lipids** contain carbon, hydrogen, oxygen, often phosphorus and occasionally nitrogen. Most are **non-polar** chemicals and therefore insoluble in water.
- Lipids are used for energy storage, protection and insulation.
- In living things there are two main types of lipid: **triglycerides** and **phospholipids**. Triglycerides are the familiar fats and oils. Phospholipids form cell membranes.
- Fatty acids vary in chain length, and may be **saturated** (with hydrogen) or **unsaturated**. These factors determine the properties of the triglyceride, such as its melting point and viscosity.
- Phospholipids are similar to triglycerides, but phosphoric acid replaces one of the fatty acids. They have a polar head, allowing them to form **bilayers** (membranes) in water.
- **Cholesterol** is a normal constituent of cell membranes.

PROTEINS

- **Proteins** consist of chains of **amino acids** linked by **peptide bonds**.
- There are 20 different amino acids in living things. All have a **carboxylic acid** group and an **amino group** but differ in their **R group**.
- **Fibrous proteins** often join to form large fibrils whose function is to provide strength or produce movement. **Globular proteins** – including enzymes, antibodies and some hormones – are usually individual molecules with a chemical function.
- The **primary structure** of a protein is the sequence of its amino acids.
- The **secondary structure** refers to the patterns and shapes formed within the polypeptide chain, for example, an α-helix.
- The **tertiary structure** refers to the three-dimensional shape of a polypeptide chain, which results from the interactions as the chain folds back on itself. If the protein consists of one polypeptide chain, the tertiary structure refers to the overall shape of the molecule.
- The **quaternary structure** refers to the overall three-dimensional shape of a protein that consists of more than one polypeptide chain.

NUCLEIC ACIDS

- There are two types of **nucleic acid**, **deoxyribonucleic acid** (**DNA**) and **ribonucleic acid** (**RNA**). RNA itself has three forms, **messenger RNA**, **transfer RNA** and **ribosomal RNA**.
- DNA carries the genetic code. Its structure allows it to store information, pass information on to RNA so that proteins can be made, and to copy itself, allowing the genetic code to pass to new cells.
- Messenger RNA copies the genes, to allow them to be used as templates for protein synthesis. Transfer RNA brings amino acids to the ribosome during synthesis. Ribosomal RNA is a major structural component of the ribosome.
- DNA and RNA are composed of **nucleotides**, which themselves contain a sugar, a **phosphate group** and a **nitrogen-containing base**.
- DNA contains the bases A, C, T and G. RNA contains the bases A, C, U and G. A always pairs with T (or U) and C always pairs with G.

? EXAM QUESTIONS

1 The figure shows two possible structures for glucose.

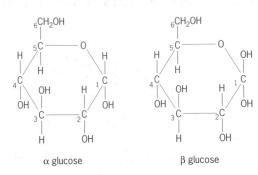

α glucose   β glucose

Both molecules α and β have the same general formula $C_6H_{12}O_6$.

a) **(i)** What is the term used to describe compounds with the same formula but different structure?
(ii) Using the information in the above diagram distinguish between α and β glucose molecules.

b) Indicate, by placing a ring around the appropriate atoms on the diagram above, which atoms are lost when the two glucose molecules join together.

c) Name the bond that is formed when the molecules join together.

d) Name the sugar that is formed after the molecules have joined.

e) Starch and cellulose are both made up from a number of glucose molecules. Suggest **two** differences between the structures of starch and cellulose.

f) Give **one** function of **each** of the molecules in cells.
(i) Starch
(ii) Cellulose.

WJEC Biology Module BI1 June 2001 Q5

2
a) Write a chemical equation to summarise how two molecules of glucose are joined to form maltose. Structural formulae are **not** expected.

b) State **two** ways by which the reaction you have shown in **(a)** could be reversed.

c) **(i)** Maltose and sucrose both have the same empirical formula, but different structural formulae. What biochemical term is used to describe this?
(ii) Maltose is a *reducing sugar* but sucrose is a *non-reducing sugar*. If you were given a solution suspected to contain a mixture of these two sugars and asked to prove their presence, describe the procedure you would use.

OCR Biology Molecules and Life March 2000 Q1

3
a) **(i)** Draw and label a diagram to show the structure of a triglyceride.
(ii) Indicate, with an X on the diagram, a site where hydrolysis takes place.

b) Explain the differences in solubility between triglycerides and the products of their hydrolysis.

c) Suggest why triglycerides release twice as much energy on oxidation compared with an equivalent mass of carbohydrates.

OCR Biology Paper 3 June 2000 Q3

4 The figure shows diagrams that represent lipid and lipid-related molecules.

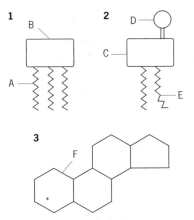

a) **(i)** Which are the **three** most common **elements** in these molecules?
(ii) Give the number of the molecule (1, 2 or 3) which contains a fourth element and name this additional element.

b) **(i)** Which molecule (1, 2 or 3) could most easily be hydrolysed to provide four molecules by boiling with alkali such as sodium hydroxide solution?
(ii) Name the products that would result from this procedure.

c) **(i)** What general name is given to molecules with a structure similar to diagram 3?
(ii) Suggest **one** important function in mammals for the type of molecule.

d) **(i)** Which **one** of the parts labelled **A** to **F** in the figure, would most likely be metabolised in order to provide energy if a person goes on a 'low calorie diet'?
(ii) State how this part is used in the metabolic pathway to provide energy.

❓ EXAM QUESTIONS

The nutritional information label on a tub of 'low fat' margarine states:
Fat … 38.0 g (per 100 g), of which, saturates = 10.6 g, unsaturates = 27.4 g.

e) **(i)** Describe briefly how molecules of **unsaturated** fats are different in structure from molecules of saturated fats.
(ii) How does this difference in chemical structure affect their physical properties?
Lipids as energy providers contain about $592\,000\ kJ\ kg^{-1}$. The equivalent figure for glycogen is $3\,800\ kJ\ kg^{-1}$.
In a human male weighing 70 kg, on average, 20% of body mass is fat stored around internal organs and under the skin.

f) **(i)** Calculate the amount of chemical energy contained in the fat.
(ii) Suggest **one** reason why humans do **not** use carbohydrates as long term energy stores.

OCR Biology Molecules and Life June 2000 Q4

5 The figure shows the structure of a lipid (triglyceride) molecule.

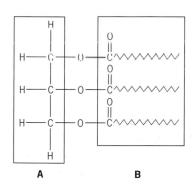

(a) Name the parts **A** and **B**.
(b) **(i)** What type of chemical reaction takes place to form the bonds between **A** and **B**?
(ii) What is the other product of this reaction besides the lipid molecule?
(c) State **one** function of this type of lipid in living organisms.

WJEC Biology Module BI1 January 2001 Q3

6 The figure represents the structure of lysozyme, a protein consisting of a single polypeptide, found in egg white.

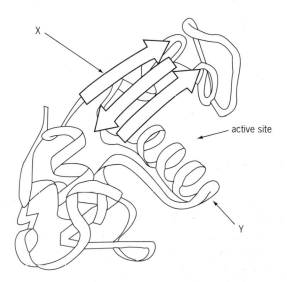

a) State the name given to the shape of this type of protein.
b) State what is meant by the primary structure of a protein.
c) In the regions labelled **X** and **Y two** different types of secondary structure are found.
(i) Identify each type of secondary structure.
(ii) State the type of bonding that is used to stabilise these structures.
d) Explain the importance of the tertiary structure of this protein to its function.

IBO Standard level November 1998 Q3

7 The following list shows some chemical compounds found in living organisms. Match the appropriate letter to one of the following descriptions. Use each letter once only.
A Glucose, **B** Cellulose, **C** Lipid, **D** Protein, **E** Starch.
(i) is destroyed by heating to 65 °C …
(ii) has a storage function in plants …
(iii) is soluble in water …
(iv) has a support function in plants …
(v) does not mix with water …

WJEC Biology Module BI1 January 2001 Q1

? EXAM QUESTIONS

8 The following table lists some features of biological compounds. Complete the table by ticking (✓) in the appropriate column(s) if the feature is found in carbohydrates, lipids or proteins. You can tick one, two or three columns for each feature.

Feature	Carbohydrate	Lipid	Protein
can be saturated or unsaturated			
contains peptide bonds			
contains the elements carbon, hydrogen and oxygen			
can contain disulphide bonds			
cellulose and glycogen are examples			

WJEC Biology Module BI1 June 2001 Q2

9

a) **(i)** In the spaces below, write the **names** of the nitrogenous bases which pair together in DNA.
… pairs with …
… pairs with …
(ii) What type of chemical bond holds these base pairs together in a DNA molecule?

b) There are two types of nitrogenous bases. Name the two types.

c) State **three** ways in which the structure of RNA differs from DNA.

OCR Biology Molecules and Life June 2000 Q1

10 The diagram shows the basic structure of amino acids:

$$R-\underset{\underset{NH_2}{|}}{\overset{\overset{COOH}{|}}{C}}-H$$

a) State what is represented in the diagram by the letter R.

b) Draw a simple diagram to show how two amino acids are linked together.

c) Amino acids are linked together to form polypeptides at special sites in the cytoplasm of both prokaryotic and eukaryotic cells. Compare the sites where polypeptides are formed in prokaryotic cells with the sites in eukaryotic cells.

IBO Standard level November 1999 Q3

11

a) The figure shows the molecular structure of a dipeptide which has been formed by joining two amino acids together.

(i) Give the formula of the chemical group present at position **X** on this molecule.
(ii) Draw a circle round the chemical bond which has been formed as the result of the condensation reaction between the two amino acids.
(iii) The table shows the chemical structure of the R-group in a number of different amino acids.

Amino acid	Structure of R-group
Alanine	CH_3
Asparagine	CH_2CONH_2
Aspartic acid	CH_2COOH
Glutamine	$(CH_2)_2CONH_2$
Serine	CH_2OH

Use the information in the table to name the two amino acids from which the dipeptide was formed.

b) The relative molecular mass (M_r) of a molecule is a measure of its size.
(i) The mean M_r of an amino acid is 110. The M_r of a particular protein is 40 000. Calculate the number of amino acids that make up this protein. Show your working.
(ii) Explain how it is possible for different proteins to have the same relative molecular mass.

AQA BYOI Processes of Life June 2000 Q3

KEY SKILLS **ASSIGNMENT**

DNA FINGERPRINTING

In 1987, Robert Melias made legal history in the UK when he was convicted of rape based on evidence obtained by DNA fingerprinting. This technique – more accurately called DNA profiling – was developed at Leicester University by Alec Jeffreys in 1984 and is proving to be an invaluable tool in forensic medicine.

The principles behind DNA profiling

Every human body cell contains 46 chromosomes. Each one consists of a single elaborately coiled piece of DNA which, if stretched out, can be as long as 5 cm. The structure of DNA can be used to identify individuals.

Between the many genes that occur along the DNA molecule are regions that code for nothing at all. Within this non-coding DNA are hypervariable regions, so called because they vary enormously in length from person to person. Hypervariable regions consist of particular base sequences called core sequences, which are repeated again and again. Different people have different numbers of repeats and so have differently sized hypervariable regions. When these are labelled and separated according to size, a pattern is produced. Each pattern is unique to each individual person and can be used as the basis of DNA profiling (Fig 3.A1).

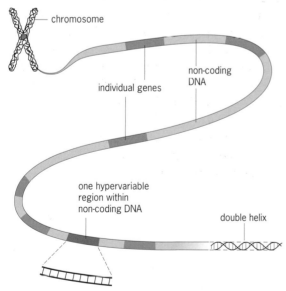

Within a hypervariable region a core sequence is repeated a certain number of times: different people have different numbers of repeats and therefore different sized hypervariable regions

Fig 3.A1 Although each chromosome contains several thousand genes, they only account for about 10 per cent of the length: the rest is non-coding DNA. Within these segments are hypervariable regions. These are unique to each individual and, when they are separated out, the pattern they form provides the basis of DNA profiling

1 **a)** What is a gene?
b) What is the name given to a fault which sometimes occurs when DNA is copied?
c) Faults accumulate in the hypervariable regions more frequently than they do in the genes. Why do you think this is?

Getting a sample of DNA

All body cells contain the same DNA, and so virtually any tissue sample can be used for DNA profiling. The amount of tissue required is very small: $0.05\,cm^3$ of blood, $0.005\,cm^3$ of semen or one hair root!

2 **a)** Which blood cells contain DNA?
b) Why do forensic scientists need a larger sample of blood than of semen?

Processing the DNA and separating the fragments

Once you have a DNA sample, the next stage is to cut the molecule using restriction enzymes, which act rather like molecular scissors (see Chapter 27). These enzymes cut DNA at specific base sequences, giving a complex mixture of DNA fragments, some of which contain the hypervariable regions.

Electrophoresis (see page 56) separates the fragments. The fragment mixture is placed in a trough in some gel. When a current is applied, the negatively charged DNA fragments move towards the positive terminal, or anode. The fragments move at different speeds according to their size: small ones move faster than larger ones. The fragments separate out into bands, as shown in Fig 3.A2 for blood.

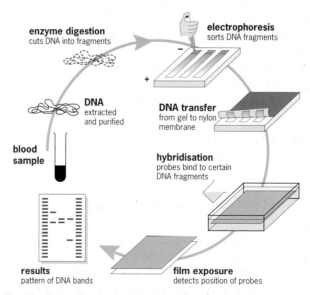

Fig 3.A2 The major steps in the preparation of a DNA profile

KEY SKILLS **ASSIGNMENT**

The bands of DNA are transferred from the gel onto a nylon membrane by Southern blotting, a process which works by capillary action. At this stage the bands are still invisible, and must be stained so that the hypervariable regions can be seen.

Labelling the fragments

Within hypervariable regions are core sequences that are common to all humans. It is the number of times the sequences are repeated which varies from person to person.

Pieces of DNA complementary to these core sequences have been isolated and are produced in bulk for use as genetic probes. They are labelled with a marker chemical, commonly the enzyme alkaline phosphatase which fluoresces (produces light) when a particular substrate is added.

When the probes, complete with enzyme, are added to the DNA sample, they attach to the core sequences, thus marking the hypervariable regions. Excess probe is washed off, substrate for the enzyme is added, and bands that contain hypervariable regions fluoresce. If the blot is exposed to an X-ray film, dark lines appear wherever bands in the blot have emitted light, forming the familiar DNA profile.

3 **a)** Why do the DNA bands fluoresce?
b) Do all the DNA bands fluoresce? Explain your answer.

Case studies

You have covered the basic theory of DNA profiling: now you can put it into practice. Here are two case studies for you to interpret.

CASE 1: WHODUNNIT?

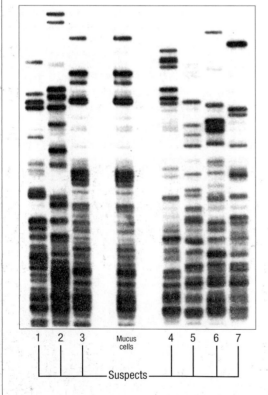

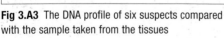

Fig 3.A3 The DNA profile of six suspects compared with the sample taken from the tissues

In 1989, a man robbed a bank at gunpoint while suffering from a particularly heavy cold. The mucus-filled tissues found on the scene provided enough DNA for a DNA profile.

Look at Fig 3.A3.

4 On the basis of the evidence, which suspect robbed the bank?

CASE 2: WHOSE BABY?

Bands in an individual's DNA profile must have come from either the mother or the father. When a child's DNA profile is compared to the mother's, it is easy to see which bands have been inherited. Any bands in the child's profile which did not come from the mother must have come from the father. Use this information to solve the next case study.

In this case study, the mother M is claiming that Mr Z, a famous rock musician, is the father of her 14-year-old daughter C. She is seeking a large maintenance settlement. Predictably, Mr Z denies all knowledge of this and his lawyer demands that a DNA profile be carried out. Fig 3.A4 shows the results.

5 **a)** Which of the daughter's DNA bands were inherited from her mother?
b) Do the remaining bands match those from Mr Z, showing that he is the father?

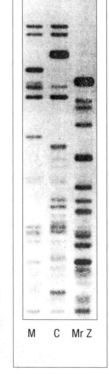

Fig 3.A4 The DNA profiles from Case Study 2, for the mother (M), child (C) and alleged father (Mr Z)

Questions for discussion

6 What would be the advantages and disadvantages of compulsory DNA testing for the whole UK population, so that the police have comprehensive files?

7 Would you like details of your DNA to be available to businesses, in the way that financial records are today? What problems could this cause?

04

Movement in and out of cells

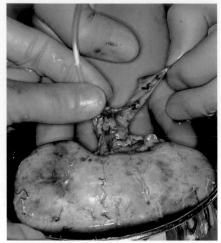

The list of people awaiting an organ transplant is growing. Improving surgery and reducing organ rejection rates has made transplants increasingly successful, leading to a greater demand for kidneys, hearts, livers and other body parts.

Surgeons face problems: an organ can become available in Dundee, but the most suitable recipient might be in Norwich. So, the organ usually has to be transported. But cells in tissues that are separated from their blood supply cannot carry out their normal metabolic functions if they are unable to exchange materials with blood. They run out of oxygen, accumulate waste and can die within minutes.

The solution is to cool the cells so that their metabolic demands are lowered. Organs packed in ice can survive a journey of several hours. During the transplant, the kidney is warmed to normal body temperature and connected without delay to the recipient's blood system.

A kidney waiting to be transplanted. The pale pink colour changes to a much darker red when the new blood supply is connected

1 THE CELL AND ITS ENVIRONMENT

Each living cell is a dynamic system that can exchange a volume of fluid several times bigger than its own volume every second! This happens only because cells are very small: if they were any larger the nutrients could not be taken up quickly enough to satisfy demand and waste could not be expelled efficiently enough to prevent poisoning of the cell. All organisms larger than about 1 mm have developed strategies to increase the exchange of materials, to meet the needs of all their cells.

Material passes into or out of cells by these basic processes:
- **diffusion** (and **facilitated diffusion**),
- **osmosis**,
- **active transport**,
- **endocytosis** and **ectocytosis**.

In this chapter, we look at each of these in detail but, since they all involve the cell surface membrane, we first take a detailed look at this important structure.

2 THE CELL SURFACE MEMBRANE

The **cell surface membrane**, often called the **cell membrane** and sometimes known as the **plasma membrane** or **plasmalemma**, is the boundary between a cell and its surroundings. It has little physical strength, but it plays a vital role in regulating the materials that pass in and out of the cell. Many of the organelles of the eukaryotic cell are also made up of membranes (Fig 4.1) with the same basic structure as the cell surface membrane.

? TEST YOURSELF

A The cell surface membrane of a human cell is about 10 nm thick. Assume that a typical human cell is about 50 μm in diameter. How thick would you make a model cell surface membrane if it had to fit around a model cell that was 5 metres across?
Remember: 1 mm = 1000 μm
               1000 μm = 1 nm

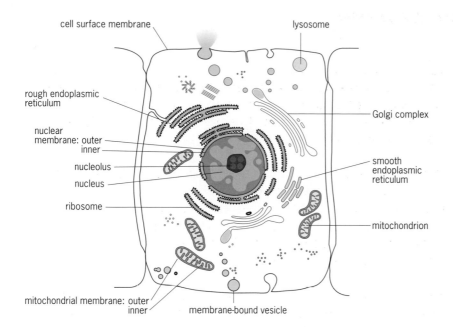

cell surface membrane

lysosome

rough endoplasmic reticulum

nuclear membrane: outer
          inner

nucleolus

nucleus

ribosome

Golgi complex

smooth endoplasmic reticulum

mitochondrion

mitochondrial membrane: outer
                 inner

membrane-bound vesicle

Fig 4.1 Whenever you draw a eukaryotic cell, nearly every line you draw represents a membrane. In this diagram of an animal cell, all the membranes are drawn as red lines

THE STRUCTURE OF THE CELL SURFACE MEMBRANE

As we saw in Chapter 1, the cell surface membrane is basically a double layer of **phospholipid** molecules about 7 to 10 nm thick (Fig 4.2). It cannot be seen with a light microscope and so its structure could not be studied directly until the electron microscope was developed. Instead, early cell biologists deduced its structure by investigating its properties.

THE FLUID MOSAIC THEORY

In 1972, aided by electron microscope studies and evidence from other techniques, Singer and Nicholson put forward the **fluid mosaic theory** (Figs 4.3 and 4.4). Today, most scientists accept it as the model that best represents the structure of living cell membranes.

phospholipid molecule

7–10 nm

head

tail

Fig 4.2 The phospholipids in a membrane are arranged tail to tail, forming a bilayer

Fig 4.3 The fluid mosaic model of the structure of a cell surface membrane. The membrane has been described as 'protein icebergs in a lipid sea'.

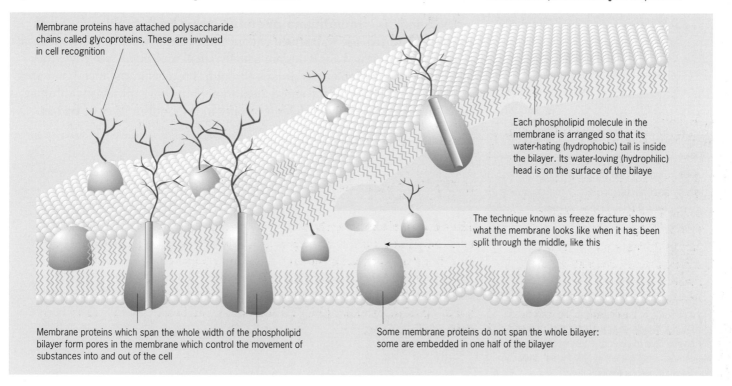

Membrane proteins have attached polysaccharide chains called glycoproteins. These are involved in cell recognition

Each phospholipid molecule in the membrane is arranged so that its water-hating (hydrophobic) tail is inside the bilayer. Its water-loving (hydrophilic) head is on the surface of the bilaye

The technique known as freeze fracture shows what the membrane looks like when it has been split through the middle, like this

Membrane proteins which span the whole width of the phospholipid bilayer form pores in the membrane which control the movement of substances into and out of the cell

Some membrane proteins do not span the whole bilayer: some are embedded in one half of the bilayer

Fig 4.4 A simplified diagram of a cell surface membrane, showing the features described in the fluid mosaic theory. This level of detail should be enough to enable you to explain the essential properties and functions of the membrane

SEE QUESTIONS 1 TO 3

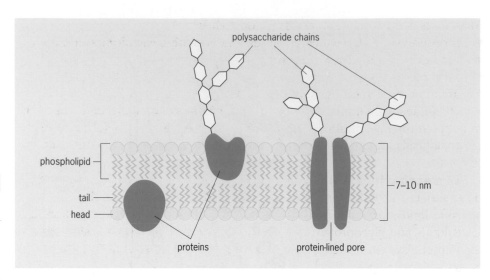

✔ REMEMBER THIS

Water is a **polar molecule** (it has regions of positive and negative charge) and is a solvent for **polar substances** such as sugars, charged ions (Na^+, Cl^-, Ca^{2+}, K^+), B and C vitamins and amino acids. Polar substances do not dissolve in lipid and so can cross a cell membrane only by going through pores.

Most fats, oils and lipids are **non-polar molecules** (they do not have charged regions) and do not dissolve in water. Other non-polar substances (such as vitamins A, D, E and K) can dissolve in lipids and so can cross cell membranes without going through pores.

? TEST YOURSELF

B Explain the terms hydrophilic and hydrophobic. How do these terms relate to a phospholipid molecule?

According to the fluid mosaic model, the cell membrane consists of a double layer of phospholipid molecules, known as a **lipid bilayer**. This is studded with proteins and other molecules. The name **fluid mosaic** is used because the bilayer is a very fluid structure (the phospholipid molecules are in constant sideways motion – a bit like table tennis balls on the surface of a stretch of water) and it contains a 'mosaic' of protein molecules.

THE CHEMICAL MAKE-UP OF CELL MEMBRANES

Cell membranes contain **phospholipids**, **proteins**, **cholesterol** and **polysaccharides**.

Phospholipids are a major constituent of cell membranes. They naturally form membranes in water because they automatically arrange themselves into a bilayer that is virtually impermeable to water and to anything that is water-soluble (see Fig 4.2). So the membrane keeps the cell contents in and it keeps everything else out – except it's not as simple as that. Some substances, mainly water-soluble chemicals, need to get in and out of cells all the time – so how does the cell manage this?

The answer is – **membrane proteins**. These protein molecules act as **hydrophilic pores**, water-filled channels that allow water-soluble chemicals to pass through (Fig 4.5). Pores are usually small and highly selective; they allow only specific molecules or ions through. Proteins in the membrane that form pores usually span the entire membrane, but other proteins with other functions can occur only in the top or only in the bottom layer of **lipids**.

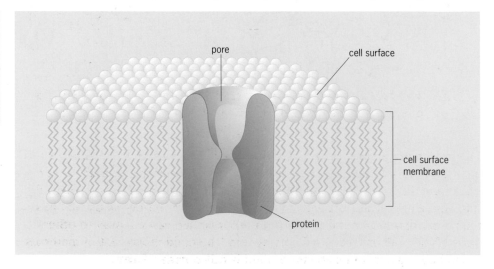

Fig 4.5 Some of the proteins in the membrane form pores. These allow particles that cannot dissolve in lipid to enter and leave the cell. Membrane pores are not simply 'holes'; they can control what passes through them

Membrane proteins have several functions. They:

- form pores through which water and water-soluble chemicals can pass,
- act as carriers in active transport,
- form receptor sites for hormones,
- are important in cell recognition.

Cell surface membranes can also contain **cholesterol** molecules, which fit in between the phospholipids (see Fig 4.15). Cholesterol is a vital constituent of animal cells, despite being notorious for its connection with heart disease (Fig 4.6).

Polysaccharides, branched polymers of simple sugars, stick out from the outer surface of some membranes like antennae (see Figs 4.3 and 4.4). They attach to lipids, forming **glycolipids**, or to proteins, forming **glycoproteins**. Glycolipids and glycoproteins help cells to recognise each other – allowing the immune system to tell the difference between body cells and invading bacteria, for example.

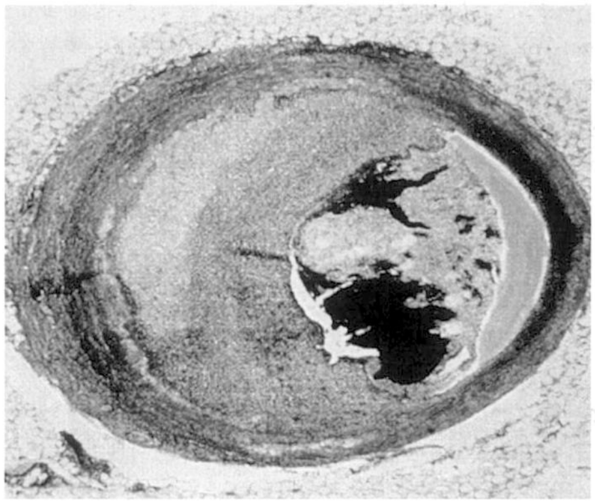

Fig 4.6 A narrowed artery. High cholesterol levels are associated with the development of coronary heart disease. The fatty deposit, atheroma, builds up under the endothelium of the blood vessel, narrowing the lumen

3 DIFFUSION

Diffusion is basically a 'mixing of molecules'. In a gas or liquid the molecules or ions move continuously, randomly bumping into each other and changing direction. In this way, particles tend to diffuse, or spread, so that they are evenly spaced rather than being concentrated in one place (Fig 4.7).

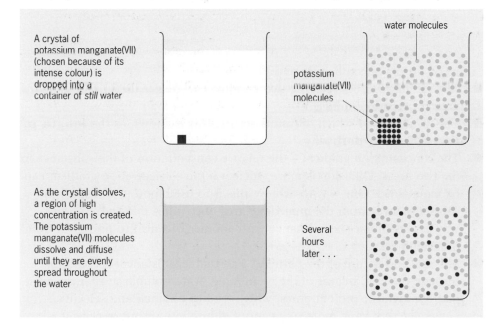

A crystal of potassium manganate(VII) (chosen because of its intense colour) is dropped into a container of *still* water

water molecules

potassium manganate(VII) molecules

As the crystal disolves, a region of high concentration is created. The potassium manganate(VII) molecules dissolve and diffuse until they are evenly spread throughout the water

Several hours later . . .

Fig 4.7 If you drop a crystal of potassium manganate(VII) (chosen for its visibility) into still water, it dissolves and the ions diffuse until evenly spread. This takes several hours (depending on temperature) but can be speeded up by stirring. Diffusion is a passive process which requires no input of energy: it does however depend on the kinetic energy (energy of movement) of the molecules in a gas or a liquid

? TEST YOURSELF

C Why does diffusion apply to particles in liquids and gases but not solids?

A useful working definition of diffusion is:

Diffusion is the movement of particles within a gas or a liquid from a region of high concentration to a region of lower concentration.

Diffusion mainly moves substances over short distances. It is too slow to move substances efficiently over distances much greater than a fraction of a millimetre. In the mammalian lung, for example, oxygen diffuses through the thin epithelium of the alveolus and into the blood, a journey of usually less than a hundredth of a millimetre (10 μm). It is then carried away to other parts of the body by the circulatory system. The rapid movement of materials around an organism in a stream of fluid is called **mass flow**.

DIFFUSION AND ENERGY

The difference in concentration of a substance between two regions is called a **concentration gradient**. Particles that are free to move have **kinetic energy**. The region of a gas or liquid with the highest concentration of particles of a particular substance has the highest kinetic energy for that substance. Particles move down a concentration gradient by diffusion, until they are spread evenly.

The most important point to remember is that diffusion is a **passive process**: it requires no input of energy from the cell. It follows that movement of a substance against a concentration gradient – **active transport** (see page 83) – requires energy.

FACTORS THAT AFFECT THE RATE OF DIFFUSION

The rate at which molecules of a substance diffuse from one region to another has a great influence on the design of cells, organs and whole organisms (Fig 4.8).

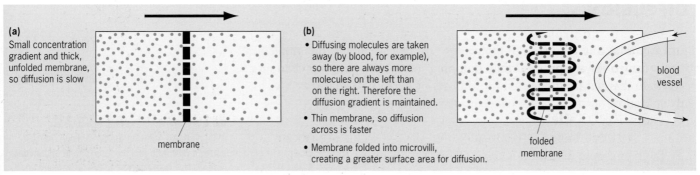

(a)
Small concentration gradient and thick, unfolded membrane, so diffusion is slow

membrane

(b)
• Diffusing molecules are taken away (by blood, for example), so there are always more molecules on the left than on the right. Therefore the diffusion gradient is maintained.
• Thin membrane, so diffusion across is faster
• Membrane folded into microvilli, creating a greater surface area for diffusion.

blood vessel

folded membrane

Fig 4.8 A schematic diagram to show some of the features which increase the rate of diffusion. An efficient exchange of materials requires a large surface area, maintenance of a diffusion gradient and a thin cell surface membrane. Increasing the temperature would also speed up diffusion, but this is not an option in living systems

Several factors affect the rate of diffusion:

● The surface area between the two regions. The greater the surface area, the greater the rate of diffusion.
● The distance over which diffusion occurs. This is known as the **length of the diffusion pathway**.
● The concentration gradient – the relative concentration of the substance in the two areas. Diffusion is more efficient if the concentration gradient can be maintained. One way to achieve this is to use blood to transport the substance away from the immediate area once it has diffused. Another way is to combine the substance with another chemical to prevent it from diffusing back.
● The size and nature of the particles. Fat-soluble substances can diffuse through the lipid bilayer of the membrane. Water-soluble substances must pass through the protein pores, which tend to be small and selective. Generally, very large molecules cannot diffuse into or out of cells at all.
● The temperature at which the process takes place. At higher temperatures, molecules have more kinetic energy and so diffuse more quickly.

CALCULATING THE RATE OF DIFFUSION

The rate at which substances diffuse can be estimated from a simple formula that takes into account the factors that affect diffusion. Rate of diffusion is proportional to:

$$\frac{\text{surface area} \times \text{concentration difference}}{\text{length of diffusion pathway}}$$

This relationship is known as **Fick's law**. For efficient diffusion, the values on the top line of this equation should be as large as possible and the value on the bottom line should be as small as possible.

DIFFUSION IN THE MAMMAL

Diffusion is not fast enough to be effective over distances greater than a few millimetres. The body of a mammal has many organs that are adapted for the exchange of materials – the lungs, intestines and kidneys are obvious examples. However, these organs are of little use without a circulatory system that transports the exchanged materials to all other areas of the body. The blood system is an example of what is known as a **mass flow system**.

The main function of this system is to bring materials to within diffusing distance of living cells, and to take away cell products. In the human body, no cell is more than a few micrometres away from a capillary (see Chapter 13).

4 FACILITATED DIFFUSION

Some substances enter and leave cells much faster than you would expect if only diffusion occurred. We now know that some membrane proteins assist, or **facilitate**, the diffusion of some substances across the cell membrane. Two types of protein are responsible for **facilitated diffusion**:

● Specific **carrier proteins** take particular substances from one side of the membrane to the other.
● **Ion channels** are proteins that open and close to control the passage of selected charged particles.

CARRIER PROTEINS

Until the 1970s, cell biologists thought carrier proteins worked by rotating within the membrane, like turnstiles. Newer research points to a different explanation (Fig 4.9). As soon as the diffusing molecule binds to the carrier protein, the protein undergoes a change in shape so that the diffusing molecule ends up at the other side of the membrane, where it is released.

Like diffusion, facilitated diffusion involves movement *down* a concentration gradient and requires no *input* of metabolic energy.

ION CHANNELS

Ion channels are proteins with a central 'hole' lined with polar groups (Fig 4.10). Ion channels facilitate the diffusion of charged particles such as Ca^{2+}, Na^+, K^+ and Cl^- ions. Many are **gated**, so can open or close. Cells use ion channels to control the movement of ionic substances between themselves and other cells, and to regulate the ionic composition of their cytoplasm.

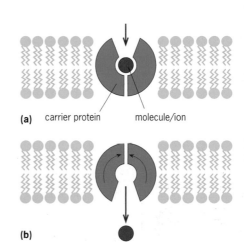

? TEST YOURSELF

E Name some organs where exchange of materials is the main function. (Note: Think about plants too.)

Fig 4.9 Facilitated diffusion using a carrier protein. The diffusing molecule interacts with the carrier protein, causing a change in shape which 'squeezes' the molecule through the channel

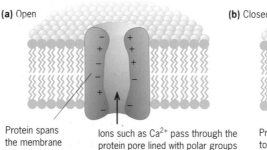

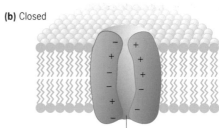

(a) Open — Protein spans the membrane — Ions such as Ca^{2+} pass through the protein pore lined with polar groups

(b) Closed — Protein changes shape, so pore becomes too to allow ions through

Fig 4.10 An ion channel. Because they are charged particles, ions cannot easily pass through the non-polar lipid bilayer. Specific membrane proteins form polar pores through which ions can pass. These channels are usually specific for one type of ion and can open and close according to the needs of the cell. When fully open, over 1 million ions per second can flow through a single channel

---

! SCIENCE FEATURE Diabetes and carrier proteins

The concentration of our blood sugar is kept relatively constant by controlling how much glucose passes from the blood into cells. After a meal, when blood glucose levels are high, the hormone insulin is released. This hormone activates a mechanism in the cell membrane that facilitates the diffusion of glucose into the cell, so lowering the blood sugar concentration.

One form of diabetes in humans is caused by a gene mutation that alters the structure of glucose carrier proteins in cell surface membranes. An affected person cannot get enough glucose into their cells. Unlike diabetes that results from the inability to make insulin (see Chapter 14), this form of the disease is difficult to treat, because injecting insulin has little effect.

A difference in concentration of ions may lead to a net positive or negative charge in a particular region. This type of concentration gradient is called an **electrochemical gradient**. Charged particles move towards regions of opposite charge. This is important in the process of transmission of a nerve impulse (see Chapter 19) and in the process by which red blood cells exchange materials with the tissues (see Chapter 13).

5 OSMOSIS

In biology, we usually talk about the diffusion of substances that are dissolved in water. But what about the water molecules themselves? Does water diffuse down a diffusion gradient? The answer is yes, and the diffusion of water is known as **osmosis**.

A working definition of osmosis is:

> **Osmosis is the diffusion of water only. It is the net movement of water molecules from a region of their higher concentration to a region of their lower concentration, through a partially permeable membrane.**

Water, water everywhere, and not a drop to drink...The concept of osmosis may be easier to grasp if you think of a real life-and-death situation. Imagine a ship-wrecked survivor floating in a dinghy on the open sea, the Sun burning down on him for 15 hours a day, no land or rescue in sight. After a few hours, he gets so thirsty, he gives in and drinks sea water. However, because sea water contains salt, its sodium chloride concentration is about three times that found in human blood. The survivor is already dehydrated, and the salt causes water to move out of his blood and into his stomach by osmosis, causing further dehydration of his body cells. Death from dehydration is more likely than if he had drunk nothing.

IMPORTANT FACTS ABOUT WATER

Understanding some of the properties of water is the key to understanding osmosis, so before going into detail about this process, we need to look at what happens when a substance dissolves in water.

As you can see from the information in the Extension box on water (see page 76), water molecules are polar – they carry a tiny positive and a tiny negative charge. In pure water, all the water molecules present are able to move around freely, not associating with each other or with anything else – think of them as 'free' water molecules. However, when a substance is placed in the water that is soluble, things change. As each of the solute molecules dissolves, it becomes charged and attracts the charges on the water molecules. It soon becomes surrounded by a shell of water molecules. When this happens, the water molecules that form the 'shell' are no longer 'free' as they were before: they have been 'tied up' by the solute molecules. This reduces the concentration of 'free' water molecules in the solution.

AN OVERVIEW OF OSMOSIS

Now think about what happens if you have two solutions, each with different concentrations of solute, on either side of a semi-permeable membrane. Fig 4.11(a) shows what happens. There are more solute molecules in solution B (on the right of the membrane) than in the solution A (on the left). In solution B, more water molecules are 'tied up' in shells around the solute molecules, so there are fewer water molecules in solution B than in solution

A. There is therefore a concentration difference between the two solutions, both of solute molecules and of water molecules.

The pores in the membrane are too small to allow solute molecules through, so their concentration has to stay the same. What can change is the concentration of 'free' water molecules on either side of the membrane. These water molecules can pass through the pores and water can diffuse down its concentration gradient. Water molecules move from the left, where there is a higher concentration of water molecules, to the right, where there is a lower concentration of water molecules (Fig 4.11b). This is osmosis.

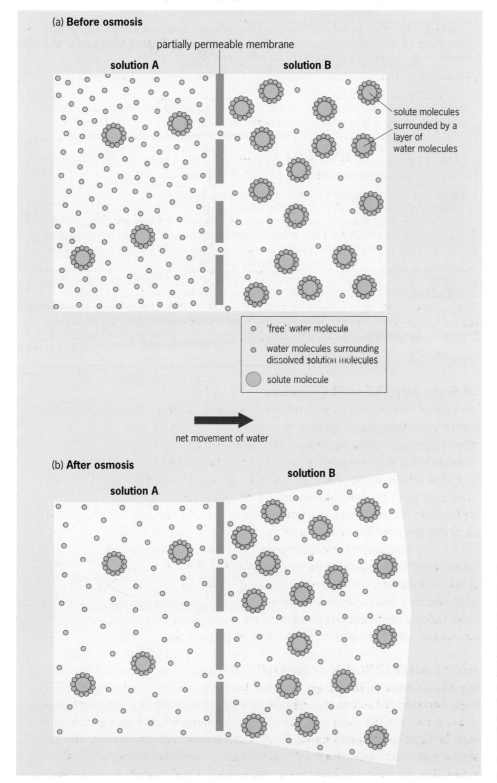

(a) **Before osmosis**

partially permeable membrane

solution A   **solution B**

solute molecules surrounded by a layer of water molecules

○ 'free' water molecule

◦ water molecules surrounding dissolved solution molecules

◯ solute molecule

net movement of water

(b) **After osmosis**

solution B

solution A

Fig 4.11(a) A 'weak' solution is one with a low concentration of solute molecules but a high concentration of water molecules. It has more 'free' water molecules than the 'concentrated' solution on the right. In osmosis, water molecules move from a region of their higher concentration to a region of their lower concentration – in this case from the solution on the left to the solution on the right
Fig 4.11(b) As osmosis occurs, the number of water molecules on either side of the membrane starts to equalise. Eventually, both solutions will have the same concentration of solute molecules relative to water. Because water has moved from left to right, the solution on the left will have a much lower volume than the one on the right

▶ EXTENSION All about water

There is no life without water: life almost certainly originated in water. Most biochemical reactions take place in solution and all organisms that live on land can do so only because they have their own internal aquatic environment. Some organisms can survive dehydration as spores or seeds, but these structures simply allow a state of dormancy, which ends when water becomes available once again. The bodies of all organisms contain a high percentage of water. Even that most complex of organs, the human brain, is 85 per cent water.

THE PROPERTIES OF WATER

All cells are tiny compartments of fluid. Living cells must exchange materials with the surroundings, and this can only take place in solution. So, as well as being full of water, cells are also surrounded by it. Water has unique physical and chemical properties, many of them significant in biology. Some of the most important are summarised in Table 4.1.

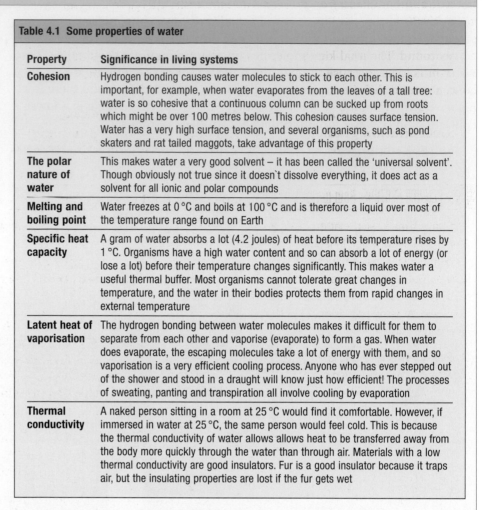

Table 4.1 Some properties of water

Property	Significance in living systems
Cohesion	Hydrogen bonding causes water molecules to stick to each other. This is important, for example, when water evaporates from the leaves of a tall tree: water is so cohesive that a continuous column can be sucked up from roots which might be over 100 metres below. This cohesion causes surface tension. Water has a very high surface tension, and several organisms, such as pond skaters and rat tailed maggots, take advantage of this property
The polar nature of water	This makes water a very good solvent – it has been called the 'universal solvent'. Though obviously not true since it doesn't dissolve everything, it does act as a solvent for all ionic and polar compounds
Melting and boiling point	Water freezes at 0 °C and boils at 100 °C and is therefore a liquid over most of the temperature range found on Earth
Specific heat capacity	A gram of water absorbs a lot (4.2 joules) of heat before its temperature rises by 1 °C. Organisms have a high water content and so can absorb a lot of energy (or lose a lot) before their temperature changes significantly. This makes water a useful thermal buffer. Most organisms cannot tolerate great changes in temperature, and the water in their bodies protects them from rapid changes in external temperature
Latent heat of vaporisation	The hydrogen bonding between water molecules makes it difficult for them to separate from each other and vaporise (evaporate) to form a gas. When water does evaporate, the escaping molecules take a lot of energy with them, and so vaporisation is a very efficient cooling process. Anyone who has ever stepped out of the shower and stood in a draught will know just how efficient! The processes of sweating, panting and transpiration all involve cooling by evaporation
Thermal conductivity	A naked person sitting in a room at 25 °C would find it comfortable. However, if immersed in water at 25 °C, the same person would feel cold. This is because the thermal conductivity of water allows allows heat to be transferred away from the body more quickly through the water than through air. Materials with a low thermal conductivity are good insulators. Fur is a good insulator because it traps air, but the insulating properties are lost if the fur gets wet

WHY IS THE WATER MOLECULE SO STICKY?

The water molecule consists of two atoms of hydrogen and one atom of oxygen. This gives water a relative molecular mass of 18. Many compounds of this size are gases:

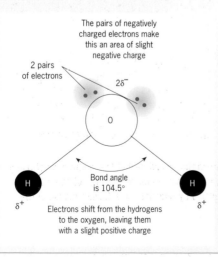

The pairs of negatively charged electrons make this an area of slight negative charge

2 pairs of electrons

$2\delta^-$

O

Bond angle is 104.5°

δ^+ δ^+

Electrons shift from the hydrogens to the oxygen, leaving them with a slight positive charge

Fig 4.12 Water is a polar molecule: it has distinct areas of positive and negative charge

water is a liquid because its molecules are 'sticky' – they cling to each other. Electrons are negatively charged. Since opposite charges attract and like charges repel, the lone pairs of electrons of the oxygen atom effectively repel the electrons of the hydrogen – oxygen bonds. This results in a 'v-shaped' molecule, as Fig 4.12 shows. The region around the oxygen atom has a slight negative charge, while the two hydrogen atoms have a slight positive charge. Molecules with different areas of positive and negative charge are said to be **polar**.

Attractive forces exist between water molecules: the positive charge on the hydrogen atoms of one molecule attracts the negative charge

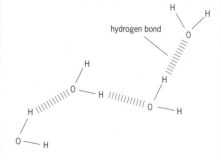

Fig 4.13 Weak attractive forces exist between the positive areas of one water molecule and the negative areas of adjacent ones. Many of the unique properties of water are due to this hydrogen bonding

on the oxygen atoms of another molecule. These attractive forces are called hydrogen bonds (Fig 4.13). In effect, each water molecule acts as a 'mini magnet'. As well as 'sticking' to each other, water molecules also cling to other charged particles.

As we saw in the section on diffusion, movement of particles occurs down a concentration gradient, *from a region of high kinetic energy to a region of lower kinetic energy*. In Fig 4.11, the total kinetic energy of the water molecules in the solution on the left is high because most of them are free to move around. The total kinetic energy of water molecules in the solution on the right is lower because fewer of them are free to move around.

So, a second definition of osmosis is:

Osmosis is the movement of water molecules down a concentration gradient of water, from a region of high kinetic energy to a region of lower kinetic energy.

! SCIENCE FEATURE How water molecules cross cell membranes

We have already said that the cell surface membrane is impermeable to water. In fact, it is not 100 per cent impermeable. A tiny amount is able to pass through a lipid bilayer. But recent studies have shown that water passes through purified phospholipid bilayers, with all the proteins removed, at rates approximately 100 to 1000 times faster than you might expect.

The exact reason for this is not known, but scientists think the constant sideways motion of the phospholipids, coupled with the flexing of the fatty acid tails, creates temporary holes (Fig 4.14)

through which the water molecules can slip. These holes appear in only one half of the membrane at a time, so the water molecules 'wait' at the halfway point until a hole appears in the other side, rather like crossing a busy road, one lane at a time.

Membranes with a high proportion of cholesterol are less permeable to water, and to simple ions such as sodium and chloride. It could be that cholesterol has a stabilising effect on the phospholipids. By slotting in between them, the cholesterol molecules minimise the creation of temporary holes (Fig 4.15).

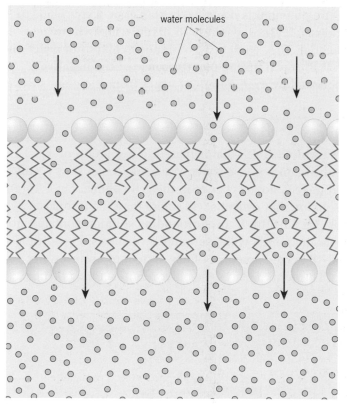

Fig 4.14 A possible explanation for the fact that lipid bilayers are permeable to water. Temporary holes, created by the random sideways movement of the lipid molecules, act as channels through which water can pass

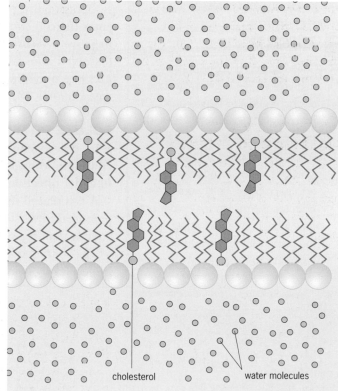

Fig 4.15 Cholesterol is an important constituent of the membranes of many types of cell. The effect of cholesterol is to reduce leakage of water and ions through the lipid bilayer, probably by reducing the sideways motion of the lipid molecules

OSMOSIS IN LIVING CELLS

All cells contain cytoplasm, a complex solution separated from its surroundings by a partially permeable membrane. So, all cells have the potential to gain or lose water by osmosis. Whether they do so or not depends on the **water potential** acting on the cell.

THE CONCEPT OF WATER POTENTIAL, Ψ

When we look at how osmosis affects cells and organisms, there are two factors to consider:

● The solute concentrations (inside and outside the cell). These are the **solute potentials**.

● Any **hydrostatic** forces. These are the **pressure potentials**.

Solute concentrations and hydrostatic forces combine to give a measurement of the overall tendency of a cell or system to gain or lose water by osmosis. We call this overall tendency the water potential.

The relationship between water potential, solute potential and pressure potential is:

$$\text{Water potential} = \text{solute potential} + \text{pressure potential}$$
$$\Psi = \Psi_s + \Psi_p$$

This tells us that the overall tendency for water to enter or leave a cell usually depends on the relative sizes of the osmotic force drawing water into the cell and the hydrostatic pressure keeping water out.

Occasionally, pressure potential is negative. In this case, water tends to enter the cell continuously. **Xylem vessels** in plants usually have a negative pressure potential. They contain a continuous column of water and dissolved minerals that is drawn up the plant because of the negative pressure potential generated by evaporation of water from the plant's leaves (see Chapter 13).

SOLUTE POTENTIAL, Ψ_s

If no other factors are involved, a cell that is placed in a solution of equal concentration will neither gain nor lose water. Water molecules pass equally in either direction and there is no *net* change in the cell volume: Fig 4.16(a).

A cell placed in a solution that contains less solute than the cytoplasm will gain water by osmosis: Fig 4.16(b). If the surrounding solution contains more solutes than the cytoplasm, the cell will lose water by osmosis: Fig 4.16(c).

The solute potential, Ψ_s, is the potential of any system to absorb water molecules by osmosis. This term replaces the older term osmotic potential,

REMEMBER THIS

Solute potential is a negative scale. For example, a relatively weak solution of sucrose may have a solute potential of –200 kPa while a more concentrated solution may have solute potential of –500 kPa. Pure water has the highest possible solute potential: zero.

REMEMBER THIS

If you have a problem with negative scales, as many people do, compare it to temperature. Of two solutions, one at –40 °C and the other at –60 °C, it is obvious which is the colder. If the two solutions were placed in contact with each other, heat would pass from the warmer to the colder until they were both about equal. It's the same with the movement of water in osmosis.

water   solute molecules   cell surface membrane

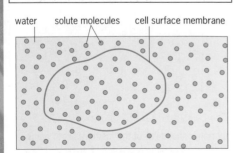

(a) When the solution surrounding the cell contains the same amount of solute as the cytoplasm (ie is isotonic), the cell neither gains nor loses water

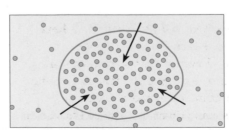

(b) When the cell's cytoplasm contains more solutes than the surroundings (ie is **hypertonic**), the cell gains water by osmosis

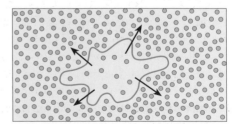

(c) When the cell's cytoplasm contains fewer solutes than the surroundings (ie is **hypotonic**), the cell loses water by osmosis

Fig 4.16 Factors that determine whether water enters or leaves a cell by osmosis

which you will, nevertheless, come across in some books. Solute potential is measured in units of pressure, **kilopascals, kPa**.

If the two solutions described above were separated by a partially permeable membrane, water would move from the more dilute solution into the more concentrated one. Water will move from a region of high water potential to a region of lower water potential. This is from the solution with the water potential of –200 kPa to the solution with the more negative value of –500 kPa.

PRESSURE POTENTIAL, Ψ_P

In addition to solute potential, we must also consider the effects of physical pressure on a cell that is subject to osmotic forces. This type of pressure is often called **hydrostatic pressure** and it arises in a cell that has a rigid cell wall, such as a plant cell. If a cell is surrounded by a less concentrated solution, water enters by osmosis. The extra water cannot increase the volume of the cell because of the rigid cell wall – instead it causes the cell contents to press against the inside of the cell wall, creating a physical pressure. A point is reached when this pressure is great enough to prevent any more water entering the cell (Fig 4.17). The physical pressure has balanced the osmotic pressure.

The **osmometer** can demonstrate the balance between osmotic and physical force (Fig 4.18). Water passes through the membrane into the sucrose solution until pressure from the weight of the column of water halts it. At this point an equilibrium is reached and the osmotic and physical forces are equal and opposite.

> ### Pressure potential, Ψ_p, describes any hydrostatic (physical) forces that act on cells.

In a plant cell, a pressure potential is created by the cellulose cell wall resisting expansion of the cell. Animal cells are also subject to pressures – those caused by the pumping of the heart, or by movement and posture that put pressure on particular areas of tissue, for example.

Since both plant and animal cells can be affected by hydrostatic pressure, the term *pressure potential* has replaced the terms **wall pressure** and **turgor pressure**, which meant exactly the same thing but referred solely to studies of cells with walls.

OSMOSIS IN PLANT CELLS

If you put a plant cell in a concentrated sucrose solution that has a lower water potential than the cell's cytoplasm and vacuole, water leaves the cell by osmosis. The vacuole shrinks and eventually the **protoplast** (the living part of the cell) becomes detached from the cell wall. The point at which the protoplast is just about to become detached is called **incipient plasmolysis**. When it has become detached, the cell is said to be **plasmolysed** (Fig 4.19).

If the plasmolysed cell is placed in solution of higher water potential (eg distilled water), it absorbs water by osmosis and begins to swell. The vacuole and cytoplasm enlarge until they begin to press against the cell wall. An

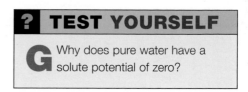

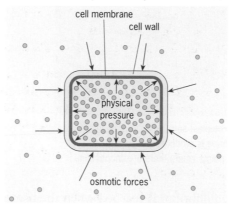
Fig 4.17 This plant cell should gain water by osmosis but water cannot get in because the rigid cell wall resists any further increase in volume

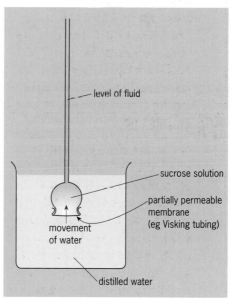
Fig 4.18 The osmometer is a simple device that is used to demonstrate osmosis and to estimate to the solute potential of a solution. The greater the solute potential, the higher the liquid rises up the tube

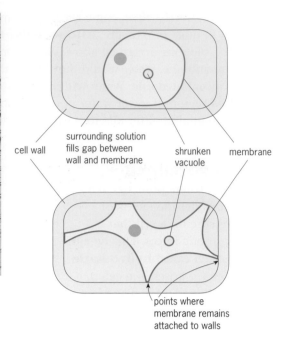

Fig 4.19 When a plant cell becomes plasmolysed, the protoplast sometimes shrinks into a sphere in the centre of the cell. At other times, a 'star' shape is produced because the cell surface membrane sticks to the inside of the wall at the points where there are connections with other cells

equilibrium is reached when the osmotic forces drawing water in are balanced by the wall pressure resisting further expansion. At this point the cell is said to be **turgid**, and the water potential is zero. In a normal healthy plant, most of the cells are turgid. In a herbaceous (non-woody) plant the straightness of the stem is due largely to **turgor pressure** (Fig 4.20).

Fig 4.20 When a plant loses more water than it gains, the solution surrounding the cells becomes hypertonic. The cells lose water by osmosis and turgor pressure is lost – the obvious consequence of this is that the plant wilts. A wilted plant can be revived if water becomes available within a short time, but a point is reached after which recovery is not possible

SEE QUESTIONS 4 AND 5

OSMOSIS IN ANIMAL CELLS

Animal cells have no cell wall, just a membrane. If you place an animal cell in distilled water, it absorbs water by osmosis and swells up. As membranes have virtually no mechanical strength, the animal cell often bursts.

Fig 4.21 shows what happens to red blood cells placed in **hypertonic**, **isotonic** and **hypotonic** solutions. When blood plasma becomes hypertonic to the cytoplasm of the red blood cells, as it does during severe dehydration, water is lost from the red blood cells, which shrink and become crinkled, or **crenated**. At the other extreme, hypotonic plasma causes red blood cells to swell and burst, leaving 'ghosts' of cell surface membrane. This process in known as **haemolysis**: literally 'blood splitting'.

❓ TEST YOURSELF

H Why is it incorrect to say that 'sea water is a hypertonic solution'?

(a)

red blood cell

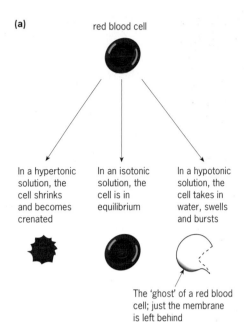

In a hypertonic solution, the cell shrinks and becomes crenated

In an isotonic solution, the cell is in equilibrium

In a hypotonic solution, the cell takes in water, swells and bursts

The 'ghost' of a red blood cell; just the membrane is left behind

(b)

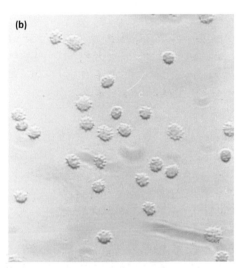

(c)

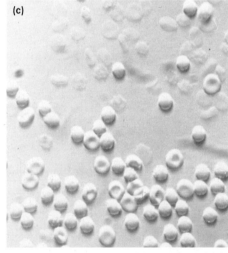

Fig 4.21(a) All animal cells are prone to the same effects as these red blood cells when placed in solutions of different concentrations
(b) Red blood cells become crenated in a hypertonic solution
(c) Red blood cells take in water and burst in a hypotonic solution (some red cells have not yet burst)

✔ **REMEMBER THIS**

The following terms are useful when describing osmosis in animal cells.
Hypertonic: a greater solute concentration.
Hypotonic: a lower solute concentration.
Isotonic: an equal solute concentration.
These are relative terms and so can only be used to compare solutions. For example, sea water is hypertonic to human blood plasma, distilled water is hypotonic to human blood plasma.

? **TEST YOURSELF**

Suggest why the marine amoeba does not posses a contractile vacuole?

! **SCIENCE FEATURE** The amoeba and its contractile vacuole

The amoeba (Fig 4.22) is a single-celled organism (a protoctist) found in freshwater and marine environments. Some species can infect the human intestine, causing amoebic dysentery.

The cytoplasm of the freshwater amoeba is hypertonic to the surrounding water, and so absorbs water by osmosis. To counteract this, the amoeba has a contractile vacuole – an organelle that fills up with water and periodically expels its contents, allowing the volume of the amoeba to remain relatively constant.

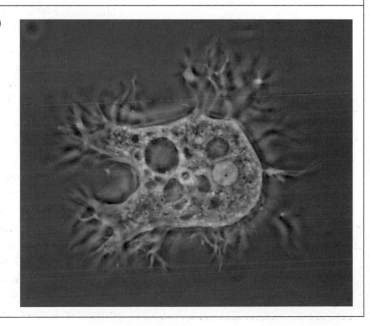

Fig 4.22 The freshwater amoeba constantly gains water by osmosis. The organism actively pumps water into the contractile vacuole. When the vacuole is full, the water is emptied to the outside, rather like a mini bladder

WATER POTENTIAL CALCULATIONS

Look at Fig 4.23. In a plasmolysed cell, the protoplast is not pressing against the cell wall, so the pressure potential is zero. The tendency for water to enter or leave the cell depends entirely on the relative solute concentrations inside and outside the cell. In this situation, water potential is equal to solute potential.

A turgid cell is in equilibrium: its volume is constant. The solute potential (a negative value) is equal to the pressure potential (a positive value), and so they cancel each other out. Thus, the water potential of a turgid cell is zero.

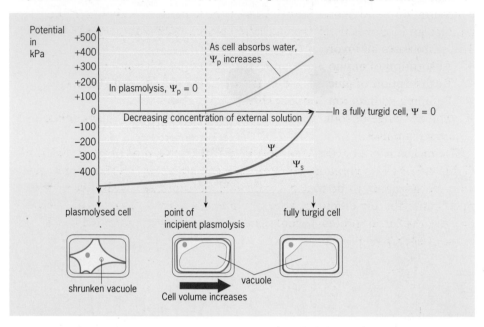

Fig 4.23 As the solution surrounding a plant cell becomes less concentrated (from left to right in the graph), the cell absorbs water until it becomes fully turgid

Now, let's look at some worked examples:

● **EXAMPLES**

Q Use the water potential equation to predict movement of water between these two cells:

Cell A Ψ_s = −500 kPa
          Ψ_p = 200 kPa
Cell B Ψ_s = −600 kPa
          Ψ_p = 100 kPa

A This question is asking you to work out the water potential of the two cells. If we put values given into the equation:

Water   =   solute   +   pressure
potential   potential   potential
Ψ = Ψ_s + Ψ_p

For cell A: −500 + 200 = −300
For cell B: −600 + 100 = −500

Now, which one is closest to zero (pure water)? Water always passes from the highest value (closest to 0) to the lowest (more negative), so in this case water will pass from A to B.

Q A turgid plant cell was found to have a Ψ_s value of −350 kPa.
a) What was the Ψ?
b) What was the Ψ_p?

A You may think that you have been given only one value and that it is impossible to work out the other two, but the clue comes in the word *turgid*. In a turgid cell, the water potential is zero, and now you know two of the values it is easy to work out the third.
a) $\Psi = 0$
b) $\Psi_s = -350$ kPa, so $\Psi_p = +350$ kPa

Q A plasmolysed cell has a Ψ of −650 kPa.
a) What is the Ψ_p?
b) What is the Ψ_s?

A Here again, you are only given one value, but you know that in a plasmolysed cell the Ψ_p is zero, because the protoplast is not pushing against the wall. In this situation the solute potential is the only force acting, so the solute potential *is* the water potential:
a) $\Psi_p = 0$
b) $-650 = \Psi_s - 0$
    $\Psi_s = -650$ kPa

6 ACTIVE TRANSPORT

We have seen that diffusion is movement of molecules down a diffusion gradient:

> **Active transport is movement of molecules *against* a concentration gradient.**

Cell membranes have evolved specialised carrier proteins that transport a particular ion or molecule across the membrane, often from a region of low concentration to a region of much higher concentration. Active transport requires an input of energy from cell respiration, usually provided by ATP. These processes all involve active transport:

- Absorption of amino acids from the gut into the blood.
- Reabsorption of glucose and other useful molecules in the kidney.
- Pumping sodium ions out of cells.
- The exchange of sodium and potassium ions which allows conduction of nerve impulses.

The mechanism of active transport is similar to that of facilitated diffusion (Table 4.2). Cell membranes contain specific carrier proteins that combine with the substance to be transported, enabling it to pass rapidly across the membrane. The key difference is that, while facilitated diffusion is a passive process, the carrier protein complex in active transport is activated by an input of energy (Fig 4.24).

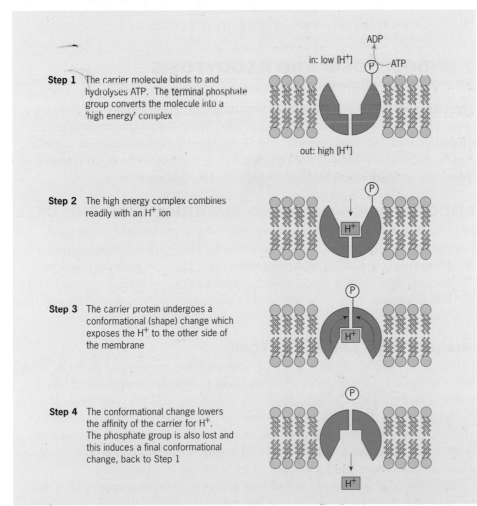

Step 1 The carrier molecule binds to and hydrolyses ATP. The terminal phosphate group converts the molecule into a 'high energy' complex

in: low [H$^+$]

ADP

P → ATP

out: high [H$^+$]

Step 2 The high energy complex combines readily with an H$^+$ ion

P

H$^+$

Step 3 The carrier protein undergoes a conformational (shape) change which exposes the H$^+$ to the other side of the membrane

P

H$^+$

Step 4 The conformational change lowers the affinity of the carrier for H$^+$. The phosphate group is also lost and this induces a final conformational change, back to Step 1

P

H$^+$

Fig 4.24 An example of an active transport mechanism. The H$^+$–ATPase pump removes hydrogen ions from a cell against a diffusion gradient. It is found, for example, in the membrane of the kidney tubule cells, where it contributes to the production of acidic urine

SEE QUESTIONS 6 AND 7

? TEST YOURSELF

J Why do respiratory poisons such as cyanide prevent active transport?

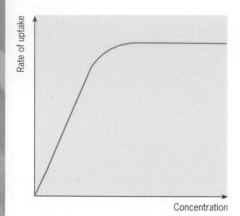

Fig 4.25 The effect of increasing concentration on rate of uptake. Note that the rate of active transport increases to the point at which all the carrier proteins are working at full capacity. If this graph was showing diffusion, it would not level off

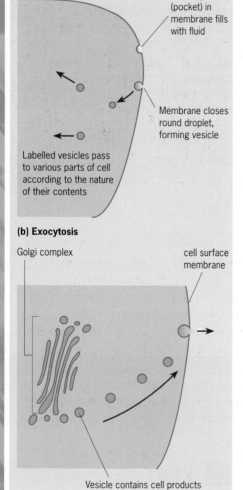

Fig 4.26 Endocytosis and exocytosis. These processes appear to be the reverse of each other, but endocytosis must be selective and the contents must be 'labelled' by the cell so that they can be processed in the correct way

Table 4.2 Comparing diffusion, facilitated diffusion and active transport

	Simple diffusion	Facilitated diffusion	Active transport
Type of membrane molecule involved	lipids	proteins	proteins
Force driving the process	concentration gradient	concentration gradient	ATP hydrolysis
Direction of transport	with concentration gradient	with concentration gradient	against concentration gradient
Specificity	non-specific	specific	specific
Saturation at high concentration of transported molecules	no	yes	yes

The rate of active transport is closely linked to the rate of respiration. We can assume that active transport is taking place if investigations show that:

● movement of particles takes place against a diffusion gradient,
● the rate of transport increases as the rate of respiration increases,
● the process is inhibited by respiratory poisons such as cyanide.

The membrane proteins responsible for facilitated diffusion and active transport are similar to enzymes. Carrier proteins have binding sites for specific chemicals, they can only work at a particular rate, becoming saturated at high concentrations of transported substance (Fig 4.25), and they can be inhibited. They do, however, differ from enzymes in one important respect: they do not alter the substances they transport.

7 ENDOCYTOSIS AND EXOCYTOSIS

The transport processes covered so far involve the movement of individual molecules or ions across the cell membrane. There are other processes, **endocytosis** and **exocytosis**, which transport larger volumes of material (solids and liquids) into or out of a cell. Fig 4.26 shows the main features of these two processes and highlights the difference between them.

ENDOCYTOSIS – BRINGING MATERIAL INTO THE CELL

In endocytosis, the cell surrounds material with a cell surface membrane, bringing it into the cell within a small vesicle. There are two main types:

● **Phagocytosis**. Cells **phagocytose** (take in) solid particles such as bacteria and old red blood cells in order to destroy them (see Chapter 1).
● **Pinocytosis**. A process identical to phagocytosis but which takes in fluid (Fig 4.26a).

PHAGOCYTOSIS AND PINOCYTOSIS

In the process of phagocytosis, a cell's cytoplasm flows round solid material. When this is completely surrounded, a segment of cell membrane is pinched off and a vesicle is formed within the cytoplasm. The membrane and the solid material inside the vesicle form a **food vacuole**. Lysosomes fuse with the vacuole membrane, allowing digestive enzymes to enter the vacuole and begin to digest the contents. Soluble products, such as amino acids and sugars, are absorbed into the cytoplasm, while undigested remains are discharged from the cell by exocytosis.

Examples of phagocytosis include the removal of foreign matter by white cells (see Fig 1.24 on page 16) and the removal of old red blood cells from circulation by the Kupffer cells of the liver.

EXOCYTOSIS – RELEASE OF MATERIAL FROM THE CELL

In exocytosis (Fig 4.26b), materials enclosed in vesicles are expelled from the cell when the vesicle membrane fuses with the cell surface membrane. When the process involves a liquid it is called reverse pinocytosis and this is how the cell secretes many of its synthesised products, such as digestive enzymes, mucus, hormones and the components of milk.

! SCIENCE FEATURE Pus

When someone has a heavy bacterial infection, such as a cut or infected sweat gland, white blood cells gather in great numbers and go into what can be described as a 'phagocytosis frenzy'. They engulf so many bacteria and dead cells that they become literally 'full'. Toxins secreted by the bacteria then kill the white cells that form pus. The lysosomes of dead white cells tend to liquefy the material surrounding them, forming a smelly, runny fluid.

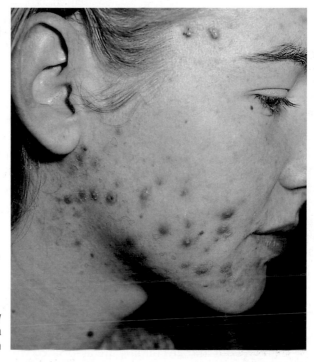

Fig 4.27 This person's acne is caused by an oversecretion of sebum, a greasy substance that blocks sebaceous glands and hair follicles, trapping bacteria underneath. Pus is formed when white cells do their best to deal with the problem

---

SUMMARY

When you have finished this chapter, you should know and understand the following:

■ **Diffusion** is the movement of particles down a concentration gradient until they are evenly distributed.

■ In order to maximise the process of diffusion, organisms have a large **surface area**, make the exchange surface as thin as possible and maintain a **diffusion gradient**.

■ **Facilitated diffusion** is diffusion enhanced by specific proteins in cell membranes.

■ **Osmosis** is the diffusion of water from a region of high water potential to a region of lower water potential.

■ **Water potential** is the tendency for water to move by osmosis. It has a negative scale. Pure water has a water potential of zero. More concentrated solutions have a more negative water potential.

■ Diffusion, facilitated diffusion and osmosis are passive processes: they do not require energy.

■ **Active transport** is the movement of particles across cell surface membranes against a diffusion gradient. This process requires energy, usually in the form of ATP. It is achieved by carrier proteins in cell membranes.

■ **Endocytosis** is the passage of droplets of fluid (**pinocytosis**) or solid particles (**phagocytosis**) into the cell by becoming enveloped in cell surface membrane.

■ **Exocytosis** is the passage of material (usually a cellular secretion) out of the cell.

? EXAM QUESTIONS

1 Give an **illustrated** account of the detailed structure and functions of the plasma membrane.

WJEC AS/A level Biology Modular Module Bl1 January 2001 Q9(a)

2 The figure below shows a section through part of the fluid mosaic model of the cell surface membrane with a Na$^+$/K$^+$ pump protein.

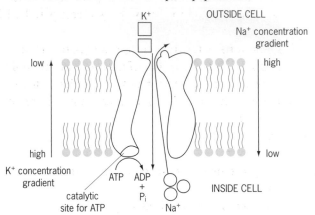

a) Explain why the cell surface membrane is described as a fluid mosaic.

b) Describe how the channel surface of the protein differs from its surface next to the phospholipid tails.

c) Explain why Na$^+$ and K$^+$ cannot pass freely across the phospholipid bilayer.

Cholesterol and glycolipids are associated with cell surface membranes.

d) Suggest **one** function of each compound in membranes.

OCR Cambridge Linear Biology paper 3 June 2000 Q2

3 The figure below represents the fluid mosaic model of a cell membrane.

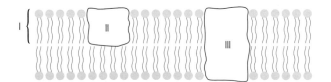

a) **(i)** State the name of the molecule labelled I.

(ii) Label the diagram to show which part of molecule I is hydrophobic and which part is hydrophilic.

b) **(i)** Identify whether molecule II is an intrinsic or an extrinsic protein.

(ii) Describe the part played by molecule III in active transport.

IBO Standard level paper 2 May 1999 Q2

4 Consider the information given below and then answer the questions which follow.

Some A level Biology students carried out an investigation involving the extraction of DNA from yeast cells. The method for extracting and purifying the DNA involved the principles of diffusion, osmosis and dialysis.

The procedure they followed is summarised below.

1. Mix yeast cells with sodium hydrogen carbonate solution in a flask and maintain at 40 °C for about 24 hours in order to autolyse (break open) the cells.

2. Filter the mixture to remove cell debris such as cell walls, retain the filtrate.

3. With the help of a syringe, transfer the filtrate into a length of cellulose (Visking) tubing which has previously been sealed at one end. When full, tie a knot in the end of the Visking tubing, which acts as a partially permeable membrane.

4. In order to concentrate and purify the filtrate, place the filled tubing in a beaker containing concentrated sucrose solution and leave for about 1 hour, until the filtrate has reduced (by exosmosis) to about half of its original volume.

5. Remove the Visking tubing from the beaker and put the concentrated filtrate into a boiling tube. Add ice-cold ethanol carefully down the side of the tube. A white, gelatinous precipitate of DNA forms at the surface.

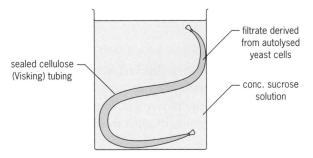

The figure shows the apparatus used.

Concentration of the filtrate occurs due to osmosis, but dialysis is occurring at the same time.

a) Explain, in molecular terms, what is happening during dialysis.

b) **(i)** Name **two** substances (molecules or ions) which would pass out of the Visking tubing during the experiment.

(ii) Name **two** types of molecule which would be retained inside the Visking tubing during the experiment.

OCR Biology Molecules and Life paper 6910 June 2000 Q2

? EXAM QUESTIONS

5

a) Define the term *osmosis*.

b) What word describes a plant cell which has gained a maximum amount of water?

c) What word is used to describe a cell where the cytoplasm has pulled away from the cell wall?

d) In what way is active transport different from osmosis?

e) What is the effect of cyanide on active transport?

WJEC Biology Module BI1 June 2001 Q1

6

a) Give **two** differences between active transport and facilitated diffusion.

b) Describe how ions are transported through a cell surface membrane by active transport.

c) The table shows the concentration of three ions inside and outside a cell of an alga that lives in pond water.

Ion	Concentration of ion in cytoplasm/mmol dm^{-3}	Concentration of ion in pond water/mmol dm^{-3}
Chloride	58.0	1.3
Potassium	93.0	0.1
Sodium	51.0	1.0

(i) By how many times have the potassium ions been concentrated in the cytoplasm compared with the pond water?

(ii) The ions enter the cell of the alga by active transport. What does the information in the table suggest about the active transport of these three ions through the cell surface membrane? Explain your answer.

AQA AS level January 2001

7 Some proteins in membranes are used for active transport.

a) List **three** other functions of proteins in membranes.

b) Explain the reasons for large numbers of vesicles often being found near to the Golgi apparatus of cells.

IBO Standard level paper 3 May 1999 Q3

KEY SKILLS **ASSIGNMENT**

THE SURFACE AREA AND VOLUME PROBLEM

Single cells exchange materials over the whole surface of their outer membrane. Multicellular organisms are larger and cannot exchange enough materials over their outside surface. In this assignment, we investigate why.

Surface area to volume ratio

The amount of material that an organism needs to exchange with its surroundings is proportional to its volume, but its ability to exchange material is proportional to its surface area. As organisms get larger, their surface area to volume ratio gets lower.

1 Imagine a cubeshaped multicellular organism that exchanges materials over its surface. (Obviously, no such organisms exist but this keeps the calculations simple.) Copy and complete the following table:

Size of each side of the organism (units)	1	2	5	10	?
Surface area (units^2) for a total of 6 sides	6	24	?	600	60 000
Volume (units^3)	1	8	125	?	?
Surface area: volume ratio	6:1	?	1.2:1	?	?

So larger organisms can only survive if they evolve ways to optimise the exchange of materials. Organisms can:

- increase their surface area.
- make membranes/barriers as thin as possible.
- maintain a diffusion gradient. As soon as materials are absorbed they are moved away, often by a circulatory system to ensure that an equilibrium is never reached.

Functional design in organisms

2 Fig 4.A1 shows examples of a simple unicellular organism called a trypanosome, the parasite responsible for sleeping sickness. Does this organism need special adaptations to increase its surface area to volume ratio? State a reason for your answer.

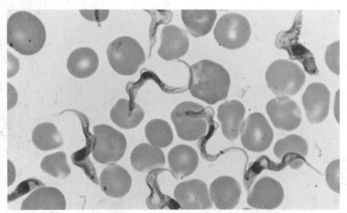

Fig 4.A1 Trypanosomes in blood

3 Two relatively simple ways of increasing the surface of area are to become flat or to become long and thin (wormlike). The tapeworm is a parasite which lives in the intestines of animals such as fish, pigs, rats and even humans.
a) Suggest why the tapeworm has no mouth, intestine or excretory system.
b) Suggest why the tapeworm could not live on land.
c) Thin shapes, such as discs and ribbons, give the greatest possible surface area to volume ratio. What shape gives the least possible surface area to volume ratio?

4 Animals that have evolved beyond the worm stage have had to develop specialised organs to increase their surface area. The nudibranc (sea-slug) in Fig 4.A2 has featherlike gills on its back to increase the surface area for gas exchange.
Suggest why some species of nudibranc constantly shake their gills.

5 The axolotl shown in Fig 4.A3 is an unusual amphibian: it retains external gills into adult life and so remains confined to water. The gill filaments are red because of the thin membranes and good blood supply that runs through them.

Suggest what would happen to the axolotl's gills on land.

Fig 4.A2 A sea slug

Fig 4.A3 Gills of axolotl

6 Like most amphibians, the frog has three gas exchange surfaces. In addition to the moist, permeable skin, the frog can exchange gases across the lining of its large mouth. The frog also uses simple lungs when oxygen demand is high.
a) Suggest why the underneath of a frog's mouth is constantly moving up and down.
b) Why must an amphibian's skin remain moist?

Find out about gas exchange in insects, fish and mammals in Chapter 12.

05

Enzymes and metabolism

Enzymes and metabolism

It is easy to see how confectioners make chocolates with hard centres: they simply pour molten chocolate over the centre and wait for it to set. But what about the soft centres? Surely it isn't possible to pour liquid chocolate over a liquid centre and still keep the shape. So, how do they make the runny, yolklike inside to chocolate eggs?

Chocolate lovers everywhere may be surprised to learn that the answer is: use an enzyme. To start with, the centre is solid and contains an enzyme and a polysaccharide. After the chocolate coating has set, the enzyme breaks down the long polysaccharide chains, turning the hard centre into the familiar runny filling. Increasingly, enzymes are being used in medicine, industry and biotechnology.

1 WHAT ARE ENZYMES?

Enzymes are complex chemicals that control reactions in living cells. They are biochemical **catalysts**, speeding up reactions that would otherwise happen too slowly to be of any use to the organism. This definition is a bit of an oversimplification. An active enzyme may speed up a particular reaction, but living organisms do not need all reactions to be going at the maximum rate all of the time. The key word is control. It is more accurate to say that enzymes interact with other molecules to produce an ordered, stable reaction system in which the products of any reaction are made when they are needed, in the amount needed.

THE ROLE OF ENZYMES IN AN ORGANISM

The metabolic pathway chart, part of which is shown in Fig 5.1, shows some of the large number of different but interconnected chemical reactions in a living cell. Charts like this are too complex to learn but they illustrate several important points:

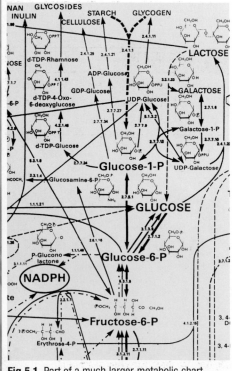

Fig 5.1 Part of a much larger metabolic chart showing some of the biochemical reactions that occur in living cells. Some reactions carry a number: this identifies the enzyme that catalyses a particular reaction

- Many of the complex chemicals that living organisms need cannot be made in a single reaction. Instead, a series of simpler reactions occurs, one reaction after another, forming a **metabolic pathway**. A single pathway may have many steps in which each chemical is converted to the next. A specific enzyme controls each reaction.

- Each individual step in the chart represents one of the simple chemical reactions. At first glance, there seem to be a great variety of reactions, but look more closely and you will see that the same types of reaction occur again and again.

- Enzymes control cell metabolism by regulating how and when reactions occur. Using this very simple pathway as an example,

$$A \rightarrow B \rightarrow C \rightarrow D$$

the final product is substance D, the chemical needed by the living organism. The pathway needs three different enzymes and, when D is no longer needed, or if too much has been produced, one of the three enzymes is 'switched off'.

ONE ENZYME, ONE REACTION, BUT WHAT A DIFFERENCE!

When carbon dioxide dissolves in water, a small proportion of the carbon dioxide molecules combine with water to form carbonic acid.

$$CO_2 + H_2O \rightarrow H_2CO_3 \rightarrow H^+ + HCO_3^-$$

It is a rather slow reaction, but, in the presence of the enzyme **carbonic anhydrase**, it is speeded up about 10^7 (ten million) times. One molecule of carbonic anhydrase can convert 600 000 molecules of carbon dioxide into carbonic acid every second. In living cells, this reaction allows other processes to occur much faster:

- In red blood cells, the enzyme speeds up the production of acid, which in turn causes oxyhaemoglobin to give up its oxygen. Without the enzyme, delivery of oxygen to the tissues would be much slower.
- In certain cells in the stomach lining, the activity of carbonic anhydrase in a pathway allows hydrochloric acid to be secreted rapidly. This creates the acidic conditions necessary for digestive enzymes in the stomach to work properly.
- In the cells of the kidney tubule, carbonic anhydrase speeds up excretion of excess acid, and so helps to maintain the pH of the body at the correct level.

2 THE CHEMICAL NATURE OF ENZYMES

Enzymes are globular proteins. They have a complex tertiary structure (see Chapter 3) in which polypeptides are folded around each other to form a roughly spherical, or **globular** shape (Fig 5.2). This overall three-dimensional shape of an enzyme molecule is very important: if it is altered, the enzyme cannot bind to its substrate and so cannot function. Enzyme shape is maintained by hydrogen bonds and ionic forces (Fig 5.3) and their function can be affected by changes in temperature and pH.

> **? TEST YOURSELF**
>
> **A** By what process are enzymes made in the cell?

> **✔ REMEMBER THIS**
>
> The active site of an enzyme is part of the enzyme molecule and not part of the substrate.

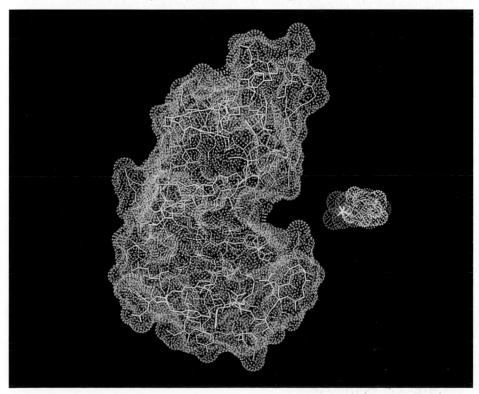

Fig 5.2 The three-dimensional shape of ribonuclease A, an enzyme that helps break up mRNA in the cytoplasm of bacteria. The **active site**, the 'pocket' into which the substrate fits, is clearly visible

An enzyme acts on a chemical known as its substrate. The name of an enzyme often comes from substituting or adding -ase in the name of the substrate, so, for example, lactose is the substrate of the enzyme lactase.

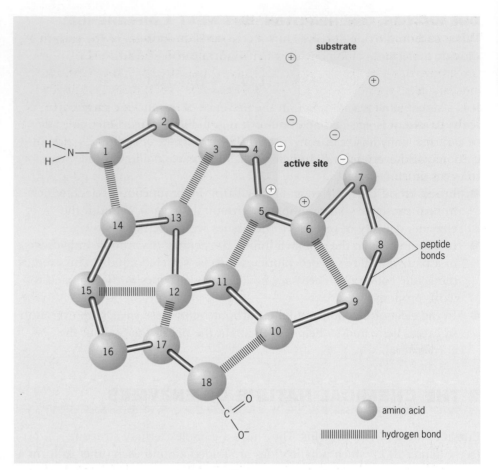

Fig 5.3 This schematic diagram shows the three-dimensional structure of an enzyme molecule. Hydrogen bonds are shown in red. The shape and the electrical charges on the substrate closely match those of the active site

SEE QUESTION 1

? **TEST** YOURSELF

B Predict substrates for the following digestive enzymes: protease, lipase, nucleotidase.

C If the enzyme amylase is specific, how can it catalyse the digestion of bread, potatoes and rice?

Enzymes have several important properties:
- Enzymes are **specific**: each one catalyses only one reaction.
- Enzymes combine with their substrates to form temporary enzyme–substrate **complexes**.
- Enzymes are not altered or used up by the reactions they catalyse, so can be used again and again.
- Enzymes work very rapidly and each has its own **turnover number** (see Table 5.1).
- Enzymes are sensitive to temperature and pH.
- Enzyme function can be slowed down or stopped by **inhibitors**.

THE SPECIFICITY OF ENZYMES

Taking digestive enzymes, you might find it difficult to believe that each enzyme catalyses only one reaction. Trypsin, for example, can begin the digestion of a wide variety of foods rich in protein: eggs, pork, chicken and soya, for example. But when you look at how trypsin works at the molecular level, you can see that this enzyme *is* **specific**. Trypsin cuts an amino acid chain at a point between two particular amino acids, arginine and lysine, and nowhere else. Most proteins have these two amino acids next to each other at some points in their polypeptide chain, and so can be partly digested by trypsin.

Several scientists in biochemistry have devised mechanisms – often called **models** – to explain how enzymes work. In coming up with ideas, they have had to take account of enzyme specificity. Two models that are used to explain how enzymes work are the **lock and key hypothesis** and the **induced fit hypothesis**.

THE LOCK AND KEY HYPOTHESIS

This idea assumes that enzyme function depends on an area on the molecule known as the **active site**, visible in Figs 5.2 and 5.3. The active site is a groove or pocket in the surface of the enzyme into which the substrate molecule fits. Typically, the active site is formed by 3 to 12 amino acids. The size, shape and chemical nature of the active site corresponds closely with that of the substrate molecule, so they fit together like a key fits into a lock (Fig 5.4) or, perhaps more realistically, like two pieces in a three-dimensional jigsaw.

Although this model helps us to understand some of the properties of enzymes, it is now generally accepted that a modified version, known as the induced fit hypothesis, better represents what happens when an enzyme catalyses a reaction.

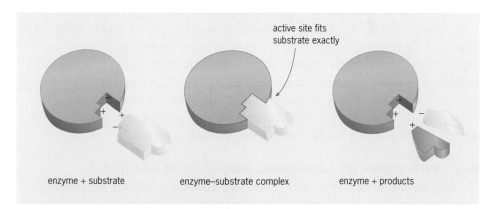

active site fits substrate exactly

enzyme + substrate   enzyme–substrate complex   enzyme + products

Fig 5.4 The lock and key hypothesis. The active site is a particular shape (the lock) into which only one substrate (the key) will fit. The enzyme and substrate combine for an instant to form an enzyme–substrate complex. The formation of this complex brings about the desired chemical reaction, converting substrate into product(s)

THE INDUCED FIT HYPOTHESIS

Experimental evidence suggests that the active site in many enzymes is not exactly the same shape as the substrate, but moulds itself around the substrate as the enzyme–substrate complex is formed (Fig 5.5). Only when the substrate binds to the enzyme is the active site the correct shape to catalyse the reaction. As the products of the reaction form, they fit the active site less well and fall away from it. Without the substrate, the enzyme reverts to its 'relaxed' state, until the next substrate molecule comes along.

Both models show why enzymes are not altered by the reactions that they catalyse: they bind to a substrate momentarily, allowing a reaction to happen, but do not themselves undergo any chemical change.

SEE QUESTIONS 2 AND 3

Fig 5.5 The induced fit hypothesis. Before substrate binding, the enzyme's active site is 'relaxed'. When the substrate binds, the active site is pulled into the correct shape by molecular interactions between the two molecules, and an enzyme–substrate complex forms. As the products fall away from the active site, the molecule becomes 'relaxed' again

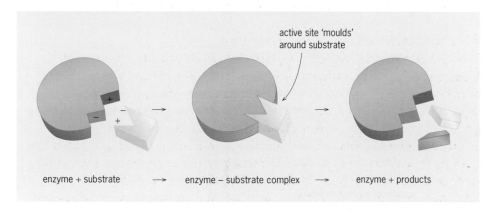

active site 'moulds' around substrate

enzyme + substrate ⟶ enzyme – substrate complex ⟶ enzyme + products

THE CHEMISTRY OF ENZYME ACTION

In biology almost all reactions are **reversible**. If they were not, there would be no recycling of molecules because once large molecules had been built up they could not be broken down again (see Chapter 36).

A simple reversible reaction can be expressed as:

$$\underset{\text{reactants}}{A + B} \rightleftharpoons \underset{\text{products}}{C + D}$$

In this reaction, the **reactants** (A and B) combine to give the **products** (C and D). If this reaction were to take place in a test tube, a proportion of the products would react to form A and B again. Eventually, an **equilibrium** would be reached in which the relative proportion of reactants and products would remain the same.

In theory, an enzyme allows a reversible reaction to reach equilibrium more quickly. An enzyme can speed the reaction in either direction; the way it does depends on whether there are more reactants or products present at the start. In living organisms, however, enzymes usually speed up reactions in one particular direction. This is because, in any organism, an enzyme acts only if the product of the reaction that it catalyses is needed. And, as the product is used as soon as it is made, equilibrium is rarely reached.

ENZYMES WORK BY LOWERING ACTIVATION ENERGY

For any chemical reaction to take place, bonds must be broken before new ones can form. The energy needed to break these bonds, and so set the reaction in motion, is the **activation energy**.

Many reactants need a large amount of energy to push them to a state where they can take part in a reaction, so many reactions take place only at high temperature. Many substances burn, for instance, but only after the

Fig 5.6 In order for a reaction to happen, activation energy must be supplied. Then the rest of the reaction proceeds, just as the boulder rolls down the hill once the energy has been supplied to push it to the top. Catalysts such as enzymes work by lowering the activation energy

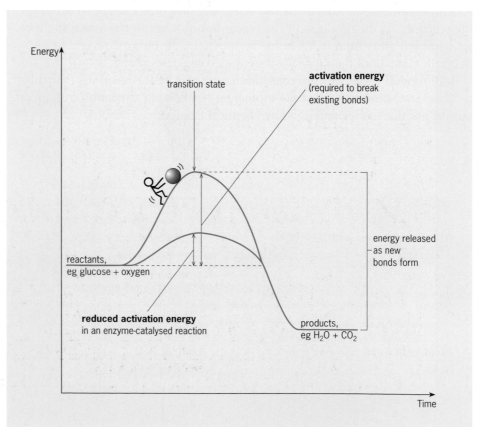
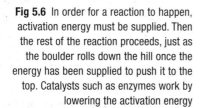

initial activation energy has been supplied, perhaps by lighting a match. In the presence of enzymes, the activation energy is greatly lowered and this allows reactions to take place at the relatively low temperatures normally found in living organisms. The 'halfway' point in a reaction is called the **transition state**, and this is represented by the top of the curve in Fig 5.6. The transition state represents the stage when the old bonds have been broken in order to allow new ones to form. Enzymes lower the activation energy by making it easier to achieve the transition state.

Enzymes speed up reactions in a number of ways:
- They can hold the substrates close together at the correct angle – this would otherwise have to occur by chance collision.
- They can position any charged groups on the active site to help the reaction to occur.
- By acid or base catalysis: the active site of the enzyme can behave as an acid or a base, donating or accepting protons (hydrogen ions, H^+) from the substrate.
- Through structural flexibility: the flexible shape of the enzyme can change during catalysis. This ensures that the substrates arc brought together in the correct sequence for the reaction.

Many metabolic reactions require energy in addition to the presence of the relevant enzyme. This energy is supplied by ATP made by cell respiration (see Chapter 8). Reactions that need energy are made to happen by coupling them with ATP breakdown, and so are called **coupled reactions**.

HOW FAST DO ENZYMES WORK?

The speed at which an enzyme works is expressed as its **turnover number**. This is usually defined as the number of substrate molecules turned into product in one minute by one molecule of enzyme. Values range from less than a hundred to many millions. Some examples are given in Table 5.1.

NAMING AND CLASSIFYING ENZYMES

Older enzyme names such as **pepsin**, **catalase** and **trypsin** give no clues about the nature of the reaction they catalyse. To cope with the rapidly expanding number of new enzymes, the International Union of Biochemistry has developed a scheme for naming and classifying enzymes. Very generally, enzymes are named be adding the suffix ase to the name of their substrate. The rest of the name attempts to indicate the nature of the reaction taking place. Alcohol dehydrogenase, for example, catalyses the removal of hydrogen from alcohol (ethanol). Further examples are given in Table 5.2.

> **? TEST YOURSELF**
>
> **D** The enzyme lysozyme helps to split bacterial cell walls. Where would you expect this enzyme to be present in the human body?

Table 5.1 Some turnover numbers

Enzyme	Turnover number
Carbonic anhydrase	36 000 000
Catalase	5 600 000
β-Galactosidase	12 000
Chymotrypsin	6 000
Lysozyme	60

Source: *Biochemical society guidance notes 3, Enzymes and their role in biotechnology.*

Table 5.2 Some common enzymes and their substrates

Enzyme	Substrate	Reaction catalysed
Maltase	maltose	hydrolysis of maltose to glucose
Amylase	starch	hydrolysis of starch to maltose
Alcohol dehydrogenase	alcohol (ethanol)	removal of hydrogen from alcohol
DNA ligase	DNA	joining together two DNA strands
RNA polymerase	nucleotides that make RNA	synthesis of mRNA on a DNA molecule
Glycogen synthetase	glucose	polymerisation of glucose into glycogen
ATPase	ATP	synthesis or splitting of ATP

? TEST YOURSELF

E People with a great deal of physical stamina have been found to have a lot of **glycogen synthetase** in their muscles. Suggest what this enzyme does.

F Suggest which type of bonds are broken when an enzyme is denatured. (Look at Fig 5.3 for help.)

✔ REMEMBER THIS

Heating denatures enzymes; it does not kill them.

✔ REMEMBER THIS

The factors that affect enzyme activity also affect the functions of the cell and, ultimately, the organism.

TYPES OF ENZYME

Although there are many different enzymes, they can be put into one of six main categories according to the type of reaction they catalyse:

- **Oxidoreductases** These catalyse oxidation and reduction (redox) reactions. In aerobic respiration, most of the cell's ATP is generated by redox reactions – see Chapter 8.
- **Transferases** These catalyse the transfer of a chemical group from one compound to another, such as the transfer of an **amino group** from an amino acid to another organic acid in the process of **transamination**.
- **Hydrolases** These catalyse **hydrolysis** (splitting by use of water) reactions. Most digestive enzymes are hydrolases.
- **Lyases** These catalyse the breakdown of molecules by reactions that do not involved hydrolysis.
- **Isomerases** These catalyse the transformation of one isomer into another, for instance the conversion of glucose 1,6 diphosphate into fructose 1,6 diphosphate. This is one of the first reactions in **glycolysis**, the first stage of respiration (Chapter 8).
- **Ligases** These catalyse the formation of bonds between compounds, often using the free energy made available from ATP hydrolysis. DNA ligase, for example, is involved in the synthesis of DNA.

FACTORS THAT AFFECT ENZYME ACTIVITY

Enzymes are proteins and their function is therefore affected by:

- Temperature
- pH
- Substrate concentration
- Enzyme concentration
- Inhibitors

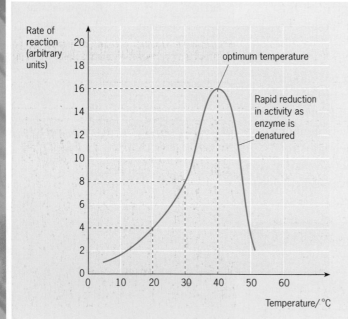

Fig 5.7(a) Up to about 40 °C, the rate of enzyme-controlled reactions increases with temperature. The optimum temperature for many enzymes is about 40 °C, although the activity of an individual enzyme may increase up to about 50 °C or beyond. However, as the temperature passes 43–44 °C, most enzymes lose their activity

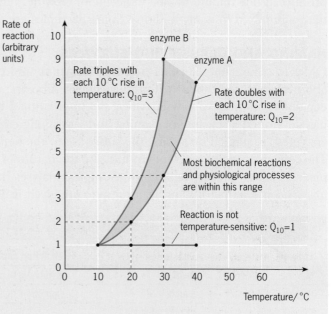

Fig 5.7(b) Not all enzymes have the same Q_{10} values. This variation is important because enzymes, even those in the same pathway, can vary significantly in their sensitivity to temperature. Outside an organism's normal temperature range, the enzymes may begin to work at different rates. This causes a metabolic imbalance that may be lethal, and may explain why some organisms die at temperatures that would seem to be relatively mild. Some Antarctic fish, for example, live in water that remains very constant at around −2 °C, and die if placed in water above 6 °C

TEMPERATURE

For a non-enzymic chemical reaction, the general rule is: the higher the temperature, the faster the reaction: Fig 5.7(a). This same rule holds true for a reaction catalysed by an enzyme, but often only up to about 40 to 45 °C. Above this temperature, enzyme molecules begin to vibrate so violently that the delicate bonds that maintain tertiary and quaternary structure are broken, irreversibly changing the shape of the molecule. When this happens the enzyme can no longer function and we say it is **denatured**.

The effect of temperature on a reaction can be expressed by the temperature coefficient, commonly known as the Q_{10}.

Where t is the chosen temperature, the formula for the Q_{10} is:

$$\frac{\text{rate of reaction at } t + 10\,°C}{\text{rate of reaction at } t\,°C}$$

Fig 5.7(b) shows how to calculate the Q_{10}. To avoid denaturing the enzyme, the values for living organisms need to fall in the range 4 to 40 °C, so we have chosen t as 20 °C.

$$Q_{10} = \frac{\text{rate at } 30\,°C}{\text{rate at } 20\,°C} = \frac{6}{3} = 2$$

In practice, most enzymes have a Q_{10} between 2 and 3. A value of 2 means that the rate of reaction doubles with a 10 °C temperature rise, 3 means that it triples.

Some organisms have enzymes that are less sensitive to heat than those found in mammals. For example, certain bacteria can survive in hot volcanic springs and deep sea hydrothermal vents (see Fig 5.8) at temperatures of over 90 °C, so their enzymes must be active at these extreme temperatures.

? TEST YOURSELF

G Egg albumen, when heated, goes solid and white. What do you think has happened to the proteins in the egg white, and why does the white not become runny again when the egg is cooled?

? TEST YOURSELF

H Why is it an advantage for humans to have a constant body temperature of around 37 °C?

SEE QUESTIONS 4 TO 6

! SCIENCE FEATURE Stable enzymes for washing powders

Enzymes are unstable, particularly at high temperature, so their commercial usefulness is limited. Many industrial processes need to take place in 'unnatural' environments, at high temperature and extremes of pH. But there are organisms that can thrive at high temperatures. Heat-loving bacteria have thermostable enzymes that are also more resistant to extremes of pH and other unfavourable conditions, such as organic solvents.

Genes that code for some thermostable enzymes have been transferred into bacteria that are easy to grow in large quantities, using recombinant DNA technology (see Chapter 30). As the bacteria that have received a particular gene multiply, the gene is translated into protein and large amounts of the enzyme are produced for commercial use.

An example of a thermostable enzyme is the alkaline protease **subtilisin**, the famous stain digester in biological washing powders. This enzyme is produced on a grand scale by the bacterium *Bacillus subtilis*, and is active in alkaline environments, and so is compatible with the other ingredients in washing powder. It is active at temperatures up to 60 °C, allowing it to be used in a wide variety of wash programmes.

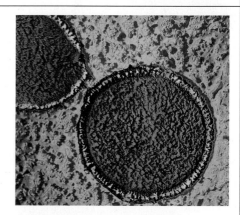

Fig 5.8 This bacterium *Staphylothermus marinus* lives near deep sea hot-water vents. It thrives at temperatures of up to 98 °C. Such heat-loving bacteria have thermostable enzymes (enzymes that are very resistant to heat damage). These have great potential in industry

? ▌**TEST YOURSELF**

▌ Chymotrypsin is found in the small intestine at pH 8. What would happen to this enzyme if it found its way into the stomach?

SEE QUESTION 7

pH

Like other proteins, enzymes are stable over a limited range of pH. Outside this range, at the extremes of pH, enzymes are denatured. Free hydrogen ions (H^+) or hydroxyl ions (OH^-) affect the charges on amino acid residues, distorting the three-dimensional shape and causing an irreversible change in the protein's tertiary structure.

Enzymes are particularly sensitive to changes in pH because of the great sensitivity of their active site. Even if a slight change in pH is not enough to denature the molecule, it may upset the delicate chemical arrangement at the active site and so stop the enzyme working (Fig 5.9).

Most enzymes are **intracellular**: they work inside cells, and their optimum pH is around 7.3 to 7.4. Most organisms have buffer systems that resist changes in pH, and many are able to excrete excess acid or alkali.

Fig 5.10 shows some enzymes that have optimums of pH that are distinctly acid or alkaline. These include digestive enzymes that normally work in the stomach or inside lysosomes. We describe enzymes that work outside cells as **extracellular**.

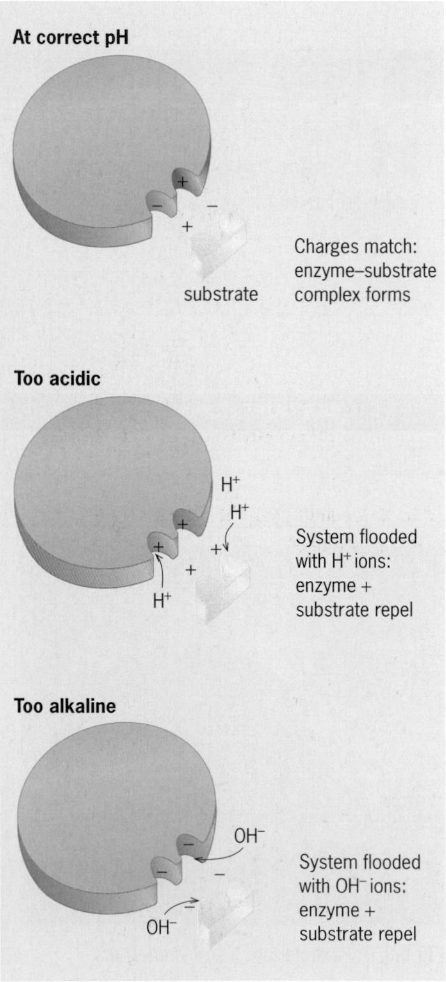

At correct pH

substrate

Charges match: enzyme–substrate complex forms

Too acidic

H^+
H^+
H^+

System flooded with H^+ ions: enzyme + substrate repel

Too alkaline

OH^-
OH^-

System flooded with OH^- ions: enzyme + substrate repel

Fig 5.9 The effect of pH on enzyme activity

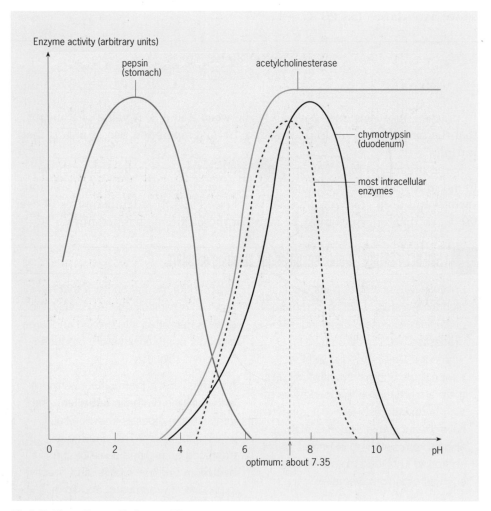

Enzyme activity (arbitrary units)

pepsin (stomach)

acetylcholinesterase

chymotrypsin (duodenum)

most intracellular enzymes

optimum: about 7.35

pH

Fig 5.10 The optimum pH of some different enzymes

SUBSTRATE CONCENTRATION

The rate of an enzyme-controlled reaction increases as the substrate concentration increases, until the enzyme is working at full capacity. At this point the enzyme molecules reach their turnover number and, assuming that all other conditions such as temperature are ideal, the only way to increase the speed of the reaction even more is to add more enzyme (Fig 5.11).

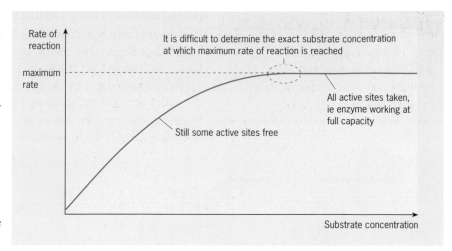

Fig **5.11** The effect of substrate concentration on the rate of enzyme action

ENZYME CONCENTRATION

In any reaction catalysed by an enzyme, the number of enzyme molecules present is much smaller than the number of substrate molecules. Look back at Table 5.1, which shows the turnover numbers of some enzymes, and you can see that one molecule of an enzyme can convert millions of substrate molecules into products every minute.

When there is an abundant supply of substrate, the rate of reaction is limited by the number of enzyme molecules. In this situation, increasing the enzyme concentration increases the rate of reaction.

INHIBITORS

Inhibitors slow down or stop enzyme action. Usually, enzyme inhibition is a natural process, a means of switching enzymes on or off when necessary. Inhibition tends to be **reversible**, and the enzyme returns to full activity once the inhibitor is removed. Many 'external' chemicals, such as drugs and poisons, can inhibit particular enzymes. This type of inhibition is often **non-reversible**. Reversible inhibitors are either **competitive** or **non-competitive**.

The different types of inhibitor are summarised in Fig 5.12.

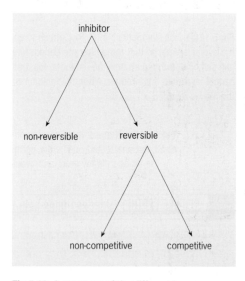

Fig **5.12** A summary of the different types of enzyme inhibitor

COMPETITIVE INHIBITORS

Competitive inhibitors compete with normal substrate molecules to occupy the active site. The inhibitor molecules must be a similar size and shape to the substrate to fit the active site, but cannot be converted into the correct product. Effectively, competitive inhibitors 'get in the way' (Fig 5.13) and reduce the number of interactions that can happen between enzyme and substrate.

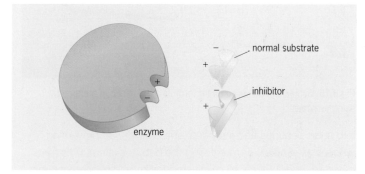

Fig 5.13(a) A competitive inhibitor fits into the active site of the enzyme, preventing the real substrate from gaining access. The inhibitor cannot be converted to the products of the reaction and so the overall rate of reaction is slowed down. If an inhibitor is present in equal concentrations to the substrate, and if both types of molecule bind to the active site equally well, the enzyme can only work at half its normal rate

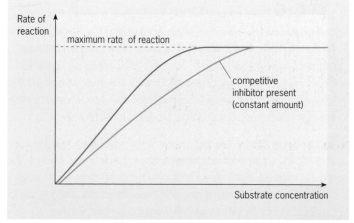

Fig 5.13(b) The effect of a competitive inhibitor on the rate of reaction. Note that the effect of the inhibitor can be overcome by adding more substrate

? **TEST YOURSELF**

J What would the graph in Fig 5.14(b) look like if a large amount of a non-competitive inhibitor was added?

NON-COMPETITIVE INHIBITORS

Non-competitive inhibitors bind to the enzyme away from the active site but change the overall shape of the molecule, modifying the active site so that it can no longer turn substrate molecules into product (Fig 5.14a).

Non-competitive inhibition has this name because there is no competition for the active site. The presence of a non-competitive inhibitor has the same effect as lowering enzyme concentration: all inhibited molecules are taken out of action completely. Fig 5.14(b) shows the effect of a non-competitive inhibitor.

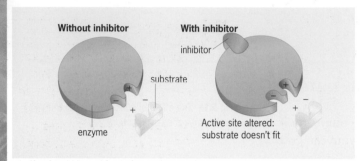

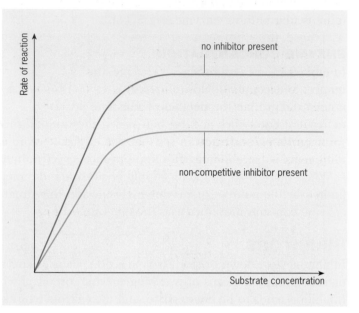

Fig 5.14(a) Non-competitive inhibitors attach to enzyme molecules and alter the overall shape, so that the active site cannot function. Although the substrate can still bind, forming an enzyme–inhibitor substrate complex, the substrate cannot be turned into product. When the inhibitor molecule is removed, normal function is restored

Fig 5.14(b) Non-competitively inhibited enzymes show no activity at all, but unaffected ones work normally. Thus, the maximum rate of reaction is lowered, as it would be if we used less enzyme

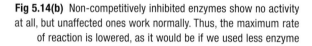

! SCIENCE FEATURE How enzymes help babies digest cow's milk

When babies are first born, their intestines don't behave at all like those of an adult. So that they can absorb and make use of important antibodies in their mother's breast milk, they have intestinal walls that have a structure that resembles a string vest. It has lots of large pores through which whole proteins can pass, without being digested into their component amino acids.

Everything is fine, as long as the baby is fed breast milk, but some mothers cannot breast feed for very good reasons. What happens then? A hundred years ago, babies that couldn't feed from their mother usually fell behind in their growth, sickened and often died. Cow's milk does not contain the same proteins as human milk, and it also has a completely different balance of minerals such as sodium. Feeding tiny babies on untreated cow's milk leads to a generalised immune response against the cow proteins and can also damage the delicate kidneys, which are not able to cope with large amounts of sodium.

Today, many babies are fed on infant formula milk (Fig 5.15) and they thrive just as well as babies that are breast fed (although the natural method is still considered the best). Enzyme technology has had a lot to do with this – enzymes have been used in the production of infant formula milk from cow's milk for over 50 years. Manufacturers used proteases to digest the milk proteins

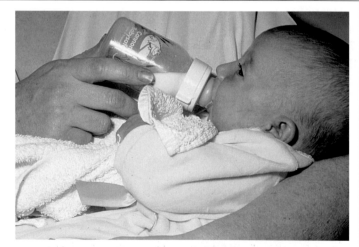

Fig 5.15 Babies can thrive on infant formula milk, which is produced using enzymes

into short peptides and individual amino acids, so preventing whole cow proteins from entering the baby's system. This 'pre-digested' cow's milk is also adjusted for fat and mineral content, and today's formula milks are as close to human milk as possible. Whether biotechnology will ever manage to add all the antibodies, cells, human enzymes and other components of naturally produced human milk remains a question for the future.

IRREVERSIBLE INHIBITORS

Irreversible inhibitors bind permanently to the enzyme, rendering it useless. For obvious reasons organisms rarely produce this type of inhibitor for their own enzymes, but they are splendid weapons to use against other organisms. A wide variety of natural toxins are irreversible inhibitors, as are many pesticides.

HOW INHIBITORS HELP TO CONTROL METABOLISM

Many metabolic pathways are self-controlling: when a substance is needed, a particular pathway is activated to produce it. When enough has been produced, the pathway is deactivated.

This happens because some enzymes in a metabolic pathway are inhibited by the end product. As Fig 5.16 shows, if too much product begins to accumulate, this inhibits one of the enzymes in the pathway. When the product is once more in short supply, the inhibition is lifted and the pathway becomes active again. This self-regulation is an example of **negative feedback**. This is a fundamental principle important in homeostasis (see Chapter 14).

> REMEMBER THIS
>
> **Cyanide** is an irreversible inhibitor of **cytochrome oxidase**, one of the enzymes involved in respiration. Organisms poisoned with cyanide die because they are deprived of ATP, their immediate energy source.

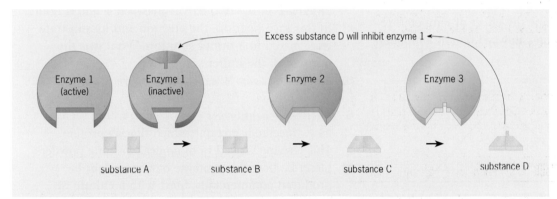

Fig 5.16 A metabolic pathway can be self-regulating by having the end product act as a non-competitive inhibitor on one of the enzymes in its production pathway

SUMMARY

After reading this chapter, you should know and understand the following:

- Enzymes are **globular** proteins with a precise, but delicate, three-dimensional shape maintained by **ionic** and **hydrogen bonds**.
- During a reaction, the substrate fits into a region on the enzyme surface called the **active site**.
- Enzymes are **specific**: each enzyme catalyses one particular reaction.
- Enzymes speed up reactions by lowering the **activation energy** needed to get the reaction started.
- Between 4 °C and 40 °C, the rate of an enzyme-controlled reaction increases between two- and threefold for every 10 °C rise in temperature. Increase in temperature beyond 40 °C usually **denatures** the enzyme. Activity is lost when its three-dimensional structure is destroyed.

- Most enzymes have an optimum pH. For **intracellular** enzymes this is usually about 7.35. **Extracellular** enzymes, such as those found in digestive juices, may have optimums of extreme pH values.
- The rate of an enzyme-controlled reaction is limited by the supply of substrate, enzyme or the enzyme cofactor.
- **Inhibitors** are substances that slow down or stop enzyme activity. They may be **reversible** or **non-reversible**. Reversible inhibitors may be **competitive** or **non-competitive**.
- Competitive inhibitors tend to be similar in structure to the substrate and compete for the active site. Non-competitive inhibitors do not bind to the active site itself but alter the shape of the enzyme so that the active site is no longer functional.

? EXAM QUESTIONS

1 Define the term 'enzyme' and relate its function and properties to its structure.

WJEC Biology Module BI1 June 2001 Q8

2 Describe the *induced fit* model of enzyme action.

IBO Standard level paper 3 November 1999 QC2

3

a) Outline how enzymes in the cytoplasm of cells are produced.

b) Compare the induced fit model of enzyme activity with the lock and key model.

c) Explain, using **one** named example, the effect of a competitive inhibitor on enzyme activity.

IBO Higher level paper 2 May 1999 Q6

4 The graph shows the results from a number of experiments to determine the effect of temperature on the rate of a reaction catalysed by a mammalian enzyme.

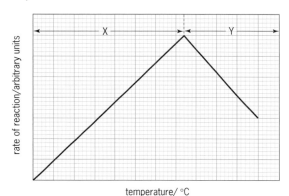

a) Add the scale and numbers on the temperature axis for a typical mammalian enzyme.

b) **(i)** Explain the shape of the curve in region **X**.
(ii) Explain the shape of the curve in region **Y**.

c) If you were carrying out this experiment state **two** factors which you would need to keep constant in order to obtain reliable results.

WJEC Biology Module BI1 January 2001 Q5

5 The total amount of product formed in an enzyme-controlled reaction was investigated at two different temperatures, 55 °C and 65 °C. The results are shown on the graph.

a) **(i)** Explain how you would calculate the rate of the reaction at 55 °C over the first 2 hours of the investigation.
(ii) Explain why the initial rate of this reaction was faster at 65 °C than it was at 55 °C.

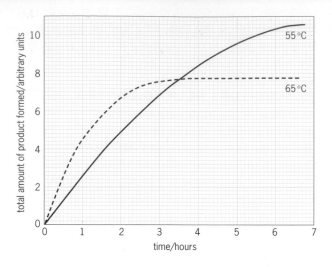

b) Use your knowledge of enzymes to explain the difference in the two curves between 4 and 6 hours.

c) In this investigation, the enzyme and its substrate were mixed in a buffer solution. What was the purpose of the buffer solution?

AQA Biology/Human Biology Molecules, Cells and Systems January 2001 Q1

6 *Bacillus licheniformis* is a bacterium that secretes an extracellular enzyme which digests protein. This enzyme is used in biological washing powders.

a) Describe how the protease enzyme might be produced commercially. Start with a culture of *Bacillus licheniformis* and a suitable medium.

b) **(i)** Biological washing powers containing protein-digesting enzymes are used to remove 'biological stains'. Suggest a 'biological stain' which could be removed by a protein-digesting enzyme.
(ii) The instructions on a packet of biological washing powder stated 'Pre-soak at 40 °C for 1 hour or leave overnight in cold water'. Explain why the recommended soaking time is longer if cold water is used.

AQA Making Use of Biology January 2001 Q3

7 One of the first enzymes to be purified and studied in the 1930s was urease, which catalyses the breakdown of urea to carbon dioxide and ammonia. The reaction may be summarised thus:

$$CO(NH_2)_2 + H_2O \xrightarrow{\text{urease}} CO_2 + 2NH_3$$

It is known that urea molecules will slowly break down to CO_2 and NH_3 in the absence of a catalyst. This spontaneous (uncatalysed) breakdown rate can be speeded up if acid (H^+ ions) is added. However,

? EXAM QUESTIONS

the rate of breakdown is very much faster in the presence of the enzyme urease. This is shown in the figure below, which is an energy profile diagram.

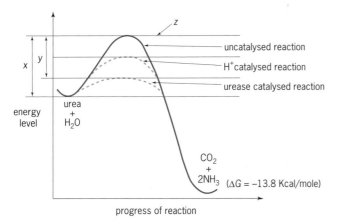

a) State, with a reason, whether this reaction is endergonic or exergonic.

b) (i) What does x represent?
(ii) What does y represent?
(iii) What does level z represent?
Experiments were performed to investigate the influence of pH on the activity or urease, using phosphate buffer solutions.
Enzyme activity is determined by measuring the amount of urea broken down to CO_2 and NH_3 per unit time. The table shows the results of the investigation.

pH	Enzyme activity (arbitrary units)
4.6	0.06
5.0	0.12
5.6	0.32
6.0	0.50
6.4	0.70
6.8	0.88
7.2	0.84
7.6	0.62
8.0	0.40
8.4	0.14

c) (i) Construct a graph of these data on graph paper.
(ii) With reference to your graph, what is the most likely optimum pH value for urease?
(iii) If the urease enzyme solution is made **more** acidic than the optimum by 1 pH unit, by what percentage is the reaction rate lowered?
(iv) Explain why a change in pH affects the activity of an enzyme.
(v) Suggest a possible mechanism, based on the results of this investigation, whereby cells could regulate enzyme activity.
(vi) State **two** factors which must be kept constant during the investigation described above.

OCR Biology Molecules and Life March 2000 Q5

KEY SKILLS **ASSIGNMENT**

ENZYMES AT WORK

Enzymes have enormous commercial potential. Since they can catalyse particular reactions at relatively low temperatures, they are more versatile and much cheaper than inorganic catalysts. Once a suitable enzyme is found, it is made on a large scale and then purified.

There is a vast worldwide demand for sweeteners, mainly for confectionery and soft drinks. Traditionally, manufacturers have used sucrose extracted from sugar beet or cane but, in recent years, **high fructose syrup**, a cheaper sugar beet product has become more common. Fructose is sweeter than sucrose, and can be made from the starch in sugar beet. About 40 million tonnes are now produced each year.

1 What sort of carbohydrate is starch?

Fructose is produced by an enzyme-catalysed process: the starch is first broken down into glucose and then glucose is converted to fructose.

2 What type of enzyme catalyses the conversion of:
a) starch to glucose?
b) glucose to fructose?

Three enzymes are involved in the production of high fructose syrups: **bacterial β-amlyase**, **fungal amyloglucosidase** and **bacterial glucose isomerase**. In the first step, the starch paste is heated to 105 °C. β-amylase is then added to reduce the **viscosity** (thickness) of the paste. The process results in a mixture of **maltodextrins** – branched sugars.

3 **a)** How do we describe an enzyme that works at high temperatures?
b) What is the advantage of higher temperatures in industrial processes?
c) Why does the action of the enzyme reduce the viscosity of the paste?

The next step is the hydrolysis of the maltodextrins to glucose. The substrate is cooled to 55–60 °C and acidified to pH 4.5. The enzyme amyloglucosidase, obtained from the fungus *Aspergillus niger*, is added. This is an **exoenzyme**: it removes the terminal glucose units from the maltodextrins.

4 **a)** What is the difference between an endoenzyme and an exoenzyme?
b) In the hydrolysis of polymers like starch, why is it an advantage to add an endoenzyme before using an exoenzyme?
c) Sketch graphs to show the likely effect of variations in temperature and pH on the activity of amyloglucosidase.

Amyloglucosidase acts on maltodextrins to produce glucose syrup. Although not as sweet as sucrose, it still has its uses in the food industry. The production of the extrasweet fructose needs the enzyme glucose isomerase, which is obtained from bacteria (*Bacillus* or *Streptomyces* sp.). This enzyme is fixed on rigid granules and packed into a column. The glucose syrup then flows between the granules and the enzyme converts the glucose into fructose.

5 What is the advantage of using a fixed (immobilised) enzyme?

The end product is high fructose syrup – a mixture of about 42 per cent fructose and 55 per cent glucose. In these proportions it has the same sweetening power as sucrose.

An alternative sweetener is **aspartame**, a dipeptide that is 180 times sweeter than sucrose. The commercial production of aspartame again involves enzymes – particularly **aspartase**. The ease and cost of aspartame production affects the demand for high fructose syrups.

6 What is the main dietary advantage of using aspartame instead of sucrose?

7 All of the enzymes mentioned in this assignment come from organisms such as yeast and bacteria. Why are these organisms particularly suitable for large-scale enzyme production?

8 Imagine that a scientist isolates a human gene that codes for an enzyme that catalyses a reaction giving a very valuable product. List, and perhaps discuss, the steps that would have to be taken in order to get this product into commercial production.

Fig 5.A1 Many soft drinks contain a high concentration of fructose syrup

Biotechnology

06 Biotechnology

Scientists in the USA have identified new strains of bacteria that could revolutionise how we deal with pollution. One 'eats' benzene, a toxic chemical found in petroleum-based fuels. Fuel spills and leaking storage tanks mean that benzene is a common contaminant of groundwater supplies and, since long-term exposure to benzene can lead to cancer in humans, it must be dealt with before it can affect soil and drinking water.

Benzene breaks down when exposed to the atmosphere but, without oxygen, this degradation process slows to a crawl. Removing the compound from airless environments is therefore difficult. However, *Dechloromonas* strain RCB and *Dechloromonas* strain JJ apparently oxidise benzene to carbon dioxide without the help of oxygen. These organisms are creating a lot of interest because they can speed up the break down of benzene in anaerobic conditions by up to 10 times. What once took 70 days now happens in 7 days.

Another bacterial species digests toluene, another common but toxic groundwater contaminant that comes from fuels, adhesives and household solvents. These bacteria also work in the absence of oxygen and research has already identified four genes that code for enzymes in the metabolic pathway responsible.

Using these organisms, or other bacteria given pollution-busting genes, to clean up contamination would have several advantages over current methods. Not only would they be cheaper, but also the bacteria could be brought to the contaminated area, which would avoid the environmental scarring caused by the removal of tonnes of contaminated soil for dumping or cleaning. The toluene-eating species also might be used to prevent pollution by treating industrial waste before it ever reaches the environment.

Culture of oil-degrading bacteria. Genetically engineered microbes may become more widely used to extract oil from the ground and valuable metals from factory wastes

1 WHAT IS BIOTECHNOLOGY?

A definition of biotechnology is quite difficult to pin down. Broadly speaking, it is the manipulation of biological organisms, systems or processes for the benefit of people, in areas such as agriculture, food production and medicine.

The oldest biotechnologies include the **fermentation process**, which has been used to make bread, beer, wine and cheese for centuries, and also **plant and animal breeding**. More recent biotechnology techniques include the **manipulation of genes** to make **transgenic** plants and animals, using **monoclonal antibodies** to target cancer cells, and using advanced chemistry to **immobilise** enzymes to control and improve manufacturing processes.

Elsewhere, you will also see biotechnology defined more narrowly – as the manipulation of DNA and genes to make organisms that are commercially useful. In this book, we consider both aspects of biotechnology. This chapter covers the wider aspects, with sections on the use of biotechnology in farming and food production, in industry and in medicine. We also consider some of the social and ethical implications of biotechnology. The manipulation of

DNA and genes is also here, particularly when it is relevant to farming and food production. However, gene technology and topics such as human cloning and the human genome project are dealt with in more detail in the section on genetics (Chapters 24–30).

2 BIOTECHNOLOGY AND FARMING

MANIPULATING FARM ANIMALS

Some people might argue that farming animals is manipulating them too much already, but biotechnology is now used increasingly as a tool to develop more productive animals. Not surprisingly, many of the developments are controversial. All have their supporters and their opponents, and in the sections that follow we have tried to present a factual balanced view. Delve into it deeper if you want, and then make up your own mind.

USING HORMONES TO INCREASE MILK PRODUCTION

In order to produce milk at all, cows need to produce a calf each year – they are constantly either pregnant or lactating, or both. With careful timing of one pregnancy after another, a dairy cow can produce a steady supply of milk. Its calf is usually taken away and hand reared, or is used for food. The quality and quantity of this supply depend on the genetic make-up of the cow, how old she is, how much food she is given and the quality of that food.

For some years now, farmers in many developed countries such as the UK and USA have been enhancing the level of **bovine somatotrophin (bST)** in dairy cows. Also known as **bovine growth hormone** (**BGH**), this is a natural protein hormone produced by the **pituitary gland** of all cattle.

Biotechnology has enabled scientists to produce a **recombinant** form of this protein, called **rbST**. This means that they have isolated the gene for rbST from cows, and then inserted it into the genome of bacteria. These are then cultured and produce the hormone is vast quantities in industrial vats. The industrial production of recombinant growth hormone is summarised in Fig 6.1.

When this readily available hormone is injected into cows on a regular basis, it reliably improves their efficiency as milk producers. The mammary glands of such dairy cows take in more nutrients from the bloodstream and produce more milk.

It is very unlikely that bST, whether recombinant or natural, has any biological effect on humans. It already exists naturally in milk, so there is just a bit more of it in the milk from rbST-treated cows. However, the human digestive juices break down the protein into the amino acids that make it up, so it disappears, except in babies who, in any case, should not be fed on cow's milk. It cannot gain entry to the human body as a complete protein. And even if it did, the hormone is **species-specific**. The form found in cows does not fit into any receptor in the human body; it is a completely different shape from human growth hormone, for example.

Cows that are given rbST increase their overall milk production between 5 and 10 per cent but they don't need to eat 5 to 10 per cent more food, so overall this is a cost-effective technique. Those in favour argue that it allows dairy farmers to be more efficient and that it enables more milk to be produced in areas where food is needed urgently. They propose that nations suffering from hunger and malnutrition could benefit from rbST.

However, those against hormone use say that milk produced by treated cows should be labelled so consumers can decide whether to drink it or not.

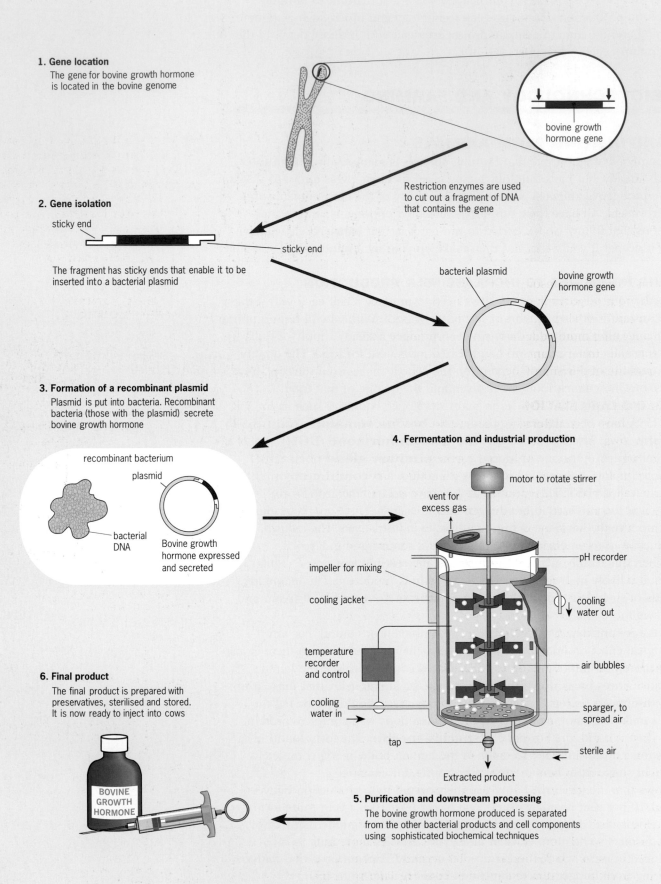

1. Gene location
The gene for bovine growth hormone is located in the bovine genome

bovine growth hormone gene

Restriction enzymes are used to cut out a fragment of DNA that contains the gene

2. Gene isolation

sticky end

sticky end

The fragment has sticky ends that enable it to be inserted into a bacterial plasmid

bacterial plasmid

bovine growth hormone gene

3. Formation of a recombinant plasmid
Plasmid is put into bacteria. Recombinant bacteria (those with the plasmid) secrete bovine growth hormone

recombinant bacterium

plasmid

bacterial DNA

Bovine growth hormone expressed and secreted

4. Fermentation and industrial production

motor to rotate stirrer

vent for excess gas

pH recorder

impeller for mixing

cooling jacket

cooling water out

temperature recorder and control

air bubbles

cooling water in

sparger, to spread air

tap

sterile air

Extracted product

6. Final product
The final product is prepared with preservatives, sterilised and stored. It is now ready to inject into cows

BOVINE GROWTH HORMONE

5. Purification and downstream processing
The bovine growth hormone produced is separated from the other bacterial products and cell components using sophisticated biochemical techniques

Fig 6.1 Steps in the production of recombinant bovine growth hormone. Note that the drawings are not to scale

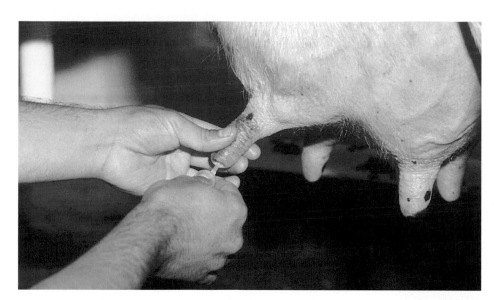

Fig 6.2 Cow's udder being injected with antibiotic

Fig 6.3 Embryo duplication allows up to 16 embryos to be obtained from one mating between a valuable female and a valuable male. The calves are carried by surrogate mothers. The cow that is the biological mother does not carry a calf; she is inseminated again soon afterwards to obtain another embryo for splitting between another set of surrogates. In this way, one pair of prime animals can be used to generate dozens of calves every year

This may lead to an unsold surplus of milk, which may need to be bought by governments to prevent farmers going out of business. They also say that cows that are given growth hormone suffer more udder infections than untreated cows, and that the greater quantity of antibiotics used to treat them then gets into the human food supply.

EMBRYO DUPLICATION

The technique of **artificial insemination** – the use of sperm from a bull or ram with characteristics that are particularly sought after in the farming world – has been used to breed 'better' animals for several years. In the last 15 years of the twentieth century reproductive technology became a lot more sophisticated and it became possible to make several female animals pregnant with identical embryos. When these are produced by mating a prime female with a prime male, the resulting embryos can be split, sometimes into two, sometimes more, and implanted into any healthy female of the same species. This means that a whole herd of prime stock animals can be produced from one breeding pair, whose embryos are carried by **surrogate mothers**.

SYNCHRONISING SHEEP BREEDING

In Britain, the changing day-length of the seasons dictates the breeding behaviour of sheep. They synchronise their reproductive cycle to produce lambs in early spring (March/April) when new grass is plentiful. This also gives the young as much time as possible to mature before the next winter. The gestation period in sheep is 21 weeks, and ewes and rams must mate in October/November to get the timing just right.

Normally, left to their own devices, sheep rely on their hormones to do this. In the autumn, the stimulus of shortening days affects the **hypothalamus**, which responds by secreting a hormone called **gonadotrophin-releasing factor**. As its name suggests, this acts on the pituitary gland to cause the release of a gonadotrophin hormone – **follicle stimulating**

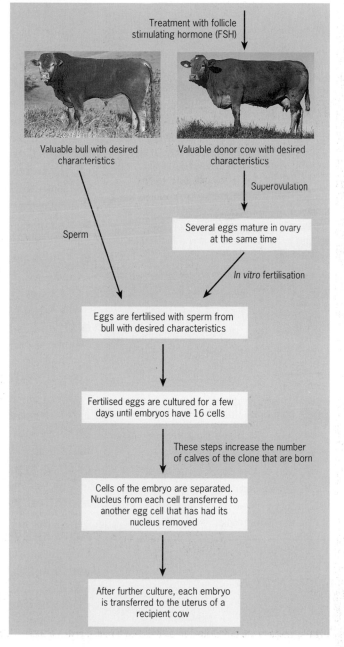

Treatment with follicle stimulating hormone (FSH)

Valuable bull with desired characteristics

Valuable donor cow with desired characteristics

Sperm

Superovulation

Several eggs mature in ovary at the same time

In vitro fertilisation

Eggs are fertilised with sperm from bull with desired characteristics

Fertilised eggs are cultured for a few days until embryos have 16 cells

These steps increase the number of calves of the clone that are born

Cells of the embryo are separated. Nucleus from each cell transferred to another egg cell that has had its nucleus removed

After further culture, each embryo is transferred to the uterus of a recipient cow

Fig 6.4 When ewes come into season, the farmers release rams into their field for a night. In the morning, all the females should have been impregnated. Just to make sure none has been missed, some farmers take the precaution of taping a block of dye to each ram's chest. This rubs off during mating. Any ewes that do not have dyed bottoms by morning have to stay with the rams for another night

hormone **(FSH)**. FSH stimulates the development of follicles in the ovaries; these secrete oestrogen, and this causes the ewe to come into season. As she begins to give off vaginal secretions that contain **pheromones**, nearby rams are aroused and mating occurs.

The problem with this for farmers of today is that the mating process, and therefore the lambing, can take place over a period of about a month. This makes it very difficult to care for all the ewes and lambs, particularly if the flock is very large. To enable farmers to concentrate their effort and resources (such as vets on standby), they try to ensure that lambing actually takes place over a 5-day period. To do this, they must ensure that most ewes mate on the same day.

Achieving this takes some extraordinary measures:

● Ewes that are to be mated must be kept away from all sexually active males during the time they are weaning their previous lambs. Sexually active males include not only rams, but male lambs and billy goats, which can secrete some of the hormones that attract the female.

● This absence of males allows the next stage to have its maximum effect. Vasectomised but thoroughly randy rams are then put with the ewes four weeks before the intended mating day to bring them into season. These rams are termed 'teasers', for obvious reasons.

● Two weeks after the introduction of the teasers, the ewes are separated again and small sponges impregnated with the hormone **progesterone** are inserted just inside the vagina of each one. The progesterone prevents ovulation in the ewes until it is removed 14 days later, at which point all the ewes should be at the same stage of their oestrous cycle.

● Sometimes, another hormone, called **pregnant mare serum gonadotrophin (PMSG)**, is injected into the ewes just after sponge removal. This slightly increases the litter size of the ewes, and may also make the timing more precise.

● 36 to 48 hours after sponge removal, ewes come into heat. Tonight's the night!

● The rams are then introduced to the ewes and, after few hours of frenzied activity, the farmer achieves his or her objective. Alternatively, the farmers use **artificial insemination** – important if they want to use a particular ram to father as many lambs as possible.

● If all has gone well, the ewes should lamb 145 days later, with some lambs born 2 days or so either side of the expected date.

USING BIOTECHNOLOGY IN CROP DEVELOPMENT

Farmers have been practising a crude form of biotechnology for centuries by **cross-breeding** plants to introduce and maintain desirable characteristics. More recently, plant technologists have bombarded plant cells or small plants with radiation that damages DNA, causing random changes that sometimes result in 'better' plants. However, the newest, truly biotechnological techniques involve **genetic engineering** and produce the **genetically modified (GM)** plants and crops that are so often in the news.

HOW ARE PLANTS GENETICALLY MODIFIED?

Genetic modification usually involves the transfer of a single gene, together with an obvious marker so that the plant breeder knows which plants have taken up the gene. The gene chosen usually codes for a protein that gives the plant a higher rate of growth, or makes it resistant to drought, pests or weedkillers, or that makes part of the plant more useful for the food industry (see the Flavr Savr tomato on page 114).

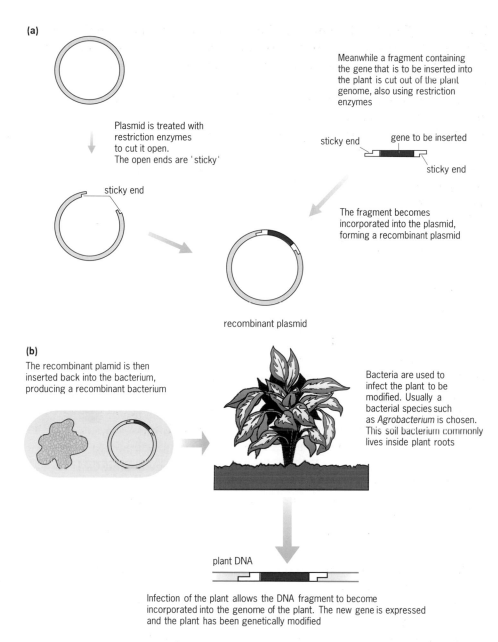

(a)

Plasmid is treated with restriction enzymes to cut it open. The open ends are 'sticky'

Meanwhile a fragment containing the gene that is to be inserted into the plant is cut out of the plant genome, also using restriction enzymes

sticky end

gene to be inserted

sticky end

sticky end

The fragment becomes incorporated into the plasmid, forming a recombinant plasmid

recombinant plasmid

(b)

The recombinant plamid is then inserted back into the bacterium, producing a recombinant bacterium

Bacteria are used to infect the plant to be modified. Usually a bacterial species such as *Agrobacterium* is chosen. This soil bacterium commonly lives inside plant roots

plant DNA

Infection of the plant allows the DNA fragment to become incorporated into the genome of the plant. The new gene is expressed and the plant has been genetically modified

Fig 6.5 (a) Production of a recombinant bacterial plasmid containing a gene for transfer into a plant (b) Bacteria such as *Agrobacterium* sp. are forced to take up the recombinant plasmid and are then introduced into the plant. The plasmid enters plant cells where it opens up to release the gene. This is then incorporated into the DNA of the plant. As the plant genome is expressed, so is the new gene. This diagram is not drawn to scale

Fig 6.5(a) shows how a gene is first incorporated into a **plasmid** (a small circular piece of DNA) and is then introduced into cells of the plant that is being modified (Fig 6.5b). Sometimes bacteria can infect a whole plant; other techniques introduce the bacteria carrying the plasmid into individual plant cells. These are then cultured to produce plant tissue and then whole plants that carry the 'foreign' gene.

THE ADVANTAGES OF GENETICALLY ENGINEERED PLANTS

The main advantage of such a specific transfer of one or perhaps a handful of genes is that the plant breeder can avoid transferring lots of other genes, which may not be identified. The only changes that will occur in the modified plants are the changes that result from the known gene. Genetic engineering is therefore more predictable and precise than 'traditional' plant breeding techniques.

AND THE DISADVANTAGES

The major problem that is preventing more widespread use of genetically modified crops and foods is public suspicion and hostility.

SEE QUESTIONS 1 AND 2

✓ REMEMBER THIS

Genetically modified plants could also be called **transgenic plants**, but this tends to be a term used more commonly for animals that have been bred to carry specific genes.

Fig 6.6 The use of genetic engineering in food production has provoked strong feelings in many people

Opponents of GM foods are particularly vocal in the UK (Fig 6.6). In other countries in Europe, and in the USA, people have accepted them more readily.

Here are some of the arguments made by people opposed to GM foods:

● GM foods are bad for your health: swapping genes from one organism to another is unnatural, and therefore dangerous. What might be the effect of eating 'foreign' genes from bacteria?

● Genes from GM plants might escape and get into other plants, creating 'superweeds'. Bacteria might pick up the plasmids that have been used to carry genes into plant cells, and then infect other plants. The genes may spread like wildfire, making weeds resistant to all known weedkillers.

● GM plants might breed with traditional varieties, producing hybrids that are no use. Organic farmers might find that crops that have taken years to establish become contaminated with genes from GM crops.

● Growing GM foods resistant to weedkillers and pesticides will increase the use of such chemicals, and insects and other useful wildlife will be lost.

● The technology behind GM crops is only available to the multinational, large agrobusinesses, not the small, independent farmer. Organic farmers may find their produce outsold by GM foods, which can be produced at a much lower cost.

EXAMPLES OF GM FOODS AND CROPS

THE POMATO

One of the earliest GM foods produced was the 'pomato', a laboratory-generated combination of two members of the *Solanaceae* family, the potato and the tomato. This could grow into a plant but could not breed. In 1994, however, researchers managed to cross the pomato **hybrid** with an ordinary potato but not without difficulty. Today, groups around the world are still trying to improve the pomato, and are also attempting to use the hybrid to transfer frost-resistance genes from the tomato plant to the potato.

Tomato is better able to withstand frost than potato. Dutch potato growers sow seed potatoes in April. If there is frost at night the leaves freeze and wither. The potato survives but yields are considerably reduced. Farmers who produce seed potatoes are most troubled by this problem. They want to sow as early as possible so that they can also harvest early in order to avoid viruses.

The plan is to cross the hybrid pomato with a potato. Each time sex cells are made through a process of **meiosis** (see Chapter 25), some of the chromosomes disappear in the hybrid. This forces single tomato chromosomes to pair up with a potato partner. The partner chromosomes then exchange parts and segments of the tomato chromosome become permanently built into a potato chromosome. However, this transfer of tomato chromosome parts to the potato does not happen as frequently as the researchers had expected, and success has so far eluded them.

? TEST YOURSELF

B Look at the list of objections made about GM crops and foods. From a scientific point of view, say which ones you think are valid, and which are not. Explain how you came to your decision.

TRANSGENIC TOMATOES COULD HAVE AN EXCITING FUTURE

Although the Flavr Savr tomato is no longer available (see the Science Feature on page 114), other transgenic tomatoes are being developed all the time. Here are a few of the developments we might see in the next few years:

- Transgenic tomatoes may be produced containing genes that help them tolerate salty growing conditions, drought, flood and extreme temperatures. One gene that regulates the plant's ability to cope with cold weather has already been identified.
- A vitamin-rich tomato has been produced that could help to prevent heart disease and cancer. This tomato has about four times the normal levels of beta-carotene, which the body uses to make vitamin A, and twice the levels of lycopene (a close cousin of beta-carotene), the compound that helps make tomatoes red. Eating lots of lycopene may reduce the risk of some cancers, particularly prostate cancer, so transgenic tomatoes may be eaten as part of a cancer prevention diet.
- Herbicide-resistant, virus-resistant and pest-resistant tomatoes are in development. *Bacillus thuringiensis* (Bt) is a naturally occurring soil bacterium that produces a protein that kills a range of common insects once it is ingested. The *Bt* gene has been isolated and inserted into crops including tomatoes.
- Tomatoes may be used as edible vaccines. Tomatoes are being engineered to carry genes that can act as vaccines for diseases such as rabies.

REMEMBER THIS

Seed potatoes may only be sold if they are certified 'virus-free'.

SCIENCE FEATURE Transgenic fish

As you will see in more detail in Chapter 30, the DNA that carries a specific gene can be introduced into an organism of the same or a different species by injecting it directly into embryo cells using a tiny glass pipette. Several species of transgenic animal have, for example, been bred to carry an extra growth hormone gene. As soon as the organism starts to develop, and this gene is expressed, it is exposed to a much higher concentration of growth hormone than usual, sometimes with fairly predictable results. Northern pike (a marine pike) with the growth hormone gene are two to six times bigger than fish without the extra gene. In one case, a fish 13 times larger than normal was created.

However, the effect that these fish have on their environment is less predictable. If this is a fish farm, they can grow as big as they can, and then be 'harvested'. However, if the fish are allowed into the open sea, there is a real danger that they could out-compete their smaller siblings and so disrupt the delicate balance of their ecosystem. Perhaps there is a case for introducing other genes as well as the growth hormone gene so that any transgenic fish are unable to breed, and therefore unable to take over.

Fig 6.7 Transgenic northern pike, *Esox lucius*

⚠ SCIENCE FEATURE The Flavr Savr tomato

The Flavr Savr tomato, the world's first transgenic tomato, was produced by an American company called Calgene in the late 1980s. It makes use of something called anti-sense technology (Fig 6.8). Basically, a gene is introduced into the tomato plant that produces messenger RNA. This blocks the transcription of a gene in the tomato that codes for an enzyme called polygalacturonase. This enzyme breaks down the pectin that normally holds cell walls together. When it is present, it causes the tomato to go soft and squishy, and the fruit rots. The messenger RNA produced by the anti-sense gene stops the enzyme being made, and the fruit stays much firmer. Other parts of the ripening process are not affected so the flavour of a ripe tomato still develops.

The major advantage of the Flavr Savr tomato was that it could be left on the vine to ripen for longer, producing tomatoes with a better flavour, but that did not become mush when transported.

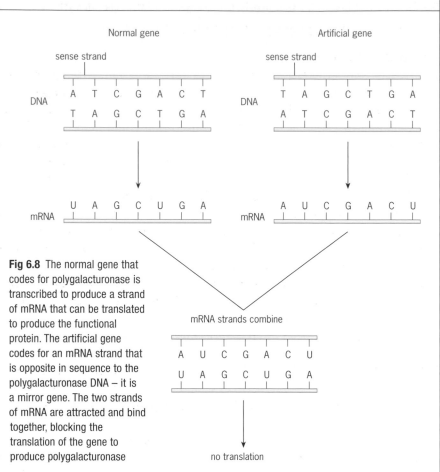

Fig 6.8 The normal gene that codes for polygalacturonase is transcribed to produce a strand of mRNA that can be translated to produce the functional protein. The artificial gene codes for an mRNA strand that is opposite in sequence to the polygalacturonase DNA – it is a mirror gene. The two strands of mRNA are attracted and bind together, blocking the translation of the gene to produce polygalacturonase

Fig 6.9 GM tomato on vine

Release of the Flavr Savr was reviewed by the Food and Drug Administration in the USA in the spring of 1994. It was found to be as safe as conventionally produced tomatoes. This is the first time a whole food produced by biotechnology had been evaluated for its public safety. However, concern still surrounded Flavr Savr's introduction to the market. People worried about whether the tomato was a danger to health, whether it would give them food allergies, whether it contained 'unnatural' toxins. It was termed a 'Frankenfood' and in 1997 it was withdrawn from the market by the company that developed it.

Fig 6.10 These tomatoes are the same age and variety. The deterioration of those on the right is normal. The tomatoes on the left have been genetically modified and stay fresher for longer

3 BIOTECHNOLOGY AND FOOD

LACTIC ACID FERMENTATION

MAKING YOGHURT

Yoghurt is basically milk (cow, goat or ewe) that is fermented by bacteria. Although it originated in Western Asia and Eastern Europe, it is now eaten all over the world. To make yoghurt, a starter culture of the lactic acid bacteria *Lactobacillus bulgaricus* and *Streptococcus thermophilus* is added to whole or skimmed milk. The bacteria multiply and ferment lactose, the disaccharide sugar in the milk, producing lactic acid. As you will know if you have ever added lemon juice to milk, acid causes the milk to curdle. Separation of the acidified milk into solid curds gives yoghurt its taste and texture.

Yoghurts can be pasteurised to destroy the fermentation bacteria but if this is not done, the result is 'live' yoghurts. Many supermarkets now sell yoghurts they describe as 'bio-yoghurts'. These have been produced in the usual way, left unpasteurised, and then extra bacteria are added. Bio-yoghurts usually contain live *Lactobacillus acidophilus* and *Bifidobacterium bifidum*. Various health claims are made for these yoghurts, but whether they actually do keep you healthy or not is still an open question (see the Assignment on page 132 of this chapter).

CHEESE-MAKING: ONE OF THE OLDEST BIOTECHNOLOGIES

Cheese-making dates back at least 5000 years and the basic method used has changed very little over that time. Three main ingredients are needed:

- Milk.
- A source of the protein-degrading enzyme **chymosin**, which causes the milk proteins to clump together. Traditionally, this was added in the form of calf **rennet** (chymosin can also be called **rennin**). Non-animal sources of this enzyme are now available and can be used in the manufacture of cheese suitable for vegetarians.
- A starter culture of lactic acid bacteria.

In many ways, cheese-making is similar to yoghurt-making. The lactic acid starter culture is added to the milk to convert the lactose to lactic acid, which helps to preserve the cheese. Chymosin causes the milk to set, producing hard cheese. Once bacterial fermentation stops, the maturation of the cheese then begins. As the bacteria die, they are digested by their own enzymes – a process called **autolysis**. This process produces more enzymes, including **peptidases**, that produce the strong flavours that we associate with a 'good' cheese. Some strains of bacteria produce carbon dioxide during maturation – this makes the holes in cheeses such as Gouda and Emmental.

A fungus of the Penicillin genus (*Penicillium roquefortii*) is used to inoculate fermented curd to produce blue cheeses such as Stilton and Danish Blue. This fungus grows through the cheese during the maturation process, producing the characteristic blue-green veins.

Fig 6.11 The pictures show the holes in Gruyere (below) and a blue cheese, Roquefort, being probed during its maturation to control the development of the blue veins (above)

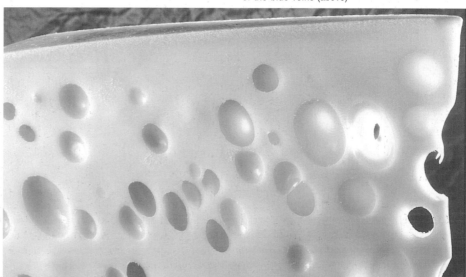

FERMENTATION USING YEAST

Yeast is an essential microorganism in the production of bread, wine and beer. Under the right conditions, yeast will ferment carbohydrates to release carbon dioxide and alcohol in a series of chemical reactions.

BREAD-MAKING

Bread is basically a mixture of flour, water and a little fat that is baked. Some breads in the world are still produced like this (pitta bread, for example) but yeast fermentation is usually used to cause the dough to rise, so that a lighter, less dense loaf can be made. As the flour, water and yeast are mixed, enzymes called amylases that are present in the flour break down the starch into the disaccharide, maltose. Other enzymes break this down into glucose, which is then fermented by the yeast to carbon dioxide and alcohol. In bread, the alcohol evaporates during baking, but the carbon dioxide bubbles become trapped in the sticky dough, causing the baked loaf to have lots of tiny pockets.

WINE AND BEER

In wine-making, the fermentation of grape juice is a complex process that involves yeasts, bacteria and filamentous fungi – but the yeasts have the starring role. During fermentation, yeasts use the sugars and other components of grape juice and convert them into ethanol, carbon dioxide and other end-products that contribute to the chemical composition and taste of wine. Fermentation stops when the yeast becomes poisoned by its own waste – the alcohol – which is why drinks made by fermentation only are no stronger than about 14 per cent alcohol by volume. Spirits such as whisky and gin, which have a much higher alcohol content, are **distilled** from a fermented mixture.

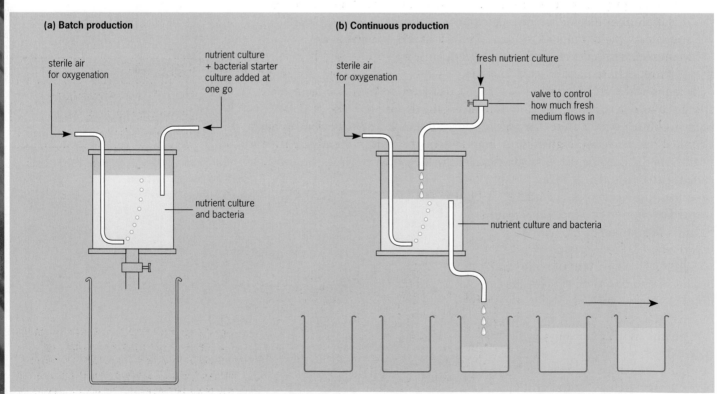

(a) Batch production

sterile air for oxygenation

nutrient culture + bacterial starter culture added at one go

nutrient culture and bacteria

(b) Continuous production

sterile air for oxygenation

fresh nutrient culture

valve to control how much fresh medium flows in

nutrient culture and bacteria

Fig 6.12 (a) Batch production: when fermentation is complete, the entire contents of the tank are released for processing. The tank is cleaned and then the whole process is repeated. Each production is a different batch
(b) Continuous production: fermentation is continuous and can carry on indefinitely as long as fresh medium is added and product is removed and the conditions for optimum growth are maintained. Continuous production runs a higher risk of contamination than batch production, so great care has to be taken to keep the incoming medium sterile, apart from the microorganism in the culture

Most wine-makers are more interested in the alcoholic by-product of yeast fermentation but a brewer producing beer needs both the alcohol and the carbon dioxide made by the yeast. Carbon dioxide gives the beer its characteristic fizziness.

The brewing industry uses barley as the source of the food that the yeast ferments to make the alcohol in beer. However, barley stores food in the form of starch – which yeast cannot use directly. In order to solve this, the brewer allows the barley grains (seeds) to germinate. During germination, enzymes in the barley convert the starch into maltose sugar, which the yeast can ferment. This process is called **malting**.

Beer production is made as efficient as possible by providing the best possible conditions for yeast to grow and ferment. This means that the temperature, oxygen supply and amount of glucose must be carefully controlled, and unwanted microorganisms must be kept out. The easiest way to make sure that this happens throughout the fermentation is to set up all the conditions at the start, together with the raw materials, and then to leave the whole system closed and untouched until the fermentation is complete. This has the disadvantage that production is only possible in batches rather than a continuous process. The main features of batch processing and continuous processing are shown in Fig 6.12.

USING OTHER MICROORGANISMS AS A FOOD SOURCE

Biotechnology has been used in the past few years to make foods that are as rich in protein as meat, but that are much cheaper. Pruteen™ is one example (Fig 6.13), mycoprotein (for example, Quorn™) is another. Both can be eaten by vegetarians. Mycoprotein has the following nutritional advantages over meat:
- It has no animal fat, little overall fat and no cholesterol.
- It has a high protein content (as high as that of skimmed milk protein).
- It is high in fibre.
- It contains useful amounts of trace elements and B vitamins.

Mycoprotein is produced by the fungus *Fusarium graminearum*, which is related to mushrooms and truffles. This fungus can use the carbohydrate present in cheap, readily available carbohydrates – wheat in the UK, potato in Ireland, and cassava, rice or sugar in tropical countries, for example. Mycoprotein contains 45 per cent protein and 13 per cent fat, a composition similar to that of grilled beef. It is also high in fibre and has a complete amino acid content.

✓ REMEMBER THIS

In exam questions on fermentation you will be expected to know about the basics of respiration, particularly glycolysis, and also be able to apply your knowledge of the properties of enzymes.

SEE QUESTIONS 3 AND 4

✓ REMEMBER THIS

Mycoprotein is made by a filamentous fungus – one that produces lots of thin strands called **hyphae**.

✓ REMEMBER THIS

Pruteen™ is a microbial protein produced by bacteria that can break down methanol.

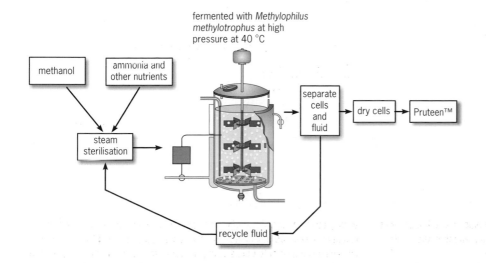

Fig 6.13 Pruteen™ production. Pruteen™ is the trade name for microbial protein produced by growing *Methylophilus methylotrophus* bacteria on methanol, which is derived from methane or natural gas. Since the product is a low-value commodity, large quantities have to be manufactured to make profits. Usually very large (3000 m³) airlift fermenters are used with a diameter of 7 m and a height of 60 m. These can be run for 100 days at a time

The industrial fermenters currently being used to manufacture mycoprotein are 40 metres high (similar in height to Nelson's Column). They run continuously for 6 weeks, after which there is a 2-week period for cleaning and preparing the fermenter for the next run. After production, the mycoprotein looks like pastry, but it is then mixed with binding agents and colours to form a product that looks like different kinds of meat.

OTHER FERMENTED FOOD PRODUCTS

SAUERKRAUT

Sauerkraut is a German delicacy that is now popular in other parts of the world, particularly the USA. It is made by fermenting cabbage that has been salted. The main bacteria responsible for the fermentation are lactic acid bacteria. *Leuconostoc mesenteroides* ferments the cabbage juices to produce lactic acid, acetic acid, carbon dioxide and ethanol. Two others, *Lactobacillus plantarum*, which produces lactic acid only, and *Lactobacillus brevis*, are used to complete the process.

TOFU, TVP AND SOY SAUCE

Biotechnology is also used to process soya beans to give the following food products:

- **Textured vegetable protein (TVP)** – defatted soya flour that has been processed and dried to give a substance with a sponge-like texture that resembles meat. Soya beans are **dehulled** (their skins are taken off) and their oil is extracted before they are ground into flour. This flour is then mixed with water to remove soluble carbohydrate and the residue is textured by either **spinning** or **extrusion**. Extrusion involves passing heated soya residue from a high-pressure area to a reduced pressure area through a nozzle – this causes the soya protein to expand. The soya protein is then dehydrated and may be either cut into small chunks or ground into granules.

- **Tofu** – soya bean curd made from coagulated soya milk. Soya beans are soaked, crushed and heated to produce soya milk to which a coagulating agent such as calcium sulphate or calcium chloride is added. The resulting soya curd is then pressed to give tofu. Tofu is sometimes known as soya cheese, and is sold as blocks packaged in water.

- **Tempeh** – a fermented soya bean paste made by inoculating cooked soya beans with the fungus *Rhizopus oligosporous*. This forms a mycelium holding the soya beans together and is responsible for the black specks in tempeh. Tempeh has a chewy texture and distinctive flavour and can be used as a meat substitute in recipes. It may be deep-fried, shallow-fried, baked or steamed.

- **Miso** – a fermented condiment made from soya beans, grain (rice or barley), salt and water. Miso production involves steaming polished rice, which is then inoculated with the fungus *Aspergillus oryzae* and left to ferment to give an end-product called **koji**. Koji is then mixed with soya beans that have been heated and extruded to form strands, together with salt and water. This is then left to ferment in large vats. Miso varies widely in flavour, colour, texture and aroma. It is used to give flavour to soups, stews, casseroles and sauces.

- **Soy sauce** – true soy sauce, called **shoyu**, is made by fermenting soya beans with cracked roasted wheat, salt and water. **Tamari** is similar but slightly stronger and made without wheat (and so is gluten-free).

Fig 6.14 Soya products: dried soya beans, soya milk (blue bowl) and two types of tofu or bean curd – dried tofu (upper right) and fresh tofu (brown bowl)

Fermentation for shoyu and tamari takes about one year. Much of the soy sauce available in supermarkets is not true soy sauce but is made by chemical hydrolysis from defatted soya flour, caramel colouring and corn syrup without any fermentation process.

4 BIOTECHNOLOGY IN INDUSTRY

Enzymes were first used on a large scale by the textile industry. Thread used for weaving is protected by a coating of starch paste – useful during the weaving process to make the thread more 'slippy' so that weaving was easier, but totally useless afterwards. The starch used to be removed by acids, alkalis or oxidising agents but these often discoloured the cloth. In the early part of the twentieth century, the industry started using enzyme extracts from malt, and later from an animal pancreas. A thermostable bacterial amylase was discovered in 1917, but its use in the mass production of cloth took a few years to perfect. At around the same time, the leather industry also started to experiment with enzymes to reduce costs. Protease enzymes began to replace calcium hydroxide and sodium sulphide for removing animal hair from skins.

These early attempts at using enzymes seem crude by today's standards and it took until after 1945 for large-scale production and use of enzymes to be commonplace in industry.

ENZYMES IN THE INDUSTRIES OF TODAY

The ability of enzymes to catalyse specific chemical reactions at body temperature makes them very useful tools in the commercial world. Table 6.1 outlines some of industrial applications of enzymes. This is a rapidly changing field, and new applications of enzyme technology appear all the time.

REMEMBER THIS

Thermostable enzymes, those that work outside the normal physiological range of the human body, are particularly useful in industry. Many have been discovered in bacteria and other organisms that live in the extreme environments of deep ocean vents. The enzymes in these organisms are not denatured at high temperatures – and so can be used to catalyse industrial processes at temperatures approaching 100 °C. An application of thermostable enzymes in washing powders is in Chapter 5 (page 97).

Table 6.1 Some applications of enzymes			(Source: Biochemical Society Guidance Notes 3, Enzymes and their role in biotechnology)
Enzyme	**Reaction**	**Source of enzyme**	**Application**
Industrial applications			
α-amylase	breaks down starch	bacteria	converts starch to glucose in the food industry
glucose isomerase	converts glucose to fructose	fungi	production of high-fructose syrups
proteases	digest protein	bacteria	washing powder
rennin	clots milk protein	animal stomach linings; bacteria	cheese-making
catalase	splits hydrogen peroxide into $H_2O + O_2$	bacteria; animal livers	turns latex into foam rubber by producing gas
β-galactosidase	hydrolyses lactose	fungi	in dairy industry, hydrolyses lactose in milk or whey
Medical applications			
L-asparaginase	removes L-asparagine from tissues – this nutrient is needed for tumour growth	bacteria (*E. coli*)	cancer chemotherapy – particularly leukaemia
urokinase	breaks down blood clots	human urine	removes blood clots, eg in heart disease patients
Analytical applications			
glucose oxidase	oxidises glucose	fungi	used to test for blood glucose, eg in Clinistix™, used by people with diabetes
luciferase	produces light	marine bacteria; fireflies	binds to particular chemicals indicating their presence, eg used to detect bacterial contamination of food
Manipulative applications			
lysozyme	breaks 1–4 glycosidic bonds	hen egg white	disrupts bacterial cell walls
endonucleases	break DNA into fragments	bacteria	used in genetic manipulation techniques, eg gene transfer, DNA fingerprinting

COMMERCIAL PRODUCTION OF ENZYMES

Table 6.2 shows some important industrial enzymes and their sources. As you can see, the majority of enzymes used in industry are **extracellular**. This type of enzyme is easier to isolate, as it is secreted into the growth medium by the source organism, and can be extracted relatively easily. **Intracellular** enzymes are more tricky, as their removal requires the cultured cells to be broken open, and then the enzyme must be separated from all the other proteins and other molecules. Fig 6.15 provides a flow chart to show how a protease for use in a stain remover is produced on a large scale from a bacterial culture.

Table 6.2 Some important industrial enzymes and their sources

	Enzyme	Source	Intracellular/ extracellular	Industry/ industrial use
Animal enzymes	catalase	liver	I	food
	chymotrypsin	pancreas	E	leather
	lipase	pancreas	E	food
	rennet	abomasum	E	cheese
	trypsin	pancreas	E	leather
Plant enzymes	actinidin	kiwi fruit	E	food
	α-amylase	malted barley	E	brewing
	β-amylase	malted barley	E	brewing
	bromelain	pineapple latex	E	brewing
	β-glucanase	malted barley	E	brewing
	ficin	fig latex	E	food
	lipoxygenase	soya beans	I	food
	papain	pawpaw latex	E	meat
Bacterial enzymes	α-amylase	*Bacillus*	E	starch
	β-amylase	*Bacillus*	E	starch
	asparaginase	*Escherichia coli*	I	health
	glucose isomerase	*Bacillus*	I	fructose syrup
	penicillin amidase	*Bacillus*	I	pharmaceutical
	protease	*Bacillus*	E	detergent
	pullulanase	*Klebsiella*	E	starch
Fungal enzymes	α-amylase	*Aspergillus*	E	baking
	aminoacylase	*Aspergillus*	I	pharmaceutical
	glucoamylase	*Aspergillus*	E	starch
	catalase	*Aspergillus*	I	food
	cellulase	*Trichoderma*	E	waste
	dextranase	*Penicillium*	E	food
	glucose oxidase	*Aspergillus*	I	food
	lactase	*Aspergillus*	E	dairy
	lipase	*Rhizopus*	E	food
	rennet	*Mucor miehei*	E	cheese
	pectinase	*Aspergillus*	E	drinks
	pectin lyase	*Aspergillus*	E	drinks
	protease	*Aspergillus*	E	baking
	raffinase	*Mortierella*	I	food
Yeast enzymes	invertase	*Saccharomyces*	I/E	confectionery
	lactase	*Kluyveromyces*	I/E	dairy
	lipase	*Candida*	E	food
	raffinase	*Saccharomyces*	I	food

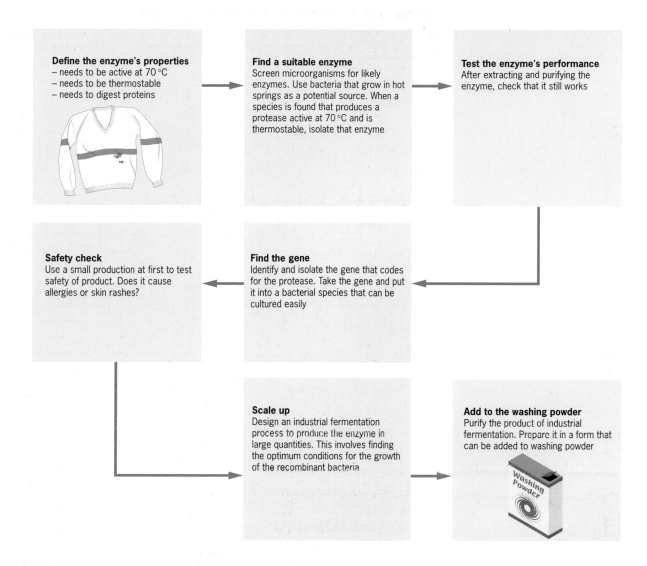

Define the enzyme's properties
– needs to be active at 70 °C
– needs to be thermostable
– needs to digest proteins

Find a suitable enzyme
Screen microorganisms for likely enzymes. Use bacteria that grow in hot springs as a potential source. When a species is found that produces a protease active at 70 °C and is thermostable, isolate that enzyme

Test the enzyme's performance
After extracting and purifying the enzyme, check that it still works

Safety check
Use a small production at first to test safety of product. Does it cause allergies or skin rashes?

Find the gene
Identify and isolate the gene that codes for the protease. Take the gene and put it into a bacterial species that can be cultured easily

Scale up
Design an industrial fermentation process to produce the enzyme in large quantities. This involves finding the optimum conditions for the growth of the recombinant bacteria

Add to the washing powder
Purify the product of industrial fermentation. Prepare it in a form that can be added to washing powder

IMMOBILISED ENZYMES

In an industrial process controlled by an enzyme, the enzyme is usually the component that costs the most. Like all catalysts, enzymes can be used several times, so it is a shame to waste enzyme. In a process in which the enzyme is simply added to the substances that you want to react, a lot of enzyme is wasted because not all of it can be removed from the products.

To solve this, biotechnologists have developed methods to **immobilise** enzymes (see also Fig 6.16):

- **Cross-linkage**: enzyme molecules can be linked together to make much larger molecular structures that are easy to extract from the products. The enzyme can also be chemically attached to a supporting material.

- **Entrapment**: the enzyme can be mixed with ingredients that form a gel. When the gelling reaction is complete, the enzyme becomes trapped. Gaps in the gel are large enough to let the substrate in, but not let the enzyme out. Enzymes can also be trapped in a compartment behind a semipermeable membrane that, again, lets reactants in, but doesn't allow the enzyme molecules out.

- **Adsorption**: the enzyme can be stuck temporarily to various surfaces – usually this method is only used for experimental studies, not full-scale industrial manufacture.

Fig 6.15 Production of protease from *Bacillus* bacteria on a large scale for use as a stain remover

SEE QUESTION 5

Fig 6.16 Immobilising enzymes

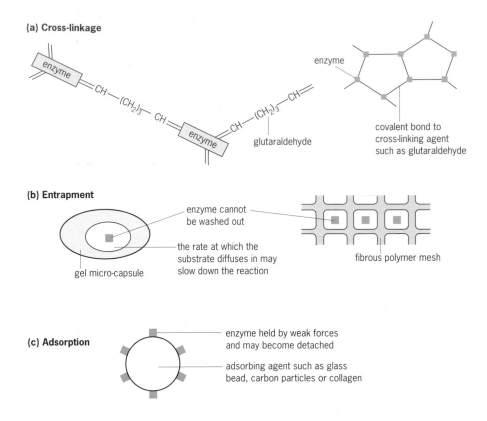

(a) Cross-linkage

enzyme

glutaraldehyde

enzyme

covalent bond to cross-linking agent such as glutaraldehyde

(b) Entrapment

enzyme cannot be washed out

the rate at which the substrate diffuses in may slow down the reaction

gel micro-capsule

fibrous polymer mesh

(c) Adsorption

enzyme held by weak forces and may become detached

adsorbing agent such as glass bead, carbon particles or collagen

The first major commercial use of immobilised enzymes took place in the 1970s, when immobilised **glucose isomerase** was first used to make high-fructose syrups from starches. Immobilised enzymes are now widely used in the food industry and the pharmaceutical industry. Immobilised **penicillin amidase** is, for example, used to prepare different forms of penicillin from the type produced by fungi.

Enzyme immobilisation is also important in **biosensors** (see page 124). These often require immobilisation of an enzyme into an electrode tip or into a paper strip (see the Science Feature on the Clinistix™ test).

BIOFUELS

Biotechnology is also used industrially to produce fuel. A biofuel is defined as a gaseous, liquid or solid fuel that contains energy that has come from a biological source. An example of a biofuel is rapeseed oil, which can be used in place of diesel fuel in modified engines. The methyl ester of this oil, **rapeseed methyl ester (RME)**, can be used in unmodified diesel engines and is sometimes known as **biodiesel**. Other biofuels include **biogas** and **gasohol**.

GASOHOL

Gasohol is produced by fermentation of sugar cane by the yeast *Saccharomyces cerevisiae* (commonly known as 'baker's yeast'). It has been produced on a large scale in Brazil, where it is manufactured in large distilleries and provides an alternative source of energy for motor vehicles. In Brazil all cars are adapted so as to be able to use gasohol or a mixture of gasohol and petrol.

Fig 6.17 Filling up with petrol made from sugar

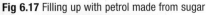

Fig 6.18 A biogas digester in India

BIOGAS

Biogas is a mixture of methane and carbon dioxide that is produced by the anaerobic decomposition of such waste materials as domestic, industrial and agricultural sewage. The decomposition is carried out by **methanogenic bacteria**, anaerobes that respire to produce methane, the main component of biogas. Methane is collected and used as an energy source for domestic processes, such as heating, cooking and lighting. The production of biogas is carried out in special **digesters**, which are widely used in China and India.

SCIENCE FEATURE The Clinistix™ test for glucose in urine

Enzymes are both specific and sensitive: this makes them ideal for use in analysis, which often involves very small samples. One such application is in the analysis of glucose – an important technique in both medicine and industry. You may have heard of Clinistix™ (Fig 6.19) – the sticks used to test for the presence of glucose in urine. Clinistix™ contain two enzymes: glucose oxidase and peroxidase.

Glucose oxidase catalyses the following reaction:

glucose + oxygen → gluconic acid + H_2O_2 (hydrogen peroxide)

In a simple, visible test, the production of peroxide is coupled to the production of a coloured dye or chromagen.

The second enzyme, peroxidase, catalyses the following reaction:

$$DH_2 + H_2O_2 \rightarrow 2H_2O + D$$

D stands for the chromagen, a colourless hydrogen donor. When it loses its hydrogen it becomes coloured. The intensity of colour indicates the amount of glucose present.

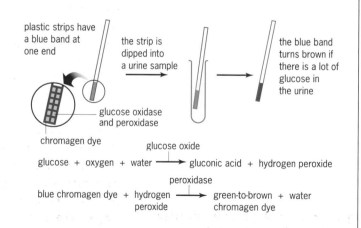

plastic strips have a blue band at one end

the strip is dipped into a urine sample

the blue band turns brown if there is a lot of glucose in the urine

glucose oxidase and peroxidase

chromagen dye

$$\text{glucose} + \text{oxygen} + \text{water} \xrightarrow{\text{glucose oxide}} \text{gluconic acid} + \text{hydrogen peroxide}$$

$$\text{blue chromagen dye} + \text{hydrogen peroxide} \xrightarrow{\text{peroxidase}} \text{green-to-brown chromagen dye} + \text{water}$$

Fig 6.19 The urine glucose test for diabetes. Glucose oxidase, peroxidase and DH_2 are fixed on a cellulose fibre pad. When this is added to a sample of urine, the colour reaction gives a quantitative measure of glucose

SEE ALSO THE ASSIGNMENT IN CHAPTER 14 – LIVING WITH DIABETES.

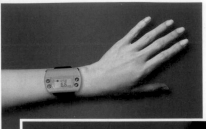

Fig 6.20 The Glucowatch® Biographer

SEE QUESTION 6

Fig 6.21 How the Glucowatch® Biographer detects the level of glucose in the blood

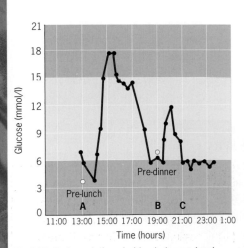

Fig 6.22 Daily variations in blood glucose level in a person with diabetes

5 BIOTECHNOLOGY IN MEDICINE

BIOSENSORS

Biosensors are being developed for many applications in medicine. One such device that is already available is the Glucowatch® Biographer (Fig 6.20), a watch that is able to measure blood glucose levels in diabetics without puncturing the skin to take a blood sample.

The watch looks like an ordinary watch; it is worn on the wrist and it has an LCD dial. however, people with diabetes use it to measure their blood glucose level, not the time.

HOW THE GLUCOSE BIOSENSOR WORKS

The back of the Glucowatch® Biographer that comes into contact with the skin has two circular flat plates, as shown in Fig 6.21. The two plates generate a low electric current that is directed into the skin. This causes positive and negative ions to move out of the skin, pulling glucose molecules from the blood with them. The glucose collects in gel discs that contain glucose oxidase. This enzyme catalyses a reaction that generates a small electrical signal. The size of the signal is proportional to the glucose level in the blood. The Glucowatch® Biographer uses this signal to calculate a reading, which is then displayed.

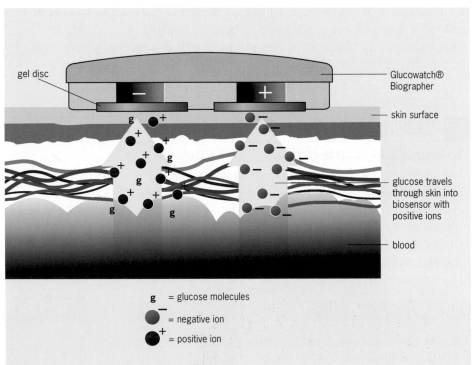

g = glucose molecules

− = negative ion

+ = positive ion

HOW IS THE BIOSENSOR USED?

Usually, people with diabetes who control their condition by insulin injections measure their blood glucose using the standard finger-prick method. They do this three or four times a day; before each meal and last thing at night. The disadvantage of this is shown in Fig 6.22. The curve shows how the blood glucose level actually changes during a 12-hour period. This can be detected by the Glucowatch® Biographer because it takes readings every 20 minutes, rather than by much less frequent finger-prick tests. These show the glucose levels at points A, B and C, lulling the person into thinking that their blood glucose is stable.

MONOCLONAL ANTIBODIES

Monoclonal antibodies are antibodies that can be produced in large amounts and that bind only to one, very specific target. This is usually part of a molecule, not even a whole molecule.

HOW ARE MONOCLONAL ANTIBODIES PRODUCED?

The method used to create monoclonal antibodies is called **hybridoma technology**. It was first devised in the 1980s by Cesar Milstein, working in laboratories in Cambridge, UK. It basically involves fusing a cell that produces antibodies with one that is **immortal**. Fig 6.23 summarises the process of making a monoclonal antibody.

The first step in the production of a mouse monoclonal antibody is to immunise a mouse with the molecule against which we want the monoclonal antibody to react. White blood cells in the mouse produce a range of different antibodies, directed against different parts of the molecule.

In the mouse, cells called B lymphocytes that are responsible for making these antibodies are isolated from the mouse's spleen. These cells are then fused with tumour cells that divide continuously and rapidly. This creates a single, large cell called a **hybridoma**. These hybridomas show properties of both 'parent' cells; they produce antibodies and they divide continuously.

One hybridoma can divide to produce a whole clone of identical hybridomas, all secreting the same antibody molecule. Once a clone has been established and is growing well in culture, the antibody produced by this collection of cells is called a **monoclonal antibody**.

MONOCLONAL ANTIBODIES USED IN DIAGNOSIS

Monoclonal antibodies can be linked to a radioactive isotope and, in this form, they have been used for diagnosing and monitoring disease since the early 1980s. Some forms of cancer can be diagnosed using monoclonal antibodies directed against molecules found only on the surface of cancer cells. When injected into patients, such a monoclonal antibody will carry its radioactive marker straight to the tumour, where it can be detected using **CT (computerised tomography)** scanning (Fig 6.24).

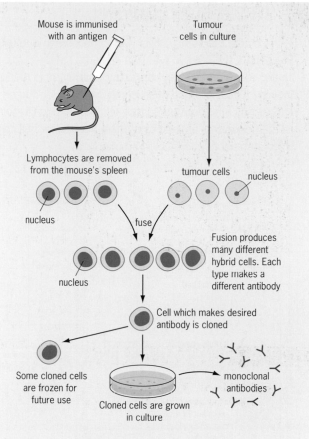

Fig 6.23 How a monoclonal antibody is made. Tumour cells, which divide without control, are fused with lymphocytes that produce antibody. The hybrid cell produced shares the properties of the two cells that formed it. It divides endlessly (it is immortal) and it produces an antibody. When one cell is then used to form a whole culture, all the cells are clones. They all produce the same antibody, a **monoclonal** antibody

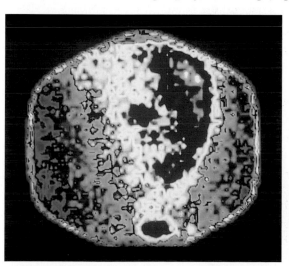

Fig 6.24 A cancer of the colon. Radioactivity from the increased concentration of monoclonal antibody in the colonic cancer cells is shown by the red, pink and white areas at the centre-right

 **REMEMBER THIS**

Monoclonal antibodies also form the basis of modern pregnancy testing kits (see Science Feature box overleaf).

SCIENCE FEATURE : The pregnancy test

By the time a pregnant woman would have been due for her next period, the embryo will have implanted and would be starting to secrete human chorionic gonadotrophin (hCG) Enough of this hormone passes into the urine for it to be detected by a modern pregnancy testing kit. This kit makes use of monoclonal antibodies that are specific for hCG and which are attached to a coloured chemical.

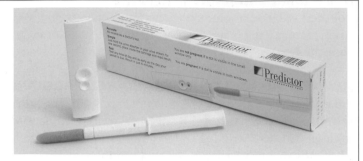

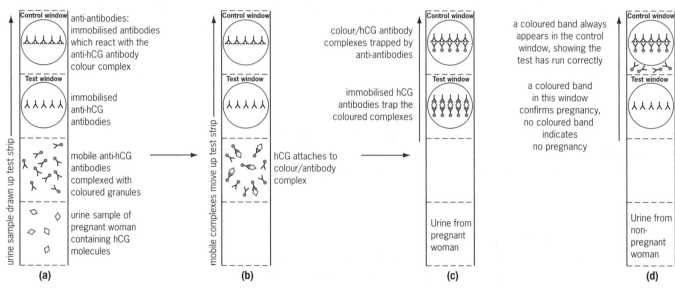

Fig 6.25 How a home pregnancy test kit works for a pregnant woman (a, b and c) and for a woman who is not pregnant (d)
(a) Urine is drawn up the test strip by capillary action
(b) hCG attached to the colour/antibody complex moves up the test strip

(c) In the urine of a pregnant woman, hCG attached to the colour/antibody complexes moves up to dock with the immobilised (fixed) hCG antibodies in the test window. All the complexes are stopped at one place, forming a visible coloured band and confirming pregnancy

(d) For the urine of a non-pregnant woman, the colour/antibody complexes have no attached hCG, so all the complexes move past the immobilised hCG antibodies in the test window and attach to the anti-antibodies in the control window, showing that the test has worked

TREATING DISEASE WITH MONOCLONAL ANTIBODIES

Many different diseases can now be treated with monoclonal antibodies. The antibody is made against a key molecule on the diseased cell, such as a protein only expressed by cancer cells, and is then linked with a drug molecule, or a radioactive isotope. The monoclonal antibody is injected into the patient and it 'homes in' on the diseased cells, binding to them and releasing its toxic passenger.

A few examples of how monoclonal antibodies are used to treat disease are shown in Table 6.3.

Table 6.3 Treatment with monoclonal antibodies

Disease	Monoclonal Ab	How it works
Non-Hodgkin's lymphoma	Rituximab	this antibody binds to molecules of B cells and kills them directly
Non-Hodgkin's lymphoma	^{131}I Tositumomab	also binds to molecules on B cells, but also carries a radioactive isotope that it carries to the tumour
Breast cancer	Trastuzumab	targets a protein found in one-third of women with breast cancer, binds to it, and slows down tumour growth
Cardiovascular disease	Abciximab	targets clotting proteins to prevent clot formation in patients undergoing angioplasty
Crohn's disease	Infliximab	binds to a protein that causes inflammation in the bowel, preventing it from causing further damage
Rheumatoid arthritis	Etanercept	binds to a protein in the body that causes inflammation, preventing it from causing further damage

RECOMBINANT PROTEINS AND THEIR USE IN MEDICINE

RECOMBINANT HUMAN INSULIN

We saw earlier in this chapter that bovine growth hormone can now be produced in bacteria as a recombinant protein. Other proteins of medical importance can also be manufactured in large quantities in this way. An important example is human insulin.

ANIMAL VERSUS HUMAN INSULIN

The daily doses of insulin that people with diabetes use to manage their glucose levels can come from several sources, including the pancreas of a pig or cow. Because pig insulin is nearly identical to human insulin, pigs used to be the most common source. However, some people reacted badly to 'foreign' insulin from pigs; researchers argued that they might be better off with human insulin.

Pharmaceutical companies now use genetic engineering to produce large quantities of human insulin for people around the world.

The common bacterium *Escherichia coli* (*E. coli*) is used to produce insulin using recombinant DNA technology. The gene for producing insulin is inserted into DNA of *E. coli*. These genetically engineered bacteria are turned into tiny insulin-producing factories. When grown in bulk in large **industrial fermenters**, they can easily produce insulin in the vast amounts needed to supply people with diabetes worldwide.

Human insulin is now widespread, and the use of purified pig insulin and insulin from other animal sources is being phased out. However, some people are unhappy about that. A few have had bad reactions to human insulin, although they had no problems with animal insulin. They are arguing that both products should be made available, so that they can have a choice if needed.

REMEMBER THIS

The hormones insulin and glucagon work antagonistically to control blood glucose levels. Insulin is produced if blood glucose rises beyond the normal range; it lowers blood glucose by making cells in the liver and muscles more permeable to glucose. Glucagon is produced when blood glucose falls; it activates enzymes inside cells to convert glycogen into glucose. (See Chapter 14 Homeostasis)

SEE QUESTION 7

SCIENCE FEATURE Farming or pharming?

A new and highly experimental use of genetic engineering is 'pharming'. This involves farming, but its objective is not to produce food. The idea is to genetically engineer crop plants and animals usually used for food so that they can produce medicines. The technique is difficult and expensive because it requires a transgenic organism that is bred to carry a human gene that makes a particular protein – it is possible, for example, to genetically engineer a cow to produce human haemoglobin in its milk. Once transgenic embryos have been produced in the laboratory, these are then implanted into surrogate mothers and carried to term, resulting in transgenic cows that secrete human haemoglobin in their milk.

Milk from a cow is relatively easy to obtain in large amounts; the secreted haemoglobin is then extracted and purified. This process has been used to produce other blood components, human growth hormone and large quantities of human proteins needed for research.

Fig 6.26
Transgenic sheep. These sheep have a human gene which makes them produce α-1-antitrypsin (AAT) in their milk. People who lack AAT suffer from emphysema

SUMMARY

- Biotechnology is defined generally as the **manipulation** of biological organisms, systems or processes for the benefit of people, in areas such as agriculture, food production and medicine.
- Biotechnology is widely used in agriculture. For example, bovine growth hormone injections are used to increase milk production in cows; artificial insemination of a prime cow using semen from a prime bull is used to generate multiple embryos carried by surrogate mothers (ordinary cows); hormones are used to control oestrus and mating in sheep to synchronise lambing in a large herd.
- The technology that creates **transgenic animals** is advancing rapidly. Some animals are bred with genes that make them more productive for the food industry (fish with an extra growth hormone gene); other animals are given genes that allow them to produce human proteins used in medicine.
- Biotechnology is also used to **modify plants** used in food production. Genetically modified foods offer both advantages and disadvantages. Some pressure groups and some ordinary people in the UK are very suspicious of the development and use of GM foods. Others are in favour but don't know anything about them. Knowledge, evaluation and caution are important.

- Biotechnology is not new. The process of **fermentation** has been used to make yoghurt, cheese, beer, wine and bread for thousands of years.
- New biotechnology techniques are used to make edible protein from fungus (Quorn™) and bacteria (Pruteen™).
- Other **fermented food products** include sauerkraut, and products made from soya beans (eg tempeh and soy sauce).
- **Enzymes** are used in industry, particularly in the food and manufacturing sectors. Enzymes allow chemical processes to occur quickly and at relatively low temperatures, so are useful for keeping costs down.
- Most of the enzymes used in industry are extracellular. These are secreted by bacteria into the growth medium and are easy to separate and purify. Some intracellular enzymes are used but extracting them involves breaking up the bacterial cells first.
- Enzymes are often immobilised so they can be reclaimed and reused. The main methods of immobilising enzymes are **cross-linkages**, **entrapment** and **adsorption**.
- Immobilised enzymes are important in **biosensors** and in **diagnostic tests**.

? EXAM QUESTIONS

1 The following diagram is a simplified representation of a technique used in genetic engineering. Look carefully at the diagram and answer the following questions.

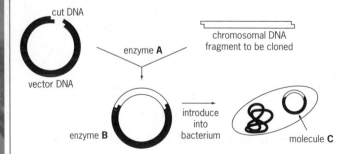

a) Give the name of enzyme **A** and enzyme **B**.

b) Suggest why the same enzyme (enzyme **A**) is used to cut both the vector DNA and the chromosomal DNA fragment.

c) What is the function of enzyme **B**?

d) What name is normally given to the molecule **C**?

e) Describe **two** ways in which the chromosomal DNA fragment can be obtained.

f) What can be added to molecule **C** to aid in selection and identification of the cells that have successfully taken up the fragment?

g) Suggest a possible medical use for this technology.

WJEC Biology Module BI1 June 2001 Q7

2 The bacterium *Bacillus thuringiensis* has a gene on its plasmid coding for the production of a protein toxic to many insect pests. The bacterial cell can be made to synthesise copies of these plasmids which can be introduced into tomato cells as shown below. Tomato plants grown from these cells are resistant to a wide range of insect pests.

a) (i) Name the chemical that makes up the bacterial plasmid.

(ii) What structure has been removed from the treated plant cell in stage 2 which would otherwise prevent uptake of the plasmid?

? EXAM QUESTIONS

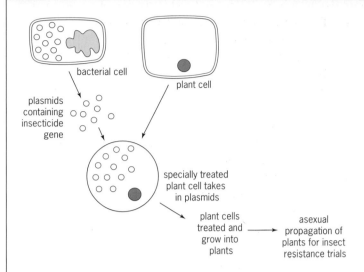

(iii) Why is it important that the entire process only involves asexual reproduction?

(iv) Suggest the **two** main advantages to farming of introducing a gene for this toxic protein into the plant.

b) What important factor would need to be determined so that genetically engineered tomatoes could be grown commercially?

WJEC Biology Module BI1 January 2001 Q7

3 Yeast cells produce ethanol under certain conditions.

a) State the conditions that cause yeast to produce ethanol.

b) Outline how glucose is converted into ethanol in yeast.

IBO Standard level Paper 3 November 2000 QC2

4 At the start of the brewing process, a small 'starter culture' of yeast is added to the fermenter under aerobic conditions. The yeast cells feed, grow and divide, leading to a growth in the yeast population.

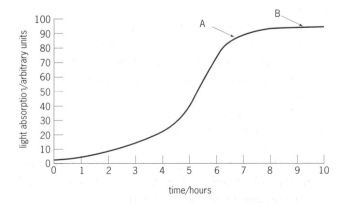

When the population of yeast is large, air is no longer allowed into the fermenter, the conditions become anaerobic and the yeast cells stop growing and dividing but are still alive. Ethanol is produced under these conditions. The growth of the yeast population in a fermenter under aerobic conditions was measured using turbidimetry. The results are shown on the graph.

a) Explain why the rate of population growth decreases after 6 hours.

b) As a result of this experiment, it was decided to turn off the air supply in future fermentations at Time **A**, rather than waiting until Time **B**. Suggest **one** commercial reason for this decision.

c) Explain why turbidimetry is **not** the best method for finding the concentration of live yeast cells in the fermenter at Time **B**.

AQA Microorganisms and Biotechnology June 2000 Q6

5

a) In industrial processes enzymes are often immobilised. Give **two** advantages of immobilising enzymes and for each explain how the industrial process is improved.

b) Chymotrypsin is an enzyme that breaks down proteins. It can be isolated from microorganisms and used industrially. Experiments were carried out on the effects of temperature on the activity of chymotrypsin. In one experiment, the enzyme was in solution, and in another it was immobilised on silica beads.

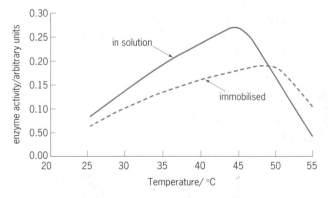

(i) Suggest reasons for the differences in the results of the experiments.

(ii) In an industrial process chymotrypsin is used at 45 °C. Suggest whether chymotrypsin is better used in solution or immobilised at this temperature. Explain the reason for your answer.

AQA Microorganisms and Biotechnology March 2000 Q7

❓ EXAM QUESTIONS

6 The figure below shows a possible structure for a biosensor that uses immobilised enzymes to detect glucose.

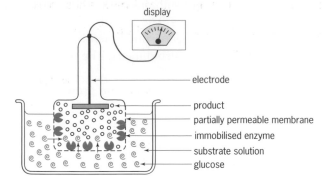

a) What are immobilised enzymes?

b) Suggest what happens when the glucose in the solution reaches the immobilised enzymes.

c) Suggest the form in which the information about the concentration of glucose is transmitted to the display.

d) **(i)** Why should a buffer be added to the substrate solution?

   (ii) Suggest **one** other variable that should be kept constant.

e) Competitive (X) and non-competitive (Y) inhibitors can be used to slow down or to stop enzyme controlled reactions. Label the following figure to suggest where these inhibitors are most likely to act. Use X for the competitive inhibitor and Y for the non-competitive inhibitor.

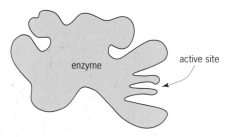

f) Describe how each of these inhibitors works.
   X...............   Y...............

g) Give an example of a competitive inhibitor.

WJEC Biology Module BI1 June 2001 Q5

7 In 1987 an attempt was made to produce human insulin from genetically engineered yeast. Genes for human insulin were inserted into small loops of DNA called plasmids. These plasmids were then used to try to carry the insulin genes into yeast cells.

a) Describe how an insulin gene could be removed from human DNA and inserted into the plasmid DNA.

b) Explain why a microorganism such as yeast might be particularly useful for producing insulin on an industrial scale.

c) The yeast used to produce insulin was a mutant strain that did not have the gene for a enzyme needed by the yeast in respiration. The missing enzyme is called triose phosphate isomerase. The DNA of the plasmids, however, did contain the gene for triose phosphate isomerase. Explain the importance for insulin production of using plasmids that have the gene for triose phosphate isomerase and a mutant yeast that does not have this gene.

AQA Microorganisms and Biotechnology March 2000 Q9

KEY SKILLS **ASSIGNMENT**

YOGHURT:
IS IT ALL IT'S MADE OUT TO BE?

People have been making yoghurt for thousands of years. It is easy to make at home, with very little equipment – just put some warm milk in a vacuum flask with a couple of tablespoons of live yoghurt, and leave it overnight. In the past, people probably did the same thing, but used a jug by the fire. Today, there is a multi-million pound international industry that has built up around yoghurt-making, which is done on a huge scale.

Fig 6.A1 outlines the major steps in the production of a commercial yoghurt.

Fig 6.A1

Production

1   Stages in yoghurt making
Mixing the ingredients
Milk, sugar, glucose and filtered
sterilised water are mixed together

2   Sterilisation
The mixture is heated to destroy any bacteria
that are not needed

3   Fermentation
The sterilised mixture is transferred to a fermentation
tank that may hold 6000 litres. The transfer is carried
out under sterilised conditions via a system of pipes
and valves that are closed to the external environment.
Bacterial starter cultures are added and the contents
of the tank are kept at about 37 °C for several hours
to several days

Post-production processing

4   Flavouring
If flavoured yoghurt is being made,
flavourings are added here

5   Storage and dilution
The concentrated yoghurt mixture is stored in
another large tank that is cooled to 2 °C. Sterile
water is used to dilute the yoghurt to the correct
consistency before packaging

6   Packaging
The finished yoghurt is put into the cartons
or bottles, sealed, labelled and stored below 6 °C.
It is then distributed for sale

KEY SKILLS **ASSIGNMENT**

1 **a)** What is the essential raw material used in the initial mix during yoghurt manufacture?
b) Give two reasons why the mixture is heat-treated.

2 **a)** Two species of bacteria that can be used in yoghurt production are *Lactobacillus bulgaricus* and *Streptococcus thermophilus*. Describe two ways by which you could distinguish between different species of bacteria such as these.
b) Why are the organisms incubated around 37 °C?

3 Describe what happens during the formation of yoghurt. Include details of the microorganisms present.

4 How are the nutritional requirements of the bacteria in the yoghurt culture satisfied by the milk used in the process?

5 Give an example of the downstream processing of yoghurt.

Is yoghurt always good for you?

A lot of advertising claims that yoghurt is the ultimate in healthy eating. Many people think that all yoghurts are non-fattening and can be used when trying to lose weight, or at least when trying not to gain any extra kilos. Those with live bacteria are represented as a vital food for a healthy digestive system. But how much truth is there behind these two assumptions?

The following table shows different nutritional contents of two types of yoghurt; A is a fruit-flavoured yoghurt with added cream; B is made with sweetener and fruit flavouring.

Quantity per 100 g	A	B
Energy (kJ/kcal)	462/110	245/58
Protein (g)	3.7	4.1
Fat (g)	3.9	0.8
Carbohydrate (g)	15	1.9
Salt (g)	0.1	0.1
Taste	sweet and fruity	sweet and fruity
Consistency	thick and creamy	thick and creamy

A gram of fat yields 9 calories; a gram of protein yields 4 calories and a gram of carbohydrate yields 4 calories.

6 **a)** How many calories in the two different yoghurts come from the fat it contains?
b) What percentage of the total energy content is this?
c) What percentage of the total energy content of both yoghurts comes from protein? And from carbohydrate?

7 **a)** Which of these yoghurts do you consider more healthy and why?
b) How do you think these two types of yoghurt are marketed?
c) Do you think the marketing of yoghurt is sometimes misleading?
d) Is there such a thing as 'an unhealthy yoghurt'?

Live yoghurts are not heated to a high temperature to kill the bacteria present before they are packaged for sale. This could have a disadvantage in that contaminant bacteria could live and grow in the yoghurt and cause it to 'go off' in storage. However, the health advantages of live yoghurt are thought to be worth it. Manufacturers claim that the bacteria in the yoghurt populate the large intestine, and set up a correct balance of 'friendly' bacteria in the bowel.

8 **a)** Using your knowledge of digestion, and the conditions present in the large bowel, do you think this claim might be true?
b) What study would you need to carry out to determine if eating live yoghurt actually did change the bacteria in the bowel?

07

Energy and life

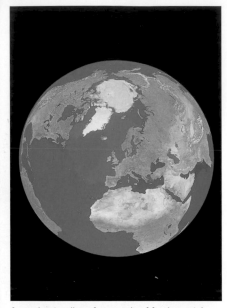

An understanding of energy is of fundamental importance in biology, from the study of individual molecules and reactions to global issues such as the use of fossil fuels, pollution, damage to the ozone layer and the 'greenhouse effect'

All living organisms need energy. As you sit and read this book, you might not think that you are using up much energy. In fact, even at rest, your body needs a constant supply of energy. Think about some of the processes going on within your body:

- Your heart is beating.
- You are breathing.
- Food is being pushed along your intestine.
- Food is being absorbed from your intestine into your blood.
- Urine is being pushed from your kidneys to your bladder.
- Nerves are taking information from your sense organs to your brain, keeping you informed of changes in your environment.
- You are thinking; your brain is processing information coming in from the sense organs, sorting out what is important and what can be ignored.
- Unless you are sitting in a sauna, you are probably losing heat to your surroundings. As a warm-blooded organism, you need to replace any heat that you lose in order to maintain a constant internal body temperature.
- Some of the molecules from food you have digested, amino acids for example, are being used to make new cell components, new cells and new tissues.

All of these processes (and many more) require energy.

1 SOLAR ENERGY, PHOTOSYNTHESIS AND RESPIRATION

This planet, and all the living things on it, must have energy and almost all of it is supplied by our nearest star, the Sun. A huge amount of sunlight reaches the surface of the Earth but only about 2 per cent of it is actually absorbed by plants. The rest of the energy is not lost; by heating up the land, air and water it prevents the planet from freezing.

In photosynthesis, plants exploit radiant energy from the Sun, using it to convert simple inorganic substances (mainly carbon dioxide and water) into larger organic molecules (glucose, starch, lipids and proteins). The plant tissues built from the products of photosynthesis form food for organisms that cannot make their own food (animals, fungi and most bacteria). Some of these become food for other organisms. Ultimately, therefore, the products of photosynthesis provide the energy that all living things need for growth, repair, movement and many other life processes. This energy is made available to living things through the process of **cell respiration**.

We will go on to look at cell respiration in detail in the next chapter but, first, there are some key concepts that we need to understand before we can study the biochemistry of the process.

Fig 7.1 Photosynthesis by blue-green bacteria was responsible for the appearance of oxygen in the Earth's atmosphere 2 billion years ago, and photosynthesis by plants maintains the oxygen level at 20 per cent of the atmosphere today

2 WHY DO LIVING THINGS NEED ENERGY?

Living things need a constant supply of energy to maintain their life processes. Let's look at the main ones.

GROWTH AND REPAIR OF CELLS AND TISSUES

Energy is required for the biochemical reactions that build large organic molecules from simpler ones. For example, energy is needed to build proteins from amino acids.

ACTIVE TRANSPORT

Energy is required to move some substances in or out of cells. The transport of nitrate from the soil into the roots of a plant, and the transport of sugars and amino acids from an animal's small intestine into its blood are both examples of active transport.

Active transport often takes place against a diffusion gradient (see page 83) and it allows organisms to control their internal environment more efficiently.

MOVEMENT

Generally speaking, all organisms move, and movement requires energy. However, movement can be considered on several levels:

- within cells, eg chromosomes separating,
- whole cells, eg white cells engulfing bacteria, sperm swimming (see Fig 7.2),
- tissues, eg muscles contracting,
- whole organs, eg a heart beating,
- parts or whole organisms, eg talking, walking.

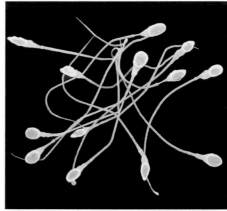

Fig 7.2 Movement requires energy. To swim far enough up into the uterus to reach an egg, a sperm cell needs enough energy to enable it to swim over 7500 times its own length – the equivalent of a 10 kilometre swim for an adult human

TEMPERATURE CONTROL

Warm-blooded animals, **endotherms**, use large amounts of heat energy to keep their body temperature constant. Humans use around 70 per cent of the energy from respiration to maintain the body at a constant 37 °C. Many smaller endotherms such as hummingbirds and shrews lose heat more rapidly because of their large ratio of surface area to volume (see also Chapter 12). Almost all of their energy of respiration is used to regulate body temperature.

Fig 7.3 The shrew has a particularly high metabolic rate because it loses heat rapidly and must replace what is lost

? TEST YOURSELF

B Give three reasons why we need energy when we are asleep.

MAKING ELECTRICITY AND LIGHT

Some organisms are able to use the energy in food to produce light or electricity. Fig 7.4 shows a glow-worm producing light at night. Fig 7.5 shows an electric eel, which uses rapid discharges of electricity to stun its prey.

Fig 7.4 The glow-worm, really a type of beetle, is one of the few organisms that can use the energy in chemicals to produce light. The male, known as the firefly, uses light to signal to females. The wingless females – the glow-worms – reply in the same way. In order to attract a partner of the right species, males of a particular species will 'flash' in a particular pattern.

Fig 7.5 The electric eel can deliver enough electricity to kill other fish swimming nearby, or to stun anyone unlucky enough to step on it

3 HOW DO LIVING THINGS GET THE ENERGY THEY NEED?

In order to respire and so release energy, organisms need a supply of food. Food can be obtained in two ways:

● Some organisms can make their own food. Organisms that can do this are called **autotrophs** (meaning 'self-feeders'). They are able to use energy from the surroundings to make their food. There are two kinds of autotroph, the **photoautotroph** and the **chemoautotroph**. Photoautotrophs, such as green plants, can photosynthesise, using sunlight to make their food. Chemoautotrophs, all of which are bacteria, can synthesise their food using energy made available from chemical reactions other than those involved in photosynthesis. Bacteria responsible for some stages of nitrogen recycling are chemoautotrophs, see Chapter 34.

● Some organisms need to obtain ready-made food because they cannot make their own. Such organisms are called **heterotrophs**. Heterotrophs must obtain their food, either directly or indirectly, from organisms that can synthesise food (usually plants). They either eat their food (ingesting it into a gut, as most animals do) or digest the food first and then absorb the nutrients, as decomposers, mainly fungi and bacteria, do. See Fig 7.6.

? TEST YOURSELF

C Heterotrophs must eat autotrophs; true or false? Explain.

Fig 7.6 All of these organisms have something in common – they are heterotrophs. They cannot make their own food and must take it, ready made, from an outside source. As you can see, there are many different ways of doing this

Fig 7.7 Sequoias are the largest trees on Earth. Apart from water, much of their weight is due to organic compounds such as cellulose, lignin and starch

NUTRITION IS COVERED IN MORE DETAIL IN CHAPTER 10.

? TEST YOURSELF

D Look at Fig 7.7 and answer the following:

a) Where did the plant get the carbon from to make the organic compounds that make up its enormous bulk?

b) Where did it get the energy from to make the organic compounds?

c) If the tree were burned, an enormous amount of energy would be released. Where would most of the energy go and which compounds would be produced?

4 ENERGY FLOW: WHERE DOES IT COME FROM, WHERE DOES IT GO?

Fig 7.8 The basic idea of energy flow. At the start of any food chain there must be an organism capable of making food – usually a green plant. From there, the energy is passed up the chain within food molecules. Some of this energy is eventually used to make the cells, tissues and organs of heterotrophic organisms

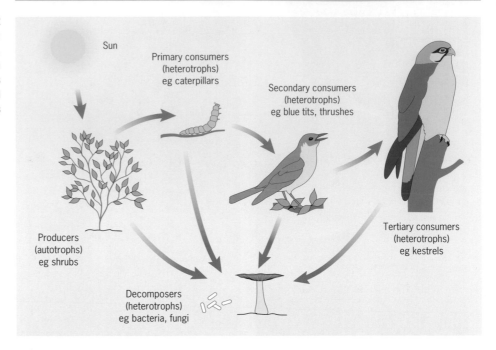

Fig 7.9

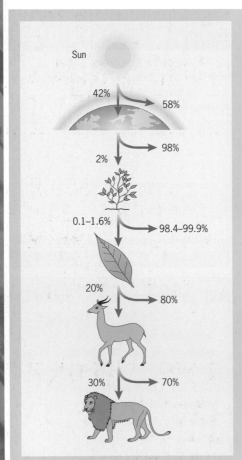

Of the light energy from the Sun that reaches the Earth, only about 42 per cent actually gets through to the surface of the planet. The rest is absorbed or reflected by the atmosphere, a process that warms the atmosphere.

Of the light energy reaching the Earth's surface, about 2 per cent is absorbed by plants. The rest goes to produce the heat that warms up the land and the oceans.

Some of the energy that reaches plants cannot be used because of its wavelength. For example, plants cannot use green light because they reflect this back – which is why they look green. Photosynthesis, like all energy transfers, is inefficient and some energy is lost in the process. Only 0.1–0.6 per cent of the energy received is ever incorporated into plant tissue during photosynthesis – and even some of this is lost because the plant itself must respire.

Of the plant tissue eaten by herbivores, only about 20 per cent (at most) is incorporated into the tissues of the animal. Most of the rest is lost as heat as the animal respires.

Of the animal tissue eaten by a carnivore as meat, only a maximum of 30 per cent ever becomes incorporated into the body of the carnivore. Most of the rest is lost as heat.

Clearly, as so little energy is transferred at each point in a food chain, there is a relatively small amount of energy available for animals at the top of the chain, usually large carnivores. For this reason, such animals are quite rare.

To understand further how energy is lost, think of your own diet. You consume large amounts of food but very little of it becomes incorporated into your body. When you compare your own body mass to the mass of all the food you have ever eaten (imagine it in a pile), you can see that a lot of energy must have been lost. The missing energy was released as heat during respiration – some of this heat was used to keep your body temperature at a constant 37 °C. So much energy is needed because you are constantly losing heat to the air, your clothes and the objects you touch. However, many people struggling against weight gain may wish the process were even more inefficient.

5 ENERGY FROM FOOD: A HUMAN PERSPECTIVE

The recommended minimum daily intake of energy varies according to a person's age, sex, occupation and/or level of activity. Fig 7.10 shows a variety of energy requirements for different people. These figures are the total requirement for all of an individual's metabolic processes for an average 24-hour period. The energy requirement must be provided by the food that they eat and is released for use by the process of cell respiration.

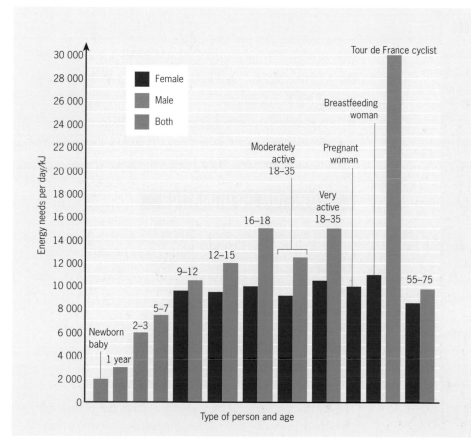

Fig 7.10 The approximate energy needs of different people

If the total amount of food eaten every day contains more energy than the body needs, the excess is stored as glycogen or lipid, as shown in Fig 7.11(a). If the food eaten contains less energy than is needed, the body makes up the difference by respiring these stored materials. When the lipid stores get low the body begins to respire an increasing amount of protein, and the actual fabric of the body starts to be used up, as illustrated by Fig 7.11(b).

? TEST YOURSELF

E When will you not lose heat to the surroundings?

? TEST YOURSELF

F Look at Fig 7.10.
a) How do the figures given explain why people tend to have more of a weight problem as they get older?
b) Compare the figures given for a moderately active adult woman of childbearing age and a pregnant woman. How much extra energy is needed in pregnancy?
c) Note down the figure for the energy requirement of a breastfeeding woman and explain it in terms of two other figures from the table.

Table 7.1 The approximate energy content of the three main food types

Food type	Energy content
1 g of carbohydrate	17 kJ
1 g of lipid	38 kJ
1 g of protein	20 kJ

? TEST YOURSELF

G Using the information in Fig 7.10 and Table 7.1, work out how many grams of
a) carbohydrate,
b) lipid
is needed by an adult male to provide his daily energy requirement.

Fig 7.11 (a) When the body takes in more energy from food than it needs, the excess is stored as fat under the skin and around the internal organs

Fig 7.11 (b) When the body does not get enough energy from food, it will be forced to respire stored fat and then protein. The very fabric of the body will start disappearing

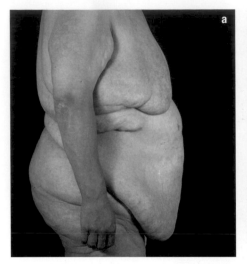
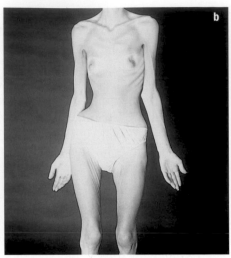

Of course, nutrition is not just about energy requirements. In theory, you could get all of your energy requirement by eating several chocolate bars every day. While you would avoid starvation, chocolate would not provide you with the correct balance of other substances your body needs to remain healthy – protein, vitamins, minerals and fibre.

DIET IS COVERED IN MORE DETAIL IN CHAPTER 10.

SCIENCE FEATURE Measuring the energy in food

The SI unit of energy is the joule (J) and the amount of energy in food is measured in joules. When we talk of the energy requirement of people, or of the energy content of food, we talk about kilojoules.

$$1000 \text{ J} = 1 \text{ kJ}$$

The energy value of food can be estimated by burning a known amount of the food in oxygen and measuring the energy released as heat. This process is called **calorimetry**. The information on energy content given on food packaging has been calculated in this way.

Fig 7.12(a) shows one very simple method of measuring the energy contained in a food sample. The sample is weighed and set alight. The energy released as the sample burns is used to heat a known volume of water. We can then calculate the total amount of energy contained in the food sample since we know that 4.18 joules of energy will raise the temperature of 1 cm^3 of water by 1 °C.

Unfortunately, although this method is simple, it is not very reliable:

- A lot of the energy escapes and does not go into heating the water. In particular, heat is lost to the surroundings.

- Food samples are very difficult to burn. There is usually a lot of unburned material left at the end of the experiment.

The apparatus shown in Fig 7.12(b) is much more accurate.

NUTRITION INFORMATION		Typical value per 100g	Per 40g Serving with 125ml of Semi-Skimmed Milk
ENERGY	kJ	1350	800*
	kcal	310	180
PROTEIN	g	8	8
CARBOHYDRATE	g	66	33
(of which sugars)	g	(31)	(19)
(starch)	g	(35)	(14)
FAT	g	1.5	2.5 *
(of which saturates)	g	(0.3)	(1.5)
FIBRE	g	13	5

Fig 7.13 Packaging for muesli shows the energy given in kilojoules and kilocalories

The calorie is an older unit for energy in food, but is still widely used. In popular nutrition guides, the calories referred to are actually kilocalories, which should be abbreviated Cal, with a capital C.

$$1000 \text{ cal} = 1 \text{ kcal} = 1 \text{ Cal}$$

A calorie is defined as the energy needed to raise the temperature of 1 cm^3 of water by 1 °C.

1 calorie = 4.18 joules
1 Calorie = 4.18 kilojoules

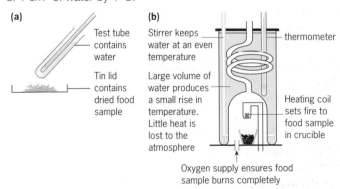

(a)
Test tube contains water
Tin lid contains dried food sample

(b)
Stirrer keeps water at an even temperature
thermometer
Large volume of water produces a small rise in temperature. Little heat is lost to the atmosphere
Heating coil sets fire to food sample in crucible
Oxygen supply ensures food sample burns completely

Fig 7.12 (a) A simple calorimeter. **(b)** A more accurate calorimeter

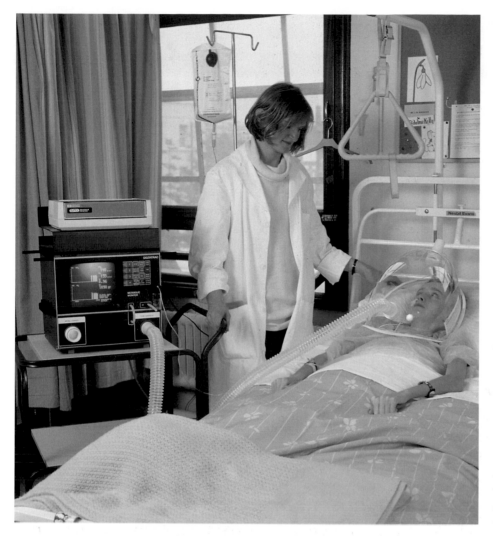

Fig 7.14 Measuring metabolic rate in a young patient with bowel disease. The plastic canopy takes exhaled air to the monitor on the left. This measures the amount of oxygen and carbon dioxide present and calculates how much energy (in kilocalories per day) the patient is using

6 BASAL METABOLIC RATE (BMR)

The BMR is the rate of respiration/metabolism the body needs to keep it 'ticking over' during periods of inactivity. For humans, BMR is defined as the rate of energy used by someone who has fasted for 12 hours and who is awake but resting, at a comfortable external temperature. The BMR is useful as it gives us a background rate against which we can compare the energy needs of activities such as swimming or walking. We can also measure and compare the BMRs of different organisms.

In warm-blooded animals the BMR is closely linked with the size of the organism and with the amount of heat that it loses. As a general rule, the smaller the animal, the higher the BMR. Chapter 8 gives more detail on the measurement of respiration and we cover thermoregulation in Chapter 15.

In practice, BMR can be measured by analysing the relative amounts of oxygen and carbon dioxide present in the exhaled air of a person under test. This allows calculation of the amount of energy that the person uses while at rest during a 24-hour period (see Fig 7.14).

Measurements of the BMR in many people of different ages has shown that BMR varies with age. In a 1-year-old child, for example, the BMR is higher, at $220 \, kJ \, m^{-2} \, h^{-1}$, than that of a young adult female at $150 \, kJ \, m^{-2} \, h^{-1}$ or a young adult male at $170 \, kJ \, m^{-2} \, h^{-1}$. As we get older, our BMR tends to decrease but the values depend on individuals and their lifestyles. Taking plenty of exercise can significantly increase BMR.

▶ EXTENSION Biology and thermodynamics

The **first law of thermodynamics** states that:

Energy cannot be created or destroyed; it can only be transferred.

This law is relevant to biology because, as we have seen, all organisms obtain their energy by transferring it from the environment in different ways.

These are the most important energy transfers in biological systems:

● Energy from sunlight becomes energy in chemicals; this occurs in photosynthesis.

● Energy in food chemicals is used to: do mechanical work, eg to move muscles; produce heat, eg to maintain body temperature; build structures within cells and tissues.

The **second law of thermodynamics** states that:

When energy is transferred, some of the energy is always lost as heat.

This law means that no energy transfer will be 100 per cent efficient, and biological processes are no exception. Photosynthesis is a complex, multistage process and a lot of sunlight energy is lost as heat rather then being incorporated into organic molecules.

During exercise, the transfer of energy from chemicals to do mechanical work within the muscles is efficient. Quite a lot of the energy is lost as heat; some is needed to keep us warm but any excess must be lost. This explains why we go red and sweat during and after strenuous activity.

Cell respiration

Cell respiration

Famine is still a fact of life for people in many parts of the world

In humans, as in all organisms, obtaining energy from food is a priority that cannot be ignored. Tragic scenes from parts of the world devastated by famine show the terrible consequences of malnutrition and starvation.

We obtain most of our energy by respiring the food that we have eaten recently. Carbohydrate is the primary source of energy, but we also respire small amounts of fat and protein. When the energy provided by recent meals runs out, as it does when we are asleep, for example, our bodies start to respire molecules stored in our tissues. These are replaced when we next eat, usually first thing in the morning when we *break* our *fast*.

During prolonged starvation, the body's stored carbohydrate runs out after about a day, and then the body must obtain its energy by respiring fat. An average 70 kg man will have about 10 to 15 kg of stored fat in his body. This is a large energy store, but one that is not readily available (fat can only be respired when some carbohydrate is still available). Increasingly the body is forced to turn to protein, the very fabric of the cells, for the energy it desperately needs.

The body can withstand the loss of about half of its protein only. Complete starvation, where no food at all is available, leads to death within weeks. The exact cause of death varies. It can be an infection such as pneumonia (starvation weakens the immune system) or circulatory problems due to a much reduced blood volume.

1 CELL RESPIRATION: A VITAL LIFE PROCESS

Understanding cell respiration can be a demanding task. You will probably find it useful to read Chapter 7 first: it looks at key concepts that you will need to understand before you tackle the details given in this chapter.

All living things obtain the energy they need from **respiration**, a chemical process that breaks down simple food molecules such as glucose. Animals, including humans, digest food to produce these molecules, which are absorbed into the blood and then transported round the body.

Multicellular organisms mainly respire **aerobically** (using oxygen). Oxygen is absorbed by lungs, gills or the body surface and is usually distributed around the body by the **circulatory system**. Only when food molecules and oxygen are together inside cells can the complex process of aerobic cell respiration begin.

SOME IMPORTANT DEFINITIONS

The terms 'respiration' and 'breathing' seem to be used interchangeably by many people, including scientists, and this can lead to confusion. As a basic guide, we use the following definitions.

Respiration is the process that releases the energy in organic molecules such as sugars and lipids. Respiration is a complex, multi-stage process that takes place in *all* cells of *all* organisms, *all* of the time (unless the organism is dormant). You may remember that respiration is one of the seven signs of life.

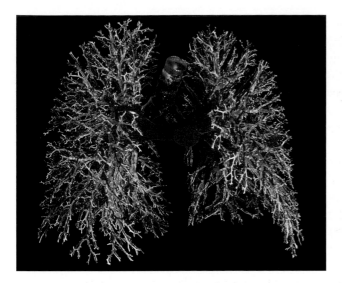

To avoid confusion with breathing, it is often called tissue respiration or cell/cellular respiration. In this book we call it **cell respiration**.

Organisms that carry out aerobic cell respiration need to take in oxygen, and they produce carbon dioxide, a waste product that must be removed. Larger organisms have had to develop specialised organs with large surface areas to increase their capacity for gas exchange (Fig 8.1). (See also Chapter 12.)

Breathing is the mechanical process that supplies oxygen to the body to drive respiration and that removes the carbon dioxide produced. Breathing *ventilates* the gas exchange surfaces.

Aerobic respiration is respiration that requires oxygen.

Most organisms respire aerobically: aerobic respiration releases a relatively large amount of energy and gives them a survival advantage over anaerobic organisms.

Anaerobic respiration is respiration without oxygen.

Some organisms, mainly bacteria, can only respire anaerobically. A few more, yeast for example, can turn to anaerobic respiration when there is no oxygen. Some animal tissues (eg muscle during strenuous exercise) and plant tissues (eg plant roots in waterlogged soil) are able to respire anaerobically if circumstances demand it.

2 AEROBIC CELL RESPIRATION

'Aerobic cell respiration' describes the cell processes that require oxygen to release energy from all types of organic molecules. In many organisms, cell respiration involves the breakdown of a variety of molecules from their food. Humans respire mainly sugars with a small percentage of amino acids and fatty acids but, when the need arises, such as during starvation, the balance can change.

> ✔ **REMEMBER THIS**
>
> Although our muscles can work anaerobically for a short time, the brain must have a continuous supply of oxygen. A lack of oxygen for just 3 minutes is enough to cause brain damage.

> ? **TEST YOURSELF**
>
> **A** Why is it important for organs involved in gas exchange to have a large surface area? What other features would you expect these organs to have?

Fig 8.2 Carnivores such as this leopard obtain much of their energy from the respiration of amino acids derived from their high-protein diet

THE AEROBIC RESPIRATION OF OTHER
ORGANIC MOLECULES AND ANAEROBIC
RESPIRATION ARE DEALT WITH LATER
(SEE PAGES 155 AND 160).

AEROBIC CELL RESPIRATION OF GLUCOSE

To make a complex topic easier to study, we shall concentrate first on the aerobic respiration of only one type of food molecule: glucose.

For any chemical reaction to take place, energy is required to break existing molecular bonds. The process of forming new bonds may either require or release energy. For there to be a *net release of free energy* (the whole point of cell respiration), the products of cell respiration must be at a lower energy level than the reactants. Chemists say the products are more **thermodynamically stable** than the reactants. The basic equation we use to sum up glucose respiration is:

$$C_6H_{12}O_6 + 6O_2 \rightarrow 6CO_2 + 6H_2O + ENERGY$$

The total energy contained in the reactants (the molecule of glucose and the six molecules of oxygen on the left) is greater than the total energy contained in the products (the six molecules of carbon dioxide and the six molecules of water on the right). The difference is the energy released by cell respiration.

In reality, glucose respiration is a sequence of many different reactions. These can produce up to 36 molecules of a chemical called **ATP** per molecule of glucose. This large amount of energy can do useful work in the cell:

$$C_6H_{12}O_6 + 6O_2 \rightarrow 6CO_2 + 6H_2O + 36ATP$$

The steps involved in the production of ATP depend heavily on redox reactions. These are reactions that involve oxidation of one reactant and reduction of another. Before going on to look at the detailed biochemistry of glucose respiration, take a look at the box on **redox reactions** on page 148, to make sure that you are comfortable with this important concept.

RESPIRATION OR COMBUSTION?

It is often said that we 'burn up' food in respiration but this is not really true. The energy contained in glucose *could* be released by setting fire to it (Fig 8.3). The glucose would be oxidised to carbon dioxide and water, but the reaction would be rapid and uncontrolled.

During the process of cell respiration, glucose and other organic molecules are broken down in small stages, some of which release energy. This step-by-step breakdown (Fig 8.4) is more gradual than combustion. The whole process of respiration is controlled by enzymes that transfer energy in food molecules to ATP, a substance that is able to power cell processes by supplying on-the-spot, instant and usable energy in controlled amounts.

Fig 8.3 The energy in food can be released by combustion, but almost all of it is lost as heat, raising the temperature of the surroundings to levels that could not be tolerated inside any living cell

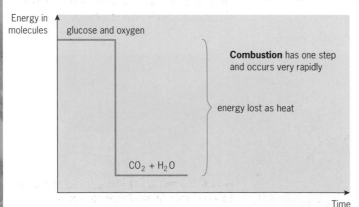

Energy in molecules

glucose and oxygen

Combustion has one step and occurs very rapidly

energy lost as heat

$CO_2 + H_2O$

Time

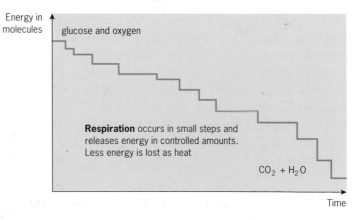

Energy in molecules

glucose and oxygen

Respiration occurs in small steps and releases energy in controlled amounts. Less energy is lost as heat

$CO_2 + H_2O$

Time

Fig 8.4 The difference between combustion and respiration. Although the difference in energy between the reactants (glucose and oxygen) and the products (CO_2 and water) is the same in both cases, the way in which the energy is released is totally different

ALL ABOUT ATP

ATP stands for **adenosine triphosphate**. ATP is a relatively small, soluble organic molecule that consists of a base (adenine), a sugar (ribose) and three inorganic phosphate (P_i) groups (Fig 8.5). ATP has a *high **free energy of hydrolysis***, which means that when ATP is **hydrolysed** (Fig 8.6), a relatively large amount of energy is released. In hydrolysis, ATP reacts with water, producing ADP and inorganic phosphate. Because of its solubility and small size (relative molecular mass 507), ATP can diffuse rapidly around cells and so can supply energy where it is needed. Metabolically active cells, such as those in muscle and in secretory tissue, synthesise and break down many ATP molecules.

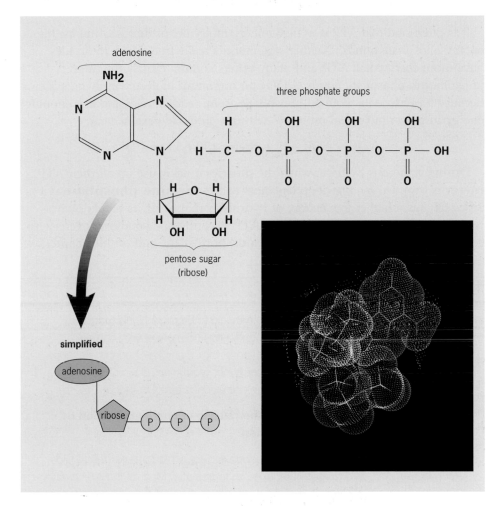

Fig 8.5 The structure of ATP. In the computer graphics image, adenosine is blue, pentose is white and the phosphate groups are red. This molecule is present in all cells and is used to provide energy. The purpose of cell respiration is to re-synthesise ATP at the same rate as it is used up

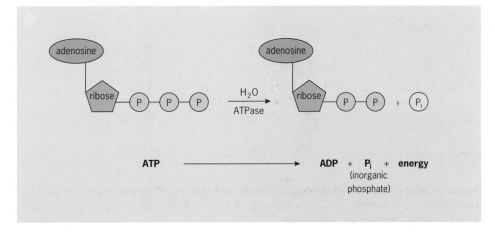

Fig 8.6 The hydrolysis (splitting) of ATP. Under the control of the enzyme ATPase, the terminal phosphate group of ATP is removed and combined with water. This reaction releases free energy which can be used to drive energy-requiring reactions such as muscular contraction

C To move our muscles, we need energy to be available 'on demand'. Why do we use ATP as an immediate energy source, instead of glucose?

Fig 8.7 For explosive bursts of effort, such as a short sprint, most of the energy comes from ATP already present in the muscles

HOW DOES ATP RELEASE ENERGY?

When ATP loses a phosphate group to become ADP, adenosine *di*phosphate, the reaction is **exergonic** (it *releases* energy). This can be used to do useful work within the cell. The same amount of energy is released when ADP loses another phosphate group to become AMP (adenosine *mono*phosphate). Less energy is released when the last phosphate group is lost.

Many of the reactions within cells are **endergonic**: they *require* energy. This energy is supplied by *coupling* them with reactions that involve ATP breakdown.

Endergonic reaction: amino acid A + amino acid B → dipeptide

Exergonic reaction: ATP → ADP + P_i + energy

It is often said that ATP is a 'high-energy molecule' or that is contains 'high-energy phosphate bonds'. Neither statement is really true. Many molecules have more energy than ATP, and many others contain exactly the same phosphate-to-phosphate bonds. ATP is an important molecule in living systems because it can lose its terminal phosphate group *readily*, releasing just enough energy to power biological processes without producing excess heat.

ATP → ADP + P_i + energy (30.6 kJ mol^{-1})

During strenuous exercise, when the muscles quickly use up all their ATP, energy is supplied by a back-up chemical called **creatine phosphate** (CP). Since CP has a higher free energy of hydrolysis than ATP, its breakdown releases enough energy to make more ATP instantly and so allows exercise to continue. See the Assignment, Energy and Sport, at the end of this chapter.

! SCIENCE FEATURE Redox reactions

Redox is short for *red*uction and *ox*idation, two chemical processes that often occur together in the same reaction.

Oxidation reactions may involve the addition of oxygen or the removal of hydrogen from molecules, but the important underlying rule is that a molecule that is oxidised *loses* electrons.

Conversely, reduction reactions involve a *gain* of electrons. As electrons carry energy, a molecule that has been reduced, and that therefore has gained one or more electrons, will usually carry more energy than the oxidised form of the same molecule.

Reduction and oxidation reactions usually occur together, because if one chemical loses electrons another must gain them (Fig 8.8).

The concept of redox reactions is a vital one in biology. The molecules that make up living things are produced, directly or indirectly, by photosynthesis. This process reduces carbon dioxide to form organic molecules (such as sugars) that contain energy. These can be later oxidised in the process of cell respiration to release energy. Lipids contain more hydrogen than carbohydrates: they have more C—H bonds than C—O bonds.

For this reason, lipids are said to be more highly reduced chemicals than carbohydrates and more energy can be released by the oxidation of their many C—H bonds. Because they contain more energy per gram than other substances, lipids are ideal storage compounds (see Chapter 3).

Cell respiration is the release of energy from food by progressive oxidation.

In cell respiration, glucose is oxidised to form carbon dioxide and oxygen is reduced to form water. Most of the ATP created during this process comes from the final stage: a series of redox reactions that occur on the mitochondrial membrane.

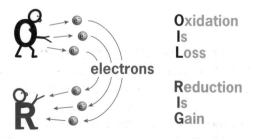

Fig 8.8 Remember – OIL RIG: oxidation is loss, reduction is gain

THE FOUR MAIN STAGES IN GLUCOSE RESPIRATION

The complete process of aerobic respiration of glucose can be divided into four distinct processes:

- **Glycolysis**; glucose is split into pyruvate.
- **Pyruvate oxidation** (the 'link' reaction); pyruvate is oxidised into acetate.
- The **Krebs cycle**; electrons are stripped off the acetate.
- The **electron transport chain**; the energy in the electrons is used to make ATP.

The way in which the four processes relate to each other is shown in Fig 8.9, and Table 8.1 gives the overall function of each stage. When learning respiration, the two compounds to look out for are ATP and NADH. The production of ATP, as we have seen, is the whole point of the process, but what is NADH?

Overall, respiration involves the removal of electrons from the glucose molecules, and then using the energy in the electrons to make ATP. NADH is an electron carrier.

$$NAD^+ \quad + \quad e^- \quad \rightarrow \quad NADH$$

$$\text{Coenzyme} + \text{electron} \rightarrow \text{reduced coenzyme}$$

NAD^+ is a coenzyme that picks up electrons from the processes that release them (glycolysis, link reaction, Krebs cycle) and delivers them to the process that converts them into ATP; the electron transport chain. You might like to think of ATP as cash and NADH as a credit note that can be turned into cash later.

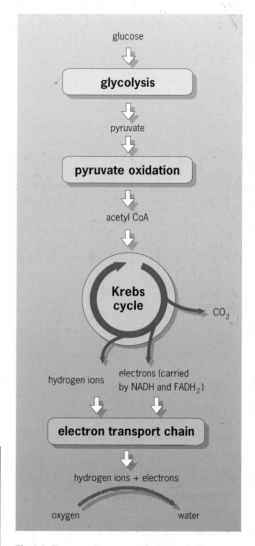

Fig 8.9 The overall process of cell respiration, showing the order of the four main stages

Stage	Site within cell	Overall process	Number of ATP molecules produced*
1 Glycolysis	cytosol	glucose is split into 2 molecules of pyruvate	2 per glucose
2 Pyruvate oxidation	matrix (inner fluid) of mitochondria	pyruvate is converted into acetyl Co-A	none
3 Krebs cycle	matrix (inner fluid) of mitochondria	acetyl Co-A drives a cycle of reactions which produces hydrogen	2 per turn, so 4 per glucose
4 Electron transport chain	inner membrane of mitochondria	hydrogen drives a series of redox reactions which release enough energy to make ATP	up to 32 per glucose

Table 8.1 The overall process of glucose breakdown. Stages 2, 3 and 4 can be thought of as the aerobic (oxygen-requiring) reactions and take place in the mitochondria

*This table indicates that, for each molecule of glucose, 38 molecules of ATP are produced. The actual figure is nearer 36 molecules. See the Example on page 155.

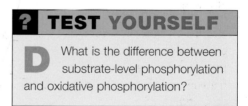

? TEST YOURSELF

D What is the difference between substrate-level phosphorylation and oxidative phosphorylation?

MAKING ATP BY PHOSPHORYLATION

In the cell, ATP is made by adding a phosphate group to a molecule of ADP. This process is called phosphorylation and it is carried out in two different ways.

Substrate-level phosphorylation is the process in which a phosphate group from a substrate molecule (a molecule other than ATP, ADP or AMP) is transferred to a molecule of ADP, giving a new molecule of ATP. This form of phosphorylation occurs in glycolysis and in the Krebs cycle.

In **oxidative phosphorylation**, ATP is synthesised using free phosphate groups. The energy required is obtained from a series of redox reactions in the electron transport chain. So whenever you see 'oxidative phosphorylation', think 'electron transport chain'.

GLYCOLYSIS

In this first stage a glucose molecule is converted, by a series of enzyme-controlled reactions, into two molecules of pyruvate, a 3-carbon compound. The process, shown in detail in Fig 8.10, yields relatively little energy (only two molecules of ATP) but this stage does not require oxygen and takes relatively little time to complete, so it can provide immediate energy.

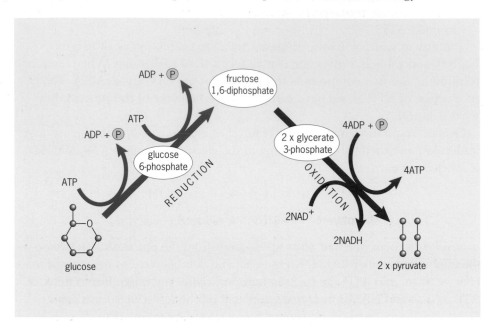

Fig 8.10 Glycolysis is a two-stage process. The 'uphill' part involves raising glucose to a higher energy level by using ATP. In the 'downhill' part, the products are oxidised, yielding two molecules of pyruvate, two molecules of reduced coenzyme and a net gain in ATP

Glycolysis takes place in the **cytosol**, the fluid part of the cytoplasm, so this basic process takes place all over the cell. The overall process consists of ten different reactions, each catalysed by a specific enzyme.

The first half of glycolysis actually uses ATP to raise the energy level of the original glucose molecules by adding phosphate groups to form **fructose 1,6-diphosphate**. This must be done if glucose is to have enough energy to complete the second half of the process.

Fructose 1,6-diphosphate is then split into two molecules of a 3-carbon compound, **glyceraldehyde 3-phosphate**. These are subsequently oxidised to pyruvate by a series of five reactions that also produce four molecules of ATP and two molecules of reduced coenzyme (NADH).

So, overall, per glucose molecule, glycolysis produces:
- two molecules of ATP (four are created but two are used up in glycolysis);
- two molecules of **NADH** (**reduced coenzyme** which later feeds electrons into the electron transport chain);
- two molecules of pyruvate (which enters the 'link' reaction if oxygen is available).

PYRUVATE OXIDATION: THE 'LINK' REACTION

This reaction links glycolysis with the Krebs cycle. It is often thought of as part of the Krebs cycle but we shall consider it to be a separate reaction.

In the presence of oxygen, pyruvate (the 3C molecule produced by glycolysis) moves from the cytosol to the mitochondrial matrix where it is oxidised into **acetate** (a 2C molecule), producing carbon dioxide as a by-product. This reaction also produces two molecules of NADH. The acetate is picked up by a carrier molecule, **coenzyme A**, and **acetyl coenzyme A** is formed.

Overall, the reaction is:

$$2 \text{ pyruvate} + 2NAD^+ + 2H_2O \rightarrow 2 \text{ acetate} + 2NADH + 2H^+ + 2H_2O$$

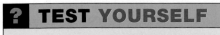

? TEST YOURSELF

E What does a reduced coenzyme have that an oxidised coenzyme does not?

SCIENCE FEATURE Pyruvate or pyruvic acid?

These terms are used interchangeably but they refer to the same substance. Fig 8.11 shows the difference: pyruvic acid in solution becomes ionised and, like all acids in solution, it loses a hydrogen ion. In the ionised state it is known as pyruvate. The same applies to other organic acids such as lactic acid and acetic acid, which become lactate and acetate respectively.

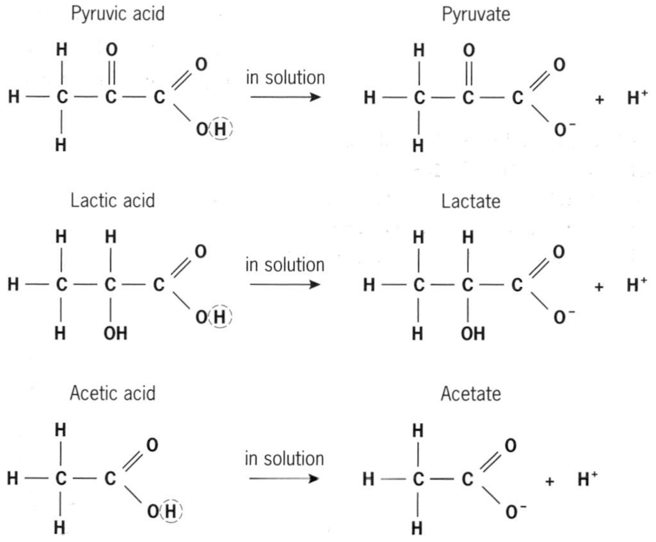

Fig 8.11 Pyruvate, lactate and acetate and their corresponding acids

KREBS CYCLE

The **Krebs cycle** is a series of reactions named after Sir Hans Krebs, the biochemist who first worked out the details. It is also known as the **citric acid cycle**, the **tricarboxylic acid cycle** or simply the **TCA cycle**. The main purpose of the Krebs cycle is to provide a continuous supply of electrons to feed into the electron transport chain.

The details of the Krebs cycle, which takes place in the matrix of the mitochondria, can be seen in Fig 8.12 on the next page. Overall, per turn, the Krebs cycle produces:

- 3 molecules of NADH;
- 1 molecule of $FADH_2$;
- 1 molecule of ATP (by substrate-level phosphorylation);
- 2 molecules of CO_2;
- 1 molecule of oxaloacetate (to allow the cycle to continue).

The Krebs cycle will 'turn' twice for every glucose molecule that enters it, so, per glucose molecule, six molecules of NADH and two molecules of $FADH_2$ are produced. NADH and $FADH_2$ are particularly important because they carry the electrons which power the next stage of glucose respiration.

F In an animal, what happens to the carbon dioxide made by the reactions in the Krebs cycle?

G Why does the Krebs cycle turn twice for every glucose molecule?

SEE QUESTIONS 1 TO 3

SCIENCE FEATURE NAD and FAD

NAD (**nicotinamide adenine dinucleotide**) and FAD (**flavine adenine dinucleotide**) are **coenzymes**, organic compounds that are catalysts for reactions. As their name implies, coenzymes work closely with enzymes. Unlike enzymes, coenzymes are not proteins. NAD and FAD carry electrons from the **electron donors** in glycolysis, pyruvate oxidation and the Krebs cycle to **electron acceptors** in the electron transport chain in the mitochondrial membrane.

When NAD and FAD accept electrons, they are reduced, becoming NADH and $FADH_2$. In this form they are called **reduced coenzymes**. Reduction of a coenzyme always requires association with a **dehydrogenase enzyme**, an enzyme that removes hydrogen from other molecules. The hydrogen removed by the dehydrogenase enzyme is split into an electron and a hydrogen ion. NAD (or FAD) accepts the electrons produced and the hydrogen ions play an important part in the electron transport chain.

In the human body, NAD is synthesised from vitamin B_3 (nicotinic acid) and FAD is made from vitamin B_2 (riboflavin): this explains why both vitamins are essential components of our diet.

Fig 8.12 The process of cell respiration. Most A-level students do not learn all the intermediate stages, but it is a good idea to look at the biochemical pathways involved to understand several important points about cell respiration:

- This is an example of a metabolic pathway. It looks complex but, like a street map, it is easy to follow once you know where you are and where you would like to go. Each individual reaction is a relatively simple step.
- Each stage is catalysed by a specific enzyme.
- You can see how the respiration of different foods (glucose, fatty acids and amino acids) is closely interconnected. The central pathway, in black, shows the complete breakdown of a molecule of glucose. The path taken by fatty acids is shown in green and the path taken by amino acids is shown in red.

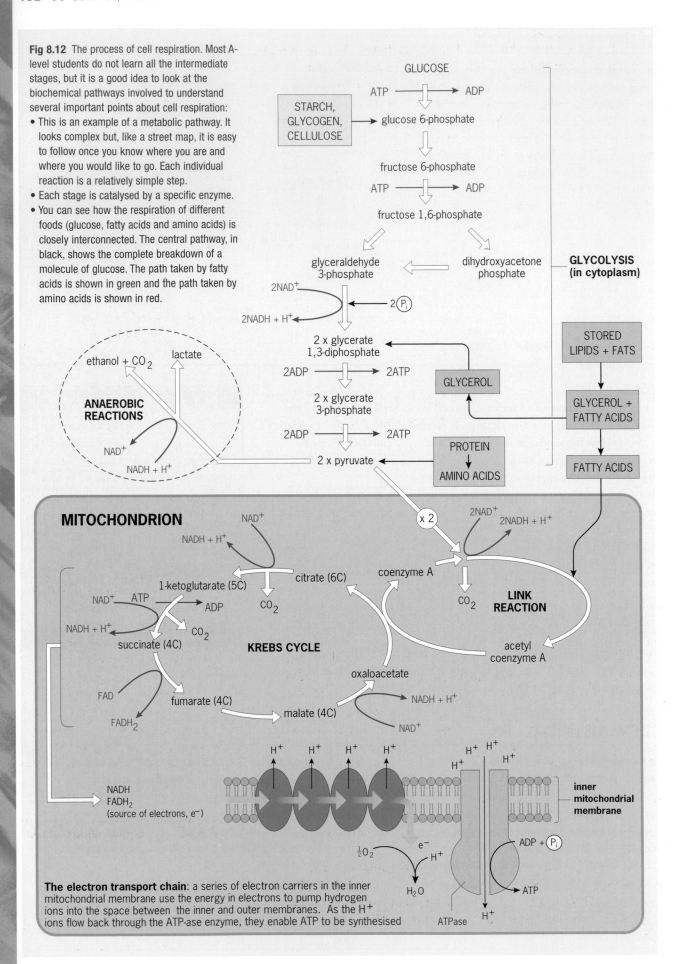

The electron transport chain: a series of electron carriers in the inner mitochondrial membrane use the energy in electrons to pump hydrogen ions into the space between the inner and outer membranes. As the H^+ ions flow back through the ATP-ase enzyme, they enable ATP to be synthesised

THE ELECTRON TRANSPORT CHAIN

This final phase of respiration produces the largest number of ATP molecules. The electron transport chain can be thought of as the 'pay day' for all the 'hard work' that has been done in the earlier parts of the process.

As Fig 8.13 shows, the electron transport chain consists of a series of carrier molecules, which will first accept an electron (thereby becoming reduced) and then lose it again (so becoming oxidised). At each of these transfers the electrons lose some energy, which can be used to power the active transport of hydrogen ions across the inner mitochondrial membrane (see Fig 8.14). This results in a high concentration of hydrogen ions in the outer mitochondrial space and a low concentration in the inner mitochondrial space.

Fig 8.13 The electrons and hydrogen ions made during the first stages of respiration are finally used to synthesise ATP

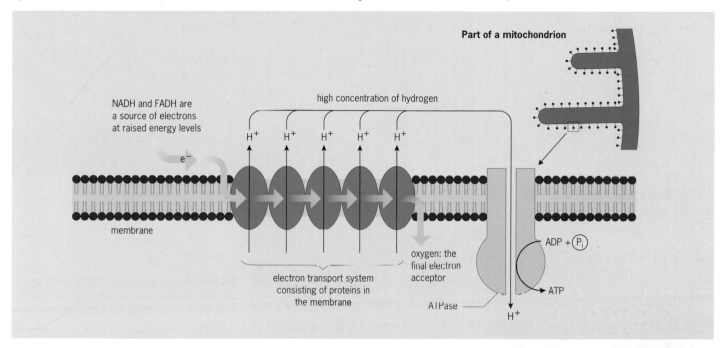

Because of the difference in concentration, hydrogen ions leak back into the inner compartment. The only route that they have is through the middle of the stalked granules – ATPase enzymes. As the stream of hydrogen ions flows down the concentration gradient, enough energy is released to allow free inorganic phosphate molecules to be added to ADP, forming new molecules of ATP. The flow of H⁺ ions is sometimes known as a **proton motive force**.

This model of ATP synthesis by **oxidative phosphorylation** is called the **chemiosmotic hypothesis** and was first proposed by Peter Mitchell in 1961. Mitchell received the Nobel Prize for Chemistry in 1978.

The end-product of the electron transport chain is low-energy electrons and hydrogen ions that combine with oxygen to form water. Although this reaction is at the end of the process, it is a key one: it is why most organisms need oxygen. If there is no oxygen to mop up the electrons and hydrogen ions from the electron transport chain, the pathway cannot be completed. This is a disaster for the cell as the intermediate compounds of glucose respiration build up and the cell loses its ability to make most of the ATP it needs.

Looking at the separate processes involved in the breakdown of one molecule of glucose makes it easy to lose track of the overall purpose of glucose respiration: ATP production. The following example gives you the chance to analyse the contribution to ATP production made by each of the different stages.

? TEST YOURSELF

H Muscle cells and cells that manufacture and secrete large amounts of hormones have more mitochondria than, say, a skin cell. Can you say why?

I The function of a kidney tubule cell is to move substances by active transport. Suggest why these cells are packed with mitochondria.

SEE QUESTION 4

Fig 8.14 Mitochondria are the organelles that house the enzymes, substrates and other chemicals associated with aerobic respiration. Pyruvate oxidation and the Krebs cycle take place in the matrix (fluid) centre of the organelle, while the electron transport chain takes place in the inner membrane itself

(a) Micrograph of a pancreatic cell: the large numbers of mitochondria suggest that this is a metabolically active cell. In this case ATP is needed to synthesise the digestive enzymes (proteins)

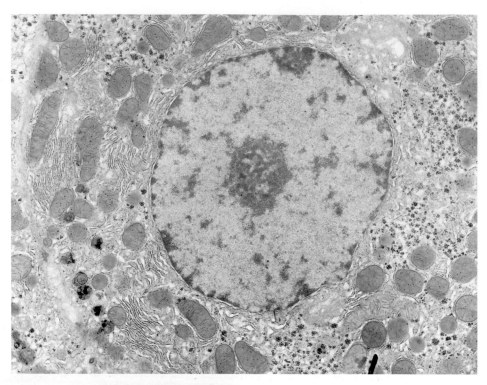

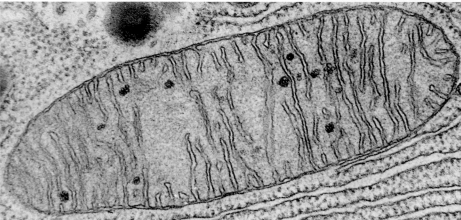

Fig 8.14 (b) Micrograph of a single mitochondrion. Generally, the more metabolically active the cell, the more folds (cristae) there are on the inner membrane

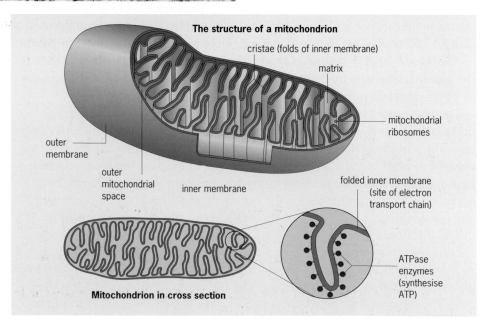

Fig 8.14 (c) 3D illustration of a mitochondrion

EXAMPLE

Q How many ATP molecules are produced, per molecule of glucose, in aerobic cell respiration?

A ATP can be made in one of two ways, by **substrate-level phosphorylation** and by **oxidative phosphorylation** (see page 149). We shall take each process in turn.

Substrate-level phosphorylation

Glycolysis and the Krebs cycle produce ATP in this way. We can complete the following table:

Stage	ATP produced
Glycolysis	2 (2 used but 4 made)
Pyruvate oxidation	0
The Krebs cycle	2 (1 per turn)
Total	4

Oxidative phosphorylation

This is the production of ATP by the electron transport chain. It relies on a continuous supply of 'high-energy' electrons from the preceding processes. Before we can calculate the final ATP total, we must calculate the number of electrons being provided by NADH and FADH$_2$.

We can draw up a second table, as follows. Remember, pyruvate oxidation and the Krebs cycle occur twice for every glucose molecule put into the system.

Stage	NADH produced	FADH$_2$ produced
Glycolysis	2	0
Pyruvate oxidation	2	0
The Krebs cycle	6 (3 per turn)	2 (1 per turn)
Total per glucose	10	2

We know that:

3 ATP molecules are produced for every molecule of NADH fed into the chain,

and that:

2 ATP molecules are produced for every molecule of FADH$_2$ fed into the chain.

We can now estimate the final ATP total:

10 NAD molecules produce 3 ATPs, giving a total of	**30**
2 FADH molecules produce 2 ATPs, giving a total of	**4**
Therefore, the total number of ATP molecules produced by oxidative phosphorylation is	**34**
The total number of ATP molecules produced by substrate-level phosphorylation is	**4**

The grand total per glucose molecule is therefore 38 molecules of ATP,

most of which are produced by the electron transport chain.

In practice, the figure is usually 36 molecules of ATP. Because the mitochondrial membrane is impermeable to NADH, the 2 molecules of NADH made during glycolysis cannot carry their electrons directly into the electron transport chain.

To get over this, most cells have a shuttle system in which the electrons are released by the NADH and passed across the mitochondrial membrane where they are picked up by FADH$_2$. As we saw earlier, FADH$_2$ produces only 2 ATPs compared with the 3 made by NADH, thus reducing the final total by 2.

This is the ATP total under *ideal* conditions. Several other factors can also reduce ATP production even further, but these are beyond the scope of this book.

AEROBIC CELL RESPIRATION OF OTHER FUELS

So far, we have studied the respiration of glucose, but many organic molecules can be fed in the same central pathway to create ATP.

LIPID BREAKDOWN

The lipid stores in the body can be used when carbohydrate is in short supply. Stored triglycerides are broken down into **glycerol** and **fatty acids**. The latter are further broken down to give several 2-carbon acetyl fragments by a process called **beta-oxidation**. These fragments are combined with coenzyme A and so enter the Krebs cycle and, eventually, the electron transport chain. Glycerol is converted into **dihydroxyacetone phosphate** so that it can be used as a fuel in glycolysis.

PROTEIN BREAKDOWN

Adult humans need up to 60 g of protein per day. We cannot store protein, so excess **amino acids**, the building blocks of proteins, are degraded in the liver by the process of **deamination**. Enzymes in the liver separate the amine group from the rest of the molecule, leaving an **organic acid**. The amine group is released as free **ammonia**, which enters the **ornithine cycle** (see Chapter 14) to be incorporated into **urea** and excreted from the body in the urine. The organic acids are fed into the Krebs cycle and are respired (see Fig 8.12).

3 THE RESPIRATORY QUOTIENT (RQ)

The **respiratory quotient**, or **RQ**, is a ratio of gas exchange, worked out by comparing oxygen uptake with carbon dioxide production. The RQ can be used to determine the type of food being respired. The basic equation for the respiration of glucose:

$$C_6H_{12}O_6 + 6O_2 \rightarrow 6CO_2 + 6H_2O + energy$$

shows that, for every glucose molecule respired, six molecules of oxygen are required and six molecules of CO_2 are produced. Equal numbers of gas molecules occupy the same volume and so the volume of oxygen taken up by an organism is equal to the volume of CO_2 produced.

> ✓ **REMEMBER THIS**
>
> Lipids can fuel aerobic exercise but not glycolysis. Aerobic exercise is therefore a good way to burn excess fat (see Assignment at the end of this chapter).

▶ **EXTENSION The origins of respiration: some speculation**

Even the simplest life form needs to release the energy locked in organic molecules and it is thought that glycolysis was one of the first metabolic pathways to evolve.

Since there was very little free oxygen at the time, aerobic respiration was impossible. It seems reasonable to assume that the process of glycolysis, as practised by early, bacteria-like organisms (Fig 8.15), used up much of the available food, giving the organisms a major problem: that of energy supply. Solving this problem proved to be a massive evolutionary step: the development of photosynthesis. Not only did photosynthesis provide a new supply, but it also released free oxygen as a by-product. The oxygen in the atmosphere is due to photosynthesis.

The build-up of oxygen in the atmosphere kick-started another evolutionary leap: the appearance of organisms that could respire aerobically. This probably happened fairly quickly because of the massive survival advantage gained by organisms that respire aerobically to produce the much larger amounts of ATP.

It is at this stage that prokaryotic organisms, which could respire aerobically, may have developed a close association with large cells, so forming primitive eukaryotic cells. Find out more about the **endosymbiont theory** on page 22.

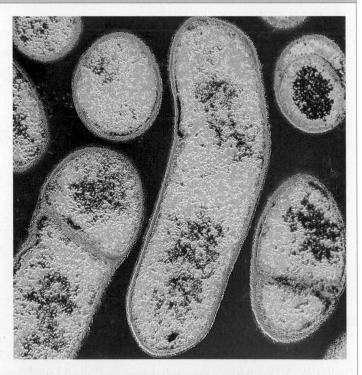

Fig 8.15 *Clostridium botulinum*, the anaerobic bacterium that causes botulism (serious food poisoning), is found in food such as meat and fish that have been badly canned. Anaerobic bacteria such as these are believed to have been the only organisms that could have survived in the oxygen-poor environment 3500 million years ago

WORKING OUT THE RESPIRATORY QUOTIENT

The RQ is worked out using the simple formula:

$$RQ = \frac{\text{volume of } CO_2 \text{ given out}}{\text{volume of } O_2 \text{ taken in}}$$

For the glucose equation, the RQ is: $6 \div 6 = 1$.

USING THE RESPIRATORY QUOTIENT

If we measure the volumes of oxygen and carbon dioxide that are exchanged by an organism and find that the RQ is 1, we can infer that the organism is respiring mainly carbohydrate. Different substrates will produce different RQ values:

- Carbohydrate respiration gives an RQ of 1.
- Protein respiration gives an RQ of 0.8–0.9.
- Lipid respiration gives an RQ of 0.7.
- Fermentation gives an RQ of infinity.

The problem with trying to work out what substrate is being respired by looking at RQ values is that many organisms, including humans, respire more than one substrate. In humans, carbohydrate is the main fuel for respiration, but a small amount of protein is also respired. The RQ figure obtained is therefore variable.

USING A RESPIROMETER

The rate of aerobic respiration can be estimated by measuring how much oxygen is taken up or how much heat is produced. Gas exchange, including oxygen uptake, is measured using a sealed chamber called a **respirometer**.

A SIMPLE RESPIROMETER

A simple respirometer is shown in Fig 8.16. Organisms placed inside exchange gases and so alter the composition of the air around them.

? TEST YOURSELF

J **a)** Why is there a range of values for the respiratory quotient of proteins?
b) Why is the respiratory quotient for fermentation infinity?

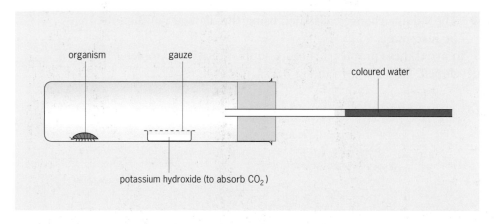

organism   gauze   coloured water

potassium hydroxide (to absorb CO₂)

Fig 8.16 A simple respirometer

To measure gas exchange due to a respiring organism, we could remove the air after a fixed time and analyse it. A simpler method is to place a carbon dioxide absorber such as sodium hydroxide in the respirometer chamber. Over a set time, a volume of oxygen is used up. Carbon dioxide is produced but is absorbed by the sodium hydroxide. So, overall, the volume of gas falls by the volume of oxygen used. The gas pressure inside the chamber falls and coloured liquid is drawn along the tube towards the chamber. The volume of oxygen used up equals the volume of the liquid moved in the fixed time. The following example shows how the respiration rate is calculated from readings taken from the tube.

⬤ **EXAMPLE**

Q a) When measuring the gas exchange of an organism with a simple respirometer, how do you calculate the volume of oxygen used?

b) In what units do you express the rate of respiration?

A a) The position of the liquid along the tube is recorded at the start and end of the set time and the distance the liquid moves is calculated. The volume of liquid (equal to the change in oxygen volume) is a cylinder inside the tube.

Therefore:

$$\text{volume of oxygen used up (in cm}^3) = \pi \times r^2 \times h$$

where r = radius or diameter/2 (in cm) and h = distance moved by the liquid (in cm).

b) The rate of respiration is given as the volume of oxygen used per organism (or per unit mass, eg per gram) per unit time; for example, as cubic centimetres of oxygen per gram per hour. So:

$$\text{rate of respiration} = x \text{ cm}^3 \text{ O}_2 \text{ g}^{-1} \text{ h}^{-1}$$

A MORE COMPLEX RESPIROMETER

The simple respirometer shows the principles of respirometry, but has these limitations:

- Changes in temperature or atmospheric pressure make gases expand or contract. These changes affect the distance the fluid moves, so the measured distance is not just due to the change in oxygen volume.
- The chemicals used might alter the composition of the gases in the chamber.
- It is difficult to restart the experiment without taking the apparatus apart.
- The volume change calculated using the diameter of the tube may be inaccurate.

To avoid these problems, a more sophisticated respirometer has been developed. This is shown in Fig 8.17.

Fig 8.17 A more complex respirometer. The left tube contains the organisms while the right tube acts as a control. In this way, any changes in fluid level that are not due to the organisms, such as changes in room temperature and pressure, are balanced out by the control tube. The experiment can be set up again using the syringe, which will also accurately measure the changes in volume due to the respiring organisms

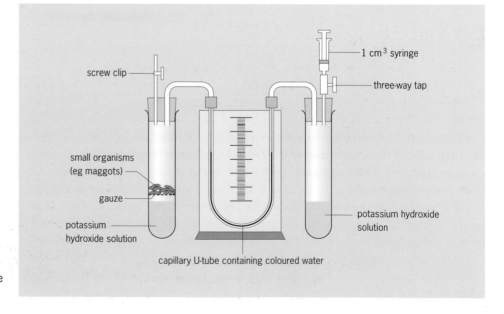

screw clip

1 cm³ syringe

three-way tap

small organisms (eg maggots)

gauze

potassium hydroxide solution

potassium hydroxide solution

capillary U-tube containing coloured water

This apparatus works on the same principle as the simple respirometer, but the syringe allows the volume change to be measured directly. After a set time, the syringe can be pulled up until the dye goes back to its starting point. This is a neat way to find the exact volume of oxygen used without having to work out the volume of a cylinder, and it allows the experimenter to move the experiment back to the start at the same time.

● EXAMPLE

Using the respirometer
Q The respirometer in Fig 8.17 was set up with 50 g of blowfly (bluebottle) larvae and left for 1 hour. The fluid moved up the left side-arm of the U-tube by 5 cm. It was moved back to its original level by drawing 2 cm³ of air into the syringe.

a) What is the purpose of the syringe and why does it give this respirometer an advantage over the simple respirometer?

b) If there were 10 larvae, what is the respiratory rate per organism?

A a) The syringe allows movement in the tube to be measured directly as a volume. This removes the need to calculate the volume of the cylinder to find out how much liquid is in the tube.

b) The volume of oxygen used up = 2 cm³. Per larva:
respiratory rate = 2/10 = 0.2 cm³ O_2 h⁻¹

Calculating the respiratory quotient
Q When the experiment was repeated using water instead of KOH, the fluid moved up 1 cm, not 5 cm.

a) What does this tell us about the volume of carbon dioxide produced?

b) What is the respiratory quotient for the larvae?

A a) Carbon dioxide was not being absorbed, and its volume is one unit less than the oxygen volume. So:

$$CO_2 \text{ out} = 4 \text{ units}$$
$$O_2 \text{ in} = 5 \text{ units}$$

b) $\dfrac{CO_2 \text{ out}}{O_2 \text{ in}} = \dfrac{4}{5} = 0.8$

Blowfly diet
Q a) Assuming the larvae are only respiring one substrate, suggest which one.

b) From what you know about the diet of blowfly larvae, does this seem reasonable?

A a) Proteins

b) Yes: they consume a large amount of protein in rotting meat.

Respirometer design
Q How does the addition of the tube on the right (known as a thermobar) help to overcome the problem of changes in atmospheric pressure and temperature?

A Any pressure changes in the tube on the left are mirrored and cancelled out by those in the right tube.

! SCIENCE FEATURE Cyanide!

Cyanide is a famous poison, often featured in crime fiction and films (Fig 8.18). It is a very effective toxin because it inhibits one of the enzymes of the electron transport chain. This stops the flow of electrons and effectively prevents aerobic respiration, the source of most of the body's ATP. Without ATP, the muscles go into spasm and the body experiences convulsions. Spasm of the muscles that control breathing usually leads to death by asphyxiation (lack of oxygen to the brain).

Fig 8.18 'A faint smell of bitter almonds and the victim lay dead, the back arched, the eyes staring.'

RESPIRATION, ENZYMES AND TEMPERATURE

Like all metabolic reactions, respiration is controlled by enzymes. The Q10 law (page 97) states that the rate of an enzyme-catalysed reaction will approximately double with every 10 °C rise in temperature between 4 and 40 °C. So, with a temperature increase of 10 °C, much more ATP becomes available to power other metabolic processes.

A dramatic example is seen in **ectothermic** animals such as reptiles, which may have wide variations in body temperature during a 24-hour period (Fig 8.19). Their rate of respiration, and therefore their metabolic rate, is roughly twice as fast at 30 °C than it is at 20 °C. Such organisms are only active when their body temperature allows them to react and move quickly.

4 ANAEROBIC RESPIRATION

Anaerobic respiration does not require oxygen, so it can take place in oxygen-poor environments such as stagnant water or deep soil. It can even occur in parts of an aerobic organism that are starved of oxygen, such as muscles during strenuous exercise.

In anaerobic respiration, glucose is broken down into pyruvate but, because no oxygen is available, the pyruvate cannot be broken down any further. Instead of entering the link reaction and the Krebs cycle, the pyruvate is converted to a waste product. The nature of the waste product depends on the organism. Fig 8.20 shows the waste products of anaerobic respiration in different organisms.

Fig 8.19 Reptiles are ectotherms: they cannot control their body temperature in the way in which mammals and birds do. This lizard must absorb sunlight in order to increase its body temperature to the level that allows its enzymes to function at their optimum efficiency

? TEST YOURSELF

K Why do some plants produce cyanide in their leaves?

L If the rate of respiration of a lizard is 400 arbitrary units at 40 °C, what will the rate be at 10 °C?

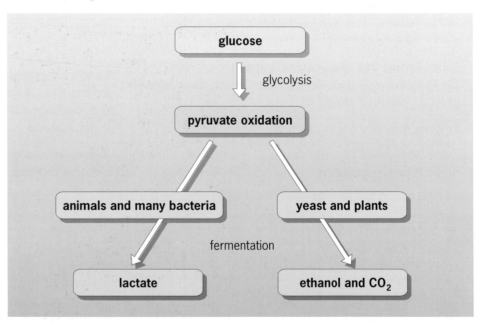

Fig 8.20 The main problem with anaerobic respiration is the build-up of NADH and the shortage of NAD^+. To solve this, H^+ is added to pyruvate to make lactate or ethanol and carbon dioxide. Fermentation is glycolysis plus the reduction of pyruvate to either lactate or ethanol and carbon dioxide

Generally speaking, animals and some bacteria convert pyruvate into **lactate** by a simple reduction reaction. In contrast, plants and fungi such as yeast convert pyruvate to **ethanol** (alcohol) and carbon dioxide.

In animals, vigorous exercise leads to anaerobic respiration in the muscles and the resulting build-up of lactate causes **fatigue**: see the Assignment at the end of the chapter.

Fig 8.21 Fermentation in different microorganisms can be used to make a variety of useful products

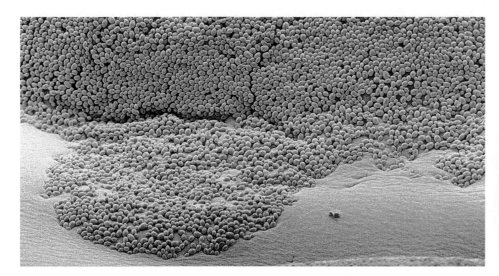

Fig 8.22 Yeast is a single-celled fungus that usually respires aerobically. When deprived of oxygen, it switches to anaerobic respiration: sugars are not completely broken down and the by-products are ethanol (alcohol) and carbon dioxide

Anaerobic respiration in microorganisms is generally known as **fermentation**. As Fig 8.21 shows, this process can give rise to many useful products. Bacteria that produce lactic acid are used in the manufacture of dairy products such as yoghurt (see page 115). The tangy taste is due to the high concentration of this organic acid. In addition to lactic acid and ethanol, other microorganisms can make solvents such as acetone and butanol.

Anaerobic respiration in yeast (Fig 8.22) is also known as **alcoholic fermentation** (see also Chapter 6). This process was discovered by accident and alcoholic drinks such as wine and beer were made for many centuries before the science behind the process was known. For fermentation, yeast simply needs a source of carbohydrate, anaerobic conditions and a suitable temperature. So wine can be made from many organic materials, such as peaches, elderberries or even potato peelings, as well as grapes (Fig 8.23).

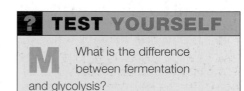

Fig 8.23 Yeast occurs naturally on the surface of many fruits. For instance, yeast is partly responsible for the bloom on a grape. Fruit stored in anaerobic conditions will therefore ferment and produce liquids with 'interesting' effects!

Since alcohol is an intermediate product in the breakdown of sugar, it contains energy that can be released by respiration. Many drinks, particularly beer (Fig 8.24), are also rich in carbohydrates that have not been turned to alcohol. Heavy drinkers become overweight because they drink large amounts as well as eating meals that, by themselves, provide enough calories to satisfy the body's energy demands.

Fig 8.24 Today, beer is brewed under controlled conditions using carefully selected strains of yeast. But the underlying process of fermentation, one of the oldest metabolic pathways known, is the same today as it was thousands of years ago

ANAEROBES AND DISEASE

Many common bacteria can only thrive in anaerobic conditions. Two such bacteria are *Clostridium tetani* and *Clostridium welchii*.

C. tetani usually thrives deep in soil, where there is little oxygen, but is present in small numbers almost everywhere. The bacteria can enter the human body through a deep cut or wound. Once inside, they multiply and produce a toxin that makes muscles go into spasm (Fig 8.25). This is **tetanus**, or lockjaw (named because of its effect on the jaw muscles), which can be fatal if it affects the muscles used for breathing.

? TEST YOURSELF

M What is the difference between fermentation and glycolysis?

C. welchii can cause **gas gangrene**. If it infects areas of dead tissue, such as frost-bitten toes or injured limbs that have a poor blood supply, it turns the tissues black (Fig 8.26). In such 'ideal' conditions, the bacteria multiply rapidly, producing a foul-smelling gas and several lethal toxins. Antibiotics help if the condition is treated early. If not, the affected limb may need to be amputated to prevent the toxins spreading further into the body.

As *C. tetani* and *C. welchii* can *only* respire anaerobically, they are known as **obligate anaerobes**. Organisms such as yeast can respire by either pathway and are known as **facultative anaerobes**. Organisms that thrive only when they are respiring aerobically are called **obligate aerobes**.

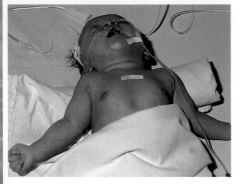

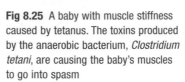

Fig 8.25 A baby with muscle stiffness caused by tetanus. The toxins produced by the anaerobic bacterium, *Clostridium tetani*, are causing the baby's muscles to go into spasm

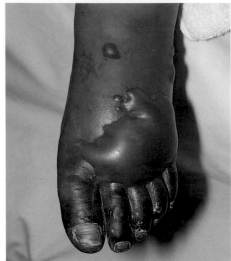

Fig 8.26 This is the dreaded gangrene. Tissue death, caused by a poor blood supply, can lead to infection by the anaerobic bacterium *Clostridium welchii*. This organism causes gas gangrene and the toxins produced can be lethal if not treated quickly

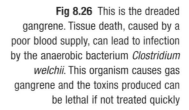

SUMMARY

By the end of this chapter, you should know and understand the following:

- **Cell respiration** is the release of energy from organic molecules. The energy released is transferred to ATP, a compound that can provide the cell with energy in small, instant, controllable amounts.

- Cell respiration consists of many reactions. Each reaction is controlled by a different enzyme. Most of the reactions occur inside the **mitochondrion**.

- Complete **aerobic respiration** of glucose has four stages: glycolysis, pyruvate oxidation, the reactions of the Krebs cycle and the reactions of the electron transport chain.

- **Glycolysis**: glucose is broken down into two molecules of pyruvate. Only two molecules of ATP are made. This happens in the cytosol of cells.

- **Pyruvate oxidation** (the 'link' reaction): in the presence of oxygen, pyruvate enters the mitochondrion and is converted into acetyl coenzyme A.

- The **Krebs cycle**: a series of reactions, fuelled by acetyl coenzyme A, which produces only two molecules of ATP but many electrons and hydrogen ions.

- The **electron transport chain**: a series of redox reactions, fuelled by the electrons and protons from the preceding three processes, as supplied by the coenzymes NAD and FAD. Electron transport releases enough energy to synthesise most of the ATP produced by aerobic respiration.

- Most food chemicals – carbohydrates, lipids and proteins – can be respired to provide energy.

- The rate of respiration can be measured as a respiratory quotient, the ratio of CO_2 evolved to O_2 absorbed. It can be calculated from measurements made using a respirometer.

- Most organisms respire aerobically (in the presence of oxygen). The substrate (eg glucose) is broken down completely to produce carbon dioxide, water and a lot of ATP.

- Some organisms and some tissues respire anaerobically. In the absence of oxygen, substrates are only partially broken down, leading to waste products such as lactate (in animals and bacteria) or ethanol and carbon dioxide (in plants and yeast).

? EXAM QUESTIONS

1 Read through and copy the account of cellular respiration given below, then complete the passage by inserting in the blank spaces the terms, or phrases, you feel are most appropriate.

The first stage in aerobic respiration is usually referred to as _____ and the site where this occurs is the _____. Glucose (a 6C molecule) is the substrate for this process but needs to be _____ in order for it to take part in the sequence of reactions. This requires the use of two molecules of _____ per molecule of glucose. Almost immediately, the resultant molecule is split into two smaller molecules of _____. This is then converted to _____ which will pass through into the central _____ region of a mitochondrion. This is where the enzymes are found which catalyse the second phase of respiration called _____.
A coenzyme called _____ is now required to link the first and second states of respiration. It is during this second stage of respiration that all of the waste _____ is removed. Also, for each original molecule of glucose, six reduced molecules of _____ are produced during this cycle. The final stage of respiration, known as the _____, takes place in the _____. Hydrogen ions are combined with _____ and _____ to form water. The iron-containing molecules called _____ are essential to this process.

OCR 6910 Molecules and Life Nov 2000 Q2

2 The diagram represents some of the biochemical reactions of aerobic respiration. Study the diagram, then answer the questions which follow.

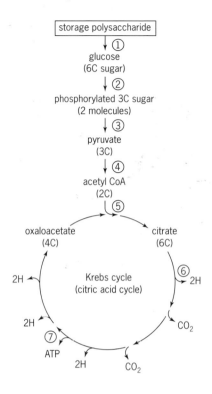

a) Although oxygen is not mentioned, state how the diagram indicates that aerobic respiration is occurring, rather than anaerobic respiration.

b) Which reaction, represented by a numbered arrow, indicates that a molecule of carbon dioxide is released at that point?

c) **(i)** At what numbered stage is ATP used?
(ii) Why is ATP needed at this stage?

d) Which numbered stage represents a coenzyme being released for recycling?

e) Which numbered stage represents the point at which a respiratory substrate enters the mitochondrion?

f) What type of enzyme would be necessary at stage 6?

g) What happens **immediately** to the removed 2H?

h) Where **precisely** would you expect to find acetyl CoA in a cell?

OCR 6910 Molecules and Life June 2000 Q7

? EXAM QUESTIONS

3 Adenosine triphosphate, ATP, provides energy within the cell.

a) Excluding respiration, state **three** different cellular activities which use ATP.

b) When ATP is used, the reaction may be summarised as:

$$ATP \rightarrow ADP + Pi + 30.6 \text{ kJ mol}^{-1}$$

Consider this reaction and the statements below, then write either T or F after each question to indicate whether the statement given is true (T) or false (F).

The reaction is exergonic.

The ADP molecule now contains two nucleotides.

The reaction is anabolic.

The ATP contains a purine base, a hexose (6C) sugar and three phosphate groups.

The ADP contains a purine base and a pentose (5C) sugar.

c) Explain precisely, **where** in the metabolic pathway, and **why**, ATP is **used** in the process of respiration.

d) State **where**, precisely, most ATP is **synthesised** in an animal cell.

OCR 6910 Molecules and Life March 2001 Q4

4 The diagram below summarises the electron-transport chain, which is a stage of cell respiration. A, B and C represent electron carriers

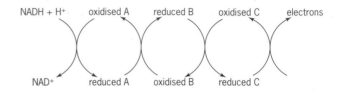

a) **(i)** Describe what happens to the electrons and hydrogen ions at the end of this stage.

(ii) State where in a mitochondrion the electron transfer chain is situated.

b) Name **one** process which produces $NADH + H^+$.

c) ATP is synthesised as a result of this stage of respiration. State **two** uses of ATP in a cell.

Edexcel B/HB1 June 1999 Q4

KEY SKILLS **ASSIGNMENT**

ENERGY AND SPORT

Muscle movement requires ATP. For prolonged exercise, muscles need a continuous supply of ATP. So, to a great extent, the stamina of athletes depends on their ability to supply their muscles with sufficient oxygen to make ATP by aerobic respiration.

In this chapter, you have studied aerobic and anaerobic respiration.

1 What is the essential difference between aerobic and anaerobic systems?

2 How many molecules of ATP are produced, per molecule of glucose:
a) from anaerobic respiration (glycolysis)?
b) from complete aerobic respiration (glycolysis, pyruvate oxidation, Krebs cycle and electron transfer chain)?

For athletes, the aerobic pathway seems to be the most important, but we must also take into account the time taken to complete the process.

Complete aerobic respiration takes between 50 and 60 seconds, but glycolysis takes only about 10 seconds to produce ATP. Glycolysis is therefore very useful to an athlete because it is fast and requires no oxygen, and still allows the muscles to work when the body cannot supply them with enough oxygen for complete aerobic respiration.

If glycolysis takes at least 10 seconds to provide ATP, how is it that we can move instantly? The answer is that there is already some ATP stored in the muscles. When we begin to move, we use this store of ATP. During strenuous exertion, such as when we lift a heavy weight, we use this store of ATP up in about 3 seconds. That leaves 7 seconds, which we get through thanks to a back-up system, a chemical called creatine phosphate (CP). CP has an even higher free energy of hydrolysis than ATP and so can be broken down to allow instant resynthesis of ATP. This allows muscular contraction to continue for another 6 to 7 seconds.

3 Why do you think that the muscles have enough ATP/CP for about 10 seconds' worth of exercise?

4 Name the chemical produced as a consequence of anaerobic respiration. Explain its effect on the muscles.

Generally, the duration of a sport or event determines the main source of ATP used. The ATP/CP system is important for events that last less than 10 seconds, the anaerobic system (glycolysis) for events between 10 and 60 seconds and the aerobic system (complete aerobic respiration) for anything longer.

5 Copy and fill in the table below, identifying the energy system that provides most of the ATP for that particular activity.

Sport/event	ATP/CP	Anaerobic	Aerobic
javelin			
200 m sprint			
800 m run			
shot put			
marathon			
100 m sprint			
400 m sprint			
5000 m run			
soccer			

Sports such as soccer, hockey and tennis use a mixture of all systems because the activity is prolonged but not constant. There are periods of intense activity such as a short sprint followed by periods of relative rest.

Fig 8.A1 Sprinting is an anaerobic activity: athletes get the energy to sprint from the ATP/CP and glycolytic systems. Note that anaerobic specialists tend to be heavily muscled

KEY SKILLS **ASSIGNMENT**

Fig 8.A2 Prolonged exercise relies on the aerobic system to provide a continual supply of ATP. Generally, fatigue sets in only when fuels such as glycogen begin to run out (see page 47 in Chapter 3 for information about glycogen). Endurance athletes like these marathon runners tend to be of slight build, with small muscles

6 **a)** Why is the 400 m (world record under 40 seconds) thought by athletes to be a particularly painful event?
b) What use would the information in the completed table (on page 165) be to someone who wanted to improve his or her fitness for a particular event?
c) Why do you think sprinters usually have large muscles but marathon runners are generally quite thin?
d) A footballer wishes to get fit for the new season, so he jogs 5 miles each day. Explain why this would not be a complete preparation and suggest the additional training he would need to do.
e) Cats in the wild hunt their prey by pouncing or by short sprints. Hunting dogs rely on stamina in a long chase. Which energy system are cats and dogs using? Explain why dogs need more regular exercise than cats.

Aerobics and weight loss

Many people want to improve their overall fitness, but can exercising help them to lose weight as well? To understand more about aerobics and weight loss, we need to think about the type of fuel being burned in respiration. The way we use respiratory fuels has been likened to the way we use money.

We all need spending power, and the quickest and most universally accepted form of payment is cash. When we have no cash, we can go to the cashpoint or bank and get some more out instantly. When our bank account is low, we have to turn to our savings. This may take a while to get at, but it provides a useful cushion. When we are completely broke, we will have to sell our belongings and property just to survive. Eventually, with no injection of cash, we can be declared bankrupt.

7 Rewrite the last paragraph, applying the analogy to human energy sources, using the terms lipid, protein, glucose, glycogen, food, dead, energy. When you have done this, either write down or discuss in a group the limitations of this analogy.

It takes time to run down the glycogen store and start to use fats. This is a shame because otherwise you could simply go for a run and return slimmer, if a little baggy.

In practice, it takes several sessions – and often a reduction in calorific intake – to bring about an improvement. Generally, exercise is very beneficial because it strengthens the heart and circulatory system and raises the metabolic rate, so you use up more calories all the time, not just during exercise.

8 Imagine you are a GP faced with a rather fat couple in their forties. Write down or discuss how you would try to persuade them to take exercise. What exercises would you recommend and what precautions should they take?

Fig 8.A3 Wild dogs chasing wildebeest

09

Photosynthesis

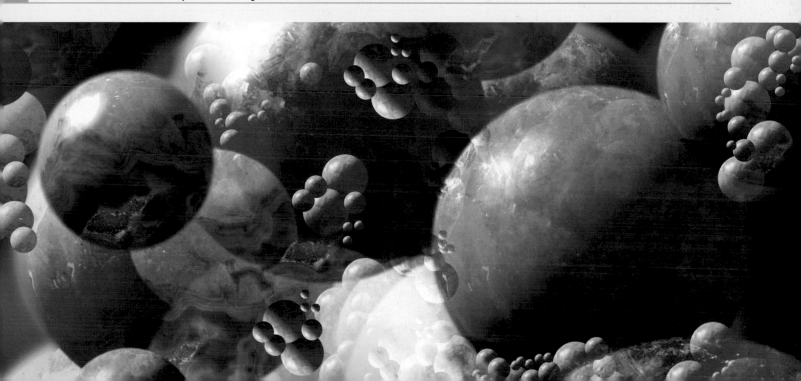

Plants are able to convert sunlight into a high-energy fuel in a complex yet very efficient process that scientists have so far been unable to duplicate. In recent years, however, progress has been made towards creating a type of artificial photosynthesis. The benefits in the long term could be huge; cheap, clean energy sources.

A team at the Massachusetts Institute of Technology in the USA has developed a protein that captures energy from a light source and stores it in the form of hydrogen gas. Creating a molecule to replace a leaf – 'photosynthesis in a beaker' – could help revive interest in the sun as a source of energy.

One possible spin-off of this technology is that in future Armed Forces operations in the field could rely on biological methods to produce electricity through photosynthesis, rather than relying on fossil fuels and bulky, heavy batteries.

Over millions of years of natural selection plants have evolved to make optimum use of the different wavelengths of light, and plants convert 98 per cent of the sunlight they receive into energy. Conversely, current artificial solar energy systems are only 10–15 per cent efficient. Using plant proteins to produce electricity – an area of research known as biological photovoltaics – could make the process much more efficient; perhaps by 40–50 per cent.

It is possible that in the future soldiers could have photosynthetic coatings on their Kevlar helmets that could produce enough energy for their electronic equipment. Other equipment and vehicles could also be covered with these solar converters. An added benefit is that the protein coatings would make whatever they coat more difficult to detect by electronic means since they would mimic the natural environment.

The light infantry? In future, helmets may by able to photosynthesise and produce enough energy to power a soldier's equipment

1 THE IMPORTANCE OF PHOTOSYNTHESIS

It is difficult to overstate the importance of photosynthesis – the process that feeds the world and replenishes the atmosphere with oxygen. Organisms that photosynthesise are able to capture the energy of the sun and use it to make carbohydrate molecules such as sugars and starches, and, indirectly, other essential compounds such lipids, proteins and nucleic acids. It is no exaggeration to say that photosynthesis turns thin air into food (Fig 9.1).

Photosynthesis is a complex process, but it can be summarised as:

$$\text{Carbon dioxide} \quad \text{and} \quad \text{water} \xrightarrow[\text{chlorophyll}]{\text{light}} \text{glucose} \quad \text{and} \quad \text{oxygen}$$

$$6CO_2 \quad + \quad 6H_2O \quad \rightarrow \quad C_6H_{12}O_6 \quad + \quad 6O_2$$

You shouldn't think of this as a chemical equation, even though it looks a lot like one. In reality, the carbon dioxide molecules never react directly with water molecules, so this is more a summary of photosynthesis, showing what goes in and what comes out.

2 PHOTOSYNTHESIS FACTS

- Organisms that can photosynthesise include higher plants, algae (including the seaweeds) and photosynthetic bacteria.
- Organisms that can photosynthesise are called **producers**, or **autotrophs** ('self-feeders').
- The process involves absorbing the light energy of the sun and converting it into chemical energy. The compounds made, such as sugars, starch, lipids and protein, contain this energy in their chemical bonds.
- Globally, photosynthesis makes billions of tonnes of organic compounds every year; one estimate puts global production at 70×10^{12} kg per year.
- Up to 50 per cent of all photosynthesis takes place in the sea, in the tiny algae known as phytoplankton (Fig 9.2).
- Virtually all ecosystems are supported by photosynthesis – it is the most important way energy gets into an ecosystem.
- Photosynthesis needs light, chlorophyll, carbon dioxide, water and a suitable temperature. The rate of photosynthesis is limited by the factor in shortest supply.
- Photosynthesis is a complex step-by-step process – each step is catalysed by a particular enzyme.

Fig 9.1 Where did this tree come from? We know that the material to make our bodies comes from the food we eat, but what about plants? This tree may weigh upwards of 3000 tonnes, but the vast majority of the organic compounds in the wood didn't come from the soil – they were made by photosynthesis using carbon dioxide from the atmosphere. A huge amount of water was also needed from the roots, but most of the dry mass comes, quite literally, from thin air

Fig 9.2 When you think of photosynthesis on a global scale you probably think of rainforests or fields of crops, but the tiny algae in the oceans produce just as much food as land plants. These are diatoms, one type of single-celled algae that are responsible for producing huge amounts of food in the world's oceans. As one of the many species that makes up phytoplankton (plant plankton) they support vast marine and freshwater ecosystems

3 PHOTOSYNTHESIS AND RESPIRATION

If you look at the overall summary for respiration, where:

Glucose and oxygen $\rightarrow$ carbon dioxide and water + energy,

you can see that photosynthesis and respiration are effectively the reverse of each other. Photosynthesis traps energy in organic molecules, and respiration releases it. Overall, photosynthesis is the reduction of carbon dioxide to form organic molecules, while respiration is the oxidation of organic molecules to form carbon dioxide. It is important to remember that plants also respire day and night, but in order for them to accumulate organic molecules and to grow, their overall rate of photosynthesis must exceed their rate of respiration.

4 WHAT HAPPENS IN A LEAF?

In most plants the leaf is an organ of photosynthesis. As Fig 9.3 shows, the leaf is perfectly adapted to maximise the process. The most actively photosynthesising cells in the leaf are the **palisade cells**, and the whole structure of the leaf is geared towards providing these cells with what they need and taking away their waste products.

Fig 9.3 Journey to the centre of a leaf. The chlorophyll pigments are housed on discs called thylakoids that are piled into grana inside chloroplasts. Most chloroplasts are found in the palisade and spongy mesophyll cells in the centre of leaves

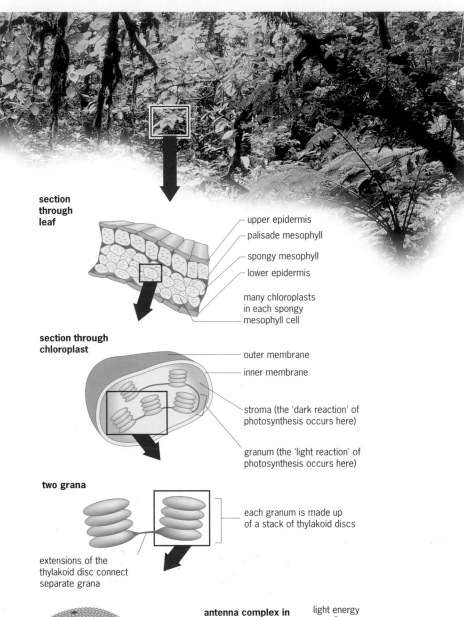

section through leaf

upper epidermis
palisade mesophyll
spongy mesophyll
lower epidermis
many chloroplasts in each spongy mesophyll cell

section through chloroplast

outer membrane
inner membrane
stroma (the 'dark reaction' of photosynthesis occurs here)
granum (the 'light reaction' of photosynthesis occurs here)

two grana

each granum is made up of a stack of thylakoid discs

extensions of the thylakoid disc connect separate grana

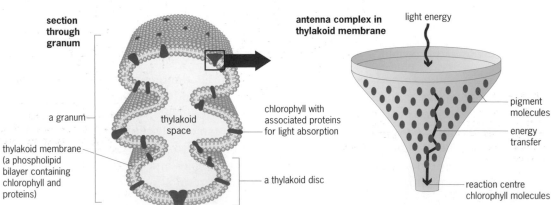

section through granum

a granum

thylakoid membrane (a phospholipid bilayer containing chlorophyll and proteins)

thylakoid space

chlorophyll with associated proteins for light absorption

a thylakoid disc

antenna complex in thylakoid membrane

light energy

pigment molecules

energy transfer

reaction centre chlorophyll molecules

Adaptations of the leaf shown in Fig 9.3:

- It is thin, so that gases can diffuse in and out quickly.
- It has a large surface area to maximise light absorption.
- It has a thin layer of wax – the **cuticle** – to reduce evaporation from the upper surface.
- It has a rich network of **xylem** fibres to deliver water and minerals, and **phloem** fibres to remove the product of photosynthesis (mainly as sucrose). This vascular tissue also gives the leaf more rigidity.
- The cells of the upper epidermis have no chloroplasts, so that more light gets through to the palisade cells.
- The palisade cells are deep and tightly packed. The chloroplasts within them can move so that all have a chance of obtaining all the light they need.
- Guard cells surround pores called **stomata** (singular = stoma; 'hole') opening in the day to allow carbon dioxide in, then closing at night to reduce water loss.
- Loosely packed **spongy mesophyll** cells create air spaces, making gas exchange more efficient between the atmosphere and the palisade cells. The air spaces also trap a little carbon dioxide, allowing some photosynthesis even when the stomata are closed.

CHLOROPLASTS

Chloroplasts are the organelles of photosynthesis. In the leaves of higher plants there can be as many as 100 or more chloroplasts in each cell, but more commonly 20–30. Their detailed structure is shown in Fig 9.4.

> **? TEST YOURSELF**
>
> **A** In order to make photosynthesis in the palisade cell as efficient as possible, what must be supplied and what must be taken away? Use the basic photosynthesis equations to help you.

THERE IS MORE ABOUT STOMATA, XYLEM AND PHLOEM IN CHAPTER 31.

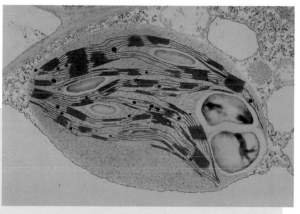

Fig 9.4 False colour transmission electron micrograph of chloroplast. The grana are connected by thin membranes; intergrana lamellae. Note the starch grains and lipid globules

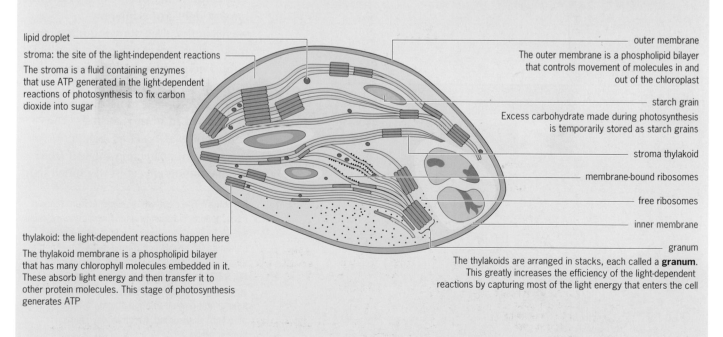

lipid droplet

stroma: the site of the light-independent reactions
The stroma is a fluid containing enzymes that use ATP generated in the light-dependent reactions of photosynthesis to fix carbon dioxide into sugar

thylakoid: the light-dependent reactions happen here
The thylakoid membrane is a phospholipid bilayer that has many chlorophyll molecules embedded in it. These absorb light energy and then transfer it to other protein molecules. This stage of photosynthesis generates ATP

outer membrane
The outer membrane is a phospholipid bilayer that controls movement of molecules in and out of the chloroplast

starch grain
Excess carbohydrate made during photosynthesis is temporarily stored as starch grains

stroma thylakoid

membrane-bound ribosomes

free ribosomes

inner membrane

granum
The thylakoids are arranged in stacks, each called a **granum**. This greatly increases the efficiency of the light-dependent reactions by capturing most of the light energy that enters the cell

REMEMBER THIS

No matter how far you take the study of biology, all you need to know about microsopic distances is these two simple facts:

- 1 mm = 1000 μm
  1 μm = 1000 nm
- 1 millimetre = 1000 micrometres
  1 micrometre = 1000 nanometres

CHLOROPLAST FACTS

- They are usually shaped like biconvex discs, about 5 μm across and 2 μm deep. The flattened shape increases the rate of the exchange of materials.
- They have a double outer membrane.
- There is an elaborate system of membranes inside the chloroplast that provides a large surface area for housing the chlorophyll pigments. The disc-shaped membranes that house the chlorophyll are called **thylakoids**, which are stacked into **grana** (singular = granum). Usually, the thylakoids are at right angles to the incoming light.
- The fluid surrounding the grana is called the **stroma**.
- Chloroplasts have their own DNA and ribosomes so they can synthesise their own proteins.
- It is thought that chloroplasts may have originated as free-living algae that became incorporated into a larger cell – see the endosymbiont theory on page 22.

REMEMBER THIS

The word *pigment* is a general name for a coloured compound. Haemoglobin and the different types of chlorophyll are all pigments.

CHLOROPHYLL AND LIGHT

In order to understand the role of chlorophyll in photosynthesis it helps to know something about the nature of light. Fig 9.5(a) shows the electromagnetic spectrum – the range of different types of radiation. Most of the radiation that reaches the Earth from the Sun falls in the region of 400–700 nm wavelength, so not surprisingly this is what our eyes have evolved to see; it is the **visible spectrum**. When stimulated by the whole 400–700 nm range we perceive white light but, as Fig 9.5(b) shows, this can be divided up into the familiar seven colours. Chlorophyll has also evolved to make use of this available light.

What we think of as chlorophyll is actually a mixture of compounds that fall into two basic types; the **chlorophylls** and the **carotenoids** (Fig 9.6). The two main chlorophylls, ***a*** and ***b***, are virtually identical molecules that absorb light mainly in the red and blue areas of the spectrum. The carotenoids, such as **β-carotene**, harvest light from different wavelengths, thus allowing plants to make much better use of the available light. Energy absorbed by the carotenoids is eventually passed on to the chlorophyll molecules before being used in photosynthesis. The carotenoids also act as a safety mechanism, protecting the leaf from the damaging effects of too much light.

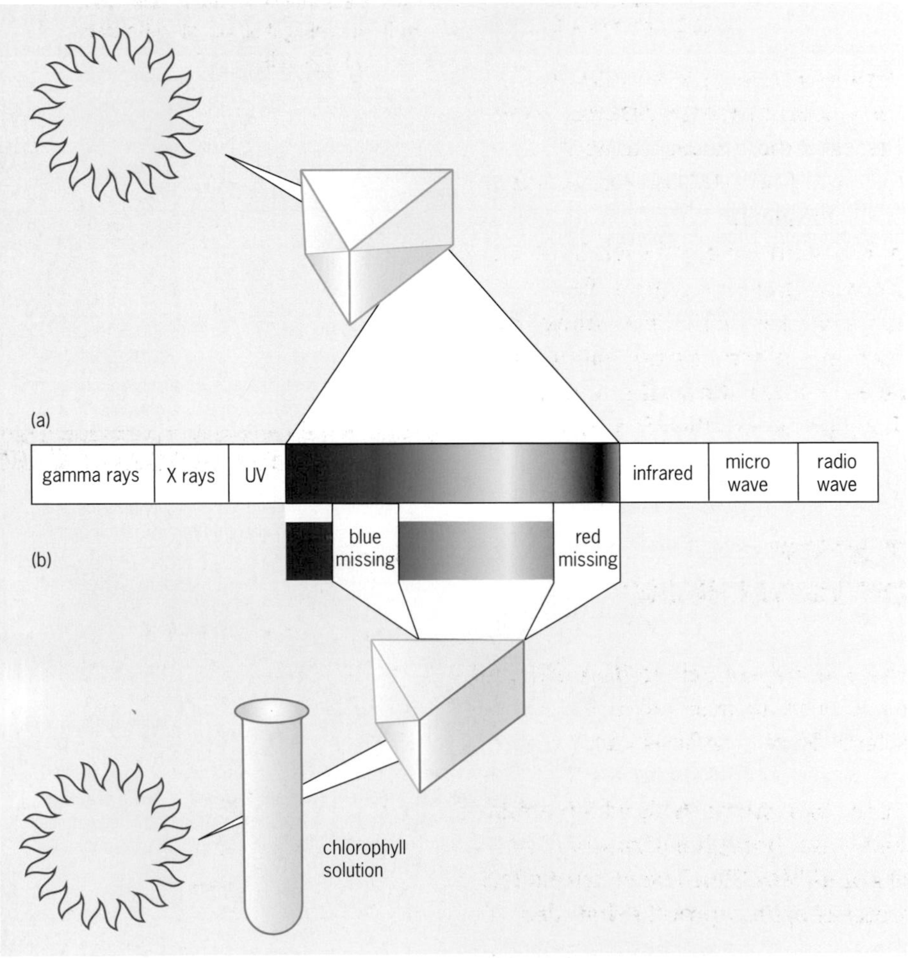
(a)
gamma rays
X rays
UV
infrared
micro wave
radio wave
(b)
blue missing
red missing
chlorophyll solution

Fig 9.5(a) The electromagnetic spectrum ranges from the long wavelength (and lowest energy) radio waves to the shortest wavelength (high-energy) gamma rays. Visible light forms just a small part of the spectrum, between ultraviolet and infared
Fig 9.5(b) White light splits into its component colours when passed through a prism. When white light is first passed through chlorophyll the red and blue light is absorbed, which is why plants appear green

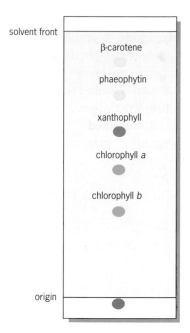

Fig 9.6 If you grind up a leaf with a little sand (to break up the cell walls) and a suitable solvent such as propanone, you can run a paper or thin layer chromatogram (see page 42 in Chapter 3) to separate out the different pigments that make up chlorophyll

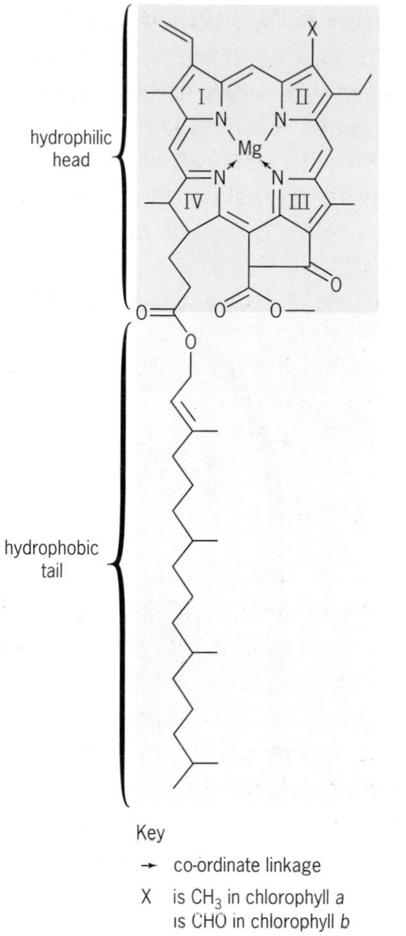

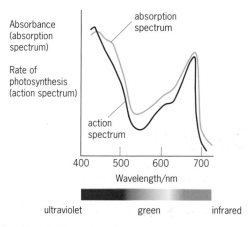

Key

→ co-ordinate linkage

X is CH_3 in chlorophyll *a*
  is CHO in chlorophyll *b*

Fig 9.7 The structure of chlorophylls *a* and *b*. The molecule consists of a head to absorb light and a tail to anchor it to the thylakoid membrane. The head is a complex *chlorin* ring structure with a magnesium ion at the centre. Different chlorophylls have slight variations on the side groups of the head, which allow different wavelengths of light to be absorbed.

Fig 9.8 The action spectrum shows the wavelengths of light that stimulate photosynthesis, while the absorption spectrum shows the wavelengths of light absorbed by isolated chlorophyll pigments

The structure of **chorophylls *a*** and ***b*** is shown in Fig 9.7. The ring structure surrounding the magnesium ion contains five smaller rings consisting of alternating single and double bonds. In reality each bond is partly double; a **conjugated system**. These bonds absorb light, so the more of them there are, the greater the light absorbance. Chlorophyll *a* absorbs strongly in the 660–700 nm range; red light. It also absorbs to a lesser extent in the blue range.

The wavelengths of light that stimulate photosynthesis are shown by an **action spectrum**, obtained by varying the wavelength of light a plant receives and then measuring the rate of photosynthesis. When the chlorophyll pigments are extracted, and their absorption at the different wavelengths is measured, we get an **absorption spectrum**. Fig 9.8 shows that the two graphs are very similar, showing that the chlorophylls are the pigments responsible for photosynthesis.

SEE QUESTIONS 1 AND 2

5 AN OVERVIEW: THE FIRST STEP TO UNDERSTANDING PHOTOSYNTHESIS

Photosynthesis is often seen as a difficult topic, but approaching it in the right way will help you make sense of it. Firstly, you need an overview or framework that you can fit the details onto later. Photosynthesis can be divided into two easy steps:

● **Step 1** The light-dependent reaction. Light hits chlorophyll, which emits two high-energy electrons. These electrons pass through a series of electron transfer reactions that make ATP and NADPH. The electrons lost by the chlorophyll are replaced when water is split, a process that also produces oxygen as a by-product.

● **Step 2** The light-independent reaction. ATP and NADPH are used to power a series of reactions known as the **Calvin cycle**. Overall, these reactions reduce carbon dioxide into the all-important carbohydrate.

For some A-level specifications, all the detail you need (and more) has been covered in the chapter so far and the summary at the end. Other specifications require the more in-depth account that follows.

Fig 9.9 An overview of photosynthesis

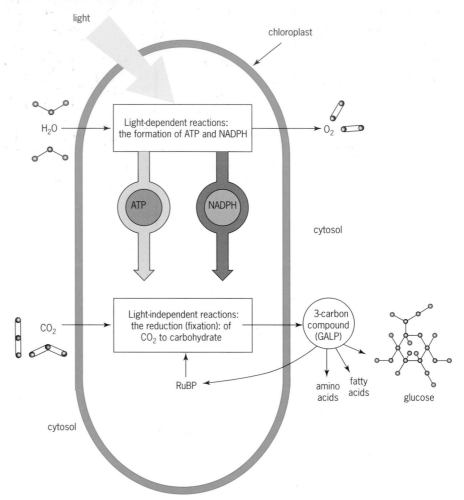

REMEMBER THIS

What is NADPH?

NADPH is a coenzyme that carries electrons. It is a powerful reducing agent, and is vital in the reduction of carbon dioxide into carbohydrate. NADPH is similar to NADH in respiration, which performs basically the same function. To avoid confusion, remember 'P for photosynthesis'; NADH in respiration, NADPH in photosynthesis (NB in reality the P stands for phosphate). While NAD^+ is reduced by an electron, NADP is reduced by an electron and a proton (ie a hydrogen atom)

REMEMBER THIS

The light-dependent reaction takes place on the thyakoid membranes, while the light-independent reaction takes place in the stroma of the chloroplasts

? TEST YOURSELF

B What is the difference between oxidation and reduction?

C Which one involves a molecule gaining energy?

6 THE LIGHT-DEPENDENT REACTION

In order to understand the light-dependent reaction, you need to be familiar with the fine structure of the chloroplast, and in particular the arrangement of the chlorophyll molecules on the thylakoid membranes. As Fig 9.3 shows, chlorophyll molecules are arranged in funnel shapes, known as **antennae complexes**. The mouth of the funnel is occupied by several different types of pigment, allowing light to be harvested across more of the visible spectrum.

When a photon of light hits any of the molecules of chlorophyll at the mouth of the complex, the electrons they contain absorb the energy and are raised from a low-energy **ground state** to an **excited state**. As the energy moves from molecule to molecule down the antenna complex, electrons are first excited, then fall back to the ground state as they pass their energy on. The energy passes on down the antenna complex until they eventually reach two chlorophyll *a* molecules at the bottom, known as the **special pair**.

The stages in the light-dependent reaction (Fig 9.11) are as follows:

● **Stage 1** A photon of light hits photosystem II and is channelled down to the reaction centre, which emits high-energy electrons.

- **Stage 2** The electrons are picked up by electron acceptors. In doing so the electron acceptor becomes reduced while the chlorophyll is oxidised. The electrons are then transferred along a series of electron carriers in the thylakoid membrane. The energy in the electrons is used to make ATP – this is **photophosphorylation**.
- **Stage 3** The electrons, having fallen to a lower energy level, are absorbed by the chlorophyll molecules in photosystem I.
- **Stage 4** A photon of light hits the reaction centre in photosystem I, which also emits two high-energy electrons. Again, the

electrons are channelled down a series of electron carriers but this time the electrons combine with H$^+$ ions to reduce NADP into NADPH.

Fig 9.10 If you shine light on pure chlorophyll it will emit high-energy electrons but without an electron acceptor they fall back to their ground state, emitting the energy as light and heat. This is why pure chlorophyll fluoresces

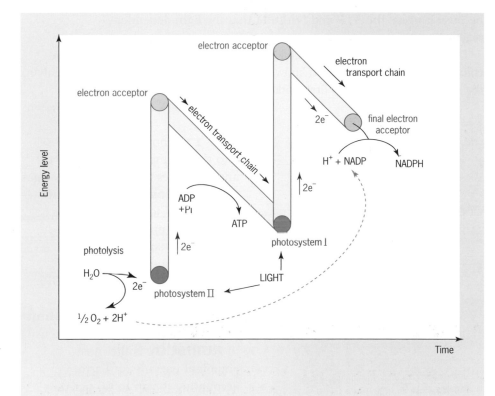

Fig 9.11 Schematic diagram of the light-dependent reaction

CYCLIC AND NON-CYCLIC PHOTOPHOSPHORYLATION

Photophosphorylation refers to the production of ATP using the energy from light. There are two methods of doing this: cyclic and non-cyclic. ATP production as described above is called **non-cyclic phosphorylation**, because the electrons are not recycled. Electrons that start off in water molecules are passed through the process until they end up in NADPH, which is then used to make carbohydrate.

However, ATP can also be made by **cyclic phosphorylation**. In this case some of the electrons from photosystem I are channelled back on to the first electron transport chain, and their energy is used to make ATP before returning to photosystem II. This process therefore makes ATP but not NADPH. No water is needed to provide electrons, and so no oxygen is produced.

SEE QUESTIONS 3 AND 4

AND WHERE DOES THE OXYGEN COME FROM?

The electrons that leave the special pair of chlorophyll molecules need to be replaced in order for the reaction to continue. Where do the electrons come from?

For photosystem I the electrons come from the electron transport chain.

For photosystem II they come from the splitting of water, a process called **photolysis**. Water is split into protons and electrons, leaving oxygen as a by-product. It is this process that has led, over billions of years, to the oxygen-rich atmosphere we have today.

$$2H_2O \rightarrow 4H^+ + 4e^- + O_2$$

7 THE LIGHT-INDEPENDENT REACTION

The light-independent reaction has in the past been referred to as the *dark reaction* but this misleading name has been dropped because it does not happen in the dark. The light-independent reaction happens in the light because it needs the ATP and NADPH that the light-dependent reaction produces.

The light-independent reaction takes place in the stroma (fluid part) of the chloroplast. The process is also known as the **Calvin cycle** after Melvin Calvin, the American scientist who first worked out the details in the 1950s.

The steps in the Calvin cycle are shown in Fig 9.12.

● **Step 1** Carbon dioxide diffuses into the stroma of the chloroplast and combines with a 5-carbon sugar called **ribulose bisphosphate** (RuBP). This forms an unstable 6-carbon compound that immediately splits into two 3-carbon compounds called **glycerate 3-phosphate** (G3P). The enzyme that catalyses this reaction, **ribulose bisphosphate carboxylase** (mercifully shortened to **RUBISCO**), is the most abundant enzyme on Earth, accounting for up to 60 per cent of the protein in many leaves.

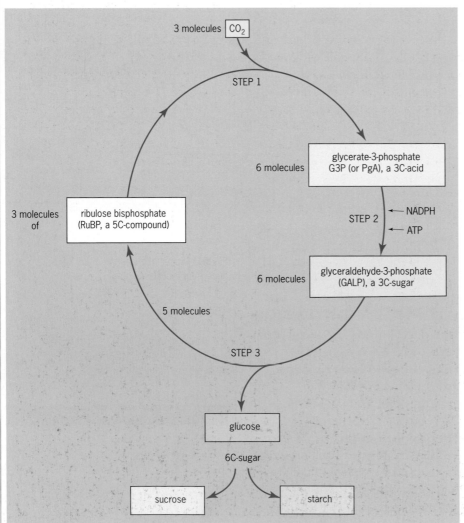

Fig 9.12 The light-independent stage of photosynthesis. See text for an explanation of each step

- **Step 2** The energy in ATP and the reducing power of NADPH are used to convert G3P into glyceraldehyde phosphate (GALP).
- **Step 3** GALP is used in two ways: five out of every six molecules are used – via a series of reactions – to make more RuBP to continue the cycle. One molecule in every six is used to make glucose.

Note that the carbon dioxide puts in one single carbon atom for every turn of the cycle. Thus the cycle must turn six times to make one glucose molecule. If the glucose were allowed to accumulate, it would lower the water potential of the cell, so it is converted into starch almost as soon as it is made, and this accounts for the large starch grains that can be seen in actively photosynthesising chloroplasts.

Plants move carbohydrate around in the form of sucrose. The plant's main storage compound, starch, is a large molecule that cannot pass across the cell membrane, so it must be converted into a smaller compound before it can pass into the phloem tissues of the plant.

SEE QUESTIONS 5 AND 6

▶ EXTENSION The discovery of the Calvin cycle

The basics of photosynthesis have been known since the nineteenth century but the pathways involved were only worked out in the 1950s. A major breakthrough, from a classic piece of experimental biology, came when Melvin Calvin and his team worked out the way in which carbon dioxide was converted into glucose. The transformation is complex, not something that can be achieved in a single chemical step. Calvin was able to work out the intermediate steps in the process that bears his name. Like most scientific research, progress was made when new investigative techniques became available. In this case it was the availability of radioactive isotopes of carbon and the techniques of two-way chromatography and autoradiography.

Calvin used the unicellular alga *Chlorella* and a unique lollipop apparatus that allowed him to give the algae carbon dioxide labelled with the isotope ^{14}C. At varying times after the introduction of the isotope he would drop the algae into hot alcohol, killing them instantly and preventing any metabolic reactions from progressing further. If the algae were killed right at the start of the experiment, he would find all the radioactivity in the carbon dioxide. If he left it too long, he would find all the radioactivity in the glucose. But by killing the algae at different times in between, he could identify the intermediates. This was achieved by a combination of chromatography and autoradiography. Calvin was able to separate out the intermediates by using two-way chromatography, then placing the chromatogram on X-ray film to show which of the intermediates contained the ^{14}C. The Calvin cycle is featured in the Assignment at the end of the chapter.

Fig 9.13 Calvin used this lollipop apparatus to figure out the basic steps on the reduction of carbon dioxide. Calvin used the alga *Chlorella* for several reasons. The small size meant that metabolic reactions could be stopped almost instantly by dropping the algae from the apparatus straight into hot alcohol. In addition, the algae could be grown in continuous batches and, as they reproduce asexually, there will be little variation in the population, producing consistent results

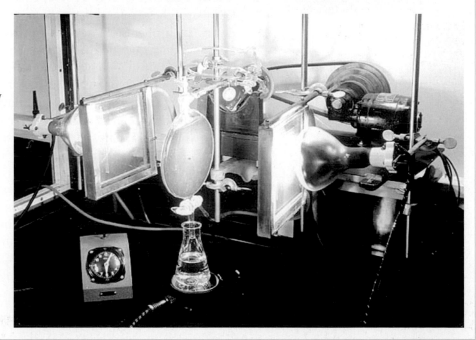

PHOTORESPIRATION

It is thought that photosynthesis evolved in an atmosphere that contained much less oxygen and more carbon dioxide than it does today. The enzyme that catalyses the reaction between RuBP and carbon dioxide, RUBISCO, can also catalyse a reaction between RuBP and oxygen – so oxygen and carbon dioxide compete for the same active site. When oxygen levels are high and carbon dioxide levels low, the oxygen reaction is favoured.

$$O_2 + RuBP \xrightarrow{\text{RUBISCO}} \text{glycolate (a 2-carbon compound)} + G3P$$

Compare this to the usual reaction of the Calvin cycle

$$CO_2 + RuBP \xrightarrow{\text{RUBISCO}} 2 \times G3P \ (2 \times \text{3-carbon compound})$$

You can see that one molecule of G3P is made instead of the usual two in the Calvin cycle. The glycolate cannot be fixed into carbohydrate, and instead passes into the mitochondria where it is respired.

Owing to its dependence on light, and the fact that it takes in oxygen and releases carbon dioxide, this process is called **photorespiration**. This process is wasteful to a plant because it means that some of the carbon dioxide that could be fixed into carbohydrate is lost. It has been estimated that photorespiration makes photosynthesis 30–40 per cent less efficient. However, some plants have overcome this problem – see the section on **C4 plants** below.

8 VARIATIONS ON THE THEME OF PHOTOSYNTHESIS

The mechanism of photosynthesis described above is known as the **C3 pathway** because the first stable product of the Calvin cycle is a 3-carbon compound, glycerate 3-phosphate. Most familiar plant species are C3 plants but there are two important variations on the theme: **CAM plants** and **C4 plants**.

CAM PLANTS DO IT IN THE DARK

C3 plants, the plants we have discussed so far, would not survive long in the desert. Their open stomata would prove to be a liability, they would rapidly lose water, dry out and die. In drier habitats some plants have overcome this problem by opening their stomata to take in carbon dioxide only at night. This drastically reduces water loss as the plants then keep their stomata tightly closed during the hours of heat and daylight. But how do they store their carbon dioxide to make it available for the light-dependent phase of photosynthesis? The key is that they use carbon dioxide to make an organic acid, which they store during the night and then break down in the day to release the essential carbon dioxide.

Plants that photosynthesise in this way are called **CAM (crassulacean acid metabolism)** plants. The carbon dioxide that enters the leaves at night is fixed into the organic acid **malate** by the enzyme phosphoenolpyruvate (PEP) carboxylase, and is then stored in the vacuoles until the daytime.

When the sun rises, the stomata of CAM plants close, preventing the loss of precious water. The malate is broken down to provide carbon dioxide according to the simple reaction

$$\text{malate} \rightarrow \text{pyruvate and } CO_2$$
$$\text{(C4)} \quad \rightarrow \quad \text{(C3)} \qquad \text{(C1)}$$

During daylight, C4 plants make their carbohydrate in the same way as 'standard' C3 plants. CAM plants are well suited to climates where there are high day and low night temperatures, especially where there is occasional drought. They tend to be native to semi-arid and tropical environments.

Overall, CAM plants overcome the problem of water loss by
● opening their stomata at night and closing them during the day;
● having fewer stomata than temperate plants;
● having a low surface area to volume ratio, which reduces evaporation.
These measures result in CAM plants losing as little as 50–100 g of water for every gram of dry mass produced, compared with 400–900 g for C3 plants.

C4 PLANTS LIKE IT HOT

In the tropics many species, including some commercially important plants such as sugar cane, maize and millet, exhibit remarkably rapid growth compared with C3 plants. They have achieved this mainly by overcoming the problem of photorespiration (see above). How have they managed to do this?

In 1966 two Australian scientists, Hatch and Slack, discovered that C4 plants were much more efficient at taking up carbon dioxide.

The key to C4 photosynthesis comes from the modified structure of the leaf. Surrounding the vascular bundles (the xylem and phloem tissue) is a layer of **bundle sheath cells** that have different chloroplasts.

The stages in C4 photosynthesis are as follows:
● **Stage 1** Carbon dioxide is fixed in the mesophyll cells. Instead of having RuBP and the enzyme RUBISCO, C4 plants use PEP (phosphoenolpyruvate) and the enzyme PEP carboxylase. This enzyme has a higher affinity for carbon dioxide than does RUBISCO, and it does not react with oxygen. The big advantage is that all the available carbon dioxide is fixed; none is lost on photorespiration. The 4-carbon compound oxaloacetate gives C4 plants their name.

$$PEP + CO_2 \xrightarrow{\text{PEP carboxylase}} \text{oxaloacetate (4C)}$$

● **Stage 2** Oxaloacetate is converted into malate (or a similar compound), which is shunted across into the chloroplasts of bundle sheath cells.

Fig 9.14 The prickly pear, *Opuntia,* is a CAM plant. This species thrives in warm, dry climates. When introduced into Australia in the 1920, this species out-competed the native plants and took over large areas of farmland until cactus moth caterpillars were introduced to control it

Fig 9.15 C4 plants such as maize show some of the fastest growth in the plant kingdom

- **Stage 3** The malate is used to make carbon dioxide, pyruvate (a 3-carbon chemical) and hydrogen, which is used to make NADPH.
- **Stage 4** The pyruvate is moved back into the mesophyll cells and used to make more PEP, while the carbon dioxide is now fixed into carbohydrate in the same was as in C3 plants, ie using RuBP and the enzyme RUBISCO. Note that in C4 plants the RUBISCO is always surrounded by a high concentration of carbon dioxide so that it is not inhibited by oxygen.

Fig 9.16 Comparisons of C3, C4 and CAM photosynthesis

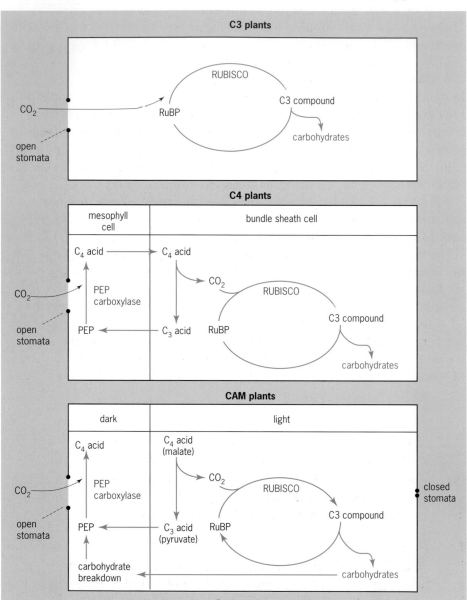

9 WHAT LIMITS PHOTOSYNTHESIS?

This is a very important question, because if commercial plant growers can increase the rate of photosynthesis, they can make plants grow faster, feed more people and/or make more money. The law of limiting factors states that the *rate of a process is limited by the factor in shortest supply*. For example, if carbon dioxide is in short supply, providing the plant with more carbon dioxide will continue to increase the rate of photosynthesis until some other factor – light intensity for example – becomes limiting. In this section we look at each limiting factor. Table 9.1 gives a brief overview of the way in which farmers can increase the growth of crops.

LIGHT INTENSITY

Light intensity is defined as the number of photons falling on a given area in a given time. Geographically, light intensity varies according to latitude – the closer to the equator, the greater the light intensity. Light intensity also varies according to the time of day, the time of year, the amount of cloud cover and whether the plant is shaded by other plants or objects. Generally, as light intensity increases so does the rate of photosynthesis until another factor – often carbon dioxide – becomes limiting. Plants have evolved to cope with different light intensities; for example plants on the forest floor have to photosynthesise in much lower light intensities than the trees above. How do they manage?

Plants respire all the time, but only photosynthesise when there is light. Photosynthesis will begin after dawn and the rate will increase until it equals the rate of respiration – this is the **compensation point**. After the compensation point photosynthesis will exceed respiration and so the plant will accumulate organic compounds. Plants that can photosynthesise in low light intensities are shade-tolerant and have a low compensation point. In contrast, shade-intolerant species have a high compensation point and so can only grow in bright environments. They cannot tolerate shading.

WATER SUPPLY

Although water is essential for photosynthesis, this represents only a tiny proportion of the water that flows through the plant in the transpiration stream (Chapter 31). A lack of water will normally cause a plant to wilt and affects all other metabolic processes long before the process of photosynthesis is affected directly.

CARBON DIOXIDE

Carbon dioxide is the commonest limiting factor when light intensity is sufficient. The atmosphere contains only about 0.035 per cent carbon dioxide, which is about 35 parts per 10 000. When the carbon dioxide levels are low and oxygen levels are high the enzyme RUBISCO can be inhibited – see the section on photorespiration and C4 plants above.

TEMPERATURE

Photosynthesis, like all metabolic processes, is temperature-dependent. As the temperature increases the kinetic energy of the molecules increases, they 'bounce around' more, so that diffusion of gases in and out of the leaf is faster, and there are more collisions between enzymes and substrates. The optimum temperature for C3 plants is about 25 °C, while for C4 plants it is about 35 °C. Above these temperatures a metabolic imbalance develops so that photosynthesis becomes less efficient, while extremes of temperature will of course denature the enzymes.

INORGANIC IONS

Plants need a variety of mineral ions from the soil in order to remain healthy – a list is given in Chapter 31. Of particular relevance to photosynthesis are nitrogen and magnesium; both are constituents of the chlorophyll molecule. Without these ions the plant develops yellow leaves, a condition known as chlorosis.

Table 9.1 How to increase the supply of limiting factors

Factor	How supply can be increased	Is it economically viable… In a glasshouse?	In a field?
Temperature	burn a fossil fuel, eg paraffin	usually	no
Carbon dioxide	burn a fossil fuel, eg paraffin	usually	no
Mineral ions	add fertiliser	usually	usually
Light	artificial light	sometimes, with a valuable crop	rarely
Water	irrigate or spray	essential	rarely in UK, essential in some climates

SUMMARY

- Photosynthesis is a process that captures the light energy of the sun and uses it to make organic molecules. Initially, glucose is made but plants use this basic carbohydrate to make the other essential organic molecules such as lipids, proteins and nucleic acids. Photosynthesis happens in chloroplasts; specialised organelles that house the chlorophyll pigments and all of the other compounds essential to the process. Photosynthesis is essentially a two-step process; the light-dependent reaction and the light-independent reaction.

THE LIGHT-DEPENDENT REACTION

- This process happens on the **thylakoids**, membranes that contain the chlorophyll molecules.
- When a photon of light hits a molecule of chlorophyll, the pigment becomes excited and emits two high-energy electrons.
- These electrons pass down an electron transport system that synthesises ATP – this is called **photophosphorylation**.
- Energy from the excited electrons is also used to split water; this is **photolysis**.
- Photolysis produces electrons to replace those lost by the chlorophyll and protons (hydrogen ions) and oxygen are made as by-products.
- Electrons are used to make NADPH (also called reduced NADP).
- ATP and NADPH are essential for the light-independent reaction, and oxygen is released into the atmosphere.

THE LIGHT-INDEPENDENT REACTION

- This process happens in the **stroma**, the fluid in the centre of the chloroplast.
- It involves the reduction of carbon dioxide by a series of reactions known as the **Calvin cycle**.
- ATP and NADPH are used to reduce the carbon dioxide. Supplies of these compounds depend on light, so, when it gets dark, the light-independent reaction finishes soon after the light-dependent reaction.
- The essential stages of the Calvin cycle are:
  - Stage 1 Carbon dioxide combines with the 5-carbon compound **ribulose bisphosphate** (**RuBP**). This reaction is catalysed by the enzyme **RUBISCO** (ribulose bisphosphate carboxylase).
  - Stage 2 This produces a highly unstable 6-carbon compound that immediately splits into two molecules of **glycerate 3-phosphate** (**G3P**).
  - Stage 3 ATP and NADPH are used to reduce the G3P into **glyceraldehyde 3-phosphate** (**GALP**).
  - Stage 4 Some of the glyceraldehyde 3-phosphate is used to make carbohydrate, but most of it is used to make more RuBP to continue the cycle. For every glucose molecule produced, five molecules are RuBP are re-synthesised.
- CAM plants save water by opening their stomata at night, storing carbon dioxide in an organic acid before releasing it for photosynthesis in the daytime.
- C4 plants avoid photorespiration, allowing them to make use of all the available carbon dioxide and therefore to grow faster.
- The rate of photosynthesis is determined by the factor in shortest supply; this could be light availability, carbon dioxide levels or temperature.

? EXAM QUESTIONS

1 **a)** Draw a curve of the action spectrum for photosynthesis on the figure right.

b) Explain the relationship between the action spectrum and the absorption spectra of photosynthetic pigments.

IBO Standard level Paper 3 November 1999 QC3

Rate of photosynthesis

? EXAM QUESTIONS

2 The results of an investigation of photosynthetic pigments in ivy leaves (*Hedera helix*) is shown below.

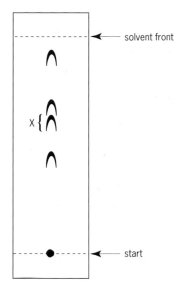

Table of R_f values

Pigment	R_f
carotene	0.95
phaeophytin	0.83
xanthophyll	0.71
chlorophyll *a*	0.65
chlorophyll *b*	0.45

a) Measure the distance travelled by spot X.

b) (i) The solvent front travelled 100 mm. Calculate R_f for spot X where
R_f = distance moved by spot/distance moved by solvent.
(ii) Identify the pigment in spot X using the table of R_f values above.
(iii) State **one** way, apart from R_f values, which could be used to identify the spot X.

c) Deduce which of the photosynthetic pigments listed in the table of R_f values was not present in *Hedera helix*.

d) Suggest **one** advantage to the plant of having more than one photosynthetic pigment.

IBO Standard level Paper 3 May 1999 QC1

3 The figure at the top of the next column represents a simplified version of the 'Z scheme' for non-cyclic photophosphorylation in a chloroplast. The letters (A) to (E) (and their dashed labelling arrows – →) refer to the questions below. Study the diagram, then answer the questions.

a) The H^+ indicated at (A) are transferred to another compound (shown on the diagram). Which compound will absorb the H^+?

b) In photosynthesis, what name is usually given to the process occurring at (B)?

c) Outline briefly what is meant at point (C) by 'PSII'.

d) Explain what happens to the OH^- seen at point (D).

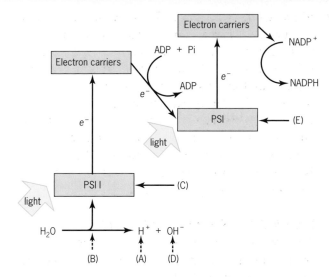

e) What is the chief difference between PSII (at point C) and PSI (at point (E)) in the way that they regain lost electrons?

f) The process shown in the diagram could be said to have three products. Name these three products.

OCR Molecules and Life March 2001 Q7

4 The diagram below is a summary of the light-dependent reaction of photosynthesis.

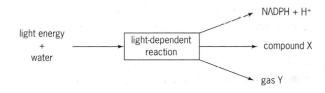

a) (i) Name the compound X.
(ii) Name gas Y and describe how it is produced.
(iii) State where in a chloroplast the light-dependent reaction takes place.

b) Name *one* pigment which is responsible for the capture of light energy during the light-dependent reaction of photosynthesis.

Edexcel B2 June 1999 Q4

? EXAM QUESTIONS

5

a) Using suitable terms or phrases from the list given below, complete the flow-diagram for the light-independent reactions of photosynthesis. Write only the **letter** of your chosen answer in the box on the diagram. **A letter may be used once only or not at all.**

A G3P (=PGA)
   (3-phosphoglycerate)
B RuBP (ribulose
   1,5-bisphosphate)
C unstable 6C compound
D 2 GALP (=3-PGAL)
E CO_2
F glucose
G several intermediate products

H NADP
I ATP
J $NADPH^+ + H^+$
K ADP + Pi
L acetyl CoA
M NAD
N starch
O chlorophyll

[Flow diagram boxes with labels: (5C), X, summary of the Calvin cycle, Y, (3C), (3C)]

b) (i) What is the name of the important enzyme required at point **X** on the diagram?
(ii) At point **Y** on the diagram, some of the product moves out from the chloroplast stroma. Where would it next be found?
(iii) Give the **letter** of a substance which could be said to provide reducing power and energy.
(iv) Give the **letter** of a substance known as the CO_2 acceptor.

OCR Molecules and Life November 2000 Q5

6 The figure shows the light-independent stages of photosynthesis.

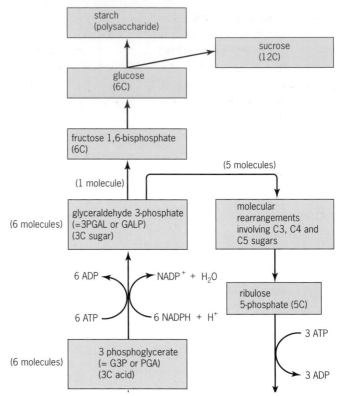

Notes: Two 3-PGAL molecules are made from 1 RUBP each time the cycle 'turns'. For every three turns of the cycle, 1 molecule of 3-PGAL leaves.

a) (i) After which famous biochemist is the cyclical sequence of reactions shown in the figure named?
(ii) Where, **precisely**, do these reactions occur in plant cells?
b) Which substance is the first carbohydrate made in photosynthesis?
c) (i) Name the enzyme that catalyses the incorporation of carbon dioxide into these biochemical processes.
(ii) Name a coenzyme involved in this cycle of reactions.
d) Calculate the percentage of the 3-PGAL produced, which is used to make the carbon dioxide-acceptor molecule.
e) Why is it important that these reactions are **cyclical**?
f) State precisely the site of production of ATP used in these reactions.
g) (i) Suggest the main function of sucrose in leaves and stems.
(ii) State **two** properties of sucrose that make it suitable for this role.

OCR Molecules and Life March 2000 Q6

KEY SKILLS **ASSIGNMENT**

INVESTIGATING PHOTOSYNTHESIS

Working out the Calvin cycle

In the 1950s, the US scientist Melvin Calvin devised a technique to investigate how plants turned carbon dioxide into glucose. Calvin was able to provide simple plants with radioactive carbon dioxide and then follow the progress of the ^{14}C in the intermediate products. Calvin used the apparatus shown in Fig 9.A1.

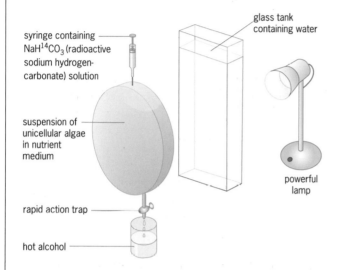

Fig 9.A1 Calvin's lollipop apparatus

1 Suggest reasons for each of the following:
   a) the flattened lollipop shape of the container with the algae;
b) the water tank;
c) the hot alcohol;
d) the $NaH^{14}CO_3$ solution.

2 Suggest two reasons why a single-celled alga was used rather than a larger plant.

3 Explain how you could obtain a sample of algae that could be used to investigate the first products of photosynthesis

4 Suggest what techniques could be used for
   a) the separation and identification of the photosynthetic products;
b) finding out which compounds contain the ^{14}C.

The results of this investigation are shown in the table below, which gives the ^{14}C content (in μmoles per cm^3) of four organic compounds (A to D) after five different periods of photosynthesis.

Compound	Time allowed for photosynthesis (in seconds)				
	5	15	60	180	600
A	0.3	2.5	6.2	10.3	7.9
B	1.0	2.0	3.1	3.2	3.2
C	0.05	0.11	0.16	1.0	1.0
D	0.01	0.02	0.08	0.17	1.7

5 **a)** Use the data in the table to place the compounds A to D in the order in which they would be formed.
b) Using your knowledge of photosynthesis, suggest a reason why the level of compound B remained steady in later samples.

Measuring the rate of photosynthesis

This part of the assignment features an experiment to investigate the effect of light intensity on the rate of photosynthesis.

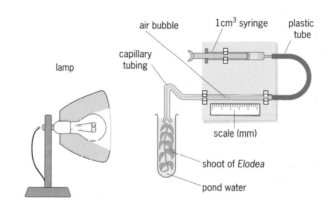

Fig 9.A2 Apparatus for measuring oxygen consumption in the aquatic plant *Elodea*

1 Why is an aquatic plant used for this experiment?

2 A problem with this apparatus is that the lamp could also heat up the water.
a) Explain how a rise in temperature will affect the rate of photosynthesis.
b) Suggest one item of equipment that could be added to the apparatus to overcome the problem.

3 Why is a small volume of sodium hydrogencarbonate usually added to the water before the experiment?

4 State two precautions that would be necessary to ensure that the results obtained by different students, using the same piece of apparatus, were comparable

KEY SKILLS **ASSIGNMENT**

The results shown in the table below were collected from six students using the same apparatus.

Distance of lamp from plant/cm	Length of bubble after 5 min/mm	Mean length of bubble after 5 min/mm
5	60, 55, 28, 62, 63, 68	
10	3, 36, 39, 33, 24, 28	
15	7, 10, 5, 6, 9, 8	
20	1, 3, 1, 2, 1, 4	

5 **a)** Identify any anomalous results in the table above.
b) Excluding such anomalous results from your calculations, find the mean bubble length for each distance.

6 **a)** What is the relationship between the distance from the lamp and the mean bubble length over the 10–20 cm distance? Express the relationship both as a simple formula and in words.
b) Do the 5 cm results fit this relationship? Explain your answer.

7 **a)** Explain how you would calculate the *volume* of gas produced *per hour* from these values.
b) Suggest one reason why the volume of oxygen given off from a plant is not a true representation of the rate of photosynthesis.

10

Nutrition

The panda's restricted diet of bamboo has contributed to its position as one of the rarest mammals in the world

Evolution has not been kind to the panda. This large, majestic bear lives on the diet of a herbivore but it evolved originally to be an omnivore. This mismatch means that, in order to maintain its bulk of 75–160 kg, the panda has to eat for 10–12 hours out of every 24. Not a bad life, you might say, but the panda has one of the most restricted diets in the animal kingdom. In those long hours of munching it has the choice of bamboo shoots or bamboo roots, which form 99 per cent of its dietary intake. Occasionally, if driven to it, the panda may eat a crocus bulb or even a small fish, but this is very much the exception.

The panda is now one of the rarest mammals in the world. There is a case to be made that its survival has not been helped by people destroying its environment and hunting it practically to extinction, but it is also at the mercy of the strange habits of the bamboo itself. This plant usually divides asexually, forming runners and shoots that spread from the base of the plant. Most of the time, the bamboo is actually invasive, growing plentifully and giving the panda the food it needs. However, every few years, the bamboo enters a stage of sexual reproduction – instead of sending out runners, it produces flowers and seeds. The problem for the panda is that as soon as it has flowered, the bamboo goes into a dormant stage called 'die-back' – it literally disappears. New growth can take up to three years to return after a widespread die-back, leaving the panda up an evolutionary creek without a paddle.

Between 1974 and 1976, bamboos over large areas of the Min Mountains of northern Sichuan, China, flowered, produced seeds, and disappeared. At least 138 pandas died of starvation during that time.

The giant panda's restricted diet is not its only problem as far as the fight for survival goes. It also breeds at a slow rate – the female is only fertile for 2–3 days every year, and so the panda birth rate can easily fall below the death rate, causing a further decline in the population.

1 WHY DO ORGANISMS NEED FOOD?

REMEMBER THIS

Inorganic substances are generally made from simple molecules. Water, mineral ions (sodium, chloride, etc) and gases such as carbon dioxide are all inorganic.

Organic substances – those based on carbon – include carbohydrates, lipids, proteins and nucleic acids; see Chapter 3.

All living things need food to obtain energy, to grow, and to repair damaged cells and tissues. This chapter is about nutrition – the process of obtaining food. There are two types of nutrition, **autotrophic nutrition** and **heterotrophic nutrition**.

Autotrophic nutrition involves taking in simple **inorganic** molecules and using them to form complex substances. Nutrition in plants is autotrophic and is commonly known as **photosynthesis**, a process in which they build up large complex molecules from carbon dioxide and water (see Chapter 9).

Heterotrophic nutrition involves taking in complex **organic** molecules, and then breaking them down into simpler molecules that are absorbed into the body. This process of breaking down food is called **digestion**, and is covered in Chapter 11.

2 HOW DO ORGANISMS GET THEIR FOOD?

Organisms that are heterotrophs can be divided further into three categories: **saprotrophs**, **parasites** or **holozoic feeders** (Fig 10.1). Saprotrophs such as fungi feed exclusively on dead and decaying material. Parasites, such as tapeworms, obtain their food from the living body of another organism. Most animals are holozoic feeders – they take in solid organic material, break the food down by digestion within their body and absorb the soluble products.

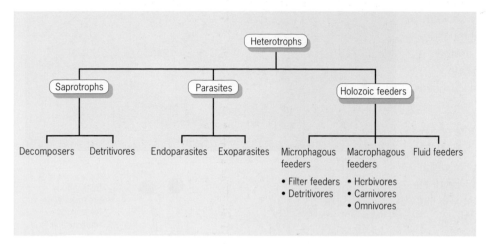

Fig 10.1 Types of heterotroph

NUTRITION IN SAPROTROPHS

Why aren't we knee deep in dead leaves, insects, birds and mice, not to mention faeces? The answer lies with a specialised group of heterotrophic organisms known as saprotrophs. The term **saprotrophic** describes organisms that feed on dead and decaying material (*sapro* – rotten).

There are two types of saprotroph: **decomposers** and **detritivores**:

Decomposers rot the dead bodies of animals and plants and so are a vital component of any ecosystem. Most decomposers are bacteria or fungi that live on or near decaying organic matter. They absorb soluble molecules, such as amino acids, organic acids and mineral salts, unchanged through their cell walls. Insoluble materials, such as starch, cellulose, lignin, fats, waxes and resins, are broken down first. Decomposers digest complex molecules by secreting powerful protease, amylase, lipase, **cellulase** and **lignase** enzymes on to their food source, digesting it externally. They then absorb the breakdown products through a permeable cell wall. Most decomposers are small or have thin structures to aid diffusion.

Detritivores, the other important saprotrophic group, can live in water and include the marine worms (Fig 10.2). These feed on the detritus (dead organic matter) at the bottom of seas or lakes. They should not be confused with decomposers; detritivores are animals with guts that actually eat the detritus.

Fig 10.2 Detritivores, such as these marine worms, feed on organic sediments on the sea bed or lake bottom

SCIENCE FEATURE Nutrition in a saprophytic fungus

Fungi, often called moulds, are saprophytes; they grow into their food, releasing enzymes to digest it extracellularly. The body of a mould such as *Aspergillus* consists of extremely thin threads called **hyphae** (Fig 10.3). The hyphae have an outer wall made of a polymer similar to cellulose, but they do not contain separate cells. The organelles, including the nuclei and mitochondria, are spread through the cytoplasm. The extensive network of hyphae forms a **mycelium**, which spreads through the food source. This mycelium may be vast. The largest single organism known is a soil fungus in a North American forest that covers about six square kilometres.

Digestion in fungi is extracellular since it takes place outside the body of the fungus. Hyphae secrete enzymes that diffuse through the cell wall and on to the food. This is comparable to the stages in the human gut in which enzymes are secreted from glands into the lumen of the stomach and the intestine. The enzymes hydrolyse the organic compounds into soluble monomers. These monomers are then absorbed into the hyphae, probably by facilitated diffusion and active transport. As they feed, the hyphae branch and grow through the decaying food material. The thin hyphae and large number of branches ensure that the mould has a large surface area to volume ratio. Thus it is well adapted for secreting enzymes over a large area and absorbing the products of digestion

SEE QUESTION 1

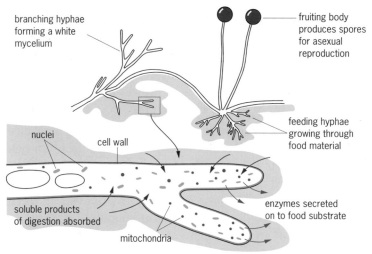

branching hyphae forming a white mycelium

fruiting body produces spores for asexual reproduction

nuclei

cell wall

feeding hyphae growing through food material

soluble products of digestion absorbed

enzymes secreted on to food substrate

mitochondria

Fig 10.3 The mould *Aspergillus* growing on a papaya fruit (top). The diagram (right) shows the main features of this saprophyte

Source: adapted from Green et al., *Biological Science 1 & 2*, Cambridge, 1990.

Fig 10.4 Tapeworms are highly adapted to living in the gut of their host

PARASITIC NUTRITION

A **parasite** is an organism that feeds from another living organism for most of its life cycle. The parasite benefits as it obtains most of its metabolic requirements, but the host is usually harmed. A successful parasite minimises this harm so that it can continue to use its host for longer periods and so that it has more chance of spreading to other hosts.

Endoparasites live inside the body of their host; **ectoparasites** live outside. Both types are adapted to live in a specific niche on or in another living body. Endoparasites usually show extreme specialisation of their organ systems, such as the digestive and sensory systems, which are very much reduced. They also tend to have complex life cycles to enable the organism to spread from host to host. Ectoparasites generally have fewer specialised features.

Tapeworms (Fig 10.4), live in the gut of their host. Once in position, they are continually bathed in predigested food and so do not need either a mouth or a digestive tract – they absorb the simple food molecules across the membrane that makes up their body surface.

Tapeworms make the most of their situation because they:
● have a body wall with a high degree of folding (microscopic folds called microtriches,
● have many mitochondria, which provide the energy required to actively transport some food products,
● produce some enzymes to aid external digestion close the body wall.
This enables the tapeworm to compete with its host for the food available.

HOLOZOIC FEEDERS

The animal kingdom consists of a huge variety of animals and, not surprisingly, they obtain food in many different ways. Broadly speaking there are three main groups: **microphagous feeders**, **macrophagous feeders** and **fluid feeders**, grouped according to the type of food eaten or the feeding method used.

MICROPHAGOUS FEEDERS

Microphagous feeders, or **microphages**, feed continuously on tiny food particles. There are two sub-types: filter feeders and deposit feeders.

Filter feeders take in organic material, often plankton, in open water. Some, such as fan worms and corals, do not move around much, preferring to wait for food to come to them. Many live on rocky shores, on the bottom of the ocean, or on a river bed. Different filter feeders have different feeding techniques but they all transfer particles of food to their digestive tract using **cilia**. A good example of a microphagous feeder for exam purposes is *Paramecium*, which can be drawn easily and has several obvious features (Fig 10.5). Mussels are another classic example.

Deposit feeders feed on rich organic material on the sea bottom. As organisms die, their bodies accumulate as sediment, or **detritus**, on the surface of the sea bed. This provides a rich source of food for detritivores such as worms and molluscs.

MACROPHAGOUS FEEDERS

Macrophagous feeders feed on larger particles of food, including animals larger than themselves. Those that feed exclusively on plant food are **herbivores**, those that consume other animals are **carnivores** and those that eat a variety of food types are **omnivores**. A **scavenger** is a carnivore that feeds on dead and rotting flesh. Technically these terms can be used to describe any animal, but in practice, they are most often used to describe mammals.

FLUID FEEDERS

Fluid feeders live on a liquid diet, which can consist of nectar (butterflies), plant phloem sap (greenfly) or blood (mosquitoes). They all have specialised mouthparts to enable them to get at their food. Sap-feeding insects such as greenfly avoid the need to break down cellulose in plant cell walls by using the plant's fluid transport systems to supply food 'on tap'. This technique can lead the insect to develop nitrogen deficiency but, to compensate, many sap-feeding insects have nitrogen-fixing bacteria in their gut.

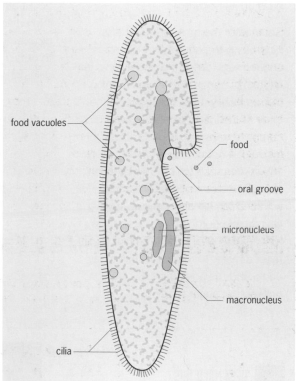

Fig 10.5 The structure of *Paramecium*

food vacuoles

food

oral groove

micronucleus

macronucleus

cilia

! SCIENCE FEATURE Nutrition in butterflies and moths

Unlike mammals and humans, who keep the same body shape throughout life, some insects have a complex life cycle in which they exist in several completely different forms. In **lepidopterous insects**, such as butterflies and moths, the first stage, the **egg**, is able to survive unfavourable conditions such as drought or cold before developing into the next stage, the **caterpillar**. The sole purpose of a caterpillar is to accumulate enough food to see it through its transformation into the sexually active adult. It needs to eat plenty of carbohydrate and protein to build up its body bulk.

Caterpillars themselves are sexless blobs; only the adult form has males and females. When the caterpillar is fat enough, it sheds its skin and forms a **pupa** or **chrysalis**. This protects the body of the insect while it undergoes a complete transformation. Studies of the inside of a pupa show that it dissolves into green mush, and from that mush the butterfly or moth develops from scratch. The adult is a sexual stage whose sole purpose is to reproduce. It needs the energy to fly around to meet a mate, to mate and then to lay eggs, and the adult generally lives on nothing but sugar solution.

✔ REMEMBER THIS

Like all insects, monarch butterflies have six jointed legs, three body parts, a pair of antennae, compound eyes, and an exoskeleton. The three body parts are the head, thorax and abdomen. The four wings and the six legs of the butterfly are attached to the thorax.

The life cycle of the monarch butterfly

The monarch butterfly (*Danaus plexippus*) is a common poisonous butterfly that is found over most of the USA, in southern Canada, Central America, most of South America, some Mediterranean countries, the Canary Islands, Australia, Hawaii, Indonesia, and many other Pacific Islands. Like other butterflies, the monarch has several distinct stages in its life cycle, going from an egg, to a caterpillar (the larva) to a pupa or chrysalis, to an adult butterfly. It takes about a month for the egg to mature into an adult. The monarch's nutritional requirements and the methods it uses to obtain food are different at each different stage.

Events in the life cycle of the monarch

A summary of the life cycle of the monarch butterfly is shown in Fig 10.6.

The egg

The monarch starts its life as a ridged, spherical **egg** only 3 mm long. The eggs are always laid singly, on the underside of milkweed leaves. The female attaches the egg to the leaf with a quick-drying glue which she secretes along with the egg. The egg hatches in about 3 to 5 days. A tiny worm-like **larva**, or **caterpillar**, emerges.

The larval stage

The caterpillar hatches from its egg and eats it. It has an urgent need for energy

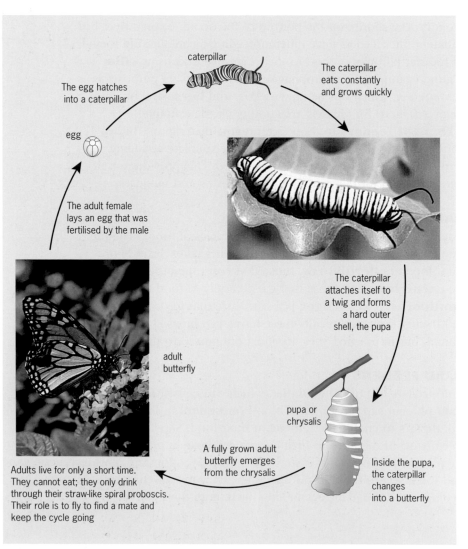

caterpillar

The egg hatches into a caterpillar

The caterpillar eats constantly and grows quickly

egg

The adult female lays an egg that was fertilised by the male

The caterpillar attaches itself to a twig and forms a hard outer shell, the pupa

adult butterfly

pupa or chrysalis

Inside the pupa, the caterpillar changes into a butterfly

A fully grown adult butterfly emerges from the chrysalis

Adults live for only a short time. They cannot eat; they only drink through their straw-like spiral proboscis. Their role is to fly to find a mate and keep the cycle going

Fig 10.6 The life cycle of a monarch butterfly

and nutrients as it needs to grow very rapidly over the next couple of weeks. After the egg has been polished off, the caterpillar starts munching through the milkweed leaves that it has hatched on to and it eats almost constantly. The milkweed leaves are poisonous and the caterpillars incorporate the milkweed toxins into their bodies in order to poison their predators.

As it grows, the caterpillar gets too big for its skin and it **moults** four times. Each time it does so, the caterpillar eats the old skin, keen not to waste valuable protein.

Pupation

When the larva is about 5 cm long, it stops eating and looks for a sheltered place to form its **pupa**. The caterpillar spins silk from its **spinneret** and attaches its hind end to a branch with the silk and small hooks in the anal prolegs. It hangs head down and moults for the last time. When the newly exposed skin dries and hardens, it takes the form of a jade-green **chrysalis**. During this stage the caterpillar turns into a butterfly as its entire body is reorganised. Ten to 14 days later the chrysalis becomes transparent and a damp butterfly emerges.

The adult butterfly

The adult butterfly pumps liquid into the wing veins to inflate them. They soon dry, and the butterfly flies off to find its first meal. It doesn't grow in the adult butterfly stage, but the flight muscles use a lot of energy. To gain energy quickly, it sucks up sugar-rich liquids through its **proboscis**, a sort of long, flexible tongue. It eats nectar from the flowers of several plants, including the milkweed on which its caterpillar fed. The adult finds a mate and produces the eggs that complete the life cycle.

Fig 10.7 Eating a monarch butterfly or caterpillar for the first time, this blue jay is unaware of the danger. However, the unpleasant after effects are remembered – this is a good example of learned behaviour in birds (see Chapter 23).

✔ REMEMBER THIS

The danger markings on the caterpillar warn potential predators – birds and small mammals that eat the caterpillar or the butterfly are immediately violently sick because the toxins cause an emetic effect (Fig 10.7). The predator remembers the experience and associates it with the distinctive markings on the monarch caterpillar and the monarch butterfly.

SEE QUESTION 2

HERBIVORES AND CARNIVORES

Mammalian herbivores and carnivores have specific features that are completely different between the two groups. These differences are summarised in Table 10.1. The main differences in the teeth and skull of the two groups are illustrated by the sheep and dog skulls (Fig 10.8).

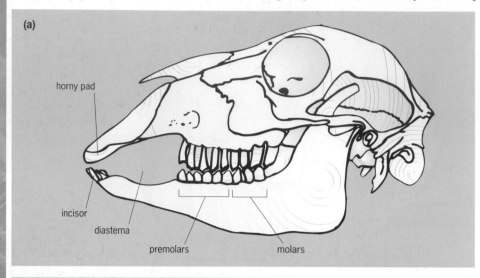

(a)

horny pad

incisor

diastema

premolars

molars

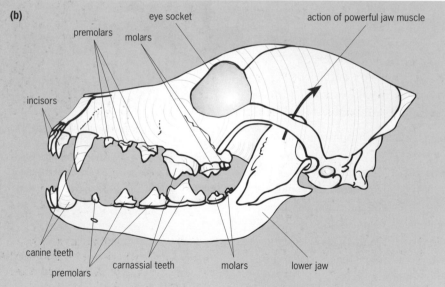

(b)

eye socket

action of powerful jaw muscle

premolars

molars

incisors

canine teeth

premolars

carnassial teeth

molars

lower jaw

Fig 10.8 Comparing the skull of a herbivore with the skull of a carnivore
(a) The skull of a sheep showing the arrangement of mouthparts and teeth
(b) The skull of a dog showing the arrangement of mouthparts and teeth

Fig 10.9 Camels need strong jaws to help them deal with tough plants

HERBIVORES

Although plant tissues are mainly water, with relatively indigestible materials such as cellulose and lignin, their roots, seeds and fruits are often rich in nutrients. Herbivores need to overcome the problems of eating these tough, fibrous foods:

● All herbivores have strong jaws, broad incisors and continuously growing ridged molars to deal with tough plants and leaves (Fig 10.9). Even grass has a high silica content and tough cellulose cell walls, which makes it difficult to break up into small pieces ready for digestion.

● The tongue in a herbivore is usually large so that it can be used to help tear off plant material, and then help break it up during chewing Herbivores that eat plants with tough defensive structures such as thorns, spines and defensive hairs have thickened mouth parts. A camel, for example, chews at acacia bushes, apparently oblivious of the thorns.

● Herbivores have specialised digestive tracts that tend to be much longer than those found in carnivores. The extra length, combined with other modifications such as the **ruminant** stomach system (see Chapter 11, page 223), ensures that the tough cellulose- and lignin-containing plant food stays inside the animal long enough to be digested.

● The bacteria usually present in the intestines of a herbivore are different from those commonly found in carnivores. Many of the bacteria produce **cellulase** enzymes that help to break down cellulose. These are found in the rumen and reticulum of a ruminant (see page 223) and in the caecum of a non-ruminant such as a rabbit.

Table 10.1 A summary of the physical adaptations in herbivores and carnivores	
Carnivores	**Herbivores**
Incisors are sharp and thin for nipping and biting	Incisors are sharp and broad for cropping
Upper incisors never absent	Upper incisors may be absent (sheep, cattle)
Canines are pointed (for grasping)	Canines are small or absent
Molars are pointed, adapted for cutting	Molars are heavily ridged for crushing and grinding
Teeth stop growing at adulthood	Teeth continue to grow to replace worn material
No diastema	Diastema present to allow efficient action of the tongue
Digestive tract short	Digestive tract long
No specialisation of stomach and caecum	Modification of stomach in ruminants with enlarged caecum and appendix

CARNIVORES

Although food of animal origin has more nutrients and energy per unit mass than plant material, it does have the habit of running off. In order to eat, the carnivore must first stalk and then catch its prey, which itself uses up a great deal of energy, particularly if a long chase is involved. Consequently, carnivores have well-developed senses and often show complex social behaviour associated with hunting. Like herbivores, carnivores have physical adaptations that enable them to deal with their food:

- Carnivores have a shorter digestive tract because their food is more readily digested and does not need to stay inside the animal for as long.
- They also have modified teeth (see Fig 10.8b and 10.10). The upper **canines** are well developed and are used to pierce and cut skin. **Carnassial teeth** have a scissor-like action: they can cut through flesh and will also crack bones. The **incisors** are sharp and can tear meat away from bone and connective tissue. Grinding teeth are small: much of the food is swallowed whole.
- Carnivores have only one stomach where food is partially digested before passing into the small intestine to be completely broken down. Soluble food products are absorbed, and tough, indigestible fragments of bone, feathers and fur pass through to the large intestine for **egestion**.
- Living exclusively on meat means than carnivores do not take in some of the water-soluble vitamins found predominantly in plant material: for example, vitamin C (although many carnivores eat the intestines of their prey as a priority, thus getting some water-soluble vitamins). Instead, carnivores synthesise their own vitamin C and other water-soluble vitamins.

Fig 10.10 Carnivores, such as this wolf, have specialised teeth, which are used to deal with a kill as quickly as possible, before rival predators and scavengers close in for their share

3 HUMAN NUTRITION

The old cliché 'we are what we eat' is perfectly true. As developing babies, we were built when molecules from the food our mothers ate were assembled according to the genes we inherited, and ever since we have been taking in the food we need to build and power our bodies.

Humans are heterotrophic organisms, or **heterotrophs**. All the food we eat comes ready-made, in the form of plants and animals.

In Chapter 3 we looked at the chemicals of life: carbohydrates, proteins, lipids and nucleic acids. This chapter looks at how we provide our bodies with these and other essential substances.

You will have heard much talk of a **balanced diet**. Many assume that to achieve a balanced diet, all you need to do is eat a little bit of everything. However, as we shall see, like many things in life, it's just not that simple.

THE CONCEPT OF A BALANCED DIET

A diet must be balanced in terms of energy and in terms of its chemical content.

Whether a person is fat, thin, or 'just right' depends on the balance of their energy intake versus their energy expenditure. Put simply, this means that if you take in more energy in the food you eat than you use up in your everyday activities, you will start to put on weight. Someone who is **obese** eats more than they need over long periods of time (look back at Fig 7.11a). If you take in less energy that you use up, you lose weight. Someone who is **starving** cannot meet their energy demands, and they start to digest their own body fabric (look back at Fig 7.11b). A diet that is balanced in terms of energy, enables us to maintain a constant, healthy weight.

It is also possible to eat a balanced diet in terms of energy, but then suffer from **malnutrition**. This is an imbalance in the chemical composition of the food we eat. If, for example, you ate only chocolate, you could easily maintain your body weight but your body would suffer from a range of deficiencies after a time – chocolate contains lots of sugar and fat, but few vitamins, minerals, complex carbohydrates and no fibre.

THE ENERGY IN FOOD

As you sit here reading this book, you might think that you are not using much energy, but even at rest, your body has a steady energy demand. Think about some of the processes going on:

- Your heart is beating.
- You are breathing.
- Food is being pushed along your intestines.
- Food is being absorbed from your intestines into your blood.
- Urine is moving from your kidneys to your bladder.
- Nerves are taking information from your sense organs to your brain, keeping you informed of changes in your environment.
- You are thinking: your brain is processing information coming in from the sense organs, sorting out what is important and what can be ignored.
- Unless you are somewhere very hot, you are probably losing heat to your surroundings. You need to replace the heat you lose in order to maintain a constant internal body temperature.
- Some of the molecules from digested food, such as amino acids, are being built up into molecules that will help to make new cells and tissues.

All these processes (and many others) require energy that is released from food molecules by respiration, a process that goes on constantly in all of our cells.

The amount of energy a person needs depends on:

- their age,
- their level of activity,
- their size,
- their genetic background (some people inherit a very high or a very low metabolic rate).

? TEST YOURSELF

A The size of an organism affects the amount of food it needs per unit of body weight. Explain.

ENERGY INTAKE AND EXPENDITURE

The energy and dietary requirements of people at different times of life are covered in detail on page 204 (see also Fig 7.10).

Ideally, we need to eat enough food to maintain an energy balance so that:

INTAKE		EXPENDITURE		
energy transfer from carbohydrates and fats	=	energy necessary for essential activities	+ energy for physical activity	+ energy for metabolising food

The energy required for the body's essential activities, that is, for the processes that take place when we are at rest, is called the **basal metabolic rate** (abbreviated to BMR) (see Section 6 in Chapter 7).

⚠ SCIENCE FEATURE Starving in the midst of plenty

In Chapter 4, we saw how someone adrift in the sea can die from dehydration even though surrounded by water. Some people in our society can also starve, even though they have access to plenty of food. Unlike the shipwreck victim, their problem is psychological rather than physiological – they have the eating disorder **anorexia nervosa**.

Anorexia usually affects teenage girls, but the number of boys affected is rising. A diet usually triggers the condition. Restricting food to lose a bit of weight gets out of control and the affected teenager becomes obsessed with food. They lose confidence as well as weight, and they often try to control the eating patterns of those around them. Some anorexics will, for example, make large meals for others, and encourage them to overeat, but eat virtually nothing themselves. Treatment does help, but an unlucky few do die from anorexia. Once the body weight of a teenage girl drops below about 7 stones (44.5 kg), her periods stop. Below 6 stones, she is unable to carry out ordinary activities like going to school, and once her weight has dropped to below 5 stones, she is in real danger of sudden death because of heart or other organ failure.

A related eating disorder, **bulimia nervosa**, also tends to affect young adults, mainly girls, and can also have devastating effects. Bulimia can also be triggered by dieting, but instead of under-eating, a bulimic will binge on food, eating vast quantities until overfull (Fig 10.11a). Often, she will then make herself vomit to get rid of the uncomfortable feeling, and because she feels shame and self-disgust (Fig 10.11b). The vomiting leads to further negative feelings, and is then followed by another bingeing session. This cycle can continue for hours, or may only happen now and again. The body weight of a bulimic does not change drastically, and if the bingeing and vomiting is carried out in secret, friends and family may not know about the problem.

Fig 10.11 (a)
Bulimic girl bingeing

Fig 10.11 (b)
Vomiting to get rid of the feeling of fullness

BALANCING THE DIET

What constitutes a balanced diet varies from animal to animal; in a big cat like a tiger, a balanced diet is a large meal of antelope every two days. Humans, because they are omnivores, need to eat a wider range of foods and the healthiest diets seem to be those based mainly on carbohydrates, with relatively low amounts of fat, some protein, and plenty of fresh fruits, vegetables and other plant products such as cereals and grains. The typical 'western' diet tends to have too much fat and protein, and too few fresh fruits and vegetables. Taking in a high proportion of fat in the diet leads to obesity because fat contains almost twice as much energy per unit mass than either carbohydrate or protein:

- 1 g carbohydrate produces in the body approximately 17 kJ;
- 1 g protein produces approximately 18 kJ;
- 1 g fat produces about 39 kJ.

CARBOHYDRATES

Carbohydrate is the most important energy provider among the food types, accounting for between 40 and 80 per cent of total energy intake of most human diets. It is also necessary for the metabolism of proteins and fats. Carbohydrates can be divided into three groups:

- **Sugars** such as sucrose and glucose – sweet foods and drinks contain large amounts. Sugars can be thought of as 'instant' fuels. The fastest way to get energy into our bodies is to have a glucose drink: the glucose does not have to be digested and so passes quickly into the bloodstream where it can be carried to the working muscles.
- **Starches**, abundant in many staple foods such as rice, potatoes, pasta and bread.
- **Non-starch polysaccharides** (NSPs) such as cellulose which collectively make up much of the fibre in our diet.

Although almost all types of carbohydrates provide similar amounts of energy, the nature of the carbohydrate can have a significant affect on our health. For instance, sugars can be classified in the following way:

- **Intrinsic sugars** are those contained within the cell walls of food, eg sugars within the cells of a banana.
- **Extrinsic sugars** that are not contained within cells, such as those found in processed foods such as sweets and soft drinks.
- **Milk sugars** occur in milk and milk products. Lactose is the primary milk sugar.

! SCIENCE FEATURE What makes for a healthy colon?

Many large studies involving thousands of people have shown that a diet low in fibre and high in red meat and fat makes it more likely that someone will develop cancer of the colon or rectum later in life. It is one of the clearest cases of a lack of a food type being associated with a particular cancer.

One large study, the European Prospective Investigation into Cancer and Nutrition (EPIC), published some of its findings in 2001. This study started in 1993, and monitored over 400 000 people across nine different countries to investigate the links between diet and cancer. It concluded that eating a diet high in fibre – at least 18 g per day – cut the risk of developing colon cancer by 40 per cent.

It is, however, difficult to explain why fibre is so important for keeping the colon healthy. It may be that a higher fibre diet increases the speed at which food passes through the alimentary canal. If food has less time to be in the colon, there is less chance that cancer-causing components in food can affect the cells in the lining.

Generally, nutritionists disapprove of eating too many simple carbohydrates, advising instead to base your diet on starchy foods. In planning a meal such as breakfast, for example, you should eat a wholegrain, unsweetened cereal, or wholemeal bread with fruit and maybe low-fat yoghurt instead of a doughnut or Danish pastry. The energy from complex carbohydrates is released slowly over a few hours and this stops you being hungry until the next meal, so you can avoid 'snacking' on more high-sugar foods. This makes sense because it keeps your blood glucose level from rising and falling rapidly – a major factor in hunger is blood glucose level. When it is low we feel hungry. Foods containing a lot of extrinsic sugars, or processed foods that allow the rapid digestion of starch, both enable glucose to be released rapidly, and it enters the blood. This triggers the body's insulin-based control mechanism, which reduces the glucose level again, causing hunger to return with a vengeance.

THE IMPORTANCE OF FIBRE

Fibre, the indigestible component of carbohydrate-rich foods, provides bulk to the digestive tract, stretching the gut wall and stimulating a faster through-put of food. Generally there is very little fibre in animal products; the vast bulk of indigestible material comes from plant tissues. If we eat a highly processed diet (one based on white bread, sweet foods, high-fat foods with little fresh fruit, vegetables or whole grains) we do not take in enough fibre. In the short term, this means chronic constipation and it increases our risk of developing gallstones and cancer of the colon later in life.

PROTEINS

The human body is made largely from protein; it accounts for more than half of the dry mass of our bodies – see Chapter 3. The proteins in our diet are digested to amino acids that then travel to our cells to be reassembled in a different order to make the proteins of the body. Proteins are needed for growth and repair of tissues, and not surprisingly are an important part of the diet of bodybuilders (Fig 10.12). Table 10.2 shows the protein content of some familiar foods.

SEE QUESTION 3

? TEST YOURSELF

C How do you test for the presence of protein in a food? (Hint – you might need to revise Chapter 3, Section 4.)

Fig 10.12 In order to make the extra muscle tissue, bodybuilders must have a high protein diet

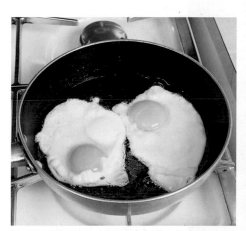

Fig 10.13 Each one of these eggs provides enough daily protein for an adult human. When proteins are heated they are denatured and often coagulate. This accounts for the change in colour and texture when an egg is cooked. However, the amino acid content is usually unchanged

Table 10.2 The protein content of some familiar foods	
Food	**Protein content/ g per 100 g**
Milk (whole, cow's)	3.2
Cheddar cheese	25.5
Beef	20.2
Chicken	17.6
Turkey	20.6
Cod	17.4
Baked beans	5.2
Red kidney beans (dry)	22.1
Peanuts	25.5
Eggs	12.3
Apples	0.4
Cornflakes	7.9
Potatoes	2.1

Source: MAFF reference book 342, *Manual of nutrition*

Table 10.3 Essential and non-essential amino acids

Essential	Non-essential
Isoleucine	Alanine
Leucine	Arginine
Lysine	Aspartic acid
Methionine	Asparagine
Phenylalanine	Cysteine
Threonine	Glutamic acid
Tryptophan	Glutamine
Valine	Glycine
	Proline
	Serine
	Tyrosine
	Histidine (essential in infants)

SEE QUESTION 4

✔ **REMEMBER THIS**

Plasma proteins are found in the blood, and one of their vital functions is to draw water back into the blood from tissue fluid. A lack of blood protein leads to the swollen abdomen seen in cases of starvation (Fig 10.14).

Although there are many different types of protein, there are only 20 different amino acids. These can be divided into two dietary groups: the **non-essential amino acids** are those that the body can make, the **essential amino acids** are those that the body cannot make (Table 10.3). These must be obtained from food. We make non-essential amino acids in the liver – see Chapter 14.

As a general rule, proteins of animal origin contain all the essential amino acids, while different plants contain only some of them. If you are a vegetarian, you need to combine proteins from different plants in the same meal so that you take in the full range of amino acids. This is not as difficult as it sounds: a simple meal of beans on wholemeal toast (pulses plus grains) contains all the essential amino acids.

It may come as a surprise to know that a healthy adult human needs only about 60 g of protein per day, about the size of a large egg (Fig 10.13). The human body cannot store protein. If we eat too much, the excess amino acids that are produced are **deaminated** in the liver. The resulting ammonia is combined with carbon dioxide to form urea, which is excreted in the urine. Although we need only a small amount of protein, this tends to be the most expensive part of the diet and is therefore often lacking (Fig 10.14).

FATS

What we usually think of as 'fat' is mainly triglyceride, but it also includes edible oils, fats, waxes and related compounds. Approximately 95 per cent of the lipids in the human diet are triglycerides (three fatty acids attached to a glycerol molecule, see Fig 3.18). The remaining 5 per cent comprise mainly cholesterol and the phospholipids. The body can synthesise many of its fatty acids from carbohydrates and proteins: only a few, the essential fatty acids, must come from food.

The fat content of most vegetables is low (seeds are an exception). In general, animal products have a higher fat content as animals store fat in their bodies – even lean meat is quite high in fat. Animal sources include mammals (beef, pork, lamb, etc) and oily fish such as herring and pilchard. Milk and milk products, such as cheese, are also rich in fats: milk contains 4.5 per cent fat by mass, and cheese is 30 per cent fat.

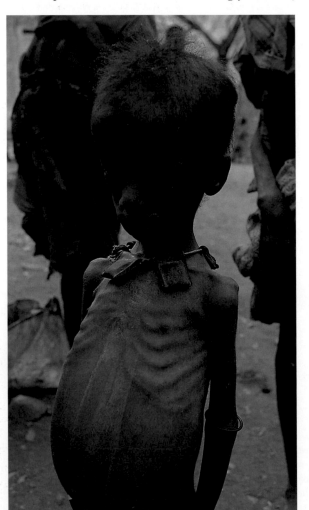

Fig 10.14 This child is suffering from kwashiorkor, a Swahili word which means 'displaced child', referring to the fact that the condition often develops after breast-feeding has stopped. A lack of protein in the blood means that water cannot be drawn back from the tissues, resulting in the swollen abdomen

VITAMINS

Vitamins are essential organic compounds needed by the body in small amounts (Table 10.4). For, example, the recommended daily amount of vitamin A for a young adult male is 1000 µg (only one thousandth of a gram). This means that the average human needs to take in only about 25 g of vitamin A (a mouthful) during an entire lifetime.

Obviously, we get most of our vitamins from our food, but there are other sources. Gut bacteria synthesise vitamin K (essential for blood clotting). In addition, humans can make vitamin A from carotene, a substance found in most plant foods, particularly carrots, and ultraviolet radiation from sunlight acts on a compound in our skin called **ergosterol**, converting it to vitamin D.

> **REMEMBER THIS**
>
> Vitamins help to regulate cell chemical processes, often through interaction with important enzyme systems. Lack of a particular vitamin in the diet leads to a **deficiency disease**.

Table 10.4 Vitamins important in the human diet

Vitamin	Name	Minimum daily requirement	Rich food source	Function	Deficiency disease
Fat-soluble vitamins					
A	Retinol	1000 µg	Fish liver oils, dairy products, liver. Most leafy vegetables and carrots contain carotene that can be converted into retinol.	Needed for healthy epithelial cells. Needed for regeneration of rhodopsin in rod cells of the eye.	Dry (keratinised) skin. Night blindness.
D	Calciferol	10 µg	Fish oils, egg yolk, butter. It can be made by the action of sunlight on skin.	Promotes absorption of calcium from intestines. Necessary for formation of normal bone and reabsorption of phosphate from urine.	Rickets in children ('soft' bones, bend easily). Painful bones in adults.
E	Tocopherol	10 µg	Vegetable oils, cereal products. Found in many foods.	In rats, formation of red blood cells, affects muscles and reproductive system. Unclear in humans. Because the vitamin is widely found in foods, deficiency is rare in humans.	Mild anaemia and sterility in rats.
K	Phylloquinone	70 µg	Fresh, dark green vegetables. Gut bacteria.	Formation of prothrombin (involved in blood clotting).	Delayed clotting time. May occur in newborn babies before the establishment of gut bacteria.
Water-soluble vitamins					
The B vitamins form a vitamin complex. They are chemically unrelated, but tend to occur together and have similar functions.					
B_1	Thiamine	1.5 mg	All plant and animal tissues contain thiamine but rich food sources are yeast, cereals, nuts, seeds, pork.	Coenzyme in cell respiration, necessary for complete release of energy from carbohydrates.	Accumulation of lactic acid and pyruvate. Beriberi (severe muscle wastage stunted growth, nerve degeneration).
B_2	Riboflavin	1.8 mg	Liver, milk, eggs green vegetables.	Coenzyme in cell respiration. Precursor of FAD.	Cracked skin, blurred vision.
B_3	Niacin	20 mg	Liver, yeast, whole cereals, beans.	Coenzyme in cell respiration. Precursor of NAD/NADP.	Pellagra (dermatitis, diarrhoea, dementia).
B_6	Pyridoxine	2 mg	Meat, fish, eggs cereal bran, some vegetables.	Interconversion of amino acids – transamination.	Dermatitis, nerve disorders.
B_{12}	Cyano-cobalamin	2 µg	Liver, milk, fish, yeast. None in plant foods.	Maturation of red blood cells in bone marrow. Maintenance of myelin sheath.	Pernicious anaemia. Nerve disorders.
	Folic acid	200 µg	Liver, raw green vegetables, yeast, gut bacteria.	Formation of nucleic acids. Formation of red blood cells.	Anaemia, especially during pregnancy.
C	Ascorbic acid	60 mg	Blackcurrants, potato, rose hips, citrus fruits.	Formation of collagen and intercellular cement.	Scurvy. Poor wound healing.

! SCIENCE FEATURE Vitamin K injection and new-born babies

Vitamin K is made in the body by helpful bacteria in the large intestine. Everybody has them, so no-one should be short of vitamin K. However, one very special group of people are at risk of vitamin K deficiency. All babies are born with sterile guts – not a bacteria in sight – and it takes about a week for the normal collection of different bugs to build up. During that first 7 days of life, this makes new babies prone to a disease called **haemmorrhagic disease of the new-born**. Vitamin K is vital for blood clotting, and a lack of vitamin K can lead to bleeding all over the body. This can occur if the skin is broken, at the navel, but also in the gut, the bowel and the brain. Brain bleeding is rare but is highly dangerous. Minor bleeding seems to happen in about one baby in 400, while more serious bleeding is very rare, about one baby in 10 000 suffers.

Fortunately, haemorrhagic disease of the new-born can be prevented very simply by giving vitamin K either by injection or by mouth on the first day of life (Fig 10.15). Recently, there has been some controversy about doing this because some studies suggested that babies given vitamin K by injection were more prone to leukaemia. Other studies have shown no such link, and most babies in the developed world now receive the vitamin K supplement.

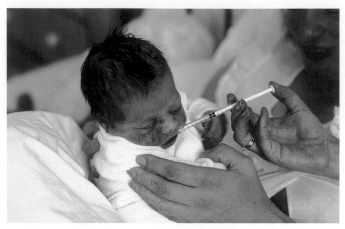

Fig 10.15 A new-born baby being given vitamin K orally (in its mouth)

Fig 10.16 Vitamin C is needed for healthy gums; the inflammation seen in this photograph is caused by vitamin C deficiency – otherwise known as scurvy

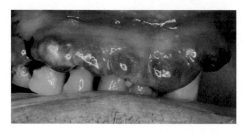

Vitamins can be divided into two groups: fat-soluble vitamins (A, D, E and K), which contain only carbon, hydrogen and oxygen, and water-soluble vitamins (vitamin C and the B complex of vitamins), most of which also contain nitrogen. Fat-soluble vitamins can be stored in the body (especially in the liver) but water-soluble ones cannot, and are excreted in the urine. Water-soluble vitamins need to be taken into the body every day, and it is virtually impossible to 'overdose' on them. Taking too much of fat-soluble vitamins can cause problems: they can build up in fatty tissues until they reach toxic levels.

MINERALS

The minerals essential to human health are listed in Table 10.5. Minerals have four functions in the body:

● They are raw materials for the construction of body tissues.
● They have a homeostatic role, providing the correct chemical environment for cells.
● They are metabolites for various cell processes.
● They are essential partners for enzymes and vitamins controlling cell activity.

Generally, the minerals in the first two categories are needed in relatively large amounts and are known as **macronutrients**. Minerals in the last two categories are needed in specific chemical reactions and in tiny amounts. For this reason they are known as **micronutrients**.

The presence of other vitamins and minerals can affect the actual amount of a mineral that the body takes up. For example, calcium, the most abundant mineral in the body, is taken up more quickly in the presence of vitamins C and D, and iron is better absorbed in the presence of vitamin C.

Table 10.5 Minerals essential in the human diet

Mineral	Major food source	Function	Total body content/g
Major minerals (macronutrients)			
Calcium	Milk, cheese, bread, watercress	Needed for muscle contraction; nerve action; blood clotting and the formation of bone	1000
Phosphorus	Cheese, eggs, roast peanuts, most foods	Bone and tooth formation; energy transfer from foods; DNA, RNA and ATP formation	780
Sulphur	Dairy produce, meat, eggs, broccoli	Formation of thiamin, keratin and coenzymes	140
Potassium	Potatoes, meat, chocolate	Muscle contraction; nerve action; active transport	140
Sodium	Any salted food, meat, eggs, milk	Muscle contraction; nerve action; active transport	100
Chlorine	Salted foods, seafood	Anion/cation balance; gastric acid formation	95
Magnesium	Meat, green vegetables	Formation of bone; form coenzymes in cell respiration	19
Trace elements (micronutrients)			
Iron	Liver, kidney, cocoa powder, watercress	Formation of haemoglobin, myoglobin and cytochromes	4.2
Fluorine	Water supplies, tea, sea food	Resistance to tooth decay	2.6
Zinc	Meat, liver, legumes	Enzyme activator; carbon dioxide transport	2.3
Copper	Liver, meat, fish	Enzyme, melanin and haemoglobin formation	0.072
Iodine	Seafood, iodised salt, fish	Constituent of thyroxine	0.013
Manganese	Tea, nuts and spices, cereals	Bone development; enzyme activator	0.012
Chromium	Meats, cereals	Uptake of glucose	>0.002
Cobalt	Meat, yeast, comfrey	Component of vitamin B_{12}; formation of red blood cells	0.0015

WATER

Just under 70 per cent of the human body is water: two-thirds of this is found within the cells and one-third outside the cells in tissue fluid and blood plasma. It is essential in the diet and can come from three main sources:

- as a drink,
- as food (lettuce, for example, contains 94 per cent water),
- as metabolic water: water is released from cellular chemical reactions, especially cell respiration.

An individual must take in enough water to balance what is lost in urine, sweating, breathing out and in faeces. Otherwise, they lose water and can become dehydrated. Severe dehydration leads to an increase in heart rate and blood urea concentration and, eventually, death.

The chemical properties of the water molecule are discussed in Chapter 4, the role of the kidneys in water balance is covered in Chapter 16.

4 A DIET FOR LIFE

The nutritional requirements of a human change according to age, sex, level of physical activity and state of health. Within groups of people there are still variations, but it is possible to say broadly what the general food and energy requirements of the groups are.

Fig 7.10 shows the energy requirements for different people at different stages of life. The Tour de France cyclist (Fig 10.17) is included as a contrast to the 'ordinary' lifestyles of people who are very active. This is one of the most demanding sporting events in the world, requiring peak output for several hours at a time, often over several days.

Fig 10.17 Tour de France cyclist being pushed to the limit

? TEST YOURSELF

F Why do experts recommend that a combination of dieting and increasing the amount of exercise is the most effective way to lose weight if you are overweight? Why not just diet or do more exercise?

Table 10.6 Composition of human milk	
Component	**Amount present per 100 ml**
Carbohydrates	
Lactose	7.3 g
Oligosaccharides	1.2 g
Proteins	
Caseins	0.2 g
Lactalbumin	0.2 g
Lactoferrin	0.2 g
Secretory IgA	0.2 g
β-Lactoglobulin	None
Milk lipids	
Triglycerides	4.0%
Phospholipids	0.04%
Minerals	
Sodium	5.0 mM
Potassium	15.0 mM
Chloride	15.0 mM
Calcium	8.0 mM
Magnesium	1.4 mM

THE FIRST YEAR OF LIFE

When a baby is first born, its mother's breasts produce a fluid called **colostrum**. This is rich in antibodies, which pass whole into the blood of the baby through large pores in its intestine. Generally, the colostrum also has about three times more protein than the breast milk produced later. There is a gradual transition between colostrum and early breast milk over the first week. As the baby suckles, it first takes in **foremilk** – a more dilute breast milk that quenches its thirst before richer, fattier **hindmilk** is produced towards the end of the feed.

Breast milk produced at about the third month of lactation has 75 calories per 100 ml; its main components are shown in Table 10.6.

During the first 4–6 months of life growth and development are rapid. Breast milk (or infant formula) contains all the nutrients required during this period; babies don't really need any solid food before the age of 4 months. As the baby grows, the fat content of breast milk decreases. After 4–6 months, milk is no longer enough and the baby starts taking in other foods – this is called **weaning** (Fig 10.18). Iron requirements increase rapidly and it is important that the diet given during weaning contains enough iron. Requirements for protein, thiamin, niacin, vitamin B$_6$, vitamin B$_{12}$, magnesium, zinc, sodium and chloride also increase between 6 and 12 months.

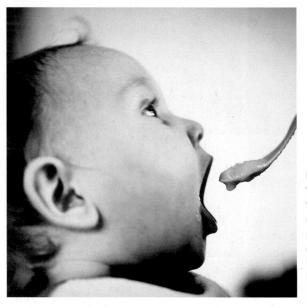

Fig 10.18 He won't stay this clean for long. Modern baby foods are well-balanced and nutritious, giving a great start in life. It's not until he reaches the age of the children on the opposite page that he'll start nagging for high-fat, high-salt or high-sugar foods. Broccoli? No chance

TODDLERS

Once babies start to toddle, walk and then run, they need a lot more energy from food. Growth and development are very rapid at this stage and, although protein requirements do not increase much, children between the ages of 1 and 3 years need more of all the vitamins, except vitamin D (this is synthesised in the skin in sunlight) and all the minerals, except zinc. Slightly lower amounts of calcium, phosphorus and iron are needed.

Children at this stage of life need energy-dense diets. The usual rules of what makes a healthy diet for an adult just do not apply. They should drink whole milk, not skimmed or semi-skimmed, and they shouldn't eat too much fibre. If the diet is too bulky, there is a possibility that a child cannot absorb enough energy from the food he or she eats. After the age of 2 years, children can switch to semi-skimmed milk, as long as the rest of the diet contains enough energy. Children shouldn't switch to completely skimmed milk until they are at least 5 years old.

CHILDHOOD

From the age of 5 until about 11, growth is still rapid and children of this age are very active (Fig 10.19):

● 5–7 years: children need more energy, protein, more of all the vitamins (except C and D) and more of all the minerals.

● 7–10 years: there is a marked increase in requirements for energy and protein. Children of this age need no more thiamin, vitamin C or vitamin A than they did at the age of 5 but their need for the other vitamins and minerals increases.

● 11–14 years: the energy needed continues to increase and young teenagers need 50 per cent more protein than they did at the age of 10. By the age of 11, the vitamin and mineral requirements for boys and girls start to differ. Boys of 11–14 need more of all the vitamins and minerals. Girls need no more thiamin, niacin, vitamin B_6, but they need more of all the minerals. When periods start, girls need more iron than boys to make up the iron lost in menstrual blood.

● 15–18 years: boys need more energy and protein and more thiamin, riboflavin, niacin, vitamins B_6, B_{12}, C and A, and more magnesium, potassium, zinc, copper, selenium and iodine. In girls, the need for energy, protein, thiamin, niacin, vitamins B_6, B_{12} and C, phosphorus, magnesium, potassium, copper, selenium and iodine all increase, but girls continue to need slightly less energy than boys of the same age. Boys and girls have the same requirement for vitamin B_{12}, folate, vitamin C, magnesium, sodium, potassium, chloride and copper. Girls still need more iron.

Fig 10.19 Young children are growing rapidly and are very active, so they require lots of energy from their food

ADULTHOOD

As young adults stop growing, their need for energy and some vitamins and minerals stabilises, and usually decreases. The energy requirements of adults depend mainly on their lifestyle – a very active person will need more energy (see Fig 10.17).

● 19–50 years: adult men and women both need less energy in their diet than when they were teenagers and their requirements for calcium and phosphorus fall as they are no longer making large amounts of new bone. There is also a reduced requirement in women for magnesium, and in men for iron. An adult's need for protein and most of the vitamins and minerals remain virtually unchanged from the teenage years.

Fig 10.20 As we age, our metabolic rate slows and we tend to become less active. This can lead to an increase in weight but elderly people often lose interest in food and eating, and can become very thin and frail

● 50+ years: energy requirements decrease gradually after the age of 50 in women and age 60 in men (Fig 10.20). Protein requirements decrease for men but increase slightly in women. Vitamin and mineral intake should continue at the same level as younger adults, for both men and women. The only exception is that once women have gone through the menopause and their periods have stopped, they need the same amount of iron as men of the same age.

THE SPECIAL CASES OF PREGNANCY AND BREAST-FEEDING

Women are now advised to take folic acid supplements to reduce the risk of spinal defects in the months when they are planning to get pregnant, and for the first 3 months of pregnancy. Despite the old saying that pregnant women are 'eating for two', a pregnant woman does not need additional energy or other food components until the last 3 months of pregnancy (Fig 10.21). Her mineral requirements do not increase at all.

Once the baby is born and the woman starts to breast-feed, she needs to take in more energy, protein, all the vitamins (except B_6), calcium, phosphorus, magnesium, zinc, copper and selenium than a new mother who does not breast-feed. The actual amount of extra energy needed depends on how much milk is produced, the size of the fat stores that have accumulated during pregnancy and how long the woman chooses to breast-feed.

Fig 10.21 Eating for two? A developing baby is fed via the placenta, an organ which takes the molecules from food directly out of the mother's blood. When the infant is born, the mother will need even more extra food in order to produce milk

EXTENSION Dietary reference values [DRVs]

In the UK, estimated requirements for particular groups of the population are based on advice given by the Committee on Medical Aspects of Food and Nutrition Policy (COMA). The COMA panel reviews scientific evidence and makes proposals that are used by the government. In 1991, the COMA report 'Dietary Reference Values for Food Energy and Nutrients for the United Kingdom' was published. This estimated the nutritional requirements of the age groups shown in Table 10.7.

Dietary reference values for energy intake exist for all different age groups, but dietary reference values for specific food groups (carbohydrates and vitamins, for example) only apply to people over 5 years old.

Source: Department of Health. Dietary Reference Values for Food Energy and Nutrients for the United Kingdom. London: HMSO, 1991.

Table 10.7 How energy requirements change during life

| | Estimated average requirements for energy | | | |
| | Megajoules per day | | Kilocalories per day | |
Age	Males	Females	Males	Females
0–3 months	2.28	2.16	545	515
4–6 months	2.89	2.69	690	645
7–9 months	3.44	3.20	825	765
10–12 months	3.85	3.61	920	865
1–3 years	5.15	4.86	1230	1165
4–6 years	7.16	6.46	1715	1545
7–10 years	8.24	7.28	1970	1740
11–14 years	9.27	7.72	2220	1845
15–18 years	11.51	8.83	2755	2110
19–50 years	10.60	8.10	2550	1940
51–59 years	10.60	8.00	2380	1900
60–64 years	9.93	7.99	2380	1900
65–74 years	9.71	7.96	2330	1900
74+ years	8.77	7.61	2100	1810
Women only pregnancy		8.9		1140

SUMMARY

When you have read and studied this chapter, you should know and understand the following:

- Food supplies us with the energy and raw materials we need for repair, growth and development of body tissues.
- Humans need a **balanced diet** which contains carbohydrate, protein, fat, **water**, **fibre**, vitamins and minerals.
- **Carbohydrate** foods (sugars and starches) should be the main source of energy in the diet, although **lipids** provide more than twice as much energy per gram.

- When intake of 'energy' foods exceeds requirements, the body will store the excess as glycogen and fat. When intake is too low, the body will turn to the lipid stores and increasingly use **protein** when the other stores run low.
- **Vitamins** are either water soluble (A, D, E and K) or fat soluble (B complex and C). Lack of particular vitamins or **minerals** in the diet can lead to **deficiency diseases**.

? EXAM QUESTIONS

1

a) Describe how saprophytic fungi obtain their nutrients.

b) You are provided with:

Petri dishes containing starch agar (starch suspended in a solid jelly);

iodine solution;

samples of five species of saprophytic fungi.

Briefly describe how you would compare the effectiveness of starch digestion by these five fungi.

AQA (NEAB) BY03 Physiology June 2000 Q4

2

a) Both mechanical and chemical digestion are involved in the breakdown of food. Explain briefly the purpose of: **(i)** mechanical digestion; **(ii)** chemical digestion.

b) The diagram shows the lateral view of the teeth and associated structures of the two different mammals.

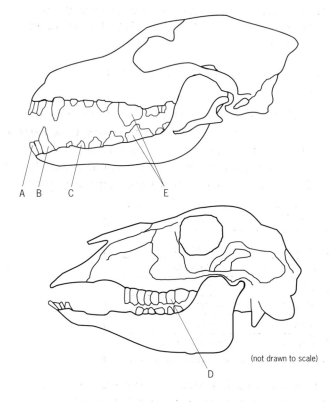

(not drawn to scale)

Name the teeth labelled **A–D**.

Describe how teeth **D** and **E** are adapted to their functions in feeding.

c) Suggest **two** features shows in the diagram, other than teeth, which adapt either mammal to its mode of nutrition.

OCR 6918 Animal Studies March 2000 Q1

3 An experiment was carried out to investigate the effects of fibre in the diet on formation of tumours (cancers) in the large intestine of rats. At the start of the experiment all of the rats received a treatment that causes tumour formation.

Immediately following the treatment the rats were divided into two groups. One group was fed a normal diet (control group). The other group was fed on a normal diet with added fibre in the form of wheat bran (experimental group).

The rats were examined every week to check for tumours in the large intestine. The results are shown below.

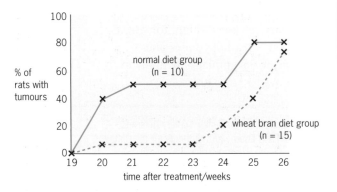

a) State the week when tumours were observed in more than 10% of the rats in: **(i)** the control group; **(ii)** the experimental group.

b) State when the difference in percentage of rats with tumours is greatest between the two groups.

c) Discuss the effectiveness of fibre in the experiment in preventing tumour formation.

IBO Biology Subsidiary Standard Level Paper 3 November 1998 Option A QA1

4

a) Define *malnutrition*.

b) Low levels in the amount of protein in the diet can result in a deficiency disorder. Suggest a reason why eating protein from only one source can cause similar disorders.

c) State **two** different sources of protein acceptable to vegans.

IBO Biology Subsidiary/Standard Level Paper 3 November 1998 QA2

KEY SKILLS **ASSIGNMENT**

OBESITY AND BODY MASS INDEX (BMI)

Obesity is a term that means 'grossly overweight' and, despite more people understanding the problems of the condition, the proportion of overweight and obese people is on the increase.

Table 10.A1 The increasing incidence of obesity

Country	Weight level	Age range	Year	Incidence/% men	Incidence/% women
England	obese (BMI >30)	16–64	1980	6	8
			1994	13	16
England	overweight (BMI 25–30)	16–64	1980	35	24
			1991	40	26
Germany	obese (BMI >30)	25–69	1985	15	17
			1990	17	19
USA	obese* (white people)	20–74	1978	24	24
			1988–91	34	34
USA	obese* (black people)	20–74	1978	26	45
			1988–91	32	49

*In the USA survey the threshold for obesity was BMI >27.8 men, >27.5 for women
Sources: Obesity Research Information Centre, London, and Dept of Health Report 46 (1994) Nutritional Aspects of Cardiovascular disease, HMSO.

1 Given the present rate of increase, what percentage of men and women in England will be obese by the year 2008?

Body mass index (BMI)

2 Obesity could be defined in terms of weight alone. What would be wrong with this?

In 1869 a Belgian astronomer Quetelet worked out that people's body mass varied not in proportion to their height, but in proportion to the square of their height.

So the formula for body mass index is:

$$\frac{mass}{height^2}$$

For example, a man who is 1.83 m tall (6 feet) and weighing 82 kg would have a BMI of

$$\frac{82}{(1.83)^2} = \frac{82}{3.35} = 24.47$$

3 Work out your own BMI.

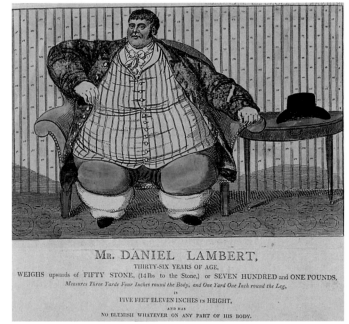

Fig 10.A1 This is Daniel Lambert (1770–1809), once thought to be the heaviest man in England. He weighed 336 kg (52 stone 11 lb) and was 180 cm (6 ft) tall

4 Work out the BMI for Daniel Lambert.

5 List some of the problems that a person as heavy as Daniel Lambert would face in their everyday life. Assume that they are alive today, not in the seventeenth century
a) What sorts of things would it be impossible for them to do?
b) Can you think of any advantages of being this weight?

Table 10.A2

BMI/kg m^{-2}	Description
less than 20	underweight
20–24.9	normal
25–29.9	overweight
30–40	obese
over 40	severely obese

The BMI is now universally used to define underweight, overweight and obesity, as shown in Table 10.A2.

KEY SKILLS **ASSIGNMENT**

The problems of obesity

There was a time when fat people were looked upon just as happy, jolly folk. Nowadays, we know the problems associated with obesity, and obese people have to deal with the intolerance of others towards their size and body image.

One look at the population of an old people's home will support the blunt statement that 'old people aren't fat and fat people aren't old'. Obesity puts a great strain on the body that more often than not leads to severe medical problems and a premature death. Consider some of the problems:

Coronary heart disease. Obesity increases the levels of triglyceride and cholesterol in the blood.

6 Suggest why obesity puts a strain on the heart. Why does it have to work harder?

Diabetes. This is the inability to control blood sugar levels. In an obese person, insulin becomes less effective than normal, a problem known as insulin resistance.

High blood pressure.

Joint problems. Excess weight puts a lot of strain on some joints.

7 Suggest which joints will be affected most.

Accidents. Obese people tend to be clumsier than less heavy people, and unable to move quickly.

Depression and suicide. These could be responses to a poor body image.

The causes of obesity

Many obese people claim that eating too much is not the problem; it's in their genes. Recent research suggests that there may be some truth in this, but we can only inherit a tendency towards obesity or 'leanness'. We must still eat the food which allows us to fulfil our potential, and studies suggest that social and environmental factors play a huge part. Children born to overweight parents tend to be given larger portions and generally follow the influence of their parents.

Lifestyle

People are generally eating more and exercising less, using the car for even the shortest journeys and taking part in leisure activities with minimal energy needs.

8 **a)** Suggest how the leisure activities of children have changed over the past few decades.
b) Imagine you are a family doctor. How would you persuade an obese couple to change their lifestyle?

Drugs

Research has shown that adipose (fat storage) tissue gives off a hormone, leptin (Greek, leptos = slender/thin) which suppresses the appetite centre of the brain. There are artificial appetite suppressants available, but their effect is only temporary. Clearly, there would be a huge market for an effective drug.

9 What properties would the perfect weight control drug have?

Surgery

Adipose tissue can be surgically removed; the process of liposuction is one of the cosmetic surgeons' biggest moneyspinners. However, all surgery carries some risk and, without a long-term change in eating and exercise habits, the problem will return.

Fig 10.A2 The United States has the highest proportion of obese people in the world

11

Digestion

Digestion

Humans are omnivores; they have a digestive system that is adapted to eat a range of foods, from both animals and plants. But that doesn't mean that all foods suit everybody. Some people suffer food allergies or are intolerant to specific foods. Lactose intolerance is well known (see page 219) but researchers are uncovering a problem with sugar digestion – an intolerance to fructose.

Everyone experiences indigestion or stomach upsets now and again, but some people have chronic digestive problems – this is commonly known as irritable bowel syndrome (or IBS), and is extremely common. Its main symptoms are abdominal discomfort, a bloated feeling and experiencing diarrhoea and then constipation for no apparent reason. In the autumn of 2001, researchers in the USA revealed that about 10 per cent of IBS cases could be explained by an intolerance to fructose. Fructose is a simple sugar found in honey and many fruits.

When people with IBS symptoms were given fructose, tests showed that over three-quarters of them produced breath that contained hydrogen and methane. This indicated that they were not metabolising fructose normally, and these abnormal by-products were being produced in the gut.

Fructose is present in many everyday foods

1 THE HUMAN DIGESTIVE SYSTEM

The overall function of the **digestive system** is to break down the larger molecules in food, such as protein and starch, into their monomers – amino acids and sugars – so that they can be absorbed easily into the body.

The **alimentary canal**, or **gut**, can be thought of as a long coiled tube that runs through the body. The food we eat is subjected to 'conveyor-belt' food processing as it passes through the different regions of the gut. A common misconception is that the food we eat is 'inside' us, when really it travels through a cavity in the middle of the body. If we cannot digest the food, it cannot be absorbed into the body and passes straight through.

The alimentary canal, or 'gut', is a muscular tube that leads from the mouth to the anus. The overall process of nutrition can be divided into several stages:

- **Ingestion:** this is taking in food. In humans food is put into the mouth and chewed. **Swallowing** takes it down through the **oesophagus** into the stomach. Food is propelled through the alimentary canal by **peristalsis**, rhythmic contractions of the gut wall.
- **Mechanical breakdown:** the breaking up of food material into smaller pieces, mainly by the action of chewing or the churning action of the stomach.
- **Digestion:** the chemical breakdown of complex food molecules such as carbohydrates, proteins and fats into simpler ones.
- **Absorption:** the passage into the bloodstream of simple food molecules such as amino acids, sugars and fatty acids, vitamins, minerals and water.
- **Egestion:** the elimination of undigested food material from the body.

2 STRUCTURE OF THE HUMAN DIGESTIVE SYSTEM

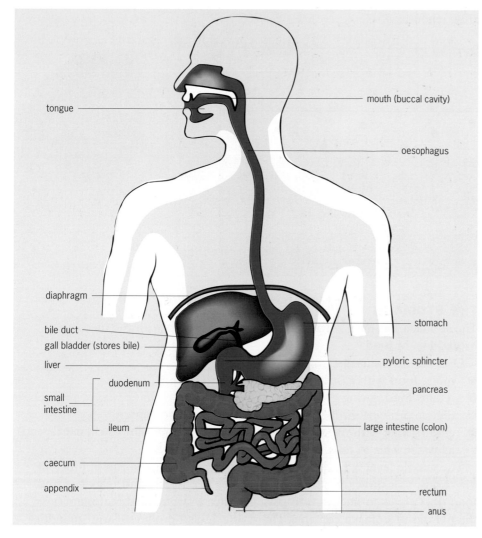

Fig 11.1 The human digestive system

Labels: tongue · mouth (buccal cavity) · oesophagus · diaphragm · stomach · bile duct · gall bladder (stores bile) · liver · pyloric sphincter · duodenum · small intestine · ileum · pancreas · large intestine (colon) · caecum · appendix · rectum · anus

The human digestive system consists of the alimentary canal and its associated glands, the salivary glands, the liver and the pancreas (Fig 11.1).

The alimentary canal begins at the mouth and ends at the anus. Between the two openings is a long convoluted tube organised into several distinct regions. The **oesophagus** carries food from the mouth to the **stomach**, a muscular bag or sac that stores food and is the first significant site of digestion.

Beyond the stomach is the **small intestine**. This has two main parts, the **duodenum** and the **ileum**. These are the main sites of digestion and absorption. The large intestine includes the **appendix**, the **colon**, whose main function is to absorb water, and the **rectum**. It ends at the **anus**.

BASIC FEATURES OF THE GUT WALL

The wall of the gut is not the same all the way along the alimentary canal. Each region has specific features, which can be related to the function of different parts of the gut.

The gut wall is divided into three main layers:

- an outer **muscle layer**, protected by a thin coating of fibres;
- a middle layer, the **submucosa**;
- an inner layer, the **mucosa**.

SEE QUESTION 1

THE OESOPHAGUS

The main function of the oesophagus is to push food from the mouth to the stomach. The structure of the wall of the oesophagus is shown in Fig 11.2. The endoscope image shows that the inside surface has large ridges, but it is fairly smooth.

Fig 11.2 The oesophagus

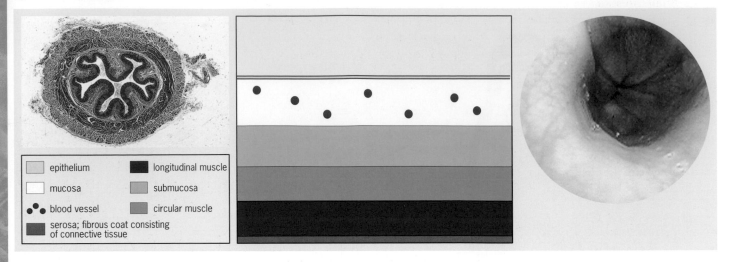

epithelium   longitudinal muscle
mucosa   submucosa
blood vessel   circular muscle
serosa; fibrous coat consisting of connective tissue

REMEMBER THIS

Note that all regions of the gut have two layers of muscle, apart from the stomach, which has three. The extra *oblique* layer runs at 45° to the other two, and helps churn the food.

THE STOMACH

The stomach has deep ridges called **rugae** that help with the mechanical breakdown of food.

Overall, the stomach:
- mixes food with gastric juice by muscular action;
- retains food, giving enzymes time to act;
- digests proteins through the action of the enzyme pepsin;
- curdles milk with the enzyme rennin, absorbs some simple chemicals such as water, salts as ions, and alcohol.

The muscle layer is thick; three different layers of muscle run in different directions. This gives the stomach the power to contract and relax to grind food up. Glands in the **gastric pits** produce a strong acid that reduces the pH of the stomach contents. The stomach also produces large quantities of **mucus** to protect its own cells from damage from the acid.

Fig. 11.3 The stomach

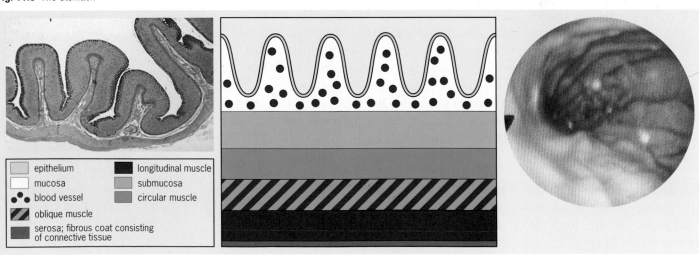

epithelium   longitudinal muscle
mucosa   submucosa
blood vessel   circular muscle
oblique muscle
serosa; fibrous coat consisting of connective tissue

THE DUODENUM AND ILEUM

The small intestine is about 5 metres long and is made up of two main parts, the duodenum and the ileum. The duodenum is the first 25 cm of the small intestine, the ileum forms the rest of it. While the duodenum is the main site of digestion, most absorption takes place in the ileum.

The small intestine:

- moves food from the stomach to the large intestine by peristalsis. Food moves through the intestine at about 1 cm per minute;
- secretes the large amounts of water necessary for most of the chemical reactions involved in digestion;
- completes the digestion of carbohydrates, proteins and fats;
- absorbs most of the small soluble food molecules produced by digestion.

Fig 11.4 The duodenum

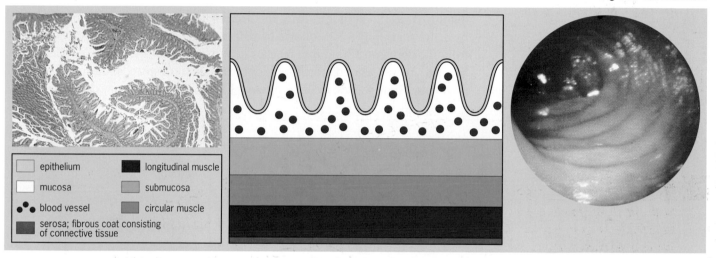

epithelium   longitudinal muscle
mucosa   submucosa
blood vessel   circular muscle
serosa; fibrous coat consisting of connective tissue

Fig 11.5 The ileum

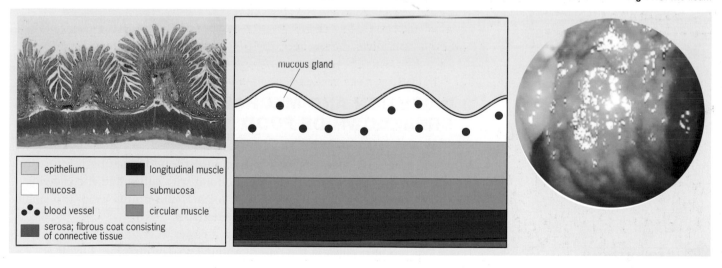

mucous gland

epithelium   longitudinal muscle
mucosa   submucosa
blood vessel   circular muscle
serosa; fibrous coat consisting of connective tissue

The pancreatic duct and bile duct join with the duodenum, releasing digestive enzymes and bile into the lumen. This neutralises stomach acid, and creates a pH that is optimal for digestion. The wall of the small intestine looks less folded than that of the stomach, but looking at it under a light microscope reveals the tiny microscopic infoldings that increase the surface area. The tiny folds are called villi; there are 4–5 million of them in the ileum. They allow intimate contact between the wall of the intestine and the products of digestion inside this part of the gut, ensuring that the absorption of digestion products is as efficient as possible.

THE COLON

The ileum joins the large intestine at the caecum, near the appendix (refer back to Fig 11.1). The **colon, rectum** and **anus** make up the final section of the digestive tract. Sodium ions leave the lumen of the colon and re-enter the blood by active transport; water follows by osmosis. The colon therefore absorbs water from the undigested food that reaches it. What is left in the colon is called **faeces**. As well as undigested food, faeces contain dead cells, bacteria and bile pigments. These wastes are then **egested** from the anus during **defecation**.

Fig 11.6 The colon

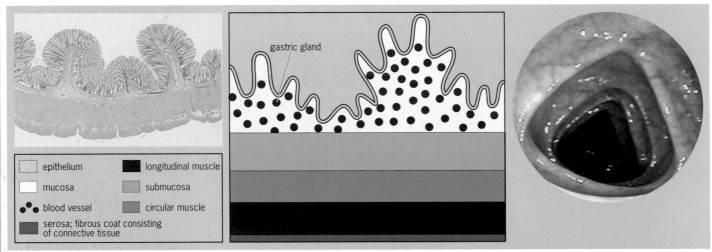

- epithelium
- mucosa
- blood vessel
- serosa; fibrous coat consisting of connective tissue
- longitudinal muscle
- submucosa
- circular muscle

gastric gland

3 INGESTION AND MECHANICAL BREAKDOWN OF FOOD

Whatever your table manners are like, you ingest food by getting it into your mouth. Your lips and tongue then mix the food with saliva and the chewing action of grinding the food between your teeth breaks up the food into small pieces. The technical term for this is **mastication**.

Adding saliva to food softens it, making it easier to swallow. Salivary glands (Fig 11.7) produce saliva at the rate of about 1 to 1.5 litres per day. It is produced constantly, but more is released when we see, smell, taste or even just think about food.

Saliva is mainly water (99.5 per cent) with some dissolved substances (0.5 per cent) including:
- **mineral salts** such as phosphates and hydrogencarbonates;
- **salivary amylase**, a starch-digesting enzyme that breaks molecules of starch into maltose;
- **mucin**, a slimy glycoprotein lubricant;
- **lysozyme**, an enzyme that kills bacteria.

Salivary amylase, the starch-splitting enzyme in saliva, begins the process of chemical breakdown. However, the speed at which most people chew and swallow their food means that salivary amylase has little chance to act, which is probably why people who lack the gene to make this enzyme are rarely aware of their 'genetic deficiency'.

SWALLOWING

Swallowing is not a voluntary action, but it is under some voluntary control because you can decide to swallow if you want to. When you are eating, the act of moving food to the back of your throat is the voluntary action but swallowing is a reflex response to food or liquid touching the back of your throat.

Theoretically, when the tongue forces food or liquid to the back of the throat, the food is able to travel in one of four directions:

- back out of the mouth;
- into the nasal cavity;
- into the trachea, or windpipe;
- into the oesophagus (where it should go).

Normally, when we swallow, the first three options are closed, and food or liquid is forced into the oesophagus. If you have ever been laughing, eating crisps and drinking a fizzy drink, you may have experienced fizzy liquid going 'up your nose' or crisps going 'down the wrong way'.

The oesophagus is about 25 cm long and 2 cm in diameter. When you swallow food, smooth muscles in the oesophageal wall contract rhythmically to propel it, by **peristalsis**, from the **pharynx** to the stomach (Fig 11.8). Elastic tissue in the walls enables the oesophagus to expand and, as in the mouth, stratified epithelium protects against friction damage.

Mechanical breakdown continues in the stomach, as the churning action breaks the food up into yet smaller pieces.

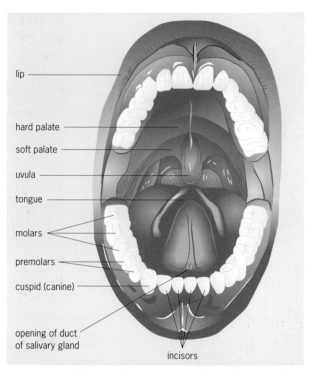

Fig 11.7 There are three pairs of salivary glands in the mouth and one pair can easily be seen. If you look under the tongue and dry the area, you will be able to see saliva oozing out

---

✔ REMEMBER THIS

The intestine, along with many other tubular organs in the body (ureters, vas deferens, uterus) are made from smooth muscle whose main function is the slow rhythmic contraction known as peristalsis.

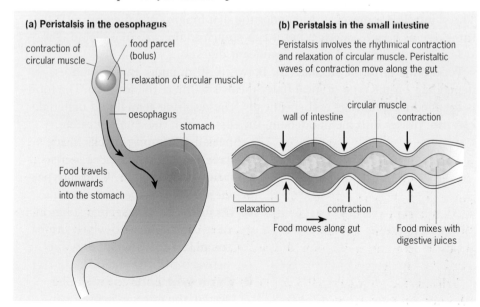

(a) Peristalsis in the oesophagus

contraction of circular muscle

food parcel (bolus)

relaxation of circular muscle

oesophagus

stomach

Food travels downwards into the stomach

(b) Peristalsis in the small intestine

Peristalsis involves the rhythmical contraction and relaxation of circular muscle. Peristaltic waves of contraction move along the gut

wall of intestine

circular muscle

contraction

relaxation   contraction

Food moves along gut

Food mixes with digestive juices

Fig 11.8 Food is propelled through the gut by peristalsis. The squeezing action of peristalsis moves semi-solid food from the mouth to the stomach in about 4 to 8 seconds. Food moves through the intestines at a rate of about 1 cm/min

4 DIGESTION

Digestion is usually thought of as the chemical breakdown of complex food molecules into simpler ones that can be absorbed and used by the body. Enzymes are key players in this process; they control all of these breakdown reactions. Table 11.1 gives an overview of the main enzymes involved in human digestion.

In the first part of this section, we look at where the different enzymes act, and what they do; in the following section we see how the release of these secretions is controlled.

Table 11.1 A summary of the main human digestive enzymes

Secretion	Enzymes produced	Site of production	Site of activity	pH	Substrate	Products
Saliva	salivary amylase	salivary glands	mouth	6.5–7.5	starch	maltose
Gastric juice	pepsin	stomach	stomach	2.0	proteins, polypeptides	
	rennin				caseinogen (milk protein)	
Pancreatic juice	trypsin	secretory cells of the pancreas (acini)	duodenum	7.0	proteins, polypeptides	short polypeptides
	chymotrypsin				proteins	polypeptides
	carboxypeptidase				polypeptides	dipeptides, amino acids
	pancreatic amylase				starch	maltose
	maltase				maltose	glucose
	sucrase				sucrose	glucose, fructose
	lactase				lactose	glucose, galactose
	lipase				fats and oils	fatty acids, glycerol

? TEST YOURSELF

A A person who has developed stomach cancer may have to have their stomach removed. Suggest what effect this will have on their eating habits.

DIGESTION IN THE STOMACH

The stomach is a muscular sac under the diaphragm. When empty, it is the same size as a large sausage, but it can stretch to the size of a melon. The stomach is the first area of any significant digestion.

Food remains in the stomach for between 30 minutes and 4 hours, depending on the type of meal (fatty meals stay there the longest). The food is digested mechanically and chemically. The resulting semi-liquid material, **chyme**, passes into the duodenum, the first part of the small intestine.

Specialised groups of cells inside **gastric pits** in the mucosa secrete **gastric juice** (Fig 11.9). Each type of cell produces a specific secretion:

- **Oxyntic cells** secrete a solution of hydrochloric acid which brings the pH of gastric fluid down to between pH 2.0 and 3.0.
- **Zymogen cells** (peptic cells) secrete the enzyme pepsinogen which is later converted into the protein-splitting enzyme, pepsin.
- **Mucous cells** secrete the mucus that protects the stomach lining from the digestive action of its own secretions.

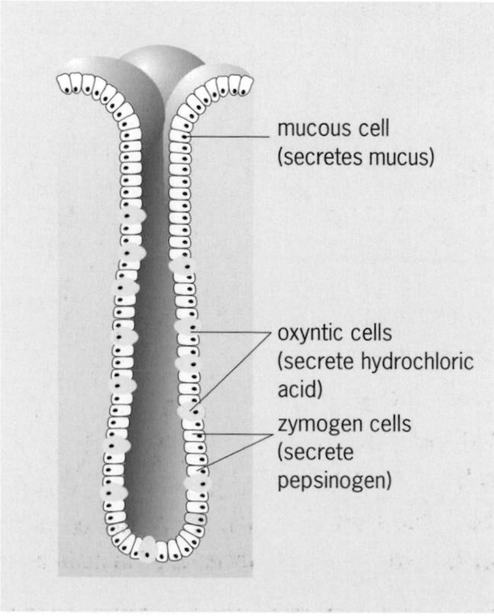

mucous cell (secretes mucus)

oxyntic cells (secrete hydrochloric acid)

zymogen cells (secrete pepsinogen)

Fig 11.9 A gastric pit from the mucosa of the stomach showing the specialised cells that secrete gastric juice

THE IMPORTANCE OF STOMACH ACID

The hydrochloric acid in gastric fluid:

- provides the optimum pH for **pepsin** and **rennin**;
- denatures proteins and helps to soften tough connective tissue in meat;
- is a strong **bactericide** (it kills bacteria) and so protects the body from some of the harmful microbes that might enter the body in food.

PROTEIN DIGESTION IN THE STOMACH

One of the main functions of the stomach is to begin to digest proteins. **Pepsin**, a powerful **endopeptidase enzyme** digests proteins. This enzyme breaks peptide bonds in the middle of the protein chain, turning protein molecules into polypeptides (see Chapter 3, page 53). Protein digestion is completed when **exopeptidase** enzymes remove amino acids from the ends of the polypeptides.

The stomach avoids digesting its own tissue by secreting pepsin in an inactive form, **pepsinogen**. This is converted to pepsin in the lumen of the stomach only after contact with hydrochloric acid. The hydrogen ions in the acid cause the pepsinogen to unfold and become pepsin, the active form of the enzyme.

MILK DIGESTION

Caseinogens, the proteins in milk, are water soluble. They are valuable nutrients (milk is the sole source of food for young mammals; see the Science Feature on this page) but, if they remained in their soluble form, they would leave the stomach before protein digestion had finished. To avoid this, the stomach produces **rennin**. This curdles milk, converting soluble caseinogen into insoluble **casein**. Like pepsin, rennin is also secreted in an inactive form, **prorennin**. Like pepsinogen, prorennin is converted to its active form by contact with stomach acid.

? TEST YOURSELF

B At what pH would you expect the human form of the enzyme pepsin to display optimum activity?

C What is the difference between an endopeptidase and an exopeptidase enzyme?

? TEST YOURSELF

D **a)** Would you expect rennin production to increase or decrease with age in mammals? Give a reason for your answer.
b) Would you expect rennin to be found in animals other than mammals?

! SCIENCE FEATURE Lactose

All mammals feed their young on milk (Fig 11.10). The main carbohydrate in breast milk is the disaccharide lactose. But why should a mother go to the trouble of combining two monosaccharides into lactose when the baby will simply have to break it down again to use it as an energy source?

The answer lies in the size of the lactose molecule. In the mother, milk is made in the mammary gland and is stored there between feeds. Since it is a disaccharide, lactose stays in the milk, rather than diffusing into the surrounding tissue as smaller monosaccharides would tend to do. In the baby, lactose remains in the gut for the same reason and is broken down only gradually to form glucose. So there is a steady absorption of glucose into the infant's blood, rather than a sudden surge at feeding time.

Several medical conditions can arise because of a failure to digest lactose. In adult life, some people stop making the enzyme lactase and so cannot break down lactose. Undigested lactose cannot be absorbed and it accumulates in the gut, encouraging the growth of bacteria that produce large amounts of carbon dioxide and lactic acid. The result is diarrhoea and wind. Affected people are said to be **lactose intolerant**.

A type of lactose intolerance that results in far more serious problems is found in people with an inherited condition called **galactosaemia**. Affected people can break down lactose into glucose and galactose but their liver cannot convert galactose to glucose. Galactose builds up to dangerous levels and the condition can be fatal.

Fig 11.10 Milk contains all of the food chemicals needed by a young mammal. The main carbohydrate in milk is lactose

DIGESTION IN THE SMALL INTESTINE

The small intestine is the major site of digestion, which is brought about mainly by enzymes and secretions that flow into the duodenum from the pancreas and liver. These neutralise stomach acid and also contain a variety of enzymes that break down proteins, carbohydrates and fats into their monomers, ready for absorption.

PANCREATIC JUICE

Pancreatic juice enters the duodenum via the pancreatic duct. Over a litre of alkaline pancreatic juice is secreted every day. The fluid contains several enzymes that are produced by secretory cells in the acini of the pancreas (see page 274, for the structure of the pancreas):

- **Trypsin**: this is a powerful **endopeptidase enzyme** that breaks down proteins into polypeptides. It is secreted in an inactive form called **trypsinogen**. Inactive trypsinogen is turned into active trypsin by the action of **enterokinase**, an enzyme released by the intestinal wall:

$$\text{trypsinogen} \xrightarrow{\text{enterokinase}} \text{trypsin}$$

- **Chymotrypsin** continues the digestion of proteins into polypeptides. This enzyme is secreted in an inactive form called chymotrypsinogen, and is converted to active chymotrypsin by the enzyme trypsin:

$$\text{chymotrypsinogen} \xrightarrow{\text{trypsin}} \text{chymotrypsin}$$

- **Carboxypeptidase** is a polypeptide-digesting enzyme, which converts polypeptides into smaller peptides and amino acids. Carboxypeptidase is also activated by trypsin in the lumen of the small intestine:

$$\text{procarboxypeptidase} \xrightarrow{\text{trypsin}} \text{carboxypeptidase}$$

- **Pancreatic amylase:** this completes the breakdown of starch to maltose started by salivary amylase in the mouth. A second amylase enzyme is needed as salivary amylase is inactivated by stomach acid.

- **Pancreatic lipase:** this continues the breakdown of fats into fatty acids and glycerol.

Hydrogencarbonate ions make pancreatic juice slightly alkaline (pH = 7.1 to 8.2), allowing it to neutralise the stomach acid and helping to create optimum conditions for the intestinal digestive enzymes.

The pancreas and liver both have other functions that are not connected directly with digestion. The pancreas produces the hormones insulin and glucagon (Chapter 14, Section 2) and the liver has many regulatory – homeostatic – functions (also Chapter 14, Section 3).

BILE

Liver cells called **hepatocytes** produce thick yellow-brown or olive green bile. This contains:
- water;
- **bile salts** (sodium glycocholate and sodium taurocholate);
- **bile pigments** (breakdown products of red blood cells, for example bilirubin) and some mucus;
- cholesterol.

Liver cells produce around 0.8 to 1.0 litre of bile daily. Secretions from individual cells pass into tiny canals called **bile cannaliculi**. These lead to the gall bladder, a small sac-like organ, which stores the bile until it is needed. Bile is released into the duodenum when the muscle wall of the gall bladder contracts. Bile reaches the duodenum through the bile duct. Bile:

- **emulsifies** fats (breaks large fat or oil droplets into an emulsion of microscopic droplets). This process massively increases the surface area available for fat-digesting enzymes;
- neutralises the (acidic) chyme from the stomach and creates the ideal pH for intestinal enzymes;
- stimulates peristalsis in the duodenum and ileum;
- allows the excretion of cholesterol, fats and bile pigments.

CONTROL OF DIGESTION

Since we do not eat food continuously, it is important that we produce digestive enzymes only when food is in the gut. If large quantities of the acids and protein-attacking enzymes were released into an empty stomach, there would be a real danger that you would digest yourself.

The release of digestive secretions is controlled by nerves and hormones working together. Fig 11.11 shows an overview of what happens.

Fig 11.11 The digestive tract is supplied by nerves from the autonomic nervous system (see Chapter 20). Particularly important is the vagus nerve, which stimulates the secretion of digestive juices. Hormones travel to the digestive tract in the blood.

Information coming into the body from food (about its appearance and smell) travels to the hypothalamus and cerebral cortex in the brain. These send signals back to the body to stimulate the salivary glands, stomach, liver and pancreas to start releasing their secretions. As food enters the stomach, gastrin is released and this encourages more gastric juice secretion. Similarly, the presence of food in the duodenum causes the release of gastrin, secretin and CCK-PZ (cholecystokinin-pancreozymin).

We do not carry on eating until we burst because food intake is controlled by hunger and satiety centres in the hypothalamus. A full stomach and high blood glucose levels stimulate the satiety centre, so that we do not feel hungry any more

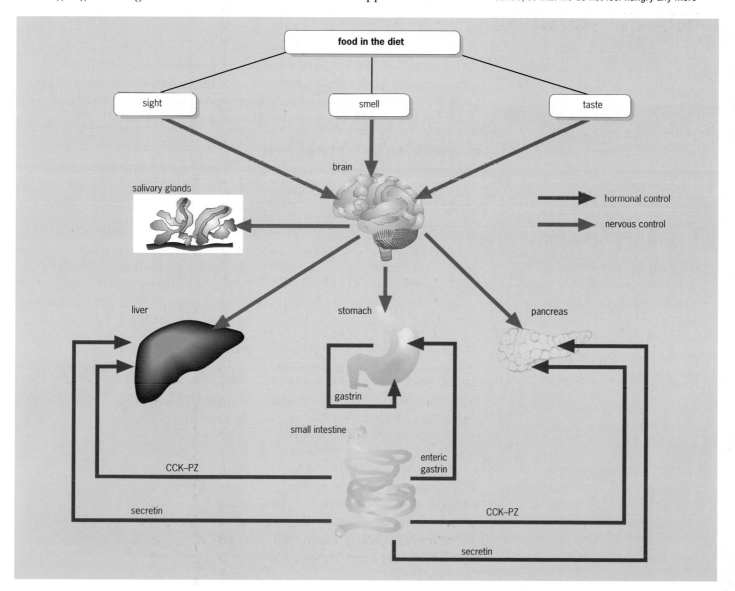

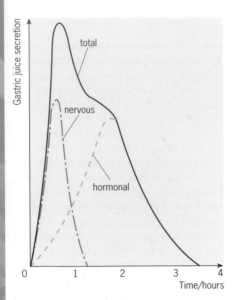

Fig 11.12 Secretion of gastric juice is partly controlled by a nerve (the vagus nerve) and partly by a hormone (gastrin). Gastrin stimulates gastric glands, causing them to release their secretions. It increases movement in the tract and relaxes the sphincters that control the movement of food into and out of the stomach

SEE QUESTION 3

CONTROL OF GASTRIC SECRETIONS

Nerves and hormones control the secretion of gastric juice in a process that has three distinct phases (Fig 11.12):

● **Nervous phase.** The sight, smell or taste of food initiates a nerve reflex in which impulses from the brain trigger gastric glands to release their secretions.

● **Gastric phase** (hormonal). Food in the stomach stimulates the lining to secrete the hormone **gastrin**. Gastrin increases gastric juice secretion through direct action on the gastric glands.

● **Intestinal phase** (hormonal). The duodenal lining, stimulated by partially digested food, produces a second hormone, **enteric gastrin**. This hormone also acts on the stomach's gastric glands, producing further small amounts of gastric juice.

Pressure sensors and chemical sensors in the stomach detect stomach stretching and the presence of chyme. The resulting nerve impulses, together with the action of gastrin, direct the stomach to start emptying its contents into the duodenum.

CONTROL OF PANCREATIC SECRETIONS

Pancreatic secretion, like gastric secretion, is controlled by both nervous and hormonal mechanisms. Impulses travel from the brain down the **vagus nerve** to trigger the secretion of pancreatic juice.

The lining of the duodenum secretes two hormones in response to acidic chyme arriving in the lumen of the duodenum. **Secretin** stimulates the pancreas and liver to secrete pancreatic juice rich in hydrogencarbonate ions. **CCK-PZ** (**cholecystokinin-pancreozymin**) stimulates the gall bladder to release bile and the pancreas to release digestive enzymes.

CONTROL OF BILE SECRETION

The hormone secretin acts together with nervous stimulation by the vagus nerve to increase the rate of bile secretion. The acidity of chyme in the duodenum and the hormone CCK-PZ stimulate the gall bladder to contract.

! SCIENCE FEATURE **Feeling sick? Blame the brain**

Lots of different things make you feel sick – eating food that is 'off', travelling on a boat or in a car or aeroplane, having a bad cold or having a serious illness such as cancer, even when the digestive system isn't affected directly. Getting rid of the stomach contents if you have eaten something bad is an understandable response by the body – but why do these other experiences cause nausea?

Travel sickness results when the inner ear is telling the brain something different from the messages being sent from the eyes. The fine hairs in the inner ear detect motion and if this is in conflict with the visual sensation of movement, you start

to feel yukky. Keeping a long-distance view on things helps – looking at near objects, such as reading a book on a long car journey, is asking for trouble.

But what about other illnesses like having a cold, or more serious illnesses such as cancer? Recently, it has become clear that, when you have an illness, your body responds by producing different chemicals that stimulate the immune system. One of these is called tumour necrosis factor. One of its functions, as its name suggests, is to clobber cancer cells, However, it also acts on neurones in the brain, in a part of the brain stem that is concerned with digestion and vomiting. When tumour necrosis factor binds to

them, these neurones cause the stomach to stop all movement. This dramatic stomach relaxation is perceived as nausea and occurs just before a bout of vomiting.

Fig 11.13

SCIENCE FEATURE Digestive differences in ruminants

Ruminant mammals such as cows, sheep and deer have a four-chambered stomach (Fig 11.14). When they eat, food enters the first two of these chambers, the rumen and the reticulum, and is reduced to a workable pulp by the addition of liquid. Bacteria and other single-celled organisms that live in these two chambers produce digestive enzymes such as **cellulase**, which help to break down plant cell walls. These microbes also contribute to the animal's metabolism by manufacturing useful organic compounds such as amino acids and fatty acids, which are absorbed directly.

A ruminant animal 'eats' its food twice. It is first broken up and partially digested in the rumen and reticulum. Later, the animal regurgitates the 'cud' into the mouth for further chewing. Eventually, the mixture of digested food and microorganisms passes to the rest of the gut during the process of regurgitation. It is then digested, releasing the nutrients it is made up from, and making them available for absorption in the cow's ileum. The use of bacteria to get the most out of nutrition-poor grass is an important adaptation of ruminants.

On its second passage down the oesophagus, food enters the **omasum** for further physical reworking, and finally reaches the **abomasum**, which is the stomach proper (it has **cardiac**, **fundic** and **pyloric** regions similar to the human stomach).

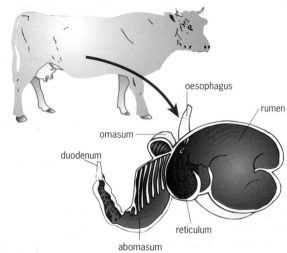

Fig 11.14 The four-chambered stomach of the cow

5 ABSORPTION OF THE PRODUCTS OF DIGESTION

Food absorption takes place mainly in the ileum of the small intestine. The small soluble molecules produced by digestion in the stomach and duodenum can be easily absorbed through the gut lining. The epithelial lining has a much greater surface area than that of an equivalent smooth-sided tube; an average 70 kg man has approximately 100 square metres of absorbing surface in his small intestine. This is possible because of three important structural features:

Fig 11.15 The absorption of the products of digestion through the lining of the intestine

- folds in the inner surface of the intestinal wall;
- movable projections (0.5–1.0 mm in length) called **villi** that are present on the folded surface of the wall;
- microscopic projections called **microvilli** on the cell surface membranes of epithelial cells that line the villi.

The epithelial cells absorb amino acids, monosaccharides, fatty acids and glycerol, water and other substances including vitamins, nucleic acids, ions and trace elements (Fig 11.15). The type of transport process used varies depending on the substance, but **diffusion**, **facilitated diffusion** and **active transport** are all involved (see Chapter 4).

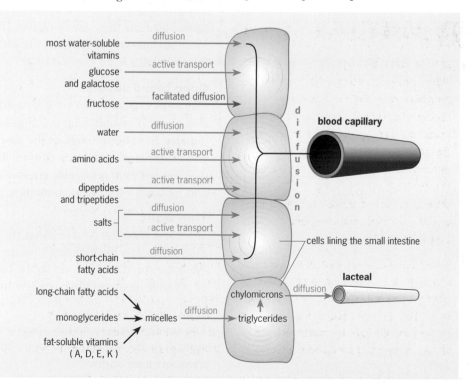

ABSORPTION OF CARBOHYDRATES

Glucose is actively transported into the cells of the small intestine and across the walls of blood capillaries. Active transport is necessary because diffusion alone would be too slow to supply the body's needs. Also, there would be diffusion out of the epithelial cells if the concentration of food in the intestines were very low. Energy for the transport is obtained from the hydrolysis of ATP.

ABSORPTION OF PROTEINS

Amino acids and small peptide molecules (di- and tripeptides) are actively transported into epithelial cells of the small intestine.

ABSORPTION OF FATS

Bile salts and lipase enzymes break up complex lipid molecules into monoglycerides and free fatty acids. The monoglycerides and some of the fatty acids combine with bile salts to form microscopic droplets called **micelles**.

⬤ EXAMPLE

Q Explain how the structure of the ileum relates to its function of absorbing the products of digestion.

A Relating structure to function is one of the basic skills a biologist must master. The structure of the small intestine is a classic example.

When faced with this question, many students will simply state a well-learned GCSE list of 'large surface area, good blood supply, thin walls, etc', and expect full marks. At A-level a little more depth is required. Why is a large surface area vital? What does a good blood supply do?

Large surface area
The basic idea behind a large surface area is that a lot of intestine is in contact with a lot of digested food molecules. These are absorbed by diffusion or active transport and so the more membrane there is, the more pores and carrier proteins there are available. The fact that the intestine is long is the most obvious way to increase the surface area, but there are also the villi and the microvilli.

Good blood supply
Like most organs which are adapted for the exchange of materials, there is also a way of maintaining the diffusion gradient so that there is always more digested food on one side of the absorbing membrane than the other, thus ensuring that absorption is continuous and as rapid as possible. As soon as the digested food molecules are absorbed into the blood, they are taken away and replaced with blood containing fewer of these molecules.

Thin walls
Fick's law (Chapter 4) states that the speed of diffusion is inversely proportional to the diffusing distance. Put simply, the thinner the membrane, the faster the diffusion, and so the barrier between food and blood is as thin as possible, being as little as a few micrometres – the thickness of the epithelial cell and the cell lining the capillary.

The small intestine also has another feature of note: the epithelial cells contain many mitochondria which make the ATP needed to power the active transport mechanisms.

Micelles are soluble in water and diffuse easily into epithelial cells, along with the remaining fatty acids. Inside the cells, the fatty acids and glycerol recombine, forming triglycerides, which acquire a protein coat (that stops them sticking together) and form particles called **chylomicrons**. It is in this state that lipids leave epithelial cells and enter, not a blood capillary this time, but a branch of the lymphatic system called the **lacteal** (see Fig 11.15).

TRANSPORT OF ABSORBED FOOD PRODUCTS

Amino acids, dipeptides and simple sugars diffuse out of epithelial cells in the small intestine, directly into blood capillaries within the villi. From here they pass to the liver via the **hepatic portal** vein. The liver processes absorbed food, converting some into storage products. Glycogen, copper, iron and vitamins A, D, E and K can all be stored. The liver breaks down other food products, including excess amino acids which are **deaminated** – their amine groups are removed. These breakdown products are then passed to the kidneys for excretion. See Chapter 14 for more about liver function and Chapter 16 for details on excretion.

The lymphatic system plays a major role in the transport of absorbed lipids. Chylomicrons that have entered the lacteals remain in suspension, giving the lymph fluid a milky-white appearance. From here, the chylomicrons move into larger lymphatic vessels that eventually drain into a large duct that empties into the blood. In the blood, these complexes are broken down into fatty acids and glycerol and enter cells to be used in the synthesis of complex lipids.

6 EGESTION

Egestion of undigested food from the body occurs after it has been through the large intestine. This is about 1.5 m long and extends from the ileum to the anus. It has four sections, the **caecum**, the **colon**, the **rectum** and the **anus**. The large intestine:
- absorbs water;
- is the site of manufacture of certain vitamins (microorganisms within the colon of some animals produce vitamin K and folic acid);
- forms and expels undigested food residue as faeces.

The caecum receives material from the ileum. In humans, the bottom of the caecum is attached to a small blind tube, the **appendix**. This tube is twisted and coiled and is about 8 cm long. The human appendix plays no part in digestion, but can become inflamed, causing appendicitis.

Much of the water that is poured on to food during its passage through the digestive tract (as saliva, gastric juice, pancreatic juice, for example) is reabsorbed. This reabsorption is vital: the digestive system pours up to 8 litres of fluid into the gut each day. If we lost all this fluid we could dehydrate. Most water absorption occurs in the small intestine, but of the litre or so that enters the large intestine, all but 100 cm^3 is reabsorbed by the colon. Minerals (as ions) can also diffuse or be actively transported into the bloodstream from this segment of the large intestine.

Diseases such as cholera and dysentery that cause severe diarrhoea can result in potentially lethal dehydration within a very short time because the toxins produced by the bacteria that cause these diseases reverse the reabsorption of water. To combat this, oral rehydration therapy can be given, a mixture of glucose and salts in water – see the Science Feature box on page 300.

The final sections of the large intestine, the rectum and anus, are concerned with the compaction of faeces and the act of **defecation**.

? TEST YOURSELF

F Why are the nutrients produced by colon bacteria (in the large intestine) of less use to us than nutrients provided by bacteria in the small intestine?

✔ REMEMBER THIS

Faeces contain some water, mineral salts, undigested food, bile pigments, products of bacterial decomposition and epithelial cells that have become detached from the digestive tract.

SUMMARY

- The human **alimentary canal** is a tube that leads from the mouth to the anus. It is associated with glands such as the **salivary glands**, **pancreas** and **liver**.
- The alimentary canal is also called the **gut** or **digestive system**. It is responsible for the **ingestion**, **digestion**, **absorption** and **egestion** of food.
- Food is broken down **mechanically** by the chewing action of teeth and by the muscular churning of the alimentary canal.
- Digestive juices are poured on to food as it travels down the alimentary canal. These soften and lubricate the food and also contain **digestive enzymes** that break the food down **chemically**.
- The muscles of the alimentary canal 'squeeze' the food – a process called **peristalsis** – and push it along from one end to the other.
- Large food molecules are broken down to their building blocks: carbohydrates are digested to form simple sugars, proteins are broken into amino acids, and fats are converted to fatty acids and glycerol.
- Digestion of food in the small intestine is due to the actions of bile and pancreatic juice, and the process is completed on the surface of the epithelial cells themselves.
- Digested food products are absorbed into the body across the wall of the alimentary canal. Most enter the bloodstream, but the products of fat breakdown enter the **lymphatic system**.
- Most digestion occurs in the **duodenum**; most absorption occurs in the **ileum**.
- A sophisticated control system ensures that digestive enzymes are released in the right place at the right time.
- Water is absorbed from undigested food that enters the colon, turning it into **faeces** that are egested from the body through the **anus**.

? EXAM QUESTIONS

1 The diagram shows a section through the stomach wall.

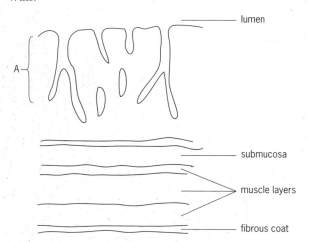

Name the part labelled **A**.

Give **two** ways in which the structure of the wall of the stomach differs from the wall of the oesophagus.

Give **one** way in which the stomach wall is adapted:

(i) to churn food;

(ii) to make food poisoning by bacteria less likely;

(iii) to prevent enzymes produced in the stomach wall digesting the surface of the stomach.

AQA Biology (B) BYB1 Core Principles June 2001 Q1

2 Read through the following passage, which refers to hormonal coordination of the functioning of the alimentary canal in humans, and then copy it out and insert the most appropriate words or words to complete the account.

The presence of partially digested food in the stomach causes the secretion of the hormone ___ which stimulates the secretion of gastric juice. On entering the duodenum, the acidic chime stimulates the secretion of ___ . This travels in the ___ and causes the release into the duodenum of an alkaline pancreatic secretion. The release of an enzyme-rich pancreatic secretion is caused by the hormone ___ which also causes contraction of the ___ .

Edexcel HB3 June 1999 Q4

3

a) Gastric juice is released by glands in the stomach lining and has an important role in digestion.

(i) Describe the processes involved in stimulating the release of gastric juice.

(ii) Describe the role of gastric juice during digestion.

b) Explain how the small intestine is adapted to its function in the absorption of the products of digestion.

AQA (NEAB) BY03 Physiology June 2000 Q9

12 Gas exchange

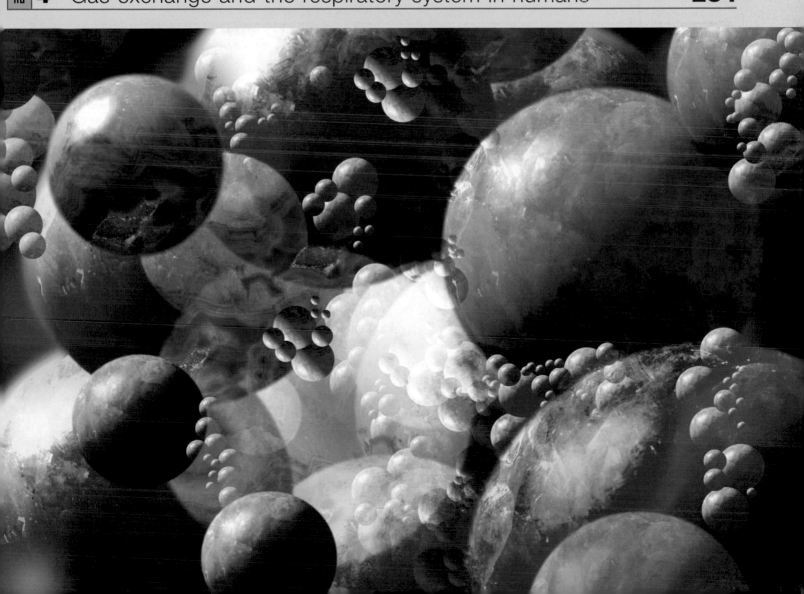

Gas exchange

When they return to the surface after a dive, whales clear their blow-hole before inhaling. This produces the familiar plume of spray

As a species we are not very good at diving without special equipment: most people cannot manage to hold their breath for more than a minute or so. But diving mammals such as dolphins, whales and seals can stay under water for much longer periods. Sperm whales, for instance, can dive for over an hour. How do they do it?

One reason is that the muscles of diving mammals contain more myoglobin (the oxygen-storage pigment that makes muscle red) and so can store more oxygen than those of non-diving mammals. Another reason is that diving mammals breathe out just before a dive. This makes them less buoyant and means there is less chance of gases in the lungs being forced into the blood at high pressure. Dissolved gases in the blood could form bubbles as the animal surfaced, causing the 'bends': a well-known problem for human divers.

During prolonged dives, diving mammals also move oxygenated blood away from organs that can survive without it for a short time (such as the skin) to those organs that need a constant supply (such as the brain and heart). The decreased flow to the skin is so effective that cuts sustained during a dive do not bleed until the animal surfaces.

1 WHY EXCHANGE GAS?

All living things need energy to carry out the processes of life. Most organisms obtain their energy from the oxidation of food in a process known as cell respiration (see Chapters 3 and 7). Cell respiration uses oxygen and produces carbon dioxide. To keep the process going, a living organism needs to obtain oxygen and expel waste carbon dioxide. This process is known as **gas exchange**.

Usually, small, simple organisms, such as those made up of only one cell, have a relatively small oxygen demand. All the oxygen they need diffuses into their body through its surface. (See Chapter 4 for an explanation of diffusion.) They expel carbon dioxide in the same way. But this strategy would not work in larger animals.

THE SURFACE AREA AND VOLUME PROBLEM

Although the amount of material an organism *needs* to exchange is proportional to its *volume*, the amount of material it *can* exchange is proportional to its *surface area*. Fig 12.1 shows that as organisms get larger, their volume increases at a faster rate than their surface area and so their need to exchange materials could quickly outstrip their ability to do so. (The Assignment in Chapter 4 should help you to understand this vital concept.)

Organisms with a diameter larger than a few millimetres can survive only if they evolve strategies to increase their surface area:volume ratio. Many species of worm have evolved a flat or cylindrical body. However, this sort of adaptation allows only a limited size increase. Much larger organisms have developed specialised organs to increase the surface area that is available for exchanging materials. Lungs, gills, intestines and kidneys are all examples of such organs.

The evolution of specialist exchange organs is linked to the development of a **transport system** (see Chapter 13). Without it, exchanged materials cannot reach other parts of the body quickly enough.

✔ REMEMBER THIS

A summary of the overall process of respiration:

glucose + oxygen

↓

carbon dioxide + water + energy
(in ATP)

✔ REMEMBER THIS

Photosynthetic organisms carry out the process of photosynthesis and respiration at the same time. If the rate of photosynthesis is greater than the rate of respiration, there is a net uptake of CO_2 and a net output of O_2.

2 RESPIRATORY ORGANS

Oxygen passes into the body of an organism by diffusion, and carbon dioxide leaves in the same way. Respiratory organs increase the rate at which gases can be exchanged. They maximise the efficiency of diffusion by having the following features in common:

- **A large surface area**. The greater the area of tissue in contact with the environment, the greater the rate of diffusion. Many aquatic animals increase the surface area for gas exchange by having projections, or **gills**. Most land animals increase surface area by having little pockets called **alveoli** in the lungs.
- **Thin, permeable walls**. These ensure that the diffusion distance is as short as possible.
- **A good blood supply**. Organisms that have a circulatory system supply their respiratory organs with plenty of blood. Blood quickly takes incoming oxygen away to a different part of the body. This ensures that a diffusion gradient is always present: there is always less oxygen in the blood flowing through the gas exchange organ than there is in the air or water.
- **Good ventilation**. Many animals increase the efficiency of gas exchange by ventilation. This is a physical pumping mechanism which continually brings fresh water or air to the gas exchange surfaces to keep the diffusion gradient as large as possible.

THE PROBLEMS OF GAS EXCHANGE IN AIR AND WATER

Animals obtain oxygen from the air or water that surrounds them. Air and water are completely different gas exchange mediums and so pose a different set of problems (Table 12.1).

Organisms that breathe air have an easier time than those that live in water. It is no coincidence that mammals and birds, the animals with the highest metabolic rate, are air breathers (Fig 12.2). They could not extract the amount of oxygen they need from water. However, a major drawback of air-breathing is water loss. The combination of a large surface area and moist membranes means that exhaled air is saturated with water vapour (Fig 12.3).

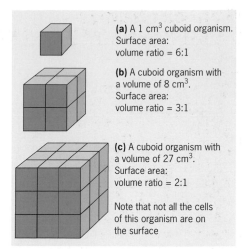

Fig 12.1 These simple shapes illustrate the problems faced by an organism as it gets larger. Think of each block as a cell and each complete cube as an organism. You can see that increasing the size of the organism lowers the surface area available to each cell. If the third 'organism' could not evolve a system to transport materials to the cell in the middle, it would not have much chance of survival

(a) A 1 cm³ cuboid organism. Surface area: volume ratio = 6:1

(b) A cuboid organism with a volume of 8 cm³. Surface area: volume ratio = 3:1

(c) A cuboid organism with a volume of 27 cm³. Surface area: volume ratio = 2:1

Note that not all the cells of this organism are on the surface

✔ REMEMBER THIS

The oxygen content of atmospheric air is remarkably constant, but the amount of oxygen dissolved in water varies considerably. Cold, turbulent water is highly oxygenated, but warm, still water contains much less oxygen. Polluted water often has very little oxygen (see Chapter 38).

Fig 12.2 There is usually about 30 times more oxygen in air than in water. Warm-blooded animals such as these harvest mice can meet their large oxygen demand only by breathing air. The oxygen consumption of this species has been measured at 2.5 cm² of O₂ per gram of body mass per hour

Fig 12.3 The air we exhale is saturated with water vapour. When you wake up on a cold day to see condensation on the windows, it is likely that much of that water came from your lungs. Loss of water in exhaled air represents a significant problem for land-using organisms – they must replace it continually to avoid dehydration

Table 12.1 Comparing air and water as breathing mediums. The differences between the two mediums affect the way in which organisms exchange gases. For example, water flows over gills as the animal moves, instead of being drawn in and out. It would take far too much energy for water dwellers to suck water in, stop it and then reverse the flow, as air-breathers do

	Water	Air	Water:air ratio	Practical consequence
Oxygen content (%)	0.7	20	1:30	far more oxygen is available from air than from water
Density (kg per litre)	1.0	0.0013	800:1	more energy is required to move water than to move air
Viscosity	1.0	0.02	50:1	air flows much faster than water
Diffusion constant for oxygen	0.000 034	11	1:300 000	oxygen can be absorbed from air far more quickly than it can from water
Diffusion constant for carbon dioxide	0.000 85	9.4	1:11 000	carbon dioxide is taken away more quickly by air than by water

LUNGS AND GILLS COMPARED

Gills usually project into the organism's external environment, see Fig 12.4(a), but, as Fig 12.4(c) shows, some organisms do have internal gills. Generally, gills are used for gas exchange only in water because the **gill filaments**, which need to be thin-walled, tend to collapse out of water. Even so, the coconut crab *Birgus latro*, in Fig 12.5, has rigid gills that do not collapse in air, and this crustacean drowns if kept under water.

Lungs are cavities that allow the environment to enter the organism – see Fig 12.4(b) and (d). In general, most animals with lungs live on land and breathe air, but, as ever, there are exceptions. Some molluscs breathe under water using lungs, but these tend to be simple pockets that supply small amounts of oxygen to meet the animal's low demand.

Air-breathing animals drown in water because, although water contains some oxygen, it does not contain enough. The effort required to move the amount of water in and out of lungs to gain sufficient oxygen would be impossibly large: they would use more oxygen than they would gain. Also, the greater density of water would burst the delicate air-sacs.

So, why do animals with gills die when removed from water? After all, there is a lot more oxygen in air. Unfortunately, the gill filaments tend to collapse out of water and the moist filaments stick together, rather like the bristles on a paint brush, drastically reducing their surface area.

Fig 12.4 Gas exchange organs

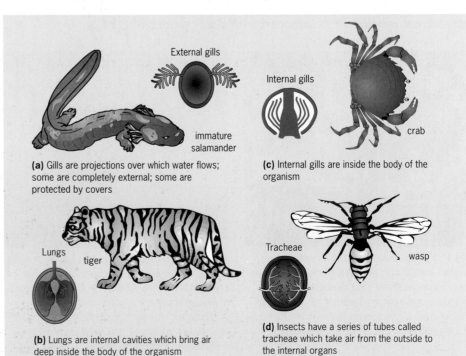

(a) Gills are projections over which water flows; some are completely external; some are protected by covers

(c) Internal gills are inside the body of the organism

(b) Lungs are internal cavities which bring air deep inside the body of the organism

(d) Insects have a series of tubes called tracheae which take air from the outside to the internal organs

Fig 12.5 The coconut crab has become adapted for life on land: it can breathe air with its gills. It gets its name from its habit of breaking open coconuts with its fearsome claws

3 GAS EXCHANGE STRATEGIES IN ANIMALS

Later in this section, we concentrate on the highly developed gas exchange organs of mammals, insects and fish. But let's look first at how some simpler organisms overcome the problem of gas exchange. Fig 12.6 shows the gas exchange strategies of several different invertebrates.

Fig 12.6 (a) This sea anemone is a cnidarian, a simple animal. It has no special respiratory organs but all of its living cells are within a millimetre or so of its surface and so each cell exchanges gases with its surroundings by diffusion

Fig 12.6 (b) This flatworm is a platyhelminth: it belongs to a phylum of animals that also includes the tapeworms. Its flat, ribbon-like shape provides a large surface area:volume ratio, so it is able to exchange sufficient gases with its environment by diffusion

Fig 12.6 (c) The long, cylindrical shape of nematode worms gives them a large surface area, further increased by the gut that passes through the middle. All active cells are very close to the external environment

Fig 12.6 (d) Larger and more complex organisms such as ragworms are also elongated, but this does not provide a large enough surface area to meet their gas exchange needs. Extensions known as parapodia further increase their surface area. These gill-like structures absorb oxygen, which the circulatory system then distributes around the body

Fig 12.6 (e) This sea slug is a mollusc. Its external gills increase its surface area

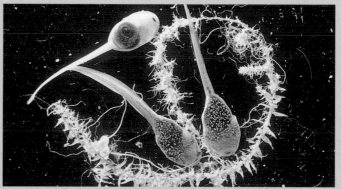

Fig 12.6 (f) Tadpoles, the larvae of amphibians, have gills. Adult frogs and toads exchange gases using a combination of simple lungs and skin diffusion

GAS EXCHANGE IN FISH

The gills of bony fish such as trout and tuna are particularly efficient structures for gas exchange in water. Fig 12.7 shows the basic structure of the gills of a typical bony fish. Each gill consists of several large **gill arches**, and each arch has two rows of gill filaments. The ends of the filaments interlock, forming a continuous mesh through which water must pass. This arrangement ensures that water passes within a fraction of a millimetre of every gill filament, allowing the maximum amount of oxygen to pass into the fish.

VENTILATION OF THE GILLS

Some types of fish (tuna, for example) have to swim forwards all the time to make sure that water flows continuously over their gills. However, most bony fish achieve a continuous flow of water over their gills even when stationary; they use a double pumping system involving the **buccal** (mouth) cavity and the opercular (gill) cavity (Fig 12.7d).

THE COUNTERCURRENT SYSTEM OF THE FISH GILL

A countercurrent system occurs when two substances flow through the same body part in opposite directions. Fig 12.7(c) shows the countercurrent system in the gill filaments of a fish. To understand the advantage of blood and water flowing in opposite directions, consider what would happen if they both flowed in the same direction. The two fluids would quickly reach equilibrium and the blood would extract much less of the oxygen available in the water. A countercurrent system ensures that there is always a diffusion gradient. Blood gathers oxygen as it flows along the lamellae, but it keeps on encountering fresh water that contains even more oxygen than the blood itself.

SEE QUESTIONS 1 AND 2

Fig 12.7 (a) A giant grouper fish having its gills 'serviced' by a cleaner wrasse. The gills are clearly visible

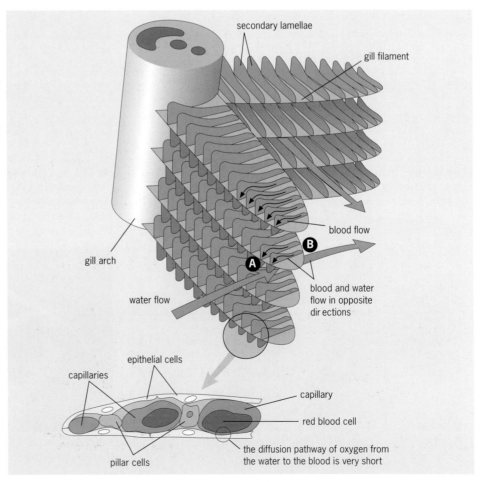

Fig 12.7 (b) The fish gill

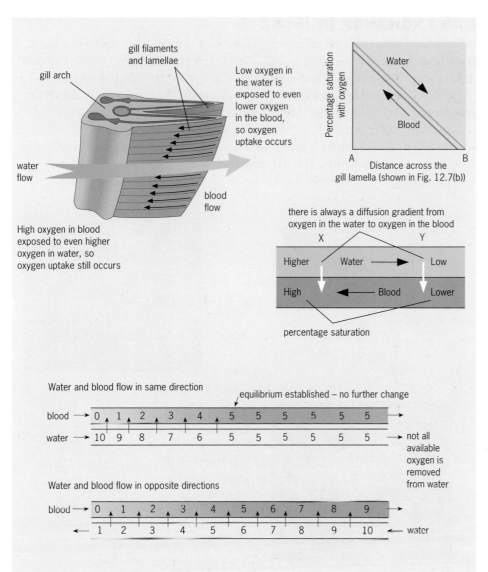

Low oxygen in the water is exposed to even lower oxygen in the blood, so oxygen uptake occurs

High oxygen in blood exposed to even higher oxygen in water, so oxygen uptake still occurs

there is always a diffusion gradient from oxygen in the water to oxygen in the blood

Water and blood flow in same direction

equilibrium established – no further change

not all available oxygen is removed from water

Water and blood flow in opposite directions

Fig 12.7 (c) The design of the gill allows water to flow over the gills in one direction. It also allows a countercurrent blood flow: the blood flowing in vessels through the gill filaments travels in an opposite direction to the water flowing over the gills. This maintains a diffusion gradient

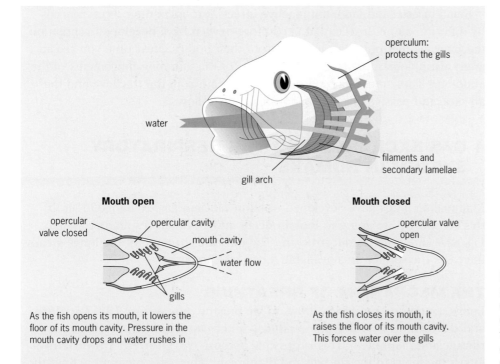

Mouth open

As the fish opens its mouth, it lowers the floor of its mouth cavity. Pressure in the mouth cavity drops and water rushes in

Mouth closed

As the fish closes its mouth, it raises the floor of its mouth cavity. This forces water over the gills

Fig 12.7 (d) Fish ventilation. In the double pumping system, water flows into the fish's open mouth. When the mouth closes, a series of valves ensures that water is forced out over the gills. Once water has passed the gills, the opercular pump expels the 'old' water and draws 'new' water in

Fig 12.8 The tracheal system in insects
(a) A generalised tracheal system of an insect. The spiracles open into the tracheoles, which form a network of tubes extending right through the body of the insect
(b) The spiracles are clearly visible on this caterpillar. Generally, spiracles open in response to a lack of oxygen or an increase in carbon dioxide. Experiments in which a tiny jet of carbon dioxide is aimed at one spiracle have shown that only that spiracle will open: each is controlled independently of all the others
(c) This micrograph of insect tracheae shows how they branch out into tiny individual tracheoles. The supportive rings of chitin are clearly visible. These spiral rings can squeeze together, as in a concertina, to expel air
(d) Individual tracheoles reach right in between cells, delivering oxygen to the mitochondria and removing carbon dioxide. Many species have air-sacs that can be inflated and deflated, like bellows, drawing air into the insect's body

SEE QUESTION 3

✓ **REMEMBER THIS**

Alveoli don't collapse when we breathe out because their surface is covered by an anti-sticking chemical called **surfactant** (see Science Feature box on page 238).

GAS EXCHANGE IN INSECTS

Insects tackle the problem of gas exchange in a unique way. An insect has a **tracheal system** consisting of tubes leading from the inside to the outside of the body. These branch and supply air directly to the internal organs.

Fig 12.8 shows the tracheal system of an insect. The individual tubes, the **tracheae**, lead to openings called **spiracles**, which can open or close to control the level of ventilation. Some of the smallest branches, the **tracheoles**, are less than 1 μm in diameter and supply oxygen to individual cells. The tiniest tracheoles, the **air capillaries**, pass within a few micrometres of every actively respiring cell.

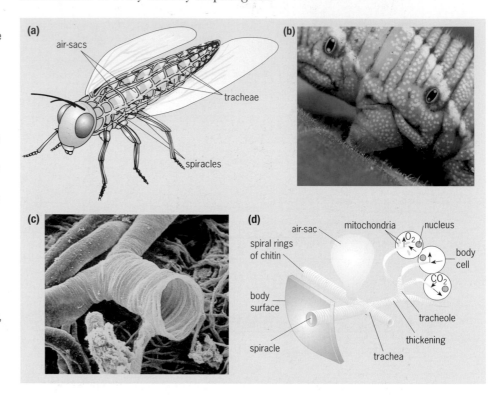

Small insects, and large but inactive insects, can meet their oxygen needs by diffusion alone. Insects that need more oxygen have developed ventilation mechanisms that pump air in and out of their bodies. Next time you see an angry wasp, look more closely and watch for the pumping movements of the abdomen. These muscular contractions expel air from the tracheae and the air-sacs, and passive inspiration automatically follows.

4 GAS EXCHANGE AND THE RESPIRATORY SYSTEM IN HUMANS

Mammalian lungs have developed to allow efficient gas exchange in air. In this section we are going to look in detail at the human respiratory system. Fig 12.9 shows its overall structure. The intricate structure of the lungs greatly increases the surface area available for gas exchange.

THE MECHANISM OF BREATHING

Lungs contain no muscle, so how do we breathe in and out? Lungs **inflate** and **deflate** because they are *expanded* and *compressed* by movements of the ribcage and diaphragm. This is possible because two **pleural membranes** attach the lungs to the ribcage and diaphragm. The outer membrane lines the

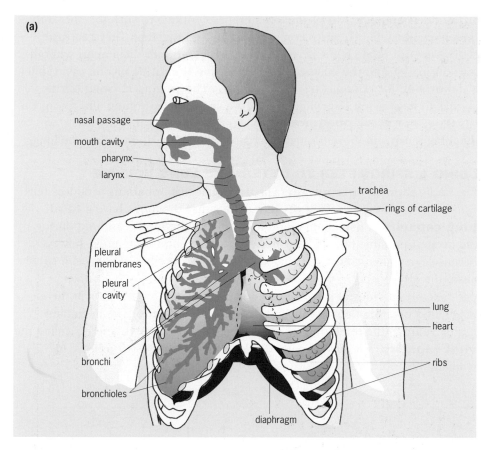

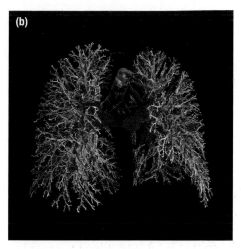

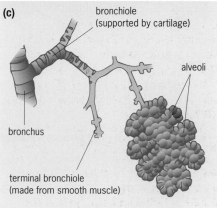

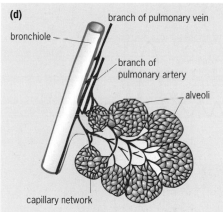

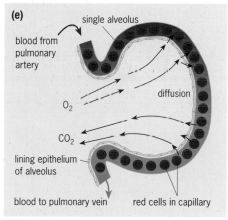

Fig 12.9 The structure of the human respiratory system

(a) The overall structure

(b) A resin cast of human lungs

(c), **(d)** and **(e)** The fine structure of the alveolus and the mechanism by which gas is exchanged

thoracic cavity; the inner membrane encloses the lungs. Between the two is a narrow space, the **pleural cavity**, filled with **pleural fluid**. This allows the pleural membranes to slide smoothly over each other during breathing and at the same time prevents them from separating. Fig 12.10 shows how we breathe in and out.

Fig 12.10 The mechanism of breathing:

(a) Inspiration (breathing in) happens when the external intercostal muscles contract and pull the ribcage upwards and outwards, away from the spinal column. At the same time, the diaphragm contracts and flattens, pushing down on the abdominal organs. These movements increase the volume and therefore lower the pressure in the thorax. As the pressure in the thorax falls below that of the atmosphere, air is forced into the lungs to equalise the pressure

(b) Expiration (breathing out) is normally a passive process: it uses no energy. When inspiration is over, the muscles in the thorax relax and breathing out follows due to a combination of gravity, the elastic recoil of the connective tissues of the lung and the pressure exerted by the abdominal organs such as the liver. Of course, we can consciously speed up expiration by forcing air out of our lungs using our internal intercostal muscles. This happens, for instance, when we blow up a balloon or play a wind instrument

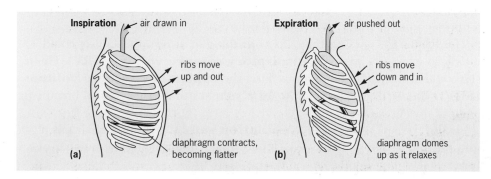

To understand why two pleural membranes continue to stick together during breathing, imagine two wet pieces of glass pressed together. They can easily slide over each other, but it is virtually impossible to pull them apart without introducing air into the middle. If air is introduced into the pleural cavity, after a stab wound for example, the lung collapses. The lung itself, being elastic, shrinks to its smallest size while air fills the space between lung and ribcage; a situation known as a **pneumothorax**. In this situation the ribs and diaphragm can still move (though it's painful) but they don't inflate the lungs.

USING A SPIROMETER TO DETERMINE LUNG VOLUME

Fig 12.11 shows a **spirometer**. The trace that this apparatus produces tells us a lot about the lungs. First of all, it shows that the lungs have a **total lung capacity**, a maximum amount of air that the lungs can hold during the deepest possible breath. We can never totally empty the lungs, however, because even when you have exhaled as much as possible, there is still some air in the alveoli and in the bronchi and tracheae (these are held open permanently by rings of cartilage). The volume of air that remains in the lungs after breathing out is called the **residual volume**. The maximum usable volume (the total lung capacity minus the residual volume) is called the **vital capacity**. The average vital capacity for men is 4.5 to 5 litres, for women, 3.5 to 4 litres.

Fig. 12.11 A spirometer and the trace it produces

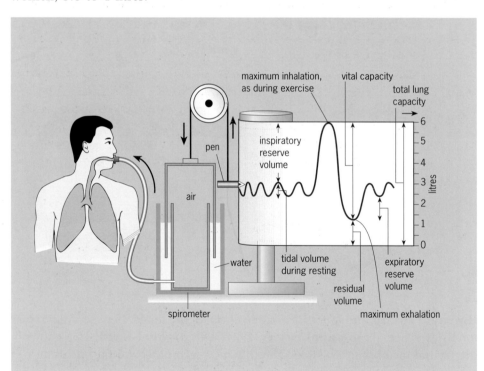

REMEMBER THIS

| Vital capacity | = | total lung capacity | − | residual volume |
| Ventilation rate | = | number of breaths taken per minute | × | tidal volume |

? TEST YOURSELF

B If the average breathing rate is 15 breaths per minute, and the tidal volume is 0.5 litre, calculate the ventilation rate.

Secondly, we can see that, during normal breathing, the volume of air that moves in and out of the lungs in each breath is called the **tidal volume**. In a normal adult at rest, this is about 0.5 litres.

Thirdly, the trace shows that, after breathing in at rest, the subject could inhale an extra 1.5 litres, the **inspiratory reserve volume (IRV)**. He or she could also breathe out another litre: the **expiratory reserve volume (ERV)**. These values represent the extra volumes of air that we can breathe in and out during exercise.

Finally, we can work out the **ventilation rate**: the volume of air taken into the lungs in one minute. We can do this by multiplying the number of breaths taken in a minute by the tidal volume.

USING A SPIROMETER TO DETERMINE OXYGEN CONSUMPTION

We can use a spirometer to estimate a person's rate of oxygen consumption. If the person rebreathes the air in the closed system of the spirometer, the composition of that air will change: oxygen levels decrease and carbon dioxide levels increase. If we include a cylinder of soda lime in the apparatus, this absorbs the carbon dioxide and so the volume of air in the spirometer decreases in volume as the oxygen is used up. Fig 12.12 shows a trace produced in this way.

To calculate the amount of oxygen used, we simply measure the volume decrease in a given time. In our example, the trace shows that in 1 minute the air volume fell from the 1600 cm^3 mark to the 1300 cm^3 mark. So, this person used up 300 cm^3 of oxygen in 1 minute.

WARNING! Re-breathing your own air can be dangerous. You should never do investigations like this without close supervision.

GAS EXCHANGE AT THE ALVEOLI

Gas exchange between air and blood occurs at the alveoli (Fig 12.13a). These tiny air-sacs create a huge surface area: 1 cm^3 of frog lung tissue has a surface area of about 20 cm^2, but the corresponding figure for a mouse lung is over 800 cm^2. Other mammals have a similar value. The total surface area of one human lung is about 100 m^2.

As we breathe in, fresh air enters the lungs and passes into individual alveoli (Fig 12.13b). Oxygen diffuses rapidly from the inhaled air through the walls of the alveoli and into the blood. Here, most of it combines with haemoglobin in the red blood cells. At the same time, carbon dioxide diffuses out of the blood and into the alveoli. It is breathed out during the next few expirations.

Table 12.2 shows the composition of atmospheric, alveolar and exhaled air. As you can see, each of the values for exhaled air is an average of the values for inhaled and alveolar air. This is because exhaled air is a mixture of the two.

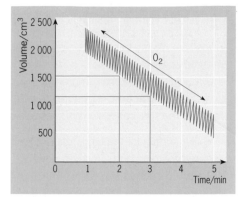

Fig. 12.12 The spirometer measures lung volume. The trace is produced on a revolving drum called a kymograph. We can use the trace to see the effects of exercise on breathing rate

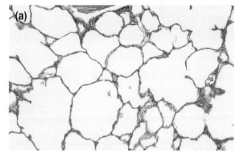

Fig 12.13 (a) In this micrograph of mammalian lung tissue, the alveoli and capillary networks that surround them can clearly be seen

Fig 12.13 (b) Gas exchange at the alveolar surface. In the human lung, the alveolar endothelium, the membrane that lines the air-sacs, is only 0.2 μm thick. This means that oxygen has to diffuse only a very short distance to enter the blood. Breathing maintains a fresh supply of air and the circulatory system takes away oxygenated blood so that constant gas exchange is possible

SEE QUESTION 4

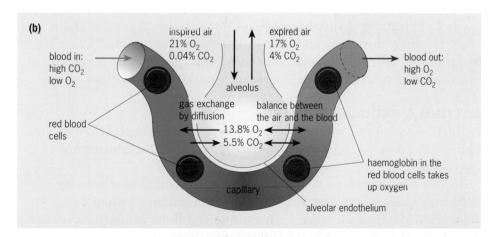

Table 12.2 The composition of inhaled, alveolar and exhaled air

| | Percentage of total volume | | | | |
	O$_2$	CO$_2$	N$_2$ + inert gases	H$_2$O vapour	Temperature (°C)
Atmospheric	21	0.04	79	variable	variable
Alveolar	13.8	5.5	80.7	saturated	37
Exhaled	17	4	79.6	saturated	37

⚠ SCIENCE FEATURE The lungs of premature babies

Alveoli are minute bubble-like air-sacs lined with moisture, and are liable to collapse because of surface tension: if their sides touch, they could stick together. To prevent this, the alveolar epithelium secretes a **surfactant**, a mixture of phospholipids, which greatly reduces the surface tension and keeps the alveoli open.

Without surfactant the lungs cannot function effectively and severe breathing problems can develop. An unborn baby does not start to secrete surfactant until about the 22nd week of pregnancy and the lungs have not accumulated enough surfactant to cope with breathing until about the 34th week.

Any babies born before this have immature lungs and suffer from a condition called **respiratory distress syndrome**. The effort needed to inhale and inflate the collapsed alveoli becomes too great and, without medical help, the baby can die from exhaustion and suffocation. Surfactant can now be made artificially. It is introduced into the lungs of premature babies to help them to continue breathing. This is a major breakthrough which means that babies as young as 23 weeks (17 weeks premature) now have more of a chance of survival.

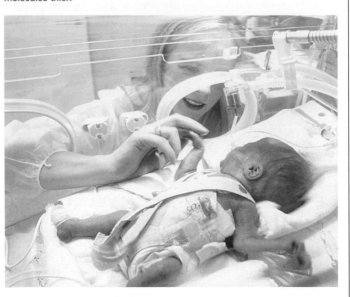

Fig 12.14 This baby was born prematurely, but survived owing to the treatment of her lungs with surfactant. Only a tiny amount (0.5 cm³) of surfactant is needed, enough to line all the alveoli with a layer one or two molecules thick

❓ TEST YOURSELF

D Why would it be impractical to breathe under water using a 6 metre snorkel?

The composition of exhaled air varies during the course of a single expiration. The first air to emerge has a very similar composition to atmospheric air because it has been nowhere near the alveoli – it has simply filled the **dead space** in the trachea and bronchi. As the exhalation continues, air that has been deep inside the alveoli is breathed out. The composition of this air has been altered by gas exchange and it contains more CO_2 and less O_2 than atmospheric air.

THE CONTROL OF BREATHING

Control of breathing is **involuntary**: we don't have to think continually about breathing in and out and our breathing rate is automatically matched to our needs. The rate changes as the brain detects the physical and chemical variations that occur in the body as we carry out different activities.

SETTING A REGULAR PATTERN

Breathing is controlled by a bundle of nerves called the **respiratory centre**. This is located in the brain in an area called the **medulla oblongata** (Fig 12.15).

Regular nerve impulses travel down efferent nerves that pass from the respiratory centre out to both the external intercostal muscles and the diaphragm. These muscles then contract, starting off inhalation. As air enters the lungs, stretch receptors in the airways start firing and feed information to the brain about how the inflation of the lungs is progressing. The more the lungs inflate, the faster the stretch receptors feed back impulses. When the lungs are inflated sufficiently, signals from the respiratory centre stop for a short time and exhalation follows automatically.

CHANGING THE BREATHING RATE TO MEET DEMAND

Ensuring the body has a constant supply of oxygen is obviously an important aspect of homeostasis (see Chapter 14) but, surprisingly, the body is relatively insensitive to falling oxygen levels. It is much more sensitive to an increase in carbon dioxide and so this is the indicator of the need for oxygen. The levels of oxygen in the arterial blood vary very little, even during exercise, but the carbon dioxide levels vary in direct proportion to the level of exertion. The heavier the exercise, the greater the carbon dioxide concentration. Lactic acid levels also increase during exercise. Any increase in carbon dioxide or lactic acid concentration in the blood lowers its pH. **Chemoreceptors** (see Fig 12.15), which are extremely sensitive to the composition of the blood that flows past them, can detect very small changes in pH.

There are three types of chemoreceptor:

● **Central receptors** in the medulla oblongata. These are sensitive to the carbon dioxide concentration in the blood that flows through this region of the brain.
● The **carotid bodies** in the wall of the **carotid arteries**.
● The **aortic bodies** on the **aortic arch**, just above the heart.

Together, the carotid and aortic bodies are described as the **peripheral chemoreceptors**. These cells sense changes in carbon dioxide and pH levels and, to a lesser extent, they are also sensitive to changes in oxygen levels.

When chemoreceptors register and change in carbon dioxide levels or pH, they send nerve impulses to the respiratory centre in the brain. This responds by sending more frequent impulses to the external intercostal muscles and diaphragm. When this happens, our ventilation rate increases: we breathe harder and faster. Heart rate also increases (see Chapter 13) and so the body automatically increases oxygen delivery at the same time as removing the extra carbon dioxide.

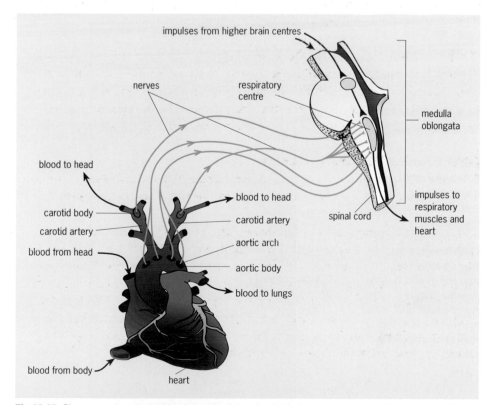

Fig 12.15 Chemoreceptors in the blood vessels above the heart are sensitive to the CO_2 content and pH of the blood. They are less sensitive to oxygen content, but will respond to very low levels

The control of breathing rate is very similar to the control of heart rate, with one important difference: we can control our breathing rate by thinking about it. This suggests that the higher, 'conscious' centres of the brain are more closely linked to the respiratory centre than they are to the cardiovascular centre. Also, research shows that pulse and ventilation rate change dramatically during exercise, even *before* the concentration of blood gases has a chance to change. It is as if the body predicts what is about to happen. How this works is not understood and is an active area of research.

THE EFFECTS OF OXYGEN DEPRIVATION

In some situations, at high altitudes for example, oxygen levels can fall without carbon dioxide levels increasing. In a rarefied or artificial atmosphere, normal breathing flushes carbon dioxide out of the blood via the lungs but there may not be enough oxygen to replace it. When this happens, the chemoreceptors often fail to register that anything is wrong, and the brain can become starved of oxygen.

! SCIENCE FEATURE Climbing Mount Everest – an explanation of partial pressures

As you study the workings of the lungs and blood in more detail, you will come across the term partial pressure. To explain this, it helps to think about someone climbing Mount Everest (height 8848 metres).

At sea level, there is a lot of air pushing down on us, and this atmospheric pressure has a value of about 100 000 pascals (Pa) or 100 kPa. We can therefore say that the barometric pressure is 100 kPa. Dry air is 20.9 per cent oxygen at sea level so the partial pressure of oxygen (P_{O_2}) is 20.9 per cent of 100, which is about 20.9 kPa.

As our mountaineer progresses up the mountain, the atmospheric pressure becomes less because there is less air pushing down on him. The partial pressure of oxygen decreases accordingly (Fig 12.16).

Oxygen passes into the lung tissues because of the different concentration in the alveolar air and the blood, and at higher altitudes the difference is smaller, making it difficult to take in enough oxygen to meet demands. At 5300 metres the P_{O_2} is only half of that at sea level, and by the time the summit is reached it is only about 8 kPa – just over one-third that at sea level. This is why most mountaineers who attempt Everest do so with pressurised oxygen containers, and to conquer the mountain without the help of additional oxygen – as has been done – is a remarkable feat.

Fig 12.16 The effect of altitude on the partial pressure of oxygen. Aircraft cabins must be pressurised and mountaineers usually need the help of pressurised oxygen canisters. The human body is able to acclimatise to lower partial pressures of oxygen by increasing the amount of haemoglobin in the blood

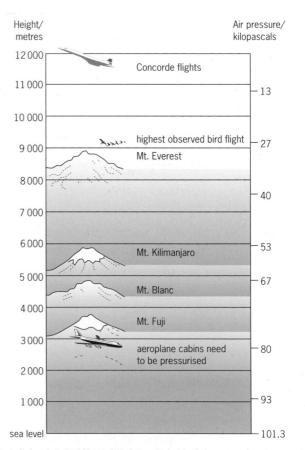

The first symptoms of oxygen starvation are feeling ridiculously happy, having impaired senses and lacking judgement. When mountaineers, fighter pilots or deep-sea divers start giggling and making stupid mistakes, it is a sure sign that they are not getting enough oxygen. It is also a signal for their colleagues to act fast, if they can, and provide them with emergency oxygen. If they don't get help quickly, they soon lapse into unconsciousness, and brain damage and death follow (Fig 12.17).

ATHLETES AND V_{O_2}(MAX)

Many physical activities – such as jogging, swimming, team sports – rely on energy released by the aerobic pathway of cell respiration (see Chapter 3). The level of performance an athlete can achieve is largely governed by how fast oxygen gets to the muscles.

The rate at which a person uses oxygen is called the $\textbf{\textit{V}}_{\textbf{O}_2}$ and is measured in terms of the volume of oxygen consumed (cm^3), per kg of body mass, per minute. The $\textbf{\textit{V}}_{\textbf{O}_2}$**(max)** is the maximum rate at which oxygen is consumed and is the amount of oxygen that can be delivered to the tissues when the lungs and heart are working as hard as possible. Athletes use a knowledge of V_{O_2}(max) in their training, as a measure of how hard they are working. A training schedule, for example, may require the athlete to work at 55 per cent of their V_{O_2}(max) for a set length of time.

The Example shows how to calculate V_{O_2} and V_{O_2}(max) for an average person. (You can work out your own values if you are able to measure your tidal volume, the number of breaths you take per minute and your weight in kilograms.) You can see that the amount of oxygen consumed during exercise increases from 4.28 to 51.42 cm^3 O$_2$ kg^{-1} min^{-1}, an increase of over 1000 per cent.

Fig 12.17 In 1875, three French physiologists decided to investigate the effects of low oxygen levels on the human body. The easiest way to get oxygen-poor air was to go up in a hot-air balloon to observe the effects of the rarefied atmosphere on each other. At first there were no obvious effects and they happily continued to throw out ballast, going up to 8000 metres. At this point they all fainted. The balloon eventually came down on its own, and one of the scientists woke up to find the other two dead

---

● EXAMPLE

Q What is the rate of oxygen consumption of a normal 70 kg adult at rest? Assume that the tidal volume of a normal adult at rest is 0.5 litres and that he or she takes 15 breaths per minute.

A We can work out the ventilation rate from the formula:

Ventilation rate $= \dfrac{\text{number of breaths}}{\text{taken per minute}} \times$ tidal volume

Ventilation rate $= 15 \times 0.5$ litres min^{-1} = 7.5 litres min^{-1}

Atmospheric air is one-fifth oxygen and from Table 12.2 we know that only one-fifth of the available oxygen is absorbed. So:

amount of oxygen used per minute

$= 7.5 \times 0.2 \times 0.2$ litres min^{-1}

$= 0.3$ litres min^{-1}

And the V_{O_2} for this person is:

$\dfrac{0.3}{70} = 0.004\,28$ litres O$_2$ kg^{-1} min^{-1},

or: 4.28 cm^3 O$_2$ kg^{-1} min^{-1}

Q What is the V_{O_2}(max) for the same adult? Assume that the volume of air taken in during each breath during strenuous exercise is 3 litres and that the number of breaths per minute increases to 30.

A Knowing that the tidal volume of the same adult during exercise is 3 litres and that he or she takes 30 breaths per minute, we can work out the ventilation rate:

Ventilation rate $= \dfrac{\text{number of breaths}}{\text{taken per minute}} \times$ tidal volume

Ventilation rate $= 30 \times 3$ litres min^{-1} = 90 litres min^{-1}

Atmospheric air is one-fifth oxygen and from Table 12.2 we know that only one-fifth of the available oxygen is absorbed. So:

amount of oxygen used per minute

$= 90 \times 0.2 \times 0.2$ litres min^{-1}

$= 3.6$ litres min^{-1}

And the V_{O_2}(max) for this person is:

$\dfrac{3.6}{70} = 0.051\,42$ litres O$_2$ kg^{-1} min^{-1},

or: 51.42 cm^3 O$_2$ kg^{-1} min^{-1}

SUMMARY

After reading this chapter, you should know and understand the following:

■ Animals must **exchange gases** to provide oxygen for cell respiration, and to take away the waste carbon dioxide that this process produces.

■ Small animals have a high **surface area:volume ratio** and gas exchange over the body surface is enough to satisfy their needs. Adopting a flat or cylindrical body is a simple strategy that increases surface area.

■ Larger organisms have **respiratory organs** that increase the surface area for gas exchange. These organs have a large surface area and thin, moist membranes. Ventilation and a good blood supply help to maintain a large diffusion gradient.

■ **Gills** are organs that are adapted to remove dissolved oxygen from water. Water flows in one direction over the gills, and the blood in the gill filaments flows in the opposite direction.

■ Insects have a **tracheal system** in which air-tubes branch out from **spiracles** on the surface of the insect and lead to individual cells.

■ **Lungs** are organs that have evolved to allow very efficient gas exchange in air. In mammals, breathing movements ventilate air-sacs called **alveoli**, bringing in a continuous supply of fresh air and flushing out waste carbon dioxide.

■ The rate of breathing is controlled by the **respiratory centre** in the **medulla oblongata** of the brain. This sets a regular breathing pattern, which is modified according to information received from **chemoreceptors**, which detect changes in the blood that indicate the body's need for oxygen.

EXAM QUESTIONS

1

a) The diagram shows part of a gill filament of a fish.

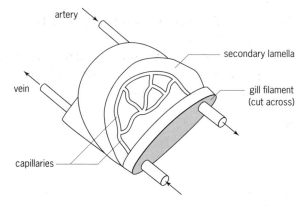

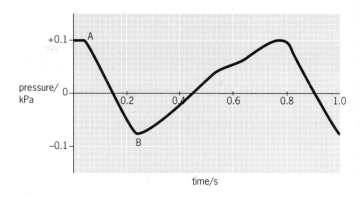

AQA Biology (B) BYB1 Core Principles June 2001 Q3

(i) Draw an arrow on the diagram to show the direction that water flows over the secondary lamella.
(ii) Explain the advantage of water flowing in this direction.

b) The graph shows the change in pressure in the mouth (buccal) cavity of a fish during ventilation of the gills.
(i) Calculate the rate of ventilation per minute. Show your working.
(ii) Explain what causes the fall in pressure between points A and B.

2 The graph shows the variation in the oxygen concentration of blood and water as they pass across a gill lamella of a fish.

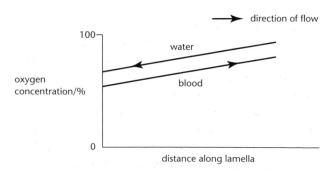

EXAM QUESTIONS

a) Use the graph to explain how the directions of flow of blood and of water increase the efficiency of the gill lamella as an exchange surface.

b) The table shows the thickness of the three layers between the water and the blood in the gill lamellae of three species of fish.

	Epithelial cells/ μm	Basement membrane/μm	Endothelial lining of capillary/μm
Dogfish	10.38	0.65	0.57
Skipjack	0.325	1.275	0.205
Mackerel	0.743	0.533	0.392

The mackerel is more active than the other two species. Explain how the figures in the table support this observation.

AQA (NEAB) BY03 Physiology June 2000 Q5

3 Gas exchange in insects involves pores in the cuticle which open into a network of tubes. These tubes have fine branches extending into all the tissues of the body.

a) Name the following: The pores in the cuticle; the network of tubes.

b) In larger insects, such as locusts, the passage of air through the tubes is helped by pumping movements of the abdomen. A student carried out an experiment to investigate the effect of carbon dioxide on the rate of these pumping movements. She set up the apparatus as shown in the diagram below.

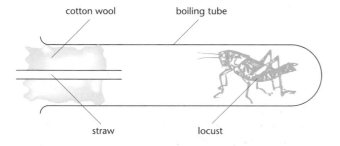

The locust was left in the boiling tube for five minutes. The number of abdominal pumping movements during one minute was counted. The student then breathed out once through the straw into the boiling tube and immediately counted the number of abdominal movements during one minute. She repeated this procedure varying the number of times she breathed out into the boiling tube. The results are shown in the table below.

Number of times student breathed into boiling tube	Number of abdominal pumping movements in one minute
0	16
1	59
2	61
3	58
4	60

(i) State why the student left the locust in the boiling tube for five minutes before she began the first count.

(ii) Describe and comment on the effect of breathing out into the boiling tube on the rate of abdominal pumping in the locust.

c) (i) During this experiment, the humidity of the air in the boiling tube may vary. Suggest how this experiment could be modified to control the humidity.

(ii) Suggest *two* factors, other than a change in carbon dioxide concentration and humidity, which may have affected the rate of abdominal pumping.

Edexcel B3 June 1999 Q6

4

a) In the lungs oxygen passes from the alveoli to the blood.
Describe and explain the features that make this process rapid and efficient.

b) The graph shows the pressure changes in the lungs during the period of one breath (inspiration + expiration) in a person at rest.

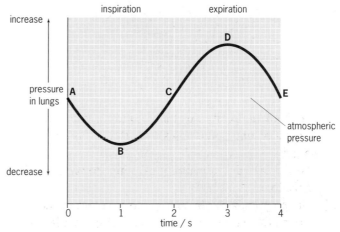

Use your knowledge of breathing to explain the changes in pressure during inspiration and expiration. The letters on the graph are to help you to refer to different parts of the curve.

AQA Biology (B) BYB1 Core Principles January 2001 Q7

KEY SKILLS **ASSIGNMENT**

COPING WITH ASTHMA

Asthma

Asthma is one of the commonest childhood ailments, affecting at least one in ten children, and many adults. The number of asthma cases has risen dramatically in recent years, and air pollution is thought to be one of the main causes.

To understand fully why asthma happens, you need to be familiar with the fine structure of the lungs – see page 235.

1 a) What is the difference in structure between the bronchi and bronchioles?
b) What type of muscle lines the bronchi and bronchioles?

What is asthma and what causes it?

Put simply, asthma is a difficulty in breathing caused by constriction of the smooth muscles of the bronchioles. Asthma sufferers find it hard to breathe because the airways leading to the alveoli become narrower, so requiring far more effort to deliver a normal amount of air to the lungs.

There are many factors that can cause asthma: an allergic reaction, exercise (especially in cold air) and lung infections. An allergy is an 'over-reaction' of the immune system to a substance that, in most other people, has no effect. The commonest allergens are house dust, fur, feathers and pollen.

2 Suggest what measures could be taken to reduce the problems caused by the house dust mite shown in Fig 12.A1.

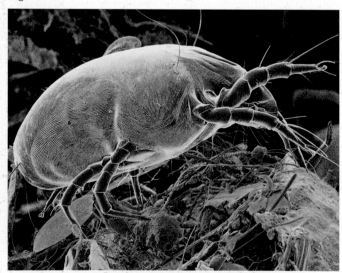

Fig 12.A1 House dust mites, *Dermatophagoides farinae*, are normal inhabitants of our mattresses. They feed on the large deposits of skin that accumulate there, and the faeces they produce can cause an allergic reaction which leads to asthma in some people

3 What advice would you give to a patient suffering from asthma caused by pollen?

Treating asthma

Several approaches are used together to help control asthma. A doctor's first priority with all but the youngest of children is to explain what is happening inside the lungs. This reduces fear and helps sufferers to come to terms with their problem. The worst physical effects of the condition are then controlled by a combination of drugs and prevention.

Generally, two types of drugs are prescribed to asthma sufferers: **bronchodilators** and **steroids**. Bronchodilators give instant relief from chest tightness and wheezing because they contain chemicals that relax the bronchiole walls. Many asthma sufferers carry an inhaler around with them. A common bronchodilator is the drug salbutamol, which is modified adrenaline. The hormone adrenaline relieves asthma, and is very useful in the treatment of severe attacks, but it is unsuitable for regular use.

4 a) What are the effects of the hormone adrenaline on the body? (You may need to look at page 410.)
b) Suggest what happens when an asthma sufferer takes an adrenaline-based drug.
c) Suggest why adrenaline is not suitable for use in long-term asthma treatment.

Steroids (see page 403) such as **Becotide**™ act by reducing the degree of inflammation of the bronchioles. Steroids are preventative: asthma sufferers take regular doses morning and night to reduce the problem of over-reaction. The amount of steroids taken is minimal but they are effective if taken according to instructions.

5 Suggest why it is an advantage to inhale steroids rather than inject them.

6 The girl in Fig 12.A2 is using an inhaler. To get the full benefit from each dose – and not just coat the inside of her mouth – she must learn the correct technique. Find out what this is.

7 How would you investigate the hypothesis that air pollution is linked with the measured increase in asthma cases?

Fig 12.A2 Using an inhaler is easy once you have mastered the right technique

13

Animal circulatory systems

During open-heart surgery, the job of the heart and lungs is performed by a heart/lung machine. This sophisticated device also has the ability to lower the core temperature of the body by about 10 °C. This reduces the body's demand for oxygen, giving the surgeons approximately twice as long to perform the operation

When the blood vessels supplying the heart muscle become narrowed by fatty deposits, the heart can become starved of oxygen. The patient will be suffering from heart pains on exercise – called angina – and is very susceptible to heart attacks. Often, the only effective remedy is surgery.

One of the available operations – sometimes known as a 'cabbage' (CABG = coronary artery bypass graft) – takes a team of two or three surgeons about five hours to complete. The overall aim is to take pieces of vein from the patient's leg and graft them from the aorta, over the blocked artery and into the heart muscle itself. But how do you sew small blood vessels onto an organ that won't keep still?

The answer is that you don't. The heart has to be stopped and this is only possible due to a heart/lung machine – a device that can take blood from the patient, oxygenate it and return it to the body at the right temperature and pressure. Once this machine has been connected the heart can be stopped, and this is done simply by pouring iced water over it. Once the heart is still the surgeon can sew on the new pieces of blood vessel. When complete, the heart is re-started by means of two electrodes placed on the heart itself.

This standard operation can greatly improve the quality of life for patients, as well as prolonging their life expectancy.

1 WHY DO ANIMALS NEED A CIRCULATORY SYSTEM?

Evolution has produced large animals, with specialised organs such as intestines, gills and lungs to increase the surface area available for exchange of materials such as food and oxygen (see Chapters 4, 11 and 12). These organs would be relatively useless without a **circulatory system**, a network of tubes filled with fluid that can deliver vital materials to all the cells of the body, and then take away their waste. In fact, in most animals, almost every cell is within a few micrometres of a branch of the circulatory system.

A circulatory system has three components:
- a *fluid* that flows in the system, carrying materials around the body;
- a *system of tubes* that carries the fluid;
- a *pump* (or pumps) that keeps the fluid moving through the tubes.

In vertebrate circulatory systems, **blood** (the fluid) is driven through **arteries**, **veins** and **capillaries** (the tubes) by the **heart** (the pump). In this chapter, we look at these three components in detail.

SINGLE AND DOUBLE CIRCULATIONS

Fish have a **single circulation**: blood passes from the heart to the gills and then slowly round the body and back to the heart in a single circuit, as shown in Fig 13.2(a). This system is adequate for fish, but animals with a higher metabolic rate need to deliver oxygen to their tissues faster than a single circulation allows.

Mammals and birds have a **double circulation**: a **pulmonary circulation** from the heart to the lungs and a **systemic circulation** from the heart to the rest of the body, as shown in Fig 13.2(b) and Fig 13.3. A double circulation must have a four-chambered heart. The right side of the heart receives deoxygenated blood from the body and pumps it to the lungs. Here, blood gains oxygen, but loses pressure. The left side of the heart receives the oxygenated blood and gives it a boost so that it can reach all the body parts quickly.

Fig 13.1 The heart of a blue whale weighs about the same as a small family car. A massive muscle, this heart pumps blood to organs over 15 metres away, down an artery large enough for a child to crawl through

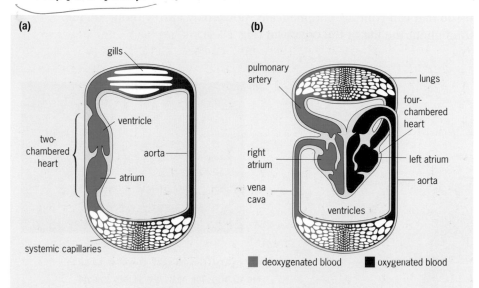

Fig 13.2(a) Fish have a single circulation and a two-chambered heart. Blood first travels to the gills, where it passes through thin capillaries, picking up oxygen but losing pressure. It continues to travel around the body of the fish, back to the heart, but more slowly because of the low pressure

Fig 13.2(b) Mammals have a double circulation and a four-chambered heart. Deoxygenated blood, coloured blue, passes into the right side of the heart and is pumped to the lungs where it picks up oxygen and releases carbon dioxide. Oxygenated blood, coloured red, returns to the heart to be pumped to all parts of the body except the lungs

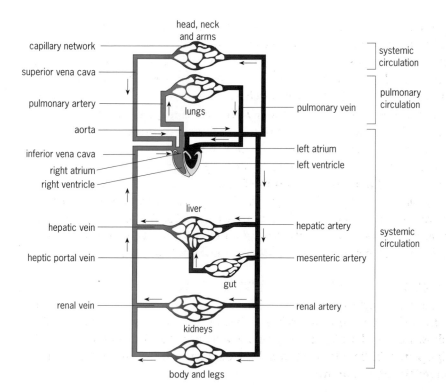

Fig 13.3 An overview of mammalian circulation: this is the system for a human. Deoxygenated blood (coloured blue) passes into the right side of the heart through the superior and inferior vena cava. The right atrium pumps the blood into the pulmonary circulation and it picks up oxygen and releases carbon dioxide as it passes through the lungs. Now oxygenated (coloured red), the blood returns to the heart. The left ventricle then pumps it around the systemic circulation and it travels to all parts of the body (except the lungs) before returning to the heart

2 THE HUMAN CIRCULATORY SYSTEM

The human heart beats over 100 000 times a day, creating the pressure to force blood through more than 80 000 kilometres (50 000 miles) of arteries, veins and capillaries. Fig 13.3 shows the extent of the human circulatory system.

STRUCTURE OF THE HEART

The mammalian heart is composed mainly of **cardiac muscle**: a specialised tissue that can contract automatically, powerfully and without fatigue, throughout the life of the organism (Fig 13.4).

(a)

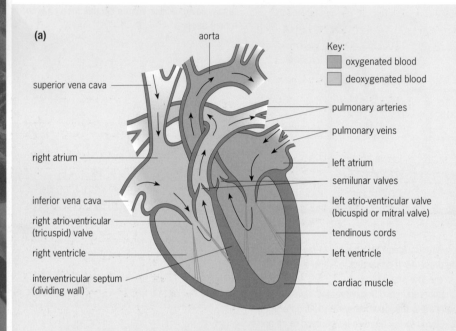

Key:
- ☐ oxygenated blood
- ☐ deoxygenated blood

aorta
superior vena cava
pulmonary arteries
pulmonary veins
right atrium
left atrium
semilunar valves
inferior vena cava
left atrio-ventricular valve (bicuspid or mitral valve)
right atrio-ventricular (tricuspid) valve
tendinous cords
right ventricle
left ventricle
interventricular septum (dividing wall)
cardiac muscle

(b)

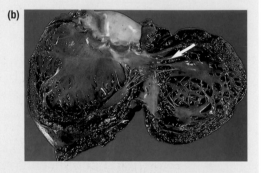

Fig 13.4(a) Longitudinal section of the heart
Fig 13.4(b) The heart is a slightly twisted, asymmetrical organ: the best way to appreciate its three-dimensional structure is to dissect a pig or sheep heart or to study a model. This is a section through the left ventricle of a human heart. The string-like cords link heart muscle to valves

(c)

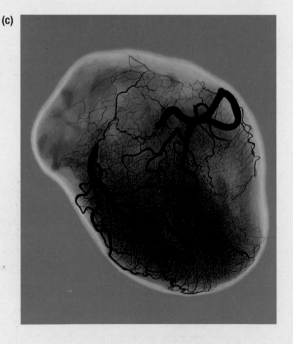

(d)

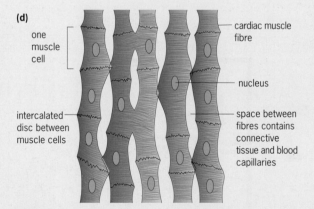

one muscle cell
cardiac muscle fibre
nucleus
intercalated disc between muscle cells
space between fibres contains connective tissue and blood capillaries

Fig 13.4(c) The coronary circulation comprises all the blood vessels that supply the heart muscle. In this photograph, these blood vessels are easy to see because they have been filled with a dye that is opaque to X-rays
Fig 13.4(d) The fine structure of cardiac muscle. The rapid spread of impulses through cardiac muscle is possible because individual cells are connected by specialised junctions called intercalated discs. This system of communication ensures that all cells 'beat' at the same time

The thickness of the walls in the different heart chambers reflects their function. The atria are thinly muscled because they pump blood only the short distance to the ventricles directly below them. The right ventricle is more heavily muscled than either of the atria because it has to force blood a much further distance to the lungs. The left ventricle has the thickest wall because it has to force blood all the way around the body.

It is important that blood flows through the heart in one direction only. Two sets of valves close to prevent backflow:

- The **atrio-ventricular valves (AV valves)**. These lie between the atria and the ventricles to prevent blood from returning to the atria when the ventricles contract. The **tricuspid valve** on the right side has three cusps, or flaps. The **bicuspid valve**, or **mitral valve**, on the left has two cusps. Both AV valves are subjected to great pressure and ultra-tough tendinous cords are needed to prevent them turning inside out – see Fig 13.4(b).
- The **semilunar valves** guard the openings to the pulmonary artery and the aorta to prevent backflow of blood into the ventricles. Both valves have three semilunar ('half moon'-shaped) cusps.

HOW THE HEART BEATS

The sequence of events in a single heartbeat is known as the cardiac cycle (Fig 13.5). The cycle involves systole, or contraction, and also diastole, or relaxation, of the atria and ventricles. The cycle has four overlapping stages:

- **Atrial systole**. Both atria contract, forcing blood into the ventricles. This stage lasts 0.1 seconds.
- **Ventricular systole**. Both ventricles contract, forcing blood through the pulmonary artery to the lungs and through the aorta to the rest of the body. This takes 0.3 seconds.
- **Atrial diastole**. The atria relax, although the ventricles are still contracted. Blood enters the atria from the large veins coming from the body. This takes about 0.7 seconds.
- **Ventricular diastole**. The ventricles relax, and become ready to fill with blood from the atria as the next cycle begins. This takes about 0.5 seconds.

Given an average heart rate of 75 beats per minute, each cycle takes 0.8 seconds.

REMEMBER THIS

Valves are simply strong flaps of tissue: they cannot move on their own. A common mistake in examinations is to claim that valves can actively control blood flow. For example, the statement: 'the AV valves close, preventing blood flow', should read 'blood begins to flow back, forcing the valve shut and preventing any further flow'.

REMEMBER THIS

The heart continues to beat when removed from the body. Individual heart muscle cells grown in culture beat on their own! Because of this, we say that the vertebrate heart is **myogenic**: the stimulus that drives it to beat originates in the muscle itself. Some invertebrate hearts are **neurogenic**: they beat only when connected to external nerves.

Fig 13.5 The timescale of the cardiac cycle in a human at rest. The heart rate is 75 beats per minute. During exercise the heart rate can increase to over 150 beats per minute, so the cycle only takes half the time

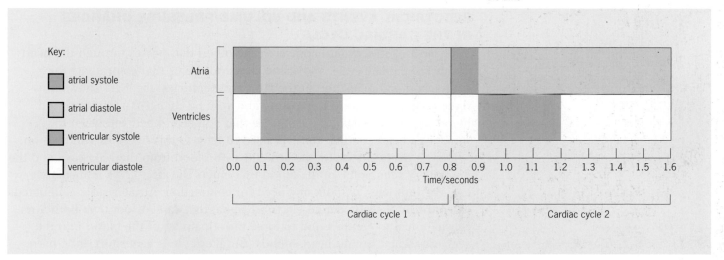

Key:

- atrial systole
- atrial diastole
- ventricular systole
- ventricular diastole

Atria

Ventricles

0.0 0.1 0.2 0.3 0.4 0.5 0.6 0.7 0.8 0.9 1.0 1.1 1.2 1.3 1.4 1.5 1.6
Time/seconds

Cardiac cycle 1   Cardiac cycle 2

CONTROL OF THE CARDIAC CYCLE

Heartbeat must be carefully controlled so that each chamber contracts only when full of blood. To achieve this, the events of the cardiac cycle are carefully co-ordinated (Fig 13.6).

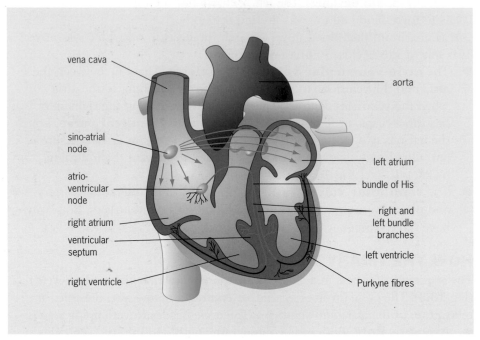

Fig 13.6 The conduction system of the human heart initiates and controls a heartbeat. Cells in the sino-atrial node act as a pacemaker, initiating impulses that spread through the walls of the atria and, after a delay, through the walls of the ventricles via the atrio-ventricular node

A single heartbeat starts with an electrical signal from a region of specialised tissue called the **sino-atrial node (SAN)**, on the wall of the right atrium. This is the 'pacemaker', or heartbeat regulator. The electrical signal spreads out over the walls of the atria, causing them to contract.

From there, the signal does not pass directly to the ventricles. If it did, the ventricles would begin to contract before they had filled with blood. Instead, the impulse is delayed slightly. A second node, the **atrio-ventricular node (AVN)** picks up the signal and channels it down the middle of the **ventricular septum** through a collection of specialised cardiac muscle fibres called the **bundle of His**. From here the signal spreads throughout the wall of the ventricles, through the **Purkyne** (or **Purkinje**) **fibres**, and the ventricles contract *after* they have filled with blood.

ELECTRICAL EVENTS AND VOLUME/PRESSURE CHANGES IN THE CARDIAC CYCLE

We have looked at what happens to the blood that passes through the heart during the cardiac cycle, and at the electrical events that co-ordinate the contraction and relaxation of the atria and ventricles. Fig 13.7 shows how these events correspond to pressure and volume changes in the heart and blood vessels, and also to the sounds that you hear when you listen to someone's heart beating. You will need to refer to this figure as you read on.

During **atrial systole**, the atria fill with blood from the vena cava and the pulmonary vein. Some of the blood that enters the atria flows immediately into the ventricles, without any need for contraction. Atrial systole is initiated when the SAN sends out an electrical signal that spreads out over both atria, causing them to contract and to force the remainder of the blood into the ventricles. There are no heart sounds. Atrial systole is very short, little more than a 'twitch'. Atrial relaxation, or **atrial diastole**, lasts for the remainder of the cycle. All the rest of the 'action' is in the ventricles.

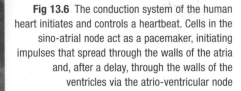

SEE QUESTIONS 1 AND 2

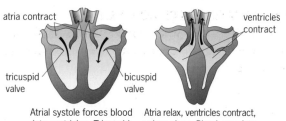

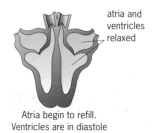

atria contract

tricuspid valve

bicuspid valve

Atrial systole forces blood into ventricles. Tricuspid and bicuspid valves open

ventricles contract

Atria relax, ventricles contract, valves close. Blood goes into aorta and pulmonary artery

atria and ventricles relaxed

Atria begin to refill. Ventricles are in diastole

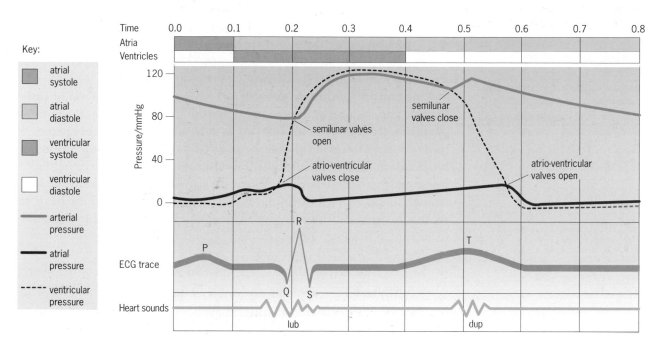

Key:

- atrial systole
- atrial diastole
- ventricular systole
- ventricular diastole
- arterial pressure
- atrial pressure
- ventricular pressure

semilunar valves open

atrio-ventricular valves close

semilunar valves close

atrio-ventricular valves open

ECG trace

P R Q S T

Heart sounds

lub dup

Ventricular systole begins when the ventricles have filled with blood. The AVN picks up the signal from the SAN and then conducts impulses down through the bundle of His and on through the Purkyne fibres in the walls of the ventricles. This stimulates the ventricles to contract. Blood is forced upwards, forcing the semilunar valves open and the AV valves shut. The closing of the AV valves causes the 'lub' of the 'lub-dup' heart sound. Pressure in the arteries rises sharply as blood is forced into them.

During **ventricular diastole**, the ventricle walls relax, arterial pressure falls and blood begins to flow back into the ventricles. This reversal of the flow causes the semilunar valves to shut, causing the second, 'dup', heart sound. Meanwhile, the atria have been filling with blood and, as the ventricles relax, blood flows from the atria into the ventricles, forcing the AV valves open again.

STROKE VOLUME AND CARDIAC OUTPUT

The volume of blood pumped by the heart during one cardiac cycle is the **stroke volume**. A typical stroke volume in an adult is about 80 cm^3: every time the heart beats, 80 cm^3 of blood is forced through the pulmonary artery to the lungs and 80 cm^3 is forced into the aorta to the body. So, the heart pumps over 8500 litres of blood per day. Stroke volume increases during exercise, and regular exercise brings about a permanent resting increase to 110 cm^3, or more.

The volume of blood pumped in one minute is called the **cardiac output**. It is calculated by multiplying the stroke volume by the heart rate and is expressed in litres of blood per minute.

Fig 13.7 The cardiac cycle. See Fig 13.12 for an explanation of the ECG trace. Many exam questions use this diagram, without labels, to test understanding of the events of one heartbeat. It is important to remember what is cause and what is effect, ie:
- Pacemaker cells initiate systole (contraction)
- The squeezing of the muscle walls reduces the volume and so increases the pressure in the chamber(s), forcing blood in a particular direction
- The direction of blood flow causes the valves to open or close. The valves ensure that blood flow through the heart is in one direction only

✓ REMEMBER THIS

Cardiac output =
stroke volume × heart rate
This is measured in litres per minute.
There are 1000 cm^3 in a litre.

? TEST YOURSELF

A Calculate the cardiac output:
a) at rest when the stroke volume is 80 cm^3 and the heart rate is 75,
b) during vigorous exercise when the stroke volume is 100 cm^3 and the heart rate is 150.

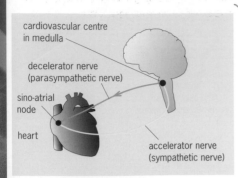

Fig 13.8 The nerves that connect the cardiovascular centre in the brain to the heart. The accelerator and decelerator nerves are part of the autonomic nervous system and act antagonistically (they have opposite effects)

CONTROL OF HEART RATE

Heart rate is modified according to the needs of the body. It increases during physical exercise to deliver extra oxygen to the tissues and to take away excess carbon dioxide. At rest, a normal adult human heart beats about 75 times per minute: during very strenuous exercise it might beat 200 times per minute. Heart rate is controlled by the SAN. The rate goes up or down when it receives information via two autonomic nerves that link the SAN with the **cardiovascular centre** in the medulla of the brain (Fig 13.8):

- A **sympathetic** or **accelerator nerve** speeds up the heart. The synapses at the end of this nerve secrete noradrenaline.
- A **parasympathetic** or **decelerator nerve**, a branch of the **vagus nerve**, slows down the heart. The synapses at the end of this nerve secrete **acetylcholine**.

A **negative feedback system** operates to control the level of carbon dioxide and, indirectly, oxygen in the blood. During exercise, the level of carbon dioxide starts to rise. This is detected by **chemoreceptors** (cells sensitive to chemical change) situated in three places: the carotid artery, the aorta and the medulla. Impulses travel down the sympathetic nerve and heart rate increases. When the carbon dioxide level drops low enough, impulses pass down the parasympathetic nerve, and the heart rate returns to normal. The control of heart rate is closely related to the control of breathing (Chapter 12).

Several factors affect heart rate:

- Secretion of **adrenaline** in response to stress, excitement or other emotions. Adrenaline is the hormone that prepares the body for action, and one of its effects is to increase heart rate (see Chapter 22).
- Movement of the limbs, as in exercise. It is thought that stretch receptors in the muscles and tendons relay information to the brain, telling the cardiovascular centre that oxygen levels will soon fall and that carbon dioxide will soon build up. This initiates signals that increase heart rate and breathing rate.
- The level of respiratory gases in the blood, as described above.
- Blood pressure. When blood pressure gets too high, a fail-safe mechanism prevents any further increase in heart rate.

SEE QUESTION 3

3 BLOOD VESSELS

There are three types of blood vessels: **arteries**, **veins** and **capillaries**. The structure of each is closely related to its function (see Fig 13.9 and Table 13.1).

Arteries carry blood away from the heart towards other organs of the body. Arteries branch into smaller **arterioles**, which branch into tiny capillaries. These are **permeable** ('leaky') vessels whose walls are one cell thick to allow exchange of materials between blood and nearby cells. Blood flows from capillaries into **venules**, which drain into larger veins.

As Fig 13.9 and Table 13.1 show, the walls of arteries and veins have the same three layers: the **tunica externa**, **tunica media** and **tunica intima** or **interna**. The relative thickness and composition of these layers vary according to the function of the vessel. The artery, which has to withstand pulses of high pressure, has an especially thick tunica media containing smooth muscle and elastic fibres. Veins have a thinner tunica media and a larger lumen to carry slower-flowing blood at low pressure. Valves in veins prevent blood flowing backwards.

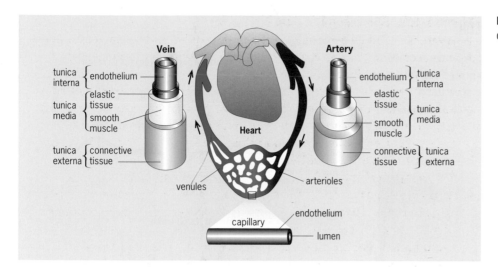

Fig 13.9 The structure of blood vessels and their distribution in the circulatory system

CAPILLARY CIRCULATION

The circulatory system keeps all cells bathed in **tissue fluid**. The tissue fluid has a reasonably constant composition because permeable capillaries allow exchange of materials between the fluid and the blood.

HOW TISSUE FLUID IS FORMED

Fig 13.10 shows what happens in living tissue. Blood from the arterioles is under high **hydrostatic pressure**. When blood enters the capillary, substances are forced out through the permeable capillary wall. The capillary walls act as filters and a proportion of all chemicals below a particular size is squeezed out, forming tissue fluid. The composition of tissue fluid is very similar to that of plasma, but tissue fluid lacks most of the large proteins that cannot pass through the capillary wall. Plasma proteins remain in the blood, where they play an important part in the drainage of tissue fluid.

> ✔ **REMEMBER THIS**
>
> ✓ All living cells are surrounded by tissue fluid. This is also known as **interstitial** or **intercellular fluid**.

> ✔ **REMEMBER THIS**
>
> Hydrostatic pressure is physical pressure. The contraction of the ventricles of the heart creates hydrostatic pressure in the blood vessels.

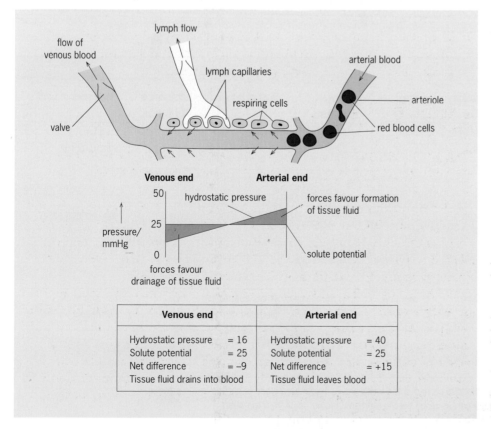

	Venous end	Arterial end
Hydrostatic pressure	= 16	= 40
Solute potential	= 25	= 25
Net difference	= −9	= +15
	Tissue fluid drains into blood	Tissue fluid leaves blood

Fig 13.10 The formation and drainage of tissue fluid. At the arterial end of the capillary, the high hydrostatic pressure forces water and small molecules out of the blood, providing the tissues with nutrients and oxygen. The hydrostatic pressure decreases as blood flows along the capillary. When it falls below the solute potential, fluid begins to drain back into the blood, taking with it wastes such as urea and carbon dioxide

⚠ SCIENCE FEATURE The electrocardiogram (ECG)

The electrical events that control the cardiac cycle are recorded by placing electrodes on certain parts of the body – Fig 13.11. A normal, healthy heartbeat produces a distinctive trace – Fig 13.12. Certain heart defects produce a modified trace and this makes the ECG a useful diagnostic tool.

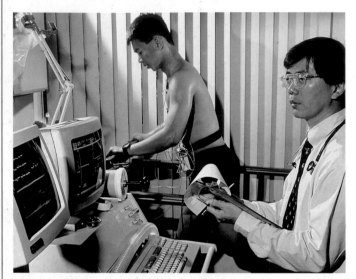

Fig 13.11 The electrocardiogram: electrodes taped to the chest pick up the electrical events of the cardiac cycle as they pass through the body. People with heart problems can now transmit their ECG trace down the telephone to a specialist who can detect problems at a distance

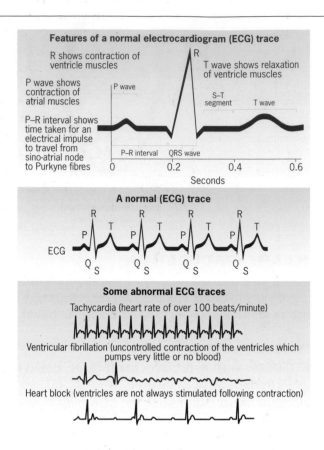

Features of a normal electrocardiogram (ECG) trace

R shows contraction of ventricle muscles

T wave shows relaxation of ventricle muscles

P wave shows contraction of atrial muscles

P–R interval shows time taken for an electrical impulse to travel from sino-atrial node to Purkyne fibres

P wave · S–T segment · T wave · P–R interval · QRS wave

A normal (ECG) trace

Some abnormal ECG traces

Tachycardia (heart rate of over 100 beats/minute)

Ventricular fibrillation (uncontrolled contraction of the ventricles which pumps very little or no blood)

Heart block (ventricles are not always stimulated following contraction)

Fig 13.12 A healthy ECG trace and some abnormal traces, showing common heart defects that can be diagnosed using the ECG

Table 13.1 A comparison of arteries, veins and capillaries							
Vessel	**Cross-section**	**Direction of flow**	**Pressure**	**Oxygen content**	**Size of lumen**	**Presence of valves**	**Properties of wall**
Arteries		away from heart	high	high*	relatively small	no	tough, powerful elastic recoil
Veins		back to heart†	low	low*	relatively large	yes	thin, distend easily
Capillaries	endothelial cell	through organs	medium	oxygen lost through wall	small, about the diameter of one red blood cell	no	one cell thick, permeable

*Except in pulmonary circulation, where oxygenated blood is carried in the veins and deoxygenated blood is carried in the arteries.
†Except portal veins, eg the hepatic portal vein, which carries blood between organs and not to or from the heart.

HOW TISSUE FLUID RETURNS TO THE BLOOD

Two main forces act on the blood as is passes along the capillary:

- The hydrostatic pressure of the blood: this tends to force water and solutes *out* of the capillary.
- The water potential of the blood: this tends to draw water *into* the capillary by osmosis.

As blood flows along a capillary, fluid passes to the tissues, and the volume and hydrostatic pressure of the blood in the capillary decreases. Since the water potential created by the large plasma proteins is relatively constant, water begins to drain back into the blood when hydrostatic pressure falls below the solute potential (see Fig 13.10). Cell waste, such as urea, and substances that have been secreted into the tissue fluid, diffuse into the capillary (Fig 13.13), contributing to the composition of venous blood.

BLOOD FLOW IN VEINS

Blood that drains into the venules is deoxygenated, under low pressure and contains many waste products and cell secretions. Several features allow blood to return to the heart:

- Valves in veins close to prevent backflow.
- Working muscles surrounding the veins squeeze blood along as they contract. The action of muscles, especially those in the legs, is so important that it has been called the 'secondary heart' or the 'venous pump' (Fig 13.14).
- Gravity helps blood to flow from organs 'above' the heart (assuming an upright position).
- The negative pressure created in the thorax during **inspiration** (breathing in) draws blood into the thorax from veins above and below.
- The action of the heart: after systole (contraction), the elastic walls of the chambers recoil, drawing blood in from the veins.

CONTROL OF BLOOD FLOW

The body often needs to alter blood flow to different areas, according to circumstances. Blood flow can be modified by **sphincters** and **shunt vessels**. A sphincter is a ring of muscle round a blood vessel that can contract, reducing the lumen size of the vessel and so reducing or preventing blood flow to a particular area. Sphincters can redirect blood into shunt vessels: these bypass a particular area. For instance, when we are cold, sphincters reduce blood flow to the skin surface, redirecting it along shunt vessels that keep the blood deeper in the body (see Chapter 15).

Blood flow can also be modified by altering the diameter of the vessel itself. Many blood vessels, particularly arterioles, contain smooth muscle fibres that can contract to reduce blood flow through the vessel: this is **vasoconstriction**. Blood flow increases when the muscle fibres relax: this is **vasodilation**. Blood pressure can be regulated by altering the degree of vasoconstriction or vasodilation.

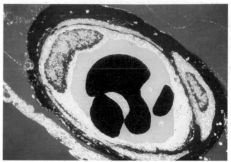

Fig 13.13 Capillary walls are about 1 μm thick. Most capillaries have a lumen that is a little larger than the diameter of a red blood cell. Note that cells in photos often seem to be strange shapes. There are two reasons for this. First, they are very flexible and will distort to squeeze along capillaries without causing blockages. Second, microscope slides take thin two-dimensional sections through three-dimensional objects, so they often give a false impression of shape

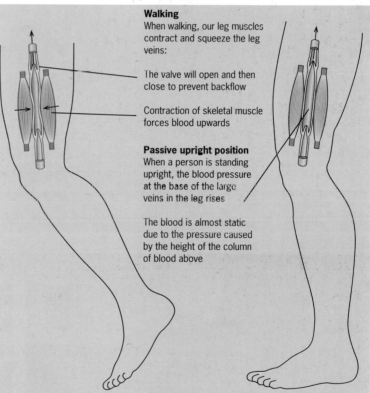

Walking
When walking, our leg muscles contract and squeeze the leg veins:

The valve will open and then close to prevent backflow

Contraction of skeletal muscle forces blood upwards

Passive upright position
When a person is standing upright, the blood pressure at the base of the large veins in the leg rises

The blood is almost static due to the pressure caused by the height of the column of blood above

Fig 13.14 The action of the leg muscles squeezes blood along veins, and the valves ensure that flow can be in only one direction, This venous pump is very important to the circulation and regularly standing still for any length of time can lead to problems. Millions of people suffer from varicose veins or haemorrhoids (piles): damaged veins whose walls have been stretched by pools of accumulated blood

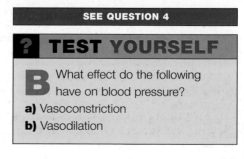

SEE QUESTION 4

? TEST YOURSELF

B What effect do the following have on blood pressure?
a) Vasoconstriction
b) Vasodilation

4 THE LYMPHATIC SYSTEM

SEE QUESTION 5

The **lymphatic system** is part of the immune system (Chapter 17) and also part of the circulatory system: it returns to the heart the small amount of tissue fluid that cannot be returned by the veins.

We can think of the lymph system as an extra set of veins (look back at Fig 13.10 and see Fig 13.15). Flow through lymphatic vessels is slow, but important (Fig 13.16).

Lymphatic flow begins in the areas round blood capillaries (**capillary beds**), where small amounts of tissue fluid drain into tiny **lymphatic capillaries** (Fig 13.17). The walls of these vessels are more permeable than blood capillaries to lipids and large molecules such as proteins. This is why lymph contains a high proportion of these substances. Many cells secrete substances that are too large to enter the blood directly, and these substances pass into the general circulation via the lymphatics. The lymph capillaries drain into larger lymph vessels that look like thin, transparent veins. These vessels have valves to prevent backflow. Lymph contains no red blood cells, and so is pale and clear. All lymph vessels flow towards the upper thorax, passing through numerous lymph nodes that filter out bacteria and cell debris.

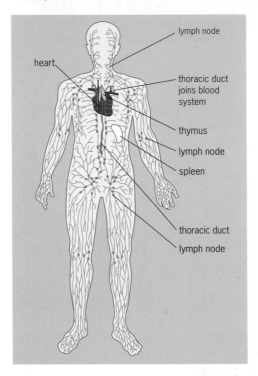

Fig 13.15 The major lymphatic vessels in the human body. The lymph system drains into the blood system in the upper thorax. A large lymph vessel, the thoracic duct, connects with the subclavian vein (sub-clavicle = under collarbone). Here lymph mixes with blood before entering the vena cava on its way to the heart

lymph node
heart
thoracic duct joins blood system
thymus
lymph node
spleen
thoracic duct
lymph node

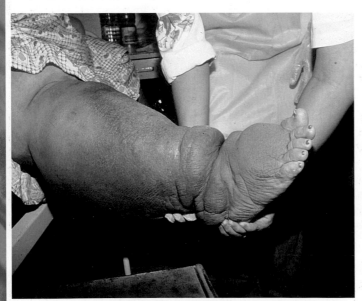

Fig 13.16 This photograph shows someone suffering from elephantiasis, a condition caused by a nematode worm, *Wuchereria bancrofti*, which blocks lymphatic vessels. Lymphatic drainage takes about 120 cm³ of fluid out of the tissues every hour. When a lymphatic vessel is blocked, the limb quickly swells until the pressure interferes with normal blood flow

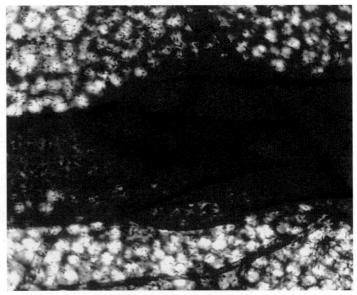

Fig 13.17 Longitudinal section through a lymphatic capillary. The cells that form the capillary walls overlap slightly, forming tiny valves through which tissue fluid can flow. This picture shows such a valve. Lymph capillaries in the villi of the intestines are called lacteals, and are important in the drainage of lipids and high molecular weight substances

5 BLOOD PRESSURE

The term **blood pressure** refers to the physical pressure of blood flowing through the arteries. This depends on the force created by the pumping of the heart, the volume of circulating blood and the size of the blood vessels. The body must keep blood pressure inside fairly strict limits. It must be high enough to force blood through all the capillaries so that all cells are well nourished, but not high enough to put the heart and blood vessels under strain: high blood pressure causes various health problems. Fig 13.18 shows how we measure blood pressure.

The **diastolic pressure** is used as an indicator for medical problems: a normal reading is between 60 and 80 mmHg, with anything over 90 usually regarded as high.

REMEMBER THIS

Have you ever stood up very quickly, and then felt faint or dizzy? This happens because the rapid change in posture has altered the body's blood distribution, causing a temporary lack of blood to the brain. Fortunately, body mechanisms quickly bring the blood flow back to normal.

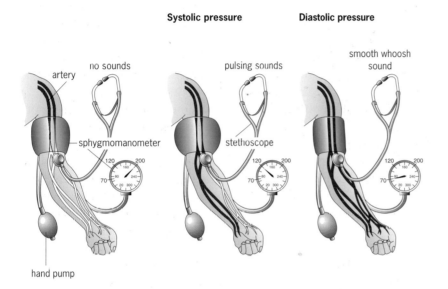

Systolic pressure   Diastolic pressure

no sounds   pulsing sounds   smooth whoosh sound

artery

sphygmomanometer   stethoscope

hand pump

Fig 13.18 Blood pressure is usually measured using a sphygmomanometer.
An inflatable cuff is placed around the upper arm and inflated until all blood flow, in or out of the arm, stops. The blood flow in the brachial artery (at the elbow) is monitored using a stethoscope.
After inflation, there is no sound, but, as air escapes from the cuff, the pressure decreases until it falls just below that created by the heart as it contracts. At this point, blood is heard spurting through the constriction in the artery. This pressure is the **systolic value**.
Pressure in the cuff continues to drop until blood can be heard flowing constantly. This is the **diastolic value** and represents the pressure to which arterial blood falls between beats

CONTROL OF BLOOD PRESSURE

A negative feedback system keeps blood pressure inside safe limits. The pressure of the blood is detected by the **carotid sinus** (Fig 13.19), a small swelling in the carotid artery. When blood pressure gets too high, the walls of the sinus expand and **stretch receptors** in the carotid artery wall send information to the cardiovascular centre in the medulla. Signals from the medulla then lower heart rate and cause vasodilation, so lowering blood pressure.

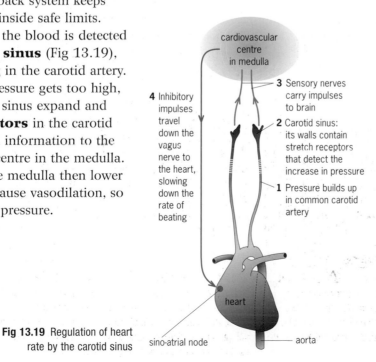

cardiovascular centre in medulla

4 Inhibitory impulses travel down the vagus nerve to the heart, slowing down the rate of beating

3 Sensory nerves carry impulses to brain

2 Carotid sinus: its walls contain stretch receptors that detect the increase in pressure

1 Pressure builds up in common carotid artery

heart

sino-atrial node   aorta

Fig 13.19 Regulation of heart rate by the carotid sinus

REMEMBER THIS

Many mechanisms work together to keep blood pressure constant. The size of blood vessels, the heart rate and the volume of circulating blood can all change to increase or decrease blood pressure, according to the needs of the body. The kidneys are particularly important because they control how much fluid we lose. A drop in blood volume and pressure causes more fluid to be retained in the blood and less to be lost in the urine (see Chapter 16).

? TEST YOURSELF

C 'Blood pressure one twenty over seventy', shouts the doctor. What does this mean?

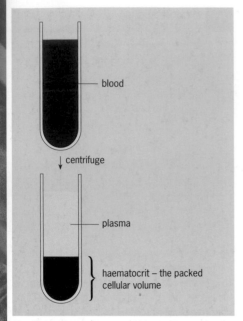

Fig 13.20 The percentage of blood taken up by cells is known as the haematocrit. A haematocrit of 39 would indicate that 39 per cent of the blood is composed of cells, mostly red cells. The average value for men is about 42 while women average 38. These values are affected by factors such as anaemia or high altitude

6 THE BLOOD

Blood is the fluid that flows through blood vessels, and the sight of blood is a sure sign that a vessel has ruptured. The loss of large amounts of blood has serious consequences: death will result if the ruptured vessels are not sealed and blood volume is not rapidly returned to normal. In an emergency, when there is no time to find out an accident victim's blood group, he or she is given fluids called **plasma expanders**. These are isotonic fluids that contain no cells but which do increase blood volume.

Blood is a complex mixture containing cells, cell fragments and a range of dissolved molecules. It does the following:

● Transports materials. Blood transports food and oxygen to respiring tissues. It also takes carbon dioxide and waste products away from respiring cells to the various organs that remove them. It carries hormones from endocrine glands to target organs.

● Distributes heat. The blood helps to maintain a constant body temperature by distributing heat from metabolically active organs, such as the liver and working muscles, to the rest of the body (see Chapter 14).

● Provides pressure. Many organs of the body depend on the physical pressure of the blood to carry out their function. For example, filtration in the kidney, formation of tissue fluid and erection of the penis all depend on blood pressure.

● Acts as a buffer. The blood contains many proteins and ions that act as buffers, keeping the pH constant by 'mopping up' any excess acid or alkali. Haemoglobin in red blood cells is an important buffer (see Chapter 16).

● Defends the body against infection (see Chapter 17).

WHAT'S IN BLOOD?

Fig 13.20 shows that centrifuged blood separates out into two distinct layers: a **cellular portion** called the **haematocrit**, and the **plasma**. The cellular portion contains red cells, white cells and platelets.

PLASMA

Plasma is the fluid part of blood. The exact composition of the plasma varies greatly (Table 13.2). For instance, in the hepatic portal vein, which leads from the intestines to the liver, the plasma is much richer in dissolved foods such as sugars, vitamins and amino acids than the plasma in a vein in another part of the body.

Table 13.2 The main constituents of plasma		
	Components	**Function**
Proteins:	albumin	osmotic balance
	antibodies	immunity
	fibrinogen	blood clotting
Salts:	sodium, potassium, chloride, hydrogencarbonate, calcium	osmotic balance, conduction of nerve impulses, carriage of CO_2, buffering, blood clotting
Products of digestion:	glucose, amino acids, fatty acids, glycerol, vitamins	nourishment of cells
Hormones:	protein, eg insulin; lipid (steroids), eg testosterone	communication
Heat	distributed around body to maintain constant body temperature	
Oxygen	vital in aerobic cell respiration	
Waste products:	urea, CO_2	none: they must be removed by excretion (see Chapter 16)

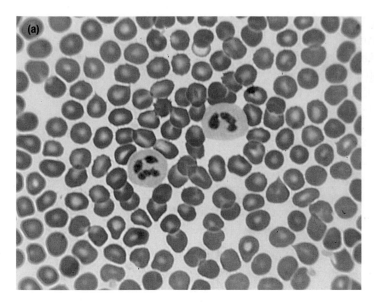

(a)

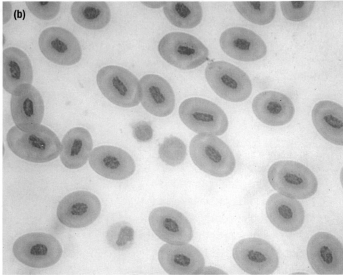

(b)

RED BLOOD CELLS

In this chapter we study red blood cells only (Fig 13.21): white cells and platelets are covered in Chapter 17.

Also known as **erythrocytes**, red cells are by far the most numerous in the blood. A single cubic millimetre of blood contains around 5 million, sometimes more. Mammalian red blood cells have no nucleus, but, as well as haemoglobin, they contain the enzymes and chemicals that allow them to function effectively. Red blood cells have one function: they carry the respiratory gases. The **haemoglobin (Hb)** they contain picks up oxygen in the lungs and swaps it for carbon dioxide in the respiring tissues.

Having the Hb inside red blood cells rather than in solution in the plasma gives several advantages:

- A much greater volume of Hb can be carried in cells than could be dissolved in plasma.
- Hb can be kept in a favourable chemical environment to allow faster loading and unloading of respiratory gases.
- Hb molecules of a particular age are kept together and can be replaced easily when old.
- Hb in cells does not increase the solute potential of the blood (free Hb would).

Red blood cells have a very regular shape, described as a biconvex disc (see Fig 13.21c). This shape is maintained by the cytoskeleton, a complex but flexible internal scaffolding made from protein fibres (see Chapter 1). When you think about what red blood cells do, it is easy to see why this is a good shape for them. They must have a large enough volume to carry useful amounts of oxygen but they also need a large enough surface area to load and unload it quickly. The biconvex shape is an efficient compromise between the maximum volume of a sphere and the maximum surface area of a flat disc.

Red blood cells are distorted when they have to squeeze through capillaries that often have lumens slightly smaller than the diameter of the red cell. Red cells rarely cause blockages because they have a flexible shape and a smooth membrane.

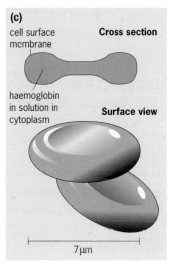

(c)

cell surface membrane

Cross section

haemoglobin in solution in cytoplasm

Surface view

7 μm

Fig 13.21(a) If you look at a drop of blood under the microscope, the most obvious feature is the mass of red blood cells. Mammalian red cells are unique as they lack a nucleus. There are also two white blood cells in this picture

Fig 13.21(b) Those of all other vertebrates, such as this example from a frog (an amphibian), are nucleated, oval in shape and generally much larger. The pink cells in the bottom left are white blood cells

Fig 13.21(c) Mammalian blood cells are biconcave discs filled with haemoglobin

✔ REMEMBER THIS

Red blood cells are made in the bone marrow of the vertebrae, ribs and pelvis by specialised **stem cells**. The turnover of red cells is very rapid: we make about 20 million red cells per second! Red blood cells circulate for about 100 to 120 days before they are destroyed in the liver by the phagocytes called **Kupffer cells**, and also in the spleen.

✓ **REMEMBER THIS**

Conjugated means 'joined'. A conjugated protein, such as haemoglobin, consists of a protein attached to a **prosthetic group** (a non-protein). The prosthetic group in haemoglobin contains iron, so we need a regular supply of iron in our diet to make this vital chemical.

✓ **REMEMBER THIS**

The concentration of dissolved oxygen is often referred to as the **partial pressure**, or **tension**. The greater the concentration of dissolved oxygen, the higher the partial pressure or tension.

SEE QUESTION 6

BLOOD AS A TRANSPORT MEDIUM

Oxygen is carried in the blood in two ways: 98 per cent travels as **oxyhaemoglobin**, the oxygen–haemoglobin complex; the remaining 2 per cent is dissolved in the plasma. Carbon dioxide is carried in three ways: as HCO_3^- ions in the plasma (70 per cent), as **carboxyhaemoglobin** (23 per cent) and in simple solution in the plasma (7 per cent).

THE STRUCTURE OF HAEMOGLOBIN

The haemoglobin molecule is a conjugated protein with a relative molecular mass of about 64 500 kilodaltons. Each Hb molecule consists of four **globin** sub-units consisting of a polypeptide chain and a prosthetic group called **haem**. At the centre of each haem is an iron ion (Fe^{2+}), which combines with oxygen. There are four haem groups, so the overall equation for the reaction is:

$$Hb \quad + \quad 4O_2 \quad \rightarrow \quad HbO_8$$
$$\text{haemoglobin} \quad + \quad \text{oxygen} \rightarrow \text{oxyhaemoglobin}$$

The polypeptide chains hold the haem groups in place and help to load oxygen. When the first haem group combines with an oxygen molecule, the haemoglobin molecule alters its shape so that the next haem group is exposed, making the loading of the subsequent oxygen molecules easier.

HAEMOGLOBIN IN ACTION

There are many substances that react readily with oxygen, but haemoglobin is one of the few that can combine with oxygen where it is abundant and then release it when the concentration falls. This property is illustrated in a graph called an **oxygen dissociation curve** (Fig 13.22). This is plotted by analysing the percentage of Hb saturated with oxygen at different concentrations of oxygen. The graph shows that Hb becomes fully saturated with oxygen at the concentrations of oxygen found in the lungs, and gives up a relatively large proportion of its oxygen in the lower oxygen concentrations that occur in the tissues.

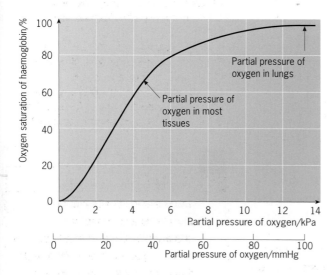

Fig 13.22 The oxygen dissociation curve. At the oxygen tension found in the lungs, Hb becomes 97–99 per cent saturated with oxygen. Surprisingly, Hb releases only about 23 per cent of its oxygen in the respiring tissues, so the blood returning in veins is still about 75 per cent saturated. This suggests that three out of four haem groups are still bound to oxygen, which allows great flexibility: if a tissue such as a working muscle becomes particularly oxygen starved, the blood can release large amounts of extra oxygen

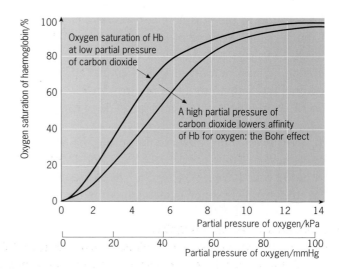

Fig 13.23 If the oxygen dissociation curve is plotted at higher CO_2 concentrations, it moves to the right, showing that haemoglobin has a reduced affinity for oxygen: this is called the Bohr effect

The dissociation curve in Fig 13.22 was plotted using mixtures of gases in which the concentration of carbon dioxide was constant. Fig 13.23 shows that, at higher concentrations of carbon dioxide, the curve moves to the right. This is a vital point: it shows that carbon dioxide lowers the **affinity** of Hb for oxygen. This means that when carbon dioxide concentration is higher, Hb does not hold on to its oxygen quite as well: Hb therefore tends to give up oxygen in areas of high carbon dioxide – such as in the respiring tissues that need it most. This lowering of affinity by carbon dioxide is called the **Bohr effect**, or the **Bohr shift**.

FETAL HAEMOGLOBIN

Before birth, a mammalian fetus must obtain oxygen from its mother via the placenta. If fetal haemoglobin had the same affinity for oxygen as the mother's haemoglobin, no transfer of oxygen to the fetus would be possible. But, as Fig 13.24 shows, fetal haemoglobin has a higher affinity for oxygen than adult Hb. It can therefore pick up oxygen in the same conditions that cause the maternal blood to release it. After birth, the baby's body makes adult Hb, which gradually replaces the fetal version.

UNLOADING OXYGEN AT THE TISSUES

The series of events that lead to Hb unloading its oxygen in the respiring tissues is shown in Fig 13.25.

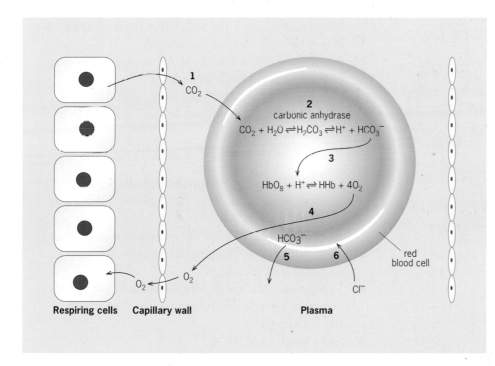

MYOGLOBIN AND MUSCLES

Mammalian muscle contains the respiratory pigment, myoglobin (myo = muscle). The myoglobin molecule is one-quarter of an Hb molecule: it consists of one polypeptide chain and one haem, and combines with one oxygen molecule. Myoglobin has a higher affinity for oxygen than haemoglobin and so it can pick up oxygen from haemoglobin and store it until the muscle becomes short of oxygen. When this happens (during exercise, for example), the oxygen tension in the muscle drops. Hb has no more oxygen to deliver and myoglobin then releases its oxygen.

Fig 13.24 The dissociation curve (above) for fetal haemoglobin is to the left of the adult version. So, at the oxygen concentrations found at the placenta there is an efficient transfer of oxygen from mother to fetus (below)

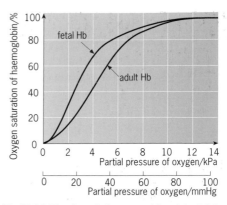

Fig 13.25 Unloading oxygen at the respiring tissues
Step 1 CO_2 diffuses from the respiring tissue, through the wall of the capillary and plasma, into the red blood cell
Step 2 Inside the cell, the enzyme carbonic anhydrase catalyses the conversion of CO_2 and water into carbonic acid
Step 3 The H^+ ions released by the carbonic acid destabilise the oxyhaemoglobin, causing each molecule to release its four O_2 molecules
Step 4 The oxygen is free to diffuse into the respiring tissues
Step 5 The accumulating HCO_3^- ions diffuse out into the plasma, leaving the inside of the red blood cell with a net positive charge
Step 6 To maintain a charge balance, Cl^- ions (the commonest negative ions in plasma) diffuse into the cell: this is the so-called chloride shift

✔ REMEMBER THIS

The presence of myoglobin in muscles is associated with endurance. Chickens, not known for their flying ability, have flight muscles that contain little myoglobin: this is why chicken breast meat is pale. In contrast, ducks and geese are accomplished long-distance fliers, and have dark meat because of the high levels of myoglobin.

SUMMARY

When you have read this chapter you should know and understand the following:

- Large organisms need a circulatory system to connect all living cells to the organs that exchange vital materials. Diffusion alone is not enough.
- Mammals have a double circulation: the **pulmonary circulation** to the lungs and the **systemic circulation** to the rest of the body.
- The mammalian **heart** is a four-chambered muscular pump made of **cardiac muscle**. This can contract powerfully and without fatigue.
- A single heartbeat, the **cardiac cycle**, consists of **atrial systole** (contraction) followed by **ventricular systole**, then **atrial** and **ventricular diastole** (relaxation).

- The heart is **myogenic**: the electrical impulses that control the cardiac cycle arise from the muscle itself.
- From the heart, blood passes into vessels in the following order: **arteries**, **arterioles**, **capillaries**, **venules**, **veins**.
- The **lymphatic system** acts as an extra set of veins, helping to drain fluid from the tissues back into the bloodstream.
- Blood consists of **red cells**, **white cells**, **platelets** and **plasma**.
- Red cells, **erythrocytes**, contain **haemoglobin**, a conjugated protein that can pick up oxygen where it is abundant (lungs) and release it where it is needed (the respiring tissues).

? EXAM QUESTIONS

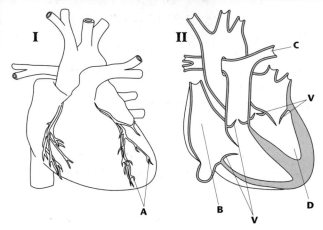

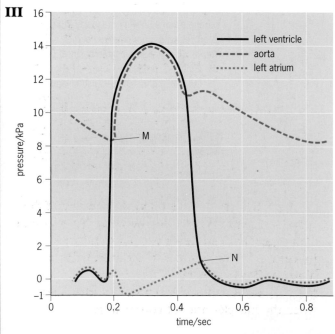

1 Diagrams I and II show the external and internal features respectively of the mammalian heart.

a) Name structures **A** to **D**.

Some people are born with heart defects, while others develop heart defects later.

b) Suggest the likely effects on the circulatory system of the following heart defects:

(i) a baby born with a hole in the wall between the left and right chambers of the heart ('hole in the heart');

(ii) valves (**V**) not working properly.

Diagram III shows the pressure changes in various parts of the circulatory system during one cardiac cycle.

At **M** and **N**, valves are either opening or closing.

c) With reference to diagram III, explain what is happening at **M** and **N**.

d) Explain why the maximum pressure in the left atrium is lower than the maximum pressure in the left ventricle.

OCR 2803/1 Transport June 2001 Q1

EXAM QUESTIONS

2 The graph below shows changes in the volume of blood in the left ventricle as the heart beats.

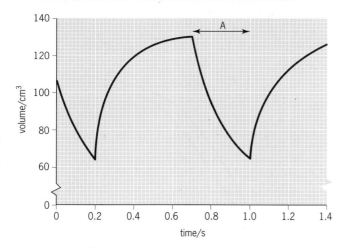

a) **(i)** The horizontal line labelled **A** on the graph shows when blood is leaving the ventricle. Explain, in terms of blood pressure, why blood does not flow back into the atrium during this period.
(ii) Draw a horizontal line on the graph, to show the period in one cardiac cycle when the muscle in the wall of the ventricle is relaxed. Label this line with the letter **B**.
b) **(i)** Draw a horizontal line on the graph to show one complete cardiac cycle. Label this line with the letter **C**.
(ii) Use line **C** to calculate the number of times the heart beats in one minute. Show your working.
c) The table shows the blood flow to different parts of the body at rest and during a period of vigorous exercise.

Part of the body	Rate of blood flow/cm^3 minute^{-1}	
	at rest	**during exercise**
Brain	750	750
Heart muscle	300	1200
Gut and liver	3000	1400
Muscle	1000	16 000
All other organs (except lungs)	1550	1550

(i) Use the figures in the table to calculate the cardiac output at rest.
(ii) Give **two** ways in which cardiac output is increased during a period of vigorous exercise.
d) Describe the parts played by the sinoatrial node (SAN) and the atrioventricular node (AVN) in controlling the heart beat.
AQA (A) BYA1 Molecules, Cells and Systems January 2001 Q8

3 The diagram below shows variations in certain features of blood vessels in different parts of the circulatory system.

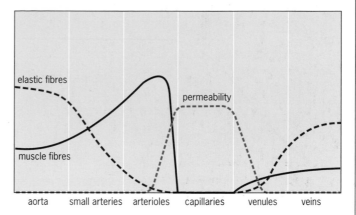

a) Describe the difference in the ratio of elastic fibres to muscle fibres between the aorta and the arterioles.
b) Explain the importance of this ratio in
(i) the aorta;
(ii) the arterioles supplying the skin.
c) Explain how the structure of the capillaries is related to their function.
AQA (NEAB) BY03 Physiology June 2000 Q6

4
a) Explain why a vein is described as an organ.
b) The diagram shows a vein passing between two muscles.

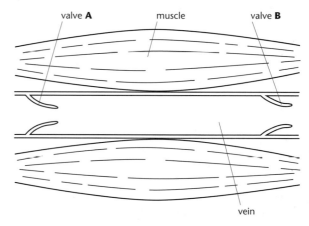

(i) When the muscles contract, the pressure of the blood in the part of the vein between valves **A** and **B** changes. Explain how this change in pressure, together with the action of the valves, helps the blood to flow to the heart.
(ii) Suggest how blood is returned to the heart when the muscles are not contracting.

c) The graph below shows the relationship between the pressure in the veins returning blood into the heart and stroke volume.

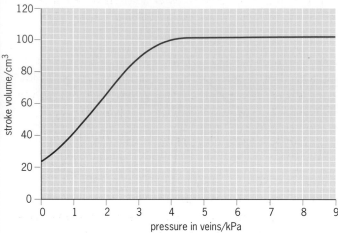

(i) Describe the relationship between stroke volume and pressure in the veins.
(ii) What information would you need, other than that in the graph, to calculate this person's cardiac output?

AQA (A) BYA1 Molecules, Cells and Systems June 2001 Q4

5

a) By using arrows and annotations complete the following diagram to show how **tissue fluid** and **lymph** are produced.

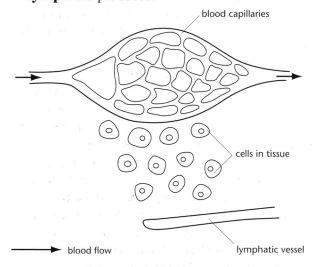

blood capillaries

cells in tissue

blood flow

lymphatic vessel

b) State **two** differences between lymph and plasma.
c) Lymph is eventually returned to the blood.
(i) Suggest **two** mechanisms by which lymph is moved through the lymphatic vessels.
(ii) Where is lymph returned to the blood?

OCR 6918 Animal Studies March 2001 Q4

6 The graph below shows oxygen dissociation curves for human haemoglobin at low and high partial pressures of carbon dioxide. The effect of high partial pressure of carbon dioxide on haemoglobin dissociation is known as the Bohr effect.

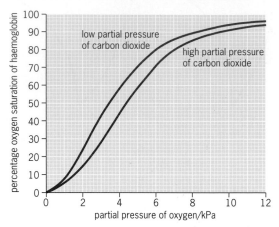

a) Explain how the shape of the dissociation curve at a low partial pressure of carbon dioxide is related to the ability of haemoglobin to transport oxygen from the lungs to respiring tissues.

b) (i) From the graph, calculate the difference in percentage oxygen saturation of haemoglobin at low and high partial pressure of carbon dioxide, at 4 kPa partial pressure of oxygen. Show your working.
(ii) Suggest why the Bohr effect is of importance to tissues when they increase their activity.

c) The respiratory pigment myoglobin is found in muscle tissue. The dissociation curve for myoglobin lies to the left of the curves for haemoglobin and it is 97% saturated with oxygen at an oxygen partial pressure of 1 kPa. Explain the function of myoglobin in muscle.

Edexcel B3 June 1999 Q7

KEY SKILLS **ASSIGNMENT 1**

HEART DISEASE

Generally, the underlying cause of heart disease is one of the following:

● Structural problems
● Inadequate coronary blood supply
● Faulty electrical conduction

Structural problems

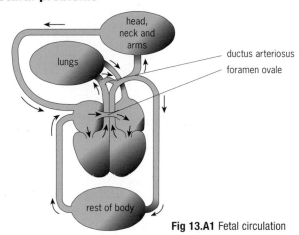

Fig 13.A1 Fetal circulation

Structural problems are defects in the anatomy of the heart, often present from birth. The diagram shows the differences in the fetal circulation before birth. Two important features are the presence of the **foramen ovale** and the **ductus arteriosus.**

The foramen ovale is a hole between the right and left atria. A valve in the vena cava directs blood through this hole so that a large proportion of venous blood passes right into the left side of the heart. The ductus arteriosus is a vessel that connects the pulmonary artery to the aorta. Its purpose is to by-pass the lungs, directing most of the blood into the aorta. The combined effect of the foramen ovale and the ductus arteriosus is to bypass the lungs.

1 Why is it advantageous to bypass the fetal lungs?

2 From which organ does the fetus obtain its oxygen
a) before birth and **b)** after birth

At birth, rapid changes in the circulation are necessary. The normal sequence of events is as follows: at birth the lungs inflate with air and the pulmonary capillaries fill with blood, greatly increasing the blood flow in the pulmonary vein. As a result, the blood flow through the foramen ovale is reversed; it starts to go from left to right. This causes a small valve in the left side of the foramen ovale to close, preventing further left to right flow. In most people, the difference in pressure between the left and right side keeps the valve permanently shut, giving the tissue a chance to bind permanently.

The ductus arteriosus closes soon after birth. The reason for this appears to be a change in the oxygen content of the blood causes a constriction of the muscular wall. This drastically reduces the blood flow and, within a few days, blood flow ceases completely.

If the foramen ovale fails to close, the baby will have a 'hole in the heart'. The severity of this depends on the size of the hole. Large holes need surgery, but many people, including athletes, have led a normal life with a small hole in the heart.

3 What problems will be caused by a hole in the heart?

Valve problems

There are two main valve problems:

● A leaky valve due to damage to one or more of the cusps
● A narrowing – or **stenosis** – of the valve opening

4 What would be the likely outcome of damage to one or more of the cusps?

5 Why would a narrowing of the valve entrance cause problems?

A faulty valve can be detected by the sound it makes – known a **heart murmur**. With leaky valves there is a lub 'swish' instead of the lub-dup, caused when some blood squirts back through the valve. A stenosed valve makes a different type of murmur caused by the narrowing of the opening. In both cases the best treatment is usually replacement with an artificial valve.

Inadequate coronary blood supply

This is the most common cause of heart disease, and also the most preventable. There is a very extensive network of vessels that supplies blood to the myocardium – the heart muscle itself. The build up of fatty deposits inside the arterial walls – known as **atherosclerosis** – will reduce and even block the blood flow through the coronary artery. This problem, known as **coronary heart disease (CHD)** or **ischaemic heart disease,** has varying degrees of severity, from mild chest pain to a full heart attack. Research has shown that the people who develop coronary heart disease always have one or more of the following factors:

● High blood cholesterol
● High blood pressure
● Cigarette smoking
● Obesity
● Lack of exercise
● Diabetes mellitus
● Genetic predisposition

KEY SKILLS **ASSIGNMENT 1**

Two common consequences of CHD are:

- **Angina pectoris**. When blood flow is reduced the heart muscle is unable to meet increase performance when the need arises. This leads to a pain called angina pectoris (commonly just 'angina'). Any extra strain on the heart – even walking upstairs – causes severe chest pains that makes the sufferer rest before they can continue.

- **Myocardial infarction** – a heart attack. When the blood flow to a particular part of the muscle is blocked, that part of the muscle will die. Infarction can cause death in several different ways. The heart may fibrillate – contract rapidly and uncontrollably. Alternatively, the cardiac output can drop below the minimum requirement. In some cases, the dead area of muscle may rupture, causing the heart to literally burst. The body can recover from a mild infarction, but extensive ones are nearly always fatal.

6 Imagine you are a GP faced with an overweight couple in their late forties. How would you persuade them to change their lifestyle?

7 Discuss the effectiveness of campaigns you have seen aimed at reducing the risk of developing heart disease. What approach would you suggest?

Faults in the conduction system

Contrary to popular belief, the heart doesn't have to stop to cause death – it just has to stop pumping blood effectively. Many heart attacks are caused by **arrhythmias** – a disruption of the normal rhythm of the heart. If the atria and/or the ventricles don't contract at the right time they won't even be able to supply the myocardium with blood.

The ECG trace is a very useful tool in detecting cardiac arrhythmias.

A common arrhythmia is a **heart block**. This results from a blockage in the conducting system – often in the AV node – and may result from a myocardial infarction as described above.

8 What will happen if the impulse created by the SAN doesn't reach the ventricles?

Coronary heart disease and other traumas such as electrocution can cause **ventricular fibrillation** – the rapid and uncontrolled contraction of the ventricles. If this is not corrected rapidly, death is inevitable. The usual way to stop ventricular fibrillation is to use a **defibrillator** – a machine that applies a very strong current across the heart for a short period of time. This works because the electric current depolarises all of the heart cells at the same time. All activity stops for 3–5 seconds, during which time the normal electrical activity has a chance to re-establish itself.

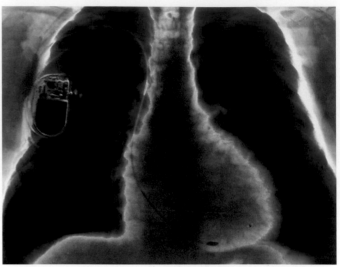

Fig 13.A2 An artificial pacemaker is basically a small box that houses small batteries and complex electronics. The device is fitted under a muscle in the upper thorax, and a wire leads down a vein into the heart. The wire ends in an electrode, seen at the base of the heart. The electrode touches the heart muscle, initiating (setting off) the cardiac cycle. New, sophisticated pacemakers can sense changes in breathing, movement and body temperature and can adjust heart rate accordingly

A patient with a permanent heart block can be fitted with an artificial **pacemaker**. These devices vary according to the nature of the fault; some will stimulate the atria, others just the ventricles. New, more sophisticated pacemakers can sense changes in breathing, body temperature and vibration, adjusting the rate accordingly.

The actual device consists of a small box that houses the lithium batteries and the complex electronics that create the impulse. The electrical impulse is delivered to the heart by means of a lead passed down the subclavian vein and the superior vena cava.

Replacement hearts: when a patient's heart becomes permanently damaged, the prospects for a long and healthy life are not good. Despite the fact that the heart has an apparently simple mechanical function, attempts to make an artificial one have defeated many of the world's best engineers. Until an artificial heart is made, the only source of an artificial heart is from a transplant.

9 What features would be needed by a perfect artificial heart?

KEY SKILLS **ASSIGNMENT 2**

ALTITUDE TRAINING

This Assignment links aspects of breathing (Chapter 12) and aspects of the circulatory system that we have covered in this chapter. So to answer the questions, you will need to refer to both chapters to find the relevant text.

Staying power

Which sport is the most demanding? Is it football? Or marathon running? It might surprise you, but these two sports are just a stroll compared with the most gruelling events that you can do. Top of the list are cross-country skiing and long-distance bike racing (such as the Tour de France).

Fig 13.A3 Cross-country skiers and long-distance cyclists top the table for maximum oxygen consumption

But how do we compare the demands of different sports and events? One way is to measure oxygen consumption. In Chapter 12 we saw that a person's oxygen consumption can be expressed in terms of V_{O_2} and $V_{O_2}(max)$.

1 **a)** Suggest what does the term V_{O_2} means.
b) What is the difference between V_{O_2} and $V_{O_2}(max)$?
c) What are the average values for V_{O_2} and $V_{O_2}(max)$, as shown in Chapter 12?

Studies show that competitors in cross-country skiing and long-distance cycling can have oxygen consumptions as high as 90 cm^3 per minute per kilogram.

In events covering many kilometres over many hours, just a few seconds' advantage can mean the difference between success and failure. Not surprisingly, there is a search for ever more effective training methods that can increase athletes' oxidative capacity – their ability to get oxygen to their muscles. For several decades, one of the favoured methods of increasing the oxygen-carrying capacity of the blood has been to train at altitude.

Altitude training

In Chapter 12, we saw that mountaineers who ascend without oxygen apparatus experience problems as the air gets thinner and the partial pressure of oxygen decreases.

2 **a)** Explain what is meant by partial pressure of oxygen.
b) Explain what happens to the partial pressure of oxygen as you go up a mountain.

The human body is able to *acclimatise*; it adapts to low oxygen levels if given time – usually several weeks. The count of red cells in the blood increases from about 5 000 000 per cubic millimetre to around 7 000 000.

3 **a)** What percentage increase in this?
b) What is the advantage of this change?

Acclimatisation

The mechanism of this acclimatisation is not completely understood, but this is what we think happens. The kidney detects the lowered oxygen levels in the blood and responds by releasing a hormone, **erythropoietin** (EPO), which targets cells in the bone marrow. In response, the bone marrow increases red cell production.

There are several ways of measuring the oxygen-carrying capacity of the blood. In addition to the number of red cells per cubic millimetre, there are the haematocrit and the amount of haemoglobin per litre. The haematocrit is the percentage of the blood taken up by cells (average values 41–42 per cent for males, about 38 per cent for females). The average amount of haemoglobin per litre is about 16 g for males and 14 g for females.

4 **a)** Write a flow diagram to summarise the mechanism of acclimatisation.
b) A person has trained at altitude and has returned to sea level. What effect will this have on his or her resting pulse rate? Explain your answer.
c) What do you think happens to the red blood cell count and resting pulse rate in the long term, if the athlete remains at sea level?

For a long time, athletes have realised the great benefit of training at altitude: if you go up into the mountains to train, the increased oxygen-carrying capacity of your blood gives you a distinct advantage over those competitors who have remained at sea level.

However, a problem comes from the fact that exercise done at altitude is not as beneficial as that done at sea level – you can't work the muscles as hard. So there's a problem; how do you 'live high and train low'?

KEY SKILLS **ASSIGNMENT 2**

There are two ways to achieve this. One is to have 'the Alps in a caravan' and live at low altitude in a mobile home with a controlled atmosphere, so that the partial pressure of oxygen is lowered to simulate high-altitude conditions.

The other solution is the highly illegal, unfair and dangerous practice of taking genetically engineered erythropoietin (RhEPO – recombinant human erythropoietin), which will mimic the effects of altitude training without the time and expense.

5 Outline the steps involved in making genetically engineered human erythropoietin. (You will need to look at Chapter 30 to answer this.)

In the past decade, the deaths of several top cyclists have been linked to RhEPO abuse and the resulting ultra-high haematocrit. The main danger comes from the increased risk of an embolism, a blood clot which circulates before becoming lodged in a vital blood vessel, such as in the heart, lungs or brain.

6 Suggest how an increased haematocrit can lead to an embolism.

7 Read the following extract and then answer the questions that follow.

> ### Human proteins are boosting performance: the evidence
>
> Genetic engineers have found a way to mass-produce EPO, which is a hormone and a front-line drug for treating anaemia. Use of EPO boosts red blood cell production and consequently blood haemoglobin levels.
>
> The increase in haemoglobin is a powerful lure for athletes in endurance events. Evidence for the use of EPO has come from comparing two cross-country skiing events at Lahti and Thunder Bay. EPO is detectable in the body for only six to eight hours but its effect lasts as long as a red blood cell – around 120 days.

(Reproduced with permission from an article in the *Daily Telegraph*)

a) Explain how the graph of Fig 13.A4(a) provides evidence that EPO has been used by some cross-country skiers.

b) Explain how increased haemoglobin concentration might lead to increased performance in endurance events.

The graph of Fig 13.A4(b) shows the oxygen dissociation curves for haemoglobin as blood passes through capillaries in the lungs and in the skeletal muscles of an athlete.

c) Explain how features of the oxygen dissociation curves for haemoglobin in the lungs and in the skeletal muscles benefit an athlete.

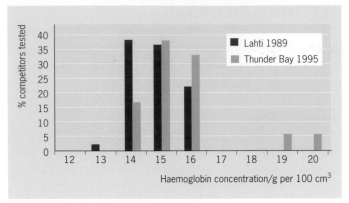

Fig 13.A4(a) Comparing the haemoglobin concentration in the blood of cross-country skiers from two events, one in 1989 and one six years later, in 1995

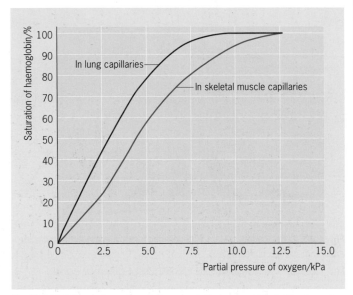

Fig 13.A4(b) The oxygen dissociation curves for haemoglobin in blood passing through capillaries in the lungs and in the skeletal muscles of an athlete

[Question 7 is an original exam question: NEAB Biology Physiology Module Test Section B March 1998 Q8]

Catching the cheats

So how can you tell when an athlete has cheated and used RhEPO? It is a natural hormone and so you would expect to find it in the blood, especially in those who have trained at altitude. The authorities have decided that for the present the only practical way to detect abuse is to measure the effect of the RhEPO, that is, the amount of haemoglobin in the blood. The assumption is that unnaturally high levels are indicators of abuse.

Recently the cycling authorities set the level at a haematocrit of 50 per cent, and the International Skiing Federation set haemoglobin levels at 18.5 grams per litre for males and 16.5 for women. Predictably, athletes have complained that it is perfectly possible to have levels this high naturally, and the debate looks set to continue.

14

Homeostasis

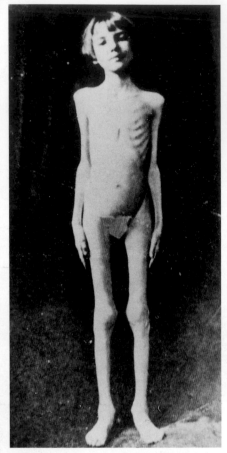
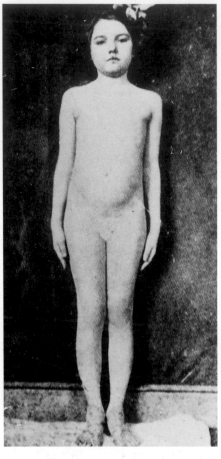

For Catherine Hayes, the discovery of insulin came just in time. Daily injections of insulin allowed the cells of her body to absorb glucose effectively, so nourishing her tissues and lowering her blood sugar

Millions of people worldwide are unable to control their blood glucose level. They have a condition known as diabetes mellitus. Early symptoms include a raging thirst, extreme fatigue and excessive urination. If sufferers are not treated, they lose weight and eventually die.

Until 1922, doctors regarded the symptoms of diabetes mellitus as a death sentence, especially when they occurred in children. Physicians confirmed a diagnosis by tasting the patient's urine to find out if it contained sugar (today, there are better methods). One noted, 'In children the disease is rapidly progressive, and may prove fatal within a few days... As a general rule, the older the patient at the onset, the slower the course.'

In 1923 a 16-year-old Canadian girl, Catherine Hayes, wrote, 'Only a year ago I was a human skeleton. Can you imagine being 5 feet 4 inches and weighing only 55 pounds [less than 4 stone; 24 kg]? My doctor had placed me on a starvation diet, the only available treatment at the time. I became too weak to engage in any physical activity, even walking, and I eventually lost most of my muscle tissue. My skin became so dry that it flaked and peeled. It was not only painful, but embarrassing, and I wondered how my school friends could bear to look at me. Many of them could not, and in the last year, I spent much of my time alone.

'However, what was even harder to bear was the knowledge that my life could end at any time. It was very difficult to live with the thought that if I took a turn for the worse I could lapse into a coma and die.

'Then came the news about the discovery of insulin...'

1 THE CONCEPT OF HOMEOSTASIS

In this part of the book we focus on the basic concepts and mechanisms of homeostasis, and then look at some examples including blood sugar, temperature (Chapter 15), the workings of the liver and kidney (Chapter 16) and the immune system (Chapter 17). Finally, we investigate exercise physiology (Chapter 18) – the short-term and long-term changes that happen to our bodies when we exercise. The fact that our muscles can work 20 times as hard and yet our body can still maintain acceptable internal conditions is one of the most impressive feats of homeostasis.

The word **homeostasis** means 'steady state'. Homeostasis describes how the body regulates its processes to keep its internal conditions as stable as possible. Homeostasis is necessary because cells, especially those of humans and other higher animals, are efficient but very demanding. To function properly they need to be bathed in tissue fluid that can provide the optimum conditions. Nutrients and oxygen must be delivered and waste needs to be removed. The concentration, temperature and pH of the fluid between cells must also be kept at levels that guarantee efficient cell functioning.

However, the phrase 'steady state' is a bit misleading. The conditions inside our bodies are not constant, but are kept within a narrow range. Some factors, such as core temperature and blood pH, fluctuate only slightly, while others, such as blood glucose, vary considerably throughout a normal day without producing any harmful effects.

In this chapter we look at the control of blood glucose and at the role of the liver. Other aspects of homeostasis are covered elsewhere in the book. Table 14.1 shows you where to find them.

Table 14.1 Homeostatic mechanisms covered in this book

Homeostatic mechanism	Covered in
Temperature	Chapter 15
Blood glucose	Chapter 14
Blood pressure	Chapter 13
Solute concentration of blood	Chapter 16
Blood pH	Chapter 16
Blood volume	Chapter 16
Blood hormone levels	Chapter 22
	Chapter 26

THE MECHANISM OF HOMEOSTASIS

When you start to study how the body controls a physiological factor such as temperature, blood glucose or blood pressure, it is important to organise your thoughts by asking the following questions:

- What conditions bring about a change in the factor being considered?
- What detects the change?
- How is the change reversed?

You will soon notice a pattern. Whenever a physiological factor changes, the body detects the change and then, by using nervous or hormonal signals, or both, it reverses the change. The extent of the correction is monitored by a system called **negative feedback** (Fig 14.1). This makes sure that, as levels return to normal, corrective mechanisms are scaled down.

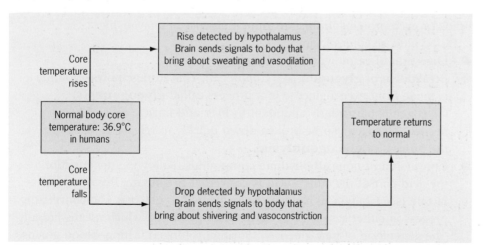

Control of body temperature illustrates this mechanism well (see Chapter 15). When we are in a very hot environment or when we have been doing strenuous exercise, our body temperature rises. The brain detects this and sends signals to the body to bring the temperature down, using various corrective mechanisms such as sweating and increased blood flow to the skin. As the body cools, the drop in temperature is monitored by the brain, which begins to send out fewer signals. Sweating then decreases.

Fig 14.1 The mechanism of homeostasis as illustrated by temperature control. All other examples follow the same general pattern

The opposite of negative feedback is **positive feedback**. In this situation, a change is amplified rather than returned to normal. Positive feedback in living systems is rare, but there are a few examples. Damage to a blood vessel causes a cascade reaction that brings about blood clotting: a few molecules of a substance become activated and each one then activates many more (see Chapter 15). The result is a tangled mesh of protein fibres that plugs the hole in the blood vessel.

Positive feedbacks also occur in abnormal situations and when normal homeostatic mechanisms get out of control. Elderly people whose sensory systems have deteriorated can suffer from hypothermia (see the Assignment at the end of Chapter 15). When they start to become cold, their body systems fail to respond, and the drop in body temperature goes uncorrected. As they grow colder, their metabolic rate decreases still further, they produce less heat and so in cold conditions they continue to cool down at an ever-increasing rate. Death occurs when their core temperature falls to about 25 °C.

HOMEOSTATIC MECHANISMS AT THE MOLECULAR LEVEL

We can also look at homeostasis at the level of molecules. In Chapter 5 we saw that enzymes control metabolic pathways, and that particular products are made, step by step, in a series of carefully controlled reactions. In a metabolic pathway, any product is made only as fast as it is needed. Excess product often prevents further quantities being made by inhibiting one of the enzymes in the pathway. When the excess has been used up, the inhibition is lifted and production continues. The process is self-regulating and so we describe it as a **homeostatic mechanism**.

2 THE CONTROL OF BLOOD GLUCOSE

In normal circumstances, we obtain most of our energy by respiring glucose. Cells therefore need a regular supply of this simple sugar. Some vital organs, notably the brain, cannot do without it even for a short time: lack of glucose can cause brain damage. Other cells and tissues, such as muscles, can respire lipids or even proteins for a short time if glucose is unavailable.

Blood glucose comes from:
- Digestion of carbohydrates in the diet.
- Breakdown of **glycogen** (see Chapter 3). This storage polysaccharide is made from excess glucose in a process called **glycogenesis**. Glycogen is particularly abundant in liver and muscle cells. When needed, glycogen can be broken down quickly to release glucose in the process of **glycogenolysis**.
- Conversion of non-carbohydrate compounds. Following deamination, the acid part of the amino acid is converted to glucose. Pyruvate and lactate (see Chapter 3) can also be converted to glucose. The conversion process in either case is called **gluconeogenesis**, which means literally 'the generation of new glucose'. During prolonged fasting, blood glucose is maintained by conversion of the body's protein and lipid stores.

The blood of a healthy person contains between 80 and 90 mg of glucose per 100 cm^3. This normal value is maintained even during prolonged fasting. But the value rises to around 120 to 140 mg per 100 cm^3 shortly after a meal, when carbohydrate digestion is in full swing. Feedback mechanisms bring the levels back to normal in about two hours. Under normal conditions, the kidney is able to reabsorb all of the blood glucose passing through it, preventing any from being lost in the urine.

REMEMBER THIS

Glucose is one of the most abundant substances in our diet. Plant material contains starch and cellulose, and meat contains glycogen. All three are glucose polymers. During digestion, both starch and glycogen are broken down into glucose, which then passes into the blood in large amounts. Humans cannot digest cellulose, and this constituent of plant material forms much of our dietary fibre.

REMEMBER THIS

The terms used to describe glucose metabolism – glycogenesis, glycogenolysis and gluconeogenesis – can be confusing. Try remembering the origins of the words that make them up:

Glyco, gluco = sugar

lysis = splitting

neo = new

genesis = generation, or formation

THE MECHANISM OF BLOOD GLUCOSE CONTROL

SEE QUESTIONS 1 TO 3

The control of blood glucose level is a good example of homeostasis. A negative feedback mechanism operates to detect and correct the level of blood glucose, maintaining it within 'safe' limits. The pancreas plays a central role in the control of blood glucose. The digestive functions of the pancreas are covered in Chapter 11 but in this section we are concerned with its **endocrine role**: how it produces the hormones **insulin** and **glucagon** to control blood glucose (look ahead to Fig 14.3).

The pancreas itself detects any change in the level of blood glucose. If blood glucose becomes too high, **β cells** in the **islets of Langerhans** respond by releasing insulin. This hormone travels to all parts of the body in the blood, but exerts an effect mainly on cells in muscles, liver and adipose (fat storage) tissue. Insulin lowers blood glucose by making cell surface membranes more permeable to glucose. It activates transport proteins in the membranes, allowing glucose to pass into cells. Insulin also activates enzymes inside the cells. Some of these enzymes convert the glucose to glycogen, others increase protein and fat synthesis.

If the levels of blood glucose get too low, **α cells** in the islets of Langerhans secrete glucagon. This hormone fits into receptor sites on cell surface membranes, and activates the enzymes inside the cells that convert glycogen to glucose. The glucose then passes out of the cells and into the blood, raising blood glucose levels.

WHEN CONTROL OF BLOOD GLUCOSE FAILS

People with the disease **diabetes mellitus** are unable to control the level of glucose in their blood. This produces a range of symptoms. Blood glucose levels can get too high because the affected person produces little or no insulin. Without insulin, glucose cannot pass into cells and remains in the blood. The solute concentration of the blood increases, interfering with effective circulation and making the individual very thirsty. The cells are starved of their main fuel and are forced to respire lipids and proteins, leading to weight loss and eventual starvation. Glucose appears in the urine because blood sugar levels are so high that the kidney cannot reabsorb it all.

Diabetes has a variety of causes and varying degrees of severity. In the UK, about 25 people in every thousand suffer from diabetes in one form or another. That means that there are over a million sufferers. About a third of these have Type I diabetes, roughly two-thirds have Type II (see below).

TYPES OF DIABETES MELLITUS

There are two main types of diabetes: **Type I** and **Type II**.

Type I diabetes, also known as 'early onset diabetes', occurs when the body cannot make insulin. This is often caused by an auto-immune reaction which attacks and destroys cells in the islets of Langerhans. This form of the disease usually appears before the age of 20, and the onset is sudden. Sufferers have this condition for the rest of their lives, but they can be treated by regular injections of insulin (Fig 14.2) matched to their glucose intake (in diet) and expenditure (in exercise). Before insulin was available, untreated Type I diabetes was usually fatal within a year of diagnosis. Most people with diabetes today need insulin injections, but they lead normal lives.

Type II diabetes, also called 'late onset diabetes', is more common than Type I, accounting for about 70 per cent of the cases in the UK. It tends to begin during middle age and is more common in those who are overweight. This form of diabetes is due to a decline in the efficiency of islet cells, or to a failure of the cell surface membranes to respond to insulin. Fortunately, in many cases it can be controlled by regulating the diet.

REMEMBER THIS

Diabetes mellitus is describes an inability to control blood glucose levels. It is not the same as diabetes insipidus. This form of diabetes is caused by a lack of anti-diuretic hormone in the body (see page 307).

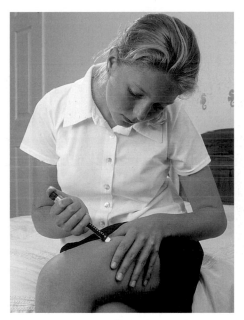

Fig 14.2 This girl has diabetes and has to inject insulin every day. For many years, insulin for human treatment was derived from cows or pigs, but both are slightly different from human insulin and can cause an immune reaction, reducing their effectiveness. Today, diabetics use human insulin produced by genetically engineered bacteria (Chapter 6)

! SCIENCE FEATURE The discovery of insulin

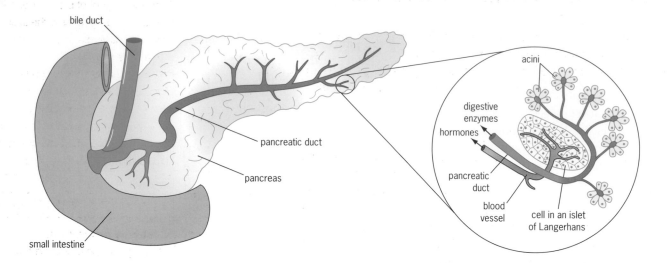

Fig 14.3 The major part of the pancreas makes a juice containing digestive enzymes, but small patches of cells, called islets of Langerhans, produce the hormones insulin and glucagon

Fig 14.4 Banting and Best. In 1923 the Nobel Prize was awarded to Banting and Macleod, but Banting shared his prize with Best, and Macleod did likewise with Collip

In the early days of science, one of the most direct ways to find out an organ's function was to remove it surgically and then observe the effects of its loss on the organism. If the symptoms could be relieved by injecting a ground-up extract of the same gland, then the organ in question was an endocrine gland – a gland that releases hormones into the bloodstream.

By the early years of this century, this effective but less-than-subtle approach had been used to clarify the role of several organs, including the thyroid gland. But, although researchers suspected that the pancreas was an endocrine gland, they ran into problems when they tried to use this method to demonstrate that the pancreas made a hormone responsible for controlling glucose metabolism. When the pancreas of a dog was removed, the animal developed the symptoms of diabetes. But injecting an extract of ground-up pancreas failed to relieve the symptoms, a result that recurred in several experiments.

Frederick Banting, a Canadian doctor, first had the idea that the digestive enzymes also made by the pancreas could be destroying any other active substances that were being made there. He persuaded the head of the physiology department at the University of Toronto, John J R Macleod, to let him have a laboratory, ten dogs and an assistant – Charles Best.

The dogs were subjected to one of two treatments. The pancreatic ducts of the dogs in group 1 were tied so that the animals could not produce pancreatic juice. Over the course of several weeks the acini, the cells that make pancreatic juice, degenerated, leaving just the islets of Langerhans functional. These dogs did not develop diabetes. The pancreases of the group 2 dogs were removed, and these animals did become diabetic.

Banting then made a pancreatic extract from the group 1 dogs and injected it into the diabetic dogs of group 2. The result was

dramatic – an instant reduction of blood sugar that could be achieved repeatedly in several different animals. There was great excitement, and then disappointment as the extracts were tried in humans and found to be too impure (they produced fever). Banting and Best enlisted the help of the biochemist James Collip. He went on to make a purer extract than had previously been possible, which could be used for people.

Some people feel that animal experiments are a bad idea, but the few dogs used in these early experiments enabled Banting's team to change a progressive and fatal illness into a chronic, manageable condition. As a direct result of these experiments, insulin became available to the world's diabetics. Today there are over 300 000 insulin-dependent diabetics in the UK, and as many as 30 million worldwide. Without insulin, they would not be alive.

3 THE LIVER PLAYS A CENTRAL ROLE IN HOMEOSTASIS

The liver receives blood from two sources: from the intestines via the hepatic portal vein, and from the hepatic artery (see Fig 14.5b). The composition of blood that flows into the liver can fluctuate greatly, depending on factors such as the timing and nature of the last meal, but the content of blood leaving the liver is remarkably constant.

THE STRUCTURE OF THE LIVER

Before we go on to look at the functions of the liver in more detail, it is important to understand its structure (see Fig 14.5 on the next page).

Blood is brought into the liver by two blood vessels. The 30 per cent that arrives in the **hepatic artery** is oxygenated blood, while the 70 per cent delivered by the **hepatic portal vein** contains relatively little oxygen but is rich in nutrients that have been absorbed from the intestines. The **hepatic vein** removes blood from the liver.

The liver consists of hundreds of thousands of **lobules**, each about 1 mm in diameter, which surround branches of the hepatic vein. Channels called **sinusoids** radiate out from each central vein. These are surrounded by rows of liver cells called **hepatocytes**. These apparently unspecialised cells perform the majority of the liver's functions, and the composition of the blood changes as it flows along each sinusoid towards the central vein.

Dotted along the sinusoids are numerous white cells called **Kupffer cells** (see Fig 14.5e). These are phagocytes (see Chapter 17) that engulf bacteria and debris. Parallel to the sinusoids are fine channels called **bile canaliculi** (singular: canaliculus). Hepatocytes secrete the constituents of bile into the canaliculi, and this drains into the gall bladder. Bile is covered in detail in Chapter 11.

THE MAIN FUNCTIONS OF THE LIVER

The thousands of different chemical functions performed by the liver contribute greatly to the overall composition of the blood. They can be grouped under a few basic headings:

- Control of blood glucose levels
- Control of amino acid levels
- Synthesis of plasma proteins
- Synthesis of fetal red blood cells
- Destruction of red blood cells
- Detoxification
- Production of bile
- Control of lipid levels
- Storage of vitamins
- Cholesterol formation

THE CONTROL OF AMINO ACID LEVELS

The human body cannot store proteins. Every day an adult needs a minimum amount (40 to 60 grams, about the weight of one egg) to provide the amino acids the body needs to repair and grow new cells. Most people take in more than this in their diet, and the liver breaks down any amino acids that are not used. Obviously, growing children and pregnant or breast-feeding mothers need more protein.

✔ **REMEMBER THIS**

Although control of blood sugar is a 'whole-body' process, the liver plays a central role. It is the first organ to receive the blood from the intestines that contains high levels of food molecules such as sugars following a meal. Liver cells are acutely sensitive to insulin and are also particularly rich in the enzymes involved in glucose metabolism.

? TEST YOURSELF

B During their experiments, Banting and Best noticed that flies congregated around the urine of the diabetic dogs. Suggest an explanation for this.

? TEST YOURSELF

C Hepatocytes have many mitochondria and microvilli. What does this suggest about their function?

Fig 14.5 Structure of the liver

(a) A liver prepared for organ transplant surgery. This liver has been cut to fit into the recipient

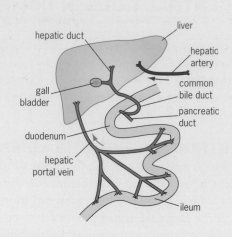

(b) The blood supply to the liver and associated organs. The liver weighs about 1.5 kg in a normal adult

(c) The liver consists of about a half to one million cylindrical blocks of cells called lobules

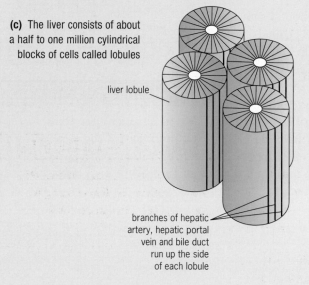

liver lobule

branches of hepatic artery, hepatic portal vein and bile duct run up the side of each lobule

Sinusoids: spaces through which blood flows as it passes from the hepatic portal vein and hepatic artery to the central vein

Kupffer cell

Bile canaliculus: tiny tube into which bile is secreted

bile duct

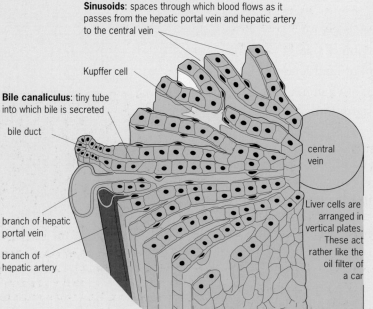

central vein

Liver cells are arranged in vertical plates. These act rather like the oil filter of a car

branch of hepatic portal vein

branch of hepatic artery

(d) A micrograph of liver lobules. Radiating sinusoids drain into the central veins

(e) The fine structure of the liver. Blood is delivered to the liver cells in branches of the hepatic artery and the hepatic portal vein. As blood flows along the sinusoids, some chemicals are removed while others are added by secretion. The liver secretes bile into the canaliculi. These drain into small branches of the bile duct

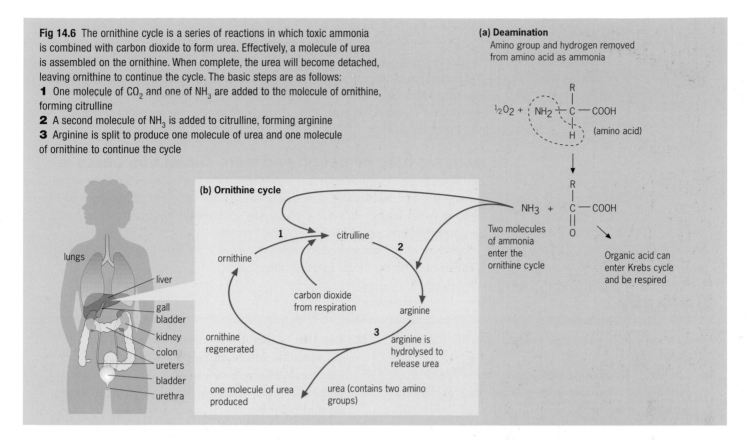

Fig 14.6 The ornithine cycle is a series of reactions in which toxic ammonia is combined with carbon dioxide to form urea. Effectively, a molecule of urea is assembled on the ornithine. When complete, the urea will become detached, leaving ornithine to continue the cycle. The basic steps are as follows:

1 One molecule of CO_2 and one of NH_3 are added to the molecule of ornithine, forming citrulline

2 A second molecule of NH_3 is added to citrulline, forming arginine

3 Arginine is split to produce one molecule of urea and one molecule of ornithine to continue the cycle

(a) Deamination
Amino group and hydrogen removed from amino acid as ammonia

Two molecules of ammonia enter the ornithine cycle

Organic acid can enter Krebs cycle and be respired

(b) Ornithine cycle

Amino acids, like many other digested foods, reach the liver via the hepatic portal vein. They can be:

- **Deaminated**. This process removes the amino group (NH_2) of an amino acid and forms ammonia (NH_3): Fig 14.6(a). The organic acid residue is usually respired, while the toxic ammonia is quickly converted into a more harmless substance, urea, via the ornithine cycle: Fig 14.6(b). The kidneys remove urea from the body (see Chapter 16).

- **Transaminated**. There are 8 essential amino acids (10 in children) that must be present in the diet. The remaining 12 are termed non-essential amino acids because they can be made in the liver by transamination. This process involves the transfer of an amino group from an amino acid to an acid (derived from carbohydrate metabolism), thereby making a new amino acid.

- Used to synthesise plasma proteins such as fibrinogen.

- Released unchanged into the general circulation (most cells need a supply of amino acids for protein synthesis).

DETOXIFICATION AND HORMONE BREAKDOWN

The liver concentrates and **detoxifies** (breaks down) many harmful chemicals. Some, such as hydrogen peroxide, are produced by the body itself. Others, alcohol and food additives for example, come from outside. Drinking large amounts of alcohol over a long period of time can cause the death of liver cells, followed by replacement with connective tissue ('scar' tissue). This is known as **cirrhosis** of the liver. (Fig 14.7)

The liver also breaks down many circulating hormones. Removal of hormones from the blood is an important aspect of the control process as it ensures that hormones act only for as long as they are needed. Insulin, for instance, is rapidly metabolised by the liver and has a half-life of about 10 to 15 minutes.

? TEST YOURSELF

D Why can't we take in a week's supply of protein in a single meal by eating one large steak or omelette?

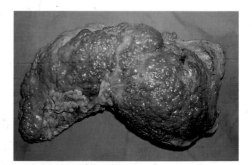

Fig 14.7 The liver of an alcoholic who developed cirrhosis of the liver. In the liver, alcohol (ethanol) is converted into ethanal by the enzyme alcohol dehydrogenase. Ethanal is toxic and seems to be responsible for much of the long-term alcoholic liver damage, although the exact mechanisms involved are unclear

STORAGE

The liver stores relatively large amounts of vitamins A, D and B_{12}, enough to supply the body for several months. Other vitamins, such as most of the B complex, and the minerals copper and iron are also stored but in smaller amounts. This high vitamin and mineral storage capacity explains why eating liver is good for you, even though some people don't find it a pleasant experience.

MANUFACTURE OF BLOOD PROTEINS AND BLOOD CELLS

The liver makes many important blood proteins, including the most abundant plasma protein, **albumin**, fibrinogen and other substances involved in blood clotting (see Chapter 17). Given a supply of vitamin K, the liver can make prothrombin and other blood clotting factors.

The liver of a fetus manufactures red blood cells, but this complex process is taken over by the bone marrow after birth.

SUMMARY

When you have studied this chapter, you should know the following:

■ The word **homeostasis** means 'steady state'. Homeostatic processes keep conditions in the body within narrow limits.

■ Homeostasis is usually maintained by **negative feedback mechanisms**. The body detects a change in a particular internal factor, such as core body temperature, and then activates a corrective mechanism to reinstate the normal level.

■ Blood glucose levels are monitored by the **islets of Langerhans** in the pancreas. If levels get too high, β cells in the islets secrete **insulin**. If blood glucose gets too low, α cells secrete **glucagon**.

■ **Insulin**, a peptide hormone, increases the permeability of cell surface membranes to glucose. It also appears to activate intracellular enzyme systems that convert glucose to glycogen, fat and protein.

■ **Glucagon** promotes the breakdown of **glycogen**, releasing more free glucose into the blood.

■ The liver is a large organ that plays a major role in controlling the composition of the blood.

? EXAM QUESTIONS

1 An experiment was carried out to determine how the level of glucose in the blood varies in a normal person. Samples of blood were taken at hourly intervals for a period of 18 hours, from 06.00 to 24.00 hours. During this period, three meals were taken: breakfast at 7.00, lunch at 13.00 and tea at 16.00.
The results are shown in the graph below.

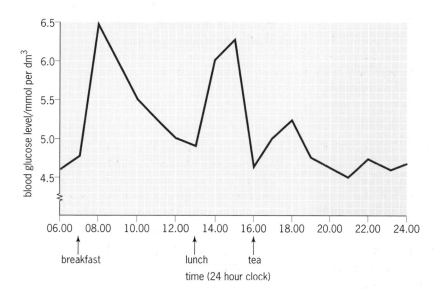

a) **(i)** Explain why the blood glucose level rises sharply between 7.00 and 8.00 hours.
(ii) Explain how the changes in blood glucose level between 08.00 and 12.00 hours are brought about.

b) Describe and explain two ways in which the changes in the blood glucose level after lunch (13.00 hours) and tea (16.00 hours) differ from those after breakfast.

c) Explain why it is important that the level of glucose in the blood is controlled.

Edexcel HB6 (Synoptic) June 1999 Q5

2 Use appropriate terms, or phrases, to complete the following passage. Each term or phrase may only be used once.

The importance of maintenance of constancy both within cells and within organisms was first appreciated by the French physiologist Claude Bernard. This maintenance of constancy is referred to as _____ . Many examples of this are carried out by the liver, for example the maintenance of constant blood glucose levels. Two other examples in the liver are the maintenance of _____ and _____ .

If the blood sugar level rises, the **β cells** in the _____ , which are found in an organ called the _____ , release a hormone called _____ . This decreases the blood sugar level by causing glucose to be converted into a polysaccharide called _____ which is stored in the liver. If the blood glucose levels fall, the hormone _____ is produced which causes this polysaccharide to be converted into glucose. Control mechanisms that involve a corrective procedure which returns levels to the norm, or set point, are referred to as examples of _____ . Control of a target organ by hormones, transported in the blood from the endocrine gland in which they are produced, is an example of chemical communication at the level of the organism. The binding of hormones with receptors on the cell surface membrane causes the release of secondary messengers within the cytosol which influence enzyme activity. This is an example of chemical communication at the _____ .

OCR 6911 Cells and Structures June 2000 Q2

3 Read the following passage.

Diabetes

Diabetes mellitus is a group of disorders that all lead to an increase in blood glucose concentration (hyperglycaemia). The two major types of diabetes mellitus are type I and type II. In type I diabetes there is a deficiency of insulin. Type I diabetes is also called insulin-dependent diabetes mellitus because regular injections of insulin are essential. It most commonly develops in people younger than age twenty.

Type II diabetes most often occurs in people who are over forty and overweight. Clinical symptoms may be mild, and the high glucose concentrations in the blood can often be controlled by diet and exercise. Some type II diabetics secrete low amounts of insulin but others have a sufficient amount or even a surplus of insulin in the blood. For these people, diabetes arises not from a shortage of insulin but because target cells become less responsive to it. Type II diabetes is therefore called non-insulin-dependent diabetes mellitus.

a) Describe how blood glucose concentration is controlled by hormones in an individual who is not affected by diabetes.

b) Suggest how diet and exercise can maintain low glucose concentrations in the blood of type II diabetics.

c) Glucose starts to appear in the urine when the blood glucose concentration exceeds about 180 mg dm^{-3}. Explain how the kidney normally prevents glucose appearing in urine.

AQA (NEAB) BY03 Physiology June 2000 Q8

KEY SKILLS **ASSIGNMENT**

LIVING WITH DIABETES

Having diabetes doesn't mean you can't live life to the full, but unlike other chronic conditions such as asthma, someone with diabetes cannot just take their medication and forget there was ever a problem. Instead, they must become experts in the management of their own condition, matching up dietary intake with insulin doses and exercise regimes in order to keep their blood glucose within definite limits.

This is an account of the way in which one person, Anna, deals with her diabetes. It should be stressed from the start that the condition affects different people in different ways. A blood sugar level that would have one person fainting on the floor may produce no symptoms in another. Bear this in mind as you read on.

1 Explain how a non-diabetic person responds to a rise in blood sugar level. Your answer should include the effect of insulin on cells.

2 What is the difference between a chronic and an acute condition?

Types of diabetes

There are two basic types of diabetes: Type I and Type II. Type I is also known as 'early onset' or 'insulin-dependent' as it appears in early life and controlling it usually requires insulin. At the root of the problem is thought to be an auto-immune reaction in which the body's antibodies destroy its own insulin-producing cells. As these cells cease to produce insulin, less insulin is released into the blood after a meal. Without sufficient insulin, most of the glucose that enters the blood from the intestines after digestion cannot get into cells. Glucose accumulates in the blood, and the cells, which are starved of their main fuel, turn to alternatives – lipid and protein.

Five of the major symptoms of Type I diabetes are:

- excessive thirst;
- excessive urinating;
- weight loss;
- glucose in the urine;
- the fruity smell of ketones on the breath (a by-product of lipid metabolism).

3 Using the information given above, suggest explanations for each of the five symptoms.

The story of Anna

Anna, who is now in her forties, was diagnosed as a diabetic at the age of 18 months. This is early, but her brother had also developed diabetes (when he was 4) and so her mother was acutely aware of the symptoms.

4 It is thought that people can inherit a susceptibility to diabetes, but it needs an environmental trigger of some sort, such as exposure to a virus. Does Anna's story support this idea?

Anna has come to terms with her diabetes and has found that management is easiest within a routine. Overall, the aim is to keep her blood glucose within the range of 4 to 11 millimoles per litre. (Non-diabetics maintain their blood glucose to within about 4 to 9 millimoles per litre.)

When she wakes up in the morning, Anna tests her blood sugar. This is done by pricking her finger and putting a drop of blood on a test strip, similar to that shown on page 123. The level of blood glucose is then obtained by either matching up the colour to the chart or by placing the strip into a machine which takes the reading automatically. Anna would expect her glucose levels to be about 3 at this time in the day. Anything less than 4 is regarded as hypoglycaemic, although Anna gets no symptoms until the value approaches 1.

5 **a)** What fraction of a mole is a millimole?
b) Suppose Anna has a blood glucose reading of 10 millimoles. If the relative molecular mass of glucose is 180, how many grams of glucose are in one litre of her blood?

After taking the reading, Anna injects herself with insulin. Two types of insulin are now available: fast-acting and slow-acting. Anna injects 2 units of fast-acting and 14 of slow-acting. Fast-acting is normal, soluble insulin and works straight away. Slow-acting insulin is attached to a retarding agent that releases the insulin slowly over the next few hours.

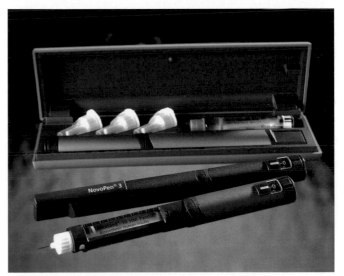

Fig 14.A1 For diabetics, there is now an alternative to syringes. Special pen-like injection devices such as this Novopen are widely available. They are very accurate and discreet to use

KEY SKILLS **ASSIGNMENT**

6 a) Suggest the advantage to diabetics of having slow-acting insulin.

b) Suggest why you must have some fast-acting insulin.

Until relatively recently the insulin was obtained from animal pancreas. This was less than perfect because the non-human insulin slowly brought about an immune reaction. Eventually, the 'foreign' insulin had little effect. With modern genetic engineering, however, it is now a relatively simple task to produce human insulin on a large scale – see Fig 14.A2.

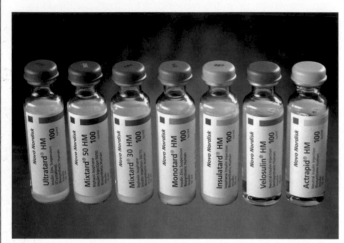

Fig 14.A2 Human insulin is made on a large scale by genetic engineering and is supplied in different doses

7 Why can't insulin be taken orally?

After breakfast, Anna goes to work. In her handbag are glucose tablets and some biscuits as well as her insulin and testing equipment. Anna has a sweet drink and a couple of biscuits at break.

8 It is frequently said that 'diabetics can't eat sugar' but this is obviously not true. Suggest why Anna needs the mid-morning biscuits.

Going 'hypo' and 'hyper'

A Type I diabetic who takes no insulin experiences the five symptoms listed on the previous page – they have gone 'hyper'. In Anna's case, the early warning signs are headaches and feeling tired or lethargic.

On the other hand, once a diabetic has taken insulin they must take care to keep their blood glucose levels up or they might go 'hypo'. For Anna, symptoms of too little blood sugar are similar to going hyper: tiredness, irritability and an inability to concentrate. Often, other people notice the symptoms first.

9 The terms hypo and hyper are short for hypoglycaemia and hyperglycaemia. What do these words mean?

10 If Anna finds that she is going hypo, suggest what she can do.

Anna has a normal lunch and another snack in mid-afternoon. At tea time she takes her blood sugar again – it is normally about 8 millimoles. After this she takes more insulin, 2 units of short-acting and 10 of slow-acting. If she is going out for a meal, and will eat more than normal, she will take an extra two units of short-acting insulin. She must also be careful what she drinks, because alcohol has a powerful effect on glucose metabolism.

Before bed time, Anna has a sweet drink, usually chocolate, and checks her blood sugar again. It needs to be around 8 millimoles because she will not eat for another 8 hours or so.

11 a) Why is it important that Anna does not go 'hypo' in the night?

b) Suggest what Anna does if her reading is, say, 6 units?

12 From what you have read of Anna's daily routine, sketch a graph of her blood glucose levels over a 24 hour period.

Type II diabetes

Many diabetics have Type II diabetes, This is also known as 'late onset' and the treatment depends on the underlying cause. In most Type II cases, the condition is due to a combination of one or more of the following:

- Muscle and fat cells do not respond to insulin.

- The body can't produce enough insulin to meet demand.

- Liver cells release too much glucose from their stores.

For many Type II diabetics, a diet and exercise plan may be enough to keep blood sugar levels within normal limits.

15

Temperature regulation

The emperor penguin (*Aptenodytes forsteri*) is over a metre tall and weighs about 35 kilograms, roughly half the weight of an average adult man. The males fast for two months while they incubate their egg and lose as much as 15 kg of body weight. Like their fathers, the chicks instinctively huddle together to reduce their heat loss

The emperor penguin lives in the Antarctic, one of the coldest places on Earth. Surprisingly, it is a warm-blooded animal and, even more surprising, it breeds in the middle of winter, when temperatures are at their lowest. At the beginning of winter, the penguins leave their feeding grounds in the sea and walk to their breeding grounds, up to 100 kilometres away. There, the female lays one egg, gives it to the male and then promptly returns to the open sea to feed.

The temperature plummets to −40 °C but the males have no choice but to stand on the ice for the next two months, incubating the egg between their feet and a fold of fat. If the egg accidentally rolls on to the ice, the male must gather it up again immediately, or the chick will die. Despite these dreadful conditions, the core body temperature of the male penguin remains at 38 °C, about a degree higher than that of most mammals.

The penguins maintain their body heat by respiring stored fat and by huddling together. Penguins are roughly cylindrical and snuggling up to their neighbours greatly reduces the surface area that is directly exposed to the freezing air. In an impressive example of social co-operation, the birds take it in turns to stand on the outside of the crowd. Scientists estimate that without this huddling behaviour, the males would not have enough fat to power the process of homeostasis that keeps them warm, let alone make the long return journey back to the open sea.

1 TEMPERATURE CONTROL AND HOMEOSTASIS

Life can exist at almost all the temperatures encountered on the Earth's surface, from the extremes of cold that occur at the polar regions to the extremes of heat that occur in hydrothermal vents. This is possible because it is the *internal*, not the *external* temperature that is important to an organism. Animals that can withstand extremes of external cold, such as the emperor penguins, do so only because they maintain their body temperature inside this 'operating range', despite the outside temperature.

WHEN THE TEMPERATURE GETS TOO HIGH

Why do animals die when the external temperature exceeds a certain limit? A common theory says that some vital enzymes become denatured, but it is probably not the whole story. It is more likely that death from excess heat is due to a metabolic imbalance, caused by enzymes working at different rates. As the temperature increases some enzymes work faster than others: some intermediate chemicals build up while some vital end-products become scarce.

WHEN THE TEMPERATURE GETS TOO LOW

Life can survive in a metabolically inactive **dormant state** at sub-zero temperatures. Sperm, eggs and even embryos kept in liquid nitrogen at −196 °C can be thawed and used successfully in infertility treatment (see Chapter 26). However, only a few organisms remain active when their body temperature drops below freezing (Fig 15.1). In most organisms, very low temperatures cause damaging ice crystals to form inside and also between cells.

REMEMBER THIS

In Chapter 5 we saw that enzymes, the chemicals that control metabolism, are temperature sensitive. Enzyme activity increases with temperature and this increases our metabolic rate. The Q10 is a value that describes the effect of a 10 °C rise in temperature on the rate of reaction. Most enzymes have Q10 values of between 2 and 3. This means that they double or triple their activity with every 10 °C rise in temperature, up to the temperature when the enzymes are denatured by heat.

Fig 15.1 This Antarctic ice-fish lives in waters where the temperature is a constant −1.8 °C. The fish has an 'antifreeze' in its body fluids. This is actually a high concentration of a glycoprotein that prevents small ice crystals from developing into large ones. If the ice-fish swims into water with a temperature greater than 6 °C, it dies

If ice crystals damage the cellular structure, how do some cells survive being frozen in liquid nitrogen? It is because it takes time for ice crystals to develop. If you immerse a sample in liquid nitrogen, it freezes so quickly that ice crystals don't form. When the cells and organelles are thawed they are able to function normally as they have not been damaged.

2 MECHANISMS OF TEMPERATURE CONTROL IN ANIMALS

Different animals control their body temperature in different ways. Many animals (fish, insects, amphibians, etc) have a body temperature that is more or less the same as their surroundings. Mammals and birds usually maintain a constant body temperature, despite the external temperature.

These two categories of animal are commonly described as cold-blooded and warm-blooded. But these terms are not really satisfactory: a cold-blooded animal, for example, is often not cold. In bright sunshine on a hot day, the core temperature of a reptile or amphibian may be higher than that of a mammal (Fig 15.2). Nor are the terms **poikilothermic** (cold-blooded) and **homoiothermic** (warm-blooded) strictly accurate. The word *poikilos* means changeable, while *homeo* means constant. However, many cold-blooded animals, especially aquatic ones, have a remarkably constant body temperature, owing to the stability of their surroundings. Hibernating mammals, on the other hand, show a dramatic drop in their core temperature during their dormancy.

Fig 15.2 This crocodile is gaping to keep cool. The external temperature is over 45 °C and so its blood is far from cold. Evaporation from its large mouth helps it to cool down

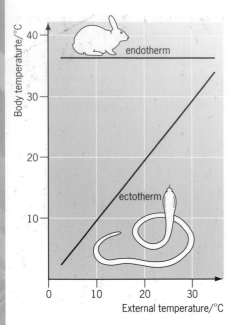

Fig 15.3 The temperature of an ectotherm tends to be the same as the surroundings. An endotherm maintains a more or less stable body temperature over the whole range of external temperatures normally encountered

To overcome these problems, we now use two other terms: **ectotherm** ('heat from outside') and **endotherm** ('heat from within'):

● **Ectothermic organisms** can control their body temperature only by changing their behaviour (Fig 15.3). All animals except mammals and birds are ectotherms.

● **Endothermic organisms** maintain a stable core body temperature using physiological and behavioural means (Fig 15.3). Mammals and birds are the only endotherms.

THERMOREGULATION IN ECTOTHERMS

Ectothermic animals **thermoregulate** (control their body temperature) with difficulty. Aquatic ecotherms, such as fish, usually have the same temperature as their surroundings. Their gills, which have a large surface area to collect oxygen, are a liability in terms of heat loss because they carry large amounts of blood to within a few micrometres of the water.

Air-breathing ectotherms such as reptiles thermoregulate more successfully. These organisms do not have the fine physiological control over their temperature practised by birds and mammals, but they manage to control their body temperature to a large extent by their behaviour.

The Peruvian mountain lizard shown in Fig 15.4 is a classic example. This reptile lives in an environment where the days are hot but the nights are very cool. After a cold night, the lizard's body temperature is well below the optimum required for activity. Its movement is sluggish and it cannot hunt or avoid predators. So, early in the morning, the lizard emerges from its burrow and basks in the sun until its body temperature reaches about 35 °C. It is then fully active and goes off in search of food. As the day wears on and gets hotter, the lizard seeks shade to avoid overheating.

Fig 15.4 (a) The Peruvian mountain lizard, sunning itself in the morning (right)
(b) The Peruvian mountain lizard comes out in the morning when the air temperature is still below freezing. By positioning itself at right angles to the Sun's rays, the body temperature of the lizard rises quickly and is soon 30 °C higher than the surrounding air. Many other species also show this behaviour

THERMOREGULATION IN ENDOTHERMS

Endotherms regulate their body temperature so that it stays at a constant level. In the section that follows, we look at the mechanisms of thermoregulation in the mammal, using the example of a human.

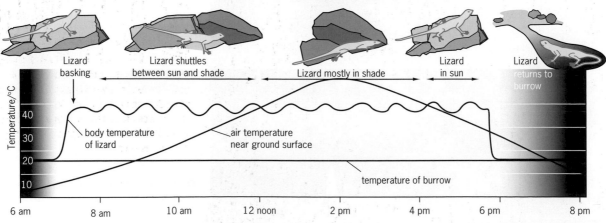

3 HUMANS AND HEAT

Some animals, such as polar bears, are obviously adapted to withstand the cold. In contrast, humans almost certainly evolved in Africa, and, as a species, we are adapted to a warm climate. We have a thin covering of insulating fat and our body hair is sparse. Most hairs are tiny and no use for insulation. We are one of the few animals to be covered in sweat glands, and our skin can make melanin, a dark pigment that blocks out harmful ultraviolet light.

It seems that we were able to spread to the colder areas of the planet only because of our ability to control our environment. We could build shelters, make clothes and control fire. These skills more than made up for the fact that we had few physical features to allow us to cope with the cold.

The core temperature of the human body remains reasonably constant at around 37 °C. It can fluctuate by a degree or so (more during fever) but it is generally very stable. Fig 15.5 is a thermal image of an adult male, showing the definition of the body core; this includes the trunk, the head and the upper part of the arms and legs.

Heat is produced inside the body as a by-product of metabolic reactions. Heat production occurs throughout the body, and is especially high in working muscles. It is often stated that the liver is a particular source of heat, but tests have shown that the temperature of blood in the hepatic vein is no higher than in other vessels, suggesting that heat production occurs more generally in the organs of the body. Since body heat is a by-product of metabolism, the amount of heat produced depends on the metabolic rate. This can be increased by doing more exercise and by the secretion of hormones such as thyroxine or adrenaline (Chapter 22).

However, the principal way we control our body temperature is by increasing or decreasing heat loss to the environment, according to our needs. In humans, the basic mechanism that underlies temperature control, or **thermoregulation**, involves part of the brain called the **hypothalamus**. This acts as a thermostat. It can detect the temperature of the blood that passes through it and, if the temperature of the blood increases or decreases even slightly, the hypothalamus initiates corrective responses such as sweating or shivering.

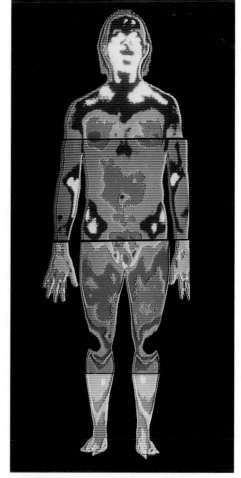

Fig 15.5 This is a thermal image of a man taken with a camera sensitive to infrared radiation. Areas of higher skin temperature look red or orange, cooler areas look blue or purple. The temperature of the skin can vary greatly, but the core temperature remains more or less stable

Fig 15.6 When it's cold, we may automatically assume a fetal position in bed to keep warm: when the arms and legs are drawn into the body this reduces the surface area that can lose heat. When it's hot we do the opposite, spreading out as much as possible

For any organism to maintain a stable body temperature, its heat loss must equal its heat gain. **Homoiotherms** such as ourselves, who live in a temperate climate, produce our own internal body heat to keep warm but we also control the amount of heat we lose to the environment.

Heat can be gained or lost in four different ways:
● conduction
● convection
● evaporation
● radiation

CONDUCTION

Conduction involves the transfer of heat between two objects that are in contact with each other. Heat is always conducted from a region of higher temperature to a region of lower temperature. When you sit on a cold seat, for example, heat is conducted from your body into the seat, until both are approximately the same temperature.

The efficiency with which a material conducts heat is called its thermal conductivity. Different materials have different conductivities. Air has a low thermal conductivity, water has a much higher one (Figs 15.7 and 15.8).

A clothed human walking in air at 15 °C could maintain body temperature comfortably. If immersed in water at that temperature, it would not be long before their core temperature dropped and hypothermia began to set in. The heat loss into the water would be much greater – water can 'draw out' heat approximately 25 to 30 times faster than air at the same temperature.

Materials with a low thermal conductivity are very good insulators. Animals with fur keep warm largely by trapping air between the hairs. Humans rely more on fatty (adipose) tissue for insulation. This has a lower thermal conductivity than other body tissues and so conducts heat more slowly to the surroundings.

Fig 15.7 When the *Titanic* sank in the North Atlantic, the water temperature was similar to that of the air – a little above freezing. However, as we now know, those who remained dry in the lifeboats had a far greater chance of survival. The high thermal conductivity of the water drew the heat out of the unfortunate victims in the sea, most of whom died from hypothermia

Fig 15.8 Most divers wear wet suits, even in tropical waters, because the body loses heat rapidly to any water that is below 29 °C. The wet suit traps a layer of water between the skin and the rubber. This warms up quickly as a result of conduction from the skin. The warm water layer cannot escape and so further heat loss is slow. For very cold water, dry suits are available. These trap a layer of air next to the skin. This provides even better thermal insulation

Extension Heat: Where does it come from? Where does it go?

CONVECTION

Convection is heat transfer due to currents of air or water. A person immersed in cold, absolutely still water would be able to heat up the water immediately next to the skin, and could reduce their heat loss to some extent by not moving.

However, this situation does not happen in real life. For a person who has fallen into the sea or a river, there are usually strong currents that continually move the water over the skin, causing water to be lost quickly by convection. Similarly, fast-moving air causes greater heat loss by convection. You have probably heard of the **wind chill factor**. A cold windy day feels a lot worse than a cold still day, even when the actual air temperature is the same.

In air, our clothes reduce heat lost by convection by trapping a layer of air next to the skin. In water, divers reduce the risk of heat loss by convection by wearing wet suits and dry suits (see Fig 15.8).

EVAPORATION

Evaporation is the change in state from a liquid to a gas. The evaporation of water uses up a large amount of energy, and this is known as the **latent heat of evaporation**. What this means for everyday life is that evaporation of water from a surface has a great cooling effect. Anyone who has ever stepped out of the bath and stood in a draught will have felt the power of this cooling. It is also why sweating is usually an effective way of losing heat when we get too hot. Even hot air – such as from a hair dryer – can have a cooling effect provided that the skin is wet so that evaporation can take place.

RADIATION

Radiation is the loss of infrared heat into the surroundings. A human sitting in a room at 20 °C radiates heat into the surrounding air, particularly from the exposed skin on the head, neck and hands. Under normal situations, at rest at comfortable room temperature, most of our heat loss is due to radiation. The heat that we radiate is in the infrared range. This is why infrared cameras can be used to search out humans and other warm-blooded animals from their colder surroundings.

In conduction, convection and evaporation, heat transfer occurs as a result of the movement of molecules. Radiation is fundamentally different: it does not depend on molecular movement. This explains why radiated heat can pass through a vacuum.

When we encounter a particularly warm or cold environment, temperature receptors in the skin inform the hypothalamus. They also stimulate the higher, **voluntary**, centres of the brain. This means that we 'feel' hot or cold and decide to do something about it such as changing position (Fig 15.6), changing our clothing or turning the heating up or down. Often, this behavioural response corrects the situation without the need for any physiological response.

Before we can look at temperature control in detail, we need to understand a little about the processes of heat transfer (see the Extension), and to study the structure of the human skin.

SCIENCE FEATURE Freezing humans

Can we keep people alive by freezing them? Can people suffering from incurable diseases be put into deep-freeze until such time as a cure is found, and then revived? The simple answer is no.

Interestingly, human life can survive sub-zero temperatures in a metabolically inactive **dormant state**. Sperm, eggs and even embryos kept in liquid nitrogen at −196 °C, can be thawed out and used successfully in infertility treatment (see Chapter 26). However, it is not possible to freeze people and bring them back to life. One of the main problems is that the formation of ice crystals destroys the cells, but a critical factor in maintaining life seems to be the timescale involved: ice crystals take time to develop. When very small samples are immersed in liquid nitrogen, the freezing process happens rapidly: all the molecules simply 'lock' in position and ice crystals do not form.

The cells and organelles are not damaged, and, when thawed, function normally. This technique, however, only works with tiny samples of tissue, which freeze instantly and entirely.

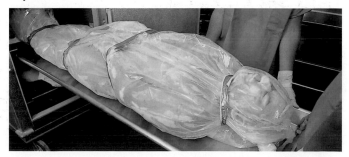

Fig 15.9 If we developed a way of instantaneously freezing all the cells of a fully developed human, could they be thawed out and live to tell the tale? Only time will tell

4 THE PROBLEMS OF TEMPERATURE EXTREMES

Humans can maintain a stable body temperature over a wide range of external temperatures. But what happens when we are no longer able to thermoregulate effectively, and our internal temperature changes? What happens to people stranded in the desert under a baking sun, or to those trapped in freezing water?

Experiments have been carried out in which a naked man is asked to sit in a room while the temperature around him is gradually raised or lowered. Fig 15.10 shows a generalised graph that plots what happens.

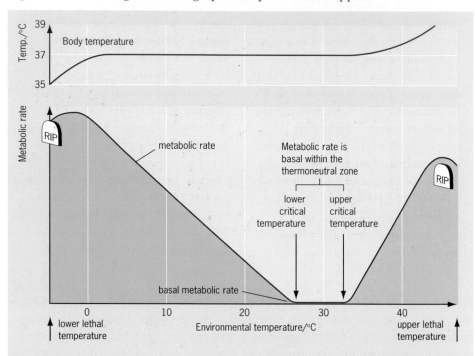

Fig 15.10 The effect of temperature on metabolic rate

Throughout the range 15 to 30 °C the naked man is able to maintain a constant core temperature. A point is reached, however, when his sweating and vasodilation are no longer able to keep him cool and he is said to have reached the **upper critical temperature**. After this, his metabolic rate starts to rise, but the experiment is always stopped at this point. If it were not, a positive feedback would begin. The man's enzymes would work faster and consequently make even more heat, raising his temperature further. At about 42 to 44 °C, the **upper lethal temperature** would be reached and he would die.

If the same type of experiment is done, but temperature is lowered, the room eventually becomes so cold that the body is no longer able to maintain a constant core temperature: the **lower critical temperature** has been reached. At this point the metabolic rate begins to rise, and this helps the situation by generating more heat. Again, the experiment is always halted at this stage but, if heat loss were allowed to continue, a **lower lethal temperature** would be reached and the subject would die. Find out more about this in the Assignment at the end of the chapter.

? TEST YOURSELF

A Why does the upper critical temperature vary with humidity?

SEE QUESTIONS 1 AND 2

5 THE STRUCTURE OF HUMAN SKIN

The skin is in direct contact with the external environment and so plays a central role in maximising or minimising heat loss. Fig 15.11 shows the structure of mammalian skin.

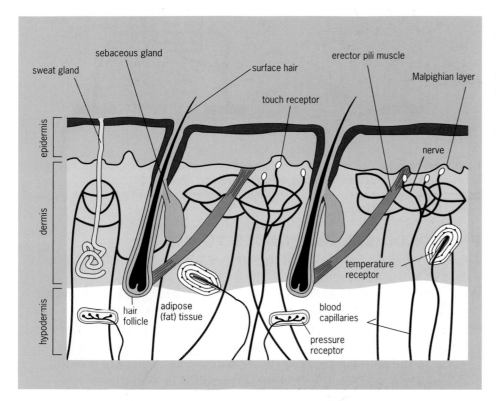

Fig 15.11 The structure of human skin. Human skin differs from that of most other mammals by being quite bare. We do not necessarily have fewer hairs, but most are tiny and no use for protection or insulation. We are also covered in sweat glands, and can lose heat by evaporation over most of our body surface. In contrast, most mammals have far fewer sweat glands, because evaporation from wet fur is not effective as a cooling mechanism

Structurally, the skin is divided into two layers, the outer **epidermis** and the **dermis** underneath. Forming the boundary between the two is the **Malpighian layer**, the function of which is to produce new epidermal cells by mitosis. These cells make the protein keratin, and once they are pushed up into the epidermis they flatten, die and dry up because they have no blood supply. We are constantly losing epidermal cells in a process called **desquamation**. House dust is mainly desquamated skin cells, and a significant proportion of the mass of our mattresses and pillows is due to accumulated skin cells (one estimate says that skin adds 1 per cent to the weight of your mattress every year).

The dermis is much thicker than the epidermis, and contains many different structures such as nerve endings, hair follicles and blood vessels, held together with elastic connective tissue. Beneath the dermis is the **hypodermis** that usually contains at least some **subcutaneous fat**. This fat storage tissue, called **adipose tissue**, determines human body shape to a large extent.

Human skin has several functions:

- It detects stimuli using cells that are sensitive to heat, cold, touch, pressure and pain.
- It prevents excessive water loss or gain.
- It plays a role in thermoregulation by adjusting heat loss according to the circumstances.
- It prevents entry of microorganisms.
- It secretes hair, fingernails and toenails.
- It allows subtle forms of communication. We secrete natural scents, **pheromones**, from modified sweat glands (see Chapter 22).

? TEST YOURSELF

B Are tattoos placed above or below the Malpighian layer? Explain.

6 THERMOREGULATION

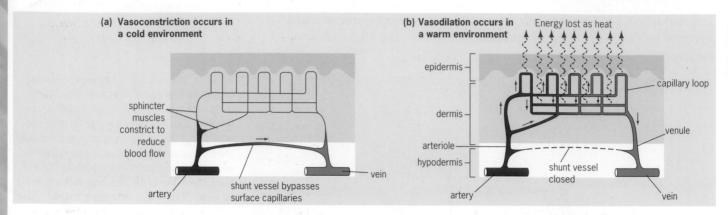

Fig 15.12 Blood flow to the skin can be altered by controlling the flow of blood in the arterioles leading to the surface capillaries

(a) When tiny sphincter muscles in the arteriole walls contract to prevent blood flow to the surface, blood is forced through shunt vessels and away from the skin, so less heat is lost

(b) When the sphincter muscles relax, the arteriole walls relax, allowing more blood to flow into the skin to increase heat loss

PHYSIOLOGICAL RESPONSES TO COLD

There are four main physiological responses when the hypothalamus detects a drop in blood temperature:

- We **shiver** when muscles contract and relax rapidly. Shivering muscles give out four or five times as much heat as resting muscles.
- **Vasoconstriction** occurs when the arterioles that lead to the capillaries in the surface layers of the skin **constrict** (narrow), so reducing blood flow to the skin (Fig 15.12a). This cuts down the amount of heat lost through the skin. Vasoconstriction is controlled by sympathetic nerves that pass from the **vasomotor centre** in the brain (Chapter 20). This centre, in turn, receives information from the hypothalamus.
- **Piloerection** means literally 'erection of hairs' and involves a reflex. In most mammals, piloerection makes the fur 'thicker', so that it traps more air to provide extra insulation. In humans, the **erector pili** muscles in the skin (see Fig 15.11) pull our tiny hairs upright, but only succeed in creating goose pimples.
- **Increased metabolic rate**. The body secretes the hormone adrenaline in response to cold. This raises metabolic rate and therefore increases heat production. People who live in cold conditions for a period of several weeks or months show a more permanent increase in metabolic rate because of the secretion of thyroxine (see Chapter 22).

PHYSIOLOGICAL RESPONSES TO HEAT

There are two main physiological responses to heat:

- **Vasodilation**. This occurs when arterioles that lead to skin capillaries dilate and shunt vessels are closed off, resulting in a greatly increased blood flow to the skin (see Fig 15.12b). As a result, more heat can be lost to the environment. Coupled with sweating, vasoconstriction is a very effective cooling mechanism.
- **Sweating**. Sweat is a salty solution made by sweat glands. Evaporation of sweat from the skin's surface leads to cooling. The efficiency of sweating depends on the humidity. In dry air, humans can tolerate temperatures of 65 °C for several hours. In humid air, however, when the sweat cannot evaporate, temperatures of only 35 °C (lower than the core body temperature) cause overheating.

SEE QUESTIONS 3 AND 4

? TEST YOURSELF

C Which of the responses to cold will be the least effective in humans? Explain your answer.

✔ REMEMBER THIS

A common misconception is that blood vessels move to the surface of the skin. Blood vessels are fixed. It is the blood flow through them that can be changed.

? TEST YOURSELF

D Why do you think that people in the tropics are recommended to take salt tablets regularly?

SUMMARY

After studying this chapter, you should know and understand the following:

- Animals are divided into two groups according to their ability to thermoregulate: **endotherms** ('heat from within') maintain a stable core temperature by behavioural and physiological means; **ectotherms** maintain a relatively steady body temperature using behaviour only.
- Humans, like all mammals, are ectotherms. We also say that mammals are warm-blooded, or **homoiothermic**. This means that they can maintain a stable core body temperature despite changes in the external temperature.
- To maintain a stable body temperature, heat loss must equal heat gain.
- Unless the climate is very hot, mammals regulate body temperature by producing heat as a result of metabolic reactions, and by controlling how much of it is released back to the environment. When mammals are too hot, they maximise heat loss; when they are cold, mammals try to lose as little heat as possible.

- Heat is transferred in four ways: **conduction**, **convection**, **evaporation** and **radiation**.
- Temperature receptors in human skin provide an early warning system for temperature change, allowing us to thermoregulate by our behaviour (altering position, clothing, etc). This often prevents the need for a physiological response.
- The **hypothalamus** can detect the temperature of the blood in the core of the body. If the blood temperature drops, the hypothalamus initiates corrective responses such as **vasoconstriction** and **shivering**. If the blood temperature increases, we respond by **vasodilation** and **sweating**. These are physiological responses.
- Extremes of both heat and cold can cause death. Death as a result of cold occurs as a result of **hypothermia**. This is common in people who have fallen into freezing water.

? EXAM QUESTIONS

1 The diagram below shows how the body reacts to environmental conditions which may lead to lowering of the body temperature.

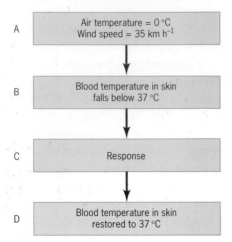

A — Air temperature = 0 °C
Wind speed = 35 km h⁻¹

B — Blood temperature in skin falls below 37 °C

C — Response

D — Blood temperature in skin restored to 37 °C

a) **(i)** Name the cooling effect caused by the conditions given in box A.
(ii) Suggest *one* response in box C which would lead to a rise in the blood temperature.
b) Explain what might happen to the exposed parts of the body if the blood temperature in the skin falls dramatically.
c) Name the condition in which the core temperature of the body may eventually fall below 35 °C.

Edexcel HB2 June 1999 Q3

2 A baboon is a large monkey. It normally regulates its body temperature in the same way as a human.
a) When a baboon's blood temperature rises, it is able to detect this rise and produce a coordinated response.
(i) Where are the receptors that detect the rise in temperature?
(ii) Describe how the response is coordinated.
b) Baboons that live in the Kalahari desert in southern Africa have to survive occasional very hot dry periods when no rain falls. During these periods, no drinking water is available.
Suggest why the body temperature of a baboon fluctuates much more during a hot dry spell when it cannot get water to drink.

AQA (NEAB) BY01 Processes of Life June 2000 Q7

3
a) Describe how blood vessels in the skin help regulate body temperature in response to hot external conditions.
b) Explain why regulation of body temperature is described as a negative feedback system.

IBO 410 Higher Paper 3 November 1998 QH2

4 Emperor penguins (*Aptenodytes forsteri*) are the only birds to live and breed through the severe Antarctic winter. In order to investigate how they survive at very low temperatures, the respiration rates of birds placed in a controlled temperature chamber were measured. The results are shown in the graph below.

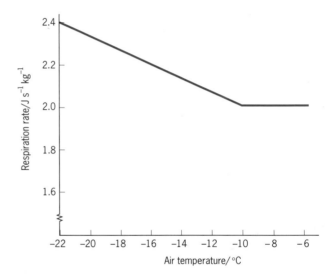

Respiration rate/J s⁻¹ kg⁻¹ (y-axis)
Air temperature/°C (x-axis)

a) **(i)** Using only the data in the graph, outline the effect of air temperature on the rate of respiration in emperor penguins.
(ii) Suggest reasons for the effect of temperature on respiration rate.
During the Antarctic winter female emperor penguins live and feed in the sea, but males have to stay on the ice to incubate the single egg that the female has laid. Throughout this time the males eat no food. After 16 weeks the eggs hatch and the females return. While the males are incubating the eggs they stand in tightly packed groups of about 3000 birds.
To investigate the reasons for standing in groups, ten male birds were taken from a colony at Pointe Geologie in Antarctica. They had already survived 4 weeks without food. They were kept for 14 more weeks without food in fenced enclosures where they could not form groups. All other conditions were kept the same as in the wild colony. The mean air temperature was minus 16.4 °C (−16.4 °C). The

EXAM QUESTIONS

composition of the captive and the wild birds' bodies was measured both before and after the 14 week period of the experiment. The results in **kilograms** are shown in the pie charts below.

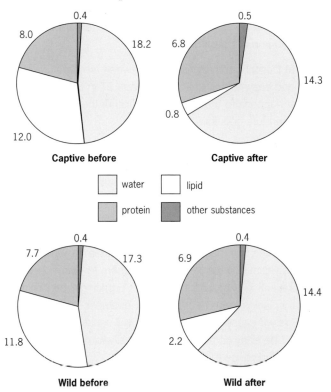

Captive before   **Captive after**

water   lipid

protein   other substances

Wild before   **Wild after**

b) Calculate the total mass loss for each group of birds: captive birds; wild birds.

c) Compare the changes in lipid content of the captive birds with those of the birds living free in the colony.

During the experiment, different processes caused both losses and gains of water in the birds. Metabolic water production was calculated from the amount of fat and other substrates oxidised. Water loss, mostly in exhaled breath, was measured by injecting labelled water and finding how quickly it disappeared from the body. Although the birds drank no water they ate some snow off the ground and this quantity was measured. The results are shown in the table below.

	Metabolic water production/g day^{-1}	Water intake from eating snow/g day^{-1}	Water loss /g day^{-1}
captive birds	144.5	16.5	206.3
wild birds	109.5	12.4	151.6

Source of data: Ancel *et al. Nature* (1997) 385, pages 304–305

d) Explain the difference in water production between the captive and the wild birds.

e) Suggest **one** reason for:
  (i) more water loss in captive birds than in the wild birds
  (ii) more snow eaten by the captive birds than the wild birds.

f) Using the information given in this question, explain the importance of forming large, tightly packed groups of male emperor penguins in Antarctica.

IBO 410 Higher Paper 2 May 1999 Q1

KEY SKILLS **ASSIGNMENT**

HYPOTHERMIA: NO ONE IS DEAD UNTIL WARM AND DEAD

The definition of hypothermia is a drop in core temperature to 35 °C or below. This condition is very common, and is regularly seen in elderly people or in people who have been immersed in cold water – such as those caught in shipwrecks or people who fall through ice.

People can be revived from the most extreme states of hypothermia, even when all the vital signs point to death. People who are cold and blue, showing no heartbeat or breathing and with fixed and dilated pupils, have been revived and have later recovered, hence the above title. Children have been 'drowned' in freezing cold water for up to one hour and still survived.

Fig 15.A1 A mountain rescue team uses an insulated bag to restore the core temperature of an accident victim

1 a) What do we mean by core temperature?
b) What temperature is normal core temperature?
c) Why are people more likely to suffer hypothermia when wet or immersed in water than when they are just out in the cold air?

THE SIGNS OF HYPOTHERMIA
There are several stages of hypothermia:

Impending hypothermia
The core temperature drops to 36 °C. The person begins to feel uncomfortable and the skin may become pale, numb and waxy. Muscles become tense and can shiver if the patient is not moving. Fatigue and signs of weakness begin to show.

Mild hypothermia
Core temperature drops below 35 °C. Uncontrolled, intense shivering begins. Victims are alert and aware of the situation and so may still be able to help themselves. Movements become clumsy and the cold causes pain and discomfort.

Moderate hypothermia
Core temperature drops below 33 °C. Shivering slows or stops, muscles begin to stiffen and mental confusion and apathy begin. Speech becomes slow and difficult to understand. Drowsiness and eccentric behaviour can follow.

Severe hypothermia
Core temperature falls below 31 °C. The skin is cold and often bluish-grey. The eyes are dilated and victim may appear drunk and deny that there is a problem. There is a gradual loss of consciousness. In extreme cases the victim can appear dead, being rigid and not breathing.

2 a) Describe the relationship between temperature and enzyme activity (refer to page 97 in Chapter 5).
b) What happens to the oxygen demand of the body as the temperature gets lower?

TREATING HYPOTHERMIA
The basic rationale behind hypothermia treatment is stop the victim's heat loss, allowing the core temperature to return to normal gradually. The basic goals of early care are to stabilise the core temperature and to prevent cardiac arrest. One of the body's first responses to cold is peripheral vasoconstriction and one of the first measures with a hypothermia victim is to wrap the victim in a blanket and allow them to warm up gently while preventing excessive heat loss. Surprisingly, heating the skin – by placing them near a fire, for example – is not a good idea. This will cause vasodilation that will cause cold blood to return to the body core, resulting in an afterdrop. This can be fatal as it can cause the heart to stop.

3 a) What is peripheral vasoconstriction, and how does it help to conserve heat?
b) Chemicals such a caffeine and alcohol often give a person a warm 'glow' and this is because they cause vasodilation. Explain why coffee or alcohol should not be given to someone suffering from hypothermia.

Fig 15.A2 Elderly people whose sensory systems have deteriorated can suffer from hypothermia. When they start to become cold, their body systems fail to respond and the drop in temperature goes uncorrected. As the temperature drops, their metabolic rate decreases and they produce less heat, and so they cool down even more quickly. Death occurs when their core temperature reaches about 25 °C

16

Excretion and water balance

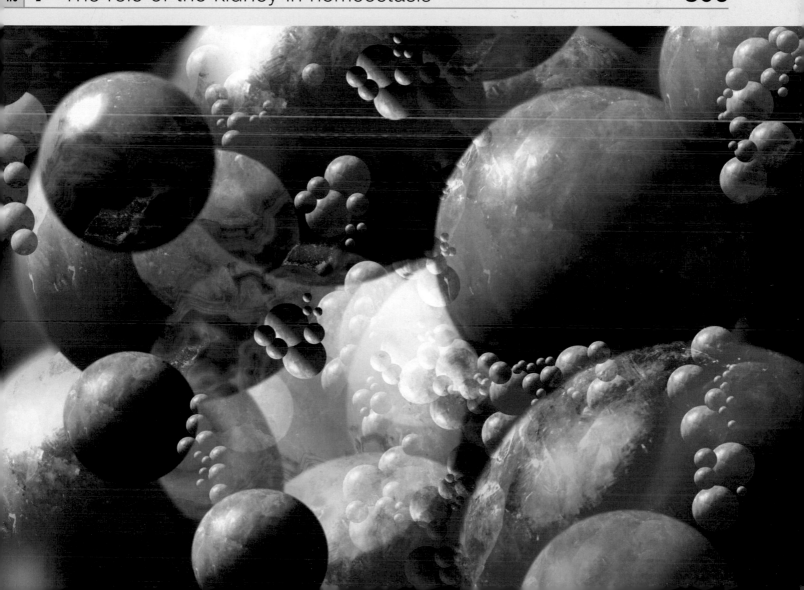

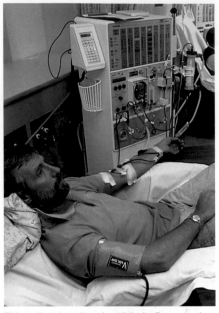

This patient is undergoing dialysis. For several hours, blood flows from the patient's forearm into the machine. Here, much of the waste, together with excess salt and water, is removed by filtration. Find out more about this process in the Assignment at the end of this chapter

If one of your kidneys were to stop working because of injury or disease, you could probably still lead a normal life. However, people whose kidneys both fail are faced with a crisis: water, urea and potassium build up rapidly in their bodies. They may continue to pass some urine, but they cannot get rid of all the waste produced by normal cell processes.

Most people suffering from kidney failure hope for a transplant, but there is a shortage of donors. Until a suitable organ becomes available, patients have to rely on dialysis to filter their blood and balance their fluid intake. As the photograph shows, dialysis is uncomfortable and inconvenient. To reduce the time they need to spend in dialysis, kidney patients must stick to the strict Giovanetti diet. This comes of something of a shock. They must limit their fluid intake to only half a litre a day, about a quarter of the amount a human adult would normally drink. They must also control their protein intake to about 30 to 40 grams per day, the amount of protein in a small egg.

But perhaps the biggest problem is the need to regulate potassium. This ion is a normal constituent of the body, but large amounts cause serious problems, including heart failure. Potassium-rich foods include citrus fruits, bananas, instant coffee, peanuts, treacle and – a big blow for many people – chocolate. In this strange diet, carbohydrates are not a priority. Few patients feel like eating anyway, and this new restricted diet makes the task of finding appetising food even more difficult.

1 WASTE AND WATER CONTROL

The human body consists mainly of water. A person weighing 65 kg contains about 40 litres of water, of which 28 litres is **intracellular** (inside cells). The rest, the **extracellular** fluid, is made up from about 9 to 10 litres of tissue fluid and 2 to 3 litres of blood plasma.

The kidneys play a major role in regulating the volume and composition of these body fluids. They excrete or conserve water and salt so that the volume and composition of the blood and body fluids remain more or less constant. This is a vital aspect of homeostasis.

The kidneys also ensure that waste products do not build up by filtering the blood. Waste is allowed to pass through and then out of the body, while important substances such as glucose are reabsorbed and conserved. The removal of metabolic waste from the blood is called **excretion**.

Urine is the end-product of all processes that occur in the kidney.

Fig 16.1 When we drink a large amount of fluid, our kidneys have the job of getting rid of the excess water, to prevent body fluids from becoming too dilute. This man is drinking beer and the alcohol it contains also affects his kidney function – more of that later

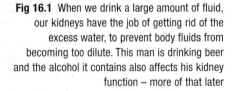

? TEST YOURSELF

A If a person has 3 litres of blood plasma, and their haematocrit (the volume occupied by cells in the blood) is 40 per cent, what is their total blood volume?

Fig 16.2 Adult humans need about 40 to 50 grams of protein per day, and so there is enough here for over a week! However, the body cannot store protein and so the daily excess is broken down in the liver. The resulting urea is excreted in the urine

Most excreted waste leaves the body in the urine, but some waste is also lost in **sweat** and in air that we breathe out. Sweat contains mainly salt and water, while **exhaled air** contains carbon dioxide and water vapour.

In this chapter we concentrate on **nitrogenous excretion**, the removal of waste compounds that contain nitrogen. The main nitrogenous compound excreted by humans is **urea**. This is made in the liver following the breakdown of excess amino acids. Enzymes in the liver remove the amine (NH_2) group from amino acids in the process of **deamination**. The ammonia formed is a highly toxic compound that must not be allowed to build up. It immediately enters a series of reactions called the **ornithine cycle**, which produce the relatively harmless compound urea:

$$2NH_3 \quad + \quad CO_2 \quad \rightarrow \quad CO(NH_2)_2 \; + \; H_2O$$
ammonia + carbon dioxide $\rightarrow$ urea + water

See Chapter 14 for more details on the ornithine cycle. Urea is carried in the blood from the liver to the kidneys to be excreted.

In this chapter we look at the role of the kidneys in homeostasis. The kidneys selectively eliminate water and solutes, such as sodium, potassium and chloride ions, so that the water and solute balance of the body is kept at the correct level. The need to balance the solute concentration of body fluids is called **osmotic regulation**, or **osmoregulation**.

Osmoregulation is the maintenance of a constant solute concentration within the body. In Chapter 4 we saw why this is important: if we place animal cells in a *hypertonic* solution, they lose water by osmosis and shrivel. If we put them in a *hypotonic* solution, they gain water and may burst. Both situations are harmful to cells, and so it is important that we maintain the solute concentration of our body fluids within narrow limits.

 REMEMBER THIS

Excretion is the removal of chemical waste from the body, waste that is produced by the metabolic processes within cells and which would be toxic if allowed to accumulate. You should not confuse excretion with **egestion** or **defecation**, the removal of undigested food and other debris from the intestine.

SEE QUESTION 1

 REMEMBER THIS

Different fluids are often compared with respect to their solute concentrations. A solution that has a higher solute concentration than another is said to be **hypertonic**. A solution that has a lower solute concentration than another is **hypotonic**. A solution that has exactly the same solute concentration is **isotonic**. So, for example, we can say that seawater is hypertonic to human blood plasma, because seawater has a much higher concentration of dissolved salts.

Most people will have experienced the unpleasant symptoms that accompany a bout of mild food poisoning due to a dodgy late-night kebab, an undercooked chicken leg at a barbecue, or those prawns that lurked in the fridge just a little too long.

Vomiting is caused by bacterial toxins that irritate the gut lining. In the intestine, the frequency of peristalsis increases and the contents move along the gut rather more rapidly than usual. This doesn't give the large intestine enough time to absorb water from the waste and the result is **diarrhoea**. Fortunately, for most of us, the symptoms are short lived.

However, both vomiting and diarrhoea can be deadly if severe and prolonged. Dysentery and cholera can rapidly lead to death by dehydration as they cause the body to lose fluid faster than it can be replaced.

Every day the average person consumes about 2 to 3 litres of fluid in one form or another, and we also pour a huge volume of fluid (over 8 litres) into our intestines in the form of digestive juices. Diarrhoea does not allow for the efficient reabsorption of these fluids. Also, vital ions such a sodium, potassium and chloride, collectively known as **electrolytes**, are lost. If untreated, this loss can lead to muscle spasms, cramps, coma and heart failure.

Oral rehydration therapy can be used to treat dehydration. This does not involve expensive drugs, simply a mixture of glucose and salt in water. In cases where the patient cannot keep anything down, the rehydration solution can be given directly into the bloodstream via a drip. This simple treatment has saved millions of lives in places where dysentery and cholera are very common.

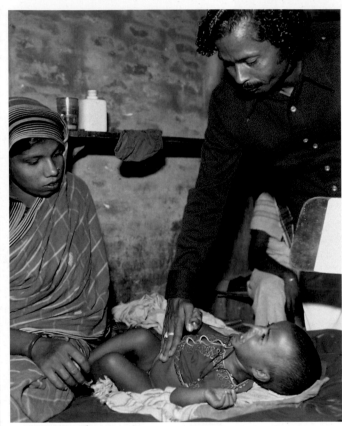

Fig 16.3 The baby in this picture has bacterial dysentery that he developed after drinking water contaminated with faeces. He risks death from dehydration if he does not receive rehydration therapy.
An oral rehydration mixture is easy to make up: one level tablespoonful each of glucose and salt dissolved in a pint of boiled and cooled water

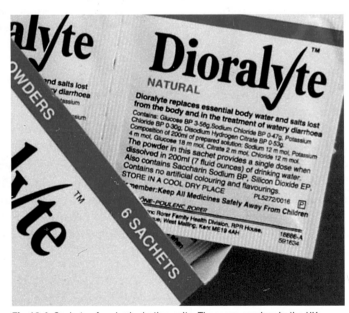

Fig 16.4 Sachets of oral rehydration salts. Those we can buy in the UK contain more than the essential ingredients

2 AN OVERVIEW OF THE HUMAN URINARY SYSTEM

 REMEMBER THIS

The urinary system is also known as the **renal system**.
If you place your hands on your hips, your thumbs show the position of your kidneys.

The human urinary system is shown in Fig 16.5. The kidneys lie at the back of the abdominal cavity, just below waist level, where they are protected to some extent by the spine and the lower part of the ribcage. Usually, the left kidney is slightly above the right.

The kidneys receive blood from the two **renal arteries** that branch off the aorta (see Chapter 13). The kidneys receive the largest blood supply of any organ, per gram of tissue. About 1200 cm^3 of blood flows to each of them every minute. Incoming blood must be at high pressure to ensure

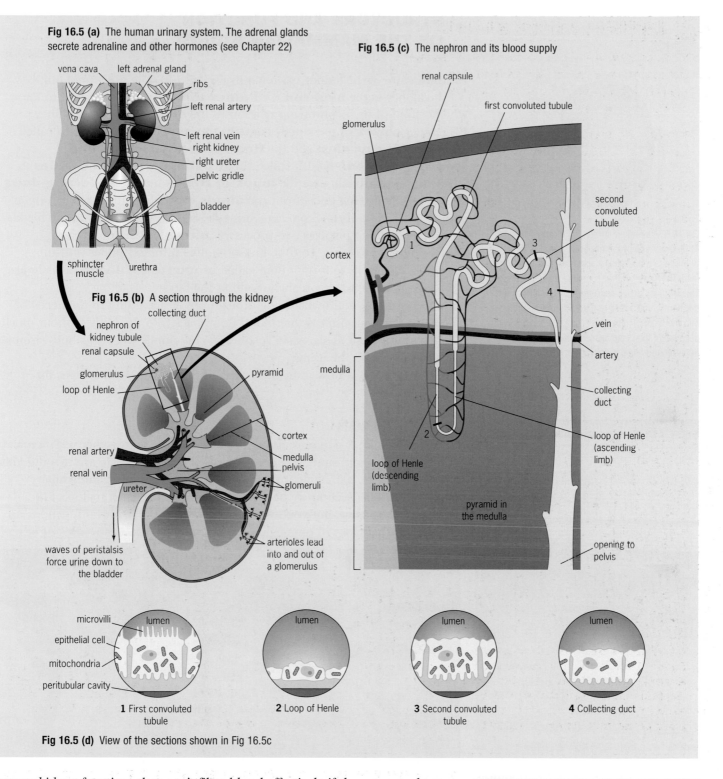

Fig 16.5 (a) The human urinary system. The adrenal glands secrete adrenaline and other hormones (see Chapter 22)

vena cava
left adrenal gland
ribs
left renal artery
left renal vein
right kidney
right ureter
pelvic girdle
bladder
sphincter muscle
urethra

Fig 16.5 (b) A section through the kidney

collecting duct
nephron of kidney tubule
renal capsule
glomerulus
loop of Henle
renal artery
renal vein
ureter
pyramid
cortex
medulla
pelvis
glomeruli
arterioles lead into and out of a glomerulus
waves of peristalsis force urine down to the bladder

Fig 16.5 (c) The nephron and its blood supply

renal capsule
glomerulus
first convoluted tubule
second convoluted tubule
cortex
medulla
vein
artery
collecting duct
loop of Henle (ascending limb)
loop of Henle (descending limb)
pyramid in the medulla
opening to pelvis

microvilli
lumen
epithelial cell
mitochondria
peritubular cavity

1 First convoluted tubule
2 Loop of Henle
3 Second convoluted tubule
4 Collecting duct

Fig 16.5 (d) View of the sections shown in Fig 16.5c

proper kidney function: they can't filter blood effectively if the pressure drops. Blood leaves the kidneys in the **renal veins**.

Urine made by the kidneys is pushed down muscular tubes, the **ureters**, by peristalsis (rhythmic muscular contractions). The ureters empty into the **bladder**, a muscular bag that stores urine until it is convenient to release it. The capacity of the human bladder varies from 400 to 700 cm³ or more. When it begins to get full, stretch receptors in the walls inform the brain of the urgency of the situation. **Urination**, or **micturition**, happens when the **sphincter muscles** relax, allowing urine to pass out of the body through the **urethra**.

? TEST YOURSELF

B A person has 5 litres of blood, and the kidneys filter 1.2 litres per minute. How many times on average does the total volume of blood in the body pass though the kidneys every hour?

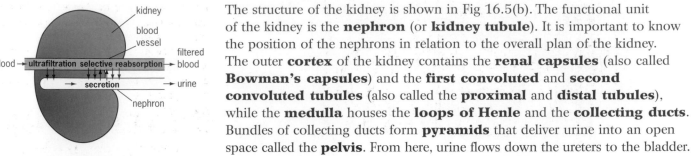

Fig 16.6 A simple summary of kidney function. At the near end of the nephron, blood is filtered to produce a fluid that is virtually identical to tissue fluid. The filtrate is then modified by active transport. This involves active secretion of substances into the filtrate and active reabsorption of substances into the blood. These processes are possible only because of the close association between the nephron and blood system (see Fig 16.5c)

SEE QUESTION 2

3 STRUCTURE AND FUNCTION OF THE MAMMALIAN KIDNEY

The structure of the kidney is shown in Fig 16.5(b). The functional unit of the kidney is the **nephron** (or **kidney tubule**). It is important to know the position of the nephrons in relation to the overall plan of the kidney. The outer **cortex** of the kidney contains the **renal capsules** (also called **Bowman's capsules**) and the **first convoluted** and **second convoluted tubules** (also called the **proximal** and **distal tubules**), while the **medulla** houses the **loops of Henle** and the **collecting ducts**. Bundles of collecting ducts form **pyramids** that deliver urine into an open space called the **pelvis**. From here, urine flows down the ureters to the bladder.

Kidney function involves two processes: **ultrafiltration** and **active transport** (Fig 16.6). Ultrafiltration is filtration under pressure: blood is 'squeezed' to form a fluid called **glomerular filtrate** (usually just '**filtrate**'). Active transport then modifies the filtrate, secreting some substances into it and reabsorbing others from it, according to the needs of the body. The end result is that blood flows back into the body without much of its harmful waste. This waste, a solution containing urea, salts and various other chemicals, is the urine.

Let's now look at how individual parts of the kidney contribute to this overall process.

THE NEPHRON

Each human kidney contains about a million nephrons, together with a maze of blood vessels and some connective tissue (Fig 16.7). There are two different types of nephron, named after their position: cortical and juxtamedullary (Fig 16.8). In this section we deal with the function of each region of the nephron in sequence, but you must remember that the nephron functions as a whole: the activities of one region are essential to the effectiveness of others.

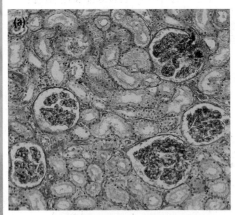

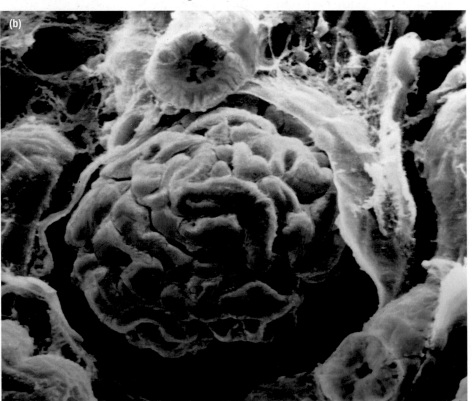

Fig 16.7 The fine structure of the kidney
(a) Under the microscope, five renal capsules (containing glomeruli) can be seen, surrounded by sections of tubules. An underwater dissection allows some of the fine nephrons to be teased out and viewed individually without the aid of a microscope. Each nephron, although only 60 μm in diameter, can be over 14 cm long when uncoiled
(b) A scanning electron micrograph of a glomerulus with part of the torn renal capsule (whitish) round it

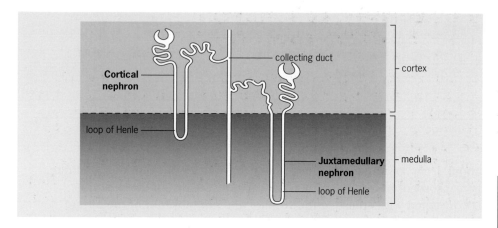

Fig 16.8 The two types of nephron.
Cortical nephrons occur mainly in the cortex.
They have a short loop of Henle that extends only
a short distance into the medulla. These are the
main functional nephrons when water is plentiful.
Juxtamedullary nephrons each have a long loop
of Henle that extends deep into the medulla.
These nephrons produce very hypertonic urine
when water is in short supply. The kidneys of
desert animals, such as the kangaroo rat, contain
mainly juxtamedullary nephrons

ULTRAFILTRATION IN THE RENAL CAPSULE

Fig 16.9 shows the **renal capsule** in detail. It is shaped rather like a
wineglass, with a central knot of blood vessels called the **glomerulus**. This
area of the nephron filters blood by ultrafiltration (filtration under pressure).
Obviously, this requires two things: a means of creating pressure and a filter.

The kidneys receive blood from the first branch off the aorta, so the blood
is already under pressure when it reaches the nephron. This pressure is
maintained and enhanced because the **afferent arteriole**, the blood vessel
that takes blood into the glomerulus, is short and has a larger diameter than
the longer **efferent arteriole** that takes blood away. This physical or
hydrostatic pressure forces blood against a filter that consists of three
layers:

- the lining or **endothelium** of the glomerular blood vessels,
- the **basement membrane**,
- the cells of the renal capsule itself.

Fig 16.9 shows how these membranes are arranged. The middle basement
membrane acts as a fine filter and is therefore mostly responsible for the
chemical composition of the filtrate. At this stage the filtrate is identical to
tissue fluid (see Chapter 13).

The rate of filtrate production is high: about 125 cm^3 per minute.
Obviously, we don't produce anything like this volume of urine or we would
be constantly in the loo and would dehydrate rapidly. On average we produce

? TEST YOURSELF

C What type of nephron would
you expect to find in the kidneys
of an otter? Explain.

✔ REMEMBER THIS

Examiners like you to talk in terms
of **water potential** when
explaining osmosis and water
movement. Water potential is a
measure of the tendency of a
system to absorb water. This is a
negative scale – pure water has a
water potential of zero, and the
higher the solute concentration the
lower the water potential. See
Chapter 4 for details.

Fig 16.9 The fine structure of the renal capsule,
a region of the kidney adapted for ultrafiltration
of the blood. Note the difference in size between
the afferent and efferent arteriole. The blood is
filtered through three layers of cells: the endothelium
of the capillaries, the basement membrane and
the capsule wall

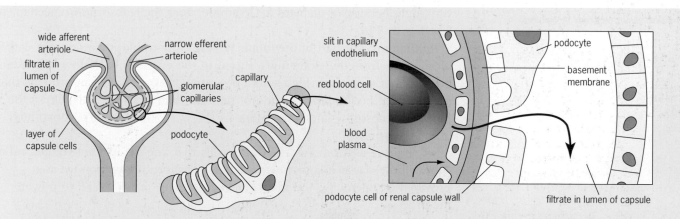

The walls of capillaries in the renal capsule are
much more permeable than those of normal
capillaries: the cells do not fit tightly together, but
have thin slits in between, through which all the
constituents of the plasma can pass

The renal capsule is lined with unique cells
called **podocytes** ('foot-cells'). These cells, like
those of the capillary, do not fit tightly together,
but form a network of slits that fit over the
capillary

Between these two relatively coarse filters is the
continuous basement membrane. This finer filter
prevents the passage of all molecules with a
relative molecular mass greater than about
68 000 kilodaltons, so the larger molecules
(mainly proteins) remain in the blood

Table 16.1 Summary of the forces acting in the renal capsule

Force acting	Opposes or encourages filtrate formation	Approximate value/kPa
Hydrostatic pressure of blood	encourages	8.0
Hydrostatic pressure of filtrate in capsule	opposes	−2.4
Solute concentration of blood	opposes	−4.3
Overall filtration pressure	encourages	1.3

? TEST YOURSELF

D If an individual lost a large amount of blood in an accident, what effect would this have on kidney function?

about 1 cm³ of urine per minute. The rest, over 99 per cent of the filtrate, is reabsorbed. In fact, after the renal capsule, the rest of the nephron is concerned with adjusting the volume and composition of the filtrate. Necessary substances are reabsorbed; toxic compounds and excess solutes and water are removed.

Several forces act on the fluids in the renal capsule, opposing or encouraging the filtration process. The high hydrostatic pressure of blood is the dominant force but it is opposed by the hydrostatic pressure and solute concentration of the filtrate (Table 16.1).

THE FIRST CONVOLUTED TUBULE

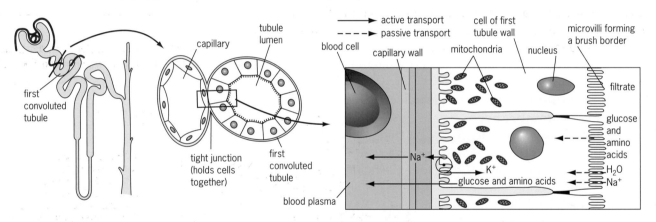

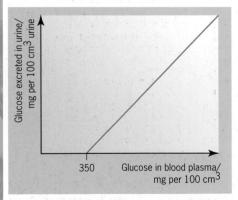

Fig 16.10 Most of the filtrate formed in the renal capsule is reabsorbed from the first convoluted tubule. Amino acids, glucose and sodium are removed from the tubule by active transport, and water follows passively by osmosis

In the **first convoluted tubule** (Fig 16.10), many solutes such as glucose and amino acids are totally reabsorbed into the blood by active transport. Normal urine should not contain glucose. Only when blood glucose becomes very high, because of diabetes for example, does the reabsorption mechanism fail (Fig 16.11).

In addition, this part of the tubule is very close to blood vessels that carry blood away from the glomerulus. Blood in these vessels has a low hydrostatic pressure but a relatively high solute potential due to the plasma proteins that remain there because they could not pass through the filter. This allows the blood to reabsorb a large percentage of the water from the first convoluted tubule by osmosis.

As you can see from Fig 16.10, the cells that line the first convoluted tubule show all the classic adaptations to active transport: a large surface area, provided by microvilli, and many mitochondria to provide ATP to power the process.

Fig 16.11 Normally the kidneys are able to reabsorb all of the glucose into the blood. Only when blood sugar exceeds a threshold of about 350 mg per 100 cm³ does glucose begin to appear in the urine

THE LOOP OF HENLE

The **loop of Henle** is a long U-shaped region of the nephron that descends deep into the medulla and then returns to the cortex. The loop creates a region of high solute concentration (low water potential) in the medulla. The collecting ducts pass through this region, and the osmotic gradient between the inside of the collecting duct and the outside draws water out of the duct by osmosis. Consequently, the urine becomes more and more concentrated (compared with body fluids) as it passes down the duct. So the loop is a vital adaptation that benefits humans and other land-living organisms: it allows us to get rid of waste without losing too much water.

Fig 16.12 outlines how the loop of Henle works. Fluid in the two limbs of the loop flows in opposite directions. We describe this sort of arrangement as a **countercurrent system**. As fluid travels up the ascending limb, sodium chloride (NaCl) is transported actively out of the limb into the surrounding area. This causes water to pass out of the descending limb by osmosis. The net result is that the solute concentration at any one level of the loop is slightly lower in the ascending limb than in the descending limb. The longer the loop, the more chance there is for this mechanism to build up a high sodium chloride concentration. If the loop in Fig 16.12 were only half the length shown, sodium chloride would accumulate to only about 600 units.

So the longer the loop, the greater the solute concentration, and the more concentrated is the urine that is eventually produced. It should therefore be no surprise that animals living in dry desert conditions have very long loops of Henle. To accommodate this extra length, the medulla of their kidneys is very thick (Fig 16.13). Table 16.2 shows the general rule: the more need an animal has to conserve water, the thicker the medulla and the more concentrated the urine. The kangaroo rat can produce urine that is 14 times more concentrated than its blood plasma.

✔ **REMEMBER THIS**

The kidneys of fish, amphibians and reptiles do not possess loops of Henle. These animals cannot produce urine that is any more concentrated than blood plasma.

SEE QUESTION 3

? **TEST YOURSELF**

E The loop of Henle is sometimes described as a **hairpin countercurrent multiplier**. Explain this description.

F People on survival courses are taught to assess their level of dehydration by looking at the colour of their urine. Explain how they would do this.

Fig 16.12 How the loop of Henle works. The numbers refer to the solute concentration in mg per 100 cm^3

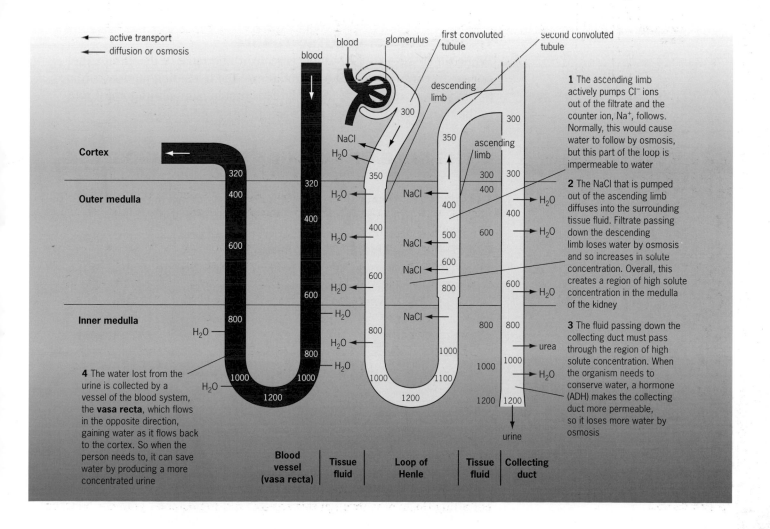

Table 16.2 The relationship between thickness of medulla (ie the length of loop of Henle) and the concentration of urine

Species	Habitat	Thickness of medulla (relative to beaver)	Temperature at which urine freezes/°C*
Beaver	freshwater	1.0	−1.0
Human	land	2.6	−2.6
Cat	land	4.2	−5.8
Kangaroo rat	desert	7.8	−10.4

*The more concentrated the urine, the lower the temperature at which it will freeze. The measurement of freezing point is a relatively easy practical method for comparing the solute concentration of urine samples.

Fig 16.13 The relative thickness of the cortex in two mammals from different environments. **(a)** The kangaroo rat, a desert dwelling animal, has a thin cortex. The deep medulla contains mainly juxtamedullary nephrons with long loops of Henle **(b)** The beaver, an aquatic animal, has a thick cortex. The relatively shallow medulla contains mainly cortical nephrons with short loops of Henle

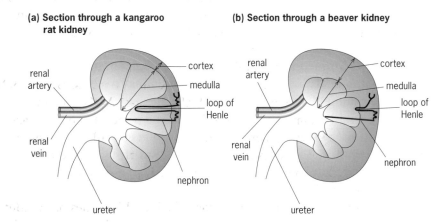

(a) Section through a kangaroo rat kidney

(b) Section through a beaver kidney

THE SECOND CONVOLUTED TUBULE

While the first convoluted tubule is reabsorbing most of the filtrate, the **second convoluted tubule** 'fine tunes' the remaining fluid, according to the immediate needs of the body. This tubule plays an important role in the regulation of pH, salt and water balance.

4 THE ROLE OF THE KIDNEY IN HOMEOSTASIS

The kidney contributes to several vital homeostatic mechanisms. One of the most important is regulation of water content and blood volume.

WATER BALANCE IN HUMANS

Table 16.3 shows a typical water balance sheet for an average person, assuming normal activity and a comfortable external temperature. On average, we get almost two-thirds of our water from drinks and a third from food.

Table 16.3 The water balance sheet for a 24-hour period

Water gain	Volume/cm³	Water loss	Volume/cm³
Food and drink	2100	through skin	350
Metabolic water	200	sweat	100
		in breath	350
		urine	1400
		faeces	100
Total	2300	Total	2300

We obtain a small but important proportion of water from metabolic reactions, notably cell respiration.

Some of the water loss shown in Table 16.3 is unavoidable. Metabolic waste must be removed in solution, and so some water loss in urine is inevitable. Similarly, water is always lost from the lungs as we breathe out. A significant amount of water is also lost by diffusion through our skin (this is not the same as sweating).

THE MECHANISM OF WATER BALANCE

Like most homeostatic mechanisms, maintenance of water balance involves a negative feedback loop that consists of a **detector** and a **correction mechanism** (see Chapter 14).

A part of the brain, the **hypothalamus**, contains **osmoreceptor cells** that are sensitive to the solute concentration of the blood. When the solute concentration rises, indicating that water loss has exceeded intake, the hypothalamus responds in two ways:

- it stimulates the thirst centre in the brain;
- it stimulates the pituitary gland to release **anti-diuretic hormone** (**ADH**).

Fig 16.14 summarises the mechanism of ADH action. ADH acts on the kidney to reduce the volume of urine produced. It achieves this by increasing the permeability of the second convoluted tubule and the collecting duct to water. The action of ADH causes more water to leave the tubule and re-enter the blood. Much more concentrated urine is produced and vital water is conserved.

Conversely, when fluid intake exceeds loss, the blood becomes more dilute. When the hypothalamus detects this, it reduces ADH production. The action of ADH on the kidneys lessens, resulting in less water reabsorption and the production of larger volumes of dilute, or **insipid**, urine.

People with the disease **diabetes insipidus** cannot produce ADH because they have a faulty pituitary gland. Once known as the 'pissing evil', this condition results in the constant production of dilute urine, leaving the sufferer permanently thirsty and unable to venture very far from the toilet. Today, it can be treated by giving extracted or synthesised ADH.

? **TEST YOURSELF**

Which areas of the nephron carry out active transport?

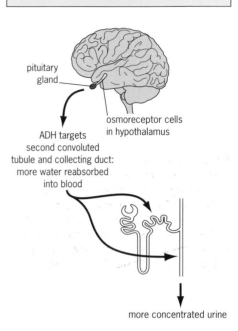

pituitary gland

osmoreceptor cells in hypothalamus

ADH targets second convoluted tubule and collecting duct: more water reabsorbed into blood

more concentrated urine

Fig 16.14 When the solute concentration of the blood rises, the osmoreceptor cells in the hypothalamus stimulate secretion of ADH from specialised nerve cells in the posterior lobe of the pituitary gland. ADH makes the second convoluted tubule and the collecting duct more permeable to water, and allows more water to pass from the filtrate and into the blood. In this way, the blood becomes more dilute and blood volume increases

SEE QUESTION 3

! **SCIENCE FEATURE** Hangovers!

Sooner or later, most people experience the unpleasant 'morning after' feeling which tends to follow a bout of drinking too much alcohol. Many hangover symptoms are due to dehydration, rather than to the toxic effects of the alcohol or other ingredients. Research has shown that alcohol inhibits the production of ADH, causing water that the body needs to be lost in the urine. Many of the symptoms disappear when the body is rehydrated.

Fig 16.15 Had he known about the dehydrating effects of alcohol, this man could have minimised his headache by having a long drink of water before he went to bed (assuming, of course, that he could find the tap)

CONTROL OF BLOOD VOLUME AND PRESSURE

Since it regulates water reabsorption, ADH also regulates blood volume. A drop in blood volume leads to a drop in blood pressure that is detected by **stretch receptors** in the walls of the aorta and carotid arteries. Impulses from these detectors pass to the hypothalamus, which then triggers the secretion of more ADH. This acts on the kidneys and causes them to retain more water, so increasing blood pressure.

ACID/BASE BALANCE

The body must maintain a relatively constant pH. As we saw in Chapter 5, enzymes are very sensitive to changes in pH, and even a small change can inhibit their activity and so have serious consequences for an organism.

Under normal circumstances, the human body is engaged in a constant battle against accumulating acid, although this depends on several factors, including diet. All of the following activities tend to lower blood pH:

● Cell respiration: this produces carbon dioxide, which dissolves in the plasma to form carbonic acid.
● Strenuous exercise: this produces lactic acid.
● Digestion: breakdown of certain foods, such as meat, eggs and cheese, result in the formation of acid products.

Generally, the body is able to regulate the pH of blood plasma and tissue fluid in the range 7.3–7.45 by using buffers and by secreting acids into the urine.

BUFFERS

A buffer is a chemical, or combination of chemicals, that can resist change in pH by mopping up any excess acid or base. The main buffers important in living organisms are:

● Blood proteins, such as **haemoglobin** and **albumin**.
● Inorganic buffer systems in body fluids, such as the hydrogencarbonate (bicarbonate) system:

$$CO_2 + H_2O \rightleftharpoons H_2CO_3 \rightleftharpoons H^+ + HCO_3^-$$

This system acts as a buffer because any extra acid tends to shift the equilibrium to the left, so the extra hydrogen ions are neutralised by the hydrogencarbonate ions.

The sodium hydrogen phosphate system in plasma:

$$Na_2PO_4 + H^+ \rightarrow NaHPO_4 + Na^+$$

$$NaHPO_4 + H^+ \rightarrow H_2PO_4 + Na^+$$

This shows that, in the presence of extra hydrogen ions, the equilibrium changes so that hydrogen ions are swapped for sodium ions that do not affect pH.

SECRETION OF ACID

Most people produce slightly acidic urine of between pH 5 and 6. Active transport mechanisms in the kidney enable the body to excrete excess acid but still retain the vital buffer chemicals mentioned above. If the blood becomes too alkaline, the kidney can reabsorb hydrogen ions and secrete basic ions such as ammonia.

SALT BALANCE: THE CONTROL OF SODIUM CHLORIDE LEVELS

The mechanism by which salt balance is regulated is shown in Fig 16.16. The concentration of sodium ions in body fluids is controlled by the hormone **aldosterone**, which is secreted by the **adrenal cortex** of the **adrenal glands** – see Fig 16.5(a). When sodium ions are actively transported, a negative ion, usually chloride, automatically follows to maintain **electrolytic balance** (balance of positively and negatively charged ions).

When the body loses sodium ions, it also loses water by osmosis, and blood volume and blood pressure fall. This is detected by a group of receptor cells, the **juxtaglomerular complex**, situated next to the renal capsule. These cells respond to a fall in blood pressure by releasing an enzyme called **renin** (not to be confused with the digestive enzyme rennin). Renin converts a plasma protein into an active hormone called **angiotensin**. This stimulates the adrenal cortex to secrete aldosterone.

Aldosterone increases the reabsorption of sodium ions from the intestines and from the kidney. The increased salt concentration in the blood leads to a greater retention of water, so bringing blood volume and pressure back to normal.

Many western diets contain too much salt and this can pose a real problem for our salt balance. A lot of extra salt in food increases salt levels in our blood and so we retain more water to dilute it. As a result, our blood pressure increases. Many people suffering from high blood pressure, or **hypertension**, need a low-salt diet.

OSMOREGULATION IN DESERT ANIMALS

Studies of kangaroo rats kept in captivity (Fig 16.17 and Table 16.4) have shown that they can survive on dried food and no drinking water for long periods. Their ability to control their water loss is remarkable and is a classic example of how animals can adapt to a harsh environment.

This remarkable rodent is well suited to life in the desert. Not only does it avoid dehydration, but it avoids overheating without the cooling effects of evaporation, which it could not 'afford' in terms of water loss. The kangaroo rat shows the following adaptations to its dry environment:

- Its kidneys consist mainly of juxtamedullary nephrons (look back to Fig 16.8), which contain extra-long loops of Henle. These allow the animal to produce very concentrated urine and so it can lose its metabolic waste without losing significant amounts of water.
- It spends long periods underground, where the air is cooler and more humid. This reduces its water loss by evaporation.
- Its nasal passages cool the air before it is exhaled, so that much of the water vapour condenses within its nose instead of being breathed out.

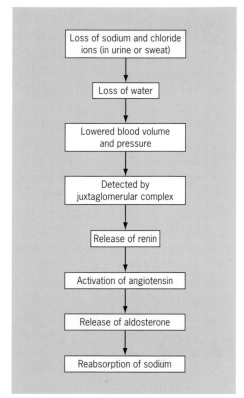

Fig 16.16 Summary flow chart to illustrate salt balance

Fig 16.17 The kangaroo rat (*Dipodomys deserti*) lives in the Arizona Desert, where rainfall is scarce and unpredictable. This mammal is able to survive on dry plant material without drinking water at all. For the study described in Table 16.4, the rats ate nothing but 100 g of barley and were kept at a constant temperature of 25 °C and a humidity of 20 per cent. The kangaroo rat 'creates' most of its water (metabolic water) by metabolic reactions, notably cell respiration. The rest is absorbed from its food. So all the kangaroo rat's water comes either directly or indirectly from its apparently dry food. Metabolism of the protein in barley produces a substantial amount of urea which must be excreted in the urine. Faeces also contain some water, but the most wasteful activity is breathing out

Water gains	cm³	Water losses	cm³
Metabolic water	54.0	urine	13.5
Water absorbed from dried food	6.0	faeces	2.6
		evaporation (mainly in breath)	43.9
Total gain	60.0	Total water loss	60.0

Table 16.4 A study into the water balance for kangaroo rats over a four-week period.

SUMMARY

After reading this chapter, you should know and understand the following:

■ The kidneys remove metabolic waste and control water and solute levels in the body. As a result of these functions, the kidney also plays a vital role in the control of blood volume and pressure.

■ Each kidney is made from around one million **nephrons**, narrow tubules closely entwined with blood vessels.

■ At one end of the nephron, the **renal capsule**, the kidney filtrate is formed by **ultrafiltration** (pressure filtration) of the blood.

■ The first filtrate has the same composition as tissue fluid. As it passes along the nephron, the composition of filtrate is altered by various active transport mechanisms that reabsorb some substances while allowing others to pass into the urine.

■ A large amount of filtrate is formed, but over 99 per cent if it is reabsorbed in the **first convoluted tubule**, mainly by active transport mechanisms and osmosis. Usually, all glucose and amino acids are reabsorbed into the blood.

■ The movement of solutes from the **loop of Henle** creates a region of high solute concentration in the medulla, through which the collecting ducts must pass. As filtrate (now **urine**) flows along the collecting ducts, water leaves it by osmosis. The resulting urine is **hypertonic** (more concentrated than body fluids).

■ The **second convoluted tubule** is involved in several homeostatic mechanisms including the regulation of salt, water and pH levels.

■ When the water potential of the blood drops (eg when we dehydrate) it is detected by the **hypothalamus** which stimulates **ADH** release from the **pituitary**. ADH makes the second convoluted tubule and the collecting duct more permeable to water, so water leaves the filtrate and enters the blood.

? EXAM QUESTIONS

1

a) Describe how urea is formed in the liver.

b) The table gives information on the nitrogenous excretory products of turtles and tortoises from different habitats.

Habitat	Percentage of total nitrogenous excretory product		
	Ammonia	Urea	Uric acid
Water	40–50	40–50	10
Land and water	12	80–90	10
Land	12	30–50	70
Desert	8	10–15	80–90

(i) Describe how the relative percentages of nitrogenous products change with water availability in the habitat.
(ii) Explain the advantage of uric acid as the main nitrogenous excretory product in the desert.

AQA (NEAB) BY03 Physiology June 2000 Q7

2   The micrograph shows a section through part of the kidney cortex as seen under the high power of a light microscope.

a) Name the structures labelled **A** to **C**.

b) Draw a simple sketch of the micrograph and show, by labelling with **X**, where ultrafiltration occurs.
Part of the kidney tubule is adapted for the reabsorption of glucose.

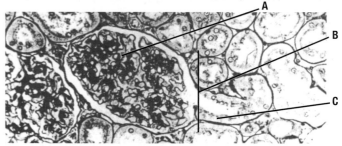

c) State **two** structural adaptations of this region of the kidney tubule and explain how each one assists reabsorption from the glomerular filtrate.
Kidney failure may be treated by use of a kidney dialysis machine.

d) **(i)** Explain how dialysis differs from ultrafiltration in the kidney.
(ii) Suggest two advantages of kidney transplantation over dialysis as a treatment for kidney failure.

OCR 9264 Biology Linear Paper 3 June 2000 Q1

3   Copy and complete the table summarising the role of each of the following hormones.

Hormone	Site of secretion	Part of kidney tubule affected	Effect of hormone
ADH			
Aldosterone			

AQA (NEAB) BY03 June 2000 Q3

KEY SKILLS **ASSIGNMENT**

TREATING KIDNEY FAILURE

In many cases of **chronic kidney failure** there is a gradual decline in kidney function, giving the patient plenty of warning. Acute renal failure is a crisis in which all kidney function effectively stops. There are many causes of **acute renal failure**, but generally they can be placed into the following categories:

- Sudden loss of large amounts of fluid (blood or tissue fluid).

- Inadequate blood flow to the kidneys.

- Bacterial infection in the kidneys.

- Effect of toxins.

- A blockage in the urinary tract caused, for example by damage to the ureter.

1 Why would a reduced blood flow to the kidneys interfere with kidney function, even though enough blood could get through to provide the kidney cells with the nutrients and oxygen they need to stay alive?

As we saw at the start of this chapter, the immediate problems of kidney failure are a build-up of fluid, urea and potassium. To minimise these problems, patients must follow the Giovanetti diet.

2 a) What are the essential features of this diet? (see Opener)
b) Suggest why protein intake needs to be limited.

Dialysis and kidney machines

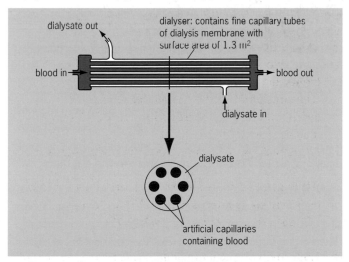

Fig 16.A1 The fine tubes seen in this dialysis filter are artificial capillaries. Blood flows through the middle of these tubes, while the dialysing fluid flows along the outside in the opposite direction. Each filter is an expensive piece of precision engineering, but it can be used for only a few dialysis sessions and must then be discarded

Dialysis is a method of separating small molecules from larger ones using a partially permeable membrane. Blood dialysis, or **haemodialysis**, separates the smaller constituents of plasma, such as urea and solutes, from the larger ones, such as proteins.

Blood is taken from the patient, usually from a vein in the forearm, and passed into the machine, where it runs through minute artificial capillaries. These are made from a partially permeable plastic that filters the blood. While blood flows inside the artificial capillaries, a special fluid, the **dialysate**, flows round the outside in the opposite direction.

In dialysis, molecules are exchanged between the blood and the dialysate. The composition of the dialysate is carefully controlled so that there is a net movement of urea, water and salts out of the blood.

3 a) By what physical process do solutes enter or leave the blood during dialysis?
b) Why do the blood and dialysate flow in opposite directions?
c) Suggest two problems that might occur if the dialysate was pure water.
d) Why must the dialysate contain glucose, amino acids and salt?
e) Why is there no urea in the dialysate?

Fig 16.A2 shows the circuit taken by the blood as it passes through the dialysis machine.

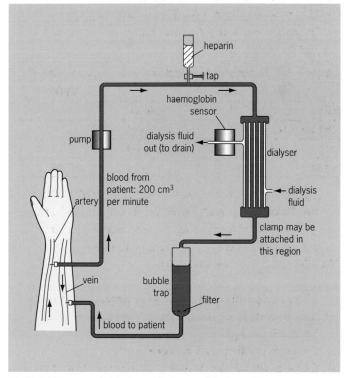

Fig 16.A2 The essential features of a haemodialysis machine

KEY SKILLS **ASSIGNMENT**

4 **a)** Calculate the volume of blood that is processed by the dialyser in four hours.

b) Why is heparin, an anti-coagulant, added to the blood?

c) Why is heparin not given in the last hour of dialysis?

d) Why is a filter included in the blood circuit?

e) Suggest why the omission of the bubble trap could prove dangerous to the patient.

f) What would a positive reading on the haemoglobin sensor indicate?

g) The dialysis fluid is maintained at approximately 40 °C. Suggest two reasons for this.

h) Excess water can be removed from the blood in a number of ways. One way is to increase the amount of glucose in the dialysing fluid. Explain how this method would work.

i) Another method involves partially clamping the blood tube at the region shown in Fig 16.A2. Explain the principle involved in this method of removing water.

CAPD

CAPD stands for **continuous ambulatory peritoneal dialysis**. In this fairly new treatment, individuals with kidney failure can use one of their own membranes, the **peritoneum**, as a dialysing membrane (Fig 16.A3). Over 3000 people in the UK are presently using this technique because of the advantages it offers over conventional dialysis.

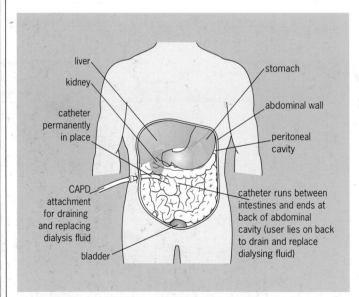

Fig 16.A3 The peritoneum is a semipermeable membrane that lines the abdominal cavity and organs such as intestines. Dialysate introduced into this cavity draws waste and excess water out of the blood

The basic principle is simple: patients have a hole, or **stoma**, made in their abdomen wall near the navel, through which a large volume of dialysing fluid is introduced through a tube or **catheter**. The patient is then free to walk around (hence the ambulatory part of the name) while dialysis occurs across the peritoneum between the blood and the dialysate. Every four to six hours the dialysate is replaced. This is a relatively simple exchange procedure that patients can carry out themselves after some basic training.

5 **a)** Suggest the biggest problem with the exchange procedure.

b) Explain why it would be no use leaving the dialysate inside the body for longer than the recommended time.

Kidney transplants

When a donor kidney becomes available, it is a relatively simple operation to transplant it into another body. Surprisingly, the old kidneys are left in place: they are rather inaccessible and so are difficult to remove, but they do no harm. The new kidney is placed in the lower abdomen. Surgeons choose this site because the new kidney can be attached easily to a large artery (the **femoral artery**, supplying the leg) and is usefully right next to the bladder.

6 What is the main problem with a transplant once it has been carried out?

Finding a suitable donor for an organ transplant is difficult. Although several hundred thousand people die each year in the UK, only a tiny fraction can provide organs for transplant. For example, accident victims can become organ donors only if their injuries have not affected the organ itself. Owing to road safety improvements, the number of serious accidents is decreasing. This is good news, but it means that the number of organs available for transplant is getting ever smaller.

A further complication is that only a minority of people carry donor cards, and permission to used body parts from recently deceased people has to be given by distressed relatives, who often say no. The problems of transplants are covered in more detail in Chapter 17, but you might want to find out more.

7 **a)** Discuss or jot down some ideas about what could be done to encourage more people to become organ donors.

b) It has been suggested that everyone should automatically be an organ donor, unless they take the decision to opt out of the scheme. Discuss whether this idea would work or not.

17

Defence against disease

Defence against disease

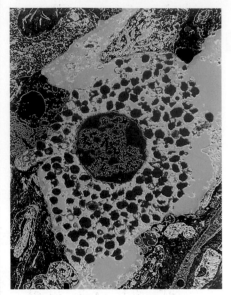

A few years ago, a major medical journal reported the tragic case of a teenage girl who died within 20 minutes of taking a health supplement containing 'royal jelly', the substance made by queen bees. Sudden death in young people usually makes headlines, particularly when it seems to be related to a trivial event such as taking a normally harmless food supplement. But similar deaths have also occurred as a result of a bee sting.

The victims all have in common a very rare and severe allergy. Just as some of us suffer from hay fever in the summer because we over-react to pollen in the air, some people over-react to the 'foreign' proteins produced by bees.

Normally, when someone is stung by a bee, the result is a painful red swelling that disappears again in a couple of days. In someone who is allergic to bee protein, the effects can be catastrophic. As soon as the poison gets into their bloodstream, many of their white cells release large amounts of histamine and other chemicals that cause severe inflammation. Fluid builds up in the tissues and the smooth muscles all contract. The whole body is affected and both blood volume and blood pressure quickly plummet. Sometimes, the airways in the lungs narrow so much that the affected person can no longer breathe. These symptoms, collectively known as anaphylactic shock, can happen very quickly and can be fatal unless immediate treatment with adrenaline and antihistamines is available.

A false-colour electron micrograph of a mast cell, a type of white cell involved in allergic reactions. The small red vesicles contain histamine, the chemical that produces many of the symptoms of allergy. Many people take antihistamines to reduce their allergy to such things as pollen, house dust or animals

1 INTRODUCING IMMUNITY

We are all surrounded by bacteria, viruses, fungi and other organisms that are capable of invading our bodies and causing disease (Fig 17.1). We are able to overcome infections by these **pathogens** (disease-causing organisms)

Fig 17.1 Spots!

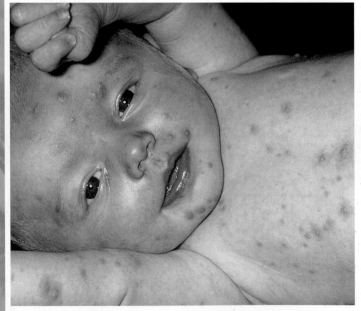

This baby boy has chicken pox, caused by the varicella-zoster virus

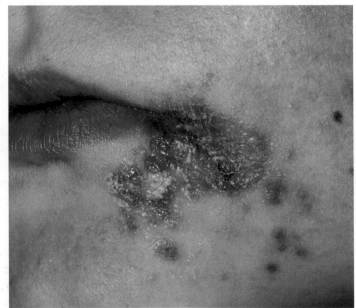

This girl has impetigo, a highly infectious bacterial infection

because we have an **immune system**. This is a complex system involving many different cells and tissues that allows us to develop **immunity** – resistance to infections.

Common pathogens include bacteria, fungi, viruses and protoctists. This last group includes microscopic parasites such as *Plasmodium*, which causes malaria, and larger parasitic animals such as tapeworms. A pathogenic organism is able to:

- break through the physical barriers of the body and enter tissues or cells;
- resist the efforts of the immune system to destroy it, long enough to multiply inside the host's body;
- get out of one host and into another;
- damage the host's tissues – either directly, or indirectly by means of **toxins** (poisons) that it releases. Some bacterial **exotoxins**, such as the one produced by *E. coli* O157, which has caused epidemics of gastroenteritis in the past few years – see page 2 – are very powerful and can be fatal.

The easiest way to start understanding the immune system is to look at its overall functions, rather than concentrating on the individual parts. Fig 17.2 summarises the main lines of defence that an organism comes up against when it tries to infect a healthy person. We look at each of these in more detail in the sections that follow.

Our study of the immune system covers:

- The ability of the immune system to distinguish between invading organisms and the tissues of its own body. This concept of **self** and **non-self** is important in determining whether the immune system keeps the body healthy, or whether it is overcome by infection.
- Problems of the immune system. In the Science Feature box on page 319, we see how the body can react against its own tissues, causing **autoimmune disease**.
- How medicine today can manipulate the immune system to provide life-saving treatments such as vaccines, blood transfusions, organ and tissue transplants and specific cancer therapy.

2 THE BODY'S BARRIERS TO INFECTION

One of the most obvious ways to avoid infection is to stop potential pathogens getting into the body in the first place. The four main strategies that the body uses are summarised below:

- **Mechanical** defence. **Nasal hairs** filter the air that is drawn into the upper airways. **Cilia**, which line the airways, sweep bacteria and other particles away from the lungs.
- **Physical** defence. The skin, made from **stratified squamous epithelium** (see Chapter 2), forms a tough, impermeable barrier that normally keeps out bacteria and viruses. The **mucous membranes** that line the entry points to the body such as the nose, eyes, mouth, airways, genital openings and anus produce fluids and/or sticky mucus. These fluids trap microorganisms and stop them attacking the cells underneath.

? TEST YOURSELF

A What conditions inside our bodies make it ideal for the growth of microorganisms?

✔ REMEMBER THIS

Pathogens are disease-causing organisms. Such diseases are communicable or infectious: they can pass from one person to another. Other diseases, such as diabetes and cancer are non-communicable.

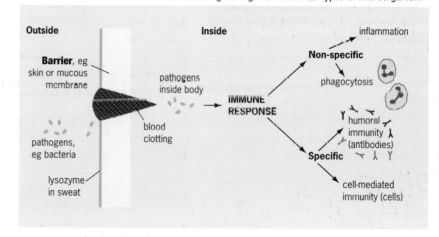

Fig 17.2 An overview of the body's defences: non-specific responses are general responses to damage. They include inflammation and phagocytosis of debris. Specific responses are targeted against individual types of microorganism

SEE QUESTIONS 1 AND 2

? TEST YOURSELF

B Smoking leads to paralysis of the cilia that line the airways. Suggest why this causes problems.

- **Chemical** defence. Fluids such as sweat, saliva and tears contain chemicals that create harsh environments for microorganisms. Sweat contains **lactic acid** and the enzyme **lysozyme**, both of which slow down bacterial growth. Stomach acid kills many microorganisms that manage to get that far. When we are injured, blood clots at the injury site, sealing the breach to prevent entry of bacteria (see the next section).
- **Biological** defence. Normally, a vast number of non-pathogenic bacteria live on the skin and mucous membranes. These do not harm the body but they out-compete pathogenic bacteria, preventing them from gaining a foothold from which to launch a full-scale infection.

3 BLOOD CELLS AND DEFENCE AGAINST DISEASE

BLOOD CLOTTING

Whenever blood vessels break, blood leaks out and **clots** (Figs 17.3 and 17.4). We can see this happening when we cut ourselves, but blood can also clot deep inside the body. Clotting enables the body to avoid blood loss and, at the surface, to prevent infection. It is important that blood clots only when it should, because when a blood clot, a **thrombus**, blocks a vital blood vessel, it can cause a fatal heart attack or stroke.

Fig 17.3 Blood clotting is a cascade reaction: a complex set of chemical reactions that occur one after the other

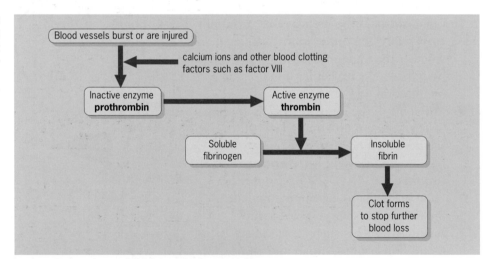

Fig 17.3 outlines the major steps involved in the control of blood clotting. When blood vessels are injured or burst, a **cascade reaction** is initiated. The activation of one molecule leads to the activation of many more. In the final steps of the process, an inactive enzyme, **prothrombin**, is converted to active **thrombin**. This form of the enzyme converts soluble **fibrinogen** into insoluble **fibrin**. The fibrin fibres form a mesh that traps red blood cells and forms a clot (Fig 17.4).

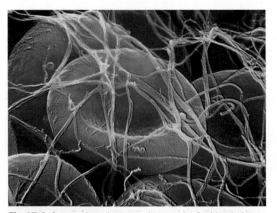

Fig 17.4 A scanning electron micrograph of a blood clot. You can see red blood cells trapped in a fine mesh of fibrin. Blood clotting is a complex process that is set off when blood vessel walls are damaged and finishes when soluble fibrinogen is converted into an insoluble fibrin mesh

If any of the factors in the cascade are missing, the blood cannot clot. This occurs in the condition **haemophilia**, a sex-linked genetic disease in which the sufferer (usually a male) cannot make factor VIII. The genetics of haemophilia are discussed in Chapter 28.

CLASSIFICATION OF WHITE CELLS

Fig 17.5 is a blood smear showing some white cells – **leucocytes** – along with many red blood cells. (The red cells outnumber the white by about 700 to 1.) White cells are made in the bone marrow and are found throughout the body. They can move and are able to squeeze between cells, passing freely in and out of the circulation. Individual types of white cell are classified according to their appearance, origin or function (Table 17.1).

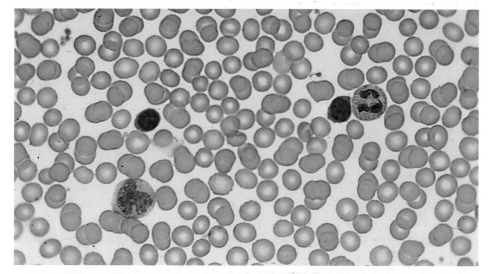

Fig 17.5 In this blood smear, the white cells stand out because their nuclei are stained. The numerous red cells, which have no nuclei, do not take up the purple stain

Table 17.1 Different types of white cell. Neutrophils, eosinophils and basophils have a granular cytoplasm and are therefore called granulocytes. The other types are called agranulocytes

Cell type	Diagram	How to recognise them	Relative abundance %	Function
Neutrophils		lobed nucleus	57	phagocytosis
Lymphocytes		large round nucleus; little cytoplasm	33	specific immunity; B cells make antibodies. T cells involved in cell-mediated immunity
Monocytes		large, kidney-shaped nucleus	6	phagocytosis; monocytes develop into macrophages: general 'rubbish collecting' cells
Eosinophils		stain red with eosin	3.5	associated with allergy
Basophils		stain with basic dye	0.5	release chemicals such as histamine that are responsible for inflammation

4 THE NON-SPECIFIC IMMUNE RESPONSE

When the body is damaged by cuts, scratches or burns, or is attacked by a pathogenic organism that manages to breach its defences, it produces a **non-specific immune response**. It is called a non-specific response because it occurs in response to tissue damage itself, not to the cause of the damage.

INFLAMMATION

Inflammation is a rapid reaction to tissue damage. Whether it is in response to a cut, insect bite or a heavy blow such as a sport injury, the classic signs of inflammation are always the same:

● **Redness**: blood vessels dilate, increasing blood flow to the area.
● **Heat**: also caused by the extra blood flow.
● **Swelling**: the extra blood forces more tissue fluid into damaged tissues.
● **Pain**: swollen tissues press on receptors and nerves. Also, chemicals produced by cells in the area stimulate the nerves.

   Inflammation (Fig 17.6) is triggered by damaged cells. Ruptured cells and some white cells (**mast cells** and **basophils**) release 'alarm' chemicals such as **histamine**. These substances dilate blood vessels and the increased blood flow leads to the classic signs of inflammation. The 'alarm' chemicals also attract white cells that remove bacteria and debris by **phagocytosis**.

WHY INFLAMMATION IS USEFUL

Inflammation prevents the spread of infection and speeds up the healing process. It also provides a way of telling the rest of the immune system what is going on. When microorganisms are phagocytosed, fragments of their cells (particularly molecules that were originally on their surface) are processed by the phagocytes. Some of these surface molecules, which we call **antigens**, allow the specific immune system to recognise and remember the type of microorganism that has tried to invade the body.

   Antigens stimulate the specific immune system to produce cells and chemicals that bind specifically to that antigen, and to no others. Find out more about specific immunity in next section of this chapter.

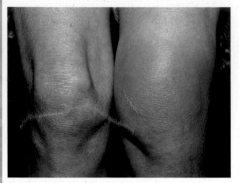

Fig 17.6 The familiar effects of inflammation: reddened skin, swollen tissues and tenderness

✔ REMEMBER THIS

An antigen is a molecule or part of a molecule, for example a protein, that is detected by the specific immune system as 'foreign': not part of the host's body. The immune system responds to an antigen by producing a very specific protein, an antibody, which reacts with that antigen, and that antigen only.

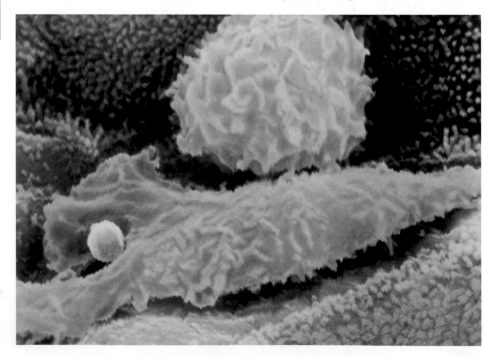

Fig 17.7 This phagocyte (a neutrophil) is engulfing a speck of dust. These white cells occur in the bloodstream and throughout the tissues. Each phagocyte (which is spherical when not active) can ingest between 5 and 25 bacteria before the toxic products of breakdown kill the cell. At sites of infection there may be many dead, liquefied white cells that, together with dead bacteria and other debris, form pus

! **SCIENCE FEATURE** Autoimmune disease: when the body attacks its own tissues

In people who have an autoimmune disease, the mechanism that enables the immune system to tell what is self and what is non-self breaks down. T and B cells (types of white cell) begin to attack the body's own cells and tissues. This is the underlying cause of **multiple sclerosis, insulin-dependent diabetes, myasthenia gravis** and **rheumatoid arthritis**.

In multiple sclerosis, T cells attack the myelin sheath around nerves. This severely limits nerve function, resulting in loss of movement and sometimes in blindness.

In insulin-dependent diabetes, the body makes antibodies that destroy the β cells in the islets of Langerhans in the pancreas. This means that the pancreas becomes unable to produce insulin and the affected person can no longer control their blood glucose.

In myasthenia gravis, the body makes antibodies that attack the motor end plates, the specialised synapses that connect motor nerves to muscles. If the motor end plates become damaged, the muscles cannot contract. The first symptom of this disease, which affects one in 30 000 people (mainly female), is rapid tiring during exertion. The muscles become progressively more unresponsive and the affected person can have difficulty breathing.

People with rheumatoid arthritis suffer from swollen and deformed joints (Fig 17.8).

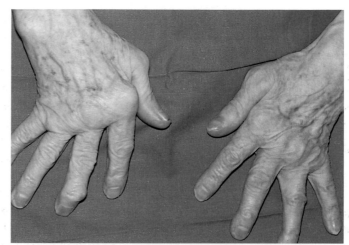

Fig 17.8 This person suffers from rheumatoid arthritis. The cartilage at the joints has been attacked by the immune system, causing swollen and painful joints

PHAGOCYTOSIS

Neutrophils are the commonest type of white cell. Together with monocytes, they are known as **phagocytes** because of their ability to 'eat' pathogens by phagocytosis (Fig 17.7). In this process, the white cell engulfs the pathogen, takes it into a vacuole inside its cytoplasm and then digests it with **lytic** enzymes (see Chapter 4).

5 SPECIFIC IMMUNITY

The specific immune system protects the body from 'invasion' by microorganisms and parasites and also makes sure that the body's defences do not turn on its own tissues. The specific immune response is made up from two different systems that co-operate closely:

- **Humoral immunity**, also called **antibody-mediated immunity**, involves only chemicals: no cells are directly involved. The chemicals, called **antibodies**, attack bacteria and viruses before they get inside body cells. They also react with toxins and other soluble 'foreign' proteins. Antibodies are produced by white cells called **B lymphocytes**, or **B cells**.

SEE QUESTION 5

- **Cell-mediated immunity**, as the name suggests, involves cells that attack 'foreign' organisms directly. Activated **T lymphocytes**, or **T cells**, kill some microorganisms, but they mostly attack infected body cells. The body uses cell-mediated immunity to deal with multicellular parasites, fungi, cancer cells and, rather unhelpfully, tissue transplants.

HUMORAL AND CELL-MEDIATED IMMUNITY

Fig 17.9 summarises the differences between humoral and cell-mediated immunity and Fig 17.10 shows how antigens from pathogens stimulate antibody production by B cells.

Fig 17.9 Lymphocytes are made in the bone marrow by stem cells. Two distinct types of lymphocytes develop, T lymphocytes and B lymphocytes (T cells and B cells). Both T and B cells migrate from the bone marrow to the lymph glands, but the T cells go via the Thymus while the B cells go straight from the Bone marrow

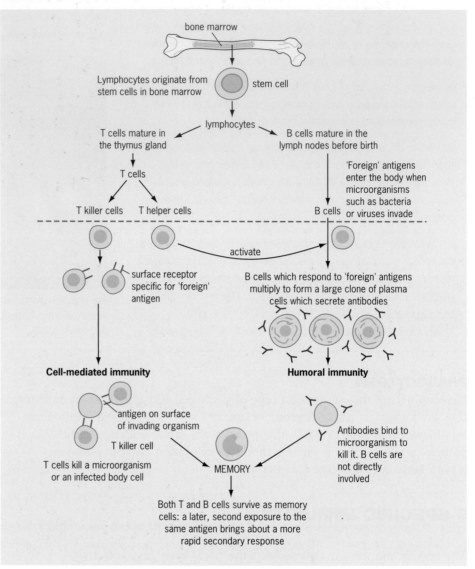

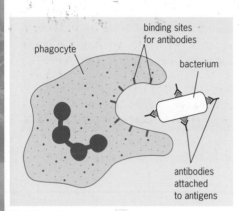

Fig 17.10 Pathogens such as bacteria are covered in antigens, which stimulate B cells to produce antibodies. Many antibodies coat the pathogen, labelling it as foreign to stimulate attack by phagocytes

B CELLS AND HUMORAL IMMUNITY

B cells are produced in the **bone marrow** and are distributed throughout the body in the **lymph nodes**. B cells respond to the 'foreign' antigens of a pathogen by producing specific antibodies. Antibodies are complex proteins that are released into the blood and carried to the site of infection. B cells do not fight pathogens directly.

An antibody, or **immunoglobulin**, is a Y-shaped protein molecule that is made by a B lymphocyte in response to a particular antigen. Fig 17.11 shows the overall structure of an antibody molecule. Antibodies interact with the antigen and render it harmless.

When a pathogen tries to invade the body for the first time, each of its antigens activates one B cell, which divides rapidly to produce a large population of cells. All the new cells are identical (we say they are **clones**) and they all secrete antibodies specific for the invading pathogen. When the infection is over, most of the newly made B cells die: their job is done. We describe this sequence of events as a **primary immune response**.

 REMEMBER THIS

A clone is a set of genetically identical individuals. In immunology, a clone refers to a population of B cells, all of which can produce the same antibody because they are genetically identical.

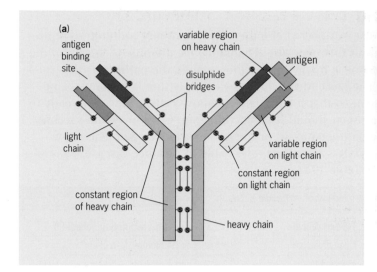

(a)

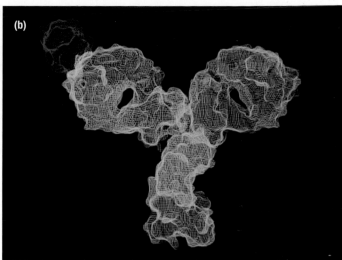

Fig 17.11 (a) Antibodies are Y-shaped molecules consisting of four polypeptide chains, two heavy and two light. Much of the molecule is constant but the tips of the Y are variable and match precisely part of a particular antigen molecule

Fig 17.11 (b) A computer-generated 3-D image of an antibody molecule (green) bound to an antigen (red)

So that the body can respond more quickly next time, some of the activated B cells persist in the body for several years. These **memory cells** 'remember' what the pathogen is like and, if it tries to invade again, they divide rapidly to produce an even greater number of active B cells, all capable of secreting specific antibody. This response is called a **secondary immune response** and is very much quicker and more effective than the primary response (Fig 17.12).

This ability of the immune system is central to **vaccination**. A vaccine stimulates the body to produce a primary immune response to a particular pathogen, without becoming infected by it. A subsequent booster produces a secondary response. Later, if the pathogen tries to invade, the body can mount a very fast response and the person does not become ill.

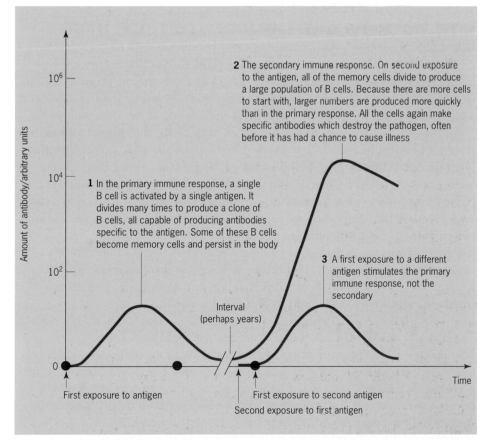

2 The secondary immune response. On second exposure to the antigen, all of the memory cells divide to produce a large population of B cells. Because there are more cells to start with, larger numbers are produced more quickly than in the primary response. All the cells again make specific antibodies which destroy the pathogen, often before it has had a chance to cause illness

1 In the primary immune response, a single B cell is activated by a single antigen. It divides many times to produce a clone of B cells, all capable of producing antibodies specific to the antigen. Some of these B cells become memory cells and persist in the body

3 A first exposure to a different antigen stimulates the primary immune response, not the secondary

Fig 17.12 The immune system can remember antigens

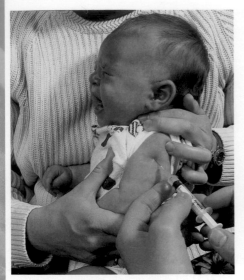

Fig 17.13 Vaccinations are an effective way of stimulating the body's own defences so that we need not suffer the infectious diseases, such as measles, mumps and whooping cough, that used to be a common feature of childhood

VACCINATIONS

Several infectious diseases overwhelm the normal primary immune response and so can be fatal on first exposure. Thankfully we are able to speed up the specific immune response by giving vaccines against the pathogens that cause them (Fig 17.13). The basic idea behind a vaccine is that it contains some form of the pathogen, so that it stimulates memory cells to develop, ready to destroy the real pathogen should it be encountered. Obviously, the vaccine can't simply be the pathogen itself, or the toxins it makes. Somehow, the vaccine must be made less **virulent** – less able to produce disease. Examples of this are shown in Table 17.2.

Table 17.2 Recommended vaccination schedule in the UK

Vaccination	Type of vaccine	Age due (UK vaccination schedule)
Diphtheria	killed organism	2, 3 and 4 months, 3–5 years
Tetanus	modified toxin	2, 3 and 4 months, 3–5 years
Whooping cough	killed organism	2, 3 and 4 months, 3–5 years
Polio	live, non-virulent	2, 3 and 4 months, 3–5 years
Haemophilus influenzae type B (Hib) for meningitis	purified bacterial capsule	2, 3 and 4 months, 3–5 years
Measles	live, non-virulent	12–18 months, 3–5 years
Mumps	live, non-virulent	12–18 months, 3–5 years
Rubella	live non-virulent	12–18 months, 3–5 years
BCG (bacille Calmette–Guérin for tuberculosis)	live, non-virulent	10–14 years
Hepatitis B	genetically engineered antigens	for people at risk, eg health professionals
Meningococcal group C conjugate vaccine	purified bacterial components of *Neisseria meningitidis* group C	2, 3 and 4 months, but also for all young adults (under 25)

HOW MOTHERS GIVE IMMUNITY TO THEIR BABIES

When babies are born, they emerge from the protective environment of their mother's uterus. They are exposed to many potential pathogens. Healthy babies have a fully functional immune system and can mount primary immune responses to many different antigens straight away. However, babies also get a bit of help from their mother, who gives them some of her own immunity.

Antibodies pass across the placenta and, after birth, the supply continues through breast milk. Babies have very porous intestines that can absorb these large proteins directly into the bloodstream without digesting them. These large pores close by the age of one year. We call this kind of immunity – passed from one person to another – **passive immunity**. It does not last long, because the antibodies are broken down within a few days, but it can help a baby to fight off common pathogens.

Passive immunity is also used to treat some types of poisoning, such as snakebites. **Antiserum**, blood that contains antibodies specific to a particular snake venom, is produced in horses, purified and then given to people who have been bitten by a snake.

T CELLS AND CELL-MEDIATED IMMUNITY

Like B cells, T cells respond to specific antigens. When a pathogen first infects the body, each individual antigen stimulates a single T cell. This divides to form a **clone**, in the same way that B cells do. Some of the **activated T cells** become **memory cells** and persist in the body, ready to mount a secondary response if the pathogen attacks again. The others, however, do not produce antibodies. They develop further to become one of three types of T cell:

- **Helper T cells** are so called because they *help* with, or rather control, the rest of the specific immune response. They tell B cells to divide and then to produce antibodies, they activate the two other sorts of T cell (see below), and they activate macrophages, telling them to get ready to phagocytose pathogens and debris.
- **Killer T cells** attack infected body cells and the cells of some larger pathogens (eg parasites) directly. The two cells face each other, membrane-to-membrane, and the killer T cell punches holes in its opponent. The infected cell or parasite loses cytoplasm and dies.
- **Suppressor T cells** are a sort of safety cut-out mechanism. When the immune response becomes excessive, or when the infection has been dealt with successfully, these T cells damp down the immune response. Obviously, this is a good idea: if the body continued to make antibody and stimulate more and more T and B cells to divide, even when there was no need, this could damage the body and would be, at best, a waste of resources.

ALLERGIES

An **allergy** or **hypersensitivity** is an over-reaction to the presence of a normally harmless substance called an **allergen**. Some common allergens are pollen, house dust mites, animal fur and feathers, fungal spores, insect bites and penicillin.

The commonest symptoms of an allergy are sore eyes, runny nose, sneezing and asthma. Many of these symptoms result from an inflammation of the mucous membranes, caused by **mast cells**, which release chemicals such as histamine. Many anti-allergy treatments suppress mast cells or neutralise histamine – chemists sell many **antihistamines** in the pollen season. For more information about asthma see the Assignment in Chapter 12.

6 BLOOD TRANSFUSIONS AND ORGAN TRANSPLANTS

In the past 100 years, advances in modern medicine have led to the technology and knowledge that allow doctors to give one person's cells, tissues and organs to another person. One of the main objectives in developing these life-saving treatments has been to overcome the natural reaction of the immune system to destroy transplanted cells and tissue, which it 'sees' as non-self.

BLOOD GROUPING AND TRANSFUSIONS

Towards the end of the nineteenth century, medical scientists realised that accidents were often fatal simply due to blood loss. Losing large volumes of blood sent victims into shock and they died, even though their injuries were otherwise not too severe. Many women, in particular, often died when they lost blood when giving birth. Different doctors tried transfusing blood from a healthy person into the injured person, to try to restore their blood volume.

Sometimes this worked and sometimes it didn't. When it failed, the results were disastrous. We now know that if the **blood groups** of the **donor** (the person giving the blood) and the **recipient** (the person receiving it) are different, the recipient's immune system reacts against the donated blood, producing a massive and deadly immune response.

In the early 1900s, an Austrian scientist, Karl Landsteiner, discovered that red blood cells from different people had different sets of antigens on their surface. The entire human population can be placed into one of four main blood groups: **A**, **B**, **AB** and **O**, according to the antigens on their red blood cells.

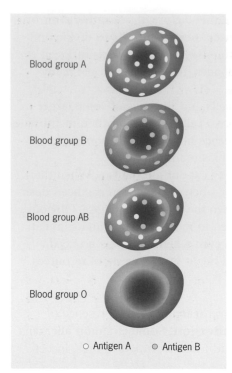

Fig 17.14 People can be placed in four groups according to which red blood cell antigens they have

Blood group A

Blood group B

Blood group AB

Blood group O

○ Antigen A ● Antigen B

From Fig 17.14, you can see that people with blood group O have no antigens on the surface of their red blood cells. Blood from these people cannot cause an adverse immune response if it is transfused into people from the other three blood groups. We say that people with blood group O are **universal donors**. By the same reasoning, people with blood group AB cannot have any antibodies to either of the blood group antigens and so cannot mount an immune response if they are given blood by people in any of the other three groups. We say that people with AB blood are **universal recipients**.

Blood transfusions are dangerous when the recipient has antibodies in their blood that react with antigens present on the surface of red cells in the donor blood (Fig 17.15). So, for example, a person of blood group A cannot donate blood to someone with group B. The recipient in this case has anti-A antibodies which react with the A antigens on the red blood cells of the donor blood.

Fig 17.15 We can find out what blood group someone has by mixing a sample of their blood with anti-A and anti-B antibodies. Blood from people with group A antigens agglutinates (forms clumps) with anti-A, blood from people with group B antigens agglutinates with anti-B. AB blood agglutinates with both, O with neither.
In the diagram, the agglutination reaction produces a granular appearance as red cells clump together. In the smooth, dark samples, agglutination has not taken place

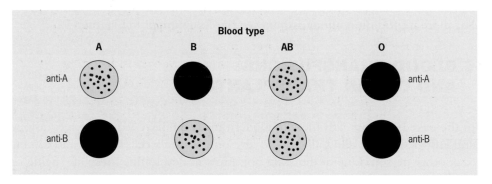

Table 17.3 gives a full list of possible transfusions.

Table 17.3 Which transfusions are possible?

Blood group	Antigens present on red blood cells	Antibodies present in blood	Can donate blood to	Can receive blood from
A	A	anti-B	A, AB	A, O
B	B	anti-A	B, AB	B, O
AB	AB	none	AB	A, B, AB, O
O	none	anti-A, anti-B	A, B, AB, O	O

TRANSPLANTS AND GRAFTS

Transplantation involves taking cells, tissues or organs from one individual and placing them into another individual. Since the first organ transplant operations in the 1950s, patients have received hearts, lungs, skin, corneas, kidneys and various other organs. The success rate of these operations has improved steadily.

The biggest problem facing most transplant recipients is that of **rejection**. The cells of transplanted tissue are covered in antigens that stimulate the specific immune response of the recipient, notably the cell-mediated response brought about by T cells. Symptoms of rejection include the degeneration of blood vessels in the transplanted organ and the destruction of whole cells followed by their replacement with 'scar' tissue. A patient whose transplant is rejected becomes seriously ill: they lose the function of the organ concerned and the massive immune response puts an incredible amount of stress on their already weakened body.

Rejection can be minimised by tissue typing or by using immunosuppressive drugs.

TISSUE TYPING

Like red blood cells, other body cells also have surface antigens that are different in different people. Body cells, such as cells in the kidney or liver, have many antigens that differ, and so matching the tissue type of the donor and recipient is much more complex than matching blood groups. In practice, the cells from two people are never identical, so the aim is to match people who have as few differences as possible.

IMMUNOSUPPRESSION

If the recipient of a transplant has a fully functional immune system, even a closely matched organ will be rejected to some extent. To counter this, transplant patients are given immunosuppressive drugs. The most effective is currently **cyclosporin**. This chemical, isolated from a fungus, inhibits the action of T cells and so has a profound effect on cell-mediated immunity. Since the introduction of cyclosporin in the early 1980s, the success rates of some transplants have risen to over 90 per cent.

The problem with immunosuppressive drugs is, of course, that they reduce the body's ability to fight off infection. Transplant patients are more prone to infection from normally harmless microorganisms, particularly viruses. Although the dose of immunosuppressive drugs can be reduced after a few months if there is little sign of rejection, transplantees must continue to take them for life.

7 ANTIBIOTICS AND THE TREATMENT OF INFECTION

Throughout human history, millions of people have died from bacterial infections. Today, we still suffer from such infections but we now have antibiotics, safe and effective drugs that can be used against the bacteria that cause disease.

The first antibiotics developed in the 1930s and 1940s were chemicals that one microorganism produced to kill another. For example, penicillin, the first antibiotic to be produced in large enough quantities to be used in combating bacterial infections, is produced by a type of fungus. Since then, antibiotics have been extracted from a range of unusual sources including snowdrop bulbs and toad skin; yet others can be synthesised.

? TEST YOURSELF

H What is the difference between agglutination and blood clotting?

SEE QUESTION 6

? TEST YOURSELF

I Suggest why, although penicillin is effective against a wide range of bacterial diseases, it is no use for treating diseases caused by fungi.

? TEST YOURSELF

J A man gets a sore throat so he goes to the doctor who prescribes antibiotics. The man takes half the course and feels better, so he stores the rest for next time, thinking he can save himself a trip. What is the problem with this behaviour?

HOW ANTIBIOTICS WORK

Antibiotics target processes such as protein synthesis that are subtly different in microorganisms compared with the cells of humans and other animals. Different antibiotics have different modes of action (Fig 17.16).

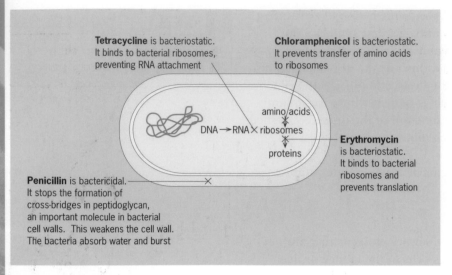

Tetracycline is bacteriostatic. It binds to bacterial ribosomes, preventing RNA attachment

Chloramphenicol is bacteriostatic. It prevents transfer of amino acids to ribosomes

DNA → RNA × ribosomes

amino acids

proteins

Erythromycin is bacteriostatic. It binds to bacterial ribosomes and prevents translation

Penicillin is bactericidal. It stops the formation of cross-bridges in peptidoglycan, an important molecule in bacterial cell walls. This weakens the cell wall. The bacteria absorb water and burst

Fig 17.16 Some antibiotics kill the microorganisms they target. These are said to be bactericidal. Bacteriostatic antibiotics do not kill microorganisms; they prevent them multiplying and give the immune system a better chance to overcome the infection

---

! **SCIENCE FEATURE** Producing penicillin on a large scale

Penicillin is produced commercially using a strain of fungus called *Penicillium chrysogenum*. This is an aerobic organism that needs a high level of oxygen to grow in culture. For this reason, it is grown commercially in fermenters with a relatively small volume – 40–200 dm³. These are easier to aerate effectively than much larger fermentation tanks. (For general information about commercial fermentation, see Chapter 6.)

The commercial production of penicillin is shown in Fig 17.17. The fungus produces most penicillin at temperatures between 25 and 27 °C. During the first 40 hours or so, the fungus is growing in the culture and increasing in biomass. Once it has reached a certain concentration in the culture, it starts to produce large amounts of penicillin, but this continues for only a short time before the fungus starts to run out of nutrients and begins to die. To get the most out of each fermentation, new nutrients are added after 40 hours to stimulate a further increase in biomass. 20–40 per cent of the culture is removed, to be processed so the penicillin can be purified, and new culture medium replaces it. This process, called **batch fill and draw**, can be carried out up to ten times before the fermentation tank needs to be cleaned out completely.

Fig 17.17 Commercial production of penicillin. The process involves several stages; scaling up of the original fungal culture, fermentation, batch fill and draw, filtration, and then three steps of downstream processing. The final product is 99.5 per cent pure

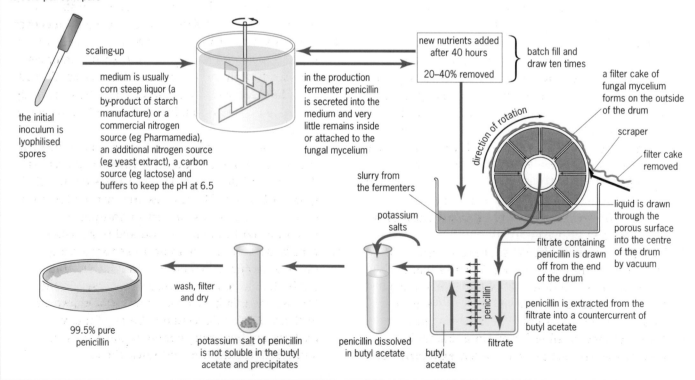

scaling-up

the initial inoculum is lyophilised spores

medium is usually corn steep liquor (a by-product of starch manufacture) or a commercial nitrogen source (eg Pharmamedia), an additional nitrogen source (eg yeast extract), a carbon source (eg lactose) and buffers to keep the pH at 6.5

new nutrients added after 40 hours

20–40% removed

batch fill and draw ten times

in the production fermenter penicillin is secreted into the medium and very little remains inside or attached to the fungal mycelium

direction of rotation

a filter cake of fungal mycelium forms on the outside of the drum

scraper

filter cake removed

slurry from the fermenters

liquid is drawn through the porous surface into the centre of the drum by vacuum

potassium salts

filtrate containing penicillin is drawn off from the end of the drum

penicillin

penicillin is extracted from the filtrate into a countercurrent of butyl acetate

wash, filter and dry

99.5% pure penicillin

potassium salt of penicillin is not soluble in the butyl acetate and precipitates

penicillin dissolved in butyl acetate

butyl acetate

filtrate

⚠ SCIENCE FEATURE Antibiotic resistance

It has been said that the most dangerous place for you if you are sick is in hospital! This, of course, is a great overstatement, but many bacteria that exist in hospitals are resistant to common antibiotics, and the problem is getting worse.

Antibiotics were undoubtedly one of the major medical advances of the twentieth century. They saved the lives of millions of people. But nothing is perfect and there is no such thing as a wonder drug. Since the 1940s when antibiotics were first used widely, resistant strains of bacteria have arisen that can cause infections that are very difficult to treat. How did this happen?

Bacteria, like all living organisms, vary. In any population of bacteria, the vast majority can be killed by an antibiotic such as penicillin. However, there might be one that has the gene that allows it to produce the enzyme penicillinase. This enzyme breaks up penicillin molecules and bacteria that produce it are resistant to the action of penicillin. If such a penicillin-resistant bug can evade the body's defences long enough to reproduce, in a day or two – bang – a new infection that is even more difficult to treat.

Penicillin resistance has become common in the past few years and bacteria have also become resistant to several other

antibiotics. We now recognise that over-prescribing has played a role, and doctors are now much more reluctant to give antibiotics for every cough and cold. Medicine now faces the challenge of staying one step ahead to develop new antibiotics faster than bacteria evolve resistance.

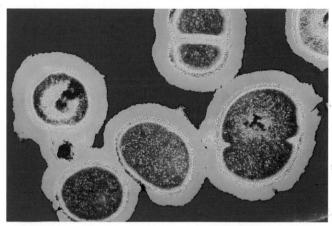

Fig 17.18 Cells of this highly antibiotic-resistant strain of the bacterium *Staphylococcus aureus* are seen dividing. In hospitals, it infects the wounds of patients, causes internal abscesses and produces boils by entering the skin through hair follicles

SUMMARY

After studying this chapter, you should know and understand the following:

- We are surrounded by potential **pathogens** (disease-causing organisms) such as bacteria and viruses which can cause disease if allowed to enter and multiply inside the body.
- The body's defence system consists of barriers to keep microorganisms out, and mechanisms to detect and destroy those that do enter. Together, these mechanisms are the **immune response**.
- The immune system is capable of non-specific responses and specific responses. Non-specific mechanisms (**inflammation** and **phagocytosis**) occur in response to any invading microorganism. Specific mechanisms allow the body to recognise and fight individual types of microorganism. There are two specific responses: **cell-mediated immunity** and **humoral immunity**.
- **Lymphocytes** (white cells) are responsible for the specific immune response. **T cells** mature in the thymus gland and **B cells** come directly from bone marrow. Both are able to recognise foreign antigens that come from pathogens.
- B cells are responsible for humoral immunity: they secrete specific proteins called **antibodies**.

- T cells are responsible for cell-mediated immunity and help to control the overall specific response. **Killer T cells** attack pathogens or infected cells. **Helper T cells** activate B cells, telling them to secrete antibodies. **Suppressor T cells** damp down the immune response when the infection is over.
- **Auto-immune diseases** occur when the body's immune system attacks its own tissues. Examples include rheumatoid arthritis, multiple sclerosis and myasthenia gravis. **Allergies** occur when the body 'over-reacts' to a normally harmless substance such as pollen.
- When cells or tissues are taken from one individual and given to another, they are often recognised as 'foreign' and destroyed. This process is often called **graft rejection**. To minimise the chance of rejection in blood transfusions and organ transplants, it is important to match blood and tissue types.
- **Antibiotics** are important drugs for treating infections caused by bacteria. In the time since antibiotics were first developed and used in the 1940s, bacteria have evolved to become resistant to them. Antibiotic-resistant strains of many bacteria are becoming common and represent a problem for doctors and researchers.

❓ EXAM QUESTIONS

1 Outline the role of each of the following in body defence against disease:

a) skin

b) mucous membranes

c) neutrophils

d) T cells and B cells

e) proteins.

OCR 6915 Health and Disease March 2001 Q2

2

a) Explain how the defence mechanisms of the body reduce the chance of entry by a pathogen.

b) Explain how the body responds both generally and specifically to pathogens that enter the blood.

AQA (NEAB) BY08 Health and Disease June 2000 Q7

3 The diagram shows one way in which white blood cells protect the body against disease.

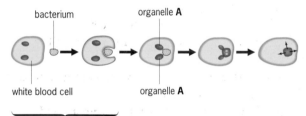

a) Describe what is happening during **Stage 1.**

b) **(i)** Name organelle **A**.

(ii) Describe the role of organelle **A** in the defence against disease.

AQA (A) BYA3 Pathogens and Disease January 2001 Q1

4 B lymphocytes are found throughout the body in the blood and in the lymphatic system. When an antigen enters the body, some of these B lymphocytes respond and change into plasma cells that secrete antibodies.

a) State the part of the body where B lymphocytes originate.

b) Explain the difference between an antigen and an antibody.

The drawing shows a plasma cell and a B lymphocyte.

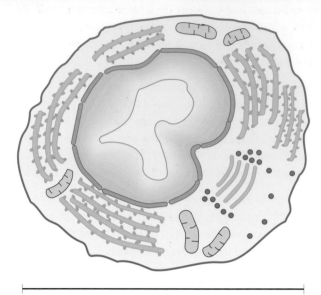

7 μm

plasma cell

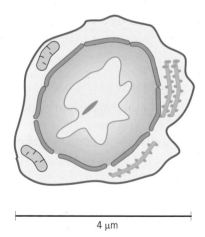

4 μm

B lymphocyte

c) With reference to the drawing,

(i) state **three** ways in which the structure of the plasma cell differs from the B lymphocyte;

(ii) explain the reasons for the differences you have described.

d) Explain why the production of antibodies is much faster when the same antigen enters the body on a second occasion.

OCR 2802 Health and Disease January 2001 Q2

5 During an immune response to a bacterial infection many different antibody molecules may be produced. The drawing shows the structure of a typical antibody molecule.

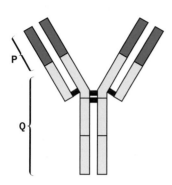

a) Name the regions of the antibody molecule labelled **P** and **Ω**.
Antibodies, like enzymes, are highly specific. This is because they are both proteins.

b) Explain how the protein nature of antibodies allows the production of many different types.
During a bacterial infection, the number of phagocytes in the blood may increase.

c) State where in the body phagocytes are made.
Below is a drawing made from an electron micrograph of a phagocyte engulfing some bacteria.

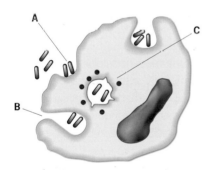

d) With reference to the figure describe the events that are occurring at **A, B,** and **C**.

e) Explain how antibodies, such as those shown in the figure, may help phagocytes engulf bacteria.

OCR 2802 Health and Disease June 2001 Q4

6 The graph shows the difference phases in the growth of a population of bacteria over a three-day period.

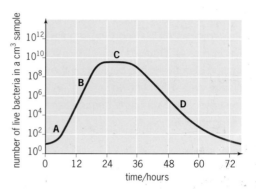

a) (i) Explain the changes in the number of live bacteria in the first 24 hours.
(ii) Suggest an explanation for the shape of the curve between 24 and 30 hours.

b) Bacteria are most susceptible to bactericidal antibiotics during phase **B**.
(i) What is a *bactericidal* antibiotic?
(ii) Suggest why bacteria are most susceptible during this phase.

c) Describe **one** way in which bacteria might produce the symptoms of disease.

d) Gonorrhoea bacteria are sexually transmitted pathogens that respond to the antibiotic penicillin. Chlamydia is another sexually transmitted disease that produces similar symptoms to gonorrhoea. Diagnosis, before treatment, is important because chlamydia does not respond to the antibiotic penicillin. A quick diagnosis is possible by the addition of monoclonal antibodies to a sample taken from the infected region.
(i) What is a monoclonal antibody?
(ii) With reference to gonorrhoea and chlamydia, suggest why monoclonal antibodies can be used to help make the diagnosis.

AQA (A) BYA3 Pathogens and Disease January 2001 Q10

KEY SKILLS **ASSIGNMENT**

AIDS AND THE HIV VIRUS

What is AIDS?

Since its discovery in the early 1980s, few scientific topics have been as controversial as AIDS. Short for Acquired Immune Deficiency Syndrome, AIDS arises when the immune system ceases to function effectively. It is caused by the human immunodeficiency virus (HIV), shown in Fig 17.A1, which infects and destroys vital cells of the immune system, notably the helper T lymphocytes.

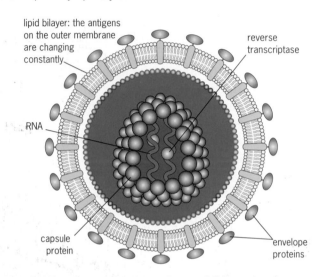

lipid bilayer: the antigens on the outer membrane are changing constantly

reverse transcriptase

RNA

capsule protein

envelope proteins

Fig 17.A1 The structure of the human immunodeficiency virus

1 HIV and AIDS are the same thing. True or false? Explain.

When the HIV enters the body, it can take many years before AIDS develops. Some people take well over 10 years to develop full-blown AIDS and some people have been infected since the mid-1980s and have not as yet developed the syndrome. Perhaps they never will.

Telltale signs that the immune system is no longer effective are the appearance of diseases that would normally cause no trouble. These include Kaposi's sarcoma (a type of skin cancer), fungal infections (eg thrush), bacterial infections (eg tuberculosis) and viral infections (eg herpes).

2 Outline briefly how viruses replicate.

HIV is a retrovirus, and contains the enzyme reverse transcriptase. Once the virus infects a cell, this enzyme allows the cell to make viral DNA from viral RNA. The foreign DNA is then inserted into the cell's own DNA where it acts as a gene, causing the manufacture of more viral RNA and, subsequently, many thousands of new viruses which eventually burst out of the cell and infect others.

3 Outline the essential difference between retroviruses and other viruses.

HIVs primarily affect a group of lymphocytes called helper T cells. These cells normally activate other lymphocytes, including B cells.

4 **a)** What is the function of B cells?
b) What will happen if B cells cannot become activated?

The transmission of HIV

Accurate knowledge of how the virus is transmitted has become a matter of basic survival for many people. Infection occurs when the virus is transferred from person to person, via body fluids. This usually occurs during sexual intercourse. The risk of someone becoming infected through sexual activity depends on the number of partners they have, the proportion of infected partners and the nature of their sexual activity.

5 What is the easiest and most usual way that a sexually active person can avoid HIV infection?

Prevention or cure?

Treating HIV infection is difficult. The viral DNA finds its way into cells in the brain, intestine and retina as well as white cells, and every time the cells divide, the viral DNA is copied. Much research is under way to develop an AIDS vaccine, but efforts have been thwarted so far by the virus's remarkable ability to vary the shape of the chemicals in its outer coat.

6 Suggest why it is difficult to make a vaccine for HIV.

Scientists agree that the most hopeful line of attack is to control the replication of HIV in the body and therefore delay the development of AIDS, perhaps indefinitely. However, this is not yet possible. The best we can do at the moment is to delay the onset of full-blown AIDS by a combination of treatments.

7 Discuss or write down details of any AIDS awareness campaigns you have seen. Which approaches do you feel are the most effective? What approach would you take if you were designing a campaign for young people?

Fig 17.A2 An AIDS information poster in Kenya

18

Exercise physiology

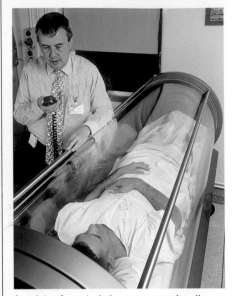

A variety of new techniques can speed up the healing process. In this controversial example a patient is being treated in a hyperbaric chamber that supplies oxygen at a pressure 2.2 to 2.4 times greater than atmospheric. This causes rapid uptake of oxygen into the blood and body tissues, speeding up the healing process in a variety of injuries. Studies suggest that this can speed up the healing process by as much as one third.

At an athletics meeting in Oxford in 1954, Roger Bannister became the first man to run a mile in under 4 minutes. In 1999, The Moroccan Hithcem El Guerrouj ran the same distance in 3 minutes 43 seconds. Genetically, humans today are no different from their predecessors. Yet year upon year, games upon games, athletes get better and world records keep tumbling. How is this being achieved?

Firstly, there are advances in science and technology. A substantial support team now surrounds top athletes: sports physicians, sports physiotherapists, sports dieticians, etc. Training can now be can be tailored to an individual's needs. Their muscle types can be analysed so that they do just the right balance of the different training methods and intensities. Training machines can isolate a particular muscle group and/or reproduce movement patterns such as a javelin throw or tennis serve. Movement can be filmed and analysed with a computer so that a biomechanics expert can target areas for improvement.

Diet can also play a big part both before and after the event. Endurance athletes prepare with a mainly carbohydrate diet so that their glycogen stores are topped up. Following an event, athletes need to rehydrate and replace lost fuels so that then can recover and begin training again as soon as possible.

Another important aspect in preparing a world-class athlete is the prevention of injury. Modern sports physiotherapists have an array of tools at their disposal to diagnose potential weaknesses. If an injury does occur, there is a variety of techniques available to speed the healing process (see the photo).

Lastly, and crucially, there has been a change from an amateur to a professional status. Roger Bannister was a doctor who fitted his athletics around his medical school training. Many of today's young athletes see their talent as a route to fame and fortune. They can devote all of their energies into excelling at their particular event or sport, knowing that success can set them up for life.

WHAT IS FITNESS?

What do we mean by **fitness**? Overall, it refers to a person's ability to perform well in their chosen activity. But there are several different aspects to the concept of 'fitness'.

There is strength and speed. Both aspects depend largely on the power in the muscles. Some men can lift 600 kilograms – about the weight of a small car – above their heads, and there are top sprinters who can cover 100 metres in less than 10 seconds – by running at a speed of about 36 kilometres per hour. A person's strength depends to some extent on the cross-sectional size of their muscles, but other factors such as the shape of their skeleton, good technique and motivation, are all important. So is skill. Some people are naturally gifted; their brain is able to control their muscles in just the right way to perform a particular activity. Of course, skill levels are further improved through coaching, training and practice.

Yet another factor is suppleness, the flexibility of the body. Often an underestimated aspect of fitness, most sports require a degree of suppleness, and performance can be improved by incorporating suppleness work into a training programme.

In this chapter we mainly focus on a fourth element of fitness: **stamina** – the body's 'staying power'. We look at the way in which the body gets its energy, and the different types of 'energy currency' that are important in different sports.

1 GEARING UP FOR EXERCISE

As you sit and read this book, your body is ticking over nicely. Oxygen demand is low, and can be met comfortably by relatively shallow breathing and low pulse rate. Blood is delivering oxygen and glucose to your cells and waste products are being taken away. The levels of these chemicals remain relatively constant, and homeostasis, that all-important *steady state* (see Chapter 14), is being maintained with relatively little fuss. There is, literally, 'no sweat'.

All of this is rudely interrupted when you begin to exercise (Fig 18.1). The metabolic rate of the muscles increases by up to 20 times, or by 2000 per cent. To fuel this frantic activity, and to maintain some sort of stability, your body must adapt. The overall response of the body to exercise is an excellent example of how the different systems work together to carry on maintaining homeostasis.

Fig 18.1 This is Naoka Takahashi who can run a marathon in a time of 2h 23min 14s. Throughout the 26 miles of this race, her body is able to keep her muscles supplied with oxygen and fuel (glucose and lipid), while ensuring that the waste products are eliminated before they build up. This is a magnificent feat of homeostasis

HOMEOSTASIS AND EXERCISE PHYSIOLOGY

Exercise physiology is the study of the responses of the body to exercise. The effects are familiar: in the short term we sweat, pant and go red, while in the long term we 'get fit', improving our muscle tone, strength, stamina and general well-being. This chapter builds on your knowledge of the human body, pulling together many different areas to show you that homeostasis is a whole-body process involving many different systems. Table 18.1 summarises the relevant topics that are covered elsewhere in the book.

Before we look at the ways in which the body responds to exercise, we need to set the scene by looking at some basic principles:

- The process of **cell respiration** (see Chapter 8) releases the energy in organic molecules such as glucose and lipids, and transfers the energy to a chemical called **adenosine triphosphate**, ATP (Fig 18.2).
- ATP provides the energy for muscular contraction; it allows the fibres to slide over each other (see Chapter 21).
- When ATP splits by **hydrolysis** into ADP and P_i (phosphate), energy is released (Fig 18.3).
- The purpose of cell respiration is therefore to resynthesise ATP from ADP and phosphate.

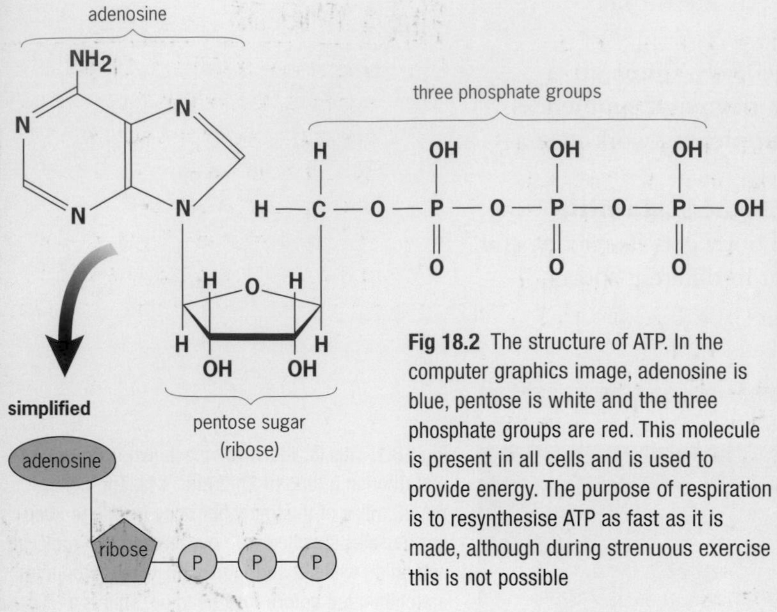

adenosine

three phosphate groups

Fig 18.2 The structure of ATP. In the computer graphics image, adenosine is blue, pentose is white and the three phosphate groups are red. This molecule is present in all cells and is used to provide energy. The purpose of respiration is to resynthesise ATP as fast as it is made, although during strenuous exercise this is not possible

simplified

adenosine

ribose

pentose sugar (ribose)

REMEMBER THIS

ATP is a relatively small, simple molecule that is generated inside the mitochondria of cells. It can diffuse rapidly in and out of the cells and tissues, and moves to places where it is needed, such as muscle fibres. ATP delivers instant energy in small, usable amounts.

Fig 18.3 When the ATP molecule is split (hydrolysed), the terminal phosphate is removed and combined with water. This reaction releases energy, which is used to drive energy-requiring reactions such as muscular contraction

- ATP splitting is a **coupled** reaction. When we exercise, ATP hydrolysis is coupled with muscular contraction.
- When ATP is split, the some of the energy is used to power the muscle but, as no energy transfer is 100 per cent efficient, some is always lost as heat. This is why vigorous exercise produces large amounts of heat that must escape from the body.

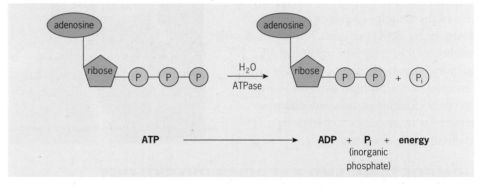

$$ATP \longrightarrow ADP + P_i + energy$$
(inorganic phosphate)

Table 18.1 Aspects of exercise physiology in this book

Topic	Found in
Homeostasis and negative feedback systems	Chapter 14
The cardiac cycle and its control	Chapter 13
Breathing an its control	Chapter 12
Respiration – the production of ATP	Chapter 8
Temperature control	Chapter 15
Movement of muscle/muscle types	Chapter 21
Electrocardiograms	Chapter 13
Artificial pacemakers	Chapter 13
Heart defects and their repair	Chapter 13
Blood transfusions	Chapter 13
Dehydration and water balance	Chapters 11 and 16
Diet and energy content of food	Chapter 10

ENERGY AND EXERCISE

If the movement of muscles requires ATP, it follows that the ability of an athlete to move his or her muscles for any length of time requires a continued supply of this essential chemical. Muscles have three sources of ATP:

- **The ATP already present**. This provides instant energy and allows us move on demand. When we contract our muscles as hard and as fast as possible, we use ATP far faster than it can possibly be made. So we rely on the ATP that has accumulated during periods of relative rest. During maximum effort there is only enough ATP for about three seconds, but there is a back-up chemical, **creatine phosphate**, CP. The energy in CP can be used to instantly resynthesise more ATP, allowing maximal exercise to continue for up to 10 seconds. This is called the **ATP/CP** system, or the **alactic anaerobic system**. Alactic means there is no build-up of **lactic acid**, or more accurately, **lactate ions**, and anaerobic means that this system does not require oxygen. All events that require explosive bursts of energy, such as weightlifting or short sprints, use the alactic anaerobic system.

- **The ATP provided by glycolysis, the first phase of respiration**. In Chapter 8 (Respiration) you can see that glycolysis provides two ATP molecules per glucose molecule. This might not seem a lot compared with the 36 or so available from complete respiration, but it has two big advantages: it is relatively quick and it does not need oxygen. Thanks to this system, exercise can continue at near-maximum levels for up to one minute. However, there is a price to pay – the accumulation of lactate ions. Lactate lowers the pH in the muscles, causing fatigue and interfering with enzymes in muscle cells. The body can tolerate only limited levels of lactate. This system is known as the **glycolytic** or **lactic anaerobic system**, and is the main energy source for events that last between 10 and 60 seconds, such as the 400 metres hurdles (Fig 18.4).

- **The ATP provided by aerobic respiration**. This is the complete breakdown of glucose, when each molecule yields about 36 molecules of ATP. The problem is that this system takes time to provide energy, and then it still has its limits. However, provided that the level of activity stays within those limits, the aerobic system can fuel exercise for a couple of hours of more. This is the main energy system for many sports: all those that last for longer than one minute. Prolonged exercise that increases heart rate and breathing rate, but that is sustainable for hours rather than minutes, is commonly known as aerobics (Fig 18.5).

Fig 18.4 Sprinters and hurdlers rely on strength and are usually heavily muscled. To power their muscles during short races they rely on the ATP/CP system for up to about 10 seconds, and for the rest of the race the glycolytic system provides the energy

Fig 18.5 A familiar sight all over the western world, sustained aerobic exercise can strengthen the heart, lungs and circulation. As an added bonus, it also uses lipid (fat) as its main fuel, so aerobics are a vital part of many weight-loss programmes

THE ENERGY CONTINUUM

From what we have learned so far, it might appear that there are three separate energy systems. But, as we all know, we don't have to stop exercising after 10 seconds and then again after 1 minute to wait for the next energy system to cut in and give us some ATP. All three energy systems blend smoothly together, one taking over from the other. This phenomenon is known as the **energy continuum** (Fig 18.6).

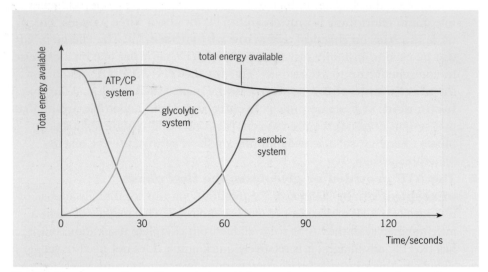

Fig 18.6 The energy continuum. The graph shows that one system gradually takes over from another, allowing us to move our muscles continuously

FUELS FOR EXERCISE

Respiration is the release of the energy contained in organic molecules such as glucose. All foods – carbohydrates, lipids or proteins – can be respired when the need arises. As a rough guide, the body uses the following fuels:

- **Glucose**. This supplies the normal, everyday energy needs when we are eating regularly (ie not dieting or starving). Anaerobic exercise always depends on glucose, because lipids cannot be used to fuel glycolysis.
- **Lipids**. Most lipid is stored in fat-storage, or **adipose** tissue. It takes a while to mobilise the lipid stores, but once the process has begun, this reservoir can fuel the body for as long as it lasts (for some of us almost indefinitely). Lipids cannot be used for short-term exercise, but they can be used to fuel aerobic exercise. This is why doing an aerobics class regularly can help to get rid of stored fat, if the exercise is done at the right intensity.
- **Protein**. The body usually gets around 5 per cent of its energy from protein. The proportion increases significantly only during starvation, when there is no other fuel available. Recent research suggests that protein can be used to fuel endurance events, but to what extent remains uncertain.

Overall, the main fuel used during exercise depends on the intensity of the activity. If you take running as an example, the main fuel for a long, gentle jog is lipid. But, as the intensity increases, so does the proportion of energy that comes from glucose. A summary of the energy systems used in exercise is shown in Table 18.2.

Table 18.2 A summary of the energy systems used in exercise

Energy system:	Anaerobic alactic	Anaerobic lactic	Aerobic alactic
Other names	ATP/CP; phosphate battery	glycolytic; lactic system	aerobic system
Energy comes from	ATP/CP already in muscles	glycolysis	complete respiration; glycolysis, Krebs cycle, ET chain
Nature of energy supply	allows maximum strength but is short lived	allows exercise to continue at near-maximum for up to 1 minute	allows long-term exercise at lower intensity
Timescale	up to 10 seconds	10 seconds to 1 or 2 minutes	anything longer than 1 to 2 minutes
By-product	no lactate	lactate	no lactate; carbon dioxide
Activity	explosive 'full strength'	longer sprint	endurance events such as cycling and marathons
Examples	sprints up to 100 metres; weightlifting and throwing events	200 to 400 metre sprints	800 metres, marathons
Training needed	to improve speed or strength	to improve lactate tolerance	to increase endurance capacity and to strengthen cardiovascular system

SCIENCE FEATURE Diet for athletes

It has long been known that the correct diet can improve an athlete's performance, and in recent years the diets of sports stars such as footballers have been increasingly in the spotlight (Fig 18.7).

The best approach is a **holistic**, or *whole-body* one. The most dedicated athletes pay close attention to their diet, taking in the right amount and types of food at the right time. In addition, they ensure that they have the right amount of sleep and minimise their alcohol intake, especially before important events.

How do you plan a diet to maximise performance? A widely used technique is known as **carbo-loading** or **glycogen loading**. About six days before an event, the athlete goes on to a low-carbohydrate, high-protein diet. This depletes the body's glycogen reserves. Then, for the three days preceding the event the athlete goes on a high-carbohydrate diet, eating 8 to 18 grams of carbohydrate per kilogram of body weight per day. This significantly increases the amount of glycogen stored in the muscles, and this obviously helps during endurance events. It is of little help, however, if the event lasts for less than 90 minutes, and the effect wears off with repetition. Generally, athletes should aim for a diet in which 60 to 70 per cent of their energy comes from carbohydrate, increasing this figure in the day or two before an important event.

Fig 18.7 The prematch meal of footballers has traditionally been steak, but in recent years much more attention has been paid to their diet. Nutritionists recommend an easily digested high-carbohydrate diet, such as pasta, along with a little low-fat protein such as chicken or fish. This will leave the athletes feeling neither hungry nor bloated, and will top up their glycogen reserves

! SCIENCE FEATURE Dehydration and isotonic drinks

Long-term exercise leads to prolonged sweating, and this can easily dehydrate an athlete. By 'dehydrate' we mean that the blood and body fluids become too concentrated. In the correct biological jargon we say that *the water potential is lowered*. The practical consequence of this is that the blood becomes more viscous and cannot flow as easily. In addition, sweating loses vital ions such as sodium, potassium and chloride, collectively known as electrolytes. A loss of electrolytes can lead to muscular cramps and a greatly reduced performance.

But, it is not too serious; all the athlete needs to do is to replace the lost water, electrolytes and glucose. This is easy enough if the athlete happens to be at home, but dehydration can strike when you are three-quarters of the way through a sporting event. Getting a drink that can do the trick takes a bit of planning.

The fastest way to combat dehydration is to drink pure water. This way, water is absorbed as fast as possible by osmosis. However, if you add glucose and salts to the drink you lower the water potential and therefore slow down the water uptake. In fact, if the drink becomes too concentrated it actually makes matters worse by drawing water out of the blood, in the same way as drinking seawater would.

The best compromise is to take an isotonic drink, a mixture that has the same water potential as body fluids. This way, all three components are absorbed as rapidly as possible. So do you have to spend money on expensive isotonic drinks? The simple answer is no. It is easy to make up an isotonic mixture on the same principle as the **oral rehydration therapy** given

to people suffering from diarrhoeal diseases (see page 300). For events lasting less than 90 minutes it is debatable whether isotonic drinks are any benefit at all. The fluid can be replaced by that cheapest of drinks, water, and the electrolytes and glucose are replaced later by any sensible meal.

Fig 18.8 Isotonic drinks contain a solution of electrolytes and glucose or short-chain polysaccharides, which can be easily digested and absorbed into the blood

2 SHORT-TERM RESPONSES TO EXERCISE

Exercise places great demands on the body. Think about the changes that occur when we exercise:

- oxygen levels fall,
- carbon dioxide and lactate levels increase,
- body temperature increases,
- blood glucose and glycogen levels fall,
- fluid and electrolytes (salts) are lost as we sweat.

All this presents a great challenge to the homeostatic mechanisms of the body. In an attempt to maintain some sort of stability, the body responds by:

- **Increasing heartbeat and breathing rate**. The increased respiration of the muscles raises the carbon dioxide level in the blood, and consequently lowers the pH (carbon dioxide is an acidic gas). These changes are detected by sensitive cells – **chemoreceptors** – situated in the **carotid** and **aortic bodies** in major blood vessels (see Chapter 12). In turn, these receptors inform the cardiovascular and respiratory centre in the brain, which responds by increasing the heart rate and the ventilation rate. The result is that gas exchange increases, as does the delivery of oxygen to the tissues and the removal of carbon dioxide.
- **Increased blood flow to the skin**. Muscle movement creates heat that must be lost. By increasing the diameter of the blood vessels that carry blood to the skin surface – a process called **peripheral vasodilation** – the blood can push heat to the body's surface where it can be more easily lost to the environment. This is why we 'go red' during and after exercise.
- **Increased sweating**. Sweat glands secrete a salty solution that evaporates from the skin, taking heat with it. Coupled with vasodilation,

✔ REMEMBER THIS

Ventilation rate is the amount of air taken in by our lungs in one minute. It can be calculated as depth of each breath (the tidal volume) × number of breaths per minute. At rest, this could be about 0.5 litres × 12 breaths = 6 litres per minute.

sweating is a very efficient means of removing excess heat (see Chapter 15 for more about heat transfer). The problem with excess sweating, of course, is that we also lose water and salts. If these are not replaced, the body becomes dehydrated and an athlete's performance can be impaired.

● **Increased mobilisation of glycogen**. In order to keep the muscles supplied with glucose, the glycogen stores in the muscles and the liver start to be mobilised. Glycogen is a highly branched polymer of glucose, which breaks down rapidly to release glucose (see Chapter 3). After prolonged exercise, glycogen stores are replenished from carbohydrate in the diet.

SEE QUESTION 1

THE RECOVERY PROCESS

After exercise, following the short-term changes listed above, the body does not immediately return to normal. Each system gradually returns to resting levels. Generally, the fitter the individual, more quickly the resting state is achieved.

Fig 18.9 shows the pulse rate of an athlete before, during and after a session on an exercise bike. Following exercise, pulse rate follows a classic pattern; a rapid initial fall followed by a slower return to normal. The shape of the curve results from two processes: the recharging of the ATP/CP system – this is known as the **alactacid component** of recovery – and the removal of lactate, called the **lactacid component** of recovery. Both occur because the body was not able to deliver enough oxygen to keep up with demand. This shortfall is known as the **oxygen debt** and is repaid after exercise stops. On a longer term basis, the body must also restore its glycogen levels. After a particularly gruelling exercise session, restoring the glycogen can take up to 48 hours and is not associated with raised heart or breathing rate.

We shall look at each of these components in turn.

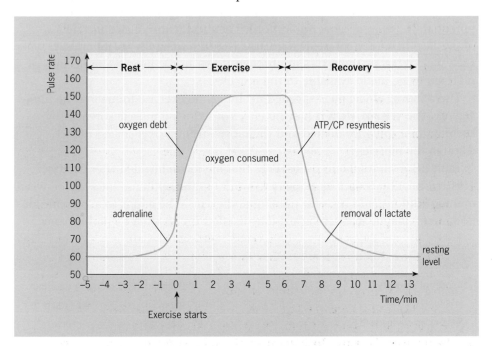

Fig 18.9 The effect of pulse rate on exercise. Note that the resting pulse is just under 60, normal for a trained athlete. The increase in heart rate just before exercise starts is due to anticipation, caused by the release of adrenaline: the more important the event, the more noticeable is the effect of adrenaline. When exercise starts, oxygen demand is greater than supply, and an oxygen debt builds up. This is paid off after the activity has finished

THE ALACTACID OXYGEN DEBT COMPONENT

This is the recharging of the ATP/CP system and takes one to two minutes to complete. Even fit people can't sprint continually, but after a short 'burst' it takes only a matter of seconds before there is enough ATP/CP to allow them to sprint again. In sports such as football, rugby, tennis or hockey, there are bursts of intense activity followed by periods of relative rest, during which the ATP/CP stores are restored by normal aerobic respiration.

Fig 18.10 This person has decided to improve his fitness by a little gentle jogging. He is comfortable at walking pace, but as soon as he starts to jog, he passes his anaerobic threshold, meaning that his muscles do not get oxygen quickly enough and lactate begins to accumulate, making his muscles ache. He will have to slow down until the lactate levels fall, then he will be able to run again. However, if he maintains regular exercise, he can raise his anaerobic threshold, allowing him to run for longer and/or faster without discomfort

THE LACTACID OXYGEN DEBT COMPONENT

This is the removal of the lactate ions that accumulated during the exercise. This acidic chemical lowers the pH of the blood, interfering with the enzymes in aerobic respiration and so reducing ATP supply to muscles. A significant build-up is painful, and there is only so much muscle fatigue we can take before we simply have to stop (Fig 18.10). Interestingly, one of the main effects of training is to increase our tolerance to lactate, so that we can continue to exercise despite higher lactate levels.

Lactate builds up in the muscles and diffuses into the blood. It then has five possible fates:

- 65 per cent is oxidised to carbon dioxide and water,
- 20 per cent is converted to glycogen,
- 10 per cent is converted to protein,
- 5 per cent is converted to glucose,
- a trace is excreted in urine and sweat.

The list clearly shows that the anaerobic system is rather wasteful in terms of energy. Only two ATP molecules are made per glucose, instead of a potential 36 (as in aerobic respiration). The production of two lactate molecules therefore represents a great waste of potentially useful energy: when each pair of lactate molecules are broken down in the liver, 34 molecules of ATP fail to be made available to the muscles.

The removal of lactate is speeded up by gentle exercise following the main activity. The **warm down** has the effect of reducing muscle soreness by keeping the capillaries dilated and therefore flushing oxygenated blood through the muscles.

Table 18.3 shows the recovery times that are recommended after exercise. This information is used by trainers to work out the optimum intervals between training sessions, and between training and events. The aim is to avoid chronic (long-term) fatigue that can have a drastic effect on an athlete's performance.

Table 18.3 Recommended recovery times after exhaustive exercise		
Aspect of recovery	**Recommended recovery time/min**	
	minimum	**maximum**
Restoration of muscle ATP and CP	2 minutes	3 minutes
Repayment of alactacid oxygen debt	3 minutes	5 minutes
Restoration of oxygen myoglobin	1 minutes	2 minutes
Restoration of muscle glycogen (after prolonged exercise)	10 hours	48 hours
Removal of lactic acid from muscle and blood	1 hour *	2 hours*
Repayment of lactacid oxygen debt	30 minutes	1 hour

* = speeded up by a warm down, ie if the muscles are kept working gently, the lactate is removed more quickly

3 LONG-TERM RESPONSES TO EXERCISE

If we exercise regularly, our bodies adapt and we 'get fit'. We feel better, look better and are able to cope easily with exercise that a few months earlier would have had us gasping in a heap on the floor. A remarkable feature of the human body is its ability to respond to exercise, making us more able to cope with our chosen activity.

SO WHAT IS ACTUALLY HAPPENING?

Observable long-term responses to exercise include changes to the heart, lungs and muscles, although the extent of the effects depends on the type of exercise done. For training to have an observable benefit, it must be above a certain intensity. Muscles must be **overloaded** before they begin to **adapt**.

For instance, if you were going to train for a rugby team, a brisk walk would be totally useless, because it would not overload your muscles. Rugby, like many sports, uses a combination of all the energy systems, and the training should reflect this balance. You would need to do endurance (cardiovascular) work as well as short and long sprints, along with training (eg with weights) that would exercise all of the relevant muscle groups. As the muscles adapt, the intensity of training must be increased continually, to ensure muscles are still overloaded. This is the concept of progressive **resistance** and ensures that the body continues to adapt and improve.

SEE QUESTION 5

CHANGES TO THE HEART

The heart responds to exercise like any other muscle; it enlarges, although the nature of the enlargement depends on the type of exercise done. Generally, there is an increase in the size of the **myocardium** (all the heart muscle) and therefore an increase in the size of the chambers. Consequently the **stroke volume** – the volume of blood pumped with each beat – increases. As a general guide, an untrained person has a stroke volume of about 90 cm^3, increasing to over 120 cm^3 after training.

When the heart can pump more blood per beat, it does not have to beat as often when the body is at rest. This is why getting fitter causes a decrease in resting pulse. This can go down from about 70 in the average untrained person to less than 50. Some of the world's top endurance athletes have a resting pulse of about 35 beats per minute.

Research also shows that exercise actually increases the strength of the blood vessels, allowing them to withstand higher pressures and reducing the risk of atherosclerosis (hardening of the arteries) later in life.

CHANGES TO THE LUNGS

The strength of the respiratory muscles (internal and external intercostals and the diaphragm) is increased, allowing a greater volume of air to be forced in and out. This means that **ventilation rate** improves. The total lung volume (the vital capacity) and surface area of the alveoli also increases, resulting in a greatly improved rate of gas exchange. Overall, the improvements in the circulation and gas exchange systems results in an increased V_{O_2} **(max)** (see Chapter 12 for a revision of these important terms).

REMEMBER THIS

Muscular contraction and the different muscle types are covered in Chapter 21.

REMEMBER THIS

A lot of damage can be done by over-training, and the effects of this can be worse than undertraining. Over-training, especially in young athletes who are still growing, can damage muscles, joints, ligaments and tendons. In the long term, the immune system can be damaged, leaving athletes susceptible to infection. For this reason training is usually cyclical. Intense periods of training are used to reach peak fitness for an important event, while at other times, the athlete simply maintains fitness.

CHANGES TO THE MUSCLES

The way in which the muscles adapt to exercise depends on the type of muscle fibres involved and the nature of the exercise. Here are some of the improvements that can occur:

- As a general rule, repetitive exercise with moderate resistance (such as aerobics, or light weights) improves muscle tone and stamina. Training with greater resistance (such as heavy weights) brings about an increase in muscle size, a process known as **hypertrophy**. Muscular hypertrophy involves an increase in the cross-sectional size of existing muscles. This is due to an increase in the number of myofibrils, the sarcoplasmic volume, and in the amount of connective tissue (tendons and ligaments).

- A general improvement in performance occurs because of an improved coordination of the motor units. A significant aspect of improving skill involves the antagonistic muscles. Studies have shown that as we practise particular movements, such as a tennis serve, we get smoother and more efficient. Not only do we train the active muscle to work better, we also learn to completely suppress the antagonistic muscle, preventing it from interfering with the action.

- With training, muscles increase their **oxidative capacity** – their capacity for respiration. This is achieved by an increase in the number of mitochondria in the muscle cells, an increased supply of ATP and CP and a rise in the quantity of the enzymes involved in respiration. The ability of the muscles to store glycogen, the amount of myoglobin and the ability to use lipid as an energy store are also increased.

- The density of the capillaries running through muscles increases in response to exercise, a process called **capillarisation**. This allows a more efficient exchange between working muscles and the blood.

FOR ASSIGNMENTS ON EXERCISE PHYSIOLOGY SEE CHAPTER 8 AND CHAPTER 21

SUMMARY

By the end of this chapter you should know and understand the following:

- Exercise places great metabolic demands on the body. Homeostatic mechanisms must cope with a much greater oxygen demand as well as a build-up of carbon dioxide and heat. In the long term, there may be a significant loss of water and salts (electrolytes) through sweating.

- Movement of muscles requires **adenosine triphosphate**, **ATP**, the chemical that supplies the muscles with the energy released in respiration.

- There are three sources of ATP: these comprise three energy systems. The most immediate is the **ATP/CP**, or **alactic anaerobic** system: ATP already present in the muscles is backed up by creatine phosphate, which is used to make more ATP instantly. The second system that kicks into action is the **glycolytic** or **lactic anaerobic** system: ATP is supplied by glycolysis and this system results in a build-up of lactate ions. The third is the **aerobic** system in which energy comes from the complete oxidation of glucose or lipid.

- The duration of an event determines the type of energy system used: up to 10 seconds, the ATP/CP system; from 10 to 60 seconds, the glycolytic system; and more than 60 seconds, the aerobic system.

- In the short term, the body responds to exercise by maintaining homeostasis. There is an increase in heartbeat and ventilation rate, coupled with vasodilation and sweating.

- During exercise we can build up an **oxygen debt**. This is paid off after exercise has finished, or during rest periods within the activity. This is why pulse and ventilation rate remain higher than normal even when activity has stopped. Extra oxygen is needed to replace the ATP/CP stores and to oxidise accumulated lactate.

- The long-term responses of the body to exercise are generally those that improve the body's fitness. These include strengthening of the cardiovascular system (heart and blood vessels) and making the lungs and muscles more efficient. The overall result is a body that is better able to cope with the chosen activity.

❓ EXAM QUESTIONS

Parts of questions marked * require a knowledge of training not covered in this chapter.

1 An investigation was carried out into effects of exercise on the heart rate and breathing rate of a student. His heart rate and breathing rate were first measured at rest. He then cycled at 5 km h^{-1}, using an exercise bicycle, for exactly two minutes. His breathing rate and heart rate were measured immediately after this exercise. After resting for five minutes, he cycled at 10 km hr^{-1} for two minutes, then measurements of heart rate and breathing rate were taken as before. This procedure was repeated for cycling speeds of 15, 20 and 25 km h^{-1}. The results are shown in the table below.

Cycling speed / km h^{-1}	Heart rate / beats min^{-1}	Breathing rate / breaths min^{-1}
0 (rest)	64	12
5	70	13
10	80	14
15	100	17
20	140	20
25	180	27

a) Describe the relationship between cycling speed and heart rate.
b) **(i)** Describe and explain the changes that occurred in breathing rate during this investigation.
   (ii) The volume of air breathed per minute is referred to as the *minute volume*. Suggest changes that are likely to occur in the student's minute volume during this investigation.
c) Suggest why the student rested for five minutes between each period of exercise.
d) During this investigation, the pH of the student's blood was found to have decreased from 7.4 (at rest) to 7.2. Explain how this change occurred.
Edexcel HB6 synoptic June 1999 Q4

2 Some athletes find that they recover faster after strenuous exercise by continuing to exercise slowly. This is sometimes called 'warming down'. An investigation was carried out to see if this was due to the removal of lactate from the blood.
A cyclist exercised vigorously on an exercise bicycle for six minutes and then rested for 34 minutes. When fully recovered, the same cyclist repeated the procedure, but this time cycled at a slow speed following the six minutes of fast cycling. Blood samples were taken at intervals and analysed for the concentration of lactate. The results are shown in the graph.

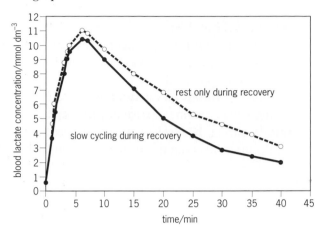

a) With reference to the graph,
   (i) describe the changes in lactate concentration in the blood when the cyclist exercised and then rested during recovery.
   (ii) explain the changes you have described in **(i)**;
   (iii) suggest how cycling at slow speed during the 'warming down' period helps to lower the concentration of lactate in the blood more quickly than resting completely.
   Trained athletes can exercise at much higher levels than untrained people before their blood lactate concentration starts to increase. This is due to changes in the cardiovascular system and in muscles.
b) Describe the changes that occur in the cardiovascular system and in muscles **during training**.
OCR 2802 Human Health and Disease June 2001 Q2

3
a) Define the term **anaerobic metabolism**.
b) The figure shows the relationship between oxygen consumption and time, before, during and after maximal exercise.

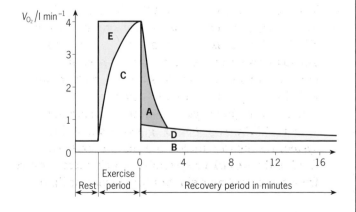

❓ EXAM QUESTIONS

Using the letters A, B, C, D and E which label each section, identify:

(i) the lactacid oxygen debt recovery;

(ii) the oxygen debt;

(iii) the alactacid oxygen debt recovery.

c) Describe the process of **ATP production** which restores the oxygen debt.

d) A student performs an interval training session during which the rate of **muscle phosphagen levels** during the recovery period was recorded. The results from this training session are given in the table below.

Recovery time/seconds	Muscle phosphagen restored
10	10%
30	50%
60	75%
90	87%
120	93%
150	97%
180	99%
210	101%
240	102%

(i) Using the results shown in the table, plot a graph of recovery time against the percentage of muscle phosphagen restored.

(ii) What resting interval would you recommend for a full recovery?

(iii) What would be the effect of restarting the exercise after 30 seconds?

e) (i) Part of the recovery mechanism after anaerobic exercise involves **myoglobin**. Explain the function of myoglobin during the recovery process.

(ii) Explain the importance of the cool down in the assistance of lactacid oxygen debt recovery and in the avoidance of muscle soreness.

(iii)* How could information on oxygen debt recovery be of use to an athlete and coach in the design of training sessions?

AEB Physical Education Paper 2 Section A Physiology of Exercise June 1995 Q1

4 An investigation was carried out to find the effect of exercise on the concentration of lactate in the blood. Four treadmills were placed at different gradients, ranging from 2 per cent to 14 per cent. A different athlete ran at a constant speed on each treadmill for five minutes. Three minutes after each run a blood sample was taken from each athlete

and the concentration of lactate was measured. The graph shows the results.

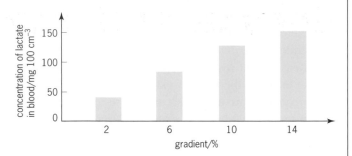

a) Name the process that produces lactate during exercise.

b) Describe the relationship between the intensity of exercise and the concentration of lactate in the blood.

c) Suggest **two** ways in which the design of the investigation could be improved to give more reliable results.

d) Suggest why the blood samples used to measure lactate concentrations were not taken from the athletes until three minutes after completion of a run.

e) The athlete running on the treadmill at a gradient of 14 per cent suffered from muscle fatigue. Explain what causes muscle fatigue.

AQA (B) BYB3 Physiology and Transport June 2001 Q3

5 During exercise, physiological changes take place within the body.

a) (i) What is the difference between a short-term physiological response and a long-term physiological adaptation? Give examples to support your answer.

(ii) If an athlete completed a 12-week programme of intense aerobic training, what physiological adaptations would you expect to take place within the cardiovascular system?

b)* The Multi-stage fitness test is commonly used by athletes to assess their level of endurance. State what this test measures and discuss the advantages and disadvantages of this method of assessment.

c) Myoglobin is found in the sarcoplasm of the muscle cell. Explain why an increase in myoglobin is beneficial to an endurance athlete.

UCLES Physical Education Paper 2 Optional Topics Section A: Scientific topics, Topic 4: Exercise Physiology June 1998 Q1

19

Nerves and impulses

19 Nerves and impulses

The Amazonian Indians, not being analytical chemists, assess the strength of their preparation by how long it takes a monkey to fall out of the tree. 'One-tree curare' is the most powerful; the monkey only escapes to the next tree before the paralysing toxin brings it tumbling to the forest floor. 'Two-tree curare' is obviously weaker and 'three-tree curare' is the weakest preparation that is useful. After that, the monkey travels too far and, although it still dies from the action of the poison, it's usually impossible to find

Some Amazonian Indians have a very effective way of hunting monkeys and other animals. They tip their blowpipe arrows in a preparation made from the bark of one of the local trees. The substance, known as curare, is a powerful neurotoxin.

For many centuries, the exact nature of curare was a mystery. In 1814, in an attempt to confirm his suspicions, the explorer Charles Waterton injected a donkey with curare. Within a few minutes, the donkey appeared dead. Waterton cut a small hole in her throat and inflated her lungs with a pair of bellows. According to his account, the donkey regained consciousness and looked around. This crude method of artificial respiration was continued for a couple of hours until the effects of the curare had worn off.

Analysis of curare has since shown that it prevents nerve impulses getting through to the muscles, causing paralysis. In addition to affecting the movement of skeletal muscles, curare interferes with heartbeat and breathing, although the exact effects depend on the dosage and whether the curare is taken orally or by injection. If heartbeat and breathing are severely affected, the result is usually fatal.

In 1939, the active ingredient in curare was isolated, and in 1943 doctors started to use it as an anaesthetic. This gift from the rainforest has proved to be a vital tool in surgery, relaxing muscles and generally making the surgeon's job a lot easier. It is also useful for treating conditions in which the muscles go into spasm – polio, tetanus, epilepsy and cholera. Today, synthetic analogues of the active ingredient in curare, such as d-tubocurarine, are used widely in medicine.

1 IT'S ALL ABOUT COMMUNICATION

The body of a mammal is incredibly complex. To ensure that the different parts work together effectively, there needs to be communication to register changes in internal and external conditions. In this section of chapters we look at the role of hormonal and nervous communication.

The nervous system is made up of specialised cells that allow the different parts of the body to communicate. These cells are called neurones (Fig 19.1). They carry information from one part of the body to another. We concentrate on the structure and function of these remarkable cells in this chapter.

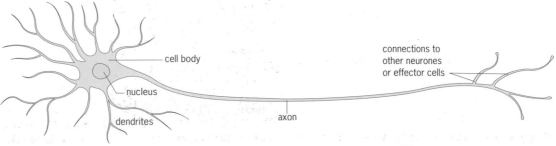

Fig 19.1 A simple diagram of a neurone

OVERVIEW OF THE NERVOUS SYSTEM

The nervous system as a whole ensures that the body responds appropriately to the external conditions at any given time. It means we can do the following:

- Gather information. Sense organs called **receptors** detect **stimuli** from the internal and the external environment (see Chapter 23).
- Transmit sensory information to the **central nervous system** by means of the **sensory nerves** (see Chapter 20).
- Co-ordinate information. Incoming information travels to the brain via the spinal cord. The brain then decides what to do. Decisions are often based on memory, the result of our past experience.
- Transmit the information to the **effectors**: the muscles and glands. Impulses pass from the central nervous system to the effectors via **motor nerves** (Fig 19.2).

The structure of the mammalian/human nervous system is covered in detail in Chapter 20, the structure of muscles and their role in movement is covered in Chapter 21 and hormonal communication is the topic of Chapter 22. This section of chapters finishes with a look at senses and behaviour (Chapter 23).

HOW INFORMATION TRAVELS AROUND THE BODY

Information passes along **neurones** in the form of electrical signals called **nerve impulses**. A nerve impulse, known as an **action potential**, is not a message, nor is it an electrical current (a flow of electrons). It is more a change in ion balance in the nerve cell, which spreads rapidly from one end to the other, like the fire travelling along a burning fuse. It is little more than an electrical 'blip' – but the brain can make sense of the blips because they vary in frequency and arrive down specific nerves (see Chapter 20).

So what happens when the nerve impulse reaches the end of the neurone? It connects with other neurones at junctions called **synapses**. A nerve impulse crosses a synapse usually by means of a **chemical transmitter**. The whole of the nervous system therefore communicates by a mixture of electrical and chemical signals. This allows information to travel around an organism with far greater speed and precision than if only chemical signals were used.

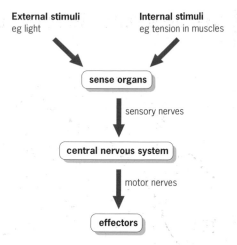

Fig 19.2 Flow diagram summarising nervous system function

✔ **REMEMBER THIS**

The entire human body is made up from just four basic types of tissue (see Chapter 2). Epithelial tissue forms coverings and linings, connective tissue holds other tissues together, muscle tissue has the ability to contract and nervous tissue is excitable – it can transmit impulses. All of our organs are formed from these four basic tissue types.

? **TEST YOURSELF**

A What is the difference between a neurone and a nerve?

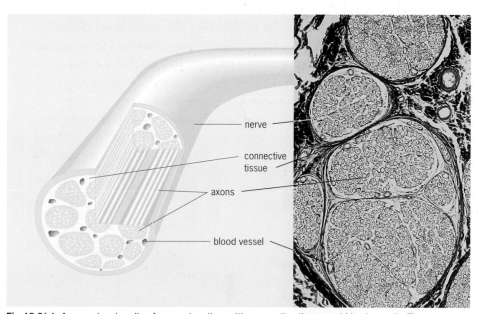

Fig 19.3(a) A nerve is a bundle of axons, together with connective tissue and blood vessels. The micrograph shows a cross section of a nerve

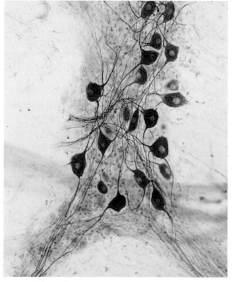

Fig 19.3(b) The brown threads (axons) and their cell bodies (see Fig 19.8) form bundles that make up nerves

NERVES

Neurones rarely act alone. They are bundled together into larger, visible structures called **nerves** (Fig 19.3). Nerves form a complex network throughout the body. Sensory nerves take information received from receptors, or sense organs, on the outer parts of the body – the eyes, ears, tongue and nose and touch receptors in the skin – into the central nervous system. Processing of that information happens in the brain, and then effector nerves take the information from the brain to effectors such as muscles and glands to cause the body to take some action. Think what happens if someone puts, say, a glass of water in front of you. Sensory nerves take the information from the eyes to the brain. If you are thirsty, the brain may then send information down effector nerves to your arm and hand and you would reach for the glass, pick it up and drink. If you aren't thirsty, information might travel down effector nerves to the muscles in your face and throat for you to say 'No thanks'.

In the reflex arc (see Chapter 20), the sensory and effector nerves are linked by a neurone in the spinal cord, rather than by neurones in the brain.

2 HOW DO NEURONES CARRY INFORMATION?

Neurones have two properties that enable them to carry information. They are **excitable** – that is, they can detect and respond to stimuli – and they are **conductive** – that is, they can transmit a signal from one end to the other. Before we look at how a neurone transmits information, let's find out what is going on in the neurone before information arrives.

THE NEURONE AT REST

At any given moment a neurone needs to be ready to conduct impulses. This state of readiness is called the **resting potential**. At this point, the axon membrane is **polarised**: the fluid on the inside is negatively charged with respect to the outside. This difference in charge, about −70 mV, results from an unequal distribution of ions known as an **electrochemical gradient**.

THE RESTING POTENTIAL

What causes the resting potential, and how is it maintained? Like most cells, neurones have selectively permeable membranes that contain specialist proteins. Some of these proteins act as passive ion channels; others are active transport mechanisms that pump ions. The steps that establish a resting potential are shown in Fig 19.4.

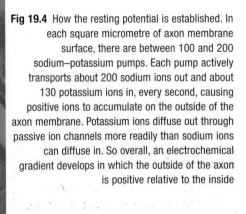

Fig 19.4 How the resting potential is established. In each square micrometre of axon membrane surface, there are between 100 and 200 sodium–potassium pumps. Each pump actively transports about 200 sodium ions out and about 130 potassium ions in, every second, causing positive ions to accumulate on the outside of the axon membrane. Potassium ions diffuse out through passive ion channels more readily than sodium ions can diffuse in. So overall, an electrochemical gradient develops in which the outside of the axon is positive relative to the inside

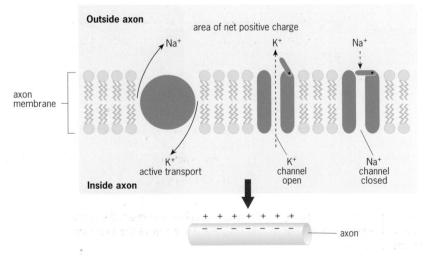

The first important thing that happens is that an active transport mechanism in the neurone membrane swaps sodium ions for potassium; Na$^+$ ions are moved out, while K$^+$ ions go inwards. On its own, this mechanism simply swaps one positively charged ion for another and so does not create the resting potential.

The second important point is that a neurone also has passive Na$^+$ channels and K$^+$ channels. In a resting neurone the K$^+$ channels are generally open while the Na$^+$ channels stay shut. Thus potassium is able to diffuse out faster than sodium can diffuse in, and diffusion via the passive ion channels is unable to balance the active transport mechanism. Positive ions accumulate on the outside of the membrane and this then becomes positively charged compared to the inside.

Overall, the resting potential results from an active transport mechanism, which uses the energy in ATP to create an ionic imbalance. If the active transport mechanism failed, the potential would quickly be lost, and the neurone would no longer be excitable.

THE NEURONE IN ACTION

So what exactly is a nerve impulse? In short, it is a temporary reversal of the resting potential; a change that spreads rapidly along the axon. A schematic diagram explaining how this happens is shown in Fig 19.5. Fig 19.6 shows an oscilloscope trace of an **action potential**, sometimes referred to as a 'spike' because of its shape.

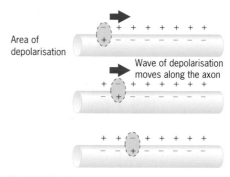

Fig 19.5 The basic concept of the nerve impulse – a wave of depolarisation which spreads along the axon. The active transport mechanism immediately re-establishes the resting potential as soon as the action potential has passed

SEE QUESTION 1

✓ REMEMBER THIS

Once the action potential starts, the wave of electrical activity travels, or is **propagated**, along the axon at great speed.

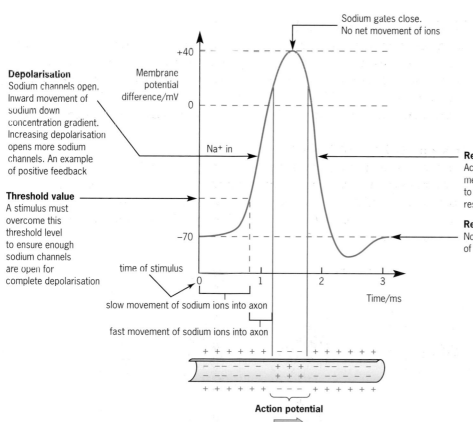

Fig 19.6 An action potential as recorded by an oscilloscope, and how it relates to the events in the axon. This is a key exam diagram that you should use when you are aiming to explain the basic sections of the trace

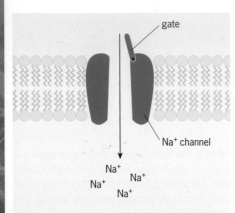

Rapid influx of Na$^+$ ions reverses resting potential

Fig 19.7 A diagrammatic model to show an ion channel in the axon membrane. When the Na$^+$ channels open the rapid influx of ions causes the action potential by reversing the resting potential

When the Na$^+$ channels close and K$^+$ channels open, this allows a rapid outflow of K$^+$ and repolarises the membrane.

In the action potential region, only about one in a million Na$^+$ or K$^+$ moves through channels

DEPOLARISATION

When a stimulus reaches the resting neurone, the Na$^+$ ion channels open at the site of stimulation. The ions move in relatively slowly at first, making the membrane potential less negative. Then, at a **threshold** of about –50 mV, a much more rapid inrush of Na$^+$ ions is triggered, and the inside suddenly becomes positive (+40 mV) with respect to the outside. As the electrical charge across the membrane is reversed, we say that the membrane becomes **depolarised**. Depolarisation initiates a nerve impulse.

THE ACTION POTENTIAL

Why does the area of depolarisation flow along the axon? When a patch of the axon membrane depolarises, a flow of current is produced locally as electrically charged ions move. This current stimulates the next patch of membrane, where Na$^+$ channels also open, as shown in Fig 19.7. In this way the action potential moves along the axon in the direction away from the cell body.

The change in voltage (the amplitude) of the action potential is the same at all points along the axon. It does not increase or fall off as the signal travels. After a very brief moment – about 1 millisecond – the Na$^+$ channels close and the resting potential is established once again. The Na$^+$ channels cannot re-open until after the resting potential has been re-established.

THE REFRACTORY PERIOD

Nerves conduct messages by 'firing' repeated action potentials along the nerve fibres. The time delay between one action potential and the next is called the **refractory period**. This has two phases:

● The **absolute refractory period**. During this time, immediately after the sodium channels close, no further impulse can be conducted.
● The **relative refractory period**. During this time, the membrane begins to recover and becomes increasingly responsive. It is possible to initiate another action potential provided that the stimulus is greater than normal.

The refractory period imposes a limit on the frequency of nerve firing. Large nerve fibres recover in 1 millisecond and could theoretically propagate 1000 impulses per second. Small fibres take longer to recover – about 4 milliseconds – and so could propagate about 250 impulses per second.

The refractory period ensures that each action potential is separated from the next, with no overlapping of signals. We can think of the information the signal conveys as coded information. The refractory period also ensures that a nerve impulse flows in one direction only: the wave of depolarisation can only move away from the refractory region, towards the axon terminal, and therefore onwards to the next neurone in the pathway.

THE SPEED OF AN ACTION POTENTIAL

The speed at which a nerve impulse or action potential travels is known as its **conduction velocity**. In human nerve fibres, values range from 1 to 3 metres per second in unmyelinated fibres, and between 3 and 120 metres per second in myelinated fibres (see below for what myelin is and what it does). In general, conduction velocity depends on the following factors:

● **Axon diameter**. The larger the axon, the faster it conducts.
● **Myelination of the neurone**. A nerve impulse travels faster in a myelinated nerve than in an unmyelinated nerve.
● **Number of synapses involved**. Communication between neurones across the tiny gaps at the synapses involves chemical release and a brief time delay. The greater the number of synapses in a series of neurones, the slower the conduction velocity.

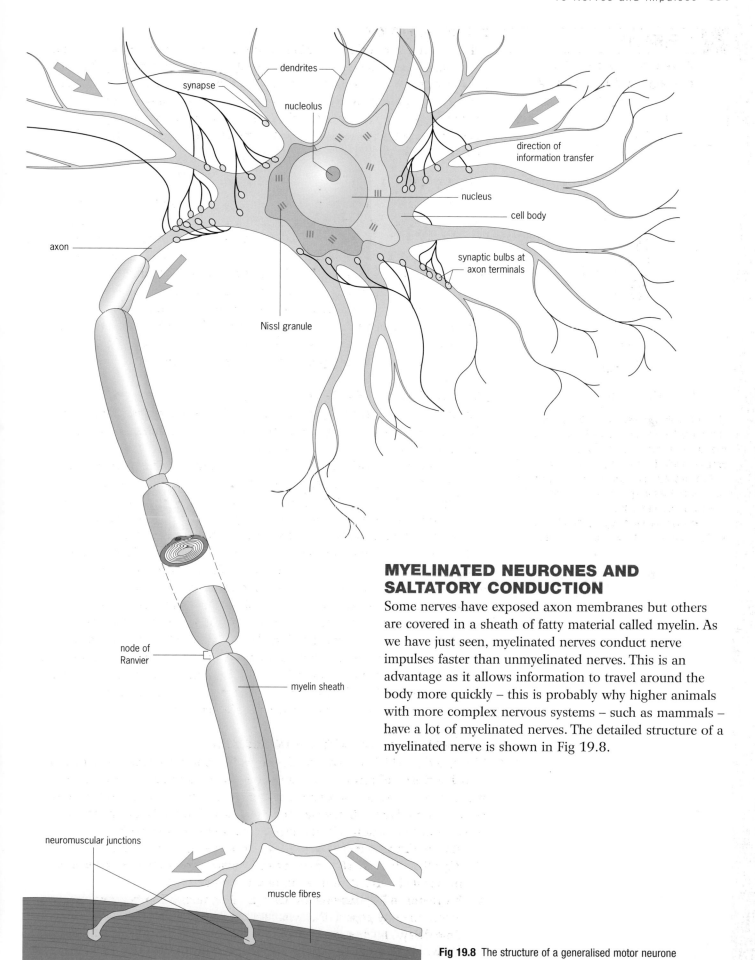

direction of
information transfer

dendrites

synapse

nucleolus

nucleus

cell body

axon

synaptic bulbs at
axon terminals

Nissl granule

node of
Ranvier

myelin sheath

neuromuscular junctions

muscle fibres

MYELINATED NEURONES AND SALTATORY CONDUCTION

Some nerves have exposed axon membranes but others are covered in a sheath of fatty material called myelin. As we have just seen, myelinated nerves conduct nerve impulses faster than unmyelinated nerves. This is an advantage as it allows information to travel around the body more quickly – this is probably why higher animals with more complex nervous systems – such as mammals – have a lot of myelinated nerves. The detailed structure of a myelinated nerve is shown in Fig 19.8.

Fig 19.8 The structure of a generalised motor neurone

SCIENCE FEATURE Myelin and multiple sclerosis

Multiple sclerosis is a chronic, often disabling disease of the central nervous system that is caused by a loss of myelin around neurones. Symptoms may be mild, such as numbness in the limbs, or severe – paralysis or loss of vision. Most people with MS are diagnosed between the ages of 20 and 40 but the unpredictable physical and emotional effects can be lifelong.

What starts MS off remains something of a mystery. In some cases, the immune system starts to recognise myelin as a 'foreign' protein and T cells (see Chapter 17) become active and start to destroy the myelin sheaths around many neurones. The resulting demyelination causes disruption to the transmission of nerve impulses within the brain and around the body. Some research suggests that this autoimmune response is provoked when a virus that has proteins similar in shape to myelin infects the body. As the body reacts against the virus, it also reacts against its own myelin.

One possible treatment for MS is a substance called copolymer-1, which is a random polymer of basic amino acids. This is structurally similar to myelin basic protein and researchers hope that it can 'mop up' some of the immune cells and antibodies that would otherwise damage myelin. People with MS would need to have an injection of copolymer-1 every day but it has been shown to help and it decreases the frequency of relapses.

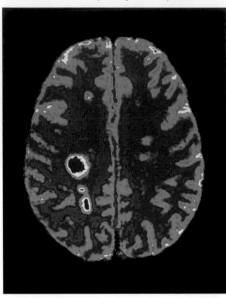

Fig 19.9 A coloured magnetic resonance image (MRI) scan of an axial section through the brain of a person with MS. The lesions due to MS are shown in red and yellow at the left and lower left

REMEMBER THIS

The pale, creamy colour of myelinated nerves is due to the fatty (lipid) nature of the myelin that surrounds them.

Fig 19.10(a) Schwann cell growing round an axon

THE IMPORTANCE OF MYELIN

But what is myelin and how does it form around neurones? Specialised cells called **Schwann cells** wrap themselves round the axons of some neurones as they develop in a growing embryo (Fig 19.10a) The Schwann cells form a thick, lipid-rich insulating layer called the myelin sheath (Fig 19.10b). This insulates the axon electrically, rather like the plastic layer round a copper wire in an electrical flex. Neurones with myelin sheaths are said to be myelinated.

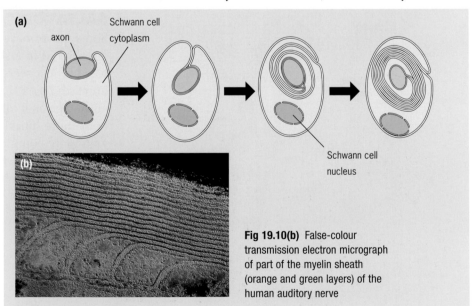

(a) axon — Schwann cell cytoplasm

Schwann cell nucleus

Fig 19.10(b) False-colour transmission electron micrograph of part of the myelin sheath (orange and green layers) of the human auditory nerve

TEST YOURSELF

D What effect does a myelin sheath have on conduction velocity?

SALTATORY CONDUCTION

Schwann cells cover most of the axon, but leave bare sections between the cells called **nodes of Ranvier** (see Fig 19.8). Nerve impulses that travel along a myelinated axon 'jump' between these gaps by **saltatory conduction** (Fig 19.11).

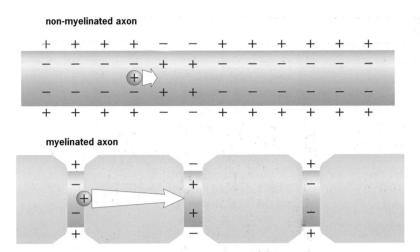

non-myelinated axon

myelinated axon

Fig 19.11 Myelin and conduction speed. Top: the impulse passes along the nerve fibre by each section of membrane depolarising the next, so transmission is relatively slow. Bottom: the impulse jumps from one node to the next, so only the membrane at the nodes is depolarised. This leads to rapid transmission

THE ADVANTAGES OF SALTATORY CONDUCTION

Saltatory conduction has two advantages:

- The conduction of a nerve impulse is fast. In human unmyelinated fibres, nerve impulses travel at 1 to 3 metres per second. Myelinated fibres conduct at speeds of up to 120 metres per second.
- Metabolically, saltatory conduction is quite economical, because fewer ions move across the membrane, so the ion pumps need less energy to restore the ionic balance.

3 COMMUNICATION BETWEEN NEURONES: THE SYNAPSE

When an action potential reaches the end of an axon it is passed on to the next neurone, or on to an effector cell such as a muscle or gland. The axon of one neurone does not usually make direct contact with the cell body of the next; the two cells are separated by a gap called a synapse (Figs 19.12 and 19.13).

> ✔ **REMEMBER THIS**
>
> In myelinated nerves, impulses can travel at up to 120 metres per second – the equivalent of a person running the length of a football pitch in one second, ten times quicker than the best sprinters can run.

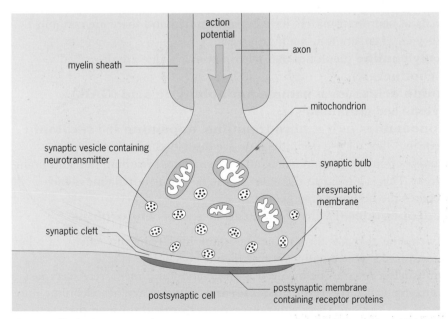

Fig 19.12 The basic structure of a synapse

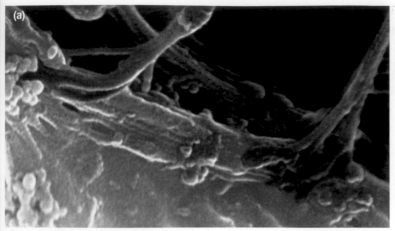

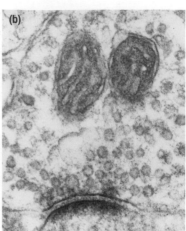

Fig 19.13(a) False-colour scanning electron micrograph of the junction sites (synapses) between nerve fibres (purple) and a neurone cell body (yellow)

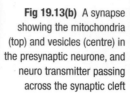

Fig 19.13(b) A synapse showing the mitochondria (top) and vesicles (centre) in the presynaptic neurone, and neuro transmitter passing across the synaptic cleft

REMEMBER THIS

The cell that carries a signal towards a synapse is a **presynaptic cell**; the cell carrying the signal away from the synapse is a **postsynaptic cell**. Presynaptic cells are always neurones but postsynaptic cells can be either neurones or effector cells.

REMEMBER THIS

Many drugs and poisons exert their effect because they interfere with the functioning of synapses – see the Assignment at the end of this chapter.

REMEMBER THIS

In some people, release of too much noradrenaline causes the heart to race. One way of treating this is to use drugs known as beta-blockers. These drugs have molecular shapes similar to noradrenaline.

THE MAIN FEATURES OF A CHEMICAL SYNAPSE

The axon terminal of a presynaptic neurone is swollen, and is often called the **synaptic bulb** or **synaptic knob** (see Fig 19.12). It meets the cell body of the next axon, leaving a gap or **synaptic cleft** of about 20 nm. This is small – you would have to split a hair 250 times to get it to fit sideways into this gap. Even so, synapses have a high electrical resistance, and this gap is too big to allow the action potential to simply jump from one neurone to the next.

HOW DO SYNAPSES WORK?

So how does the action potential get across? It is relayed by chemicals that diffuse across the gap and initiate an action potential in the neurone at the other side. The synaptic bulb contains many mitochondria, which provide energy for the manufacture of chemicals called **neurotransmitters**. Neurotransmitters are small molecules and they can diffuse easily across the synaptic cleft. Synaptic vesicles are temporary vacuoles (membrane-bound spheres) that store neurotransmitter chemicals, the most common being **acetylcholine**. Synapses that have acetylcholine as their transmitter are called **cholinergic synapses**.

TYPES OF NEUROTRANSMITTER

Hundreds of neurotransmitters have been identified and there are certainly more to find. There are four main groups:

● **Acetylcholine** (neurones that release acetylcholine are described as **cholinergic**).
● **Amino acids** such as **gamma-aminobutyric acid (GABA)**, **glycine** and **glutamate**.
● **Monoamines** such as **noradrenaline**, **dopamine** and **serotonin** (neurones that release noradrenaline are described as **adrenergic**). Synapses that use noradrenaline affect heart rate, breathing rate and brain activity. This is similar to the effect of the hormone adrenaline, which prepares the body for emergencies.
● **Neuropeptides** (chains of amino acids) such as **endorphins**.

Acetylcholine also acts throughout the brain, modifying the activity of other neurotransmitters. Nerve pathways in which acetylcholine is a neurotransmitter seem to be involved in motivation and memory. A very fast-acting enzyme called **acetylcholinesterase** breaks down acetylcholine into acetic acid and choline. These substances are reabsorbed through the presynaptic membrane. ATP energy from mitochondria is used to provide the energy to resynthesise acetylcholine, which is then returned to the vesicles. The chemicals in some 'nerve gases' work by inhibiting acetylcholinesterase.

TRANSMISSION AT A SYNAPSE – A DETAILED EXPLANATION

As you read the next section, follow the stages of chemical transmission at a synapse numbered in Fig 19.14.

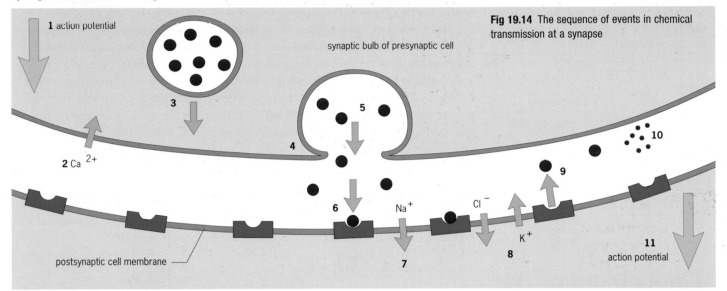

Fig 19.14 The sequence of events in chemical transmission at a synapse

1 An action potential arrives at the synaptic bulb.
2 Calcium channels open in the **presynaptic membrane**. As the Ca^{2+} ion concentration inside the bulb is lower than outside, Ca^{2+} ions rush in.
3 As the Ca^{2+} concentration increases, synaptic vesicles move towards the membrane.
4 The vesicles fuse with the membrane, releasing neurotransmitter into the synaptic cleft.
5 The short journey across the synapse takes about a millisecond, longer than an electrical signal takes to travel the same distance. This time is therefore called the **synaptic delay**.
6 At the **postsynaptic cell**, the neurotransmitter binds to receptors on the postsynaptic cell surface membrane.
7 Some neurotransmitters open sodium channels in the postsynaptic membrane, causing an inflow of Na^+ ions. This creates an **excitatory postsynaptic potential** (EPSP) in the membrane. This potential lasts for only a few milliseconds and can travel only a short distance, but it makes the membrane more receptive to other incoming signals.
8 Other neurotransmitters open chloride and potassium channels, causing Cl^- ions to flow into the cell and K^+ ions to flow out. This creates the **inhibitory postsynaptic potential** (IPSP) that makes the postsynaptic membrane less receptive to incoming signals. If more EPSPs than IPSPs are produced in the postsynaptic membrane, the change in potential can exceed the threshold potential needed to create a new action potential.
9 Once the neurotransmitter has acted on the postsynaptic membrane, it is immediately broken down by an enzyme on the postsynaptic membrane. If the transmitter remained, it could continue to stimulate the neurone, even without new impulses coming from the presynaptic cell.
10 The enzyme acetylcholinesterase splits acetylcholine into choline and ethanoic acid. These components then diffuse back into the presynaptic membrane, when they are resynthesised to acetylcholine using the ATP from the mitochondria.
11 An action potential is set up in the postsynaptic cell.

? TEST YOURSELF

E The synaptic cleft has a high electrical resistance but is very narrow. Suggest reasons for these two observations.

F Why can transmission occur only one way across synapses?

SEE QUESTION 2

✔ REMEMBER THIS

Receptor binding can also lead to the formation a second messenger (a transmitter substance) such as cyclic AMP (cAMP). This also changes the ionic permeability of the membrane, but it has a longer-lasting metabolic effect on the ion channels. Such long-term changes to brain neurones are thought to underlie memory.

SYNAPSES IN ACTION: FACILITATION AND SUMMATION

Does the arrival of an impulse at a synapse mean that an action potential is always generated on the postsynaptic neurone? The answer is no, because it would lead to chaos; all neurones would automatically connect to others. Synapses therefore have a vital role in **information processing**. Transmission of information across synapses is graded. They can amplify or damp down the information they receive. In many cases, they will not transmit it at all.

A neurone can be fed information by both excitatory synapses that produce EPSPs and inhibitory synapses that produce IPSPs (Fig 19.15). Whether the cell develops an action potential is determined by the sum of all the excitatory and inhibitory synapses at any particular moment. Put simply, impulses arriving at some synapses will 'excite' the cell, while others will 'calm it down'. Whether or not a neurone generates an action potential depends on the balance of the two types.

Facilitation occurs when many EPSPs reach the same area of the postsynaptic membrane at the same time, making it more likely that the threshold potential will be reached. When this happens, an action potential is propagated.

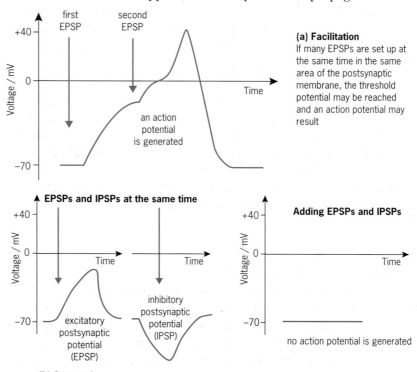

(a) Facilitation
If many EPSPs are set up at the same time in the same area of the postsynaptic membrane, the threshold potential may be reached and an action potential may result

(b) Summation
If the same number of EPSPs as IPSPs are set up at the same time in the same area of the postsynaptic membrane, the two potentials cancel each other out

Fig 19.15 (a) Facilitation and **(b)** summation

Imagine a synapse discharging its transmitter on to a postsynaptic neurone. This will set up an EPSP, but if it is not big enough to reach the threshold, no action potential is generated. However, if other synapses discharge their transmitter at the same time, or shortly after, the EPSPs will add up, or **summate**, until an action potential is generated. Generally, there are two types of summation:

● **Temporal summation** – summation of two or more impulses that arrive rapidly one after the other down the same neurone.
● **Spatial summation** – summation of two or more impulses arriving down different neurones at the same time (spatial = related in space). One neurone can receive information from many others – this is synaptic convergence. It follows that the arrival of one excitatory impulse will leave the neurone more responsive to another one. This is known as facilitation, and results from the summation of two or more synapses discharging their transmitter substance at the same time.

As a simple example of this idea, imagine the touch receptors from one area of skin feeding into one sensory neurone. An action impulse down just one receptor is almost certainly an insignificant stimulus, and can be ignored. It will not create an EPSP large enough to generate an action potential in the sensory nerve. However, if several touch receptors are stimulated at the same time, they will summate and produce a sensory impulse.

WHY HAVE SYNAPSES?

Synapses are important because they allow the transfer of information in nerve networks to be controlled. Synapses:

- allow information to pass from one neurone to another;
- help ensure that a nerve impulse travels in one direction only;
- allow the next neurone to be excited or inhibited;
- can amplify a signal (make it stronger);
- protect nerve networks by not firing when over-stimulated. When this happens the synapse is said to be fatigued. Over-stimulation might damage muscle or gland tissue;
- can filter out low-level stimuli. For example, you fail to notice the sound of a clock ticking because synapses are 'filtering out' the signal of sound;
- aid information-processing by the action of summation (adding together the effect of all impulses received, see page 356);
- are modifiable and can form a physical basis for memory.

Overall, the significance of synapses cannot be over-emphasised. They allow us to select particular neural pathways. The process of learning is largely one of educating the synapses. People can play the violin or piano, or play tennis, because their synapses allow their brains to co-ordinate their senses and muscles in the right way. Your memories, too, have a basis in synapses choosing specific pathways. If you are asked, 'What's the capital of France?' your synapses will (we hope) select a pathway of neurones in your brain which will lead you to the answer 'Paris'.

A SPECIAL SORT OF SYNAPSE: THE NEUROMUSCULAR JUNCTION

When a motor neurone terminates on a muscle, it branches into many specialised synapses called **neuromuscular junctions**. Fig 19.16 shows a typical neuromuscular junction. These structures are wider than ordinary synapses and come into close contact with the surface membrane of the muscle, the sarcolemma. The area of sarcolemma in contact with the synapse is called the motor end-plate, and contains acetylcholine receptor sites. When an action potential arrives at a neuromuscular junction, vesicles of acetylcholine are released in the usual way. The transmitter changes the permeability of the motor end-plate to Na^+ and K^+, creating an end-plate potential (EPP) which results in an action potential passing along the sarcolemma. This impulse brings about the contraction of muscle fibres in that area.

REMEMBER THIS

Temporal = related in time.
Spatial = related in space.

SEE QUESTION 3

REMEMBER THIS

Facilitation is not a result of temporal summation, which is simply the accumulation of EPSPs because impulses arrive before the preceding EPSPs have died down.

SEE QUESTION 4

REMEMBER THIS

Chapter 21 gives more detail about muscular contraction, and the Opener and Assignment of this chapter illustrate how chemicals can affect the neuromuscular junction.

SEE QUESTION 5

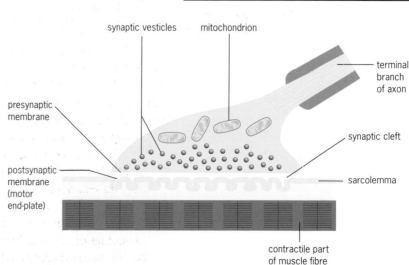

Fig 19.16 Studies have shown that each vesicle contains about 10 000 acetylcholine molecules, and that about 100 vesicles need to be released before an action potential (leading to a muscle twitch) can be generated

4 DRUGS AND SYNAPSES

Transmitters are released in tiny amounts: only 500–1000 molecules from each synaptic knob are required to transmit an impulse. So, drugs that affect transmitters or their binding sites can have powerful effects when given in fairly small doses. Some chemicals, many of them from plants, have a dramatic effect on the nervous system.

NICOTINE AND ATROPINE

Nicotine is a substance found in tobacco. It has a similar shaped molecule to acetylcholine and so competes with acetylcholine to bind with its receptors. Once the nicotine has bound to the receptor, it opens the sodium channel that forms part of the receptor, causing nerve impulses to be generated. The effect of nicotine is covered in the Science feature on the next page.

Atropine also binds to acetylcholine receptors, but does not open the sodium channels. It blocks the receptors, and prevents acetylcholine binding to them. When this happens in motor neurones, it causes muscle paralysis. The binding of nicotine and atropine is summarised in Fig 19.17.

Fig 19.17 Drugs and transmitter binding sites

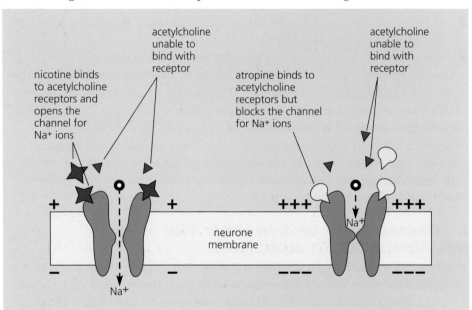

DRUGS AND NORADRENALINE

Beta-blockers have a similar shape to noradrenaline and so compete with it for receptors on the postsynaptic membrane. Amphetamines and cocaine also affect noradrenaline synapses, but they work in a different way. They prevent the reabsorption of noradrenaline from the synaptic gap. So, noradrenaline remains in the gap and the neurone keeps on firing. One of the effects of amphetamines is to make a person feel energetic and carefree, which is why they are often known as speed. Before the harmful effects were known, amphetamines were given to pupils with poor attention spans, to help them concentrate.

Amphetamines are psychologically addictive. Users become dependent on the drug to avoid the 'down' feeling they often experience when the drug's effect wears off. This dependence can lead a user to turn to stronger stimulants such as cocaine, or to larger doses of amphetamines to maintain a 'high'.

People who abruptly stop using amphetamines often experience the physical signs of addiction, such as fatigue, long periods of sleep, irritability and depression. How severe and prolonged these withdrawal symptoms are depends on the degree of abuse.

SCIENCE FEATURE Nicotine — the drug that mimics acetylcholine

Do you wonder why people continue to smoke, even though they know the increased risk of lung cancer and heart disease associated with their habit? The answer lies partly with one of the components of tobacco: nicotine.

Nicotine is very addictive. It affects the brains of smokers, making them feel less stressed, better able to concentrate and less likely to eat sweet foods. Smokers become tolerant to nicotine over time, needing to smoke more to achieve the same effects. But how does nicotine cause addiction?

Studies carried out in the early 1980s using nicotine labelled with a radioactive tracer showed that it is taken into the brain very rapidly. Once there, it binds tightly to acetylcholine receptors, fooling postsynaptic cells into 'thinking' they are being stimulated. It also binds to other receptors which normally accept another neurotransmitter, dopamine.

The action of nicotine on both types of receptor in specific areas of the brain causes long-lasting changes to cell connections and may explain why it is addictive. We know that dopamine receptors, in particular, are involved in addictions to other substances such as amphetamines and cocaine.

Because nicotine is addictive but not carcinogenic (it is the chemicals in tobacco tar that have been shown to cause cancer),

smokers keen to kick the habit can get help. They can buy skin patches and gum that deliver nicotine to the brain, but not tar to the lungs.

Although patches and gum do help smokers to cut down or stop smoking altogether, they are only a partial solution. Nicotine affects the acetylcholine receptors in the parasympathetic nervous system that are involved with the constriction of blood vessels. Over time, circulatory problems and heart disease can result, and it is best to avoid these effects altogether.

Fig 19.18
Transdermal patches can be used almost anywhere on the skin. They slowly release substances such as nicotine into the bloodstream

CURARE

Curare is used as a muscle relaxant in abdominal surgery. It competes with acetylcholine at the neuromuscular junctions, preventing the nerve impulses that cause the abdominal muscles to contract. South American Indians first used curare as a poison. They used it mainly to catch animals for food by smearing arrow tips with it. In a poisoned animal it first affects the muscles of the toes, ears and eyes, then those of the neck and limbs, and, finally, those involved in respiration. The leads to death caused by respiratory paralysis.

PROZAC AND TRANQUILLISERS

Serotonin is a neurotransmitter normally active in the brain. Some forms of depression are caused by a reduced concentration of serotonin in the brain. The antidepressant drug Prozac is a known as a serotonin re-uptake inhibitor. Prozac alleviates depression because it competes with serotonin for the 'active' sites on the proteins that reabsorb serotonin, leading to a higher concentration of serotonin at synapses in the brain. It also competes for the active sites on the enzymes that break down serotonin at synapses.

Tranquillisers are drugs that reduce tension. Benzodiazepine tranquillisers, such as Valium, work by increasing the binding of inhibitory transmitters in the brain. Inhibitory transmitters hyperpolarise rather than depolarise the membrane of the next neurone. This makes the next neurone less excitable. Valium reduces stress and anxiety, but it can be addictive.

SUMMARY

When you have read this chapter you should know and understand the following:

■ The basic unit of the nervous system is the nerve cell, or neurone. This is a specialised cell with **dendrites** that take impulses into the cell body, and greatly elongated axons that take impulses away.

■ The axon membrane is able to use an active transport mechanism to establish a **resting potential**. This is an electrical charge across the membrane caused by an unequal distribution of ions.

■ The nerve impulse itself, called the **action potential**, is a momentary reversal in the resting potential, caused by a sudden rush of sodium ions into the axon. The action potential spreads rapidly along the axon.

■ The action potential lasts for only a millisecond or so, after which the resting potential is re-established. When an action potential has passed, there is a brief period of time – the **refractory period** – during which it is impossible to generate another action potential.

■ The nerve impulse passes from one nerve to another (or from a neurone to a muscle) by means of **synapses**.

■ Synapses are vital in selecting some neural pathways and not others. As such, synapses play a vital role in memory, skill and co-ordination of the body's activities.

■ Transmission across synapses occurs when a chemical, the neurotransmitter, is released by the **presynaptic membrane**. This diffuses across the gap and changes the permeability of the **postsynaptic membrane**, generating an **excitatory postsynaptic potential** (**EPSP**). If the EPSPs are sufficiently large, an action potential is generated in the next neurone.

■ **Summation** means that the effect of several action potentials can add up to produce transmission at a particular synapse.

■ Many drugs and poisons work by affecting synaptic transmission.

? EXAM QUESTIONS

1

a) Describe how the resting potential is maintained across the cell surface membrane of a neurone.

b) The graph below shows the change in potential difference across the cell surface membrane of a neurone during the propagation of a nerve impulse.

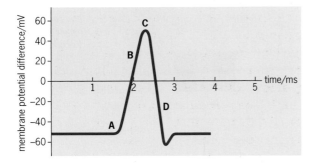

Explain, in terms of ion movements, the change in potential difference which takes place between:
(i) points **A** and **B**; **(ii)** points **C** and **D**.

c) In an investigation into action potentials, neurones were placed in different solutions. The concentration of sodium ions in solution **X** was the same as in tissue fluid. The concentration in solution **Y** was only 30% of that found in tissue fluid. The graph below shows the change in potential difference across the axon membranes after stimulation.

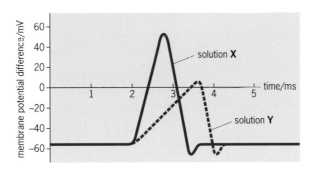

(i) Explain the difference in the size and shape of the curve obtained for the neurone placed in solution Y.

(ii) Explain why it was not possible to produce a series of action potentials when a respiratory inhibitor was added to solution X.

AQA (NEAB) BY04 Biological Basis of Behaviour March 2001 Q4

2 The graph shows changes in concentration of sodium ions inside the axon of a large neurone. The axon was stimulated at the points indicated on the graph. Between **A** and **B** on the graph the neurone was treated with dinitrophenol. This prevents the production of ATP.

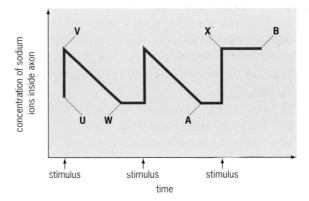

a) Explain the changes in sodium ion concentration
  (i) between **U** and **V**;
  (ii) between **V** and **W**.

b) Explain why the concentration of sodium ions did not change between **X** and **B**.

AQA (NEAB) BY04 Biological Basis of Behaviour June 2000 Q5

? EXAM QUESTIONS

3 The membrane potential of a neurone leading from a stretch receptor in a muscle was measured before, during and after a period in which the muscle was stretched. The results are shown in the graph.

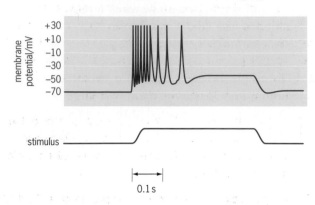

a) Explain what causes the initial increase in membrane potential.

b) How do these results illustrate the 'all-or-nothing' signal?

c) **(i)** For how long was the muscle stretched?
(ii) Describe and explain the effect on the neurone of keeping the muscle stretched for this period.
(iii) Suggest **one** biological advantage of this effect on the neurone.

AQA (NEAB) BY04 Biological Basis of Behaviour March 1999 Q6

4

a) Describe the structures of a sensory neurone and of a motor neurone and explain their roles in a reflex arc.

b) Explain how a nerve impulse is transmitted across a synapse.

OCR 9264/3 Biology Linear Paper 3 June 2000 Q9

5 The diagram shows a neuromuscular junction.

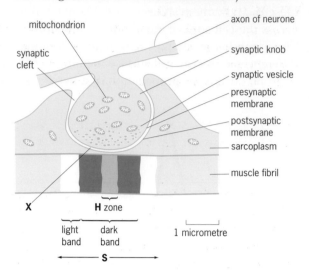

a) **(i)** Name the neurotransmitter that is released from the synaptic knob.
(ii) By what process does this neurotransmitter cross the synaptic cleft?
(iii) Calculate the width of the synaptic cleft at **X**. Show your working.

b) Suggest **two** functions of the energy released by the mitochondria in the synaptic knob.

c) Describe how the appearance of the section of the muscle fibril labelled **S** would change when the fibril is stimulated by the neurotransmitter.

AQA (NEAB) BY04 Biological Basis of Behaviour March 2000 Q3

KEY SKILLS **ASSIGNMENT**

TOXINS AND DRUGS

A huge variety of chemicals have an effect on the human body, and many act by interfering with the ways nerves or synapses work. In particular, they affect the transmission of impulses at synapses. In this Assignment we take a brief look at toxins and drugs.

1 Write a simple flow diagram outlining the sequence of events at a normal synapse.

Toxins

A toxin, by definition, is a poisonous substance of biological origin. The curare featured in the Opener is a toxin. In nature, animals and plants produce a huge variety of chemicals that organisms have developed to use against each other. Various species of spiders, frogs, jellyfish, snakes, molluscs, fish and plants – to name but a few – all make toxins that they use either to protect themselves or to kill prey.

Snake venom is of particular interest to neuroscientists. Venomous snakes have evolved poison glands as modifications of their salivary glands. Not only does the toxin subdue or kill the victim, preventing the snake from the serious injuries that might result from a violent struggle, some of them also start breaking down the prey's tissues. So when the snake gets round to eating it, the dead animal doesn't take so much digesting.

Fig 19.A1 The krait (*Bungarus multicinctus*). The krait's venom contains a powerful neurotoxin, bungarotoxin. Cobras produce similar substances in their venom

One of the first neurotoxic snake venoms was isolated from the krait (Fig 19.A1). In 1963, a small polypeptide consisting of about 70 amino acids was purified from the whole venom and was given the name bungarotoxin. An antibody to bungarotoxin was made and then attached to a fluorescent 'tag' so that its destination could be followed in the body of an animal injected with the krait's venom. The bungarotoxin was found almost exclusively at the neuromuscular junctions where it blocked the acetylcholine receptor sites on the postsynaptic membrane.

2 Predict the symptoms of krait bite. Explain your answer.

Bacterial toxins

Why is it that some bacteria cause disease while others are harmless? One of the key features is that pathogenic (disease-causing) bacteria produce an exotoxin that interferes with the host's metabolism in some way. A few of these toxins affect the nervous system. Botulism, a particularly nasty if rare form of food poisoning, is caused by the bacterium *Clostridium botulinum*. Botulinum toxin prevents the release of acetylcholine from the presynaptic membrane.

3 The symptoms of botulism include a weakness of the facial muscles, dilation of the pupils and difficulty in swallowing and breathing. Suggest why the botulinum toxin produces these effects.

Most people in the UK are vaccinated against tetanus (Fig 19.A2). Also known as lockjaw because it causes muscle paralysis in the face and neck, tetanus results from infection from the bacterium *Clostridium tetani*. The tetanus toxin that this bacterium produces prevents the release of the substance at the neuromuscular junction that breaks down neurotransmitter. As a result, the effects of acetylcholine persist much longer than required.

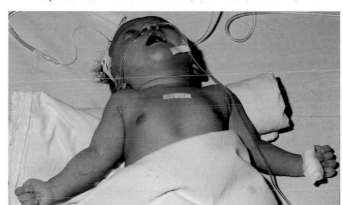

Fig 19.A2 A baby with muscle stiffness caused by tetanus. The toxins produced by the bacterium are causing the baby's muscles to go into spasm

4 What is the essential difference between the actions of tetanus toxin and botulinum toxin?

Drugs

Something that we touched on in this chapter, and will go on and study in more detail in the next, is the brain. This organ is incredibly complex. It contains billions of nerve cells, each with thousands of connections to other cells, allowing for an infinite variety of pathways. There are many different transmitter substances in the brain that mediate synaptic transmission through these pathways. It should be no surprise that chemicals that interfere with neurotransmitters in the brain often produce sensory effects such as hallucinations.

KEY SKILLS **ASSIGNMENT**

These effects can occur as a side-effect of a legal, prescription drug that could be used to reduce pain, to induce sleep, or to reduce tension or depression. There are also substances that, since they have been recognised to alter the way the brain processes information, have been abused. Alcohol and nicotine both act on the brain and can be abused, but they are legal. So-called 'recreational' drugs are illegal and often highly addictive. They interact with the brain and central nervous system and can have extremely dangerous side-effects.

5 Table 19.A1 includes the main classes of abused drugs, their mode of action and the common names of some individual drugs. However, they have been mixed up. Copy the table, matching up the correct examples and descriptions.

Table 19.A1 The main classes of abused drugs, the effect they have and some examples		
Group	**Effect**	**Examples**
Depressants	interfere with brain function, distort perception and induce a dream-like state	a variety of preparations of the plants *Cannabis sativa* and *Cannabis indica*. AKA hashish, marijuana, tac
Stimulants	similar to alcohol, they sedate the central nervous system	LSD, mescaline, magic mushrooms
Hallucinogens	pain killers	cocaine, ecstasy, amphetamine
Cannabis	keep the user awake, alert, excited	a variety of anti-anxiety drugs, sleeping tablets including benzodiazepines (eg Valium)
Analgesics	produce a relaxed, mild euphoria	morphine, heroin, methadone

Becoming dependent on drugs

If illegal drugs had a well-understood mode of action, were very short-lived in the body, had no side-effects and did not induce dependency or addiction, few people would worry. There are people who argue that drugs, if used 'sensibly', are not dangerous. However, the evidence is overwhelmingly against this.

With repeated use, many drugs can lead to addiction. This may be psychological, or physical. Psychological dependence

is the most widespread. The person develops a strong desire for the state induced by the drug – elation, euphoria, stimulation, sedation, hallucination – and the state becomes much more desirable than normality. In many cases, the person experiences tolerance to the drug. Any attempt to kick the habit can result in severe symptoms of depression.

6 How is a drug user most likely to respond when they start to become tolerant to the drug they are abusing?

Some drugs can, if used repeatedly, produce physiological changes and, after a time, the body cannot function 'normally' if the drug is not taken. Often a drug mimics one of the body's own products and this is no longer made as long as the drug is being taken. Therefore, when the drug is stopped, the body cannot respond immediately and start making the required product again. This situation occurs with heroin, which stops the body making its own endorphins. This is physical dependence and an attempt to stop will lead to dangerous and painful withdrawal symptoms.

7 Find out what some of the common withdrawal symptoms are.

8 Controversially, some drugs – notably alcohol and tobacco – are legal, while those listed in Table 19.A1 are not, or are available only with a prescription. What are the arguments for and against the legalisation of these drugs?

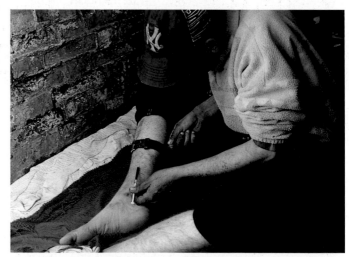

Fig 19.A3 Abuse of drugs brings many dangers. Drugs bought on the street might be mixed with a variety of other substances that can be lethal, and the sharing of needles can spread infections such as AIDS and hepatitis

20

Nervous systems

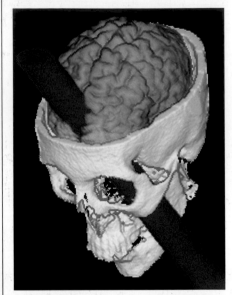

Ouch! Early knowledge about brain function was pieced together with the help of accidents such as that suffered by Phineas Gage. The metal rod missed the vital blood vessels and the crucial areas of the brain that keep us alive. It did, however, damage the frontal lobes, and the subsequent changes in Gage gave us important clues about the function of this part of the brain

The human brain is an organ of great mystery. In addition to being unbelievably complex, it is very difficult to investigate experimentally. We now know that different areas of the brain have particular functions, but how did we find out? Some of the earliest clues came from accidents in which people suffered damage to a particular part of the brain.

Consider the tale of Phineas Gage, a US railroad worker. He was a popular and reliable man, polite and responsible, and he had been made a foreman. In September 1848, he was jamming a stick of dynamite into a hole using a tamping iron. A spark from the iron ignited the dynamite, and the metal rod came out of the hole a lot faster than it went in.

The rod entered Gage's face below his left cheekbone, passing through the eyeball and through the top of his skull before landing several metres away. Gage fell back in a heap, as you would, but remarkably he did not die. He did not even pass out. He was driven by oxcart to a local physician, John Harlow. As the doctor stuck his fingers into the holes in Gage's head until the tips met, Gage asked when he would be ready to return to work. Within a couple of months he had recovered physically, but was no longer himself. Instead of being a gentle, honest, conscientious worker, Gage became 'a foul-mouthed and ill-mannered liar'.

He lived on as something of a celebrity for another 13 years before dying from an epileptic fit. On hearing of the death, Harlow managed to get Gage's body donated to medical research. He believed that the changes in Gage's behaviour had been caused by the damage to the frontal lobes of the brain. 'The equilibrium … between his intellectual faculties and animal propensities seems to have been destroyed', Harlow wrote.

Some 130 years on, scientists were able to use computer modelling to trace the damage done to Gage's brain. The metal rod missed the areas associated with language and motor function, but destroyed the ventromedial region on the left side of his frontal lobe. This is what made Gage so antisocial. People who have had tumours in this area of the brain have undergone the same sort of transformation.

1 WHAT IS A NERVOUS SYSTEM?

All organisms are sensitive to changes in their surroundings. Even the simplest single-celled organisms can detect and respond to stimuli. Some bacteria, for example, move towards areas of high oxygen concentration using their whip-like flagella, and an amoeba can 'swim' away from areas of high salt concentration. In these examples the single cell must act as both receptor and effector; the stimulus is detected and the response brought about in the same place.

In multicellular organisms different cells become specialised for different roles. Even in the simplest animals such as the **cnidarians** (hydra, jellyfish and sea anemones) there are **neurones** connecting receptors and effectors. These animals have a **nerve net** made up of interconnected neurones (Fig 20.1) but there is nothing that looks remotely like a brain.

 REMEMBER THIS

Receptors detect stimuli and **effectors** bring about a response. They are connected by nerve cells or neurones. For more about the structure and function of neurones see the previous chapter.

Fig 20.1 Not the most sophisticated of creatures, this jellyfish is one of the simplest organisms that is actually classed as an animal. Despite the fact that it has very few specialised sense organs and no head or brain to speak of, it does have a network of nerve cells that allows responses and co-ordinated swimming movements

Fig 20.2 At the other end of the scale from the jellyfish, the octopus is the most intelligent invertebrate. It has well-developed eyes and a brain that allows it to solve problems. An octopus can be made to work for its dinner, by giving it shrimps inside a Perspex box with a catch. The animal quickly learns to open the catch. In one recorded incident, an octopus climbed out of its tank, walked across a dry table and climbed into a tank that contained a crab – its favoured prey. Having been out to dinner, the animal went home again

IF YOU WANT TO GET A HEAD GET A BRAIN

There is a clear and predictable pattern in animal evolution; nervous systems become more complex and sophisticated. In the next 'step up' from the jellyfish stage animals began to move in a particular direction, sense organs became concentrated around the mouth and mouthparts at the anterior (front) end, and a distinct head developed. Behind the sense organs a mass of nerves developed to process all the incoming information. This process of 'head development' is called **cephalisation.**

In many invertebrates such as annelid worms and arthropods the mass of neurones that develops behind the head is called a **ganglion**, and there can be many more ganglia along the length of the body (Fig 20.3). These ganglia are little more than nerve exchanges, where a particular stimulus leads to what is generally a fixed response. There is little capacity for learning or memory, though it would be wrong to think that the behaviour of these organisms is completely inflexible.

In the course of evolution the ganglia become less numerous and congregate towards the head end. As these ganglia become more elaborate, and gain more control over the whole organism, they become what we call brains. The annelid worms, molluscs and insects all have ganglia rather then brains. It is well-documented 'fact' that if you cut off the head of some insects, eg cockroaches, they will continue to live until they starve to death. This doesn't work with vertebrates – a physiological fact that the French used with great effect when they dismissed their monarchy in 1789.

Vertebrates generally have more sophisticated nervous systems than invertebrates, though there is some overlap. In many ways the octopus could be said to be more intelligent than the average fish. The brain of a fish is little more than a swelling at the front end of the spinal cord. It is divided into three areas, the **forebrain**, **midbrain** and **hindbrain**. In fish these three areas are associated with the senses of smell, sight and balance, respectively.

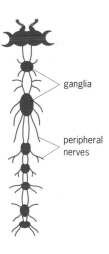

ganglia

peripheral nerves

Fig 20.3 We know that insects are capable of amazingly complex social behaviour, but do they have brains? Actually, no – an insect's nervous system consists of a ventral nerve cord and a ganglion around the oesophagus, together with the peripheral nerves

? TEST YOURSELF

A How would you define intelligence?

The mammalian brain shows the same three areas as the fish early in embryonic life, but the frontal areas become so enlarged (Fig 20.4) that it is difficult to recognise them as the same organ. In this chapter we focus on the nervous system in humans (Fig 20.5).

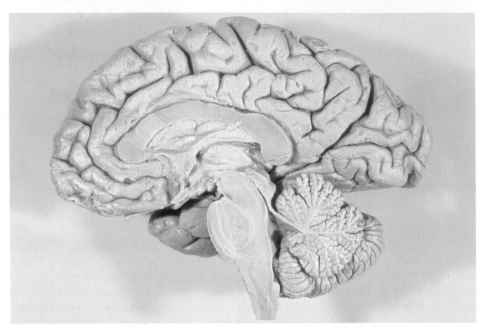

Fig 20.4 The forebrain of the human brain, which consists mainly of the huge cerebral hemispheres, is responsible for our capacity for conscious thought, memory, emotion and language

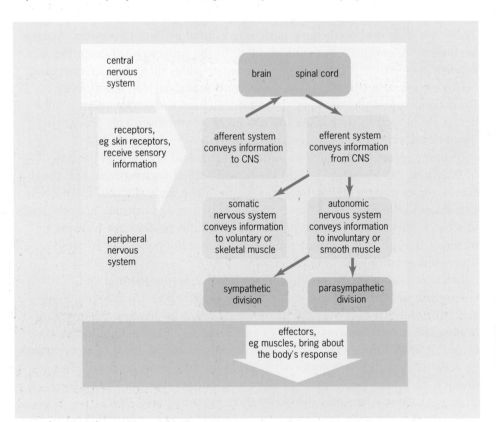

Fig 20.5 Overview of nervous system organisation

2 THE HUMAN NERVOUS SYSTEM

GENERAL ORGANISATION

The human nervous system can be divided into two parts: the **central nervous system** (CNS) and the **peripheral nervous system** (Fig 20.5). The CNS consists of the **spinal cord**, which is protected by the vertebral column, and the **brain**, which is enclosed within the bony shell of the **cranium**. The peripheral nervous system brings information from the sense organs into the CNS, and then relays information out to the structures that bring about responses: the muscles and glands. The peripheral nervous system is divided into the **somatic** and **autonomic** nervous systems. The somatic system is responsible mainly for controlling voluntary actions, such as walking and talking, while the autonomic system keeps the involuntary body functions such as heartbeat, blood pressure and breathing ticking over. The autonomic system is further divided into the **sympathetic** system and the **parasympathetic** system.

THE SPINAL CORD

The spinal cord starts at the base of the brain and ends at the first **lumbar vertebra** (this is roughly at waist level). The spinal cord is enclosed within the **vertebral column** and has a diameter of about 5 millimetres (Fig 20.6). Further protection is provided by three layers of tough membranes called **spinal meninges**, and by the **cerebrospinal fluid** that cushions the cord, acting as a shock-absorber. Each vertebra has an opening on its right and left sides to let spinal nerves pass through. These nerves extend into the body, forming the **peripheral nervous system**.

> **✔ REMEMBER THIS**
>
> What is grey matter? Different areas of nerve tissue in the central nervous system are different colours. Grey matter contains nerve cell bodies; their nuclei are responsible for the grey colour. White matter consists largely of myelinated fibres (their fatty sheaths are creamy-white).

SEE QUESTION 1

Fig 20.6 Two images of the spinal cord showing a slipped, or prolapsed, disc

Fig 20.6(a) This model illustrates the basic structure of the spinal cord. The delicate nerve cord itself runs down the middle of a channel in the centre of the vertebra. Between each bony vertebral disc is a shock-absorbing intervertebral disc made of cartilage. When the outer layer of the disc breaks or ruptures, part of the cartilage can push against the nerve cord, causing pain, numbness and – in severe cases – partial paralysis

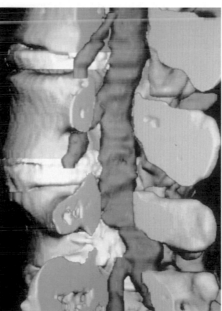

Fig 20.6(b) A coloured 3D computed tomography (CT) scan image of the same problem. The slipped disc (yellow, lower centre) can be seen pushing against the nerve cord (purple). Compare this with the two normal discs above

> **✔ REMEMBER THIS**
>
> Meningitis is a potentially fatal disease characterised by inflammation of the meninges – the protective membranes that surround the brain and spine.

THE REFLEX ARC

A reflex arc is the simplest example of co-ordination, so is a good place to start a study of how the nervous system works. The key feature of a reflex is that a particular stimulus leads to a fixed response – this is very rapid and can't be controlled because it does not pass through the conscious parts of the brain. An important feature of reflex arcs is that they contain as few synapses as possible. This speeds up the response and in many cases – such as the blinking reflex – avoids danger or minimises damage.

THE KNEE-JERK REFLEX

Fig 20.7(b) shows the knee jerk – a classic example of a reflex arc. This is a postural reflex, one of the many mechanisms we have to maintain our position and body control without having to constantly think about fine adjustments. Most reflexes involve either the constant fine adjustments to posture or have evolved to avoid danger.

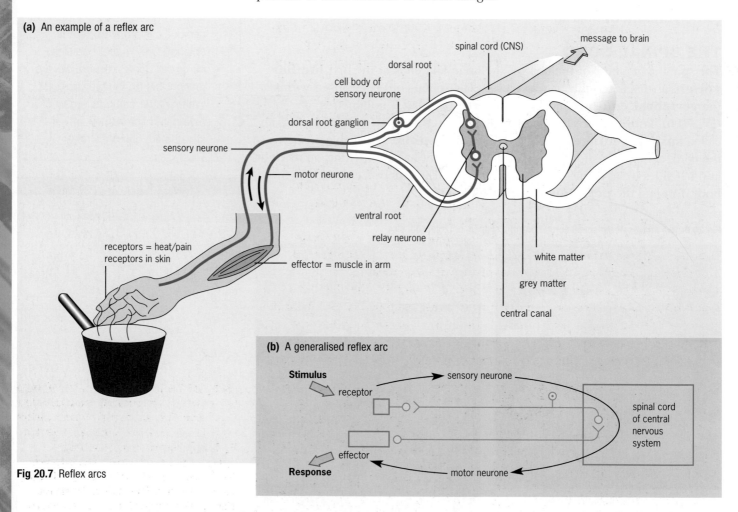

(a) An example of a reflex arc

message to brain

spinal cord (CNS)

dorsal root

cell body of sensory neurone

dorsal root ganglion

sensory neurone

motor neurone

ventral root

relay neurone

white matter

grey matter

central canal

receptors = heat/pain receptors in skin

effector = muscle in arm

(b) A generalised reflex arc

Stimulus

receptor

sensory neurone

spinal cord of central nervous system

effector

Response

motor neurone

Fig 20.7 Reflex arcs

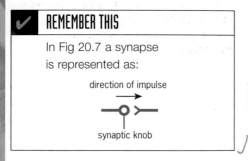

✔ REMEMBER THIS

In Fig 20.7 a synapse is represented as:

direction of impulse

synaptic knob

In the knee-jerk example, the reflex is initiated by the stretch receptor in the patellar tendon – a tap on this tendon just below the knee has the same effect as the knee bending. Nerve impulses pass up the sensory neurone and into the spine. Here, the sensory neurones synapse with motor neurones and the nerve impulses pass straight out of the spine in the motor nerve, where they pass to the thigh muscle (the quadriceps). Contraction of the quadriceps straightens the leg. Impulses will also pass from the sensory nerves up the spine to the brain, but we are conscious of the stimulus only after the response has been initiated.

OTHER REFLEXES

Most reflexes contain more synapses than the knee-jerk example. Blinking when a foreign object enters the eye and the withdrawal reflex, or pulling your hand away from a hot pan (Fig 20.7a), involves a circuit containing sensory receptors, sensory neurones, spinal relay neurones (interneurones), motor neurones and effector muscles. The principle is the same: the heat of the pan stimulates receptors, and impulses pass along sensory neurones towards the spinal cord. Here, instead of making synapses with motor neurones, they pass their signals on to relay neurones. These connect with motor neurones that cause muscles to contract, moving your hand away from the pan.

You know that you have touched the hot pan because some impulses do travel to the brain, but the movement that is part of the reflex is involuntary, not under your conscious control. It is more difficult to persuade people that the swearing that accompanies such an event is also involuntary.

THE BRAIN

The human brain weighs about 1.5 kilograms, is 85 per cent water and has the consistency of thick blancmange (Fig 20.8). It is, however, the most complex material known. It copes with a huge amount of information from the various senses, deciding what is important and what can be ignored. It stores thousands and thousands of memories for decades, and can sort them into chronological order. It allows us to control the complex functions of the body, while at the same time allowing us to maintain posture, read, write and talk. Can we make a computer to do all of this? Not a chance.

The brain is the organ that makes humans human. We owe our success to this remarkable mass of nerves which takes over 20 years to mature, and which allows us to make tools and use language to an extent that far surpasses our nearest relative, the chimpanzee.

The brain of a human makes up about one-fiftieth of your body's mass. Its delicate tissues are protected by the skull or cranium and by the **cranial meninges**, membranes that are continuous with the spinal meninges. Cerebrospinal fluid bathes the outside of the brain and fills the chambers – the **ventricles**. Twelve pairs of cranial nerves innervate (supply nerves to) various regions of the head. The human brain is thought to contain ten thousand million (10^{10}) neurones. Each neurone may be in contact with a thousand other cells, providing an immense number of different communication routes.

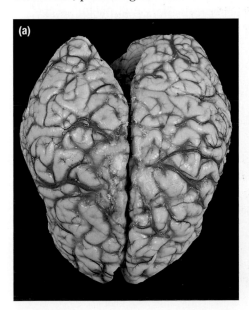
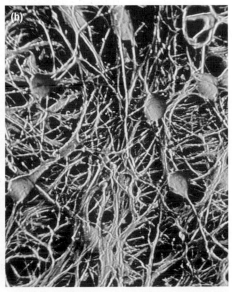

Fig 20.8(a) The most obvious feature of the human brain (seen here from above) is the two huge cerebral hemispheres. These are the site of higher conscious functions such as memory, language and emotion

Fig 20.8(b) False-colour scanning electron micrograph (SEM) of neurones (nerve cells) from the cerebral cortex – the outer, heavily folded, grey matter of the brain. Have you ever thought about thinking? It's a strange idea, but our thoughts are pathways of nervous activity between the neurones of the brain. This is what makes us what we are, to the extent that there will never be a brain transplant, it would be a body transplant

THE AREAS OF THE BRAIN

The brain receives a vast number of impulses from receptors both inside and outside the body (see Table 23.1). It maintains basic involuntary body 'housekeeping' functions such as heart rate, breathing rate and temperature control. It also co-ordinates the semi-automatic muscular actions of the body such as swallowing, and it initiates and controls voluntary activities such as walking and running. The human brain is the site of higher mental functions such as reasoning, emotion and personality.

Like all vertebrate brains, the human brain consists of three parts, a hindbrain, midbrain and forebrain. However, our brains have evolved and enlarged to such an extent that the basic three-part pattern is only noticeable in the early stages of development of the embryo.

The **hindbrain** has three distinct structures. The **medulla oblongata** is a swollen portion at the bottom of the brain stem that houses vital centres controlling heart rate, breathing and blood supply. The **cerebellum** (see below) controls body movement and maintains balance, and the **reticular activating system** (a collection of neurones in the centre of the brain stem) filters incoming stimuli and controls wakefulness and sleep.

As in other animals, the **midbrain** in humans links the forebrain to the hindbrain. Our emotions, which are located in the forebrain, can affect basic functions of the hindbrain such as control of blood vessel diameter, heart rate and sweating. When we are worried about something – exam results for example – the forebrain interprets this as stress and brings about the release of adrenaline, a hormone that prepares the body for action – see Chapter 22.

The **forebrain** has two main parts, the **cerebrum** and a region containing the **thalamus** and the **hypothalamus**.

HYPOTHALAMUS

The hypothalamus is a key area. It receives a huge amount of internal and external sensory information and acts as a co-ordinating centre between the nervous system and the endocrine (hormonal) system. The hypothalamus is responsible for sensations such as hunger and thirst. It also helps control the autonomic nervous system since it regulates body temperature (see Chapter 15) and the balance of water and salts in the blood (see Chapter 14). It is linked directly to the pituitary gland by means of blood vessels and nerves. Generally, the hypothalamus controls the release of hormones from the pituitary, including the antidiuretic hormone (which controls water reabsorption in the kidneys), growth hormone and some reproductive hormones. We deal with the endocrine system in more detail in Chapter 22. The thalamus (see Fig 20.9) directs sensory information from the sense organs to the correct part of the cerebral cortex.

REMEMBER THIS

The **brain stem** is the medulla oblongata plus the pons plus the midbrain.

SEE QUESTION 2

Fig 20.9 Main regions of the brain

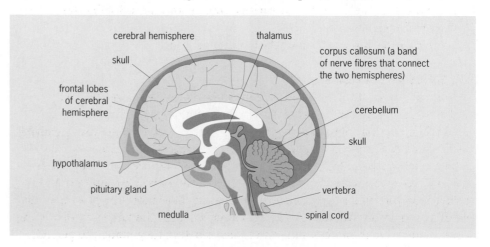

CEREBRUM

The cerebrum is made up of two large **cerebral hemispheres** (Fig 20.9). They have a thin outer layer, the cortex, which is thrown into many folds with fissures (grooves) between. The cortex is the surface layer of the hemispheres (Fig 20.10), and most of our conscious thought takes place here. The cerebral cortex carries out the 'higher' mental activities of reasoning, and is regarded as the site of personality and emotion and a sense of 'self'. This is what sets humans apart from other species.

Different areas of the cerebral hemispheres are associated with different sensory and motor functions – these are mapped out in Fig 20.11. The cerebrum of mammals is very large compared with the forebrains of other vertebrates and is certainly the dominant feature of the human brain. Folds in the surface of the cerebrum increase the surface area for centres of control, where incoming nerve impulses are interpreted, or integrated, in the light of information already stored in the brain. Overall, the function of the cerebral cortex can be summarised in three stages:

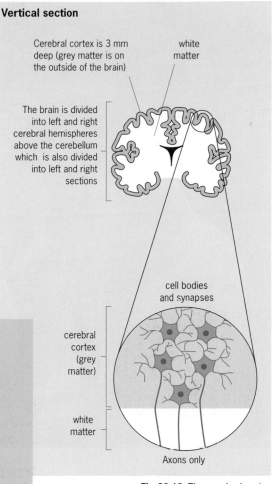

Vertical section

Cerebral cortex is 3 mm deep (grey matter is on the outside of the brain)

white matter

The brain is divided into left and right cerebral hemispheres above the cerebellum which is also divided into left and right sections

cerebral cortex (grey matter)

cell bodies and synapses

white matter

Axons only

Fig 20.10 The cerebral cortex

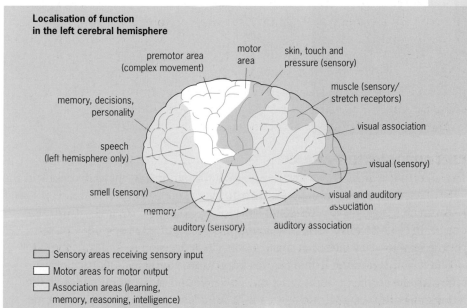

Localisation of function in the left cerebral hemisphere

premotor area (complex movement)

motor area

skin, touch and pressure (sensory)

muscle (sensory/ stretch receptors)

memory, decisions, personality

visual association

speech (left hemisphere only)

visual (sensory)

smell (sensory)

visual and auditory association

memory

auditory (sensory)

auditory association

- Sensory areas receiving sensory input
- Motor areas for motor output
- Association areas (learning, memory, reasoning, intelligence)

Fig 20.11 The key areas of the human cortex

- The **sensory areas** receive sensory information.
- The information is processed and interpreted in **association areas,** which decide on an appropriate response.
- Responses are initiated in the **motor areas** and modified by the **cerebellum**.

In this section we look at each of these areas in turn.

THE SENSORY AREAS

The sensory areas of the brain receive incoming impulses from the vast number of receptors that provide us with information about the world outside. The external receptors might tell us that it is cold, dark, raining and that the skin is getting wet. Your brain has to decide what to do.

The model shown in Fig 20.12 shows how our brains 'see' our body. There are many receptors in the skin but as you will know, some parts are more sensitive than others. The lips, tongue and fingertips have more receptors per unit area of skin than anywhere else. Fig 20.12 reflects this; the largest parts are the most sensitive. We think they ran out of plastic before they made the genitals.

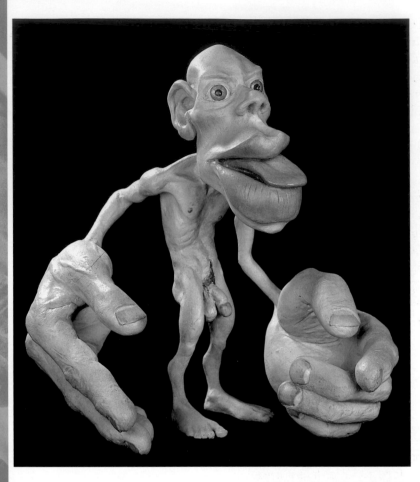

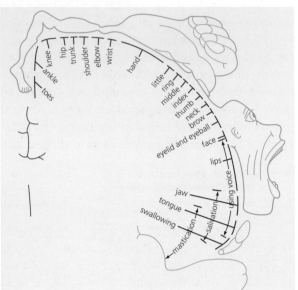

Fig 20.12 No, you won't meet him on a dark night. Some parts of the body contain more receptors than others, and consequently a larger area of brain is devoted to them. In this distorted model the man's anatomy is proportional to the area in the brain that deals with the sensory information. Thus the hands, mouth and tongue are huge – they have many receptors per square centimetre – while other areas such as the torso and legs are relatively small. The eyes, however, are not in proportion in this model. These take up more of the human brain than the rest of the body put together
Fig 20.13 The different areas of the body are also mapped out on the motor area of the cortex

THE ASSOCIATION AREAS

The association areas *interpret* the sensory information and make decisions in the light of previous experience, ie our **memory**. If you are cold and wet, you may see a friend's house and your visual association areas will compare it with the images of houses stored away. Once you have recognised it, impulses pass to your frontal lobes, which then make a conscious decision to knock at the door.

The way information from the eyes is processed gives us a particularly impressive example of brain function. The millions of rods and cones in the retinas of both eyes generate a stream of impulses that pass down the optic nerves to the visual cortex at the back of the brain. The different information from the two eyes is processed in the **visual association area** into a single image in which we have a perception of depth as well as colour and detail.

THE MOTOR AREAS

The motor areas generate the impulses that cause the muscles to contract. This simple statement hides the fact that the process is incredibly complex and requires such a degree of computation that we are unlikely to be able to make lifelike 'android' robots for a long time to come.

Muscular contraction is not a matter of all or nothing. If it were, muscles would be either relaxed or fully contracted. Movement, if possible at all, would be very jerky. However, we can move our muscles with a great deal of precision, creating just the right tension for the action we are performing.

This is made possible by the process of **proprioception**, in which the brain receives constant feedback about the degree of contraction of the muscles and the position of the joints in relation to each other. Vital information also comes from the inner ear about our position with respect to gravity, direction of movement, acceleration and deceleration (see page 425).

The process of movement control and postural maintenance is a matter of countless fine adjustments to alter the tension in the muscles. Interestingly, the left side of the body is controlled by the right motor area, and vice versa (Fig 20.14).

SEE QUESTION 3

B A woman has a stroke and as a result is unable to move the left side of her body properly. Where did the stroke probably occur?

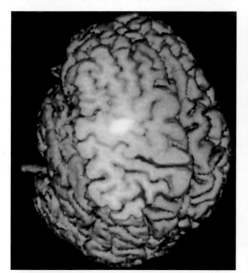

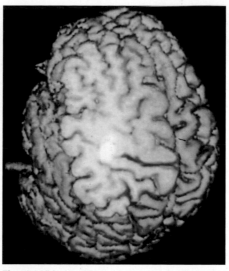

Fig 20.14 Two computer images of the left side of the human brain. These pictures were created by a process called magnetoencephalography, a technique that measures magnetic fields generated by neurone activity in the brain

Fig 20.14(a) Just before the subject moves her right index finger, neurones in the motor area can be seen to 'fire'

Fig 20.14(b) 40 milliseconds later, the hand sensory area processes feedback from the muscles in the finger, so that ongoing modifications can be made.

There are three key components to the process of moving:
● selection of a **motor programme**;
● initiation of a motor programme – the muscles contract;
● modification of movement in the light of feedback.

Given the complexity of the task it is not surprising that several different areas of the brain are involved in the control of movement.

The process starts with the **basal ganglia**; a collection of structures lying deep in the forebrain. These are of fundamental importance because they select and initiate the motor programme. Fig 20.15 shows the pathway taken by motor impulses before they reach the muscles. From the basal ganglia impulses pass to the motor cortex.

The motor cortex contains two major regions, the **primary motor cortex** and the **supplementary motor cortex.** The primary motor complex contains neurones that send impulses to skeletal muscles along nerve fibres passing down the brain stem and spinal cord.

The cerebellum is a complex structure lying on the posterior surface of the brain stem. Its function is to co-ordinate ongoing movements, thus producing smooth flowing movements. The cerebellum uses information from proprioceptors to decide whether the body's actions are going to plan or need modifying.

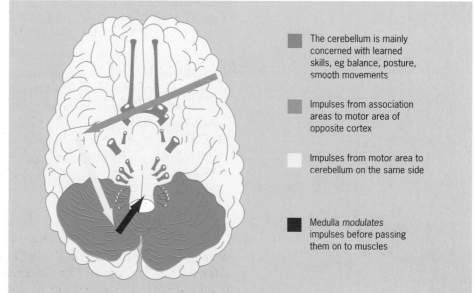

The cerebellum is mainly concerned with learned skills, eg balance, posture, smooth movements

Impulses from association areas to motor area of opposite cortex

Impulses from motor area to cerebellum on the same side

Medulla *modulates* impulses before passing them on to muscles

Fig 20.15 Motor pathways and the cerebral cortex

⚠ SCIENCE FEATURE Dyslexia and the cerebellum

Dyslexia is a very common condition that affects the learning ability of both children and adults. Commonly over-simplified as 'word blindness', dyslexia is a complex condition that involves an inability to concentrate and difficulty in co-ordinating, along with reading and writing problems. Dyslexics can become very frustrated and distressed with school and learning generally, to the extent that some become suicidal. One survey suggested that 70 per cent of convicted criminals have some sort of learning difficulty.

Now a new area of research is focusing on the cerebellum, an area previously thought to be solely concerned with movement, not with the higher intellectual processes. However, when dyslexics are given exercises to improve their balance and co-ordination, dramatic improvements are seen in their dyslexia within six weeks. Treatment is resulting in improvement time after time, to the extent that even the most sceptical of researchers are having to take notice. Centres offering this treatment are springing up all over the UK.

Fig 20.16 Exercises that improve balance and co-ordination may offer new hope for dyslexics

PARKINSON'S DISEASE

Parkinson's disease is a disease in which the death of a small number of cells in the basal ganglia leads to an inability to select and initiate patterns of movement. The exact symptoms vary, but generally Parkinson's sufferers have an increased rigidity in their muscles, usually accompanied by a tremor. This leads to slowness of movement, poor balance and speech problems. This shows the importance of the basal ganglia – without them the body is incapable of normal movement even though most of the motor system is intact.

Quite what triggers the onset of Parkinson's is still not clear. It appears to be due to a genetic predisposition and an as yet unidentified environmental trigger. What is known is that the underlying cause is an inability to produce **dopamine**, a neurotransmitter that has a number of functions, including enabling us to move smoothly and normally.

Some relief from Parkinson's has been achieved by treatment with **levadopa,** a precursor that is transformed into active dopamine in the brain. Dopamine reduces the symptoms – often spectacularly so – but is not a cure. Trials involving transplants of dopamine-producing cells from fetal or animal sources also show promise, but it is still early days.

EEGS AND BRAIN STEM DEATH

In 1929 the German scientist Berger discovered that the brain produces electrical activity that can be measured by electrodes placed on the scalp. The machine that measures brainwaves is called an **electroencephalogram**, or EEG. It was discovered that the patterns of EEG activity change in different situations, particularly with levels of consciousness. The four basic types of brainwaves, alpha, beta, theta and delta, are shown in Fig 20.17(b). The absence of β waves or any electrical activity from the brain of a patient indicates brain stem death. The absence of heartbeat, breathing movements and electrical brain activity is a clinical definition of death.

EEGs tell us little about how the brain works, but they are useful for locating and diagnosing the various types of epilepsy, sleep disorders and brain tumours.

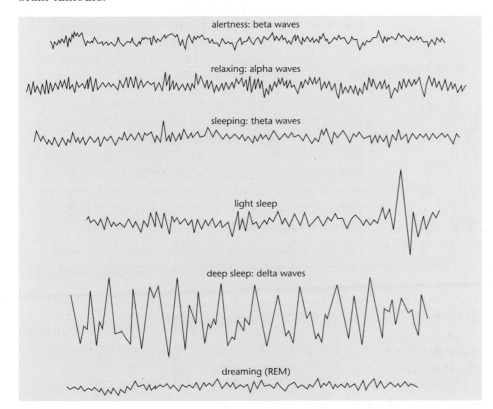

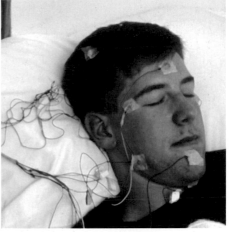

Fig 20.17(a) Person wired up to EEG machine

Fig 20.17(b) EEG traces of the different brain waves

THE PERIPHERAL NERVOUS SYSTEM

The nerves of the peripheral nervous system behave like major road systems, carrying traffic in and out of the central nervous system. **Afferent nerves**, also called **sensory nerves**, carry information from sensory receptors into the CNS. **Efferent nerves**, also called **motor nerves**, carry information from the CNS out to effector organs. The efferent system can be further subdivided into the **somatic** and **autonomic** systems. These differ in their function, rather than their structure or position in the body.

THE SOMATIC NERVOUS SYSTEM

The somatic nervous system contains both afferent and efferent nerves. It receives and processes information from receptors in the skin, voluntary muscles, tendons, joints, eyes, tongue, nose and ears, giving an organism the sensations of touch, pain, heat, cold, balance, sight, taste, smell and sound. It also controls voluntary actions such as the movement of arms and legs.

 REMEMBER THIS

Afferent means 'incoming' while efferent means 'outgoing'. You can refer to 'afferent nerves' or 'efferent blood vessels', for example.

! SCIENCE FEATURE Epilepsy and the two sides of the brain

Epilepsy is a common disorder of the brain. Symptoms range in severity from a mild loss of concentration, known as an absence or petit mal, to full-blown convulsive fits (grand mal) in which the subject blacks out and falls to the floor. These can be dangerous if the sufferer lashes out – injuring himself and others – or bites his own tongue.

The underlying cause of epilepsy is random, uncontrolled activity of some cells in the brain. This chaotic activity in both sensory and motor nerves causes patients to see and hear a variety of strange things, such as flashing lights and bells, while muscles jerk uncontrollably.

A diagnosis of epilepsy can be confirmed and studied using an EEG machine (see Fig 20.17a). Fig 20.18 shows a trace from a person during an epileptic seizure: you can compare it with the normal reading in Fig 20.17(b).

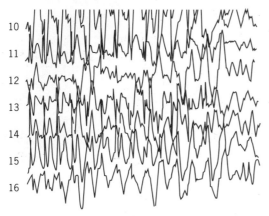

Fig 20.18 EEG traces taken during an epileptic seizure

Epilepsy can often be controlled successfully by drugs. However, in extreme cases, the condition is treated by brain surgery, and one such operation has given us a fascinating insight into the workings of the brain. The cerebral hemispheres have been described as two separate brains, and in order to work effectively as a whole, the two halves must communicate. The bridge between the two halves is known as the **corpus callosum** (Fig 20.19). Neuroscientists discovered that the corpus callosum was involved in the spread of epileptic seizures.

Fig 20.19 The corpus callosum is a broad, thick mass of nerve fibres connecting the right and left cerebral hemispheres. The left eye supplies information to the right hemisphere, and vice versa. The corpus callosum allows the two sides of the brain to communicate

In a seemingly drastic operation, surgeons sever most of the corpus callosum. This often causes the seizures to be less intense and dangerous. However, there are other amazing consequences. Initially, subjects appear to be perfectly normal: they can talk and read, and have no problems in recognising the world around them. However, if they close their right eye, and are given a familiar object such as a comb, they cannot put a name to it. Open the other eye, however, and, 'Ah, it's a comb!'

The same happens with words. If a word such as 'TIGER' is looked at it with the left eye only, the patient can't read it. If they open the right eye, they can read the word immediately. This is because the left eye supplies information to the right side of the brain, and that is not where the language centre is situated. The right eye supplies information to the left side of the brain, to the language-processing neurones.

From studies of split-brain patients, and other studies, it appears that different sides of the brain have different functions. The left hemisphere contains the language centre, and the three R's – reading, writing and 'rithmetic. The right side, in contrast, is responsible for our imagination and sense of humour. It can also appreciate form, geometry and music. It cannot, however, put words to things. If the right hemisphere needs a word, it has to put in a call to the left side, via the corpus callosum.

Split-brain patients do not experience the symptoms forever. Within a few months the right hemisphere develops more language skills and can function on its own. It has even been suggested that split-brain patients could read two books simultaneously, one with each eye!

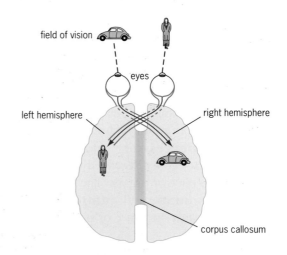

THE AUTONOMIC NERVOUS SYSTEM (ANS)

The ANS consists of two sets of involuntary nerves that generally act antagonistically – they have opposing effects. The system is entirely motor, made up of efferent nerves only. It does not carry sensory information: feedback from muscles and glands travels via the somatic system.

The ANS controls basic 'housekeeping' functions such as heart rate, breathing, digestion and blood flow. Heart rate, for example, can increase or decrease. This and many functions like it are controlled by the two branches of the ANS, the sympathetic and the parasympathetic systems (Fig 20.20). These two sets of nerves have antagonistic (opposite) actions. So, for example, while the sympathetic system increases heart rate, the parasympathetic system lowers it. Generally, the sympathetic system has a stimulatory effect, and prepares the body for action, while the parasympathetic system returns body functions to normal.

Fig 20.20 The autonomic nervous system, showing the antagonistic (opposing) effects of the sympathetic and parasympathetic division.
Autonomic neurones, like others in the nervous system, release chemical transmitter substances where they communicate with other cells. Autonomic neurones can synapse with other neurones within the ANS or with effectors such as muscles and glands. Neurones are classified as cholinergic or adrenergic on the basis of the transmitter they release

REMEMBER THIS

There is a parallel between the nervous and muscular systems of the human body. In the same way as we have voluntary and involuntary components of our nervous system, we also have skeletal (striped) muscle to deal with voluntary muscle contraction, and visceral (smooth) muscle to deal with involuntary muscle contraction. The different muscle types are described in Chapter 21.

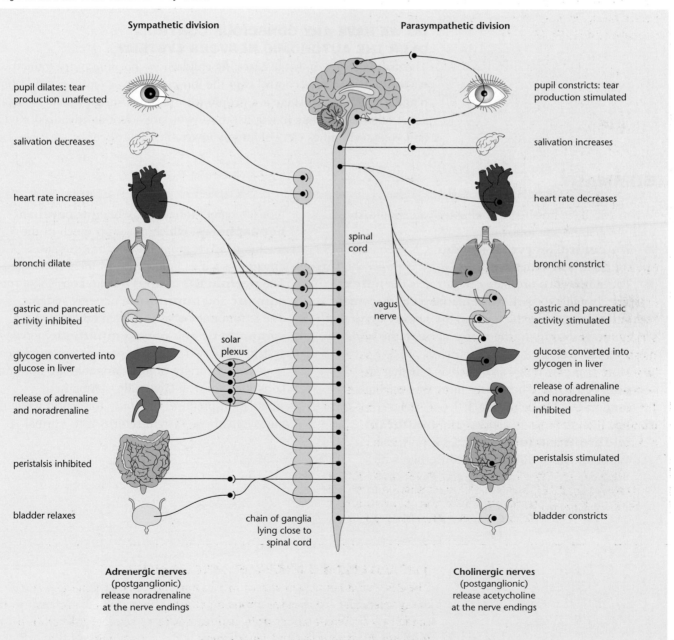

Sympathetic division

pupil dilates: tear production unaffected

salivation decreases

heart rate increases

bronchi dilate

gastric and pancreatic activity inhibited

glycogen converted into glucose in liver

release of adrenaline and noradrenaline

peristalsis inhibited

bladder relaxes

solar plexus

spinal cord

chain of ganglia lying close to spinal cord

Adrenergic nerves
(postganglionic)
release noradrenaline
at the nerve endings

Parasympathetic division

pupil constricts: tear production stimulated

salivation increases

heart rate decreases

vagus nerve

bronchi constrict

gastric and pancreatic activity stimulated

glucose converted into glycogen in liver

release of adrenaline and noradrenaline inhibited

peristalsis stimulated

bladder constricts

Cholinergic nerves
(postganglionic)
release acetycholine
at the nerve endings

Normally, the activity of both systems is balanced. But if the body is stressed, then the 'fight or flight' reactions of the sympathetic nervous system take over, causing an increase in heart rate, faster breathing, an increase in blood pressure and an increase in blood sugar level. This makes the body ready for sudden strenuous activity. When the emergency is over, the parasympathetic system takes over. It decreases the heart and breathing rates and diverts blood supply back to 'housekeeping' activities such as digestion and food absorption. The actions of the parasympathetic nervous system have been described as 'feed or breed' because parasympathetic stimulation leads to increased blood flow and peristalsis in the intestines, and sexual responses such as gaining an erection.

The autonomic system was originally thought to be independent of the rest of the nervous system, hence the term autonomic, meaning 'on its own' or 'self-governing'. Now we appreciate that it is not autonomous, but is regulated by areas within the central nervous system, including the hypothalamus, cerebral cortex and the medulla oblongata.

SEE QUESTIONS 4 AND 5

DO WE HAVE ANY CONSCIOUS CONTROL OVER THE AUTONOMIC NERVOUS SYSTEM?

The answer is yes, in some cases. As children we became potty trained when we learned conscious control over the muscular valves (sphincters) in our bladder and rectum. There are people who have learnt conscious control over some other autonomic functions; those who are adept at advanced meditation and yoga techniques can voluntarily lower their heartbeat.

SUMMARY

When you have finished this chapter you should know and understand the following:

- The **central nervous system** (CNS) consists of the brain and spinal cord.
- The simplest co-ordinated response is the r**eflex arc**. In the knee-jerk reaction, the pathway consists of the receptor in the tendon, the sensory nerve that connects directly to the motor nerve in the spine, and the effector, the thigh muscle.
- More complex reflex arcs such as blinking are **polysynaptic**. These have more intermediate connections inside the CNS.
- The human brain is divided into **hindbrain**, **midbrain** and **forebrain**. Generally, the midbrain and hindbrain are concerned with 'housekeeping' functions, while the **cerebral hemispheres**, which make up much of the forebrain, are responsible for the conscious functions such as language and memory.
- The **peripheral** nervous system consists of the **somatic** and **autonomic** nervous systems.
- The autonomic nervous system consists of the **sympathetic** and **parasympathetic** nervous systems. The two sets of nerves act antagonistically to control a variety of functions that are not under conscious control. Generally, sympathetic stimulation prepares the body for action while the parasympathetic system returns it to normal.

❓ EXAM QUESTIONS

1 The diagram below shows a transverse section of the spinal cord of a small mammal.

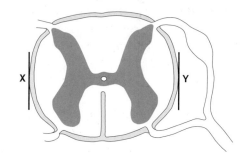

a) Name the substance present in the central canal.

b) The magnification of the diagram is ×25. Calculate the actual diameter of the spinal cord at the position between **X** and **Y**.

c) On the diagram, draw and label the neurones involved in a simple spinal reflex.

Edexcel B3 June 1999 Q2

2 The diagram shows a vertical section through a human brain.

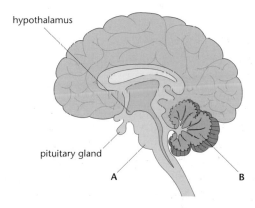

a) Give the name and one function of the part of the brain labelled:
(i) A;
(ii) B.

b) The hypothalamus monitors the blood. It is also involved in homeostasis.
(i) Name one of the features of the blood monitored by the hypothalamus.
(ii) Explain what is meant by homeostasis.

AQA (NEAB) BY04 Biological Basis of Behaviour March 2001 Q1

3 The diagram shows a vertical section through a human brain.

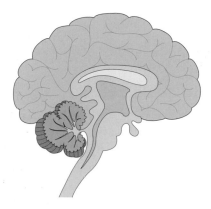

a) (i) A person drinks a glass of lemonade. Name the parts of the brain to which the following descriptions would apply.

P receives impulses from touch receptors in the lips.

Q enables the person to coordinate the movements necessary to drink from the glass.

R contains receptors that respond to a change in concentration of the blood plasma.

(ii) Use the appropriate letter and a guideline to show the position of each of parts, P, Q and R on the diagram of the brain.

b) When the lemonade touches receptors in the throat, a swallowing reflex occurs. What are the effectors in the reflex action?

AQA (NEAB) BY04 Biological Basis of Behaviour March 1999 Q1

? EXAM QUESTIONS

4

a) State the name of the neurotransmitter substance normally produced at the effector in:
(i) the sympathetic nervous system;
(ii) the parasympathetic nervous system.

b) Atropine is a drug that depresses the activity of the parasympathetic nervous system. Predict the effect of atropine on the heart rate.

c) Identify **two** other possible effects of atropine.

IBO Biology Higher Level Paper 3 November 1998 QE2

5 The sinoatrial node in the heart controls the heartbeat. The rate of heartbeat is coordinated by the brain.

a) Name:
(i) the **two** types of nerves that connect the brain to the sinoatrial node;
(ii) the part of the brain where these nerves originate.

b) Suggest:
(i) why it is necessary to have two separate nerve connections from the brain to the sinoatrial node;
(ii) how the two nerve connections are able to have different effects on the sinoatrial node.

AQA (NEAB) BY04 Biological Basis of Behaviour June 1999 Q1

21

Support and movement

Support and movement

Astronauts must often live and work in zero gravity. Weightlessness is not a restful state and it can put great stress on body systems, particularly the skeletal system.

Many changes occur in zero gravity. With less work to do, body muscles begin to break down and the bones start to lose calcium at an increased rate. The number of red blood cells in the body falls and there are dramatic shifts in fluid distribution: the face becomes puffy and the legs become thinner (astronauts call this condition 'bird's legs'). The lack of gravity also affects the spine. Without the constant downward force, the spine lengthens by as much as eight millimetres. This can lead to blocked nerves and back pain. Often, astronauts also lose their touch sensitivity.

NASA scientists are busy looking at forms of treatment and exercise that might prevent bone breakdown. Such information might also have practical benefits closer to home. By studying and finding ways to slow the accelerated changes produced in space, it may be possible to develop new treatments for bone diseases, such as osteoporosis which occurs when bones become brittle due to a loss of calcium.

It might look fun, but in zero gravity the body is under a lot of stress

✔ REMEMBER THIS

Movement is a feature of all living things and occurs at all levels. Atoms move within molecules and cell contents move inside cytoplasm. Plants move when they tilt their leaves towards the Sun and when their flowers open and close. The human rib cage moves up and down as we breathe.

1 SUPPORT, MOVEMENT AND LOCOMOTION IN LIVING ORGANISMS

The bodies of most multicellular organisms need some form of support: body tissue is soft and collapsible and so needs to be held in a rigid frame. Land-living animals need more support than those that live in water.

Support in animals is usually provided by a **skeleton**. The skeletons of animals are of two basic types. **Fluid** or **hydrostatic** skeletons are used by a wide range of soft-bodied invertebrates such as earthworms, flatworms and jellyfish (Fig 21.1a). Arthropods, echinoderms and vertebrates have **rigid skeletons** (Fig 21.1b and c).

A rigid skeleton supports important body organs, enabling them to work efficiently (they cannot operate properly when squashed), it protects the internal structures from damage and it allows the animal to **locomote**, or move from place to place.

Locomotion is different from movement. It refers not to a simple shifting of the body parts, but to the movement of the *whole* organism from place to place.

Fig 21.1(a) Jellyfish have a hydrostatic skeleton

Fig 21.1(b) Many invertebrates, including these lobsters, have an exoskeleton

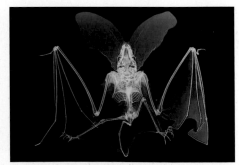

Fig 21.1(c) This X-ray shows the bony endoskeleton of a long-eared bat

Fig 21.2(a) An adult cheetah can run at speeds approaching 120 kilometres per hour

Fig 21.2(b) The bacterium *Rhizobium* fixes nitrogen in the root nodules of plants such as peas and beans (see Chapter 36). It has a rigid flagellum which turns like a corkscrew, propelling the organism through its environment

Bacteria, for example, move from place to place using flagellae – Fig 21.2(b). Many protoctists (single-celled organisms) move around by altering their cell shape – Fig 21.2(c). But the most dramatic and most efficient forms of movement result from muscular contraction and occur in higher animals – Fig 21.2(a).

Getting enough food and finding a mate are major priorities for most animals. Animals need to locomote to do both. Animals may also move to escape from predators or harmful stimuli, to disperse themselves through a particular environment, to avoid competition from other individuals, or to move to conditions that are more favourable (into the shade on a hot day, for example).

In this chapter we first describe the features of different types of animal skeletons and then look at how each type of skeleton is used in locomotion.

Fig 21.2(c) Amoebae locomote by forming pseudopodia

2 HYDROSTATIC SKELETONS

A hydrostatic skeleton consists of an enclosed, fluid-filled cavity, which provides support. Water is difficult to compress and so, when it fills body cavities such as the **coelom** of the earthworm (Fig 21.3), it provides a fluid mass against which muscles can contract.

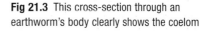

epidermis

circular muscle

longitudinal muscle

chaetae

chaeta muscles

intestine

fluid-filled coelom

ventral nerve cord

Fig 21.3 This cross-section through an earthworm's body clearly shows the coelom

LOCOMOTION IN ANIMALS THAT HAVE A HYDROSTATIC SKELETON

Animals such as earthworms, leeches, caterpillars, snails and slugs locomote using a hydrostatic skeleton. Earthworms are burrowing animals that need sustained and powerful locomotion. They can locomote because their muscles, attached to the body wall, pull against the fixed volume of fluid in the coelom.

> ✔ **REMEMBER THIS**
>
> The supporting properties of water are also seen in animals that have other types of skeleton. In mammals, **amniotic fluid**, the fluid inside the uterus, surrounds and supports the developing fetus; and the **vitreous humour**, the jelly-like material inside the eyeball, supports the structures inside the eye.

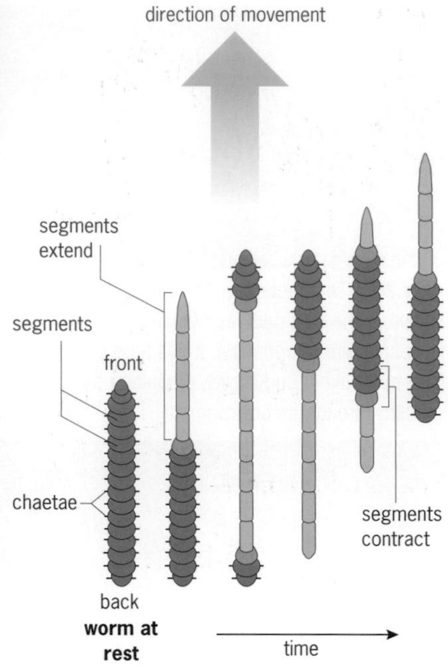

direction of movement

segments
extend

segments

front

chaetae

back

worm at rest

time

segments
contract

Fig 21.4 Earthworm locomotion. Contraction of circular muscle causes the extension of groups of individual segments. Tough bristles (chaetae) then anchor the body and longitudinal muscle contraction shortens the body segments at the front, pulling the rest of the body forwards. Waves of contraction flow backwards away from the direction of movement

The muscles are arranged in blocks and each enclosed segment of the earthworm's body is controlled separately. Contraction of the **longitudinal muscles**, which run along the length of each segment, shortens that segment. Contraction of the **circular muscles**, which run around the circumference of each segment, extends it.

To move forward, the earthworm extends the segments at the front of its body (Fig 21.4). This part of the body moves forwards, the front segments anchor themselves to the ground using bristles, or **chaetae**, and then these segments contract. The following segments 'catch up' with the front section by relaxing, anchoring and then contracting in the same way.

Other animals that have hydrostatic skeletons move in a very similar way. In leeches, the whole body acts as a single hydrostatic system (Fig 21.5).

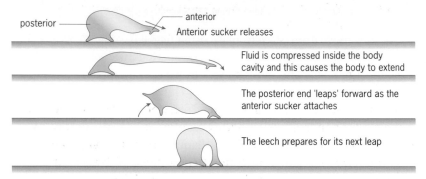

posterior

anterior

Anterior sucker releases

Fluid is compressed inside the body cavity and this causes the body to extend

The posterior end 'leaps' forward as the anterior sucker attaches

The leech prepares for its next leap

Fig 21.5 The typical form of locomotion seen in leeches and caterpillars. In the leech, anterior and posterior suckers anchor the ends of the animal. Contraction and relaxation of body muscles compress fluids in the body spaces

3 EXOSKELETONS

Exoskeletons, literally 'skeletons on the outside', occur mainly in arthropods such as crustaceans. Arthropods have a hardened outer 'skin' that protects and supports the body (Fig 21.6). This outer covering, the **cuticle**, is made up of a rigid polysaccharide compound called **chitin** (see page 47). In land-living arthropod groups such as insects, spiders, centipedes and millipedes, this cuticle is covered by a layer of wax which provides a waterproof outer coat. The cuticle of crustaceans such as crabs and lobsters does not have a waxy layer (understandable in these mainly aquatic animals) and contains calcium salts for greater strength. The cuticle is not completely smooth. Internal folds add to its strength and ridges provide sites to which muscles attach.

Fig 21.6 An arthropod exoskeleton is a tough, protective covering. The outer cuticle of an arthropod skeleton consists of chitin and proteins arranged as a series of plates joined by thin membranes, rather like a medieval suit of armour protected by chain mail at the joints

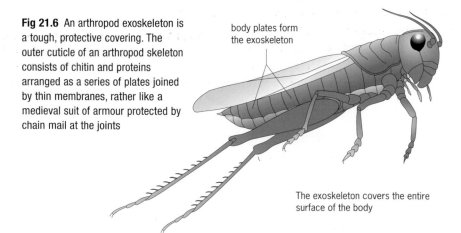

body plates form the exoskeleton

The exoskeleton covers the entire surface of the body

HOW INSECTS LOCOMOTE USING THEIR EXOSKELETON

The arthropod leg (Fig 21.7) is basically a jointed hollow cylinder. Tendons attach muscles to projections on the inside of the exoskeleton and transmit the pulling force of the muscles to the hardened, chitinous cuticle. Areas of soft cuticle form flexible joints that allow the limbs to bend.

Insects walk like us. They bend and straighten each of their six legs in turn, while moving their body weight forwards. We can stand on one leg because we've got big feet, but an insect, which has small feet, needs to keep at least three of them on the ground. Muscles act across a joint working in pairs, one to **flex**, or bend the limb, the other to **extend**, or straighten it. Two sets of muscles that oppose each other in this way are called **antagonistic muscles**. Vertebrate muscles also work antagonistically (see page 390).

Most of the propulsive force needed for forward movement is generated by large muscles at the top of the leg.

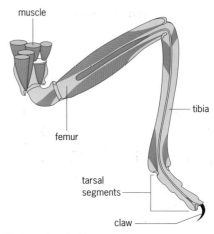

Fig 21.7 A typical insect leg

4 ENDOSKELETONS

An endoskeleton is a rigid structure that is inside the body, enclosed by soft tissues. Although we are more familiar with vertebrate endoskeletons, some invertebrates also have internal skeletons. The soft bodies of sponges are supported by sharp, mineral rods called **spicules**, and starfish and sea urchins are supported by small internal plates called **ossicles**.

Vertebrate skeletons are made up of **bone** and **cartilage**, which are specialised connective tissues (see Chapter 2). Some vertebrate skeletons, such as those of the cartilaginous fish, are made almost entirely of cartilage. In a developing vertebrate with a bony skeleton, cartilage appears first and is then replaced by bone. We call this process **ossification**.

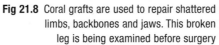

SCIENCE FEATURE Coral: a substitute for bone?

Mending damaged bones is a major part of a surgeon's work, yet acceptable substitutes for bone are hard to find. It is possible to use bone from another part of the patient's body but only small amounts can be used. Artificial substitutes run the risk of being rejected by the body's immune system (see Chapter 17).

A few species of coral have a porous structure similar to that of bone. When grafted into the body, the honeycomb texture of coral provides the conditions necessary for new blood vessels to grow into it, and this promotes new bone growth. In addition, coral is tough, carries no risk of infection and is unlikely to be rejected by the body.

'Liquid bone' can also be made from coral. This is based on a calcium-rich solution and a phosphate-rich solution. The surgeon mixes the compounds in a little acid before applying it, rather like toothpaste, to the site of a fracture. Within 12 minutes it has solidified, and after one hour the 'new' bone is as hard as real bone.

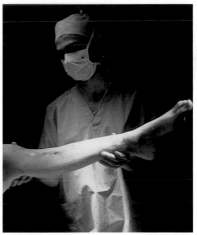

Fig 21.8 Coral grafts are used to repair shattered limbs, backbones and jaws. This broken leg is being examined before surgery

THE HUMAN SKELETON

The human skeleton has several functions:

- It acts as a framework that supports soft tissues.
- It allows free movement through the action of muscles across joints.
- It protects delicate organs and structures such as brain and lungs.
- It forms red blood cells in the bone marrow.
- It stores and releases minerals from bone tissue.

Fig 21.9 shows the structure and features of the human skeleton.

Fig 21.9 The human skeleton contains a total of 206 bones. It is divided into two parts: an axial skeleton that comprises the skull and vertebral column and an appendicular skeleton that is made up of the limbs and limb girdles.

All vertebrates show skeletal modifications related to their lifestyle. Since humans are bipedal (they walk on two legs), the hips and lower spine take most of the weight of the body and so the pelvic girdle (the hip) is larger and less flexible than the pectoral girdle (the shoulder)

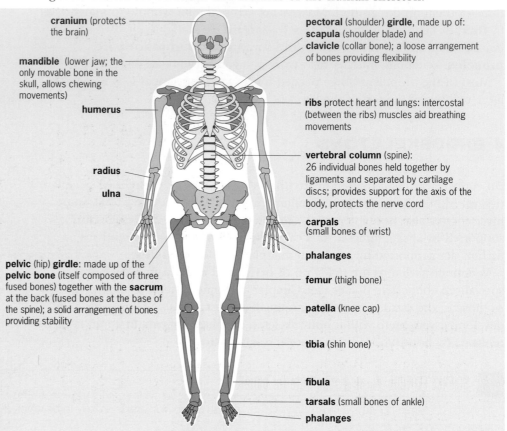

cranium (protects the brain)

mandible (lower jaw; the only movable bone in the skull, allows chewing movements)

humerus

radius

ulna

pelvic (hip) girdle: made up of the pelvic bone (itself composed of three fused bones) together with the sacrum at the back (fused bones at the base of the spine); a solid arrangement of bones providing stability

pectoral (shoulder) girdle, made up of: scapula (shoulder blade) and clavicle (collar bone); a loose arrangement of bones providing flexibility

ribs protect heart and lungs: intercostal (between the ribs) muscles aid breathing movements

vertebral column (spine): 26 individual bones held together by ligaments and separated by cartilage discs; provides support for the axis of the body, protects the nerve cord

carpals (small bones of wrist)

phalanges

femur (thigh bone)

patella (knee cap)

tibia (shin bone)

fibula

tarsals (small bones of ankle)

phalanges

Parts of the skeleton	Number of bones
Axial skeleton:	
skull	29
spine	26
rib cage	25
Total	**80**
Appendicular skeleton:	
pectoral girdle	4
pelvic girdle	2
arms	60
legs	60
Total	**126**

? TEST YOURSELF

A Look at Fig 21.9. Why are the bones of the legs thicker than the corresponding bones in the arms?

B Which structure is protected by the vertebral column or backbone?

THE STRUCTURE AND PROPERTIES OF HUMAN BONE

Bone is one of the hardest tissues in the human body and is second only to cartilage in its ability to absorb stress. It is made up of bone cells called **osteocytes**, which sit in a bone **matrix**. The matrix consists of inorganic matter (mainly compounds of calcium and phosphorus) interwoven with **collagen fibres** (see page 57).

Bone needs to be tough and resilient. These properties are provided by two types of bone tissue: **compact bone** and **spongy bone** (Fig 21.10). **Compact bone** is deposited in sheets called **lamellae** arranged as cylinders inside cylinders. Nerves and blood vessels run in a central canal. The compacted structure of the lamellae gives immense strength. The lamellae of **spongy bone** are arranged in a criss-cross pattern, forming a spongy honeycomb. This structure has excellent shock-absorbing properties.

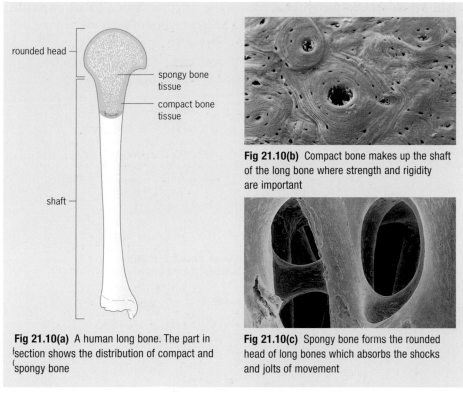

Fig 21.10(b) Compact bone makes up the shaft of the long bone where strength and rigidity are important

Fig 21.10(a) A human long bone. The part in section shows the distribution of compact and spongy bone

Fig 21.10(c) Spongy bone forms the rounded head of long bones which absorbs the shocks and jolts of movement

JOINTS

As in other vertebrates, the human skeleton is jointed. A joint is simply a place in the body where two bones meet. Most joints are **movable** but some, such as between the bones making up the **skull**, the **sacrum** (base of the spine) and the **pelvis**, are **immovable**, or **fused**. Elastic **ligaments** bind bones together, while tough inelastic **tendons** attach muscles to bone (Fig 21.11). The human knee joint is seen in Fig 21.12. Internally, the joint is lubricated by a viscous fluid, **synovial fluid**, secreted by the **synovial membrane**, the membrane that lines the joint.

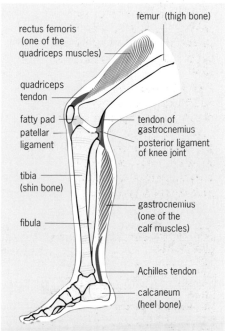

Fig 21.11(a) A diagram of the human leg showing the position of the main ligaments (green) and tendons (blue)

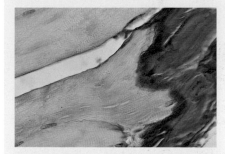

Fig 21.11(b) This micrograph shows a slice of tendon (yellow) connected to a piece of muscle (red). Tendons and ligaments look similar but their composition is different: tendons contain mainly **collagen fibres** (see page 57); ligaments are made up of fibres formed from the protein **elastin**

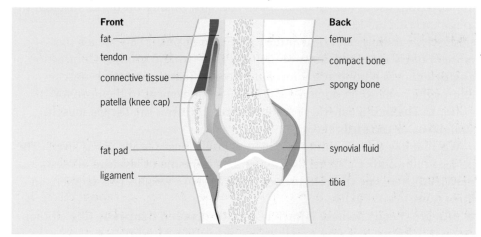

Fig 21.12 The features of a typical joint can be seen clearly in the knee joint

SEE QUESTION 1

The ligaments and the synovial membrane together form a **joint capsule** which surrounds the end of the bones. Different types of joint allow for different kinds of movement (Fig 21.13).

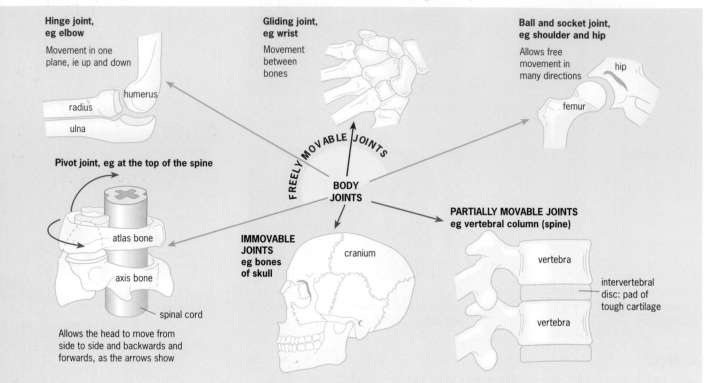

Fig 21.13 There are many different types of joint in the vertebrate skeleton. They are classified both by their shape and by their mobility

? TEST YOURSELF

C Why are ligaments elastic?

D Why does the tendon need to be inelastic?

HOW THE HUMAN BODY MOVES

A simple action, such as tapping your finger on the table, involves the skeletal system, the muscular system and the nervous system. This voluntary action begins as a stimulus in the cerebrum of the brain. Motor nerves carry impulses to muscles and cause them to contract, pulling on bones through the tough inelastic tendons.

Muscles can only pull, or contract; they cannot push. This means that muscles rarely act alone: most of the time they work in groups. Contraction of a muscle moves a bone at a joint, but a second (**antagonistic**) muscle returns the bone to its original position. Take movement in the arm, for example. The biceps muscle bends or **flexes** the arm; the triceps muscle straightens or **extends** the arm.

We all know that machines can help us to lift heavy loads using levers. The bones in our bodies also act as **levers**. The principle of leverage allows a muscle to overcome the **resistance** supplied by a weight. This happens in three ways:

- An arm bends because a force is applied between the **pivot** (the elbow) and the resistance (the weight to be lifted) (Fig 21.14a).
- We stand on tiptoes by pivoting the toes on the ground and using our calf muscles to raise the weight at the ankles. This is the wheelbarrow principle (Fig 21.14b).
- When standing or sitting, we can raise our head because muscles at the back of the neck tip the skull at the pivot at the top of the spine. This is the seesaw principle (Fig 21.14c).

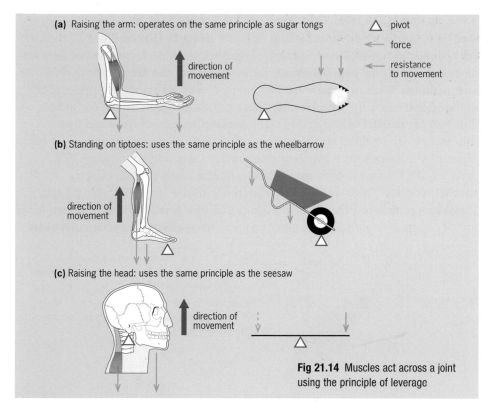

(a) Raising the arm: operates on the same principle as sugar tongs

△ pivot
← force
← resistance to movement

direction of movement

(b) Standing on tiptoes: uses the same principle as the wheelbarrow

direction of movement

(c) Raising the head: uses the same principle as the seesaw

direction of movement

Fig 21.14 Muscles act across a joint using the principle of leverage

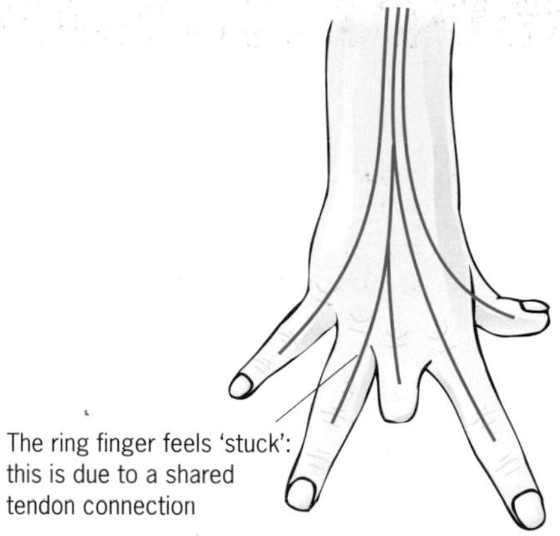

The ring finger feels 'stuck': this is due to a shared tendon connection

Fig 21.15 The stuck finger investigation

The 'stuck finger' test demonstrates the effect of muscles working in groups (Fig 21.15). Each of your fingers has a separate tendon connecting it to muscles in the forearm. Try it. Curl up your middle finger and place the other four fingers on a hard surface as shown. Now lift up the other fingers one by one. You will find that you can lift all of your fingers except the ring finger. This is because these two fingers share the same tendon connection.

5 MUSCLES AND MOVEMENT

The bones of the human skeleton provide a basic system of levers and joints that makes the skeleton potentially movable, but neither levers nor joints can move without muscles. Skeletal muscle (also called striped or striated muscle) provides the main source of power for human locomotion. In this section we look at the structure and function of skeletal muscle.

THE PROPERTIES OF SKELETAL MUSCLE

Like other sorts of muscle, skeletal muscle has three basic properties:

- **Excitability**: it can receive and respond to a stimulus.
- **Extensibility**: it can be stretched or it can contract.
- **Elasticity**: it can return to its original shape after it has contracted.

Skeletal muscle is also described as **voluntary muscle** because we can contract it when we want to. Smooth muscle and cardiac muscle are **involuntary muscle**: we cannot consciously control its contraction.

Skeletal muscles contract and relax, moving bones at joints in the skeleton. This allows for movements such as walking, running and waving, and fine movements such as those needed to use tools. The action of skeletal muscles also helps to maintain the position of the body when we are sitting or standing, even though there is no obvious movement. The contractions of skeletal muscle also produce heat, and much of this is used to keep the body temperature at a normal level (see Chapter 15).

> ✔ **REMEMBER THIS**
>
> There are three types of muscle.
> **Skeletal muscles** are responsible for whole-body movement.
> **Smooth muscle** is responsible for automatic movements such as those involved in peristalsis.
> **Cardiac muscle** is found only in the heart. The properties of the different muscle types are described in Chapter 2 and cardiac muscle is covered in detail in Chapter 13.

> ? **TEST YOURSELF**
>
> **E** What type of muscle fibre (skeletal, cardiac or smooth) is found in the following structures?
> **a)** stomach
> **b)** aorta
> **c)** biceps muscle
> **d)** left ventricle of heart
> **e)** face

THE STRUCTURE OF SKELETAL MUSCLE

Skeletal muscle is made of specialised cells, the **muscle fibres** (Fig 21.16). Each cell is long and thin and contains several nuclei. Each fibre is surrounded by a cell membrane called the **sarcolemma**; the cytoplasm of the fibre is the **sarcoplasm**. The sarcoplasm contains many mitochondria and a large number of thread-like fibres, the **myofibrils**. These run along the length of the muscle fibre and are parallel to one another. The **sarcoplasmic reticulum** that surrounds each myofibril consists of a network of tubes that contain calcium ions. These play a major role in bringing about muscle movement (see page 393).

As Fig 21.17 shows, each **myofibril** contains many threads called **myofilaments**. Thick myofilaments are made of the large-molecule protein **myosin**, the thin myofilaments are composed of a smaller protein, **actin**. Actin myofilaments also contain two other proteins: **troponin** and **tropomyosin**.

SEE QUESTIONS 2 AND 3

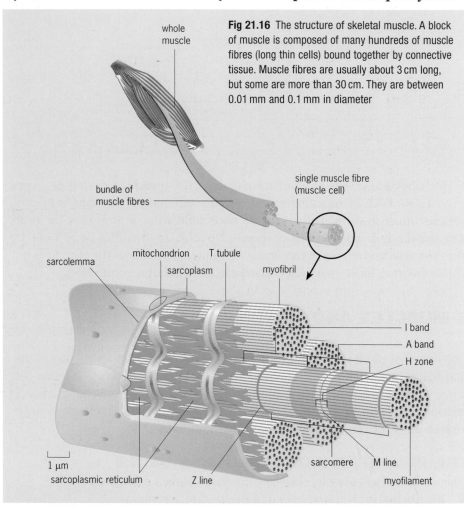

Fig 21.16 The structure of skeletal muscle. A block of muscle is composed of many hundreds of muscle fibres (long thin cells) bound together by connective tissue. Muscle fibres are usually about 3 cm long, but some are more than 30 cm. They are between 0.01 mm and 0.1 mm in diameter

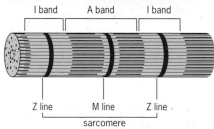

Fig 21.17 The ultrastructure of the myofibril. Each myofibril contains myosin and actin strands. Where the two overlap, the myofibril looks dark: this region is called the A band. In the regions where only actin is present, the myofibril looks lighter: this region is called the I band. At the centre of each I band is the dark Z line. At the centre of each A band is the dark M line. A bands and I bands alternate and so give skeletal muscle its striped appearance.

The sarcomere, the basic unit of muscle contraction, is the section of muscle fibre between one Z line and the next

THE SLIDING FILAMENT THEORY OF MUSCLE CONTRACTION

The relationship between the ultrastructure of skeletal muscle and its ability to bring about movement is explained by the **sliding filament theory of muscle contraction** (Fig 21.18).

As we have seen, the actin and myosin filaments lie parallel to each other. When the muscle is at rest, the 'heads' that stick out from the myosin filament cannot attach to the actin filaments that are alongside them. Movement occurs when the heads move towards the actin filament, attach to it and act as hooks to pull the actin myofilaments past them. As the actin myofilaments from opposite ends of the sarcomere are pulled towards each other, the sarcomere becomes shorter. This happens along the whole length of each myofibril and causes the muscle fibre to contract. Later, as the actin myofilaments become detached from the myosin heads, both filaments return to their original position and the muscle relaxes.

THE ROLE OF CALCIUM IONS AND ATP IN MUSCLE CONTRACTION

Contraction of skeletal muscle does not simply occur 'out of the blue'. It must be initiated by a stimulus in the form of a nerve impulse (see Chapter 19) which arrives at the muscle fibre from the nervous system. So when a muscle contracts and then relaxes, the following events take place (refer again to Fig 21.18):

● The brain makes a decision that the body is to move. Nerve impulses travel from the brain, along the spinal cord and through the motor nerves, towards the muscle.

● The electrical signal reaches the **neuromuscular junction**, a specialised synapse. The arriving signal causes the release of acetylcholine (see Chapter 19). This chemical is a neurotransmitter: it travels across the gap between the end of the nerve and the muscle, and binds to receptors on the **motor end-plate**.

● This binding causes an electrical change in the muscle fibre which triggers the release of calcium ions from the sarcoplasmic reticulum into the myofilament.

● Calcium ions bind to troponin and tropomyosin, the proteins that are closely associated with actin filaments. This changes the three-dimensional shape of the troponin–tropomyosin–actin complex, revealing parts of the actin filament that were previously hidden. These are the active sites to which the heads of the myosin filaments will attach.

● The calcium ions also act directly on myosin, activating it so that it splits ATP into ADP and inorganic phosphate, releasing energy. This energy is used to move the heads of the myosin filaments towards the newly exposed binding sites on the actin filaments.

● As cross-bridges form between the actin and myosin filaments, the myosin heads tilt, pulling the actin myofilaments past them. As the myofilaments slide, the heads detach from one site and attach to the next. As many as 100 such attachments can occur every second.

● As the actin myofilaments from opposite ends of the sarcomere move towards each other, the muscle fibre contracts.

● When the stimulation of the muscle fibre stops (when the brain decides the body should stop moving), the calcium ions are pumped back into the sarcoplasmic reticulum.

● As the level of calcium ions falls, troponin and tropomyosin move back to their original positions, blocking once again the active sites available for myosin attachment. The myosin heads can no longer attach to the actin filament, and both sets of filaments return to their original position. When the antagonistic muscle contracts, the sarcomeres in the first muscle return to their normal length and the muscle relaxes.

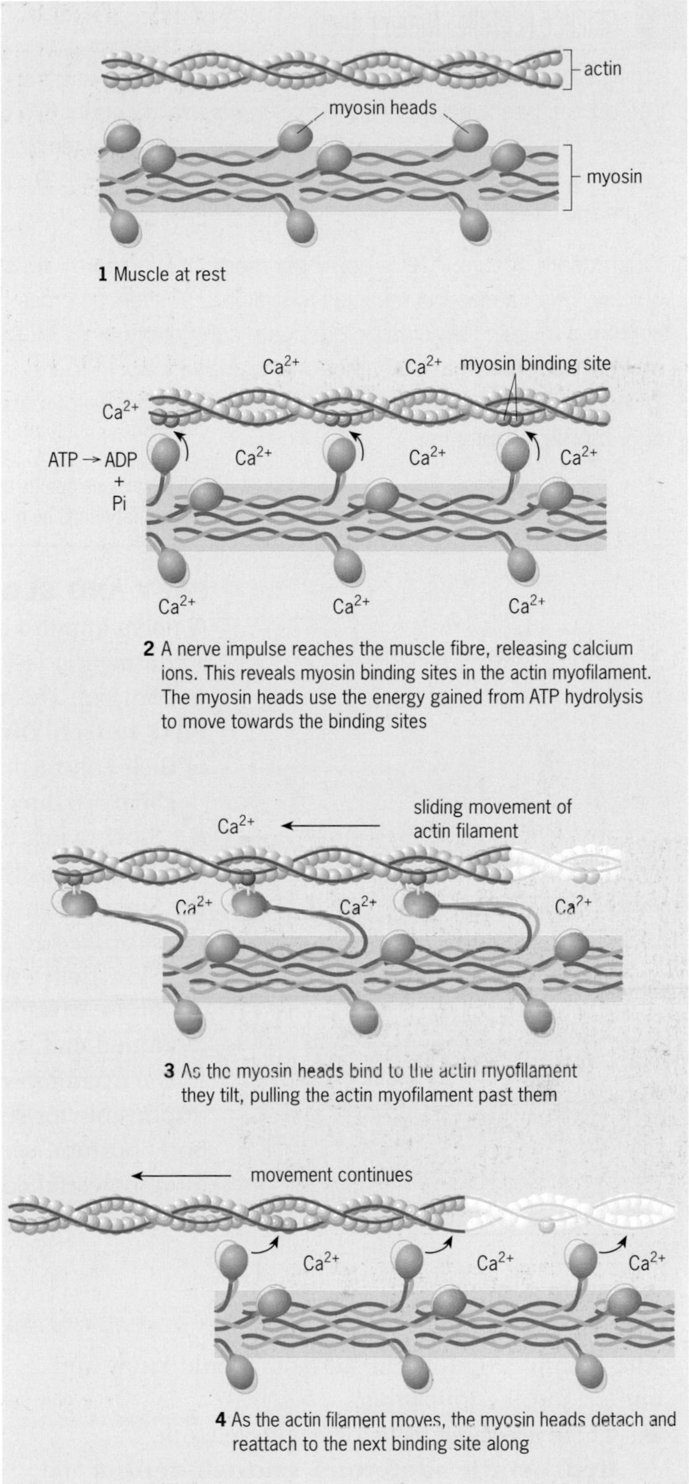

1 Muscle at rest

2 A nerve impulse reaches the muscle fibre, releasing calcium ions. This reveals myosin binding sites in the actin myofilament. The myosin heads use the energy gained from ATP hydrolysis to move towards the binding sites

3 As the myosin heads bind to the actin myofilament they tilt, pulling the actin myofilament past them

4 As the actin filament moves, the myosin heads detach and reattach to the next binding site along

Fig 21.18 The sliding filament theory to explain muscle contraction

---

! **SCIENCE FEATURE Time of death**

When a body is discovered, a pathologist investigates and tries to estimate how long the person has been dead. This information can be important in a murder enquiry. The pathologist assesses the time of death, partly by taking the internal temperature, to see how much the body has cooled from the normal body temperature of 37 °C, and partly by looking at the state of the muscles.

When death occurs, ATP is no longer made. It is a short-lived chemical and so it runs out fairly quickly. This causes the muscles to lock into position as cross-bridges that formed between actin and myosin filaments before death can no longer be broken. This condition, **rigor mortis**, happens in all body muscles. It appears about four hours after death and lasts about 24 hours. After this time, muscle proteins are destroyed by enzymes within the cells and so rigor mortis disappears.

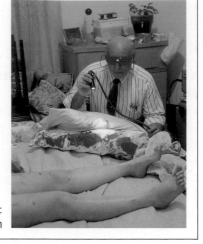

Fig 21.19 All suspicious deaths are investigated by a police pathologist: one of the first things he or she does is estimate the time of death

---

FAST AND SLOW MUSCLE FIBRES

A nerve impulse is the trigger for muscle contraction, but the length of time a contraction lasts depends on how long calcium ions remain in the sarcoplasm. This time is different in different types of skeletal muscle fibres. **Fast twitch fibres** and **slow twitch fibres** are classified on the basis of their contraction times.

There are three important differences between fast and slow fibres:

● Slow twitch fibres have less sarcoplasmic reticulum than fast twitch fibres. This means that calcium ions remain in their sarcoplasm longer.

● Slow twitch fibres have more mitochondria which provide ATP for *sustained* contraction.

● Slow twitch fibres have significantly more myoglobin than fast twitch fibres. Myoglobin has a higher affinity for oxygen than the haemoglobin in blood and so is particularly efficient at extracting oxygen from the blood.

The overall result of these differences is that slow twitch fibres are responsible for sustained muscle contraction, such as that which maintains body posture, while fast twitch fibres are responsible for the shorter-acting but more powerful contractions important in locomotion (see the Assignment at the end of this chapter).

SUMMARY

After completing this chapter you should know and understand the following:

■ There are three basic types of skeleton: **hydrostatic skeletons**, **endoskeletons** and **exoskeletons**. Invertebrates such as earthworms have hydrostatic skeletons, arthropods and crustaceans have exoskeletons, and vertebrates, including humans, have endoskeletons.

■ Skeletons support and protect the body and allow animals to **locomote** (move from place to place).

■ The human skeleton is made of **bones**, **tendons** and **ligaments**. The human skeleton is movable because different bones meet at **joints**, across which **muscles** are attached. Because muscles can

pull but not push, muscles often work in pairs call **antagonistic pairs**.

■ Muscles are **excitable**, **extensible** and **elastic**. **Skeletal muscle**, also called **striped**, **striated** or **voluntary muscle**, is responsible for human locomotion, maintaining body posture and generating body heat.

■ Skeletal muscle is made of **muscle fibres** which contain bunches of **myofibrils** which in turn contain **myofilaments**. Thick **myosin** filaments lie alongside thin **actin** filaments.

■ The **sliding filament hypothesis** explains muscle contraction at the molecular level.

? EXAM QUESTIONS

1

The diagram below shows a section through a human synovial joint.

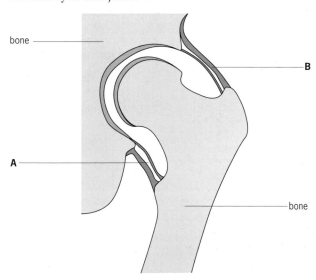

a) **(i)** Name the type of synovial joint shown in the diagram.
(ii) State **one** location of this type of joint in the human skeleton.
(iii) Describe the range of movement permitted at this type of joint.
b) Describe **one** function for each of the structures labelled **A** and **B**.
c) One form of arthritis, which is common in elderly people, causes the joints to become painful so that movement is difficult.
Suggest how the joints of a person affected by arthritis would differ in appearance from that shown in the diagram.

Edexcel HB3 June 1999 Q3

2 The diagram below shows part of a myofibril as seen through an electron microscope.

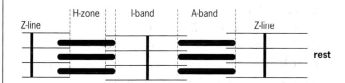

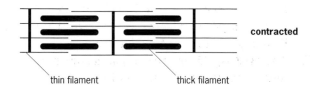

a) Name the main protein present in
(i) the thick filaments;
(ii) the thin filaments.
b) Describe the mechanism that brings about the change in position of the filaments when the myofibril contracts.

AQA (NEAB) BY04 Biological Basis of Behaviour June 2000 Q4

3
a) The diagram shows some of the muscles in a human leg.
(i) Describe what happens to the leg when: muscle **A** contracts; muscle **B** contracts.
(ii) Explain why muscles such as **C** and **D** are arranged in pairs.

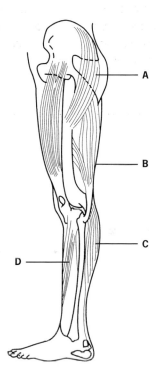

b) The photograph shows part of a muscle fibre.

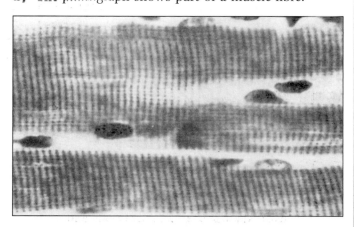

Reproduced from *An Atlas of Histology*, W.H. Freeman and B. Bracegirdle by permission of Heinemann Ltd.

Give **three** differences in structure between a muscle fibre and an epithelial cell from the lining of the small intestine.

AQA (NEAB) BY04 Biological Basis of Behaviour March 2000 Q1

KEY SKILLS **ASSIGNMENT**

MUSCLE FIBRES AND ATHLETICS

This Assignment follows on from the Chapter 8 Assignment **Energy systems and sport**. You may need to revise this to answer all the questions here.

The different types of muscle fibre

You may have noticed that athletes are highly specialised for their particular event. Marathon runners can run great distances without fatigue, but they lack the power to run as fast as the sprinters. In contrast, sprinters fatigue far too quickly to run long distances. These features are in part due to the type of muscle fibres they have inherited, and partly to the training they have done. Research has shown that there are two basic types of muscle fibre: **fast twitch** and **slow twitch**. Fast twitch fibres can be further subdivided, giving us the three different muscle types:

Slow twitch oxidative (type I) fibres. These contain large amount of myoglobin, many mitochondria and a dense capillary network. They split ATP slowly and therefore can only contract slowly. These fibres are red and have a high resistance to fatigue.

Fast twitch oxidative (type IIA) fibres. These show similar features to slow twitch fibres: a lot of myoglobin, many mitochondria and blood capillaries; but they are able to hydrolyse ATP far more quickly and therefore to contract rapidly. They are relatively resistant to fatigue, but not as resistant as type I fibres.

Fast twitch glycolytic (type IIB fibres). These fibres have a relatively low myoglobin content, few mitochondria and few capillaries. They contain large amounts of glycogen which provides fuel for anaerobic ATP production via the process of glycolysis. They contract rapidly but fatigue quickly, owing to the build-up of lactic acid. Type IIB fibres appear whiter than the other two types.

1 Summarise the above information in the form of a table.

2 Movement of muscles requires energy which is provided by one or more of three energy systems: (i) the ATP/CP system, (ii) glycolysis and (iii) the aerobic system. Outline the essential features of these systems.

3 a) What is the function of myoglobin? (See page 394 for a reminder.)
b) Suggest why type IIB fibres have very little myoglobin.

Muscles are made up of motor units: a motor unit is a bundle of fibres controlled by a single motor nerve. Analysis of muscle tissue by biopsy shows that all the muscle fibres in a single motor unit are of the same muscle type, but there can be different muscle types in different units within the same muscle. The various muscles of the body have different proportions of the three types, according to their function.

4 Why is it an advantage to have all three types of muscle fibre in the same muscle?

The distribution of the different muscle types in the body

For the rest of this Assignment we will consider the two main types of muscle fibre: slow and fast twitch.

A central principle in physiology is that structure is closely related to function and this is illustrated clearly by the distribution of the muscle types in the body. For instance, the postural muscles – such as those in the neck and back – have a high proportion of slow twitch fibres. In contrast, the muscles that bring about fast, explosive movements such as running, jumping and throwing are packed with fast twitch fibres.

5 a) Where in the body would you expect the muscles to have a high proportion of fast twitch fibres?
b) Explain why slow twitch fibres are suitable for postural muscles.

The distribution of different fibre types in different athletes

The muscle fibre types of a selection of age-matched athletes were analysed, and the results are summarised in Table 21.A1. These are average values: there is considerable variation between different athletes.

Table 21.A1 Analysis of the average proportions muscle fibre types in selected muscles in different male athletes

Type of athlete	Approx % fast twitch	Approx % slow twitch
Marathon runner	18	82
Swimmer	25	75
Cyclist	40	60
800 metre runner	52	48
Untrained person	55	45
Sprinter and jumper	62	38

6 What does 'age-matched' mean?

7 Explain in general terms how the proportion of fast and slow twitch fibres is related to the nature of an athlete's chosen activity.

8 What do you think the information about untrained people tells us?

22

Hormones

The European fire bug *Pyrrhocoris apterus*. Fire bugs grow through five nymph stages. Normally, as the fifth nymph sheds its skin, it emerges as a fully mature, reproductive adult

We usually think of hormones as complex chemicals that occur in the human body. But hormones control all sorts of processes in all sorts of organisms. The vital part that one hormone plays in the life cycle of an insect, the European fire bug, was discovered by accident. In the 1950s, a young Czech scientist, Karel Sláma, travelled from Europe to Harvard University, carrying a small number of European fire bugs. This insect develops in five nymph stages, shedding its skin each time it grows. After the last moulting, it becomes an adult, capable of reproduction. When he arrived, Sláma was astonished to find that his bugs had not changed into mature adults: they had moulted more than five times and had grown into abnormally large nymphs.

The paper lining the insects' cages seemed to be responsible. Cages lined with *The Times* or the science journal *Nature* produced normal insects but those lined with *The New York Times* or *Scientific American* produced large nymphs. Sláma his colleagues later showed that wood pulp from balsam fir, used to make the paper for some North American publications, contains a chemical that mimics insect juvenile hormone. This hormone prevents development from nymph to adult. But why should a tree mimic an insect hormone? Could it be a defence mechanism against aphids? Aphids that attack the balsam fir absorb the hormone and are then unable to mature into reproductive forms. Their colonisation attempt fails, and the tree remains free from aphids.

1 HORMONES ARE CHEMICAL SIGNALS

Chemical signals are produced by most living organisms. In plants they control processes such as growth and flowering (see Chapter 32). Animals use a wide variety of chemical signals to control and co-ordinate many aspects of their everyday life (Fig 22.1). In mammals, as Fig 22.2 shows, there are three main types of chemical signals:

Fig 22.1 Chemical signals occur in many organisms

Sexual reproduction in most organisms is controlled by several hormones. In humans, hormones control gamete production, puberty, pregnancy, birth and lactation

This dog is searching for olfactory clues: airborne chemicals produced by frightened prey or given off by a female dog to let potential mates know that she is 'on heat'

This dragonfly began life as a water-living nymph. Because its cuticle was not flexible, it shed its outer skin to grow bigger. The timing of these moultings is controlled by hormones. One, juvenile hormone, prevents metamorphosis into the adult form. Another, ecdysone, is a moulting hormone which helps the old cuticle to split

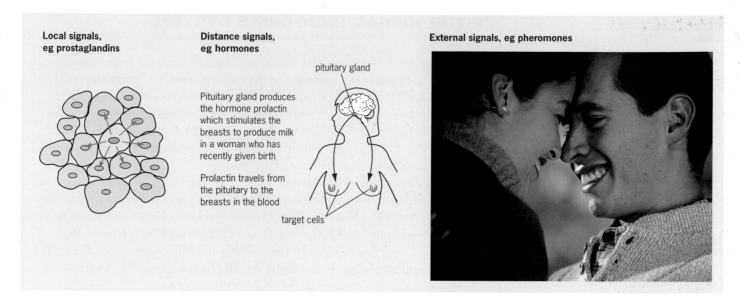

Local signals, eg prostaglandins

Distance signals, eg hormones

pituitary gland

Pituitary gland produces the hormone prolactin which stimulates the breasts to produce milk in a woman who has recently given birth

Prolactin travels from the pituitary to the breasts in the blood

target cells

External signals, eg pheromones

- Locally acting chemicals such as **endorphins, prostaglandins** and **histamines** (Fig 22.3). These act on cells that are close to their site of production.
- **Hormones.** These are produced by special tissues called **endocrine glands**. They pass into the blood and act away from their site of production. Hormones affect **target cells** and **target organs** in the body.
- **Pheromones** are chemical messengers that allow external communication: simple exchanges of information between people. Generally, pheromones are not an important means of communication in humans, despite the claims of some newspaper reports and advertisements.

Fig 22.2 A summary of the different types of chemical signalling that exist within and between organisms. Local signals include neurotransmitters such as acetylcholine (see Chapter 19) and local messengers such as prostaglandins, endorphins and histamines. Hormones allow communication between different organs, via the blood. Externally produced messengers allow communication between organisms

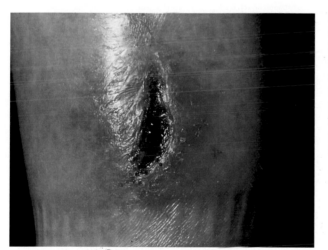

Fig 22.3 Histamine is a local messenger which is released by injured cells. It acts on the cells around an injury like the deep cut shown above, causing blood vessels near to the injury to dilate, bringing more blood into the area. This is an important part of the inflammation response (see Chapter 17)

In this chapter we look at how chemical signals control many life processes in animals, concentrating on the systems that occur in humans. We focus mainly on hormones. The definition of a hormone in mammals is:

A chemical messenger that is produced by cells or tissues of the endocrine system and that travels through the blood to act on target cells or target organs, before being broken down.

The main part of the chapter describes the human endocrine system and we finish with a brief overview of prostaglandins and endorphins and the effect of pheromones on animal behaviour.

2 THE HUMAN ENDOCRINE SYSTEM

In humans, more than a dozen tissues and organs produce hormones. Some, including the **pituitary gland**, the **thyroid gland**, the **parathyroid glands** and the **adrenal glands,** are endocrine specialists: their major function is to secrete one or more hormones. Others, such as the pancreas, ovaries and testes, secrete hormones in addition to their other functions. Together, these glands make up the **endocrine system**, shown in Fig 22.4.

The endocrine system has four main functions:

● It maintains **homeostasis**, the balance of the body, by making sure the concentration of many different substances in body fluids are kept at the correct level. The control of blood sugar level, blood pH and water balance provides examples of homeostasis (see Chapters 14, 15, 16 and 18).

● It works with the nervous system to help the body respond to stress. The release of adrenaline in the **fight or flight** response is an example of this (see the Assignment at the end of this chapter).

● It controls the body's rate of growth.

● It controls sexual development and reproduction (see Chapter 26).

We know of more than 50 human hormones, and we divide them into two groups according to their origin: those made from *fatty acids* and those made from *amino acids*. The first group, which includes **steroids** such as oestrogen and progesterone, are *lipid-soluble*. Hormones based on **amino acids** are **modified amino acids**, **peptides** or **glycoproteins**: they are *water-soluble* and include insulin and adrenaline. Table 22.1 lists most of the hormones produced by the endocrine system and shows where they act and what they do. You might find it useful as a revision aid.

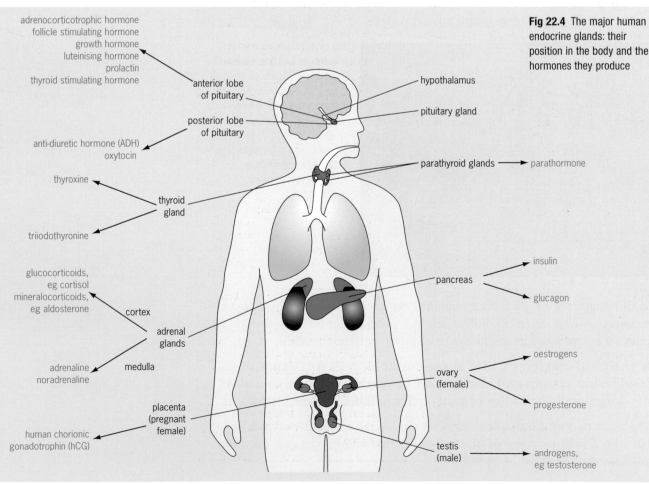

Fig 22.4 The major human endocrine glands: their position in the body and the hormones they produce

adrenocorticotrophic hormone
follicle stimulating hormone
growth hormone
luteinising hormone
prolactin
thyroid stimulating hormone
→ anterior lobe of pituitary

anti-diuretic hormone (ADH)
oxytocin
→ posterior lobe of pituitary

thyroxine
triiodothyronine
→ thyroid gland

glucocorticoids, eg cortisol
mineralocorticoids, eg aldosterone
→ cortex

adrenaline
noradrenaline
→ medulla

→ adrenal glands

human chorionic gonadotrophin (hCG)
→ placenta (pregnant female)

hypothalamus
pituitary gland

parathyroid glands → parathormone

pancreas → insulin
pancreas → glucagon

ovary (female) → oestrogens
ovary (female) → progesterone

testis (male) → androgens, eg testosterone

Table 18.1 Human endocrine glands and their principal secretions. You do not need to learn all of this, but you should find it useful for reference as you read the rest of the chapter

Gland	Hormone	Chemical structure	Effect
Pituitary 1 Posterior lobe	Anti-diuretic hormone (ADH)	Peptide	Reduces amount of water lost in urine. Raises blood pressure by constricting arterioles
	Oxytocin	Peptide	Contraction of smooth muscle during childbirth. Stimulates secretion of milk from mammary glands
2 Anterior lobe	Adrenocorticotrophic hormone (ACTH)	Peptide	Stimulates production and release of hormones from adrenal cortex
	Follicle stimulating hormone (FSH) also called ICSH in males	Glycoprotein	Controls the development of follicles in the ovary, and sperm cells in the testis
	Growth hormone	Protein	Promotes growth (especially of skeleton and muscles). Affects body metabolism
	Luteinising hormone (LH)	Glycoprotein	Stimulates ovulation and formation of the corpus luteum (stimulates testosterone production in males)
	Prolactin	Protein	Stimulates milk production and release during pregnancy
	Thyroid stimulating hormone (TSH)	Glycoprotein	Stimulates growth of thyroid gland; synthesis and production of thyroid hormones
Thyroid	Thyroxine + Triiodothyronine	Iodine-containing amino acids	Thyroxine Increases rate of cell metabolism, controls aspects of growth and development and controls basal metabolic rate (BMR)
Parathyroid	Parathormone	Peptide	Raises blood calcium levels by stimulating release of calcium from bone
Pancreas (Islets of Langerhans)	Insulin (produced by the β cells)	Protein	Lowers blood glucose levels by making cell membranes more permeable to glucose, increases glycogen storage in liver
	Glucagon (produced by the α cells)	Peptide	Raises blood sugar levels by stimulating glycogen breakdown in the liver
Adrenal I Cortex	Glucocorticoids, eg cortisol	Steroids	In response to stress, raises blood glucose
	Mineralocorticoids, eg aldosterone	Steroids	Concerned with water retention. Increases reabsorption of sodium chloride in kidneys, so important in control of blood volume and pressure
2 Medulla	Adrenaline	Modified amino acid	'Fear, flight and fight' reactions. Prepares the body for heightened activity. Mimics effects of the autonomic nervous system
	Noradrenaline	Modified amino acid	As adrenaline
Gonads 1 Ovary (follicle)	Oestrogens	Steroids	Female sex characteristics, rebuilding of uterus lining after menstruation, inhibits FSH
2 Ovary (corpus)	Progesterone	Steroids	Stimulates maturation of uterus lining, maintains pregnancy, inhibits FSH
3 Testis	Androgens, eg testosterone	Steroids	Support sperm production. Important in the development of male secondary sexual characteristics
Placenta	Human chorionic gonadotrophin (hCG)	Steroid	Causes corpus luteum to secrete progesterone, thus maintaining pregnancy

adrenal gland

main blood vessels

kidney

ureter

capsule

adrenal medulla

This secretes adrenaline

The adrenal cortex is made up of 3 layers of cells

These secrete glucocorticoids such as cortisone, mineralocorticoids such as aldosterone, and gonadocorticoids such as the sex hormones

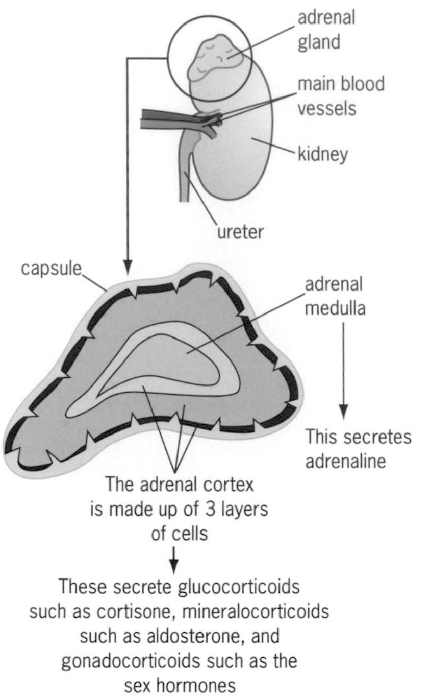

Fig 22.5 The adrenal glands sit on top of the kidneys. They consist of the adrenal medulla, the tissue in the centre of the gland which secretes adrenaline, and an outer layer called the adrenal cortex

SEE QUESTION 3

✔ REMEMBER THIS

All endocrine glands contain some of the hormone they produce, at any given time. Although the hormone thyroxine is released continuously, most are not. **Negative feedback loops** control the release of hormones from many endocrine glands so that homeostasis is maintained. See Chapter 14.

Fig 22.6 Control of metabolic rate involves a negative feedback loop

MAKING AND RELEASING HORMONES

Endocrine glands are tissues and organs that produce and release hormones. These glands have no ducts. The hormone is simply secreted into the blood and then travels to target organs or cells elsewhere in the body. A good example of an endocrine gland is the **adrenal gland** (Fig 22.5).

Hormones are formed inside endocrine cells. Some amino acid-based hormones, such as insulin and glucagon, are produced directly from a copy of a gene by the processes of transcription and translation (see Chapter 27). Most hormones, including the steroids, are made as an end-product of a series of chemical reactions. Each reaction makes minor changes to precursor molecules, until, eventually, the active hormone is produced. For the steroid hormones progesterone, oestrogen and testosterone, the precursor is cholesterol (see Chapter 3).

THYROXINE: AN EXAMPLE OF NEGATIVE FEEDBACK

Metabolic rate is the rate at which all cells in the body carry out their biochemical reactions. It is a vital whole-body function that must be controlled within very strict limits. The hypothalamus in the brain detects even a small decrease in metabolic rate and responds by releasing more **thyrotropin releasing hormone** (Fig 22.6). This acts on the pituitary gland, causing it to release more **thyroid stimulating hormone**. This passes to the thyroid gland, which responds by secreting more **thyroxine**, the hormone that acts on individual cells to increase metabolic rate. As soon as metabolic rate gets back to normal levels, the hypothalamus responds by releasing less thyrotropin releasing hormone and, in a healthy person, homeostasis is maintained.

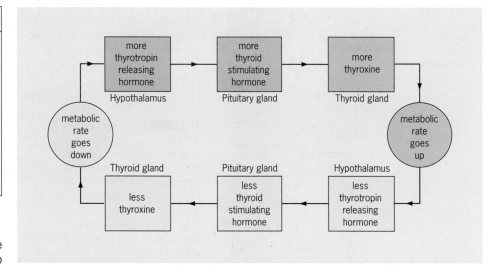

HOW HORMONES AFFECT TARGET CELLS

Hormones exert their effects in two different ways. Peptide hormones travel in the blood to all parts of the body. They do not, however, affect all cells in the body. Target cells or organs have specific proteins on their surface that act as receptor sites. The hormone fits into these sites as a key fits a lock, or like an enzyme fits its substrate (see Chapter 5). Once in place, the hormone–receptor complex brings about changes inside the cell, although the hormone itself never enters the cytoplasm. In contrast, steroid hormones pass easily into cells because they can get through the lipid bilayer. They bind to a specific receptor molecule inside the cytoplasm: this complex then causes biochemical changes inside the cell.

PEPTIDE HORMONES AND SECOND MESSENGERS

Hormones based on amino acids are water-soluble and vary in size from **thyroxine** (just 2 modified amino acids) to **growth hormone**, a protein made up of 190 amino acids. Because they are not lipid-soluble, they cannot get into cells through the lipid cell surface membrane. Instead, they act as first messengers, binding to target receptors on the outside of the cell surface membrane.

As Fig 22.7(a) shows, this binding event activates an enzyme, **adenylate cyclase**, which is located on the inside surface of the membrane. The activated enzyme converts ATP into a substance called **cyclic AMP**. This is the **second messenger** that moves about inside the cell, causing biochemical changes. The final effect on cell function depends on the type of hormone and the type of target cell involved.

STEROID HORMONES

Steroids are small lipid-soluble molecules that can get through cell membranes easily. They enter cells and bind with **target receptors** inside the cytoplasm (Fig 22.7b). Cortisol, progesterone, oestrogen and testosterone all act in this way.

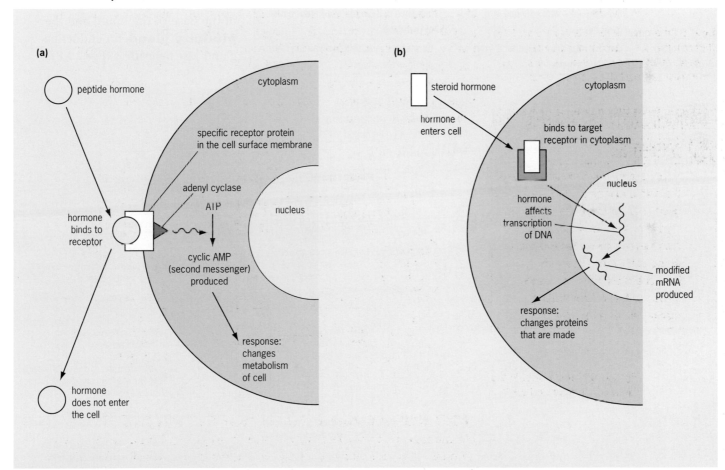

Fig 22.7(a) Water-soluble hormones work via a second messenger which is formed inside the cell when they bind to receptors on the cell surface

Fig 22.7(b) When steroid hormones bind to a receptor inside the cytoplasm, they form a complex similar to an enzyme–substrate complex. This enters the nucleus of the cell and binds directly to the DNA, interfering with the cell's ability to read some of its genes. A steroid hormone can either switch on or switch off protein synthesis of particular genes. The overall effect of hormone action is different for each steroid hormone (see Table 22.1)

3 COMPARING THE NERVOUS AND ENDOCRINE SYSTEMS

Table 18.2 Differences between the nervous system and the endocrine system in humans

Property	System	
	Nervous system	**Endocrine system**
Nature of signal	Nerve impulses are **electrical** signals, transfer of information across synapses, is chemical (see Chapter 19)	All hormones are **chemical** signals
Size of signal	**Frequency modulated** – determined by the frequency of nerve impulses sent along a nerve fibre, and the number of nerve fibres being stimulated	**Amplitude modulated** – determined by the concentration of hormone
Speed of signal	**Rapid**. Human nerves conduct nerve impulses at speeds ranging between 0.7 metres per second and 120 metres per second	**Usually slower**, by comparison. The release of insulin from the pancreas in response to a rise in blood sugar level takes several minutes
Effect in the body	**Localised** effect – each individual neurone links with only one or a few cells	More **general** effect – hormones can influence cells in many different parts of the body
Capacity for modification	**Can be modified** by learning from previous experience	**Cannot be modified**: no learning from previous experience

Hormones do not act independently. They work together with other hormones and also with the nervous system. The nervous system and the endocrine system both control and co-ordinate the function of different parts of the body and they both rely on chemical messengers, but they have obvious differences (Table 22.2). Despite these differences, we now realise that the link between the endocrine system and the nervous system in humans is more definite than once thought. The main physical link between the two systems is between the **hypothalamus**, part of the base of the brain, and the **pituitary gland**, an endocrine gland just beneath it (Fig 22.8).

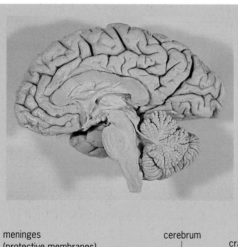

Fig 22.8(a) The position of the hypothalamus and pituitary gland in the brain can be identified from Fig 22.8(b)

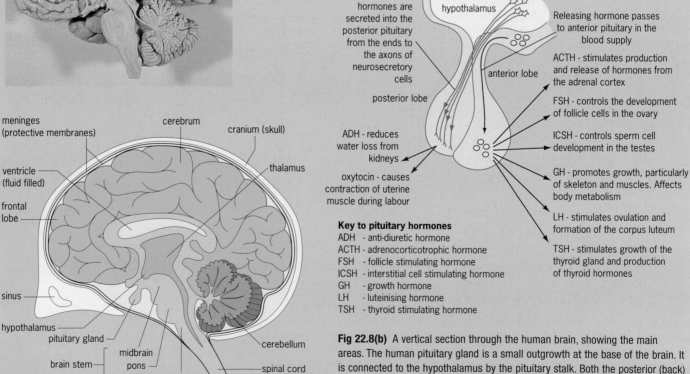

Key to pituitary hormones
ADH - anti-diuretic hormone
ACTH - adrenocorticotrophic hormone
FSH   - follicle stimulating hormone
ICSH - interstitial cell stimulating hormone
GH   - growth hormone
LH   - luteinising hormone
TSH   - thyroid stimulating hormone

Fig 22.8(b) A vertical section through the human brain, showing the main areas. The human pituitary gland is a small outgrowth at the base of the brain. It is connected to the hypothalamus by the pituitary stalk. Both the posterior (back) and anterior (front) lobes secrete hormones

! SCIENCE FEATURE Gigantism and acromegaly

Gigantism

Robert Wadlow, the tallest man who ever lived, had a condition called gigantism; he probably had a tumour that produced large amounts of growth hormone. The growth plates in his long bones didn't fuse as they normally do in the late teens.

Acromegaly

Robert grew to be a touch under 9 feet tall (2.7 metres) and had size 36 feet (US sizes). Born in Alton, Illinois, USA, Wadlow reached the height of over 7 feet at the age of 13 and continued to grow throughout his short life of 22 years. He might have continued to grow, but he died prematurely following an infected foot wound.

Fig 22.9 Robert Wadlow made it into the record books as the tallest man who has ever lived. There might have been people who have grown taller – stories exist of men over 9 feet tall – but their existence, unlike Wadlow's, cannot be proved. Today, such extremes in size are rare because conditions can be diagnosed early and treated effectively

But what happens if someone develops a tumour that produces growth hormone later in life, after their growth plates have closed over? Tumours of the pituitary gland that cause excess production of growth hormone are rare, but they do happen. They cause a condition called acromegaly. The name acromegaly comes from the Greek words for 'extremities' and 'enlargement' and reflects one of its most common symptoms, the abnormal growth of the hands and feet. Soft-tissue swelling of the hands and feet is often an early feature, with patients noticing a change in ring or shoe size. Gradually, bony changes alter the patient's facial features: the brow and lower jaw protrude, the nasal bone enlarges and spacing of the teeth increases. So, growth still occurs, but this does not lead to a change in height.

The disease does cause problems and is fatal if not treated. Overgrowth of bone and cartilage often leads to arthritis. When tissue thickens, it may trap nerves, causing numbness and weakness of the hands. Other symptoms of acromegaly include thick, coarse, oily, smelly skin; skin tags; enlarged lips, nose and tongue; deepening of the voice due to enlarged sinuses and vocal cords; snoring due to upper airway obstruction; excessive sweating, fatigue and weakness; headaches; impaired vision; abnormalities of the menstrual cycle and sometimes breast discharge in women; and impotence in men. Body organs, including the liver, spleen, kidneys and heart, can also enlarge. This is bad enough, but in the longer term, the most serious health consequences are diabetes mellitus, high blood pressure and increased risk of cardiovascular disease. Patients with acromegaly are also at increased risk for polyps of the colon that can develop into cancer.

Acromegaly is difficult to diagnose as the changes it brings about happen very slowly and can be mistaken for changes that result from the normal ageing process. However, once it is diagnosed, it can be treated and people with acromegaly can avoid the changes in their facial features that were common until about 25 years ago, when treatment became available. The first treatment is surgery to remove the pituitary tumour – this reduces the production of growth hormone significantly, but perhaps not down to normal levels. The slight excess must then be 'mopped up' using drug treatment. The main drugs in use are called octreotide and lanreotide. These are analogues of the hormone somatostatin – this dampens down the production of growth hormone by the pituitary. Early forms of the drug had to be injected under the skin every 8 hours but longer-acting preparations that last a month are now available. After surgery, a patient is given the short-acting form to judge which dose is most suitable for them, and then they move on to the once-a-month injections, which need to continue for the rest of their lives.

SEE QUESTION 4

THE HYPOTHALAMUS AND PITUITARY: PARTNERS IN COMMUNICATION AND CONTROL

The pituitary has been called the 'master gland' because it controls the activity of most of the other endocrine glands, but it is itself actually under the control of the hypothalamus. The range of hormones produced by the pituitary and the way the hypothalamus controls pituitary function are shown in Fig 22.8(b).

The human pituitary gland has a **posterior lobe** and an **anterior lobe**. The **pituitary stalk** connects the posterior lobe to the brain. This direct physical link allows nervous communication between the pituitary and the hypothalamus. A rich network of blood vessels also links the hypothalamus with the anterior lobe, allowing hormonal signals to pass from the brain to the pituitary.

NEUROSECRETORY CELLS AND THE POSTERIOR PITUITARY

A **neurosecretory cell** is a modified neurone that synthesises a hormone. The hormone is produced in the cell body and is then transported down the axon and released into the blood through the synapse, which terminates on a capillary. The neurosecretory cells that arise in the hypothalamus and end in the posterior pituitary make two hormones: **anti-diuretic hormone** (**ADH**) and **oxytocin**. When the hypothalamus is stimulated by appropriate nerve impulses, it releases these hormones into the posterior pituitary. From here they pass into the blood and then travel around the body. ADH controls the reabsorption of water by the kidneys (see Chapter 16); oxytocin causes contraction of the uterus during childbirth (see Chapter 26).

HORMONAL CONTROL OF THE ANTERIOR PITUITARY

Blood vessels that connect the hypothalamus with the anterior pituitary carry hormones and chemical messengers. These either stimulate or inhibit the release of pituitary hormones, including growth hormone, prolactin, the gonadotrophins, adrenocorticotrophic hormone and thyroid stimulating hormone (see Fig 22.4).

NERVOUS AND HORMONAL CONTROL OF DIGESTIVE SECRETIONS

The body uses a combination of nerve communication and hormones to control the secretion of digestive juice to coincide with the presence of food in particular areas of the gut. Below is a brief summary of how this happens: for more detail see Chapter 11.

The taste or smell of food encourages the brain to signal the salivary glands and stomach to release saliva and stomach secretions. Food leaving the stomach stimulates the vagus nerve, which then triggers the release of bile and pancreatic juice. Digestive secretions can also be controlled by hormones. The hormone gastrin, produced by the stomach wall, travels in the bloodstream but exerts its effect locally, stimulating the production of both pepsinogen and hydrochloric acid.

Secretin and cholecystokinin control pancreatic and liver secretions. Both are formed by cells in the duodenal wall. Secretin causes the release of sodium hydrogencarbonate in the pancreas. In the liver, it increases the rate of bile formation. Cholecystokinin triggers the release of pancreatic enzymes such as lipase and trypsinogen.

THE ACTION OF ADRENALINE

As the nervous system and the endocrine system are closely linked, we should not be surprised that hormones can affect the way that we feel emotionally. The classic example of this is the **fear, fight or flight response** (see the Assignment at the end of this chapter).

At the molecular level, adrenaline binds to the outside of a target cell and activates the enzyme **adenyl cyclase**. This converts ATP to cyclic AMP. Cyclic AMP, is a second messenger which sets off a **cascade**, a chain of reactions that convert glycogen to glucose.

As adrenaline binds to receptor molecules on the surface of a cell, changes in the membrane cause the production of **G proteins**. For every one molecule of adrenaline that binds, 10 molecules of G protein are produced. Each of those 10 molecules of G protein catalyses the production of 10 molecules of adenyl cyclase. Each molecule of adenyl cyclase stimulates the production of 10 molecules of cyclic AMP. This **amplification** effect continues along a chain of reactions that eventually breaks down glycogen into glucose.

4 PROSTAGLANDINS

Prostaglandins are a group of hormone-like compounds. They were originally thought to be produced by the prostate gland – hence the term prostaglandins. They are not true hormones as they are not produced by endocrine glands. Most mammalian cells synthesise prostaglandins, which act locally on surrounding cells.

Prostaglandins control cell metabolism, probably by modifying levels of cyclic AMP inside the cell, and are involved in a wide range of activities including blood clotting, inflammation and smooth muscle activity. Prostaglandins are extremely powerful: one billionth of a gram produces measurable effects.

5 ENDORPHINS

Endorphins are one of several morphine-like chemical messengers produced in the brain (endorphin = endogenous morphine). These polypeptide molecules mimic the effects of drugs such as morphine and heroin: like prostaglandins, they are involved in a wide range of activities, but their most important role seems to be in the management of pain. It is thought that endorphins bind to pain receptors and so block the sensation of pain. Long-distance runners who run until they collapse do so because abnormally high levels of endorphin block out the acute pain and discomfort.

Endorphins also seem to affect the 'pleasure centres' of the brain. Stimulation of these areas provides the intense feelings associated with orgasm. Scientists studying the chemistry of pleasure have shown that pleasurable sensations begin when the hypothalamus releases serotonin. This stimulates the release of endorphins, which turns on the supply of dopamine and simultaneously turns off the supply of GABA (an amino acid that suppresses dopamine). Dopamine stimulates the pleasure centre directly.

Drugs such as heroin work by mimicking endorphins. The brain of an addict stops making endorphins and so withdrawal from heroin results in a sudden lack of endorphin and a build-up of GABA, producing unpleasant withdrawal symptoms.

REMEMBER THIS

The process of amplification seen in the enzyme cascade, which results in the fear, fight or flight response, allows one molecule of adrenaline to stimulate the production of millions of glucose molecules.

REMEMBER THIS

Prostoglandin pessaries can be used to set off labour in a pregnant woman whose baby is overdue. However, semen is rich in prostaglandins, and some experts recommend to expectant couples the natural alternative of having frequent sex. There is a sound physiological rationale behind this, but, from a practical viewpoint, it is rather difficult advice to follow.

REMEMBER THIS

The prostaglandins in human semen cause muscles of the uterus and oviduct to contract during female orgasm, helping the sperm on their journey towards the ovum.

6 PHEROMONES AND BEHAVIOUR

Pheromones, sometimes called **ecto-hormones**, are substances that organisms release into the environment to communicate with members of the same species. Pheromones are small, volatile molecules that spread easily in the environment. They are active in very small amounts: the pheromone of the female gypsy moth causes a response in the antennae of the male moth at concentrations as low as one in a thousand million million molecules (Fig 22.10).

Pheromones are usually classified according to the type of response they produce:

● **Alarm pheromones** are produced by bees and ants when attacked. They excite other insects of the same species to swarm around the attacker.

● **Sex attractants** are released by moths, rats and possibly humans to attract members of the opposite sex. Humans do not have a particularly good sense of smell but there is some evidence that very young babies can recognise their mother by her characteristic smell. Some people also think that sexual partners can also recognise each other using their sense of smell, but there is less evidence to back this up.

● **Trail substances** are produced by ants to show other ants where to find sources of food.

● The **queen substance** is produced by the queen bee within a hive to suppress the production of other queens.

Fig 22.10 Sex attractants such as the pheromone bombykol are released into the air in minute quantities by female moths. The male is extremely sensitive to these pheromones, because his antennae are large and crammed with many specialist receptors. This male has about 17 000 receptors in each antenna that respond only to bombykol

SUMMARY

After reading this chapter, you should know and understand the following:

■ There are three basic types of chemical signal: **local signals** such as histamine, prostaglandins and endorphins; **hormones** such as insulin and adrenaline; external signals or **pheromones**.

■ The human **endocrine system** has four main functions. It controls growth, sexual development and **fear, flight and fight reactions**. It maintains body homeostasis.

■ Endocrine glands lack a duct: their secretions are delivered directly into the bloodstream.

■ Hormone levels are usually controlled by **negative feedback loops**.

■ There are two basic types of hormone: **peptide hormones** which are derived from amino acids (eg insulin) and **steroid hormones** which are made from fatty acids (eg oestrogen).

■ Peptide hormones bind to receptors on the outer membrane of target cells and achieve their effect via a **second messenger** such as cyclic AMP. Steroid hormones pass straight through the cell membranes and bind to target receptors inside the cytoplasm.

■ There is a close link between the nervous and endocrine systems, shown by the way in which, in the brain, the **hypothalamus** interacts with the **pituitary gland**.

■ Adrenaline, the hormone that causes the classic fear, flight and fight response, acts at the molecular level to bring about the production of millions of glucose molecules.

■ **Prostaglandins** and **endorphins** are local messengers which affect many different types of cell. Prostaglandins are best known for their effects on the female reproductive system: endorphins are chemicals that influence our perception of pleasure and pain.

? EXAM QUESTIONS

1

a) Given below is a table of mammalian hormones and a list of hormonal functions **A–H.** Copy the table. From the list, choose function of each hormone and place the appropriate letter **A–H** in the table.

Hormone	Function
insulin	
anti-diuretic hormone	
glucagon	
oxytocin	
thyroxine	
progesterone	

List of hormonal functions

A Lowers the resting potential across the membranes of the muscle fibres in the wall of the uterus

B Causes hair follicles to produce hair in the lower abdominal regions of females

C Stimulates the hydrolysis of glycogen to glucose in the liver

D Increases the metabolic rate by affecting the rate of protein synthesis

E Inhibits the secretion of follicle-stimulating hormone during pregnancy

F Increases the rate of contraction of the cells in the sinoatrial node

G Renders the walls of certain areas of nephrons permeable to water

H Increases the ability of skeletal muscle to use glucose as a respiratory substrate

b) State three differences between hormones and enzymes.

c) State two differences between control using the endocrine system and control using the nervous system.

OCR 6918 Animal Studies March 2000 Q2

Refer to Chapter 14 page 273 to help you answer this question.

2 Read the following passage about hormones involved in the regulation of blood glucose levels, then add the most appropriate word or words to complete the passage.

When blood glucose levels rise above normal, ––– is released from the ––– cells of islets of Langerhans in the –––. This lowers the blood glucose levels by increasing the uptake of glucose into cells and promotes the conversion of glucose into –––. Other cells in the islets of Langerhans secrete ––– which raises the level of blood glucose.

Edexcel B2 January 2000 Q1

3 Thyroxine is a hormone secreted by the thyroid gland and it controls the basal metabolic rate. The diagram shows how the secretion of thyroxine is controlled.

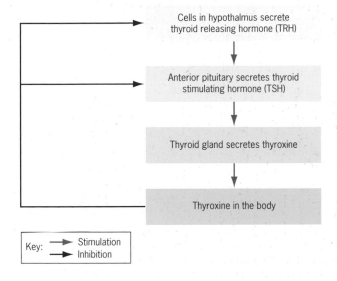

Key: → Stimulation → Inhibition

a) Explain how the control of thyroxine secretion illustrates the principle of negative feedback.

b) In the formation of thyroxine, a protein-digesting enzyme detaches amino acid molecules from a globular protein called thyroglobulin. Thyroxine is formed from an amino acid molecule joined to iodine atoms. Name each of the following:
(i) The organelle where thyroglobulin is manufactured.
(ii) The type of bond that is broken when the amino acid is detached from thyroglobulin.
(iii) The process by which thyroxine is secreted from a cell in the thyroid gland.

c) Give three ways in which hormonal control differs from nervous control.

Edexcel B6 January 2000 Q5

4 Complete the table summarising the role of each of the following hormones.

Hormone	Site of secretion	Part of kidney tubule affected	Effect of hormone
ADH			
Aldosterone			

AQA (NEAB) BY03 Physiology June 2000 Q2

KEY SKILLS **ASSIGNMENT**

ADRENALINE AND STRESS

When we are nervous, familiar symptoms appear. We get butterflies in the stomach, our pulse increases, our hands become cold and clammy. This happens because our body is preparing for action by secreting **adrenaline**. This hormone:

● increases heartbeat,
● increases ventilation rate,
● causes vasoconstriction, resulting in redirection of blood from intestines and skin to brain and muscles,
● increases metabolic rate,
● dilates pupils.

1 **a)** Which organs secrete adrenaline?
b) What is the overall effect of adrenaline on the body?
c) Use your answer to 1b to explain why adrenaline leads to an increased blood pressure.

The effects of adrenaline are very similar to the effects of the sympathetic nervous system (see Chapter 20). For example, secretion of adrenaline speeds up the heart, as does direct stimulation by a sympathetic nerve.

Fig 22.A1 Adrenaline helps this mountain biker to concentrate and compete to the best of her abilities. Her heart is pounding, blood pressure is raised and all of the capillaries in the muscles are open to maximise performance

The difference is that adrenaline produces a general state of readiness for action, while the sympathetic nervous system is used for fine control. If your body needs to increase the heart rate a little, but does not need the other effects of adrenaline, it does so via the sympathetic nerves.

The effects of stress

Fig 22.A2 There are many stressful lifestyles: being a busy executive is just one. Stress happens when adrenaline is continually released, ready for action that is not taken. This leads to raised blood pressure which can increase the risk of heart disease or stroke

It is difficult to define stress. What is a stressful situation to some people may be sheer enjoyment to others. Certainly, some stimulation is desirable – it makes us feel more alive.

In the short term, the responses of the body to stress have a beneficial effect and leave the individual better able to cope with the crisis if and when it comes. In the long term, however, the effects can be far from beneficial. Continued secretion of adrenaline and general sympathetic stimulation bombard the body with stress chemicals. Table 22.A1 outlines some of the common symptoms of stress.

2 **a)** List some activities/sports that people enjoy because they stimulate release of adrenaline.
b) List some occupations or lifestyles that are commonly associated with stress.

3 **a)** Discuss or write down some ideas that might explain why there is more stress-related illness today than in previous generations.
b) Discuss or write down some strategies that people might use to minimise stress.

Table 18.A1 The effects of stress on bodily functions			
Normal state	**Adrenaline response**	**Short-term effect**	**Long-term effect (stress)**
Brain: normal blood flow	blood flow increases	think more clearly	headaches and migraines
Muscles: normal blood flow	blood flow increases	improved performance	muscular tension and pain
Heart: normal pulse and blood pressure	output and pressure increase	improved performance	hypertension and chest pain
Intestines: normal blood flow and peristalsis	blood flow decreases, peristalsis increases	slower digestion	abdominal pain and diarrhoea
General biochemistry: normal rate of oxygen use. Glucose and lipids liberated	oxygen, glucose and lipid use increases	more energy available quickly	rapid tiredness

23

Senses and behaviour

Many dolphins and porpoises die entangled in fishing nets each year

Dolphins, porpoises and toothed whales make sense of their environment by producing a variety of squeaks, whistles and clicks at high frequencies and then detecting echoes from nearby objects. Though sophisticated, this system can be fooled by human inventions such as fishing nets. These are too fine to reflect the sonar signals that dolphins send out and are therefore 'invisible'.

Understanding more about how dolphins' sensory systems work has enabled scientists to devise a way to allow fishermen to carry on their activities without being a threat. Apparently, dolphins can detect shoals of fish by picking up the sonar reflections from their swim bladders. When fishing nets are fitted with plastic devices that look like small rugby balls, dolphins come to an abrupt halt some way from the net, firing rapid and repeated sonar at it. The plastic balls seem to 'look' a bit like swim bladders, but are not similar enough to attract the dolphins.

When tested in the open sea, the initial curiosity of the dolphins turned to suspicion: they avoided the nets completely and, after turning back briefly to bombard the back of the net with more sonar, they then resumed their normal activities. Scientists hope to get more funding to do larger-scale tests. If they are successful, dolphin deaths from fishing net accidents could become a thing of the past.

1 RESPONDING TO THE ENVIRONMENT

All living organisms are **sensitive**. Whole organisms and individual cells respond to physical and chemical changes in their environment. Bacteria, fungi and plants respond to their environment. Plants, for example, grow towards the light. Single-celled protoctists such as amoeba, paramecium and euglena (Fig 23.1) react to touch, heat and light by simply moving closer or further away from the source. But it is in the animal kingdom that we find the widest range of body senses and the most sophisticated behavioural responses.

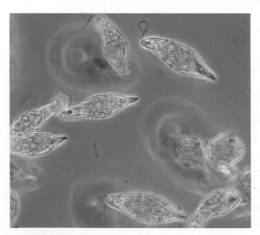

In this chapter we look at sensory systems in animals, concentrating on the human body senses, as illustrated by the Pacinian corpuscle, the eye and ear and then we go on to look at animal behaviour.

Fig 23.1 *Euglena gracilis*, a single-celled protoctist, is capable of hunting for food and eating prey, and it can also use light to build up sugars by the process of photosynthesis. It usually swims towards light in its environment

SENSORY SYSTEMS IN ANIMALS

Animals have a nervous system (see Chapter 20) which receives information from the environment, decides on the best course of action and then signals the body to respond. Humans sense the environment through the body senses of **touch**, **taste**, **sight**, **hearing** and **smell**. We make sense of our surroundings by using our **sense organs** to gather information and relay it to the central nervous system. Table 23.1 shows the huge amount of sensory information that reaches the brain. Fortunately, we deal with most of it at a subconscious level, which is just as well, since otherwise we would have no time for thoughts. Overall, the brain copes with a huge range of sensory information and uses it to decide on appropriate courses of action.

Table 23.1 Summary of the sensory information that goes to the brain	
External stimuli	**Internal stimuli**
Sound: volume, pitch, harmonics	Blood pressure
Vision: colour, shape, intensity, movement	Solute concentration of the blood: do we need a drink?
Touch: light or heavy pressure?	Internal pain: ulcer, wind, headache?
Texture: sandpaper or velvet?	Tension in muscles: posture?
Temperature: hot or cold?	Fatigue in muscles
Pain: mild or severe?	Tension in bladder or rectum
Movement: direction, acceleration or deceleration	
Direction of gravity	
Smell: which one of thousands of different chemicals?	
Taste: salt, sweet, sour or bitter?	

Before going on to look at sensory systems in more detail, you will find it useful to become familiar with some important terms:

- A **stimulus** is a physical event, usually some form of energy change, that we can detect with our receptors. Examples include **mechanical stimuli** such as pressure, touch and movement of air, **thermal stimuli** such as heat and cold, **light** and **chemical stimuli** such as taste, smell and the concentration of carbon dioxide or oxygen in the blood. There are many factors in our environment that are not stimuli – we cannot detect them because we do not have the appropriate receptors. We do not sense radio waves or high-pitched dog whistles, for example.

- A **sensation** refers to a general state of awareness of a stimulus. Aristotle first identified the five senses of hearing, sight, taste, smell and touch, but we now know that there are actually many more senses than this. The skin can experience the sensations of light touch or deep pressure, heat, cold and pain. There are also the general senses of balance and body movement. Internally, the body can sense changes in blood pressure and blood levels of chemicals; and we experience hunger and thirst when we need food or fluids.

- **Perception** is the interpretation of stimuli by the brain. Perception of a stimulus is different from just sensing that it is there – it allows us to assess the significance of the stimulus. Many stimuli can be ignored, such as the sensations from your clothes, to allow us to concentrate on more urgent stimuli (like the information in this book).

REMEMBER THIS

Perception involves reception *and* interpretation of stimulus information.

Fig 23.2 The head of this ragworm, an active hunter, contains many receptors. It has well-developed eyes, chemoreceptor cells and large palps to enable it to find its prey. The ragworm withdraws into its burrow for safety, guiding itself into position using tactile (touch) cells and sensitive tentacles called cirri

SEE QUESTION 1

? TEST YOURSELF

A Look at the receptors in Fig 23.3. What type of receptors are they? Classify them according to stimulus, location and complexity, as in Table 23.2.

2 FEATURES OF SENSORY SYSTEMS

A sensory system must allow the following processes to take place:

- **Transduction**. This is the process by which a stimulus initiates a nerve impulse. Stimuli vary widely in their nature; they can be physical, light, heat or chemical, and so on. All, however, result in the generation of a nerve impulse (see Chapter 19). For example, when physical pressure is applied to the skin, sensory cells inside specialised receptors in the skin depolarise, and the physical stimulus is transduced, or changed, into the electrochemical event known as the action potential.
- **Transmission**. A nerve impulse is transmitted along a sensory neurone into the central nervous system (CNS) (see Chapter 20). The brain recognises the strength of a stimulus not by the size of the impulse but by its frequency. The stronger the stimulus, the higher the frequency of impulses.
- **Information processing**. The CNS processes information that it receives from several parts of the body and makes a decision about the best course of action to take. We have superb memories compared with other animals and so we are able to combine new sensory information with past experience before deciding on an appropriate course of action.

RECEPTORS

A receptor is part of the nervous system that has become adapted to receive stimuli. Some receptors are modified neurones while others are separate cells connected to neurones. Some receptors, such as those in the skin, occur alone or in small groups. Others are concentrated within a specialised sense organ such as the eye or ear.

Sensory receptors are vital. They sense the external environment and they also provide information to the CNS about the body's internal environment. Receptors that detect changes in the outside world are positioned at or near the body surface, usually at the front end (Fig 23.2). These include the specialised sense organs, the ear and eye, the smell receptors of the nose and the taste receptors of the tongue. An animal's head is basically a set of sense organs in front of a mass of nerves (the brain) that is capable of processing all the incoming information. In contrast, receptors that respond to internal stimulation, such as stretch receptors in blood vessels or muscles, are found deep inside the animal's body.

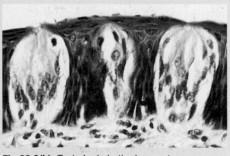

Fig 23.3(a) Sensory hairs on the mouth-parts of a blowfly

Fig 23.3(b) Taste buds in the human tongue

TYPES OF RECEPTOR

There are many different types of receptor in the animal world (Fig 23.3) Receptors can be classified in three ways: by the type of stimulus they respond to, by their location within the body, or by their level of complexity (Table 23.2).

There are also two basic types of receptor, and both can be any of the classes of receptor described in Table 23.2. Simple or primary receptors consist of a single neurone whose axon carries a nerve impulse into the CNS. The temperature and pressure receptors in the skin are both examples of **primary receptors**.

Table 23.2 Classification of receptors

Receptor name	Action	Example
Classified by stimulus		
Photoreceptor	responds to light	eye
Chemoreceptor	responds to chemicals	taste bud
Thermoreceptor	responds to changes in temperature	temperature receptors in skin
Mechanoreceptor	responds to mechanical (touch) stimuli	nerve fibres around hairs
Baroreceptor	responds to changes in pressure	baroreceptors in carotid artery
Classified by location		
Interoceptor	responds to stimuli within the body	baroreceptors in carotid artery
Exteroceptor	responds to stimuli outside the body	eye and ear
Proprioceptor	responds to mechanical stimuli conveying information about body position	muscle spindle
Classified by complexity		
General senses	single cells or small groups of cells	temperature receptors in skin
Special senses	complex sense organs	eye and ear

? TEST YOURSELF

B Look at Table 23.2. What type of receptors are the photoreceptors of the eye: are they exteroceptors, proprioceptors or interoceptors?

Secondary receptors receive information from a cell that is not part of the nervous system and then pass the information on to a nerve cell for transmission to the CNS. Taste buds in the tongue are secondary receptors.

THE PACINIAN CORPUSCLE

When people shake hands, there is a subtle exchange of information. A firm, confident handshake communicates friendliness but 'I'll take no nonsense', while a feeble, cold, clammy, loose grip says 'I don't really want to be here'. This is only possible because of the simplest receptor in the human body – the **Pacinian corpuscle**. This is a **mechanoreceptor** that responds to pressure or any kind of mechanical stimulus.

The corpuscle itself is a nerve ending surrounded by several concentric capsules of connective tissue, separated by a viscous (stiff) gel. In its resting state, the Pacinian corpuscle in cross section resembles a dartboard with the nerve ending at the bull's eye (Fig 23.4). Touching the hands or feet makes this epidermal receptor deform into an oval cross section including the nerve ending. The viscous gel then moves and allows the nerve ending to resume its normal shape. If the pressure is released, the corpuscle as a whole returns to its original shape and the nerve ending is again deformed, signalling the end of the hand or foot pressure. The viscous gel flows back and everything returns back to the normal resting stage. The nerve sends the signal to the brain, alerting it to the exact location of the pressure.

✔ REMEMBER THIS

The Pacinian corpuscle is named after Filippo Pacini, a nineteenth century Italian anatomist.

✔ REMEMBER THIS

Pacinian corpuscles can detect vibrations and touch in terms of frequency, duration and intensity.

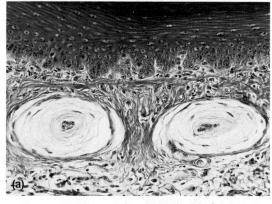

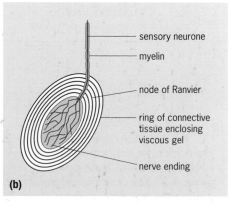

Fig 23.4(a) A micrograph of two Pacinian corpuscles (touch receptors) in human skin. They are abundant at fingertips. In this touch receptor, pressure is detected by the end of the neurone at the centre of the connective tissue capsule which is layered like an onion. When the neurone is stimulated, action potentials are propagated along the neurone. The brain interprets these impulses as pressure on that part of the body

(b) A simple diagram of a cross section through a Pacinian corpuscle

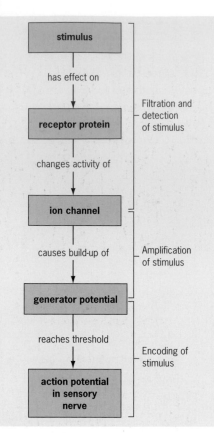

Fig 23.5 Flow chart summarising the function of receptor cells

---

✔ **REMEMBER THIS**

A stimulus alters the electrical properties of membrane of the receptor cell. The charge on the membrane is reversed: it goes from –70 mV to +40 mV. We say that it has depolarised. When a membrane depolarises there is the formation of a nerve impulse, or moving action potential, that is transmitted along a nerve fibre. Find out more about action potentials in Chapter 19.

HOW DO RECEPTORS WORK?

How do cells detect stimuli? How, for instance, can we distinguish between a light and a heavy touch? The answer lies in the fact that primary receptors have proteins in their cell surface membranes that are sensitive to the type of stimulus. Whatever the stimulus, transduction follows the same basic pattern, which is summarised in Fig 23.5. A receptor protein is activated by a specific stimulus. In turn, the protein opens or closes specific ion channels in the cell surface membrane. This change in ion movement causes depolarisation of the membrane; this is the **generator potential**.

If the depolarisation goes beyond the threshold level for that cell, it triggers an action potential (see Chapter 19). All receptors therefore act as **biological transducers**: they create action potentials in neurones in response to physical and chemical stimuli from the environment.

The greater the stimulation of the sense cell, the greater the depolarisation. Once the generator potential exceeds the threshold level, an action potential is produced and is transmitted along the axon. It is important to remember that all action potentials are of the same magnitude – you can't have powerful and weak nerve impulses. The body is able to tell the difference between strong and weak stimuli by the frequency of the action potential created.

Receptors are very sensitive – they have to be in order to detect small changes in the environment. The human nose can detect ethanoic acid (acetic acid, the acid in vinegar) at a concentration of 5×10^{11} molecules per litre of air; a dog's nose, though, can detect the same smell at a concentration of about 2 million times less, 2×10^5 molecules per litre.

Body senses also **amplify** the incoming signal. A stimulus is often quite weak – just a few photons of light when viewing a distant star for example, or a few molecules of a chemical. But the action potential travelling from the eye to the brain, for example, has about 100 000 times as much energy as the few photons of light that triggered it.

The endocrine system also shows amplification. One molecule of the hormone adrenaline, for example, can release millions of glucose molecules in the cascade effect (see Chapter 22).

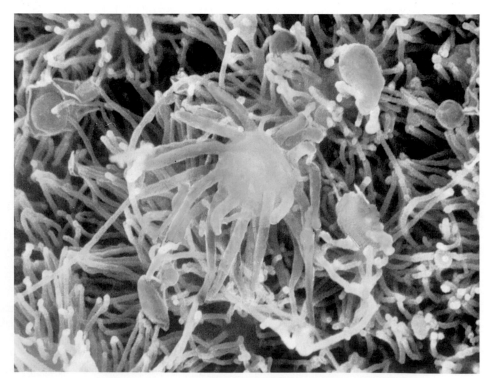

Fig 23.6 A coloured scanning electron micrograph of the surface of the olfactory (smell) epithelium of the rat. At the centre (blue) is the ending of an olfactory receptor cell process. This small swelling gives rise to a number of long cilia (projections). These cilia are thought to be the sites of interaction between odiferous substances and the receptor cells. There are many more cilia in the rat (over 20) than in human cells (12), indicating the rat's greater sense of smell

3 VISION

The eyes are literally our window on the world (Fig 23.7). This most sensitive and intricate of sense organs has fascinated both scientists and poets for centuries. In this section we look at the basic structure of the eye before going on to study focusing and the correction of defects. Finally, we look at vision at the molecular level.

Fig 23.7 Mammalian eyes are contained in the orbit, a bony socket of the skull. They are protected by the bone and also by a pad of fat at the rear of each eyeball. Other features of the eye are also protective:
• The eyebrows prevent sweat running from the forehead into the eye
• The eyelashes prevent the entry of airborne particles such as flying insects, often adding to the efficiency of the eye blink reflex
• Tears constantly bathe the surface of the eye, removing the surface film of dust and dirt and killing bacteria
• The conjunctiva, a thin, transparent membrane which covers the cornea, also protects the eye from airborne debris

STRUCTURE OF THE HUMAN EYE

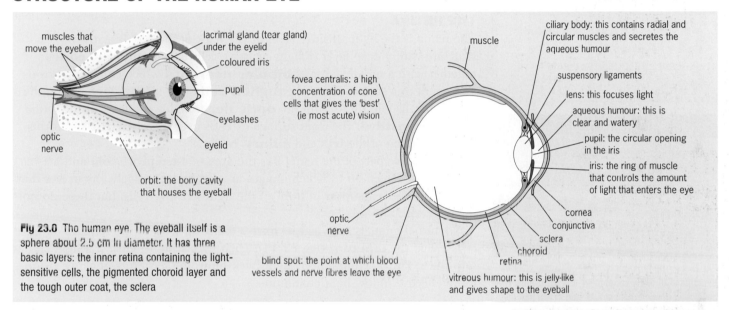

Fig 23.8 The human eye. The eyeball itself is a sphere about 2.5 cm in diameter. It has three basic layers: the inner retina containing the light-sensitive cells, the pigmented choroid layer and the tough outer coat, the sclera

The eye is an important and delicate structure. The **orbits**, bony sockets in the skull, protect the eyes from physical damage, and a pad of fat behind the **eyeball** helps to cushion it from shocks. The other protective features of the eye are shown in Fig 23.8.

The main body of the eye is divided into two parts by a **biconvex lens**. This lens contains transparent **lens fibres** (long thin cells that have lost their nuclei) together with an elastic **lens capsule** made of glycoprotein. The lens is flatter at the front than at the back and is soft and slightly yellow. With age it becomes flatter, yellower, harder and less elastic; it can also become cloudy (Fig 23.9). There is no blood in the lens – substances diffuse in and out of it from the surrounding fluid.

The eyeball has three cavities: the region between the cornea and the iris, the region between the iris and the lens, and the cavity that fills all the space behind the lens. The first two cavities are filled with a fluid called the **aqueous humour**, which is thin and watery. The third cavity, the largest (it takes up about 80 per cent of the volume of the eyeball) is filled with a thicker fluid called the **vitreous humour**. This transparent gel has the consistency of egg white and contains 99 per cent water and 1 per cent collagen fibres and hyaluronic acid. The vitreous humour preserves the spherical shape of the eyeball and helps to support the **retina**.

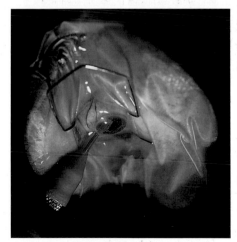

Fig 23.9 As people get older, the lens in their eyes can become cloudy, interfering with normal vision. In this routine operation, an ultrasound probe is being used to break up a cloudy lens before a plastic lens is fitted

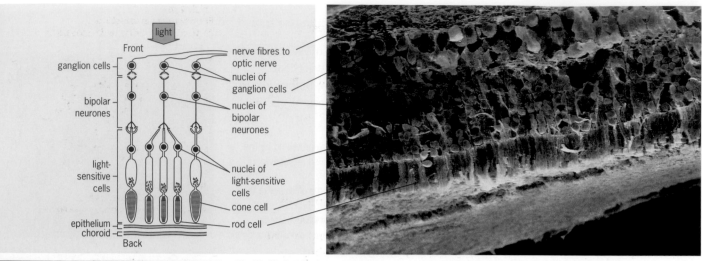

Fig 23.10 Detailed structure of the retina showing photoreceptors and their connections

C The eye is often compared with a camera. Which part of the camera corresponds to the retina of the eye?

Fig 23.11 The tapetum is a light-reflecting layer in the choroid of many vertebrates. It is particularly well developed in nocturnal animals, and in fish that live in deep water. In hoofed mammals such as elk, the tapetum contains glistening fibres of connective tissue. In cats and other carnivores, it contains shiny crystals of guanine

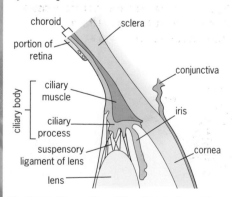

Fig 23.12 The position of the ciliary body and the suspensory ligaments in the eye. When we focus on objects that are distant from us, the ciliary muscles relax, increasing the tension on the suspensory ligaments. This pulls the lens into a thinner and flatter shape. When we look at objects closer to us, the ciliary muscles contract, the tension on the ciliary ligaments decreases and the lens becomes fatter

THE RETINA

The retina is the light-sensitive layer at the back of the eye. It is a complex structure with a deep layer of light-sensitive cells called **rods** and **cones**, together with a middle layer of **bipolar neurones** which connect the rods and cones to a surface layer of **ganglion cells** (Fig 23.10). Fibres from the ganglion cells join up to form the **optic nerve**. Only the back of the retina is **photosensitive**. The front surface extends over the **choroid** and forms the inner lining of the **iris** and **ciliary body**.

The sensitive part of the retina has the area of a ten pence coin and acts as a projection screen on to which images are directed. Strangely, the nerves that come from the retina pass in front of the light-sensitive cells, but these do not interfere with vision.

Nocturnal animals such as cats have a shiny backing to the retina, a layer of cells called the **tapetum** (Fig 23.11). Their eyes seem to glow if a light shines into them. The tapetum reflects light back on to the receptors, making vision more effective in low light conditions.

THE CHOROID

The **choroid** contains many blood vessels. Blood supplies the cells of the eye with nutrients and oxygen and removes waste. The choroid is dark-coloured because of a high concentration of the pigment melanin in its cells. These pigmented layers prevent internal reflection of light rays and so prevent us seeing a confused and blurred image.

At the front of the eye, the choroid expands around the edge of the lens to form the ciliary body. The smooth muscles in the ciliary body alter the shape of the lens (Fig 23.12). The lining of the ciliary body secretes aqueous humour, the fluid that fills the front of the eye.

The iris is also an extension of the choroid, partially covering the lens and leaving a round opening in the centre, the pupil. Its function is to control the amount of light entering the eye. The iris contains **radial** and **circular muscles** that can alter pupil size. Iris muscles are controlled by the autonomic nervous system (see Chapter 20).

THE SCLERA

The **sclera** forms the 'white' of the eye. It is a tough external coat mainly of collagen fibres. Six exterior muscles attached to the sclera enable the eye to look up, down and side-to-side. The central one-sixth of the sclera is colourless and transparent, forming a clear window, the cornea.

HOW THE EYE WORKS

The process of seeing is complex and involves five different stages:

- Light enters the eye.
- An image is focused on the retina.
- Energy in the light that makes up the image is transduced into an electrical signal.
- Nerve impulses carry information about the image into the brain.
- The brain decodes the information and perceives the image.

Let's look at each of the stages in more detail.

LIGHT ENTERS THE EYE

Light enters through the cornea and then passes through the pupil, the aqueous humour, the lens and the vitreous humour, before it reaches the retina. To operate well in conditions of different light intensity, the eye must control how much light reaches the retina. It does this by changing the diameter of the pupil. In bright light, the pupil **constricts** (gets smaller) to prevent overstimulation of the retina and the perception of 'dazzle'. In dim light, the pupil **dilates** (gets wider), letting in more light.

The size of the pupil is controlled by the muscles in the iris (Fig 23.13). These are themselves controlled by the autonomic nervous system and so pupil adjustment is a reflex response, not under the conscious control of the brain.

FOCUSING LIGHT ON THE RETINA

If you do not focus a camera correctly, the picture you take is blurred. Similarly, the eye must focus light to produce a high-resolution image on the retina. The eye focuses an image by refracting or bending light using the cornea and the lens, forming an upside down or inverted image on the retina (look ahead to Fig 23.15).

Because of the composition and curvature of the cornea, most of the refraction of light occurs in this structure. The lens is important in the fine focusing of light onto the retina. The cornea is fixed but the lens is adjustable, allowing **accommodation**, a reflex that makes the eye focus on objects that are at different distances from the eye.

ACCOMMODATION

Generally, when we open our eyes in the morning, they are not focused on nearby objects. At rest, the ciliary muscles have relaxed, pulling the lens flat. In this state we focus on distant objects.

When we focus on an object that is only a metre or so away, the ciliary muscles of the eye contract involuntarily (Fig 23.14). This reduces the tension

Bright light
Relaxation of radial muscles and contraction of circular muscles cause pupil to constrict

Dim light
Contraction of radial muscles and relaxation of circular muscles cause pupil to dilate

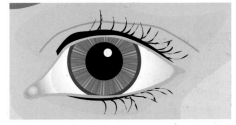

Fig 23.13 The eye is prevented from being dazzled by a speedy reflex that makes the pupil smaller. To achieve this **constriction** of the pupil, radial muscles in the iris relax and circular muscles contract. In dim light, the opposite happens: radial muscles contract and circular muscles relax, causing the pupil to enlarge

SEE QUESTION 2

Fig 23.14 The eye can focus on near and distant objects by altering the shape of the lens. This process is called accommodation

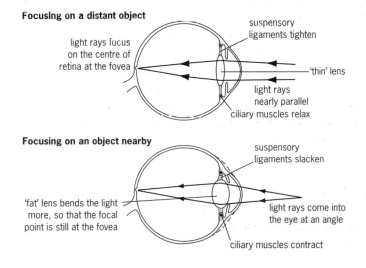

Focusing on a distant object
light rays focus on the centre of retina at the fovea
suspensory ligaments tighten
'thin' lens
light rays nearly parallel
ciliary muscles relax

Focusing on an object nearby
'fat' lens bends the light more, so that the focal point is still at the fovea
suspensory ligaments slacken
light rays come into the eye at an angle
ciliary muscles contract

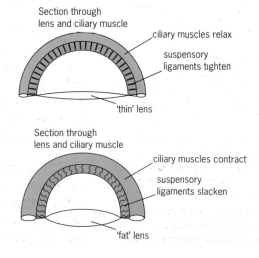

Section through lens and ciliary muscle
ciliary muscles relax
suspensory ligaments tighten
'thin' lens

Section through lens and ciliary muscle
ciliary muscles contract
suspensory ligaments slacken
'fat' lens

on the ligaments that hold the lens in place, and the lens becomes 'fatter'. The focal length changes, and the image is focused. When we then look at an object much further away, the ciliary muscles relax, again automatically. The tension on the ligaments supporting the lens increases, and the lens becomes 'thinner' allowing our vision to **accommodate**.

PROBLEMS WITH THE LENS SYSTEM

In some individuals the lens and cornea do not focus correctly (Fig 23.15). If you are short sighted, a condition known as **myopia**, you focus light from an object in front of the retina. This produces a blurred image. Short sight can result from an elongated eyeball or a thickened lens. It is corrected by using a **diverging** or **concave** lens.

If you are long sighted you are said to have **hypermetropia**, and you focus light behind the retina. Again, objects appear blurred. Long sightedness results from a shortened eyeball or a lens that is too thin, and is corrected using a **converging** or **convex** lens.

A more complicated visual defect is **astigmatism**. In astigmatism, the surface of the cornea is irregular, the object appears blurred because some of the light rays are focused and others are not (this is similar to the distortion produced by a wavy pane of glass). Astigmatism is corrected using a **cylindrical** lens that bends light rays in one plane only.

? TEST YOURSELF

D Imagine you are standing at a bus stop reading a book. What happens **(a)** to the ciliary muscle, **(b)** to the lens and **(c)** to the pupil of your eye as you look up to see the bus approaching in the distance?

Short sight (myopia)

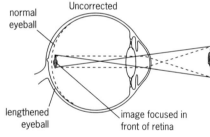

Long sight (hypermetropia)

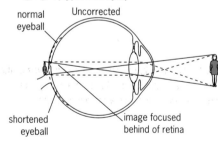

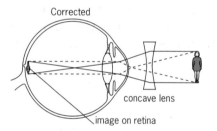

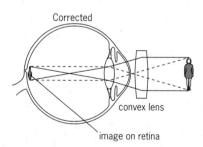

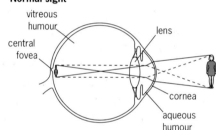

Normal sight

Astigmatism

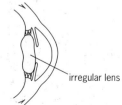

Fig 23.15 In some people, an image forms in front of the retina (this is myopia) or behind the retina (this is hypermetropia). In these cases the eyeball is either too long or too short. Astigmatism can result from an irregular cornea or an irregular lens

BINOCULAR VISION

Humans and some other mammals, notably primates and predators such as cats, have **binocular vision** (Fig 23.16). We have eyes at the front of our heads, so the field of vision seen by each eye is similar but not identical. Normally, the brain is able to compute the incoming signals so that we see only one image.

Fig 23.16 Primates have both binocular and colour vision. Binocular vision allows animals to judge distance. This is vital when jumping from branch to branch. It is thought that colour vision allows food to be chosen with greater accuracy; many fruits use colour to signal when they are ripe. Humans, being primates, owe our vision to our ancestry

Binocular vision gives us a big advantage. When two sets of signals are decoded by the brain, we get the impression of distance and depth and of objects being three-dimensional. This **stereoscopic vision** allows animals to judge distance and depth accurately. It is easy to see why primates need binocular vision; when living in the trees, any animal that cannot judge distance is likely to fall sooner rather than later.

Predators often have their two eyes placed centrally to maximise this overlap of retinal images (Fig 23.17). It gives them excellent stereoscopic vision to judge distances accurately and locate their prey. Prey species, on the other hand, tend to locate their eyes to the side of the head. They have reduced stereoscopic vision but an all-round vision that allows them to spot nearby predators, even if they try to sneak up from behind.

Fig 23.17 This Sri Lankan Leopard has eyes at the front of its head so it can focus on its prey. This deer has eyes on the side to spot predators

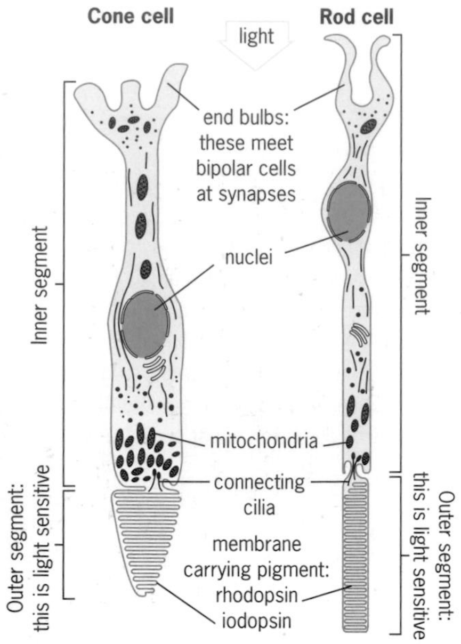

Cone cell   light   **Rod cell**

end bulbs: these meet bipolar cells at synapses

Inner segment

nuclei

Inner segment

mitochondria

connecting cilia

membrane carrying pigment: rhodopsin iodopsin

Outer segment: this is light sensitive

Outer segment: this is light sensitive

Fig 23.18 Rods and cones: the two types of light-sensitive cell in the human eye

? TEST YOURSELF

E It has long been known that a deficiency of vitamin A causes the condition of night blindness. Give an explanation for this.

? TEST YOURSELF

F Which cones are stimulated by **(a)** blue light at 490 nm, **(b)** violet light below 440 nm? (Look at Fig 23.19.)

VISION AT THE MOLECULAR LEVEL

When light reaches the back of the retina it is focused on to the **photoreceptors**, cells that are specialised to detect light. These cells contain pigment molecules that are bleached by light. Bleaching of the pigment results in a generator potential and, when the threshold is reached, action potentials are created and nerve impulses begin to travel towards the brain.

The two photoreceptors in the human eye are the rods and cones (Fig 23.18 and Table 23.3).

Table 23.3 Comparing rods and cones		
Feature	**Rod cells**	**Cone cells**
Shape	rod-shaped outer segment	cone-shaped outer segment
Connections	many rods connect with one bipolar neurone	only a single cone cell per bipolar cell
Visual acuity	low	high
Visual pigment(s)	contain rhodopsin (no colour vision)	contain three types of iodopsin, responding to red, blue and green light
Frequency	120 million per retina – 20 times more common than cones	7 million per retina – 20 times less common than rods
Distribution	found evenly all over retina	found all over retina but much more concentrated in the centre, particularly the yellow spot or fovea centralis
Sensitivity	sensitive to low light intensities (used in dim light)	sensitive to high light intensities (used in bright light)
Overall function	vision in poor light	colour vision and detailed vision in bright light

PHOTORECEPTION IN ROD CELLS

Rods contain a reddish-purple compound called **rhodopsin**, or visual purple. This consists of two joined compounds – **opsin**, a protein, and **retinal**, a light-absorbing compound derived from vitamin A. When retinal is exposed to light and absorbs light energy, it changes shape (it is actually converted into an isomer) and this causes the two compounds of rhodopsin to break apart. The free opsin acts as an enzyme which sets in motion a series of biochemical events that leads to the hyperpolarisation of the rod cell surface membrane, that is, it becomes more negative. This generates action potentials in the connecting nerve cells. Nerve impulses then pass to the brain for decoding. In the absence of further light stimulation, the retinal molecule goes back to its original shape and then recombines with opsin to form rhodopsin. It is worth remembering that this resynthesis process takes time.

Rhodopsin is very sensitive to light and so rods can function effectively in dim light. In bright light our rhodopsin is broken down quicker than it can be re-formed, so that most of it is bleached for most of the time. In this state we are said to be light adapted. If we enter a dark room from a brightly lit one we cannot see anything because much of our rhodopsin is bleached. However, as we resynthesise the pigment and become dark adapted, things become clearer and we are able to see in dim light, even if it is only in shades of grey.

PHOTORECEPTION IN CONE CELLS

The function of the cones is to see colour and detail, although they can only do this in bright light. In dim light they do not function at all. Cone cells contain the pigment **iodopsin**, a photosensitive pigment made up of **photopsin**, a different protein from that found in rods, and retinal. The events that occur in cone cells stimulated by light are basically the same as those that occur in rod cells. There are, however, three different types of cone.

Each one contains a slightly different pigment and responds to a different wavelength of light. One responds to red light, one to green light and one to blue light, so allowing us to see in colour (Fig 23.19).

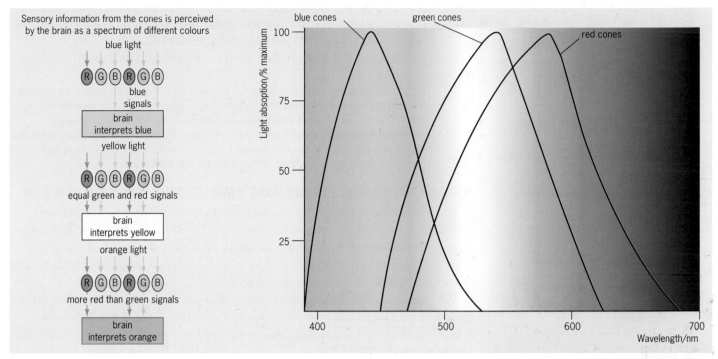

Fig 23.19(a) There are three types of cone responding (optimally) to three wavelengths of light. The colour we perceive depends largely on the proportion of the stimulation of the three types. If a single type of cone is missing, the individual will not be able to distinguish between certain colours – red–green is the commonest – and is said to be colour blind

The colour that we 'see', or more accurately **perceive** (interpret in our brain), depends on which cones are stimulated. When all the cones are fully stimulated we see 'white'. When very few are stimulated we see black. Stimulation of separate types produces red, green or blue, and all colours in between are produced by combinations of different levels of stimulation of the three types together. This model of how three different types of cone can produce the range of colour vision that we experience is known as the **trichromatic theory of colour vision**.

VISUAL ACUITY: THE ABILITY TO SEE DETAIL

From what distance can you read a newspaper? Can you read the bottom line of the standard optician's eye-test board? Tests such as these are a measure of **visual acuity**, which is defined as the ability to resolve detail. For instance, two lines such as these ══════════ will obviously appear as two lines, but if you prop the book up and walk backwards, a point will be reached when they look like one grey line.

An interesting feature of visual acuity is that we only have a very narrow field of accurate vision. Focus on any particular word on this page and you see it in detail, but those on either side are unclear. We see only a very small proportion of our field of vision in detail – and that is the part that is focused on the fovea, and therefore that stimulates the cone cells.

Overall, rods are more sensitive than cones. Rods can function in poor light, but they cannot perceive detail; you can't read small print in dim light. The cones have a higher visual acuity than the rods, but they need bright light. Why is there such a difference? The answer lies largely in a phenomenon known as **retinal convergence**.

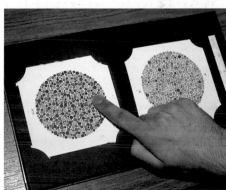

Fig 23.19(b) Colour blindness can be diagnosed by means of Ishihara tests. A person with normal vision will see the pink and red loop, while a red–green colour-blind person will not.

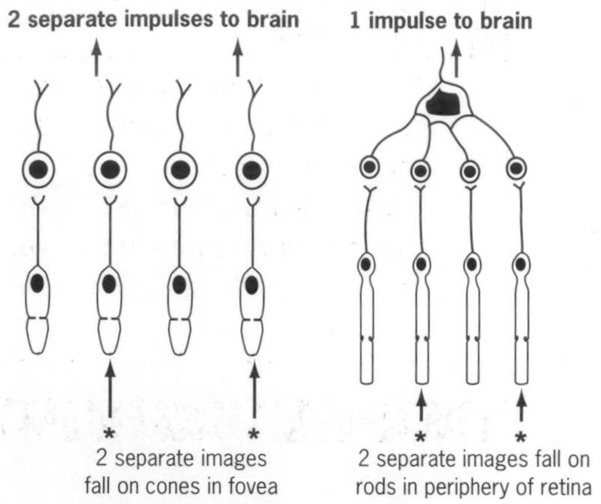

2 separate impulses to brain   **1 impulse to brain**

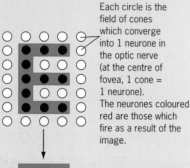

2 separate images fall on cones in fovea

2 separate images fall on rods in periphery of retina

RETINAL CONVERGENCE

Figs 23.20 and 23.21 illustrate retinal convergence. There are over 120 million rods and about 7 million cones, but only about 1 million neurones in the optic nerve. Clearly, each rod and each cone cannot all have their own neurone. In fact, many rods converge into one neurone, while only a few cones are associated with each neurone. In the centre of the fovea it is thought that each cone actually has its own neurone.

Fig 23.20 The concept of retinal convergence. Generally, the further out you go from the centre of the fovea to the edge of the retina, the greater the degree of convergence. Two separate stimuli (such as two dots or lines) may stimulate two separate cones, but not the one in between. Therefore, the brain will see these as two separate images. The same two stimuli falling on the rods will not stimulate separate neurones, because a group of rods all feed into the same neurone. In this case, the brain will perceive only one stimulus, and see only one line or dot

TRANSMISSION OF NERVE IMPULSES TO THE BRAIN

If you look back at Fig 23.10, you can see that the photoreceptor cells are linked to a set of nerve cells in the retina called bipolar cells. These link with a second type of nerve cell called ganglion cells. It is the axons of ganglion cells that take information from the eyes to the brain. Ganglion cell fibres are bundled together to form the optic nerve.

We cannot see an object whose image falls on the retina at the point where axons of ganglion cells leave to form the optic nerve. Since it contains no receptor cells, any light striking this small area is not sensed; hence its name, the blind spot (Fig 23.22).

Fig 23.22 To demonstrate the blind spot, hold your book at arm's length with the two symbols straight in front of your eyes. Close your left eye and concentrate on the cross with your right eye. Bring the book slowly towards you. Keep looking at the cross on the left: eventually, as the dot falls on the blind spot of your right retina, the image of the dot on the right will disappear

(a) When image falls on cones in fovea

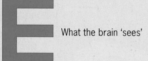

Each circle is the field of cones which converge into 1 neurone in the optic nerve (at the centre of fovea, 1 cone = 1 neurone). The neurones coloured red are those which fire as a result of the image.

What the brain 'sees'

(b) When same image falls on rods in periphery of retina

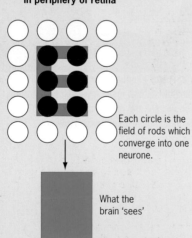

Each circle is the field of rods which converge into one neurone.

What the brain 'sees'

Fig 23.21 Why can we see in detail only when we look straight at something? Taking the letter E as an example, it can be seen that if the image falls on the cones it will stimulate some neurones but not others. However, if it falls on the rods it will simply stimulate lots of adjacent neurones, and all we will see is a splodge, not a clear letter E

SEE QUESTION 3

HOW THE BRAIN PERCEIVES AN IMAGE

Action potentials do not differ very much from one another and so there are only two possible ways in which stimuli going into the brain can be coded. One is the rate at which action potentials arrive and the other is their destination in the brain.

The rate at which sensory neurones fire tells the brain the strength of the stimulus (see page 350): specific neurones tell the brain which part of the retina has been stimulated. The brain analyses this information to assess the nature of the original stimulus. This process, called **visual processing**, is immensely complex.

Specific areas of the cerebral cortex are associated with different sensory functions (see Chapter 20). The **visual cortex** in the **occipital lobes** deals with visual information (Fig 23.23). However, body senses must not be considered in isolation. Sensory and motor systems interact within the control regions of the CNS, and sensory information is processed at various points along the nerve pathway before it reaches the brain. In the brain it is processed further with information from other senses and stored memories.

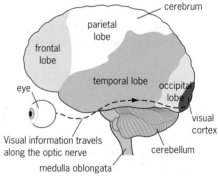

Fig 23.23 The location of the visual cortex in the human brain

4 HEARING AND BALANCE

The ear is a miniature **receiver**, **amplifier** and **signal-processing system**. It is divided into three parts – an **outer ear**, **middle ear** and **inner ear** – and is connected to the brain by the **auditory nerve** (Fig 23.24).

Sound waves pass through the outer and middle ear and into the inner ear through the **oval window**. Pressure on the oval window squashes the fluid in the inner ear, and compression waves travel through the canals of the **cochlea**. The round window eventually receives these movements of fluid and dampens them down (this prevents them from being reflected back into the cochlea) by bulging into the middle ear. Sensitive hair cells in the cochlea fire off signals, sending action potentials along the auditory nerve to the brain.

Fig 23.24 The structure and function of the human ear

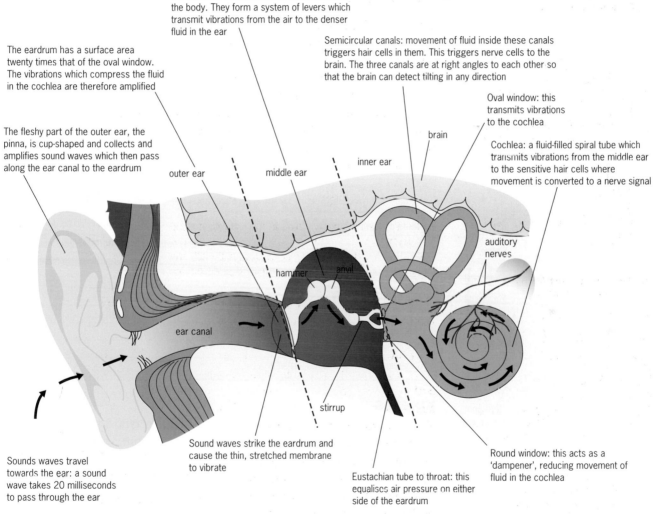

Ear ossicles: these are the smallest bones in the body. They form a system of levers which transmit vibrations from the air to the denser fluid in the ear

Semicircular canals: movement of fluid inside these canals triggers hair cells in them. This triggers nerve cells to the brain. The three canals are at right angles to each other so that the brain can detect tilting in any direction

Oval window: this transmits vibrations to the cochlea

The eardrum has a surface area twenty times that of the oval window. The vibrations which compress the fluid in the cochlea are therefore amplified

brain

Cochlea: a fluid-filled spiral tube which transmits vibrations from the middle ear to the sensitive hair cells where movement is converted to a nerve signal

The fleshy part of the outer ear, the pinna, is cup-shaped and collects and amplifies sound waves which then pass along the ear canal to the eardrum

outer ear   middle ear   inner ear

auditory nerves

hammer   anvil

ear canal

stirrup

Sounds waves travel towards the ear: a sound wave takes 20 milliseconds to pass through the ear

Sound waves strike the eardrum and cause the thin, stretched membrane to vibrate

Eustachian tube to throat: this equalises air pressure on either side of the eardrum

Round window: this acts as a 'dampener', reducing movement of fluid in the cochlea

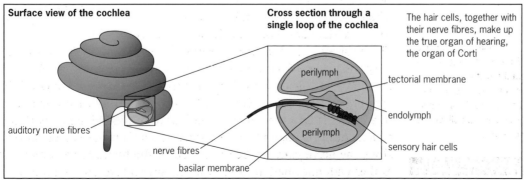

Surface view of the cochlea

Cross section through a single loop of the cochlea

The hair cells, together with their nerve fibres, make up the true organ of hearing, the organ of Corti

perilymph

tectorial membrane

endolymph

perilymph

auditory nerve fibres

nerve fibres

basilar membrane

sensory hair cells

5 BEHAVIOUR

SIMPLE BEHAVIOURS

One of the simplest types of behaviour is **orientation**. This enables organisms to move in order to find favourable environments and to avoid unfavourable ones. Animals such as insects show two types of simple orientation:

● **Taxes**: movements in a specific direction that are directed by a stimulus such as light or food.
● **Kineses**: more random movements that are not directed by a stimulus.

Taxes and kineses are usually regarded as **reflexes**. Reflexes also occur in higher animals – including mammals and humans; these are covered in more detail in Chapter 20.

TAXES

In many orientation behaviours, the animal moves directly towards or directly away from a stimulus. Maggots (fly larvae) have very simple eyes and can respond to the direction of light (Fig 23.25). Honey bees move towards a scent which is associated with food. These types of directional response are called taxes (singular: taxis).

A positive taxis involves movement towards a stimulus, a negative taxis involves movement away from the stimulus. As Table 23.4 shows, taxes are also classified according to the type of stimulus involved (light, water, gravity) or to the type of behaviour shown (straight-line responses, turning movements).

> ✓ **REMEMBER THIS**
>
> A reflex is a rapid, automatic respose to a specific stimulus. It is a built-in response: it does not have to be learned (see Chapter 20).

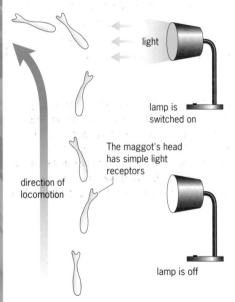

Maggot shows avoidance reaction when light is turned on

light

lamp is switched on

direction of locomotion

The maggot's head has simple light receptors

lamp is off

The maggot has to move its head from side to side to compare light levels on either side of its body

Fig 23.25 The maggot has simple eyes that can distinguish changes in light intensity but it cannot tell from which direction the light is coming. It moves its head from side to side to judge on which side the light is brightest

SEE QUESTION 5

Table 23.4 Summary of some common taxes. These are normally classified either by the type of stimulus that triggers them or by the type of response they elicit		
Name of taxis	**Type of stimulus**	**Type of response**
Telotaxis		fixing on a stimulus without scanning from side to side
Klinotaxis		moving head repeatedly from left to right to judge stimulus intensity
Geotaxis	gravity	
Thermotaxis	heat	
Phototaxis	light	
Anemotaxis	air currents	
Rheotaxis	water currents	
Chemotaxis	chemicals, eg salts	

KINESES

Woodlice are crustaceans, the group that includes crabs, shrimps and lobsters. Woodlice are the only truly terrestrial crustaceans, but are prone to desiccation (drying out). Thus they need dark, damp conditions where they are also safe from predators and can find rotting vegetation to eat.

If placed in dry, light conditions, woodlice will move rapidly (by their standards), changing directions frequently. They don't move in any particular direction, which is why this behaviour is classed as a kinesis. They can't see or use any other senses to locate dark and damp places, but their behaviour means that sooner or later they have a good chance of finding the conditions they require. When they find dark, damp conditions, their movements are much reduced. Some common types of kineses are shown in Table 23.5.

Table 23.5 Some common kineses. Like taxes, these are normally classified either by the type of stimulus that triggers them or by the type of response they elicit		
Name of kinesis	**Type of stimulus**	**Type of response**
Photokinesis	light intensity	
Chemokinesis	chemical gradients	
Hygrokinesis	humidity levels	
Klinokinesis		turns faster when stimulus is more intense
Orthokinesis		moves with greater speed when stimulus is more intense

MORE COMPLEX BEHAVIOURS

We can divide all animal behaviour into two categories: **instinctive behaviour,** or **innate behaviour,** which is inherited, and **learned behaviour**, which is modified by experience.

INSTINCT

There are many examples where a piece of behaviour appears without any previous experience. A herring gull chick automatically pecks at the red spot on its parent's bill to get it to regurgitate food. The digger wasp, *Ammophila*, builds a nest in the sand and immobilises a caterpillar using precisely the right amount of poison injected into exactly the right spot, without any prior knowledge. We describe this sort of behaviour as instinctive.

Charles Darwin was the first to propose a definition of instinct in terms of animal behaviour. Natural selection, he argued, applies just as much to behaviour as it does to body parts and so behaviour can be inherited. Those animals that do not behave in the right way (eg finding the right food or looking for a mate) do not pass on their genes to the next generation.

But the classical, ethological view of instinct was put forward by Konrad Lorenz and Niko Tinbergen. They said that **motivational energy** – the drives and urges such as hunger, thirst and sex – controlled complex sequences of behaviour, which they called **fixed action patterns** (Fig 23.26). These behaviours, which include nest-building, courtship and parental behaviour, are automatic and occur in response to specific stimuli called **releasers**.

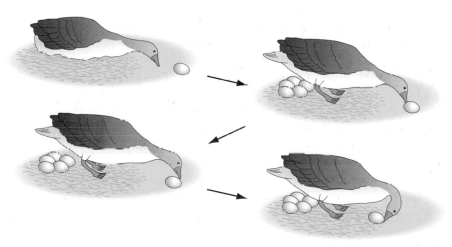

Fig 23.26 Egg rolling behaviour in the greylag goose is a good example of a fixed action pattern. The parent bird retrieves an egg that has been moved out of the nest in a set sequence of movements. If the egg is taken away in the middle of the retrieval, the goose carries on moving an 'imaginary' egg, until the set sequence is complete

LEARNED BEHAVIOUR

Learning is behaviour that is modified by experience. A 3-day-old chick eventually learns to peck at food, not at stones on the ground, and a bee learns to visit those flowers with most nectar. Such behaviours are useful because they allow animals to respond to changing conditions.

Some features of the environment, such as gravity and seasonal changes, are highly predictable. An animal does not need to learn how to respond to these situations: to do so would waste energy that could be used for other things. Animals learn:

- to avoid predators;
- to avoid harmful environments;
- to find food;
- to avoid harmful food;
- to find a suitable mate;
- to recognise important individuals within the group;
- to find their way home.

LEARNED RESPONSES

Fig 23.27 gives a simple classification of learned responses. The following section looks in more detail at some of these responses.

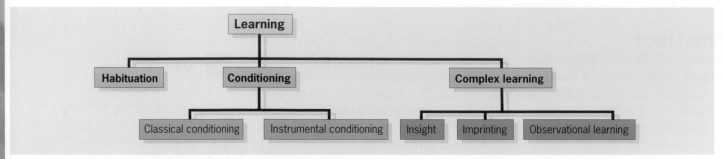

Fig 23.27 Learned responses in animals

HABITUATION

Habituation can be described as the process whereby an animal becomes accustomed to non-threatening environmental stimuli and learns to ignore them. This learning process occurs with most effect in the first few weeks of life; it is part of the socialisation and survival of most mammals. Wolf cubs, for example, must learn to recognise other wolves in their pack as non-threatening, but still be aware of the danger of wolves from other packs or from other animals such as snakes. They become habituated to other members of the pack when this response is at its most potent. After a few weeks the response is reduced, just as they start to venture out beyond the safety of their own parents and group. They cannot become habituated as easily and so are wary and fearful of things they have not come across before.

Habituation is also important for domestic animals, such as the dog. This can be illustrated by these three examples of what can happen to a puppy in its first 10 weeks of life, and how this affects how it develops into an adult:

- A puppy is kept apart from its mother, given adequate food and shelter, but little contact with other dogs, or with humans. At 6 months, a dog like this would be fearful of people and new situations, and would not interact well with other dogs, even its own siblings.
- A puppy is kept with its mother and siblings for 10 weeks, with regular human contact and food and shelter. The dog at 6 months is likely to be well adjusted; able to interact well with other dogs, and be a good family pet.
- A puppy exposed to cruel treatment, poor conditions and physical violence in its first few weeks tends to become withdrawn and unpredictable. It does not have a fearful response when further mistreatment occurs, and becomes withdrawn. It may attack a person making a friendly approach.

These are examples only, but indicate the importance of habituation in early life. Manipulation of this response can, however, be useful in some circumstances. Dogs bred for future use as police or army dogs are often exposed to loud bangs during their first weeks of life; this can enable the dog to become 'bomb-proof' as an adult dog, able to cope with loud, frightening conditions such as a riot, without panicking or endangering the safety of the public.

CONDITIONING

Conditioning is another type of learning. An animal is able to detect a stimulus and then to predict what is likely to happen, simply because it has happened several times before. The animal associates two events that occur together. For example, cattle respond to the farmer entering the field in a tractor, but ignore other people with other vehicles. They associate the farmer and tractor with food.

In his original experiments on conditioning, the Russian physiologist Ivan Pavlov (1849–1936) recorded saliva produced by hungry dogs that were given food. Pavlov delivered a neutral stimulus, the sound of a bell, at the same time as the food. The dogs began to connect the neutral stimulus with food and, after a short time, the animals would salivate at the sound of the bell, whether food appeared or not. We now call this type of learning **classical conditioning** and the response it produces a **conditioned reflex**. It is found in all complex animals.

Since Pavlov's initial research, many different physiological responses such as fear, dislike, love, hunger and sleepiness have been conditioned in response to a wide variety of neutral stimuli.

The other type of conditioning, **operant conditioning**, was first described by B. F. Skinner. Operant conditioning is learning as a result of consequences. Humans, and other animals, tend to prefer behaviours that are followed by some type of **reinforcement**, and they avoid behaviours that are followed by some type of **punishment**. Reinforcement can be either positive – getting something good – or negative – losing something bad. Reinforcement can occur only sometimes in connection with a particular behaviour (partial reinforcement), or every time that behaviour is followed (continuous reinforcement). Continuous reinforcement is generally the most effective at bringing about learning.

Punishment can be effective in reducing unwanted behaviours, particularly if it is administered soon after the unwanted behaviour, consistently, and either not too weakly or extremely in relation to the behaviour (ie the punishment must fit the crime). However, too much reliance on punishment as a tool to modify behaviour can be counterproductive in the long run because it doesn't teach new behaviours (it only eliminates old ones), it models aggression and it creates a negative association between the person doing the punishing and the punishment itself.

COMPLEX LEARNING

Not all learning is the result of conditioning or habituation. When a banana is left just out of reach outside its cage, a chimp might use a nearby bamboo pole to pull in the fruit. Or, if a banana is tied to the branch of a tree, again out of reach, the chimp may eventually stack boxes one on top of another and climb up to reach it. This behaviour appears to involve reasoning and we call it **insight learning**. If one chimp watches the behaviour of another, and then copies it, this is **observational learning** (see page 431).

It is well known that birds and mammals form an attachment to people or animals that they come into contact with shortly after hatching or birth. This behaviour is called **imprinting** and it allows young animals to learn the characteristics of its parents. This is particularly important where young animals need to learn quickly who their parents are. Young ducks, for example, follow their parents to the water soon after hatching.

SOCIAL BEHAVIOUR

There is safety in numbers. Schools of fish are less vulnerable to predators than single fish: large numbers tend to confuse a predator. Living in a group may also give benefits in terms of feeding success – for example, flocks of birds eat better than single individuals because group members can 'share' tasks such as keeping a lookout when feeding. Birds such as geese are also more likely to find food when there are many pairs of eyes to look for it. We call this type of group behaviour **social behaviour**.

REMEMBER THIS

Classical conditioning occurs when an animal learns to associate a 'neutral' stimulus with an important one: the dog associates the bell ringing with the arrival of food.

Instrumental conditioning occurs when an animal learns to associate an action with a 'punishment' or a 'reward': a bird learns that eating a particular butterfly makes it sick.

? TEST YOURSELF

H A cow touches an electric fence and gets a shock. Thereafter it avoids the fence. Is this an example of classical or instrumental conditioning?

SEE QUESTIONS 6 AND 7

! **SCIENCE FEATURE** The work of Wolfgang Kohler and Jane Goodhall

Kohler's chimps

A hundred years ago, people thought that only human beings made and used tools but we now know that many animals, including birds, make and use tools. One of the first scientists to show this was Wolfgang Kohler, a psychologist who studied learning in chimpanzees. Kohler, who trained at the University of Berlin, was working at a primate research facility maintained by the Prussian Academy of Sciences in the Canary Islands when the First World War broke out. Marooned there, he had at his disposal a large outdoor pen and nine chimpanzees of various ages. Kohler decided to make the best of this situation, and set up a series of experiments to test his theory that chimps were capable of insight learning, just as humans are.

Kohler worked mainly with four chimps; Chica, Grande, Konsul and Sultan. He placed a chimp in an enclosed play area. Somewhere out of reach he placed a prize, such as a bunch of bananas. To get the

Fig 23.28(a) Sultan making a tool

bananas, the chimp would have to use an object as a tool. The objects in the play area included sticks of different lengths and wooden boxes. He discovered that chimpanzees were very good at using tools. They used sticks as rakes to pull in bananas, and as clubs to bring down fruit that was hanging overhead. Sometimes they stood long sticks on end and quickly climbed up the 7 metres and grabbed the bananas before the stick fell over. The chimpanzees also learned to use boxes as stepladders, dragging them under the hung bananas and even stacking several boxes on top of one another. Kohler concluded that his chimps were showing insight learning – they were looking at a problem, at what was available to solve it, and then thinking up a strategy before carrying it out.

Fig 23.28(b) Grande making a 4-storey tower

Fig 23.28(c) Jane Goodall

Jane Goodall

Another scientist, Jane Goodall, has since observed chimpanzees making and using tools in the wild. As a young woman, Goodall worked on a project in Africa, under the direction of Louis Leakey, the famous archaeologist and anthropologist. As part of this study, she observed a chimp that picked a blade of grass and carefully trim the edges. The chimp then stuck the grass into a termite mound, left it there for a moment, and pulled it out. Termites swarmed over the blade of grass. He then ate the termites clinging to the grass blade; this had been made into a 'fishing rod' for termites.

In a society individuals cooperate (Fig 23.29). Animals take on individual tasks, a practice called **division of labour**, and there are complex systems of communication involving recognition of either family or other species members. Social organisation can range from the simple cooperation between a male and a female during mating to the complex societies of insects and primates.

Group living also has its costs. In a large bird colony, for example, males run the risk of females mating with other males. Females run the risk of male desertion. In mixed-species flocks such as Australian finches, there can be increased competition for food and in any large group there is always the risk of contracting disease or parasites.

Fig 23.29 Prairie dogs live in large groups in the plains of North America, in tunnels up to 30 metres long. They show complex social behaviours, in which different members specialise to do different tasks – such as sentry duty and pup-sitting. The advantages of group living for the prairie dog outweigh the disadvantages – the increased likelihood of catching fleas, and the greater chance of getting into a fight

OBSERVATIONAL LEARNING

One advantage of living in groups is that animals gain information from one another. Young animals, such as juvenile mammals and birds, learn skills from their parents by watching and observing. On the Japanese island of Koshima, macaque monkeys learned to wash dirty sweet potatoes originally provided for them by researchers in the early 1950s (Fig 23.30). This behaviour was passed on from parents to offspring and was

observed by other members of the troop. After a few months, all the monkeys were washing their potatoes in the sea. Nearly 50 years later, this colony of monkeys still wash their potatoes even though they are not given dirty potatoes any more. Scientists think they like the salty taste.

Closer to home, blue tits and great tits have learned the 'trick' of opening milk bottle tops to get at the cream inside (Fig 23.31). These birds already had the habit of hammering and pecking at nuts with their beak. It was therefore a relatively small step to use the same technique on attractive, shiny bottle tops. This behaviour did not spread randomly but seemed to spread from certain focal points, suggesting that birds learned by watching each other.

In both these examples, **observational learning** occurred. The behaviour spread rapidly through the population until it became the norm. We call this phenomenon **cultural transmission**.

HUMAN BEHAVIOUR

Consider two examples of human behaviour:

- If children see pictures of small children's faces and adult faces they prefer the adult face. A change in preference occurs at 12 to 14 years of age in girls when they express a preference for baby pictures. Boys display the same trend two years later.
- If we sit too close to a stranger in a library then we tend to use physical 'barriers' such as books or bags to keep the stranger at a distance. If this person gets too close we often move seats.

These are typical of the types of behaviour studied in human ethology – the biology of human behaviour. It asks the question 'why do we behave the way we do?'

Ethologists view human behaviour in the same way as that of other animals: they think it is adaptive and has evolved to suit its purpose. In the two examples above, there is strong evidence to link children's behaviour with child rearing and adult behaviour with protection.

One goal of human ethologists is to look for patterns of behaviour shared by all peoples of the world. These are seen in examples of behaviour which involve body movements, rather than speech. Gestures such as smiling and raising the eyebrows, and behaviours such as grooming and hugging seem to be common to all human societies.

We all think that we have a free choice in our behaviour, but do we really? We may be under the control of our genes or be shaped by our surroundings. There is evidence for both points of view.

Fig 23.30 Japanese macaques learn how to wash potatoes in the sea to remove soil. This behaviour is passed on to other members of the troop by observational learning

Fig 23.31 A great tit removes the top from a milk bottle before drinking the cream

SUMMARY

After completing this chapter you should know that:

- A **sensation** is a general state of awareness of a **stimulus**; **perception** involves the **interpretation** of **sensory information**. A **receptor** is a **biological transducer**: it converts energy from one type of system (eg chemical) to another system (eg electrical). A receptor is classified according to the type of stimulus it receives (eg photoreceptor), its location (eg interoceptor) or its level of complexity.

- The **Pacinian corpuscle** is a simple receptor in the skin that detects pressure and vibration.

- The eyeball contains three outer layers, the **sclera**, **choroid** and **retina**.

- Incoming light rays are **refracted** (bent) by the cornea and by the **lens** of the eye.

- **Accommodation** is the ability to focus the eye automatically, to see near and distant objects.

- **Rod** and **cone** cells are stimulated by the effects of light on two pigments, **rhodopsin** and **iodopsin**. Breakdown of these compounds alters the electrical properties of the cell membranes.

- There are three types of cone cell pigment. Each responds to a different range of wavelengths of light. Humans have **trichromatic** vision: our cone cells detect three colours, blue, green, red.

- **Stereoscopic vision** enables us to perceive depth and distance.

- The human ear has two main functions, to detect sound and to maintain balance.

- Ethology is the study of animal behaviour under natural conditions.

- Patterns of behaviour are controlled by the nervous system; simple behaviours include **taxes** and **kineses**.

- A taxis is an orientation response related to the direction of the stimulus. A positive taxis describes movement *towards* a stimulus; a negative taxis, movement *away* from a stimulus.

- A kinesis is an orientation behaviour where the response relates not to the direction of the stimulus, but to its intensity.

- **Instinct** or **innate behaviour** is unlearned. It has survival value.

- There are three main types of learning: **habituation** (learning not to respond), **conditioning** (associating a particular stimulus with a particular response) and **complex learning**.

- Animals live in social groups for a number of reasons including protection, food capture, reproduction, shelter. They show cooperative behaviour.

- Behaviour can be passed on from parents to offspring genetically (innate behaviour, eg **fixed action patterns**) and culturally (eg **observational learning**).

? EXAM QUESTIONS

1   Chitons are small animals that live on the seashore. When the tide is out they are found on the lower surfaces of stones. When stones are turned over the chitons move to the new lower surface as shown in the diagram.

a) Suggest **two** advantages to the chitons of this response.

b) Give **two** factors, other than light, to which the chitons might be responding.

c) A student investigated the response of chitons to light. Three covered dishes, X, Y and Z, were arranged as shown in the diagram. One half of the top of each dish was painted black, the other half was transparent. The dishes were placed on a table outside at noon on a bright cloudy day. Ten chitons were placed in the light half of each dish. The number of chitons in each half of the dishes was recorded every five minutes for the next hour.

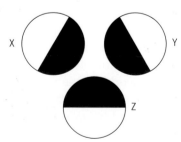

(i) Suggest why the dishes were arranged as shown in the diagram.

(ii) The results of the investigation are shown in the table.

❓ EXAM QUESTIONS

Time/ minutes	Number of chitons					
	Dish X		Dish Y		Dish Z	
	Light	Dark	Light	Dark	Light	Dark
0	10	0	10	0	10	0
5	9	1	7	3	8	2
10	6	4	7	3	7	3
15	5	5	7	3	6	4
20	3	7	5	5	5	5
25	3	7	4	6	5	5
30	3	7	3	7	5	5
35	3	7	3	7	5	5
40	2	8	3	7	4	6
45	1	9	3	7	4	6
50	1	9	2	8	4	6
55	1	9	3	7	2	8
60	1	9	1	9	1	9

What conclusions may be drawn from these results?
(iii) Explain how kinesis could account for the results shown in the table.

AQA (NEAB) BY04 Biological Basis of Behaviour June 2000 Q1

2
a) A person, who has been watching a distant view from a window, looks down and starts to read. Describe how a clear image is formed on the retina of this person's eye, and explain the changes that take place in the eye in order to produce a clear image when the person starts to read.
b) Explain the roles of the eye and brain in enabling humans to distinguish colours in a picture.

AQA (NEAB) BY04 Biological Basis of Behaviour March 1999 Q8

3 The diagrams show sections through the front of an eye during distance vision and near vision.

distance vision

near vision

a) Describe the mechanisms which bring about the changes to the lens and pupil as the eye changes from distance vision to near vision.
b) The diameter of the pupil may also change when light intensity changes. This is caused by a reflex action which involves the brain rather than the spinal chord. Describe the route taken by the nerve impulses between the receptor and the effector in this reflex action.

AQA (NEAB) BY04 Biological Basis of Behaviour June 2000 Q7

4 The following structures are all found in the mammalian ear. Explain concisely the function of each of the structures.
a) Tympanic membrane
b) Maleus, incus and stapes
c) Basement membrane (basilar membrane) in the organ of Corti
d) Tectorial membrane in the organ of Corti
e) Auditory nerve
f) Maculae in the walls of the utricle and saccule
g) Semicircular canals
h) Sensory hair cells in the cochlea

OCR 6918 Animal Studies March 2000 Q3

5 An experiment using choice chambers was arranged to determine the environmental conditions favoured by woodlice. Choice chamber A was divided to provide two environments. Choice chamber B was divided to provide four environments. In each of the choice chambers, 20 woodlice were placed at random and they were allowed to move freely. Their positions were recorded at intervals of ten minutes.

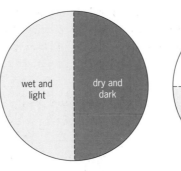
Chamber A

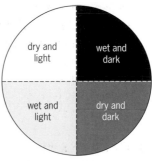
Chamber B

? EXAM QUESTIONS

The table below shows the results of the experiment.

Time/ minutes	Chamber A			Chamber B		
	wet and light	dry and dark	dry and light	wet and dark	dry and dark	wet and light
0	9	11	5	4	6	5
10	10	10	2	11	4	3
20	9	11	0	17	2	1
30	11	9	0	20	0	0

a) **(i)** Identify which two conditions are preferred by the woodlice.
(ii) State which chamber provided inconclusive results.
(iii) State the biological term for the response shown by the woodlice when they move in response to differences in humidity.

b) Predict what difference in the behaviour of the woodlice would have been observed in Chamber A and Chamber B if the experiment had continued for 3 hours. Give a reason.

IBO Biology Higher Level Paper November 1998 QE2

6 The diagram shows apparatus used to investigate behaviour in rats. When a rat presses the lever a food pellet drops into the food tray.

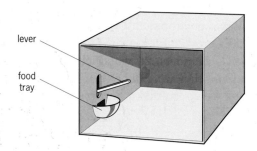

The graph shows typical results from these investigations.

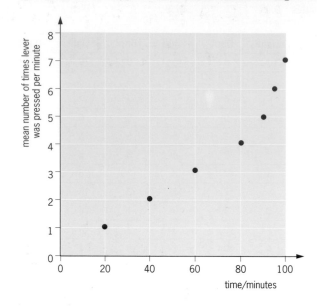

a) What type of learning is shown by the rat?
b) Explain why the rate of lever-pressings changes during the investigation.
c) Explain how similar principles may be used when training a dog.

AQA (NEAB) BY04 Biological Basis of Behaviour June 2000 Q3

7
a) **(i)** Describe one example of imprinting.
(ii) Give one biological advantage of the imprinting that you have described in part **(i)**.
(iii) Explain why imprinting is considered to be learned rather than innate behaviour.

b) Birds can be a danger to aircraft. One method used to keep them away from airfields is to play recordings of alarm calls which the birds make in response to predators. This works well for a time, but then becomes less effective.
(i) Explain why the alarm calls become less effective after a time.
(ii) Name the behavioural process involved.

AQA (NEAB) BY04 Biological Basis of Behaviour June 1999 Q2

Genetics: the basics

Genetics: the basics

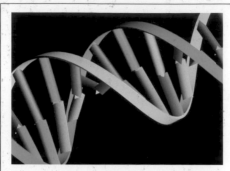

If you reproduce, your DNA may well live longer than you do. This is a computer model showing the structure of DNA. The 'backbones' of the double helix (shown here as ribbons) are made from alternating phosphate and sugar groups connected to atoms of carbon in pentose molecules. Between the two backbones are the bases: adenine, guanine, cytosine and thymine

Why are we all similar? Why are we all different? These are two of the fundamental questions on which the science of genetics is based. When the sperm from your father met the egg from your mother, a unique individual was created, different from anyone who had gone before or will ever exist again.

But would it surprise you to learn that you have quite a lot of genes in common with bacteria, lettuce, cockroaches, tapeworms and ferrets? Certain essential genes – such as those that code for the enzymes of respiration – are very similar in all organisms as they are vital to this basic life process. Such genes must have been copied faithfully, generation after generation, for the last three billion years or so. It is also intriguing to realise that, although almost all organisms die young before reproducing, not a single one of our ancestors did. Their DNA must have had what it takes to be passed on. Well done Granny and Granddad, and Great Granny and Great Granddad.

1 WHAT ARE GENES?

Genes are the 'instructions' that control everything that happens inside a cell. Physically, genes are different sized stretches of DNA that code for the manufacture of proteins such as enzymes. Enzymes organise the chemistry that goes on in our bodies and are behind all the life processes, including growth, repair and reproduction. Throughout our lives, the genes remain in the nuclei, directing the activities of each of our cells.

Children grow and develop according to the genes they have inherited from their parents. It's a harsh fact, but nothing lives forever: the inevitable fate of all organisms, often sooner rather than later, is death. However, although we die, our genes live on in our offspring.

✓ REMEMBER THIS

In organisms that reproduce sexually, a set of genes from the male combines with one from the female to form a unique individual (see Chapter 26).

Fig 24.1 The film *GATTACA* presents us with a world where the privileged classes are genetically modified at birth so that all the potentially harmful genes are deleted; 'It's you, but it's the best of you.' This isn't such an unbelievable vision of the future. What do you notice about the letters used to make the title *GATTACA*?

2 STUDYING GENETICS WITH THIS BOOK

Studying genetics means learning about DNA and what it does. In this section of the book, we divide our study of this remarkable molecule into five areas:

- The **cell cycle** is the study of how cells divide. Discovering just how a single fertilised egg can develop into an organism as complex as a human is one of the central challenges of biology. When the control of cell division goes wrong, it can lead to cancer. Find out more about mitosis and meiosis, the two types of cell division, in Chapter 25.
- **Molecular biology** is the branch of science that deals with all aspects of the DNA molecule: its structure and function. How are genes used to make products? Why are some genes active when others aren't? How can DNA copy itself? How do mutations occur? This is covered in Chapter 27.
- **Genetics** is the study of inheritance: the way in which genetic information is passed on from one generation to the next. Why are only some genes passed on? Why are some characteristics always shown while others are hidden? This is covered in Chapter 28.
- **Evolution** looks at how species change and develop with time. What is the mechanism of this change? How long does the process take? Can we observe evolution in action? What do we know about the evolution of humans? This is covered in Chapter 29.
- **Genetic engineering** is the popular name for the technology that manipulates the DNA molecule. How can we change DNA? How can we transfer DNA between organisms? What are the benefits of genetic engineering? Can we cure genetic disease? Should we pursue this technology at all? This is covered in Chapter 30.

3 THE POTENTIAL OF GENETICS

Few scientific subjects make the headlines as often as advances in genetics and DNA technology. We are witnessing an accelerating revolution that is likely to have far-reaching effects. We have mapped the complete human genome and are now embarking of the daunting task of finding out where the genes are, what they do and how they interact with each other. We may have the 'book of humans', but it's huge and written in a foreign language.

The implications of this progress are enormous:

- Two out of every three people die for reasons connected to their genes. Although only a small percentage inherit a lethal genetic disease, many more inherit a *tendency* to develop a condition such as heart disease, premature senility or cancer. We call this a **genetic predisposition**. An in-depth knowledge of the human genome will make it possible for doctors to screen babies before or soon after birth for genetic abnormalities and tendencies to develop such diseases. But this raises ethical and moral questions, not just scientific ones: would you want to know just how likely you are to develop cancer in middle age? Would you want anyone else to know – life insurance companies, potential employers, friends or partners? Who would have access to your DNA files?

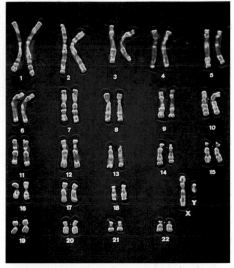

Fig 24.2 A complete set of human chromosomes. Each body cell contains two sets of 23 chromosomes and each chromosome contains up to 4000 genes. The total amount of DNA in a cell is called the genome. The human genome consists of about 30 000 genes

✓ REMEMBER THIS

Chromosomes condense and become visible during cell division, as we see in more detail in Chapter 25.

Normal cell division – mitosis – duplicates the DNA and then divides it accurately into two so that the two new cells formed are genetically identical.

Meiosis is the type of cell division that produces haploid cells such as eggs and sperm. This process halves the amount of genetic material that goes into each new cell and shuffles it at the same time so that no two sex cells are the same.

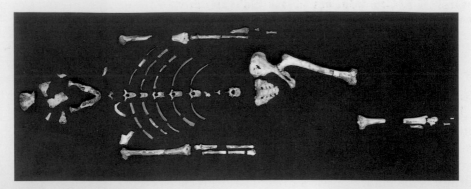

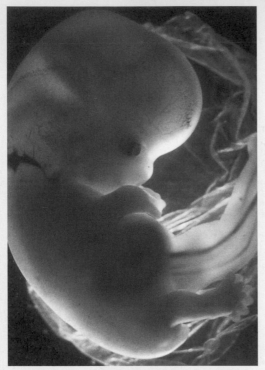

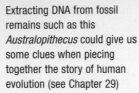

Extracting DNA from fossil remains such as this *Australopithecus* could give us some clues when piecing together the story of human evolution (see Chapter 29)

A human embryo 7 weeks after conception. We have developed the technology to clone animals as advanced as humans, but this area of genetics is an ethical minefield

Wheat has been bred selectively to produce a greater yield. Many new types are easier to harvest and are also resistant to disease

Fig 24.3 The applications of our knowledge of DNA

This fermenter houses genetically engineered bacteria which contain the human growth hormone gene. Inside the fermenter the bacteria multiply and produce large amounts of human growth hormone

Male and female sperm can be separated according to the amount of DNA they contain. The sperm are stored in liquid nitrogen until needed. Farm animals can be artificially inseminated to produce offspring of the required sex

- We can isolate genes for useful products, such as insulin and growth hormone, and place those genes into another organism. This is **recombinant DNA technology** (see Chapter 6). Bacteria already produce human insulin and other products on a large industrial scale, and it is also possible to transfer such genes into higher animals such as sheep so that they produce milk containing the required protein.

- We can isolate faulty genes, such as those which cause cystic fibrosis, and replace them with multiple copies of healthy genes. Although still at the experimental stage, this **gene therapy** holds much promise for the future.

- We could be able to halt some of the genetic causes of ageing, and allow more people to live even longer.

Only the bravest scientists would dare to speculate about where our genetic knowledge might eventually take us. It could be to a greatly improved world where suffering is reduced while quality of life and life expectancy are improved. Genetic engineering and agricultural techniques might give us food-producing species that can live in the harshest of conditions, so providing food in areas of most need.

But can we be this optimistic? There are still many things we don't know about living organisms and it is impossible to predict exactly the effects of interfering with an organism's DNA. Already, companies are trying to patent genetic material and its products, leading to hot debate about whether or not such natural products can be owned, like a brand name or an invention. Will genetic advances be used solely to make money, and therefore be available only to those who can pay?

The ethical implications of some very recent developments also need to be considered. The first mammals have now been cloned but the process is unreliable. This has led to a heated debate about whether or not we can, or should, clone humans.

4 WHAT GOES ON IN THE NUCLEUS?

The nucleus of every human cell contains DNA (Fig 24.4). Each nucleus contains a set number of long, elaborately coiled DNA molecules that condense into chromosomes. Genes, the individual instructions, are dotted along the chromosomes.

In an adult organism most cells are not dividing: the chromosomes are uncoiled, and active genes are producing proteins to control the activities of the cells. Not all genes are active at the same time: controlling which genes are active and which are not enables different cells to carry out different functions. For instance, all cells in the human body contain two copies of the insulin gene. Only in certain specialised cells in the pancreas, however, are the genes switched on and used to make insulin.

GLOSSARY OF GENETICS TERMS

Like many branches of science, genetics comes with its own jargon, seemingly designed to make some very straightforward ideas inaccessible to the average person. You will need to know the following terms:

Diploid: cells or organisms containing two sets of genes. Human body cells, for instance, are all diploid.

Haploid: cells or organisms containing one set of genes. Human sex cells (egg cells and sperm) are haploid.

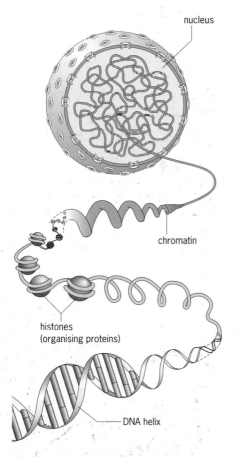

Fig 24.4 Most of the DNA in the cell is locked away in the nucleus where, in effect, it forms a reference library. The genes, which act as instructions for all cell functions, must be passed on when the organism reproduces

Gene: a length of DNA that codes for a particular polypeptide. Some proteins consist of more than one polypeptide; these are coded for by more than one gene.

Allele: one form of a gene. There may be two or more alleles of any particular gene. Diploid organisms contain two alleles, one on each pair of chromosomes. For example, in pea plants, the gene for flower colour may have a red allele and a white allele.

Genotype: the genetic make-up of an individual.

Phenotype: the observable characteristics an organism possesses. For example, in peas the genotype Aa may result in a phenotype of red flowers. Many alleles are expressed, or used, and these contribute to the characteristics of the individual. Other genes – recessive alleles in the presence of a dominant version – are not expressed and so remain hidden.

Dominant: the allele that, if present, is expressed. If the red allele in the pea plant is dominant, the pea has a red flower, even if it has a white allele as well.

Recessive: the allele that is expressed only in the absence of the dominant. In the pea, if the white allele is recessive, the plant has white flowers only if it has two white alleles.

Homozygous: when both alleles of a particular gene are the same. In the pea example, an individual would be homozygous if it had two red alleles or two white alleles. We usually give letters to denote alleles, for example A for the dominant allele and a for the recessive allele. Individuals with genotype AA or aa are homozygous.

Heterozygous: the two alleles of a gene are different. In a pea plant, a heterozygous individual could have a red allele and a white allele, so their genotype would be shown as Aa.

5 VARIATION: WHY IS EVERYBODY DIFFERENT?

The group of people in Fig 24.5 shows what we all know: that everyone is different. Height, weight, intelligence, personality, eye colour, size of feet – the list is endless. No two people have the same combination of features. Even identical twins show subtle differences. So what makes us the way we are?

Fig 24.5 Like most species, humans show great variation between different individuals

Each human baby is born with a unique set of genes – we call this its **genotype**. However, only some of these genes affect the observable characteristics of the organism – its **phenotype**. Recessive genes, for example, remain hidden, masked by dominant versions.

The other factor that contributes to the phenotype is the environment – the circumstances in which the organism is brought up affects the way in which genes are expressed. For example, humans are born with a set of genes that give us the potential to grow to a certain height. We cannot realise this potential, however, without an adequate diet. Two hundred years ago the average height was several centimetres shorter than it is now: the difference is due to improvements in nutrition and general health care.

WHY IS VARIATION NECESSARY?

The simple answer is: survival. Variation in a population has great survival value because it greatly increases the chance that at least some individuals will adapt to changing conditions. To illustrate this, consider a **clone**, a population of genetically identical individuals. All is well when conditions are favourable, but when a serious disease comes along it is very likely that all of the organisms in the population will die. In a more varied population that has developed as a result of sexual reproduction and variation, there is a much greater chance that some of the individuals could be resistant to the disease. These individuals survive to breed, ensuring that the population, and perhaps the whole species, continues to exist.

This 'survival of the fittest' is more correctly called **natural selection**. In the long term, natural selection, acting on the variation in a population, is the main driving force behind the process of **evolution** – the subject of Chapter 29.

HOW VARIATION OCCURS

Variation arises because individual genes **mutate**. **Mutations** are changes in the genetic material of an organism, often caused by faults in the copying of the DNA (see Chapter 28). Once there is some variation to work on, the differences are maximised by the process of sexual reproduction. Using humans as an example we can see how this happens.

Firstly, gametes are produced by meiosis. This is a special type of cell division (see Chapter 25) which not only produces haploid cells, but shuffles the genes so that all eggs or sperm made by a particular individual are different. In meiosis, there are two key events that produce variation:

- **Crossover during prophase I**. Blocks of gene are swapped between chromosome pairs, bringing some new genes together and separating others.
- **Independent assortment during anaphase I**. Either chromosome from each pair can pass into the gamete. As a human cell has 23 pairs of chromosomes, 2^{23} different combinations are possible.

Secondly, following meiosis, **fertilisation** joins gametes from two separate individuals to form a genetically unique zygote from which the new diploid organism develops.

REMEMBER THIS

The phenotype of an organism = its expressed genes + effects due to the environment that it is exposed to.

REMEMBER THIS

A population is a group of interbreeding individuals of a particular species in a particular place (see Chapter 29). A species usually consists of many different populations. Some endangered species become particularly vulnerable because they consist of just one isolated population, restricted to a particular area.

CONTINUOUS AND DISCONTINUOUS VARIATION

Table 24.1 Examples of continuous and discontinuous variation

Continuous	Discontinuous
Most dimensions in animals and plants such as height, weight, length of bones, number of leaves	Ability to roll tongue in humans
Tolerance to adverse conditions, such as cold, heat, dehydration	Ability to taste phenylthiourea
Human fingerprints	Flower colour in peas
	ABO blood group in humans
	Coat colour in cats

There are two main types of variation in a population: **continuous variation** and **discontinuous variation** (Table 24.1):

● Continuous variation is so-called because the factors, or variables, can be any value inside a given range. Examples of continuously varying characteristics include height and weight in humans (Fig 24.6a), and the number of leaves in plants.

● Discontinuously varying features do not display a range: they are either one thing or the other. The ABO human blood group system is a classic example of discontinuous variation: people are either A, B, AB or O. They cannot be anything in between. Height in pea plants is another example: pea plants can be either tall or dwarf (Fig 24.6b).

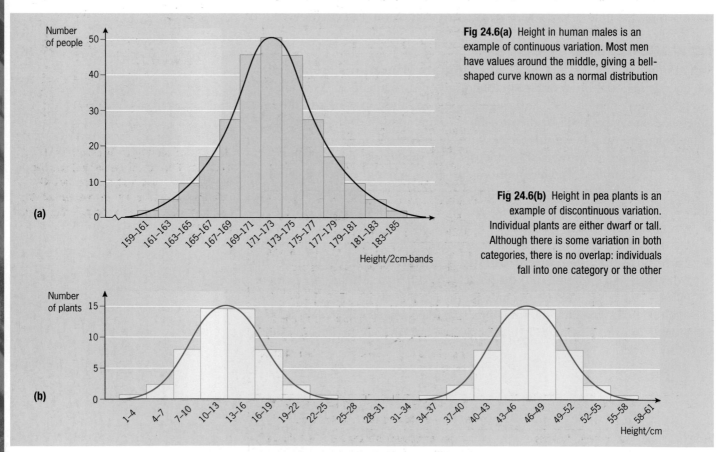

Fig 24.6(a) Height in human males is an example of continuous variation. Most men have values around the middle, giving a bell-shaped curve known as a normal distribution

Fig 24.6(b) Height in pea plants is an example of discontinuous variation. Individual plants are either dwarf or tall. Although there is some variation in both categories, there is no overlap: individuals fall into one category or the other

Often, continuous variation is **quantitative** (you can put a numerical value on it) and it is **polygenic** (controlled by several genes). Discontinuous variation is more often qualitative (you can describe it, but not measure it) and is controlled by one or only a few genes. In Chapter 28 we look at the principles of inheritance using examples of discontinuous variation.

6 A BRIEF HISTORY OF GENETICS

The history of genetics gives an insight into the changing nature of scientific progress. Early scientific discoveries and breakthroughs were few and far between, often made by individuals who observed and studied nature and put forward radical theories to explain what they had seen. Mendel (page 505) and Darwin (page 526) are classic examples.

More recent discoveries have been made by teams of dedicated people working together, using the scientific method and latest technology to build on knowledge accumulated by others. Table 24.2 outlines some of the landmarks in this most dynamic of scientific areas.

Table 24.2 Some landmark discoveries in genetics. In many of these, the scientists mentioned by name led a much larger team of scientists who worked towards the discovery described

Year	Scientist(s)	Discovery
1858	Charles Darwin, Alfred Russel Wallace	Jointly announced their theory of evolution by natural selection: 'Survival of the fittest'
1859	Charles Darwin	Published his book *The origin of species*
1866	Gregor Mendel	Published his laws of genetics following studies on pea plants. His findings were ignored
1900	Hugo de Vries, Carl Correns, Eric von Tshermak	Discovered meiosis, providing an explanation for Mendel's laws: his work is rediscovered by all three and gains recognition
1905	Nettie Stephens, Edmund Wilson	Independently discovered the principles of sex determination: XX = female, XY = male
1910	Thomas Hunt Morgan	Proposed the theories of sex linkage, mutation, linkage and chromosome maps following work on *Drosophila* (fruit fly)
1928	Fred Griffiths	Proposed that some 'transforming principle' had changed a harmless strain of bacteria into a lethal one. The hunt for DNA began
1941	G. W. Beadle, E. L. Tatum	Irradiated the bread mould *Neurospora*, producing mutations which suggested that genes code for enzymes
1944	Oswald Avery	Purified the 'transforming principle' in Griffiths' experiment, showing it to be nucleic acid (DNA)
1950	Erwin Chargaff	Discovered that the base pairing ratios in DNA were always the same, whatever the organism (Chargaff's principles: A = T, C = G)
1951	Rosalind Franklin	Obtained high-quality X-ray diffraction studies of DNA, showing that it has a helical structure
1952	A. D. Hershey, M. Chase	Used bacteriophages to show that DNA, not protein, is the material of heredity
1953	Francis Crick, Jim Watson	Built on the work of Chargaff and Franklin to work out the 3D structure of DNA

? TEST YOURSELF

B Look at Fig 24.6(b) on the previous page. A pea plant is 34 cm in height. Is it a tall or a dwarf plant?

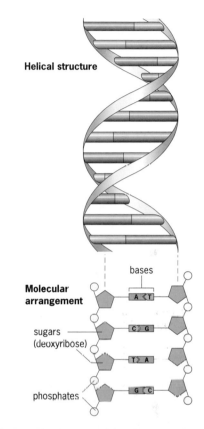

Helical structure

Molecular arrangement

bases

sugars (deoxyribose)

phosphates

Fig 24.7 The discovery of the structure and function of DNA is one of the greatest scientific discoveries of the twentieth century. The DNA molecule stores genetic information and, with the help of enzymes, makes exact copies of itself, time after time

Table 24.2 Some landmark discoveries in genetics. In many of these, the scientists mentioned by name led a much larger team of scientists who worked towards the discovery described (continued)

Year	Scientist(s)	Discovery
1958	M. Meselson, F. W. Stahl	Used radioisotopes of nitrogen to prove the semi-conservative mechanism of DNA replication
1958	Arthur Kornberg	Purified DNA polymerase from *E. coli* and used it to make DNA from nucleotides in a test tube
1961	F. Jacob, J. Monod	Propose a mechanism for switching genes 'on' or 'off': the operon hypothesis of metabolic control
1966	Marshall Nirenberg, Gobind Kharana	Cracked the genetic code: particular triplets of bases on DNA code – via mRNA – for the 20 amino acids
1970	H. O. Smith, K. W. Wilcox	Isolated the first restriction enzyme, *Hind*II, which can cut DNA
1972	Paul Berg, Herb Boyer	Produced the first recombinant DNA molecules
1973	Annie Chang, Stanley Cohen	Showed that DNA could be inserted and cloned inside a bacterium
1977	Fred Sanger	Developed a method for sequencing DNA
1977		The first genetic engineering company (Genentech) is founded, using recombinant DNA to make pharmaceuticals
1983	James Gusella	Located the gene responsible for Huntington's chorea on chromosome 4 (see Fig 24.2)
1984	Cary Mullis	Developed the polymerase chain reaction (PCR) in which minute samples of DNA can be copied, prior to analysis
1984	Alec Jeffreys	Developed DNA profiling (or 'fingerprinting')
1988		The Human Genome Project began to map the human DNA sequence
1989	Francis Collins	Identified the gene (*CFTR*) responsible for cystic fibrosis on chromosome 7
1990	French Anderson	First attempts at gene therapy: T cells of a 4-year-old girl were exposed to viruses containing working copies of her defective gene. After treatment, her immune system began working again
1994		Genetically engineered 'Flavr savr' tomatoes went on sale
1995		Transgenic sheep made to express human genes in their mammary glands, so that they produce milk containing valuable pharmaceuticals
1997	Roslin Institute, Edinburgh	First mammal cloned. Dolly the sheep fuels controversy about genetic research
1997		Dolly the cloned sheep gives birth to lambs conceived normally.
2000		Entire human genome announced
2001		First stem cells extracted from cloned human embryo Gaur, an almost extinct bovine species, cloned using a cow as a surrogate mother Pigs cloned for first time, bringing the possibility of pig organs for transplant into human patients a step closer Researchers use cow stem cells to construct a complete cow kidney in the laboratory
2001	Texas A&M University	First cat ('cc') cloned
2002		Scientists in the UK were given the go-ahead to create human embryo clones under strictly controlled conditions by a House of Lords select committee

25

Cell division

Cell division

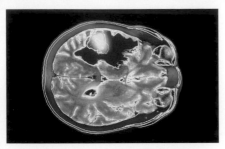

A magnetic resonance imaging (MRI) scan of a section through the brain of a 42-year-old woman. A tumour, coloured yellow, can be seen in the left hemisphere. This is a **metastatic** tumour; it has spread from another tumour elsewhere in the body

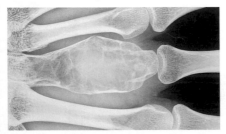

Coloured X-ray of hand showing a tumour growing on the bone on one finger

In a healthy human adult, cells divide only when they should. Some cells, such as those that line the gut, are replaced at a remarkable rate. Other cells, such as muscle cells, live longer and need to be replaced far less often. Occasionally, the systems that control cell division break down, and a cell that should be stable divides uncontrollably. Soon, a mass of tissue called a tumour forms. Tumours can be either benign or malignant.

A benign tumour is not cancerous. Cells divide within a small, confined area and the growth is often surrounded by a membrane. New cells form in the centre of the mass. This type of tumour can be dangerous if it presses on important organs, or if it grows very large. It is usually treated by surgery. Because the tumour cells are confined and do not invade other tissues, recovery and survival rates are good.

A malignant tumour is commonly known as cancer. The tumour grows at the edges, spreading out and invading the surrounding tissues (supposedly like a crab – hence the name cancer). This type of tumour is much more dangerous. Vital tissues and organs can be destroyed quickly and this can lead to death. Even if the tumour is removed from the body, actively dividing cells may have already broken off and set up secondary growths elsewhere in the body.

1 SOME BASIC PRINCIPLES

An understanding of cell division is of great importance in the study of biology. Our bodies grew and developed by cell division, and as adults we rely on the same process for maintenance and repair. When control of cell division is lost, the result can be cancer. A different type of cell division halves the chromosome number in eggs and sperm (the sex cells, or **gametes**) so that when fertilisation occurs, the resulting zygote has the normal double number.

Humans, like all animals, are eukaryotic organisms: in eukaryotic cells, DNA is organised into chromosomes and is enclosed within a double nuclear membrane. Before going on to study the mechanism of cell division in eukaryotic cells, familiarise yourself with some of the basic terms and concepts by looking at Table 25.1.

AN INTRODUCTION TO CELL DIVISION

Our bodies are made from two different types of cells; body (or somatic) cells and sex cells (or **gametes**). The nucleus of each body cell contains two sets of chromosomes and so has two sets of genes. Body cells are described as diploid (from Greek, meaning 'double number') (Fig 25.1).

Mitosis is the process of normal cell division. In mitosis, the chromosomes are copied and then divided equally between the two new daughter cells. So each mitotic division produces two cells, both diploid and each with exactly the same genes as the parent cell.

Table 25.1 Some basic terms important in cell division

Term or feature	What it is or what it means
Gene	A length of DNA that codes for the production of a particular polypeptide or protein
Genome	The name given to the full set of genes in a cell. The human genome consists of between 50 000 and 100 000 genes (the exact number is not yet known)
Diploid	Diploid cells contain two versions of every gene
Haploid	Haploid cells contain one version of every gene
Chromosome	A long single molecule of DNA, organised around proteins called histones. The largest human chromosomes contain about 4000 genes. Chromosomes exist in cells all the time but they can be seen only during cell division, when they condense and separate
Homologous chromosomes	Chromosomes exist in pairs – humans have 23 pairs. Homologous chromosomes have the same genes at the same positions, but not necessarily the same versions of each gene
Chromatid	During cell division, the DNA of a cell is replicated (copied). When the chromosomes condense, they therefore appear as double structures: each unseparated chromosome within such a pair is called a chromatid
Sister chromatid	When two chromatids are genetically identical (as in mitosis), they are called sister chromatids
Bivalent	A bivalent is a pair of homologous chromosomes that line up together, as they do during meiosis
Locus	The position of a gene on a chromosome
Transcription	The process in which a molecule of mRNA is assembled on an active gene. mRNA thus becomes a mobile copy of the gene
Translation	The process of converting the code on the mRNA into a protein. This is achieved by protein synthesis: amino acids are joined in a particular order to make a protein such as an enzyme

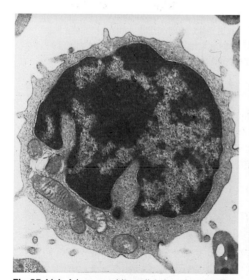

Fig 25.1(a) A human white cell (a lymphocyte). The nucleus of this cell contains all the genes necessary to make a whole new organism, a clone of the person from whom this cell was taken. Although the technology exists to clone humans, the area is fraught with huge moral, ethical and legal problems

Fig 25.1(b) Dolly the sheep, born in 1997, was the first ever mammal to be cloned from a diploid adult body cell

Fig 25.1(c) A micropipette puts the nucleus from an adult cell into a non-fertilised egg cell that has had its own nucleus removed

Fig 25.2 This human embryo is 4 days old, but the fertilised egg has already divided mitotically four times, producing a cluster of 16 cells. The cells will soon specialise to form the tissues and organs of the body

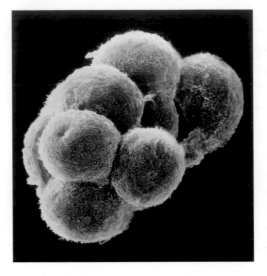

SEE QUESTION 1

Some cells in the human body are not diploid. Gametes, the sperm and eggs, contain only one copy of each gene as they have only one set of chromosomes. These cells are haploid and are produced by a special type of cell division called **meiosis**.

A male and female gamete join together at fertilisation, to form a new diploid cell called a **zygote**. This one cell divides by mitosis to produce a complete new organism (Fig 25.2). Reproduction is covered in Chapter 26.

SCIENCE FEATURE Cell turnover

As children, we grow because the cells in our bodies divide mitotically, and cell production outnumbers cell death. When we reach our adult size, our cell population stays relatively constant. The rate at which cells die and are replaced is known as **cell turnover**.

Research has shown that different tissues and organs have very different turnover rates. Brain cells, for instance, are not usually replaced, and after our 20s there is a slow but steady decline in number. This is not normally something to worry about – you won't run out. Other cells have a low turnover. Liver cells, for example, might divide only once every one to two years although, unlike the brain, they can regenerate if some cells are damaged or destroyed. Most of the high turnover cells are epithelial cells (Figs 25.3 and 25.4). These are the cells of the body that most frequently give rise to tumours. Cells of the bone marrow (Chapter 17) and cells in the testes (Fig 25.5) also divide very frequently.

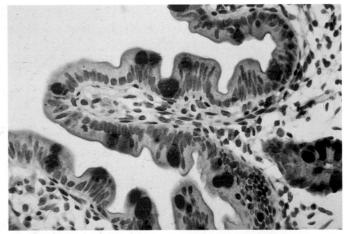

Fig 25.4 Intestinal epithelial cells have a very rapid turnover: the entire gut lining is replaced every couple of days. Much of the content of these cells is digested and reabsorbed in an efficient recycling system, but even so, gut cells form a significant proportion of faecal matter

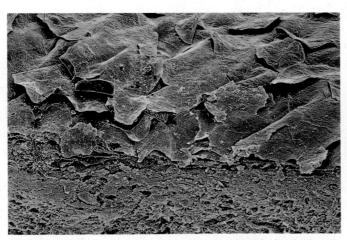

Fig 25.3 Scanning electron micrograph of human epidermis. Dead skin cells are constantly being lost and replaced by mitosis. A large proportion of household dust consists of human skin cells

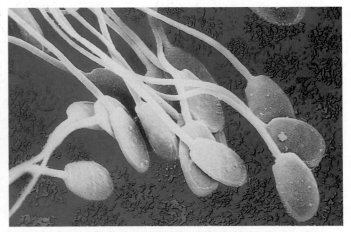

Fig 25.5 The production of sperm is a remarkably intense process. The average healthy human male produces about 1000 new sperm every second. The process of sperm production, spermatogenesis, which involves meiosis, is described in Chapter 26

2 THE CELL NUCLEUS

The DNA of a cell is contained in the nucleus. In a cell that is not actively dividing, DNA exists as **chromatin** (see Fig 3.42). This has a granular, or grainy, appearance (Fig 25.6). Separate chromosomes are present, but the DNA is so spread out that we cannot tell one from another.

When a cell is about to divide, the chromosomes condense and separate. They become visible under the light microscope as dark, rod-like structures. Each pair of chromosomes is known as a **homologous pair**. This means that they both contain genes at the same positions, or **loci**. You received one chromosome of each pair from your mother and one from your father. So, although homologous pairs contain the same genes, they do not necessarily carry the same versions of each gene. This is the key to understanding why organisms vary. Alternative forms of genes are called **alleles** (see Chapters 24 and 28).

The appearance, number and arrangement of chromosomes in the nucleus are referred to as the **karyotype**. The human karyotype consists of 23 pairs of chromosomes (Fig 25.7). All diploid human cells contain 23 pairs of chromosomes and this is often given the notation $2n = 46$. This means that a diploid cell has 46 chromosomes (a haploid cell has 23). Other organisms have different numbers of chromosomes (Table 25.2).

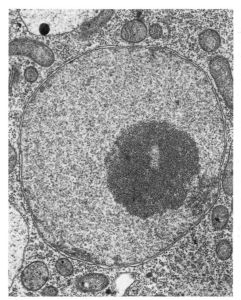

Fig 25.6 In a non-dividing cell, the chromosomes are spread out as chromatin and appear as pale granules. In this form, the DNA is used as a template to make mRNA in the process of transcription. mRNA passes to the cytoplasm and is itself used as a template for protein production in the process of translation. In this cell, the nucleolus is also clearly visible: it is the dark sphere inside the nucleus

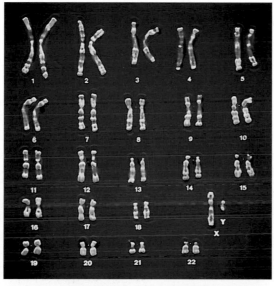

Fig 25.7 The 23 pairs of chromosomes in the human karyotype. Karyotypes are used to detect chromosomal abnormalities and to perform sex tests on athletes. The presence of XX confirms the athlete is female, the presence of XY shows the athlete is male

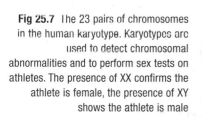
Table 25.2 The number of chromosomes in the cells of some plants and animals, compared with humans. Note that there is no connection between chromosome number and the complexity of an organism. Some ferns have hundreds of chromosomes

Species	Number of chromosomes
Penicillin mould (*Penicillium notatum*)	5
Broad bean (*Vicia faba*)	12
Lettuce (*Lactuca sativa*)	18
Yeast (*Saccharomyces cerevisiae*)	34
Cat (*Felis cattus*)	38
Human (*Homo sapiens*)	46
Potato (*Solanum tuberosum*)	48
Chimpanzee (*Pan troglodytes*)	48
Horse (*Equus caballus*)	64
Chicken (*Gallus gallus*)	78
Dog (*Canis familiaris*)	78

REMEMBER THIS

How can a potato have more chromosomes in its cells than humans do? It's quality, not quantity, that counts. All organisms have the same DNA but it's what the DNA codes for that makes the difference. When you translate potato DNA, you get a potato plant. When you translate human DNA, different proteins are made, and eventually a human being results.

3 THE STAGES OF THE CELL CYCLE

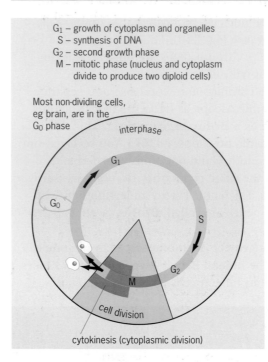

G$_1$ – growth of cytoplasm and organelles
S – synthesis of DNA
G$_2$ – second growth phase
M – mitotic phase (nucleus and cytoplasm divide to produce two diploid cells)

Most non-dividing cells, eg brain, are in the G$_0$ phase

interphase

G$_1$

G$_0$

S

G$_2$

M

cell division

cytokinesis (cytoplasmic division)

Fig 25.8 The four stages of the cell cycle: the three stages of interphase and the stage of active division in which mitosis occurs

The cell cycle is the complete sequence of events in the life of an individual diploid cell. The cycle starts and ends with cell division and consists of the stages of mitosis plus **interphase**, the interval between divisions in which the cell carries out its normal functions. Fig 25.8 shows the main stages of the life cycle of a cell.

Cells that do not divide, such as those in muscle and nerve tissue, are always in interphase: this is the normal state for a functioning cell. In interphase the DNA in the nucleus is unwound, and active genes are read to produce proteins such as enzymes.

A new cell has three options:

- It can remain stable, in interphase for many months or years. Brain and other nerve cells rarely divide, if ever.
- It can undergo mitosis within a short period of time. Skin cells and cells that line the gut all have a very rapid turnover (see the Science feature on cell turnover on page 448).
- It can undergo meiosis. Specialised germ cells in the ovary and testes include meiosis in their cell cycle and produce egg cells or sperm.

INTERPHASE

We start looking at the cell cycle at the point when a cell has just divided. The cell is now in interphase. Interphase has three distinct stages: **G$_1$**, **S** and **G$_2$**.

G$_1$ – THE FIRST GROWTH PHASE

Just after it has been produced by division of its parent, a new cell is in the early part of the first **g**rowth phase, **G$_1$**, sometimes called G$_0$. Cells that do not divide remain at this point in the cell cycle because they do not need to go further: they never replicate their DNA. 'New' cells are relatively small, with a full-sized nucleus but relatively little cytoplasm.

During G$_1$, protein synthesis starts and the volume of cytoplasm and the number of organelles increase rapidly. The process is actually quite complicated, not least because some organelles such as mitochondria (which have their own DNA) divide independently of the cell nucleus. In later G$_1$, the cell takes on more 'normal' proportions: the nucleus begins to look smaller as the surrounding cytoplasm increases in volume.

THE S PHASE

The **S** phase – DNA **s**ynthesis phase – follows G$_1$. In this phase, the cell's DNA **replicates** (copies itself). Predictably, a cell only enters the S phase if it is going to divide. The point at which DNA replication starts is called the **restriction point**. After this, the cell becomes **restricted**, or locked into an automatic sequence that moves inevitably on to cell division.

G₂ – THE SECOND GROWTH PHASE

Before the actual mitotic cell division (see below), the cell enters a second, shorter **g**rowth phase, **G₂**, in which the proteins necessary for cell division are synthesised.

MITOSIS

Mitosis is a continuous sequence of events but, for clarity, they are divided into four distinct stages: **prophase**, **metaphase**, **anaphase** and **telophase**. The stages of mitosis are illustrated in Fig 25.9. You will need to refer to this constantly as you read the text.

Fig 25.9 The stages of mitosis

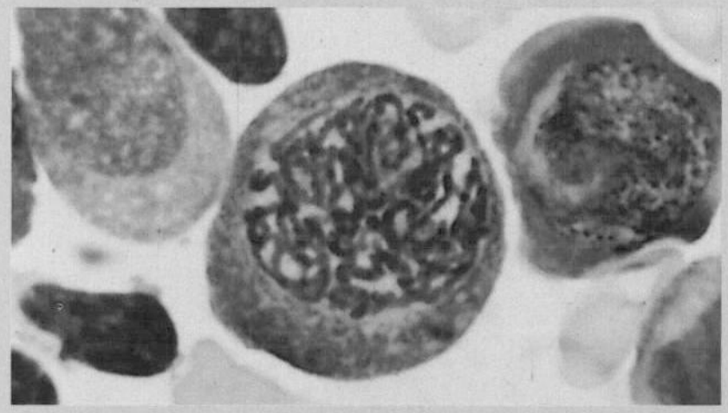

(a) **Early prophase**. The DNA has already replicated during interphase and the chromosomes begin to condense. Each chromosome is a double structure made from two genetically identical chromatids

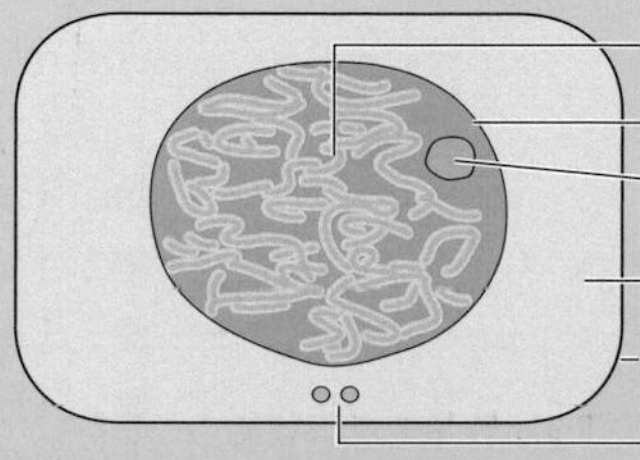

- chromatin threads
- nuclear membrane
- nucleolus
- cytoplasm
- cell surface membrane
- centrioles

(b) **Late prophase**. The nuclear membrane has disappeared and the spindle has started to develop

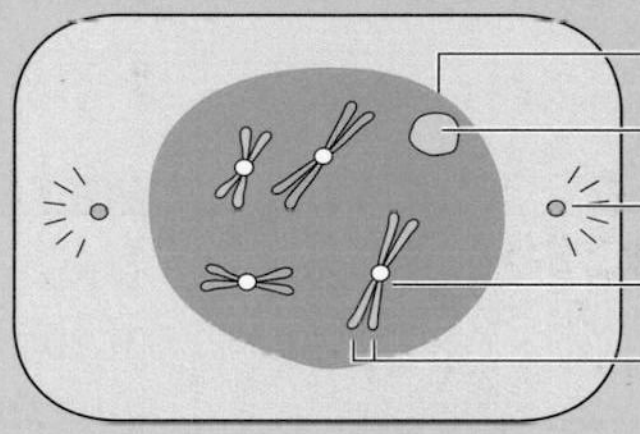

- nuclear membrane
- nucleolus
- centriole
- centromere
- pair of chromatids

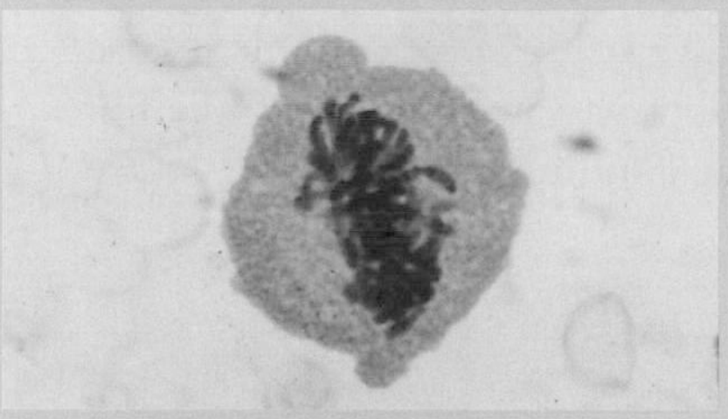

(c) **Metaphase**. The nuclear membrane has gone and the chromosomes arrange themselves on the equator (middle) of the spindle

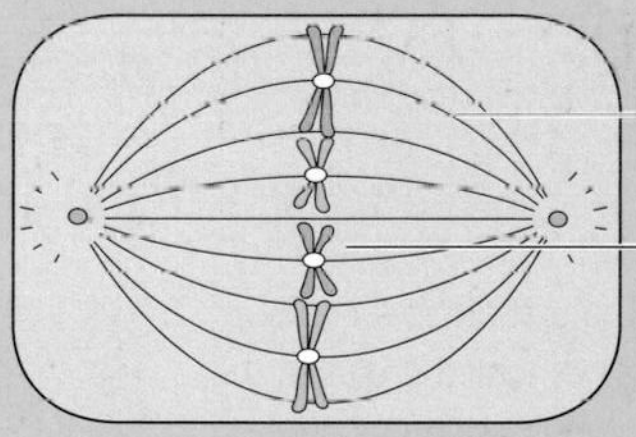

- spindle fibres (microtubules)
- centromeres on 'equator' of spindle

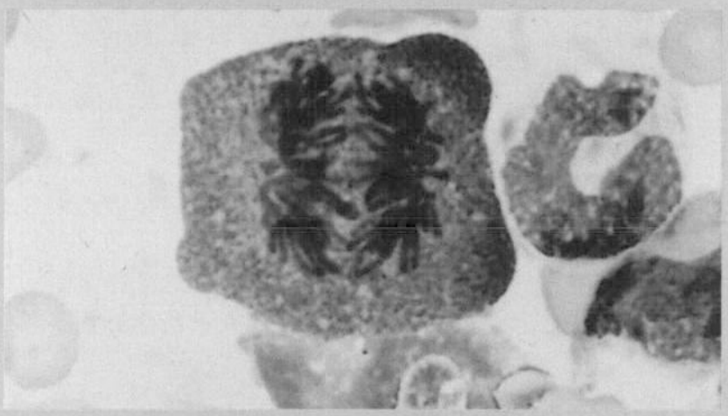

(d) **Anaphase**. The chromatids are pulled apart and move to opposite poles.

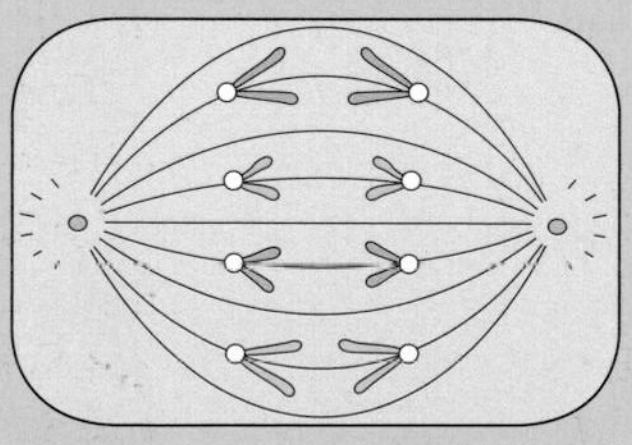

Daughter chromosomes move apart, led by their centromeres

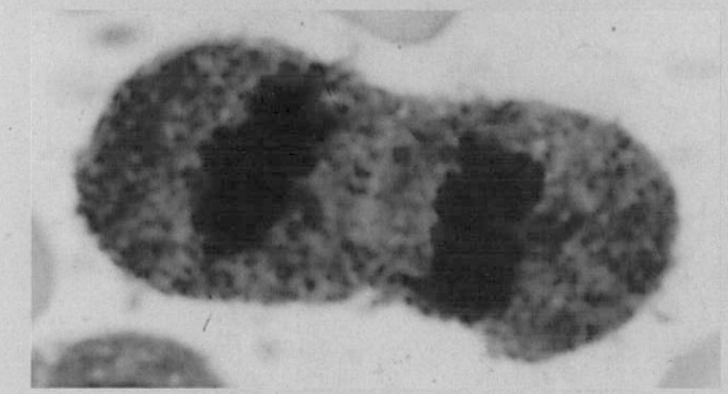

(e) **Telophase**. Cytokinesis (cytoplasmic division) is beginning. The chromosomes becomes no longer individually visible

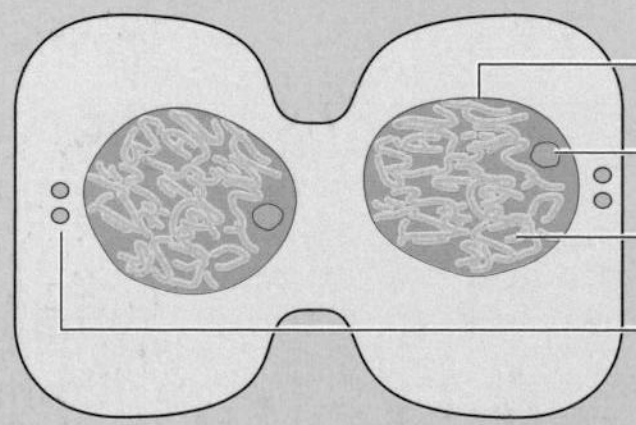

- nuclear membrane
- nucleolus
- chromatin threads
- pair of centrioles

(f) **Cytokinesis**. Cytoplasmic division is achieved

SEE QUESTIONS 2 TO 4

✓ **REMEMBER THIS**

Exam hint – Remember the mnemonic: **IPMAT** – **I**nterphase, **P**rophase, **M**etaphase, **A**naphase, **T**elophase.

The movement of chromosomes during cell division is controlled by microtubules (page xx), which form the **spindle**. In a non-dividing cell the microtubules are found as two bundles of fibres, the **centrioles**, in an area of cytoplasm known as the **centrosome**, or **microtubule organising centre (MTOC)**. During cell division the centrioles move to opposite sides of the nucleus, from where they form the spindle.

PROPHASE

Chromatin begins to condense into chromosomes: Fig 25.9(a). At this stage, each chromosome has replicated itself and now consists of two identical **chromatids**. These are known as **sister chromatids** and are joined by a **centromere**. Unless there has been a mutation (a fault in the DNA replication), these sister chromatids are genetically identical. The subsequent stages of mitosis organise and split the pairs of chromatids, so that one chromatid of each pair goes into each daughter cell.

As the chromosomes condense, other changes occur in the cell:
● The nucleolus begins to break down.
● The centrioles move to opposite sides of the nucleus.
● The centrioles begin to assemble the spindle.
● The nucleolus disappears.
● The nuclear membrane begins to break up.
From prophase onwards, most 'normal' cell activity, such as protein synthesis and secretion, is halted until division is over.

METAPHASE

The beginning of metaphase (meta = middle) is marked by the disappearance of the nuclear membrane, which breaks down into separate vesicles, moves into the surrounding cytoplasm and joins with the endoplasmic reticulum: Fig 25.9(b).

The spindle (Fig 25.10) becomes fully developed and fills the space that was occupied by the nucleus. Then the most obvious event of metaphase happens: the chromatid pairs attach themselves to individual spindle fibres and align themselves on the equator of the spindle.

Fig 25.10 The spindle is a cradle of microtubule fibres which organise the chromosomes during cell division

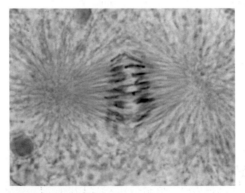
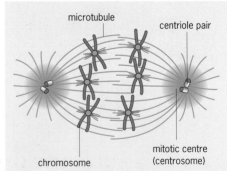

ANAPHASE

At the start of anaphase (ana = apart), as Figs 25.9(c) and 25.10 show, the chromatids are pulled apart by movements of the spindle fibres: sister chromatids are pulled to opposite poles. The newly separated chromatids are now called chromosomes, and are single structures. If you watch a film of mitosis (highly recommended for learning purposes), you will see that anaphase is the most obvious event.

TELOPHASE

In telophase (telo = final) the chromosomes reach the poles of the spindle: Fig 25.9(d). They then unravel and become indistinct, forming the familiar chromatin of interphase. The nuclear membrane re-forms and the nucleolus reappears. Soon afterwards, transcription resumes and the cell restarts protein synthesis, endocytosis and other normal cytoplasmic functions. This marks the end of mitosis.

CYTOKINESIS – DIVISION OF THE CYTOPLASM

In the events of mitosis just described, we have looked only at the splitting of the nucleus of the cell. Splitting of the cell itself is called **cytokinesis** and usually begins during anaphase. In most animal cells, microtubules form a furrow in a ring around the cell. These gradually constrict until the cells separate, as shown in Fig 25.11.

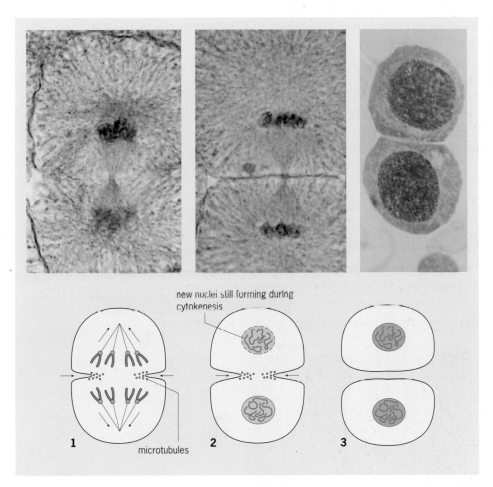

new nuclei still forming during cytokenesis

microtubules

1   2   3

Fig 25.11 Cytokinesis in animal cells. As the events of mitosis come to an end, movement of microtubules causes a constriction around the centre of the cell. Eventually the cytoplasm divides, leaving two new daughter cells

MEIOSIS: MAKING THE GAMETES

Meiosis is the type of cell division that halves the genetic material in cells. For this reason, it is known as a **reduction division**. In humans, as in all animals, meiosis makes haploid gametes (sex cells).

Meiosis is a remarkable process that produces haploid gametes and shuffles the genes, so that each gamete produced is genetically different. This is why children born to the same parents are usually not identical: they are produced by the fusion of a genetically unique sperm and egg. Identical twins are an exception because they originate from the same fertilised egg.

TEST YOURSELF

E **a)** Where in the body does meiosis occur in (i) women and (ii) men?

b) Why is meiosis sometimes referred to as reduction division?

THE STAGES OF MEIOSIS

Meiosis has two separate divisions. Both divisions have stages that are given the same names as in mitosis, but they have a number to denote whether they refer to the first or second division, for example anaphase I, prophase II.

It is important to remember here that a normal diploid cell contains two copies of each chromosome, one from the 'mother', and one from the 'father'. Just before meiosis, the cell's DNA becomes organised into chromosomes and replicates to form pairs of chromatids in preparation for division, just as it does before mitosis. So at the first stage of meiosis, the cell contains 92 (2 × 46) chromatids. The 23 chromosomes originally from the mother have been duplicated, making 23 identical pairs of chromatids, as have the 23 chromosomes originally from the father.

An overview of meiosis is shown in Fig 25.12. There is a basic difference between meiosis and mitosis: in meiosis, the homologous chromosomes pair up; in mitosis this does not happen. During the first part of meiosis, each homologous pair can swap sections of DNA, so that each pair becomes 'genetically mixed'.

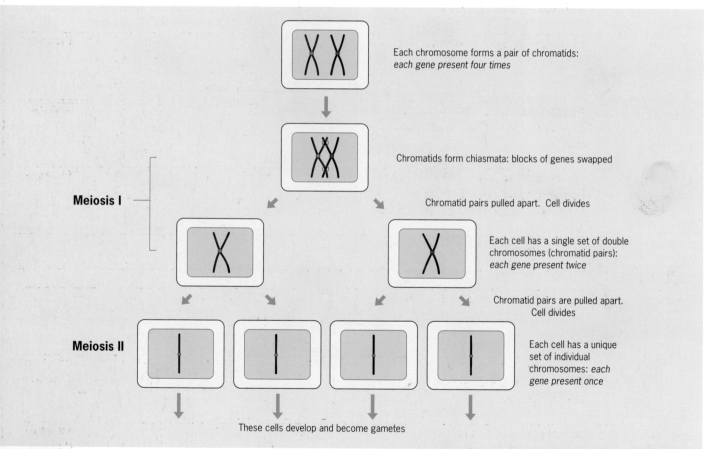

Fig 25.12 An overview of meiosis in an imaginary animal cell with just one pair of homologous chromosomes

The individual stages of meiosis are shown in Fig 25.13. In the first meiotic division (prophase I to telophase I), one of each homologous pair of 'genetically mixed' chromosomes passes into each new cell. In the second meiotic division, the individual chromatids are pulled apart so that one chromatid (now a chromosome) goes into each daughter cell. The end result of meiosis is four haploid cells, each containing a single set of chromosomes. Each cell is genetically unique.

Fig 25.13 The stages of meiosis in an imaginary animal cell showing just one of the pairs of chromosomes

Stage of meiosis First division	What is happening	What it looks like
Interphase	Just before meiosis, DNA replicates, so cells which contained two copies of each chromosome now have four. Chromosomes not yet visible	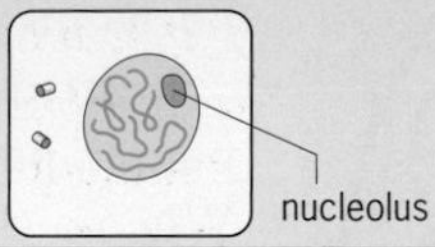 nucleolus
Early prophase I	Chromosomes become visible. Centromeres move to opposite sides of cell	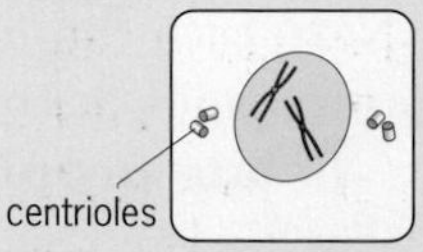 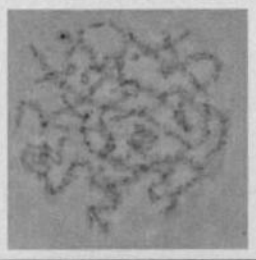 centrioles — Early prophase I
Mid prophase I	Each homologous pair of chromosomes comes together to form a bivalent	 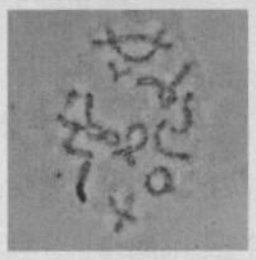 Mid prophase I
Late prophase I	Each chromosome in a bivalent forms two chromatids. Genetic mixing occurs: chiasmata, the points of crossover, are visible	
Metaphase I	The bivalents arrange themselves on the equator of the spindle	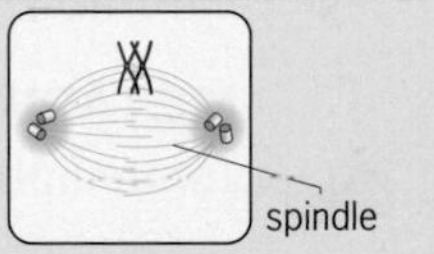 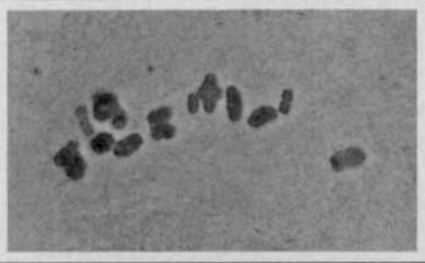 spindle — Metaphase I
Anaphase I	The chromatid pairs from each homologous chromosome split apart and move to opposite poles of the cell	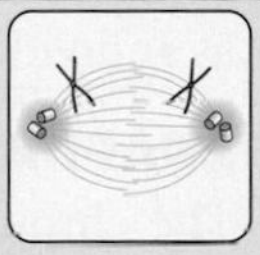 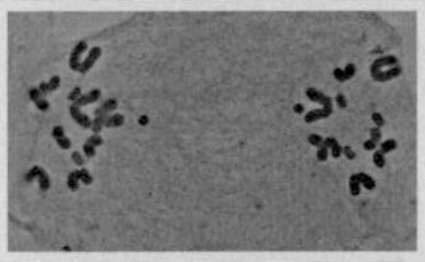 Anaphase I
Telophase I	Cytokinesis begins, two new cells form, each has two copies of each chromosome. These chromosomes are genetically different from those in the original cell	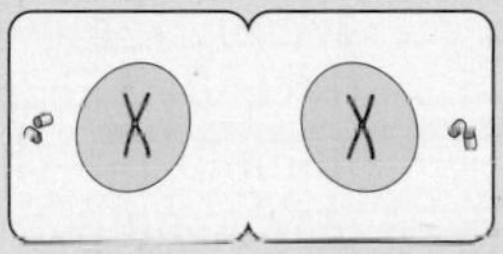 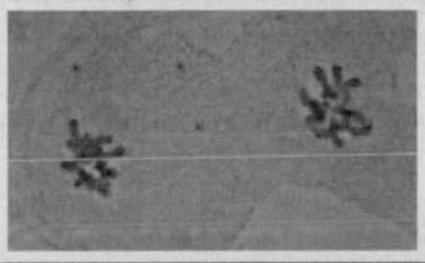 Telophase I
Interphase	A resting time (length varies between cell types)	
Second division		
Prophase II	A new spindle forms, at right angles to the first	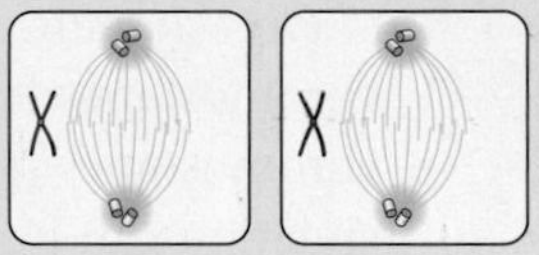
Metaphase II	Chromosomes, each of which is a pair of chromatids, align themselves on the equator of the spindle	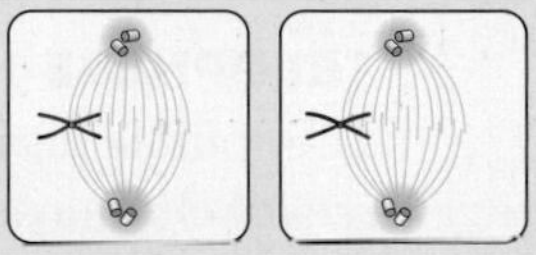 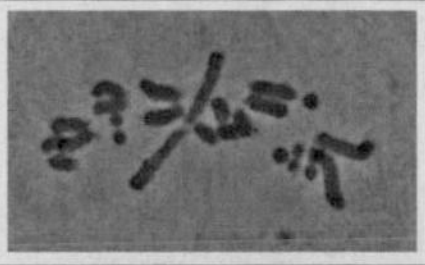 Metaphase II (1 cell only)
Anaphase II	Chromatids are pulled apart to form two chromosomes that then move to opposite poles of the cell	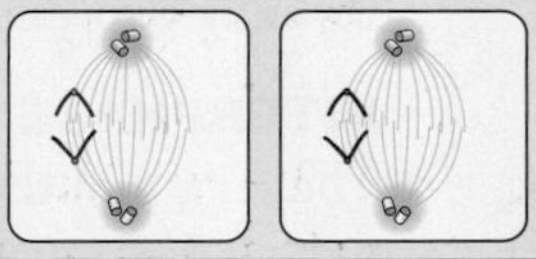  Anaphase II (1 cell only)
Telophase II	Cytokinesis begins. Four haploid cells, each with only a single chromosome, have been formed. Each chromosome is genetically different	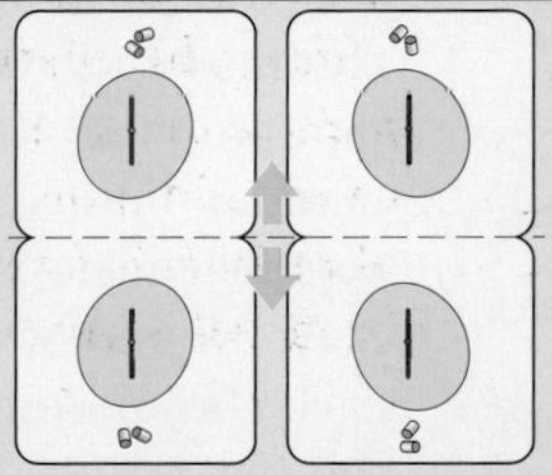 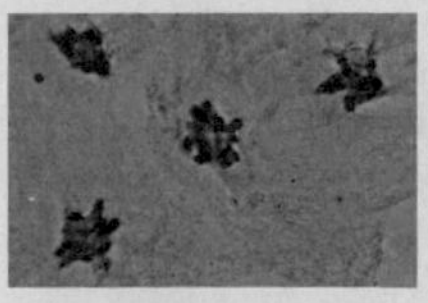 Telophase II

SEE QUESTION 5

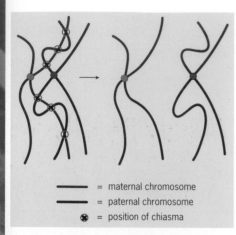

Fig 25.14 = maternal chromosome
_____ = paternal chromosome
⊗ = position of chiasma

Fig 25.14 Chiasmata form between homologous chromosomes during late prophase I of meiosis. At these points, parts of the maternal chromosome separate and join with the paternal chromosome, and vice versa. This produces new chromosomes that are genetically different from each other, and from both the maternal and paternal chromosomes from which they are derived

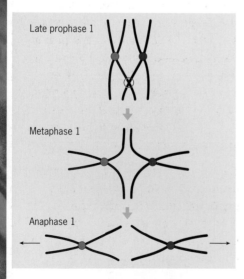

Late prophase 1

Metaphase 1

Anaphase 1

Fig 25.15 Chiasmata form during late prophase I and crossover takes place. The newly formed chromatid pairs separate during metaphase I and then the two pairs move to opposite poles of the cell during anaphase I

PROPHASE I

Prophase I is more complex than the corresponding phase in mitosis, and can be subdivided into early, middle and late.

Early prophase I starts when the chromosomes condense and the nucleolus disappears.

In **mid-prophase I**, the homologous chromosomes, one from each parent, pair up. Each pair forms a bivalent. This does not happen in mitosis. When a chromosome pair is exactly aligned, it is said to be at **synapsis**. Next, each chromosome in a pair divides into two chromatids, giving four chromatids per bivalent.

In **late prophase I**, **recombination** – or crossover – takes place (Fig 25.14). One (or both) of the chromatids of the two homologous chromosomes breaks off at certain points and fuses with a chromatid of the other chromosome in the bivalent, forming joints called **chiasmata** (the singular, *chiasma*, means 'crosspiece'). This process ensures that blocks of genes are swapped between maternal and paternal chromosomes. The position of chiasma formation varies, even within the same species, and this produces a large variety of new gene combinations.

Prophase ends with the chromatids of each bivalent (pair of homologous chromosomes) entwined and joined by chiasmata.

METAPHASE I

In metaphase I, the nuclear membrane disappears and the spindle is fully developed. The bivalents move to the equator of the spindle in the same way as individual chromosomes do during the matching phase in mitosis.

ANAPHASE I

The chromatid pairs of a bivalent are pulled apart (Fig 25.15) because of the action of spindle fibres, a process that separates the entwined chromatids. At the end of anaphase I, the chromatid pair from one of the original homologous chromosomes is positioned at one pole of the cell and the chromatid pair from the other homologous chromosome is at the other pole. Either the maternal or the paternal chromosome can pass into either cell. This is the process that allows **independent assortment of alleles** (see Chapter 28). It is another reason why meiosis increases variation.

Occasionally, this phase is not completed successfully and a pair chromosomes fails to separate. The result is that both homologous chromosomes pass into one daughter cell, the other receiving neither. This situation can lead to conditions such as Down's syndrome.

TELOPHASE I

The spindle disappears and the nuclear envelope re-forms around the two sets of chromosomes. At the same time cytokinesis separates the cytoplasm, forming two daughter cells, ready for the second meiotic division.

INTERPHASE

The length of the resting interphase between the two meiotic divisions varies widely. It is sometimes short or even non-existent. If there is no interphase, the chromosomes remain condensed and the cell passes straight from telophase I into prophase II. In human females, however, ova may remain in interphase for decades. Basic egg cells are made in a girl before birth, but they do not complete meiosis until just before ovulation, anything up to 50 years later.

At the end of meiosis I, there are two cells, each containing two copies of each chromosome on a chromatid pair. The second meiotic division separates the chromatids, so that each daughter cell formed is haploid (has one set of single chromosomes).

THE SECOND MEIOTIC DIVISION

Prophase II
For each chromosome, the chromatid pair attaches itself to the new spindle, which forms at right angles to the first.

Metaphase II
Each chromatid pair lines up on the equator of the spindle.

Anaphase II
The chromatids are pulled apart to form two chromosomes that move to opposite poles.

Telophase II
The spindle disappears, the nuclear membrane re-forms, chromosomes unravel and cytokinesis produces two separate cells.

THE END-PRODUCT OF MEIOSIS

Meiosis produces four genetically different haploid cells, known as a **tetrad**. Genetic variation has been produced in three ways:

- The homologous chromosome pairs originate in different organisms, one maternal and one paternal, and so are genetically different.
- Blocks of genes are swapped between the chromatids of homologous chromosomes as the chiasmata form during prophase I.
- Each daughter cell can receive a copy of either chromosome from a pair, and each copy may have undergone crossover and have different genes from the other three. This is called independent assortment and is covered in more detail in Chapter 28.

> **? TEST YOURSELF**
>
> **F** How many bivalents are formed in a human cell undergoing meiosis?

> **✓ REMEMBER THIS**
>
> The importance of genetic variation should become apparent as you study evolution. Find out more in Chapter 29.

SEE QUESTION 6

! SCIENCE FEATURE Down's syndrome

In the United Kingdom, two children in every 1000 are born with Down's syndrome (Fig 25.16a), a genetic condition that arises from a fault in meiosis. Individuals with this condition have an extra chromosome 21 (Fig 25.16b) and usually have physical and mental disabilities. They often have a small mouth with a normal-sized tongue (making eating and speech difficult), reduced resistance to disease and heart abnormalities.

The reason for the extra chromosome is usually a failure of the chromosome 21 pair to separate during anaphase I of meiosis (see Fig 25.13). Both chromosomes 21 pass into one daughter cell, and the other gets no copy at all. If a gamete with no chromosome 21 forms a zygote, the embryo fails to develop. But, if the gamete containing both chromosomes is fertilised, the resulting baby later develops Down's syndrome.

Fig 25.16(a) A child with Down's syndrome

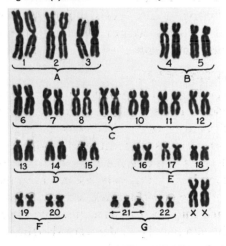

One reason for the non-separation may be the length of time the egg spends in the ovary. In females, egg cells do not complete their meiotic division until they are 'selected' for ovulation (see Chapter 26). The longer the eggs remain in the ovary, the 'stickier' the chromosomes become and the greater the probability that they will not separate during anaphase I. This explains why the chance of having a child with Down's syndrome increases as women get older. Women below the age of 25 have less than a 1 in 2000 chance, but this rises to 1 in 50 for women over 45.

The age of the father is also important: recent studies have shown that in a significant minority of cases the extra chromosome comes from the father, that is, a normal ovum is fertilised by a sperm containing two chromosomes 21.

Fig 25.16(b) The complement of chromosomes (the karyotype) of a person with Down's syndrome, showing the extra chromosome 21

LEARNING MEIOSIS

Meiosis is a complex process that many people find difficult to learn. Here are some hints that you may find useful. First, divide your study of meiosis into three sections by looking at three questions:

What is the point of meiosis?

It makes gametes by shuffling the genes to produce haploid cells that are all genetically different from either of the parents. Each fertilisation (combination of an egg and a sperm) produces an individual that is genetically unique.

What do each of the meiotic divisions achieve?

Start by learning an overview of each division: Fig 25.13 is a good place to start. In the first meiotic division, homologous chromosomes pair up, divide into chromatids, swap blocks of genes and then separate. In the second division the individual chromatids separate. The end result is four genetically different haploid cells from one diploid original.

What are the stages of the two divisions?

Once you are confident that you can answer the first two questions, you can put some flesh on the bones. Describe one stage at a time and then draw what you have just described. Try using different colours for paternal and maternal chromosomes. Alternatively, you could make some chromosomes out of modelling clay and work through the meiotic sequence yourself. Many students find this a valuable exercise.

SUMMARY

After reading this chapter, you should know and understand the following:

- Almost all the cells in the human body are **diploid**: they contain a double set of chromosomes. The only exceptions are the gametes: sex cells – egg cells and sperm – are **haploid** and contain only one set of chromosomes.

- The cell cycle consists of four stages: G_1, in which the cell grows; **S**, in which the DNA is doubled; G_2, a second growth phase, and finally **mitosis** or **meiosis**. G_1, S and G_2 are collectively known as **interphase**.

- Mitosis is normal cell division, the process by which we grow and develop. One diploid cell divides to produce two genetically identical diploid cells.

- You can remember the stages of mitosis using the mnemonic IPMAT: Interphase, Prophase, Metaphase, Anaphase, Telophase.

- Non-dividing cells are said to be in interphase. There are no chromosomes visible and the active genes in the DNA are being used to direct the activities of the cell. If the cell is going to divide, the cell prepares itself during interphase and the DNA is replicated.

- In prophase, the chromosomes condense and become visible. The nuclear membrane disappears and the **spindle** forms. The chromosomes are double: two identical chromatids are joined at the **centromere**.

- In metaphase, the chromosomes arrange themselves in the middle of the spindle (remember, meta = middle).

- In anaphase, the double chromosomes are pulled apart (remember, ana = apart).

- In telophase, the nuclear membrane re-forms around the two new sets of chromosomes and the cytoplasm divides forming two new, genetically identical cells.

- Meiosis is a special type of cell division which produces sex cells or gametes. It is also known as a 'reduction division'. One diploid cell will divide meiotically to produce four genetically different haploid cells.

❓ EXAM QUESTIONS

1 The diagram shows a sperm and an egg from the same species, drawn to the same scale.

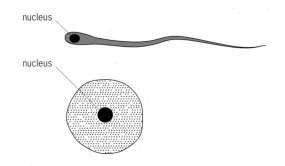

nucleus

nucleus

a) Assume that both the egg and its nucleus are spheres.
The volume of a sphere is given by $4\pi r^3/3$ where r is the radius of the sphere.
Calculate the proportion of the egg that is occupied by the nucleus.
Show your working.

b) The nucleus occupies a much smaller proportion of an egg cell than it does of a sperm.
Give **one** reason for this difference and explain its importance.

AQA (B) BYB2 Genes and Genetic Engineering June 2001 Q2

2 The photographs show stages in mitosis labelled **A–D**. The stages are in sequence with stage **A** being the earliest.

A 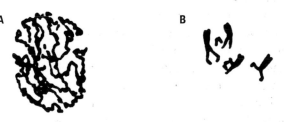 B

C 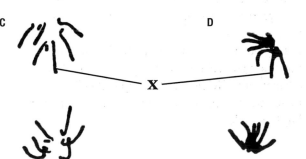 D

X

a) Describe what occurs between
(i) stage **A** and stage **B**;
(ii) stage **B** and stage **C**.

b) Describe what happens to structure **X** between the end of stage **D** and the next time stage **A** occurs.

AQA (B) BYB2 Genes and Genetic Engineering January 2001 Q1

3 Some cells divide by mitosis.
a) **(i)** Describe the products of mitosis.
(ii) State **two** processes in which mitosis occurs in humans.
The diploid number of the gorilla is 48.
b) State the number of chromosomes which would be found in the following cells of the gorilla: brain cell; epithelial cell; sperm; muscle cell.
The chart lists six of the events that take place during mitotic cell division. Each event is identified by a letter.

A	chromatids separate

B	centromeres divide

C	chromosomes become visible

D	nuclear membrane disintegrates

E	chromosomes align at equator

F	cytoplasm divides (cytokinesis)

c) list the letters shown in the chart in the order in which the events occur during a mitotic cell division. The first one is **C**.
d) State **three** factors which can increase the chances of cancerous growth.

OCR 2801 Biology Foundation January 2001 Q1

4 The drawing indicates the appearance of a chromosome at early prophase of mitosis.

A

identical chromatids

a) With reference to the drawing,
(i) name the structure labelled **A**;
(ii) explain why the two chromatids are identical.
The diagram represents the nucleus of an animal cell ($2n = 6$) at early prophase of mitosis.

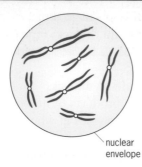

nuclear
envelope

b) Copy the diagram and shade **one** pair of homologous chromosomes.

c) Draw an annotated diagram to indicate what happens in this cell at anaphase of mitosis.

d) (i) State the number of chromosomes which would be found in a **haploid** cell in this animal.
(ii) Explain why haploid cells need to be produced during a life cycle which includes sexual reproduction.

OCR 2801 Biology Foundation June 2001 Q6

5

a) Describe the behaviour of chromosomes during meiosis.

b) Explain the similarities and differences between homologous chromosomes.

OCR 9264/3 Biology Linear Paper June 2000 Q8

6 The diagram shows the life cycle of the honey bee.

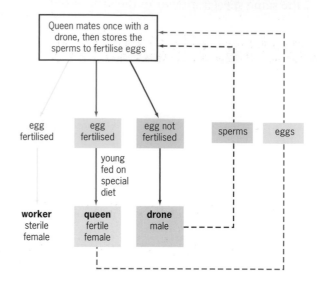

a) The queen has body cells containing 32 chromosomes.
How many chromosomes are there in the body cells of:
(i) the drone;
(ii) the workers?

b) Name the type of cell division in bees that produces:
(i) eggs;
(ii) sperms.

c) Are all the drones identical ?
Explain the reason for your answer.

AQA (B) BYB2 Genes and Genetic Engineering January 2001 Q2

26

Reproduction

Male freshwater fish, such as these sunfish, guard nests (the light-coloured circles) of their eggs against predators and other males. New genetic information by Purdue Assistant Professor Andrew DeWoody reveals that the fish are willing to eat some of their own eggs in order to survive while guarding the nest

Fig 26.1 The overwhelming urge to reproduce is illustrated by these salmon. Born in fresh water, they swim downstream to the sea where they spend several years feeding and growing. Eventually, they return to the river of their birth, often using massive amounts of energy to complete the journey upstream. Finally, they lay and fertilise their eggs and then die

The expression 'cold fish' is often used to describe someone who is unresponsive and dispassionate. But it would be difficult to apply this label to a real fish. Some freshwater fish, for example, have sex lives more colourful than the raciest characters from 'EastEnders'.

Using the genetic fingerprinting techniques usually used to investigate criminal cases (see the Assignment in Chapter 3), researchers have uncovered an unknown world of fish behaviour. Most freshwater fish reproduced by building nests and laying eggs in the spring. But, unlike birds and reptiles, it is the male fish that tends the nest. He tries to make it look as attractive as possible to encourage females to visit to lay their eggs, and then he fertilises the eggs by depositing great clouds of semen over them. He gets many females into the nest to ensure that his genes are passed on as often as possible.

The male also protects the nest once the fertilised eggs have started to develop. Unfortunately for him, other fish would come and eat the eggs if he left the nest, even for a moment, so he needs to stay with the nest during this crucial time. He can't forage for food, and, if he is in danger of starving, he will eat a few of his own fertile eggs for food, making the choice that to lose a few is better than losing the lot.

The male also has another problem. Some male fish of the same species don't build nests and tend them in the usual way. About 15 per cent of them are 'rogues' that make use of the nests of other males. There are two categories: sneakers and satellites. The sneakers are adult males disguised as immature males. They are small and weedy but their testes are three times larger than those of an ordinary male. A sort of swimming sack of sperm. A sneaker, as his name suggests, sneaks up on a nest into which a female has just laid eggs. He then deposits as much of his own sperm as possible, avoiding the nest male. The satellite is even more devious – he looks like a female fish and approaches the nest as if it wants to lay eggs; instead releasing his own sperm into the nest.

These tactics aren't that great from a genetic point of view though. Genetic fingerprinting showed that these rogue males only managed to sire 1.3 per cent of embryos in the nests they targeted. Not a huge return for the effort.

1 THE NEED TO REPRODUCE

None of us is immortal and so, in common with all other types of organism, we must reproduce in order for our species to survive. The need to reproduce, so passing genes on to the next generation to ensure the survival of the population, is one of the basic features of living organisms. In many organisms, the urge to reproduce can over-ride the urge to live (Fig 26.1).

An interesting way of looking at the whole business of reproduction is to think of organisms as disposable containers for the genes they contain. None of us will live forever, but our genes, mixed with those from other people, might survive long after we are forgotten.

Reproduction is either **sexual** or **asexual** (non-sexual). All animals reproduce sexually, and some animals also reproduce asexually in some circumstances.

2 ASEXUAL REPRODUCTION IN ANIMALS

Asexual reproduction involves no fertilisation: one individual organism produces one or more new individuals. Since there is only one parent, there is no mixing of genetic material and so the new organisms are **clones** – organisms that are genetically identical to the parent. Asexual reproduction can quickly increase the number of organisms in a population and it avoids the need to find a mate. This works well if the environment is favourable and does not change. The problem with asexual reproduction, though, is that the organisms produced are all identical and so the population cannot adapt to change.

The only variation that occurs in asexual reproduction is by mutation (see Chapter 27). This means that organisms that reproduce asexually tend to evolve more slowly, and hardly any animal species uses it as their only means of reproduction. Asexual and sexual reproduction are compared in Table 26.1. The processes of asexual and sexual reproduction in plants are covered in Chapter 33.

There are three main methods of asexual reproduction in animals:

● **Budding**. Some cnidarians, flatworms and annelids multiply by budding (Fig 26.2). A new individual simply grows out of the body of the parent.
● **Parthenogenesis**. This term means 'virgin birth' or reproduction without fertilisation. It occurs in a variety of animal types, including some insects, fish, amphibians and lizards, when unfertilised eggs develop into new individuals. See the Science feature on aphids on the next page.
● **Regeneration**. Many echinoderms such as the starfish (Fig 26.3) and some flatworms and annelids reproduce successfully by regeneration. If their bodies are broken into fragments, each one can develop into a new individual.

Fig 26.2 Budding is one type of asexual reproduction. In this *Hydra* (a cnidarian), the new individuals simply grow out of the body wall of the parent

Fig 26.3 A memorable example of the power of some organisms to regenerate was when some fishermen in the USA launched a campaign to prevent starfish from destroying their oyster beds. They collected a large number of starfish, chopped them up and threw them back into the sea. Bad move. As the picture shows, a severed arm can regenerate into a whole new organism. The fishermen caused a population explosion

Table 26.1 Comparing sexual and asexual reproduction		
Type of reproduction	**Advantages**	**Disadvantages**
Asexual	Fast – one individual can build up a population in favourable circumstances	All offspring are identical. There is less genetic variation, so evolution is slower
Sexual	Produces variation. Allows organisms to evolve and adapt to changes in their environment	Slow – requires two individuals. Individuals are vulnerable during mating and pregnancy

Aphids illustrate the pros and cons of sexual and asexual reproduction quite dramatically. These insects, which include greenfly, are pests: they cause millions of pounds of damage to crops world-wide. Although the exact details vary from species to species, a generalised aphid life cycle is as follows:

After surviving the winter on tree bark, aphid eggs hatch in spring, producing a population of wingless females. These females are born pregnant – aphid embryos produced by parthenogenesis are already developing inside them (Fig 26.4).

- The aphids build up their numbers rapidly, taking advantage of the very favourable conditions at the start of the summer – big healthy plants full of sap, no competition for food and no predators.
- The large numbers of insects – and their waste – damage host plants and so food becomes scarce. (No population can go on expanding indefinitely: there are always limiting factors.) Predators such as ladybirds become more common as they take advantage of the abundant food and, towards the end of summer, conditions generally become unfavourable to the aphids. This triggers a 'switch' and the reproductive process produces winged aphids that can fly away and colonise new plants.

- Towards the end of summer, as the days get shorter, aphids of both sexes are born. Winged males and females then reproduce sexually, mating and laying eggs that can survive the winter.

Fig 26.4 Aphids give birth to live young produced by the process of parthenogenesis. Each wingless female aphid is born with embryos of the next generation already growing inside her. This phenomenon is also known to occur in organisms as advanced as fish, amphibians and reptiles

3 SEXUAL REPRODUCTION IN ANIMALS

Sexual reproduction involves the fusion of **gametes**, or sex cells. Female sex cells, called **egg cells** or **ova**, fuse with the male gametes, called **spermatozoa** or **sperm**, in a process known as **fertilisation**. The resulting cell, the **zygote**, develops into a new individual.

In sexual reproduction, the genetic material of two different individuals is mixed and combined to produce an individual that is genetically different from either parent. This produces variation within a population (see Chapter 24). All individuals, unless they have an identical twin, are genetically unique: different from all other individuals in the group.

Most species are unisexual: individuals are either male or female. But some organisms carry both male and female sex organs in the same body: they are described as hermaphrodites. Many familiar animals such as earthworms and snails are **hermaphrodites**.

Having both types of sex organs in the same body gives an organism greater reproductive capacity, or **fecundity**. When two snails mate, each female sex organ receives sperm and both snails become pregnant and can lay eggs. One mating therefore results in twice the number of offspring. In some circumstances a hermaphrodite can fertilise itself and so start up a new population.

ANIMAL LIFE CYCLES

Animals are diploid organisms that produce haploid gametes (Fig 26.5). These fuse to produce a new diploid individual, so there is no alternation of generations (in contrast to plants – see Chapter 33).

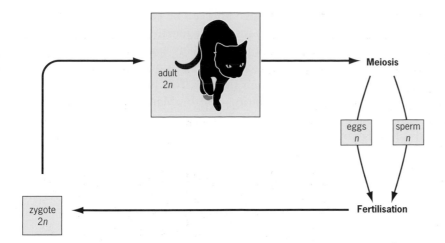

Fig 26.5 Animals have diploid life cycles. All of the body cells of this cat are diploid, except for its sex cells. These are produced by meiosis and are haploid. When a male gamete fertilises a female gamete, a diploid zygote is produced. This divides by mitosis many times. The new cells continue to divide and differentiate (cells become specialised to carry out particular functions), to produce a new individual

INTERNAL VERSUS EXTERNAL FERTILISATION

The male gametes of all animals can swim. Animals that reproduce in water, such as fish and amphibians, fertilise egg cells outside the female's body simply by releasing sperm (Fig 26.6). These swim through the water to reach the egg cells. We now know that the egg cells of many species release chemicals that attract the sperm, greatly increasing the chances of successful fertilisation.

As animals evolved and began to colonise the land, they had to overcome the problems of reproducing out of water. Internal fertilisation became a necessity. Animals developed specialised organs such as a penis and complex behaviour patterns to enable the male to introduce sperm directly into the female's body (Fig 26.7).

Animals have developed various strategies to protect the growing embryo(s) and prevent them from drying out. Some animals – birds and some reptiles, for example – produce waterproof eggs. The egg has a tough material round the fertilised ovum. Eggs are laid by the animal and provide the embryo with a self-contained, watery environment. Organisms that lay eggs are said to be oviparous (ovi = egg, parous = birth).

Fig 26.7 In terrestrial animals such as reptiles, birds and mammals, the eggs are fertilised inside the female

Fig 26.6 Common frogs: the female produces eggs which are fertilised in the water by the male's sperm

Fig 26.8 This guppy is giving birth to 'live' offspring: the eggs hatch inside the female. This process avoids the vulnerable egg stage of the life cycle

Table 26.2 The gestation period of some mammals

Species	Approx. gestation period
Hamster	18 days
Stoat	28 days
Rabbit	30 days
Cat	63 days
Lion	110 days
Human	40 weeks
Horse	11.5 months
Elephant	22 months

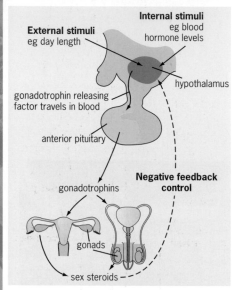

Fig 26.9 The hormonal control of reproduction. The hypothalamus can respond to external and internal nervous stimuli and is also sensitive to hormone levels in the blood. The hypothalamus controls the levels of several different circulating hormones using a negative feedback mechanism

? TEST YOURSELF

A The lioness is a big cat that weighs more than a human, yet its gestation period is only 110 days, compared with 240 days (40 weeks) in humans. Suggest a reason for this difference.

B The hypothalamus controls the level of sex hormones via a negative feedback mechanism. Suggest how the hypothalamus would respond to a fall in oestrogen levels.

✔ REMEMBER THIS

The role of hormones in controlling the timing of reproduction in sheep is covered in detail in Chapter 6.

Even so, eggs are vulnerable to damage and can be eaten by predators. A much safer place for fertilised ova is inside the female, and it makes sense to lay eggs as late as possible in development. Many species have taken this to its logical conclusion: their eggs hatch inside the mother and the young are born as immature miniatures of their parents. This process is known as ovoviviparity (ovo = egg, vivi = live, parity = birth). Many species of fish (Fig 26.8), amphibians and reptiles reproduce in this way.

Mammals also give birth to live young, but these do not hatch from eggs. In most species of mammal, the embryo grows inside a specialised organ called the uterus and is fed by a placenta. The young are born live, but their state of maturity at birth depends on the species. This reproductive process is called viviparity.

REPRODUCTION IN MAMMALS

Mammals show a great range of reproductive strategies that reflect their circumstances, but the basic life cycle of most mammals follows a general pattern. After mating, the fertilised egg cell(s) develop inside the uterus, nourished by the placenta. The **gestation period** of a mammal, the time between fertilisation and birth, often reflects the metabolic rate of the organism. It is shorter in small short-lived mammals such as mice and longer in large long-lived mammals such as elephants.

Other factors complicate the length of gestation. For example, it is also limited by the size of the fetal skull (which must fit through the female's pelvis at birth) and the mobility of the mother. In species where the mother needs speed and agility to survive, the gestation period tends to be shorter. Birth is followed by a period when the mother gives close protection and suckles her young. The length of upbringing varies greatly: it depends on factors such as degree of maturity at birth and the amount of learning required to survive.

The onset of sexual maturity marks the change from **juvenile** to **adult**. Most female mammals conceive as soon as they become sexually mature and enter their first **oestrus** (season). From then on, throughout their reproductive life, they are either pregnant, feeding young or both. A female's reproductive life is usually brought to an end only by death. This has been the harsh fact for humans, too, until fairly recently. Even now, in societies where contraception is either unavailable or not used for cultural reasons, many women reproduce continuously.

THE ROLE OF HORMONES IN MAMMALIAN REPRODUCTION

For many animals the timing of reproduction is vital. If there is a severe variation in the seasons, animals must ensure that their offspring are born when food is plentiful and conditions are mild. **Hormones** that control and synchronise reproductive events are produced by three different organs: the **hypothalamus**, the **pituitary gland** and the **gonads** (sex organs).

The hypothalamus is a major point of contact between the nervous and endocrine (hormone) systems (see Chapter 22). Internal and external stimuli reach the hypothalamus, which responds by releasing hormones. They, in turn, control the rest of the endocrine system (Fig 26.9).

Day length is a classic example of an external stimulus. When the hours of daylight reach a threshold level, perhaps indicating the arrival of spring, the hypothalamus responds by releasing a hormone called a **gonadotrophin releasing factor** (**GnRF**). This controls the activity of the nearby pituitary gland, stimulating it to release hormones called **gonadotrophins**.

Gonadotrophins have a direct effect on the gonads (the ovaries and testes). These respond by releasing **steroid sex hormones** (Chapter 3) such as **oestrogen** and **progesterone** and **testosterone**.

4 HUMAN SEXUAL REPRODUCTION

Sexual activity in humans has evolved to achieve more than just fertilisation. Although humans become sexually mature in their early to mid-teens, many societies have cultural and legal controls aimed at restricting sexual activity before the late teens. Although beyond the scope of this book, it is interesting to consider whether human sexual behaviour may have evolved to strengthen the emotional bond between a couple. This 'social cement' may make it more likely that couples will stay together and so provide a stable environment for raising children.

PUBERTY

Puberty is the time between childhood and adulthood: it marks the process of sexual maturing. We still don't know exactly what triggers the onset of puberty, but we know that, early on, GnRF is secreted from the hypothalamus (look back to Fig 26.9). GnRF travels the short distance to the anterior pituitary, where it stimulates the release of gonadotrophins, **follicle stimulating hormone** (**FSH**) in females and **interstitial cell stimulating hormone** (**ICSH**) in males (Fig 26.10). These two hormones are chemically identical, but have different names because they have different effects in the two sexes.

In girls, FSH targets the ovaries, which it stimulates to produce **oestrogen**. This steroid hormone is responsible for many of the female sex characteristics. In addition, oestrogen stimulates the ovaries to start releasing egg cells (ovulation) and this leads to the first monthly period, or **menstruation**, an event known as the **menarche**.

To start with, periods tend to be irregular and unpredictable, and may sometimes occur without ovulation. Within about a year, hormone levels have increased to the point where they stimulate the regular development of follicles. This makes periods more regular and, unfortunately for many girls, painful. Period pain is mainly due to the hormone progesterone, which causes uterine cramps.

In boys, ICSH targets the **interstitial cells** of the testes. These are embedded in the connective tissue between the seminiferous tubules – see Fig 26.12(b). ICSH stimulates these testis cells to secrete **testosterone**, the steroid hormone that stimulates development of male sex characteristics.

THE AGE OF ONSET OF PUBERTY

The average age for the onset of puberty is 12 to 13 in girls, 13 to 15 in boys. Interestingly, this is much earlier than in previous centuries. Two hundred years ago the average age of the menarche in girls was 16, around four years later than it is now. The reason for this is almost certainly the improvement in diet. A better diet enables us to grow faster and so reach the same stage of maturity at an earlier age. In girls, the proportion of body fat appears to be important: a girl who is a dedicated athlete (a gymnast for example) and has a high muscle:fat ratio, may find that the menarche is delayed. Also, girls who have started their periods but who then crash diet and lose a lot of weight may find their periods stop for a while.

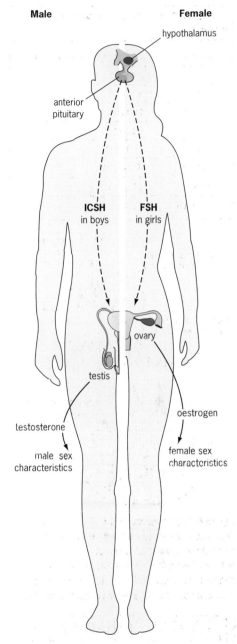

Fig 26.10 The onset of puberty is controlled by hormones.

Testosterone causes the penis, scrotum and testes to enlarge and mature. It also causes enlargement of the **larynx** (voice-box) which deepens the voice. Body hair is more extensive, with coarse hair appearing on the chest and face. Ironically, in later life, the same hormone leads to male-pattern baldness. A man's shoulders and chest tend to be large and there are often changes in facial structure, with the nose and chin becoming more prominent. The muscles tend to be better developed and generally the male has greater physical strength than the woman.

Oestrogen causes the ovaries, oviducts, uterus and vagina to mature and brings about the development of secondary sex characteristics. Breasts grow, pubic and underarm hair grows, the shape of the pelvis changes and the body fat is redistributed so that a woman tends to have wider hips and a more curved shape than a man

THE MALE REPRODUCTIVE SYSTEM

The overall structure of the male reproductive system is shown in Fig 26.12. The male system secretes testosterone, makes and stores sperm and delivers them into the female's body.

Spermatozoa are made in the testes, a pair of organs that are held in a pouch of skin called the **scrotum**. It may seem odd that such delicate and vital organs are relatively unprotected outside the body, but there is a reason for this. The process of sperm production, **spermatogenesis**, is most efficient at around $35\,^{\circ}\text{C}$, two degrees cooler than the core of the body. Men whose testes do not descend into the scrotum cannot produce healthy sperm.

THE PENIS

The penis introduces sperm into the female's body so that internal fertilisation can occur. Some animals do not have a penis – most birds and reptiles, for example – and their attempts at fertilisation are a little more haphazard. The male produces semen from his genital opening and simply rubs it on to the female's genital opening.

THE TESTES AND SPERMATOGENESIS

The production of sperm in human males is a continuous process, beginning at puberty and continuing well into old age: men in their nineties have fathered children. Spermatogenesis centres around the process of meiosis, and occurs at a remarkable rate: over a thousand human sperm are made every second. Look again at Fig 26.12(b) and you will see that each testis is composed of a series of **lobules**. They contain **seminiferous tubules**, the structures in which sperm production takes place. This process, shown in Fig 26.13, has three main phases:

- **Multiplication**. As large numbers of sperm are needed, cells of the **germinal epithelium** divide by mitosis to produce many **spermatogonia** (sometimes called sperm mother cells).
- **Growth**. The spermatogonia grow into **primary spermatocytes**. At this stage the cells are still diploid ($2n$).
- **Maturation**. The diploid primary spermatocytes undergo meiosis. After the first division they become **secondary spermatocytes** and when meiosis is complete they have become haploid **spermatids**. In the final part of the maturation process, spermatids differentiate into the familiar spermatozoa (sperm).

! SCIENCE FEATURE Is the male pill in sight at last?

Until the 20th century, there were no reliable methods of contraception. Today, there are a variety of barrier and hormonal methods but women often point out that most of them are their responsibility. Only the condom really affects the male partner – the contraceptive pill, hormone injections and the cap are all used by the female. But things might soon change.

In 2001, researchers discovered a new ion channel that only occurs in the tail of sperm. It exists nowhere else in the body. Ion channels are specialised proteins that carry ions – usually sodium, potassium or calcium ions – from one side of a cell surface membrane to the other (see Chapter 4). CatSper, the ion channel in sperm, carries calcium ions and is

necessary for the sperm to penetrate the outer coating of the ovum. If this channel could be blocked, the sperm would no longer be able to fertilise the egg. A small molecule that acted as a CatSper blocker would therefore be an ideal male contraceptive. It would be short acting, and so would need to be taken only when sex was likely to take place. Unlike the female contraceptive pill, it isn't hormonal, and so would not cause some of the side effects that have prevented male hormonal contraceptives from being developed. Researchers have tried but these tend to cause lack of libido and impotence – for the perfect contraceptive that sort of misses the point.

Because CatSper is present only in sperm tails, blocking it would also be unlikely to cause side-effects in other parts of the body. However, although biotech companies are currently working on such a male pill, several years of safety testing and clinical trials lie ahead.

Fig 26.11 The sperm is the target of a potential male contraceptive pill

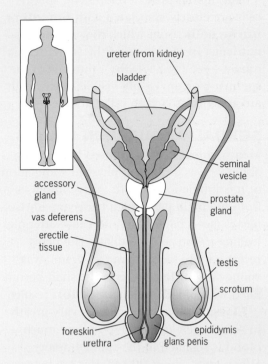

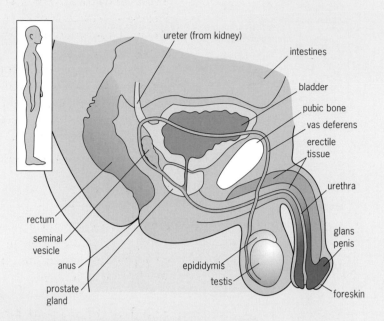

Fig 26.12(a) The main parts of the male reproductive system, seen from the front and the side

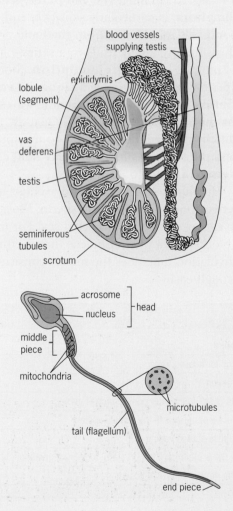

Fig 26.12(b) The microscopic structure of the testis showing a section through a seminiferous tubule. The testis contains many seminiferous tubules. Inside each of these at any time, many thousands of sperm are maturing

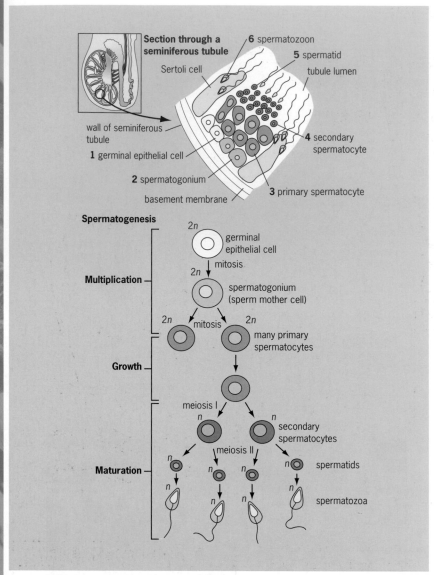

Fig 26.13 The process of spermatogenesis. Sperm cells develop as they pass from the outer wall of the seminiferous tubule to the lumen

SEE QUESTIONS 1 AND 2

Throughout their development, sperm cells are closely associated with **Sertoli** or **nurse cells**, from which they obtain nutrients. In the lumen of the seminiferous tubule – see Fig 26.12(b) – the tails of the spermatozoa are clearly visible: their heads are attached to Sertoli cells.

SEXUAL AROUSAL IN MALES

Men become sexually aroused by thinking about sex, as a result of physical stimulation or a combination of both. Nerve impulses from the brain pass down parasympathetic nerves (see Chapter 20) and cause arterioles leading to the penis to dilate. The penis receives more blood than can drain away, spongy **erectile tissue** in the shaft becomes filled with blood, and an **erection** results.

Flaccid (non-erect) penises vary greatly in size, largely depending on how much blood is retained in the spongy tissue. When erect, about 90 per cent are between 14 and 16 cm long. The end of the penis, the **glans**, is particularly sensitive, and continued stimulation from rhythmic thrusting eventually leads to a series of reflexes known as **ejaculation**. Stored spermatozoa are propelled along the **vas deferens** by powerful peristaltic waves. As they pass various accessory glands, different secretions are added to the sperm and the final ejaculate is a milky fluid called **semen** (Table 26.3).

Human males ejaculate, on average, about 5 cm³ of semen, which contains between 50 to 200 million sperm. Most sperm never get anywhere near the egg, even after unprotected sex. For most of the monthly cycle, the cervix is blocked by a plug of mucus that sperm cannot penetrate. Only at around ovulation time does the mucus consistency change, allowing sperm to pass through easily.

? TEST YOURSELF

C What do you think are the advantages of producing small numbers of non-motile female gametes and large numbers of motile male gametes?

Table 26.3 The constituents of semen		
Gland	**Secretion**	**Purpose**
Seminal vesicles	fructose	energy for spermatozoa
	mucus	lubrication
	protein	forms clots, which alter consistency of semen
	prostaglandins	stimulate peristalsis in the female system
Prostate	alkaline chemicals	neutralise acid in vagina
	clotting agent	clots the protein from the prostate, forming a gelatinous mass
Cowper's gland	clear fluid	cleans urethra prior to ejaculation

THE OVULATION PHASE

On day 12 to 13, the blood oestrogen level reaches a threshold level, which triggers the release of **luteinising hormone (LH)**, from the anterior pituitary gland. The rapid increase in LH levels triggers ovulation around day 14. The ovum lives for only 24 to 36 hours. During this time it moves only a few centimetres from the ovary.

SEE QUESTION 3

THE SECRETORY PHASE

Luteinising hormone has a second effect: it causes the Graafian follicle to develop into a **corpus luteum** ('yellow body'). The name comes from the yellow appearance of the secretory cells that develop inside the 'remains' of the Graafian follicle. The corpus luteum secretes oestrogen and progesterone. Progesterone causes spiral-shaped blood vessels to grow into the endometrium. This thickened lining begins to secrete nutrients and mucus to prepare for an embryo to be implanted. During this phase the high levels of progesterone inhibit the production of FSH. As long as progesterone levels are high, the endometrium is maintained and no new follicles are stimulated. The 'contraceptive pill' takes advantage of this inhibition (see Science feature on page 473).

If the egg cell is not fertilised, the corpus luteum lasts for about 10 to 12 days and then degenerates, ceasing to secrete progesterone. This is a key event because the inhibition of FSH is lifted. The endometrium is no longer protected and the cycle can start again.

THE MENSTRUAL PHASE

The drop in progesterone and oestrogen levels causes the uterine capillaries to rupture, and the endometrium is lost from the body through the cervix, together with some blood. Renewed secretion of FSH begins around day 2, and the cycle begins again.

? TEST YOURSELF

E Many women do not have a 28 day cycle. Although the second half of the cycle (ovulation and menstruation) is usually 14 days, the first half can vary considerably. If a woman's cycle is 35 days, on which day is she likely to ovulate?

! SCIENCE FEATURE The menopause and hormone replacement therapy

The menopause is a natural event that occurs when the ovaries stop working. For most women, this happens between the ages of 45 and 54. This time is often difficult and traumatic. The lowered levels of oestrogen and progesterone are directly responsible for the unpleasant symptoms of the menopause:

- Circulatory problems such as hot flushes and night sweats.
- Psychological problems, such as depression, anxiety and insomnia.
- Skeletal problems. Oestrogen inhibits reabsorption of bone, and after the menopause, bone loss can be as great as 7 per cent per year. This condition, osteoporosis, affects the spongy bone particularly and the sufferer is more likely to break a bone.
- Oestrogen is also thought to give women some protection against some types of heart disease. Women below menopausal age are less likely than men to have heart disease, but afterwards they catch up.

Hormone replacement therapy can help to reverse many of these symptoms and effects. The basic idea behind HRT is simple: to restore hormone levels to those of the early follicular phase of the menstrual cycle using tablets, implants or transdermal patches (Fig 26.18).

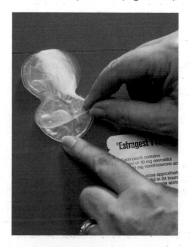

Fig 26.18 One method of administering the hormones in HRT: a transdermal patch (a patch attached to the skin) that releases controlled amounts into the bloodstream. The hormones used in HRT are extracted from the urine of pregnant mares or made artificially

THE EVENTS OF PREGNANCY

If the egg cell is fertilised, the menstrual cycle is interrupted and the female's body changes in response to the events of pregnancy.

FERTILISATION

Fertilisation is a complex sequence of events that begins when the sperm reaches the egg cell (oocyte). The events of fertilisation are shown in Fig 26.19.

It may also help you to look back at Fig 26.16. The secondary oocyte is surrounded by several layers of follicle cells, the **corona radiata**, and a layer of glycoprotein, the **zona pellucida**. Before a sperm can penetrate the layers surrounding the oocyte, it undergoes a process called **capacitation**. The actual mechanism is poorly understood, but it seems to set off the **acrosome reaction**, which allows the sperm head to enter the oocyte. The **acrosome** is a bag of digestive enzymes on the tip of the sperm head (see Fig 26.19). During the acrosome reaction the bag splits, releasing the enzymes, which digest a pathway through any remaining follicle cells and the zona pellucida.

As soon as the outer membrane of the first sperm penetrates the cell surface membrane of the oocyte, a rapid reaction occurs. Many **cortical granules** fuse with the zona pellucida, forming a **fertilisation membrane**. This reaction starts at the point of entry of sperm head and spreads rapidly over the surface of the oocyte, preventing entry of other sperm. So, only one sperm enters the diploid secondary oocyte, even though many reach it at the same time. Entry of the sperm nucleus triggers the completion of meiosis II in the female nucleus, leading to the formation of the second polar body and a haploid mature ovum.

Almost immediately afterwards, a spindle forms and the paternal and maternal chromosomes come together, forming a diploid zygote. Within 12 hours, the first mitotic division takes place. Cell division is now rapid, forming a bundle of cells called a **morula** (Latin for the blackberry it resembles). As divisions continue, this becomes a **blastocyst** and moves slowly along the Fallopian tube through the action of cilia, which create a steady current of fluid towards the uterus (see Fig 26.20).

It takes around 6 or 7 days for the embryo to complete the journey down the Fallopian tube. When it arrives at the uterus on day 21 of the cycle, the lining must be in just the right condition to accept it. For a successful pregnancy there must be exact timing between the preparation of the endometrium and the development of the embryo.

IMPLANTATION

Pregnancy begins not with fertilisation but when the embryo **implants** in the wall of the uterus. This happens about a week after fertilisation and is not always successfully completed. Many women trying to conceive may have a 'near miss', when an egg is fertilised but fails to implant.

Implantation begins when the blastocyst makes contact with the endometrium (Fig 26.20), usually on the back wall. The outer layer of the blastocyst, the **trophoblast**, causes an **inflammatory-type response** (normally a response to damage) and this causes an outgrowth of the endometrium at this point. The placenta develops where the trophoblast and the endometrium interact.

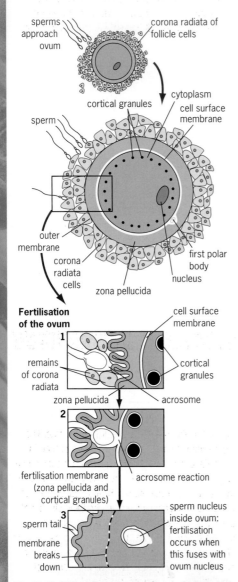

Fig 26.19 With a diameter of 100 μm, the egg cell is the largest cell in the human body. Most of the volume is taken up by inert food material that will fuel the embryo during its first few cell divisions. A spermatozoon must pass through any remaining follicle cells (the corona radiata) and the zona pellucida, before it can fertilise the egg cell. The head of the sperm enters the body of the egg cell, but the tail remains outside

If implantation is successful, the embryo begins to secrete **human chorionic gonadotrophin (hCG)**. This hormone forces the corpus luteum in the ovary to continue to secrete progesterone, thereby maintaining the endometrium and inhibiting FSH production.

The **chorion**, one of the membranes that later grows and surrounds the embryo, develops **villi** (projections) that burrow into the endometrium. These are thought to break down the mother's blood vessels, causing the chorionic villi to become bathed in maternal blood.

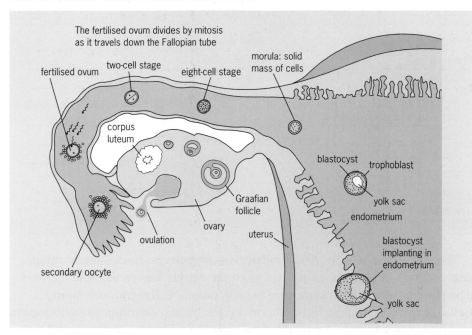

Fig 26.20 Following fertilisation, the embryo undergoes several mitotic divisions as it passes slowly down the Fallopian tube. When the embryo reaches the uterus, it implants in the endometrium

THE PLACENTA

The placenta is a temporary organ that allows the blood systems of the fetus and the mother to come into close contact, without actually mixing (Fig 26.21). The placenta allows nutrients and oxygen to pass to the fetus from the mother, and allows metabolic waste back into the mother's blood (see Table 26.5).

SEE QUESTIONS 4 TO 7

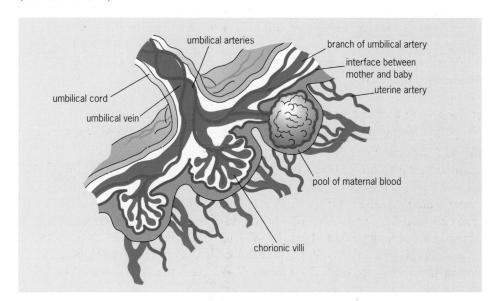

Fig 26.21 The fine structure of the placenta. The capillaries within the chorionic villi are bathed in pools (lacunae) of maternal blood. The placenta is a large disc-shaped organ that is usually attached to the back wall of the uterus. At birth, contraction of the uterine muscle causes the chorionic villi to split away from the endometrium and the placenta is delivered after the baby: hence its common name of afterbirth. Most mammals eat the placenta after the birth of their young. It is an important source of nourishment at a time of great need

✔ **REMEMBER THIS**

Pregnancy testing kits make use of the hormonal changes that occur in early pregnancy (see page 126 in Chapter 6).

The placenta is an organ adapted to maximise the exchange of materials so, as you would expect, it has a large surface area provided by the **chorionic villi**. A close look at the cells of the villi shows that the membranes are folded into microvilli and also contain many mitochondria: these two features maximise the processes of diffusion and active transport (see Chapter 4). There are many small vesicles in the cells of the villi, suggesting that substances are being absorbed by pinocytosis (see also Chapter 4).

Table 26.5 Exchange of materials across the placenta

Mother to fetus	Fetus to mother
oxygen	carbon dioxide
glucose	urea
amino acids	other waste products
lipids, fatty acids and glycerol	
vitamins	
ions; Na, Cl, K, Ca, Fe	
alcohol, nicotine, many drugs	
viruses	
antibodies	

The placenta is an important **endocrine organ**. It secretes the hormones that maintain pregnancy, taking over from the corpus luteum at about 12 weeks. The placenta secretes progesterone (which maintains the endometrium), oestrogen (which inhibits the ovulatory cycle), human chorionic gonadotrophin (see the Science feature in Chapter 6 page 126) and **human placental lactogen** (which stimulates the development of the mammary glands).

PREGNANCY, LABOUR AND BIRTH

In humans, pregnancy lasts for an average of 40 weeks. Some of the main stages of growth of a baby are shown in Fig 26.22.

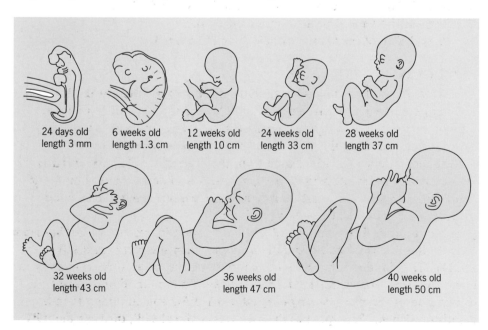

24 days old
length 3 mm

6 weeks old
length 1.3 cm

12 weeks old
length 10 cm

24 weeks old
length 33 cm

28 weeks old
length 37 cm

32 weeks old
length 43 cm

36 weeks old
length 47 cm

40 weeks old
length 50 cm

Fig 26.22 Stages in the growth of a baby in the uterus

From around the twelfth week of pregnancy, progesterone secreted by the placenta inhibits uterine contractions. The level of progesterone rises steadily until just before birth, usually around 38 weeks after conception, when it starts to fall dramatically. This lifts the inhibition of uterine contractions. The mother's anterior pituitary begins to secrete **oxytocin** and the placenta secretes prostaglandins, two hormones that actively promote contractions. Oxytocin stimulates uterine contractions at about 40 weeks. The resulting tension in the muscle and pressure on the cervix are stimuli that bring about further secretion of oxytocin, causing more powerful contractions.

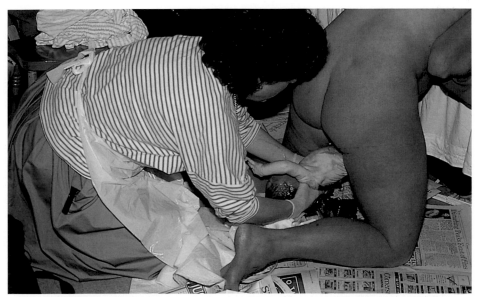

Fig 26.23 Childbirth is one of the most intense human experiences. Childbirth used to be a time of great danger to both mother and child, but advances in medicine have greatly reduced the risk. Today, in the developed world, 99 out of every 100 babies born survive beyond their first birthday (200 years ago, only 54 out of every 100 lived)

There are three stages of labour:
- Stage 1. The cervix dilates (opens) to a diameter of 10 cm.
- Stage 2. The fetus is pushed out of the uterus.
- Stage 3. The placenta and umbilical cord are expelled.

LACTATION

All mammals produce milk from specialised **mammary glands**. Humans are unique in having permanent breasts but these are normally composed only of fatty tissue. Throughout pregnancy, however, oestrogen and progesterone stimulate the development of milk-producing tissue.

After birth, the first fluid released from the breasts is called **colostrum**. This contains no fat and very little sugar, but has important antibodies that 'lend' the baby immunity until it has time to develop its own. After about three to four days, normal milk is produced.

Milk is produced constantly and, although a certain amount of leaking from time to time is usual, milk flow only happens when the baby suckles (Fig 26.24). The stimulus of sucking at the nipple causes the posterior pituitary gland to secrete oxytocin. This hormone travels in the blood and causes contraction of the muscular **myo-epithelial cells** surrounding the milk glands or alveoli. This squeezes the milk out of the alveoli, through the milk ducts and into the infant's mouth. This mechanism, called the **let-down reflex**, takes several seconds to take effect but is quite powerful. If the unsuspecting infant lets go of the nipple, it can get a jet of milk in the eye.

Throughout the period of lactation the pituitary continues to secrete **prolactin**, a hormone that maintains the milk ducts and, to some extent, inhibits ovulation. This inhibition is called **lactational anoestrus** and reduces the risk of conception so soon after birth. However, it is not always effective and breast-feeding cannot be relied on as a contraceptive.

Fig 26.24 During pregnancy the milk-producing ducts grow, substantially increasing the size of the breasts. When the new baby suckles, the physical stimulation leads to a series of hormonal events that increase the amount of milk that is made

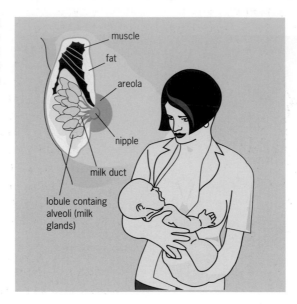

SUMMARY

After studying this chapter, you should know and understand the following:

- **Asexual reproduction** usually involves one individual. All offspring are genetically identical: they are **clones**.
- **Sexual reproduction** involves two sexes which produce **haploid gametes**. Males produce **spermatozoa** by **spermatogenesis**, females make **ova** (egg cells) by **oogenesis**. Both processes involve a special type of cell division called **meiosis**.
- At fertilisation, a new **diploid zygote** is formed. The zygote, which is genetically unique, grows and develops by a series of mitotic divisions.
- The timing of reproductive events in humans is controlled by hormones. The **hypothalamus** controls the **pituitary gland**, which releases **gonadotrophins**. These stimulate the gonads to secrete steroids: **oestrogen**, **progesterone** and **testosterone**.

- The human menstrual cycle lasts, on average, for 28 days. The cycle begins with **menstruation**, which occurs as the next ovum (egg cell) is prepared inside the ovary. By day 14 a new endometrium has grown. Ovulation takes place on day 14 and the egg cell is then available to be fertilised. If the egg cell is not fertilised the endometrium breaks down and the cycle repeats.
- In humans, the egg cell is fertilised in the Fallopian tube. The **zygote** begins mitotic divisions, becoming a **morula** and then a **blastocyst** as it moves along the tube towards the uterus. When the blastocyst has implanted in the endometrium, it is known as an **embryo**.
- The developing embryo (which is called a **fetus** after about 8 weeks) is nourished via the placenta, a temporary organ that does the job of the fetal lungs, intestines and excretory system.
- Birth is brought about by a series of hormonal changes that begin uterine contractions. In mammals, the infant is nourished by milk produced by the mother.

❓ EXAM QUESTIONS

1

a) Describe the processes involved in spermatogenesis.
b) Explain the hormonal control of puberty in boys.
c) Discuss the ethical issues of contraception.

IBO Biology Higher level Paper 2 May 1999 Q7

2 The diagram below shows some cells from a transverse section through a seminiferous tubule from a human testis.

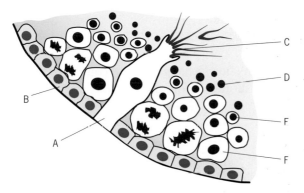

a) Give the functions of the cells labelled A and B.
b) **(i)** Give the letters of two cells which have the haploid number of chromosomes.
 (ii) Give the letters of two cells which have the diploid number of chromosomes.
c) Give two ways in which the process of oogenesis in humans differs from that of spermatogenesis.

Edexcel HB6 (Synoptic) June 1999 Q3

3

a) Copy and complete the table which gives information about some of the hormones which control the oestrous cycle.

Hormone	Site of secretion	Target organ	One effect
Luteinising hormone (LH)			
Oestrogen	ovary	uterus	stimulates growth of uterine lining
Progesterone	ovary	uterus	

b) The diagram shows the way in which hormones are involved in controlling part of the oestrous cycle.
(i) Some women only produce very small amounts of FSH. Explain why these women are infertile.
(ii) Clomiphene is a drug used to treat this type of infertility. It blocks the action of oestrogen. Explain how treatment with clomiphene could be used to stimulate production of FSH.

AQA (NEAB) BY01 Processes of Life June 2000 Q6

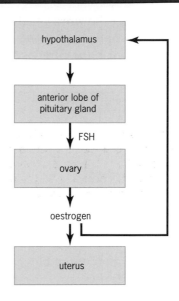

4 Give an account of the structure and functions of the placenta.

Edexcel HB3 June 1999 Q8

5 The diagram shows a simplified diagram of a placenta and fetus.

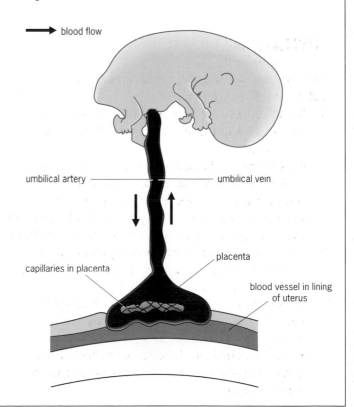

? EXAM QUESTIONS

a) Using the words, higher, lower, same, as appropriate complete the following table to compare the composition of the blood in the umbilical vein with that in the umbilical artery.

Content in umbilical vein	Compared to umbilical artery
oxygen	
urea	
glucose	
carbon dioxide	
immunoglobulins	
oestrogen	

b) On the diagram draw an arrow to show the direction of the blood flow in vessels in the lining of the uterus.

c) State four ways by which the placenta is adapted to perform its functions.

d) Suggest why the cells of the chorion, which develop into the placenta, are genetically identical to those in the embryo.

e) State how the following cells and structures are provided with nutrients.

(i) Spermatozoa in the female reproductive tract.

(ii) The developing ball of cells in the oviduct.

OCR 6918 Animal Studies March 2001 Q6

6 For each statement, write T if you consider that it is correct, and F if you consider that the statement is incorrect.

a) Both testosterone and spermatozoa are produced in the testes of mammals.

b) Fertilisation is the term used to describe the introduction of spermatozoa into the reproductive tract of the female.

c) Semen is produced by the testes.

d) Oogonia and spermatogonia are produced by mitosis.

e) Polar bodies are produced during oogenesis and during spermatogenesis.

f) Oestrogen stimulates the production of FSH.

g) LH stimulates the production of oestrogen.

h) In males FSH may be referred to as ICSH.

i) Fertilisation normally takes place in the oviducts.

j) Oxytocin is involved both in birth and in lactation.

k) A fall in oestrogen and progesterone levels in the blood causes the lining of the uterus to break down.

l) In humans the oestrous cycle is referred to as the menstrual cycle.

OCR 6918 Animal Studies March 2000 Q1

7 The graph shows changes in the concentration of some hormones during the oestrous cycle of a cow.

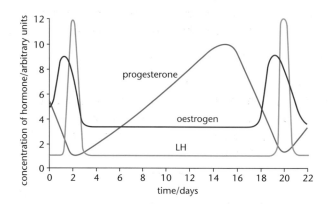

a) Between which days is fertilisation most likely to be successful? Explain your answer.

b) Did the cow become pregnant during this cycle? Give evidence from the graph for your answer.

c) **(i)** Describe how oestrus may be detected in a **named** farm animal.

(ii) Explain why it is important for a farmer to be able to detect oestrus in farm animals.

AQA (A) BYA2 Making Use of Biology January 2001 Q5

KEY SKILLS **ASSIGNMENT**

INFERTILITY TREATMENT

Nine out of ten couples who use no contraception and actively try to conceive, are successful within a year. But, one in ten couples may face the distressing problem of infertility. Many are 'sub-fertile' rather than infertile, and can be helped by the various methods described in this Assignment. However, for a few, nothing works.

In this exercise we look at the possible causes of infertility and at some of the treatments. When you have read both and worked through the questions, you should be in a position to prescribe the best course of treatment for a particular couple.

Causes of infertility

Doctors accept that a couple may need infertility treatment if they have been trying to conceive for at least a year without any success. The first objective is to establish whether either partner has an obvious problem that is preventing conception.

Female infertility may be caused by:

- Blocked Fallopian tubes.
- Altered hormone levels leading to a failure to ovulate or implant. Failure to ovulate, known as **anovulatory infertility**, is usually due to a failure to secrete the right balance of hormones.
- Cervical mucus that halts, repels or kills sperm.

1 **a)** How would a specialist find out if a woman's Fallopian tubes were blocked?
b) Why might blocked Fallopian tubes cause infertility?
c) Which hormones combine to cause ovulation?
d) What should happen to the cervical mucus around the time of ovulation?

Male infertility may be caused by:

- A low sperm count. Samples that are found to have fewer than 20 million sperm per cm^3 are said to be abnormally low.
- Production of large numbers of abnormal sperm (more than 4 per cent).
- Production of antibodies that make the sperm stick together.

2 Men who wear tight-fitting clothes, or who spend a lot of time in a hot bath, have sometimes been found to have a low sperm count. Suggest a reason for this.

Treatment methods: ovulation induction

A woman may not ovulate because the balance of hormones in her body is abnormal. To restore the hormone levels and initiate follicle development and ovulation, she can have treatment with artificial gonadotrophins or drugs that stimulate the natural secretion of gonadotrophins. One such drug, clomiphene, works by increasing FSH secretion.

After such treatment, the response of the ovaries can be followed by ultrasound, which shows how many follicles are developing in each ovary. A follicle is considered to be ready for ovulation when it reaches 17 mm in diameter. The endometrium is also checked – it should be at least 8 mm thick at this time.

When a ripe follicle is detected, ovulation can be stimulated artificially by injecting human chorionic gonadotrophin (hCG), a hormone that has a similar effect to LH. Ovulation should occur after about 36 to 48 hours, and so intercourse should be timed to coincide with this.

3 **a)** Why will increased FSH secretion help a woman who isn't ovulating regularly?
b) What is a potential problem with treatment that stimulates ovulation? Hint: this may not be a problem if the couple want a large family.

Treatment methods: In vitro fertilisation (IVF)

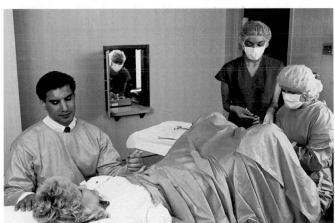

Fig 26.A1 A woman undergoing IVF treatment must endure weeks of hormone treatments, followed by the uncomfortable process of egg cell collection. Up to about 20 egg cells may be recovered. Her egg cells are fertilised by her partner's sperm '*in vitro*' and, when the tiny embryos have developed, two of these are put back into her uterus. IVF has, at best, about a 30 per cent success rate

In vitro (= 'in glass') fertilisation is what most people think of as 'test-tube baby' treatment. In IVF, the egg cells and sperm are taken from the couple and fertilised in a dish. The process of fertilisation normally takes 12 to 15 hours and after this the new embryos start to develop. Cell division is taken as a sign that fertilisation has been successful. Then, the tiny embryos (no more than balls of 8 to 16 cells at this stage – see Fig 26.A2) are placed into the uterus.

KEY SKILLS **ASSIGNMENT**

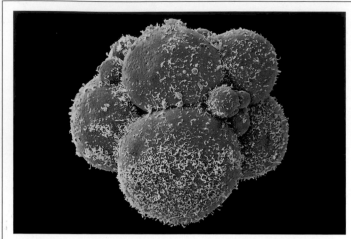

Fig 26.A2 A human embryo at the 8-cell stage. The eight large cells (red) are covered with microvilli (yellow). The smaller cells will degenerate

4 How often does an ovary normally produce an ovum?

Treatment methods: gamete intra-Fallopian transfer (GIFT)

GIFT involves stimulating the ovaries and collecting the ova in much the same way as in IVF treatment. The important difference is that in GIFT the egg cells are mixed with sperm and immediately introduced into the Fallopian tubes (Fig 26.A3), without waiting to see if fertilisation occurs. The advantage of this procedure – which would seem to be less controlled than IVF – is a 5 per cent better success rate. This is possibly because the Fallopian tube is the natural site for fertilisation: it may secrete chemicals that stimulate the process.

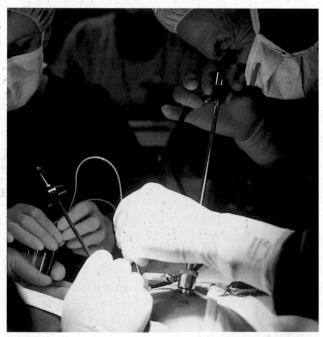

Fig 26.A3 A mixture of egg cells and sperm is being inserted into the woman's Fallopian tube through a small insertion

5 Many specialists feel that counselling couples is an important part of IVF and GIFT treatment. Think of two reasons why these treatments might cause anxiety.

Treatment methods: intra-uterine insemination (IUI)

In around 20 per cent of infertility cases, there seems to be no problem with either partner. Surprisingly, in some of these cases, intra-uterine insemination proves successful. The basic idea behind IUI is to introduce the partner's semen into the uterus.

The steps in IUI are as follows:
● Follicle development is stimulated and monitored.
● Treatment to induce ovulation is given.
● A fresh sperm sample is introduced into the uterus.

6 What advantage does IUI give over trying to conceive naturally?
After working through the Assignment so far, you should now be able to tackle the following question:

7 Copy and complete the following table, matching up the causes of infertility to suitable treatments. Remember that there may be more than one suitable treatment.
Your options are:

OI – ovulation induction

IVF – *in vitro* fertilisation

GIFT – gamete intra-Fallopian transfer

IUI – intra-uterine insemination

Problem	Possible treatments
Blocked Fallopian tubes	
Hostile cervical mucus	
Low sperm count	
Woman has antibodies against partner's sperm	
Many abnormal sperm	

8 A programme of infertility treatment can put a great strain on a relationship. Suggest reasons for this.

9 Hormone replacement therapy can be used to allow post-menopausal women to become pregnant. A donated egg can be fertilised by sperm from the woman's partner, and can lead to women as old as 60 bearing children. Discuss the advantages and disadvantages of this technology.

27

DNA, genes and chromosomes

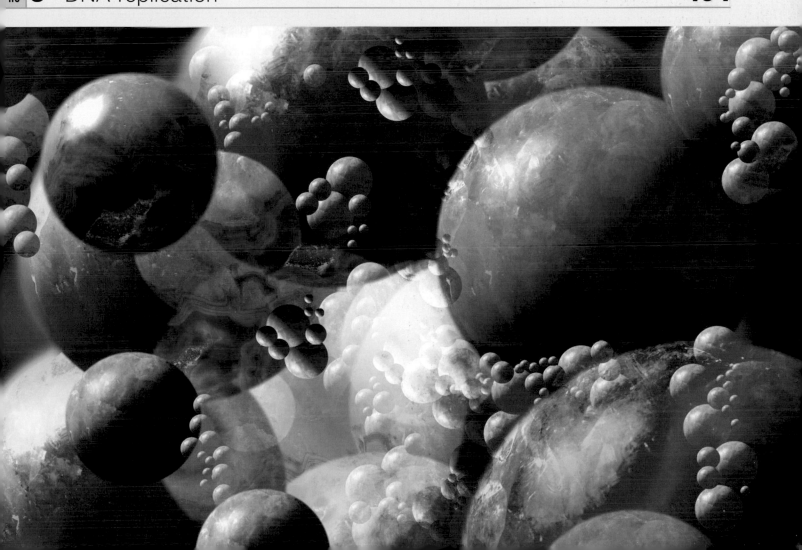

DNA, genes and chromosomes

Why do we age? Until a few years ago, scientists and doctors thought that our bodies simply wore out like machines. They now think that many aspects of ageing are genetically determined. Organisms develop and grow by cell division, and the same process happens in adults to repair and maintain their bodies. It seems likely that each of our cells has a genetic timer that counts how many times a cell divides: after a maximum number of divisions, the cell dies. But what sets this limit?

The latest research is focusing on the ends of chromosomes, the telomeres. Every time a cell divides, the telomeres shorten, and when they disappear completely, the cell is unable to divide further and its repair systems break down. Researchers have been intrigued to find that eggs and sperm are 'immortal' – they have an enzyme, telomerase, which ensures that the embryo's telomeres are complete. This is necessary to make sure that the genetic clock that produces a baby is set to zero. It might seem obvious, but babies need to be born with 'new' cells so that they have their full life expectancy, regardless of the age of their parents.

The discovery of telomerase opens up the exciting, if ethically difficult, possibility that we may be able to manipulate the ageing process. If we can use telomerase to repair the telomeres of normal body cells, it may be possible to extend potential life expectancy to 150 years, or more. The social and economic consequences of this are massive, and are bound to be the cause of much heated debate in years to come.

Imagine the implications of prolonging your life well past 100. How long would you work for? When would you retire? Could any government afford pensions and healthcare bills? What would happen to societies in which six, seven or even eight generations of the same family were alive together?

1 WHAT IS A GENE?

The nucleus of a eukaryotic cell has a set number of chromosomes – every human body cell has 46. Each chromosome is a single, elaborately coiled DNA molecule that has individual genes dotted along its length (Fig. 27.1). A gene is a length of DNA that codes for the synthesis of one polypeptide.

We used to think that a gene contained all the information needed to build a complete protein, such as an enzyme. However, many proteins consist of more than one polypeptide chain, so these molecules are coded for by more than one gene. For instance, haemoglobin molecules consist of four polypeptide chains, two alpha and two beta, so it takes two genes to make a complete haemoglobin molecule, one is copied twice for the alpha chains, the other twice for the beta chains.

A gene is a region within the DNA molecule

Fig. 27.1 Genes are sequences in the DNA molecule. The sequence of bases shown here represents one very short gene. In the genetic code, a group of three bases codes for one amino acid, so a protein consisting of 500 amino acids requires a gene of at least 1500 base pairs long, probably more. When unravelled, chromosomes are incredibly long – each one consists of up to 4000 genes and these make up only about 10 per cent of its total length (the rest is non-coding DNA)

GENES IN ACTION:
THE CENTRAL CONCEPT OF MOLECULAR BIOLOGY

Can we sum up the way genes work in one sentence? It is a tall order, but the underlying theme, or **central dogma**, of genetics is:

The DNA of a gene codes for the production of messenger RNA which, in turn, codes for the production of a polypeptide
(Fig 27.2).

In this chapter we look at how genes work, and at how they are copied to allow cell division, reproduction and growth.

After Crick and Watson worked out the structure of DNA in 1953, two vital questions remained:

- How does information get from the DNA in the nucleus to the site of protein synthesis in the cytoplasm?
- How does the base sequence of the DNA translate into the amino acid sequence in a polypeptide?

To answer the first question, Crick, Brener and Monod developed the **messenger hypothesis**. This states that a specific molecule, named **messenger RNA** (mRNA), is copied directly from the DNA sequence of the gene. mRNA is therefore a mobile copy of a gene. It can move from the nucleus to the cytoplasm. Here, ribosomes use mRNA as a template (pattern) to assemble amino acids in the correct order to make the required polypeptide.

Crick also answered the second question: he proposed that an 'adaptor' molecule existed which had two specific binding sites. At one side, the molecule fitted a specific DNA base sequence and at the other, a specific amino acid. This molecule was later discovered and named **transfer RNA** (tRNA) (see Fig 27.6 on page 491).

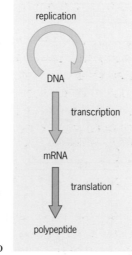

Fig 27.2 The central concept of molecular biology. The base sequence on a particular gene is copied, or transcribed, to produce a messenger RNA template. This mRNA template passes out of the nucleus and moves to the ribosomes where it is read, or translated, into a polypeptide

 REMEMBER THIS

The genome is defined as 'all the DNA sequences in an organism'. It is a complete set of genes, together with the non-coding DNA in between. The human genome is thought to consist of over 3 billion base pairs. Remarkably, the entire genome is present in every one of our body cells.

 REMEMBER THIS

The position of a gene on a chromosome is known as its **locus** (plural: loci).

2 THE GENETIC CODE

Along the DNA molecule the base sequence is continuous, but it is read in blocks of three. Each three, or triplet, is called a **codon**. Each codon codes for a particular amino acid. A particular sequence of bases therefore codes for a specific amino acid sequence. When the amino acids are assembled into a polypeptide, they interact to twist and bend the chain into its final shape. When the polypeptide chain is complete, it forms all or part of a protein which performs a vital role in the organism.

You will have noticed that there are two strands of bases in the DNA molecule. One strand, called the **template strand** or **sense strand**, contains the all-important genetic code for any particular gene. The corresponding part of the other strand is simply there to stabilise the molecule. However, in the same DNA molecule, different genes are present on different sides and the enzymes that transcribe DNA can use either side as the sense strand.

 REMEMBER THIS

Retroviruses such as the Human Immunodeficiency Virus (HIV) are an exception to the central dogma. Retroviruses contain RNA from which they make a DNA copy using the enzyme **reverse transcriptase**.

SEE QUESTIONS 1 TO 3

 REMEMBER THIS

The base pairings are always the same. In DNA, A bonds with T and G with C. In RNA, T is replaced by U, so A bonds with U, and G with C (see Chapter 3).

⚠ SCIENCE FEATURE Nucleic acids — molecules with jobs

A favourite topic for biology examiners is to ask the candidate to relate structure to function. On a molecular level, the nucleic acids are classic examples. This feature builds on your knowledge of the basic structure of nucleic acids covered in Chapter 3. The differences between DNA and RNA are shown in Table 27.1.

Table 27.1 The differences between DNA and RNA

	DNA	mRNA	tRNA
Sugar	deoxyribose	ribose	ribose
Bases	C, G, A, T	C, G, A, U	C, G, A, U
Strands	two	one	one
Size of molecule	enormous	small compared with DNA, varies with size of gene	small. constant size
Life span	long term	short term	short term
Site of action	nucleus	nucleus and cytoplasm	cytoplasm

DNA

How is the structure of DNA related to its function?

- It contains a sequence of bases that codes for a sequence of amino acids. Cells use this genetic code to make the proteins that build organisms. The base sequence is read in blocks of three, eg GAU – this is a **codon**.
- It is long, so can store a lot of information.
- It has complementary base pairing that allows replication and transcription.
- It has a stable sugar–phosphate backbone to give strength.
- It has relatively weak hydrogen bonds between the bases. This allows the two strands to come apart – again allowing replication and transcription.

Messenger RNA

mRNA is simply a mobile copy of a gene, carrying a base sequence that is complementary to – not the same as – the gene on which it was made. mRNA molecules are long, ribbon-like molecules consisting of a single strand of nucleotides. Molecules of mRNA vary in length – there are as many different ones as there are genes.

Transfer RNA

This is the adaptor molecule, connecting the genetic code to the protein. tRNA molecules have an **anticodon** on one end and an amino acid binding site at the other. The role of tRNA is to pick up a particular amino acid, bring it to the ribosome and hold it in place so it can be added to the growing polypeptide. tRNA binds to the mRNA by means of the anticodon. It could be argued that because there are 20 different amino acids, there are 20 different types of tRNA molecule. Another way of looking at it is that there are 61 different codons (64 minus 3 **stop codons**) so there are 61 different types of tRNA.

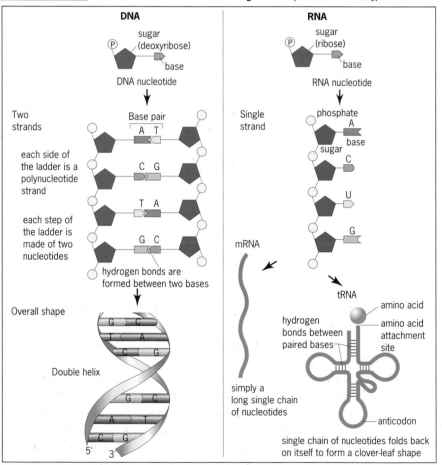

Fig 27.3 Comparison of the three types of molecule

TRANSCRIPTION

DNA can be thought of as a permanent reference library: it does not leave the nucleus and so cannot be used directly for protein synthesis. Instead, the genetic code in a gene is transferred, or **transcribed**, on to a smaller RNA molecule, **messenger RNA**, which can then pass to the ribosomes to act as a template for protein synthesis. In effect, mRNA is a mobile copy of a gene.

Before transcription can begin, the DNA helix must unwind and the two halves of the molecule must come apart, so exposing the base sequence. This process begins when the DNA is 'unzipped' by specific enzymes and the enzyme **RNA polymerase** attaches to the DNA molecule at the **initiation site**. This is a base sequence at one end of the gene which effectively says: start here. Once in place, the enzyme moves along the gene (Fig 27.4), assembling the messenger RNA molecule by adding the matching nucleotides, one at a time.

Once formed, the mRNA copy begins to peel away from the DNA. When RNA polymerase has passed a particular region of DNA, the double helix rewinds. When the whole gene has been transcribed, the complete mRNA molecule leaves the nucleus. As on a production line, RNA polymerase molecules can follow each other along the gene, assembling several copies of mRNA together.

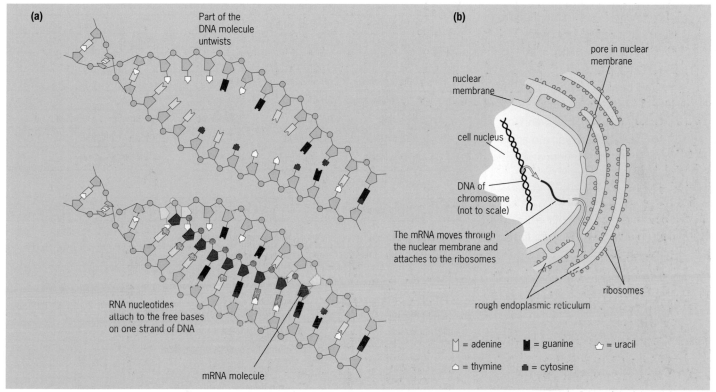

Fig 27.4(a) The process of transcription. The process is controlled by enzymes that are not shown here. Starting at the initiation site, helicase enzymes separate the two sides of the DNA molecule, and then the enzyme RNA polymerase calalyses the assembly of mRNA, one nucleotide at a time, until the gene has been transcribed
(b) The sites of transcription and translation in the cell. Transcription happens in the nucleus and involves the assembly of a molecule of messenger RNA. Translation happens on the ribosomes and involves the assembly of a protein according to the code on the mRNA

TRANSLATION

Translation occurs when the base sequence on the mRNA is used to synthesise a polypeptide. The process is called translation because the information contained in the gene – and delivered by the mRNA – is translated into a polypeptide (Table 27.2 on the next page).

Protein synthesis is one of the major activities of cells, and it is highly likely that you are already familiar with many of the organelles and chemicals involved. The roles of individual organelles in the process are listed in Table 27.3. See also Chapters 1 and 3 for background information.

Translation occurs on **ribosomes**, structures that hold all the components together as an amino acid chain is made. The mRNA strand passes out of the nucleus and attaches to the ribosome. At the point of attachment, two tRNA molecules deliver the required amino acids and hold them in position so that they can be added to the growing polypeptide.

The mechanism of translation is shown in detail in Fig 27.6. The process continues until an mRNA **stop codon** – which has no tRNA molecule – moves into the ribosome's first binding site. At this point all the components separate, releasing the completed polypeptide.

REMEMBER THIS

Genes code for the manufacture of polypeptides. Fig 3.27 on page 52 shows the importance of protein in the human body.

 REMEMBER THIS

Many people find protein synthesis a difficult topic to learn. A valuable exercise is to make the different components – either from modelling clay or just paper – and work through the events shown in Fig 27.6(b).

You will need to make a ribosome, an mRNA molecule with several codons written on it, and enough tRNA molecules and amino acids to match to all the codons on the mRNA.

Table 27.2 The genetic code: the base sequence in each mRNA triplet code translates to a particular amino acid. This same code is used by all organisms					
First base	**Second base**				**Third base**
	G	A	C	U	
G	GGG glycine	GAG glutamic acid	GCG alanine	GUG valine	G
	GGA glycine	GAA glutamic acid	GCA alanine	GUA valine	A
	GGC glycine	GAC aspartic acid	GCC alanine	GUC valine	C
	GGU glycine	GAU aspartic acid	GCU alanine	GUU valine	U
A	AGG arginine	AAG lysine	ACG threonine	AUG methionine	G
	AGA arginine	AAA lysine	ACA threonine	AUA isoleucine	A
	AGC serine	AAC asparagine	ACC threonine	AUC isoleucine	C
	AGU serine	AAU asparagine	ACU threonine	AUU isoleucine	U
C	CGG arginine	CAG glutamine	CCG proline	CUG leucine	G
	CGA arginine	CAA glutamine	CCA proline	CUA leucine	A
	CGC arginine	CAC histidine	CCC proline	CUC leucine	C
	CGU arginine	CAU histidine	CCU proline	CUU leucine	U
U	UGG tryptophan	UAG **stop**	UCG serine	UUG leucine	G
	UGA **stop**	UAA **stop**	UCA serine	UUA leucine	A
	UGC cysteine	UAC tyrosine	UCC serine	UUC phenylalanine	C
	UGU cysteine	UAU tyrosine	UCU serine	UUU phenylalanine	U

Table 27.3 The main organelles and chemicals involved in protein synthesis	
Organelle/molecule	**Role in protein synthesis**
Nucleus	houses the DNA
Nucleolus	manufactures the ribosomes
Ribosome	site of translation – where polypeptide assembly occurs
Endoplasmic reticulum	isolates, stores and transports polypeptides
Golgi apparatus	modifies and packages polypeptides
DNA	stores the genetic information
Messenger RNA	is a mobile copy of a gene on DNA
Transfer RNA	each brings a specific amino acid to the ribosomes

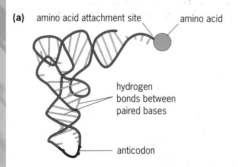

(a) amino acid attachment site   amino acid
hydrogen bonds between paired bases
anticodon

(b)
amino acid attachment site   amino acid
hydrogen bonds between paired bases
anticodon

Fig 27.5 Two ways of representing the tRNA molecule. **(a)** The 3D shape of the molecule, stabilised by hydrogen bonds as the single nucleotide chain folds back on itself
(b) A schematic diagram: each type of transfer RNA transports a specific amino acid to the ribosomes, the site of protein synthesis. At one end of the molecule is the anticodon which attaches to the corresponding codon on the messenger RNA. At the other end of the molecule is the amino acid attachment site

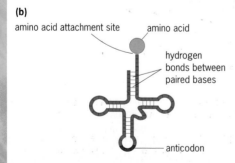

After translation, polypeptides are processed according to their final destination:

● Those that are to be exported from the cell, such as digestive enzymes, are threaded through pores in the endoplasmic reticulum to accumulate on the inside. Here they are processed and packaged before being secreted (see pages 12 and 14).

● Polypeptides that will form membrane proteins follow the same route as those for export, but they remain on the cell surface membrane rather than being released.

● Polypeptides that will be used inside the cell, such as those that form haemoglobin, remain free in the cytoplasm.

THE ROLE OF TRANSFER RNA

Transfer RNA molecules are smaller than mRNA molecules, containing only about 75 to 80 nucleotides. Each consists of a single strand of nucleic acid folded back on itself to form a 'clover-leaf' shape (Fig 27.5). Transfer RNA molecules bring specific amino acids from the cytoplasm to the ribosome so that they attach to the growing polypeptide.

Fig 27.6(a) The ingredients needed for protein synthesis

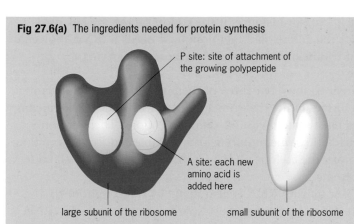

P site: site of attachment of the growing polypeptide

A site: each new amino acid is added here

large subunit of the ribosome

small subunit of the ribosome

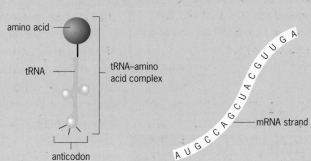

amino acid

tRNA

tRNA–amino acid complex

anticodon

mRNA strand

Fig 27.6(b) The process of protein synthesis

(1) The two subunits of the ribosome come together. The mRNA strand binds to the ribosome. Then an amino acid–tRNA complex with an anticodon complementary to the first codon on the mRNA, binds to the P site

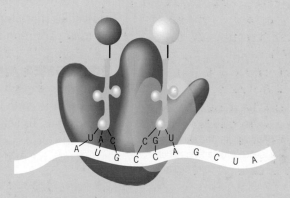

(2) In the next step, an amino acid–tRNA complex with an anticodon complementary to the next codon on the mRNA strand binds to the A site.

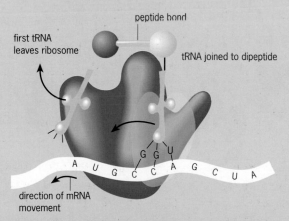

peptide bond

first tRNA leaves ribosome

tRNA joined to dipeptide

direction of mRNA movement

(3) A peptide bond forms between the two amino acids which have been brought together on the ribosome. The bond between the first amino acid and the first tRNA is broken and the tRNA molecule leaves the ribosome. The ribosome moves along the mRNA strand and the second tRNA molecule, now joined to a dipeptide, shifts across from the A site to the P site

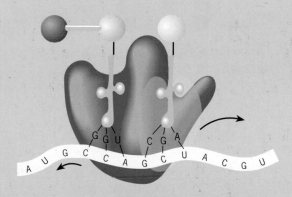

(4) A third amino acid–tRNA complex with an anticodon that matches the third codon on the mRNA strands comes into the A site

The sequence of events from **(3)** to **(4)** is repeated until the entire length of the mRNA strand has been translated. The polypeptide is then complete

The process by which an amino acid binds to a tRNA molecule is controlled by an enzyme. The process also involves the splitting of ATP. This is important because ATP gives the tRNA–amino acid complex enough energy to form a peptide bond when the amino acid is added to the growing polypeptide.

At one end of the tRNA molecule is the **anticodon**, a three-letter base sequence that matches the codon on the mRNA molecule. At the other end is a particular amino acid. So, for instance, if the codon on the mRNA reads AUG, which codes for the amino acid methionine, it needs a tRNA molecule with an anticodon of UAC which arrives at the ribosome carrying a methionine molecule.

There are 64 different codons, but three of them code for 'stop translating'. When such a codon arrives at the ribosome, no tRNA is needed. This means that tRNA molecules must match with 61 different codons. There are only 20 different amino acids, so some amino acids are translated from more than one codon. For example, the codons GGG, GGA, GGC and GGU all code for the amino acid glycine.

Like messenger RNA, transfer RNA is also made by transcription. In eukaryote DNA there are many genes whose sole function is to make tRNA. In any cell, most genes occur only once, but there are hundreds of genes that code for tRNA synthesis, to provide the vast numbers of these workhorse molecules that are needed. That tRNA genes do not code for proteins is another exception to the central dogma.

POLYSOMES AND THE RATE OF TRANSLATION

We have seen that the mRNA molecule is a long single strand of nucleotides, and that the code is translated into a polypeptide at the point of contact with a ribosome. To achieve protein synthesis at a reasonable speed, ribosomes occur in clusters, called **polysomes** or **polyribosomes**, which all translate a different bit of the mRNA at the same time (Fig 27.7).

Fig 27.7 A polysome is a cluster of ribosomes which can simultaneously translate a different section of the mRNA molecule. The situation is similar to having a message on a long ribbon, with several people reading different bits at once

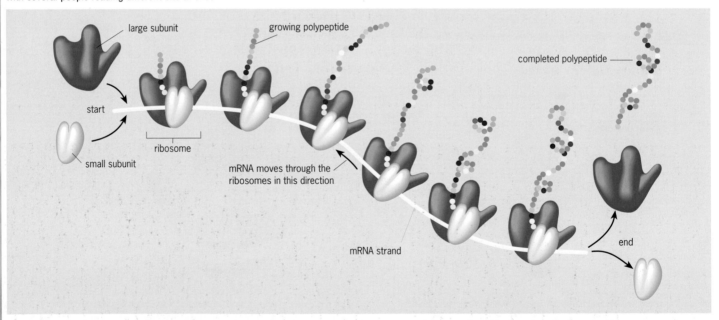

3 A JOURNEY ALONG A CHROMOSOME

If one single chromosome were unravelled, like pulling the wool out of a piece of knitting, we would be left with one long DNA strand. Although this is one single molecule, it is several *centimetres* long and contains up to four thousand genes.

How do you go about investigating such a molecule? One way is to work backwards; if you can isolate the mRNA molecules that are in the cytoplasm, you can trace their origin in the DNA.

A clever technique called **nucleic acid hybridisation** is used to pinpoint the origin of the mRNA, giving the location of the gene itself. In this technique, DNA strands are denatured: when heated to 87 °C, hydrogen bonds that hold the two strands together are broken, and the individual strands separate. If messenger RNA strands from the cytoplasm are added to the mixture, they stick to the region of DNA on which they were originally made.

Studies using this technique have unexpectedly revealed that there is some non-coding DNA in genes (Fig 27.8). Most eukaryote genes contain regions, called **introns**, which do not find their way into the mRNA molecule at all. The parts of the gene that are expressed in the mRNA are called **exons** because they are the expressed regions of the gene.

Generally, a gene consists of a DNA sequence which is present only once. The non-coding sequences, however, tend to consist of repeated or 'stuttered' sequences. Humans all have the same repeated sequence, but in different individuals it is repeated a different number of times and in different places. This is the basis of DNA profiling – see the Assignment in Chapter 3.

The function of non-coding DNA is not known. It is probably not useless: much of the repeated DNA in a chromosome occurs around the region of its centromere and may have an important role in maintaining the physical stability of the chromosome as a whole.

The DNA code can be compared to a page of writing in a book. If the letters were strung together continuously, we would find it difficult to read: the gaps between words help us to make sense of the sentences.

4 THE CONTROL OF GENE EXPRESSION

Most multicellular organisms start life as a zygote: a single fertilised cell. This cell has a complete set of genes. As the cell divides by mitosis, the DNA is copied faithfully, so each new cell also contains the same complete set of genes. How, then, do cells differentiate? How, for example, do some animal cells specialise into muscle, nerve or skin?

The answer lies in **gene expression** – how different genes are 'switched on', or **expressed**, in different cells (Fig 27.9).

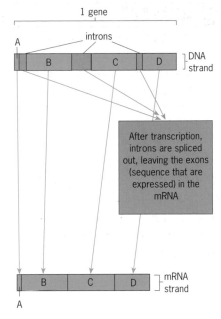

Fig 27.8 Even in a gene, there are non-coding lengths of DNA called **introns**. Only the **exons** are encoded into the mRNA molecule, and therefore expressed

Fig 27.9 A spectacular example of the power of gene expression is seen in the life cycle of some insects such as this peacock butterfly. The genes present in the fertilised egg are still present in the adult, but a different set is activated to give rise to the caterpillar and to control its metamorphosis into a pupa and then an adult – which looks like a completely different organism

✔ **REMEMBER THIS**

If we could control gene expression, we would be able to make new tissues and organs for repair and transplant. See the section on stem cells in Chapter 17 for more about the potential of controlling gene expression.

At any one time, the average cell is probably using only 1 per cent of its available genes. Some of the expressed genes are used for 'housekeeping' – they code for the proteins needed by all cells, such as the enzymes involved in respiration. In contrast, other genes are expressed only in specialised cells. The genes for making insulin, for example, are expressed only in the β cells of the islets of Langerhans in the pancreas (see Chapter 14).

Selective activation of genes is the key to the development of a complex multicellular organism and, not surprisingly, it is the subject of intense research. If we could understand how and why some genes are activated while others remain dormant, we could unlock the mystery of embryonic development – how a single fertilised egg grows and differentiates into something as complex as a human being.

5 DNA REPLICATION

Whenever a cell divides, the DNA must be copied. Otherwise, there would be no reproduction and no growth. When cells divide, each new daughter cell must have a complete set of genes. **Replication**, the mechanism of DNA copying, is therefore of fundamental importance.

Just three years after Crick and Watson published their model of DNA structure in the journal *Nature*, their theory about DNA replication was confirmed by Arthur Kornberg. He put DNA into a mixture containing all four **nucleotides** (each of the four bases attached to a phosphate and a sugar), together with the enzyme DNA polymerase, and showed that the DNA could replicate without any other factors present.

In the mixture, each DNA strand unwinds and acts as a template for the construction of a new strand. The exposed strand acts as a template on which the free nucleotides arrange themselves in exactly the same sequence as the intact strand they replace. This model is called **semi-conservative replication** because each of the resulting strands of DNA contains one strand from the original DNA and one newly synthesised strand: half has been conserved and half is new.

Strong support for the semi-conservative model of DNA replication came from the work of Meselson and Stahl (Fig 27.10). By growing bacteria with bases containing the radioisotope ^{15}N (sometimes called heavy nitrogen) and then following its progress, they showed that each new DNA strand contains half the original strand.

Fig 27.10 Meselson and Stahl's experiment to support the semi-conservative replication model of DNA replication

1 To start with, many generations of *E. coli* bacteria are grown in a medium containing nucleotides labelled with the heavy nitrogen radioisotope, ^{15}N. Eventually, almost all of the bacterial DNA contains ^{15}N, and so is denser than normal DNA. A control culture of the same bacterium is grown with 'normal' ^{14}N nucleotides for comparison

2 The bacteria grown in ^{15}N are transferred to ^{14}N medium. After 0, 20 and 40 minutes, samples of the DNA are extracted, placed in a solution of caesium chloride and centrifuged. This separates out the DNA according to its density and allowed Meselson and Stahl to distinguish between DNA containing ^{15}N and ^{14}N

3 The results confirm that, initially, all the DNA contained ^{15}N. After 20 minutes the DNA had replicated, producing two strands of ^{15}N/^{14}N hybrid DNA. After 40 minutes, the second generation strands consisted of half ^{15}N/^{14}N hybrids and half all-new ^{14}N strands

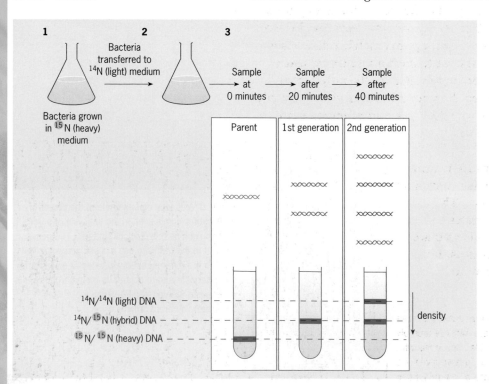

? **TEST YOURSELF**

E Predict the outcome of the third generation of DNA molecules in Meselson and Stahl's experiment. Of the eight DNA strands produced, how many would be hybrids and how many would be all ^{14}N strands?

THE MECHANISM
OF SEMI-CONSERVATIVE REPLICATION

The basic mechanism of DNA replication is shown in Fig 27.11(a). DNA is copied before the cell divides (see Chapter 25) and the basic idea is simple: the helix unwinds and each strand acts as a template for the manufacture of a new, identical strand.

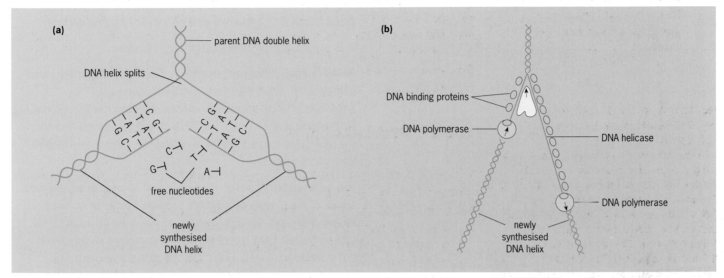

In practice, DNA replication is rather more complex than it appears, and is a good illustration of the importance of enzymes (Fig 27.11b). Two **helicases** use energy from ATP to separate the DNA strands. Next, **DNA binding proteins** attach to keep the strands separate. **DNA polymerase** enzymes then move along the exposed strands, synthesising the new strands, often in short segments. Finally, **DNA ligase** enzymes join the segments together, completing the new DNA strands.

Fig 27.11(a) When DNA replicates, the double helix unwinds and each exposed strand acts as a template for the synthesis of a new strand. This is a simplified diagram showing how free nucleotides are added to produce two new DNA helices

Fig 27.11(b) DNA replication is actually more complicated and several enzymes other than DNA polymerase are involved. DNA helicase unwinds the parent helix and then this is 'held open' by DNA binding proteins so that the DNA polymerase can gain access to the parent strand

DNA PROOFREADING

Mistakes in DNA replication occur regularly. The chances of the wrong base being added are between 1 in 10^8 and 1 in 10^{12}. This may seem low, but so many nucleotides are added that this rate means approximately one mistake occurs for every ten genes that are copied. Clearly, such a rate would cause the death of most organisms. This disaster is prevented by the polymerase enzymes themselves. They can detect when the wrong base has been inserted, and pause to correct the mistake. In this way 99.9 per cent of mistakes are corrected. Uncorrected mistakes give rise to **mutations**.

There are many different DNA repair mechanisms in cells to put right the genetic damage that inevitably accumulates during the life of an organism. The condition xeroderma pigmentosum illustrates the importance of these mechanisms: those affected are liable to get skin cancer because their cells cannot repair genetic damage caused by exposure to ultraviolet light (Fig 27.12).

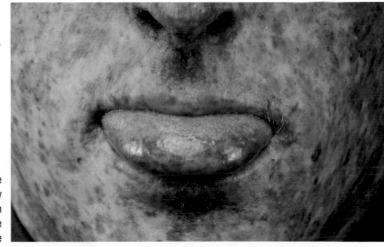

Fig 27.12 Xeroderma pigmentosum is a rare genetic disease characterised by dry, freckled skin which is extremely sensitive to sunlight. One theory suggests that individuals with this condition, who have to stay indoors, could have given rise to the vampire legends of central Europe

! SCIENCE FEATURE How do you compare the DNA of different organisms?

The technique of DNA hybridisation (Fig 27.13) can be used to compare the similarity of the DNA of different species, and is proving to be a useful tool in the search to piece together evolutionary trees.

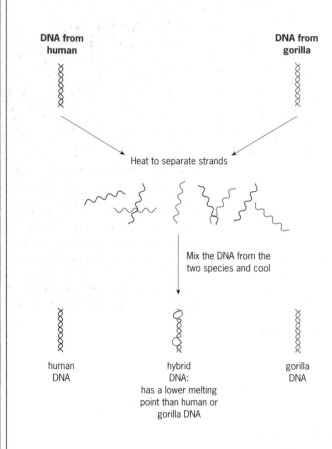

Fig 27.13 The principle of DNA hybridisation

DNA is much more stable than protein because it is held together by regular hydrogen bonds along the whole of its length. However, when DNA is heated to about 87 °C, known as the melting point of DNA, the hydrogen bonds break and the two halves of the molecule separate. When the temperature drops below 87 °C, the two strands join up: they **re-anneal**.

This is the basic process of DNA hybridisation:

DNA from two different species is cut into small fragments and heated to 87 °C so that the two strands separate. The separated strands from each species are then mixed together. As the temperature drops the DNA strands re-anneal. Some of the molecules formed will be hybrids: they contain one strand from each species.

The temperature at which the two hybrid strands re-anneal tells us about the degree of similarity between them. When two strands from the same species re-anneal, they do so at 87 °C. Closely related DNA re-anneals at temperatures slightly lower but close to 87 °C. This indicates that they have many base sequences in common and so form many hydrogen bonds along the length of the molecule. In contrast, DNA from more distantly related species must cool to much lower temperatures before they can anneal as there are fewer hydrogen bonds to hold the strands together.

For example, human DNA can be hybridised with gorilla DNA or rabbit DNA. The melting point for DNA from all three single species is 87 °C. The melting point for the human/rabbit DNA is lower than for the human/gorilla DNA. This shows that humans and gorillas have more DNA sequences in common than humans and rabbits.

POLYMORPHIC GENES

Genes that have two or more alleles are called polymorphic. Many genes have no alleles – they code for essential proteins and any mutation in them is selected against. For instance, the genes that code for the enzymes in respiration must be copied exactly time after time, or the organism dies. Some genes, however, can cope with a bit of variation. The genes that code for eye colour or hair colour are obvious examples. In humans, it has been estimated that 31.7 per cent of our genes are polymorphic.

Find out more about polymorphic genes in the Science feature on cheetahs in Chapter 29.

MUTATIONS

A mutation is a spontaneous change in the genetic material of an organism (Fig 27.14b). It usually occurs when DNA is copied and cells divide. Generally, there are two types of mutation:

● a gene mutation,
● a chromosome mutation.

Both of these may occur in **somatic cells** (normal body cells) or **germ cells** (sex cells).

Fig 27.14(a) To many people, this is the image conjured up by the word mutant. But in biology, a mutation is a fault in the copying of genetic material and is usually lethal or makes no difference to the organism. Only in rare cases will mutation give rise to new variation, and it is sure to be less spectacular than science fiction writers would have us believe

Fig 27.14(b) An example of a real-life mutation. A chance mutation in a single gene of the speckled moth *Biston betularia* produced a fortunate change from the normal speckled coloration, seen in the moth on the left, to the melanic form seen on the right. In sooty areas, this new form was better camouflaged and so had a better chance of avoiding predators. More of the black moths passed on their genes to the next generation, making the melanic form dominant in industrialised areas

Somatic cell mutations are not passed on to the next generation. Generally, organisms accumulate somatic mutations as they get older, and this is thought to be one of the causes of cancer. In contrast, when a mutation arises in an egg or sperm cell which then goes on to become a zygote, the mutation is passed on to all the cells of the new individual. Germ cell mutations alter the genome of an organism.

GENE MUTATIONS

A gene mutation is a change in the base sequence of DNA. We have already seen that when DNA replicates, two identical strands are formed. Occasionally, however, a fault can occur. In a typical gene of, say, 1400 base pairs, a change in just one of the base pairs has the potential to make the whole gene useless, and can prove lethal to the organism. A change in a single base is known as a **point mutation**.

How can such a tiny change be so disastrous? It is because a change to any base will alter a codon so that it will probably code for a different amino acid. In turn, this 'wrong' amino acid can affect the way the polypeptide chain folds, so changing the shape of the whole protein molecule. It might then be unable to function. (Remember: many genes code for enzymes or parts of enzymes whose proper functioning relies on precise shape.) Table 27.4 shows the different types of gene mutation that can occur, using words instead of codons. In all cases, the original meaning of the message is lost.

SEE QUESTION 7

? TEST YOURSELF

F From the information given in Table 27.4, which type of mutation is likely to cause the fewest changes to a protein?

Table 27.4 Different types of gene mutation	
Type of gene mutation	**Effect on code (using words rather than codons)**
Normal code	THE FAT OLD CAT SAW THE DOG
Addition (frame shift)	THE EFA TOL DCA TSA WTH EDO G
Deletion (frame shift)	THE ATO LDC ATS AWT HED OG
Substitution	THE FAT OLD BAT SAW THE DOG
Duplication	THE FAT TFA TOL DCA TSA WTH EDO G
Inversion	THE TAF OLD CAT SAW THE DOG

From Table 27.4 we can see that the two least damaging types of mutation are substitutions and inversions, because these alter only one of the seven amino acids. In some cases, if the new amino acid occurs in a non-vital part of the chain, or has similar properties to the correct one, then the mutation might not actually matter at all.

In contrast, deletions and additions are potentially much more damaging because they cause **frame shifts**: only one base may be added or lost, but this causes the whole sequence to shift along, altering all the codes. In this case, the protein made is completely different from the original and is unlikely to function as it should. The Assignment at the end of this chapter deals with sickle cell anaemia, a disease caused by a fault in just one base.

HOW COMMON ARE GENE MUTATIONS?

The rate at which genes mutate varies between species and between genes at different loci. At a rough estimate, there is an average mutation rate of about 10^{-5} per locus per generation, meaning 1 in 100 000. So, in a population of one million individuals, you could expect ten to have inherited a mutation at any particular gene. The rate of gene mutation (and chromosome mutation) can be greatly increased by environmental factors, known as **mutagens**, such as ultraviolet, X-rays, alpha and beta radiation, and chemicals such as mustard gas and cigarette smoke.

CHROMOSOME MUTATIONS

During mitosis (normal cell division) chromosomes are copied, condensed and pulled apart. The end result should be that each daughter cell receives a complete set of perfect chromosomes. But faults can occur.
A chromosome mutation occurs when the structure of a whole chromosome – or set of chromosomes – is altered in some way.

Fig 27.15 shows the four basic types of chromosome mutation: deletion, duplication, inversion and reciprocal translation.

While gene mutations normally affect just one gene, chromosome mutations involve the disruption of whole blocks of genes. For this reason, many chromosomal mutations give rise to **syndromes**, which are complex sets of symptoms with a single underlying cause.

In addition to the structural mutations outlined above, another common failure is **non-disjunction**. This occurs when pairs of chromosome fail to separate during anaphase I of meiosis. Some gametes get both chromosomes, while others receive neither. Down's syndrome, for example, occurs when a gamete containing two copies of chromosome 21 joins with a normal gamete containing one, giving the affected individual three copies of chromosome 21 (see Chapter 25).

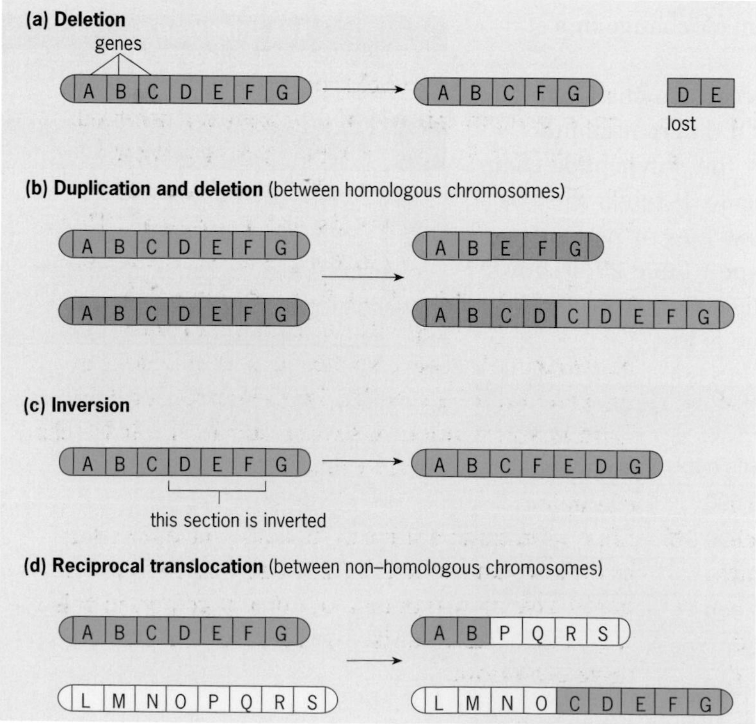

POLYPLOIDY

Occasionally, a mutation causes a whole set of chromosomes to be changed. For instance, a fault in meiosis might result in some gametes having two sets of chromosomes while others receive none. When such a diploid gamete joins with a normal one, the result is a **triploid** zygote – one with three sets of chromosomes. In another case, a fault in mitosis after

Fig 27.15 The main types of chromosome mutations **(a)** Deletion occurs when a chromosome splits and a fragment is lost as the parts rejoin. In this case, a whole set of genes is lost, and so this is often fatal **(b)** A duplication occurs when a section of chromosome is copied twice. This can occur when pairs of homologous chromosomes break at different places and then reconnect to the wrong partners. Thus, one of the partners would have two copies of a particular sequence, but the other would have neither. This can have serious consequences **(c)** An inversion results when a broken segment is reinserted in the wrong order. Any genes contained in this region will be transcribed and translated backwards, and will therefore not make the correct proteins **(d)** A reciprocal translocation occurs when two non-homologous chromosomes exchange segments. This makes the chromosome pairs of unequal size, and this can cause problems in meiosis, with daughter cells receiving the wrong chromosomes

(a) Deletion

genes

A B C D E F G → A B C F G | D E

lost

(b) Duplication and deletion (between homologous chromosomes)

A B C D E F G → A B E F G

A B C D E F G → A B C D C D E F G

(c) Inversion

A B C D E F G → A B C F E D G

this section is inverted

(d) Reciprocal translocation (between non–homologous chromosomes)

A B C D E F G → A B P Q R S

L M N O P Q R S → L M N O C D E F G

fertilisation could result in the chromosomes doubling but not separating, leaving four sets in a single **tetraploid** cell. Neither of these embryos could develop.

The possession of multiple sets of chromosomes is known as **polyploidy**. This phenomenon is very common in flowering plants and has played a vital role in their evolution. Botanists estimate that over half of the world's flowering plant species are polyploid, including some of the world's most economically important crops: sugar beet, tomatoes and tobacco.

WHY MUTATIONS ARE IMPORTANT

A mutation can have one of several effects:

- It can be lethal: an organism with a lethal mutation cannot survive and might not even develop beyond the zygote. Genes code for vital proteins such as enzymes, so many gene mutations mean that the protein is the wrong shape, and so cannot function metabolically. Chromosome mutations which involve the loss of whole blocks of genes are usually lethal.
- It can have no effect. It might occur in a non-coding part of DNA, or in a gene that is not expressed. Alternatively, it can result in a different amino acid being incorporated into the protein, but one that does not alter its ability to function.
- It can be beneficial. Occasionally, a mutation produces an improvement in phenotype. Statistically, this is a very rare event, but given the number of DNA replications and the number of individual organisms, it is bound to happen sooner or later. Beneficial mutations are hugely important because, ultimately, they are the source of all variation. Natural selection favours these 'improved' mutants (see Chapter 29).

Fig 27.14(b) on page 497 shows an example of a beneficial mutation. Following the Industrial Revolution, a mutation in the peppered moth produced a black (melanic) form that was better camouflaged against sooty backgrounds. The black moth had an advantage because it was hard for predators to see and so survived to pass on its genes to the next generation. Consequently, the mutated gene spread and the genome of the species was altered (see Chapter 29).

✔ REMEMBER THIS

Meiosis is also known as reduction division. See Chapter 25 for more detailed information about this important process.

? TEST YOURSELF

G Polyploid cells can be induced experimentally by adding the chemical colchicine, a substance extracted from the crocus. Colchicine inhibits the formation of the spindle during cell division. Suggest why colchicine results in polyploid cells.

SUMMARY

- A chromosome is a large, densely staining body consisting of a single, supercoiled DNA strand. The genes along a chromosome are found at particular points called **loci**.
- A gene is a length of DNA that codes for the synthesis of a particular polypeptide or protein. Only about 10 per cent of a chromosome consists of genes; the rest is non-coding DNA.
- The genetic code is the sequence of bases contained in the DNA molecule. The code is read in groups of three bases, called **codons**. Each codon codes for a particular amino acid.
- The central dogma (principle) of molecular biology is that genes are read by **messenger RNA** which in turn moves out into the cytoplasm where it codes for the assembly of a protein.
- At the ribosome, messenger RNA attracts **transfer RNA** molecules with the right amino acids attached.

- The amino acids are held together so that peptide bonds form; in this way the protein is assembled.
- DNA replication occurs when the two strands separate and the enzyme **DNA polymerase** assembles two new strands on the originals. This is called semi-conservative replication as each old strand is conserved as half of the new DNA.
- A fault in DNA replication leads to a **gene mutation**. The base sequence is altered so that the gene no longer codes for the correct protein. Chromosome mutations can result in whole blocks of genes being lost, or translated backwards for example.
- Most mutations are harmful, some have no effect, but a few can be beneficial. Beneficial mutations might give the organism an advantage and in this way mutations give rise to variation and hence drive evolution.

❓ EXAM QUESTIONS

1 Give an account of the structure and replication of deoxyribonucleic (DNA) [**essay question**]

Edexcel B/HB1 June 1999 Q8

2 The diagram shows part of a DNA molecule.

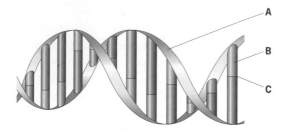

a) DNA is a large molecule made up of a large number of nucleotides. Each nucleotide has three components.

(i) Name the components that make the parts of the DNA molecule labelled **A** [two components] and **B**.

(ii) Name the type of bond found at **C**.

b) 1 kilobase of DNA is a length of a DNA molecule with 1000 base pairs.

The length of 1 kilobase is 0.34 micrometres (μm). The DNA in a human cell contains a total of 2 900 000 kilobases.

Calculate the total length of DNA in a human cell. Give your answer in millimetres (mm). Show your working.

c) Give **three** ways in which the structure of the DNA molecule enables it to carry out its functions.

AQA (B) BYB2 Genes and Genetic Engineering June 2001 Q1

3 DNA is made up of two polynucleotide strands, the sense strand and the antisense strand. Messenger RNA is transcribed from the DNA sense strand, which contains the genetic code.

a) The graph shows the number of bases found in the sense strand and the antisense strand of a short piece of DNA, and the mRNA transcribed from it.

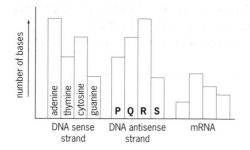

(i) Identify the base represented by each of the following letters: **P; Q; R; S**.

(ii) Explain why the total number of bases in the DNA sense strand and the total number of bases in the DNA antisense strand are the same.

(iii) Explain why the total number of bases in the DNA sense strand and the total number of bases in the mRNA are different.

b) The mRNA has a sequence of 1824 bases. How many amino acids will join to form the polypeptide chain?

c) Although DNA is double-stranded, only the sense strand determines the specific amino acid sequence of a polypeptide. Suggest a role of the antisense strand.

AQA (A) BYA3 Pathogens and Disease January 2001 Q3

4

a) Draw a labelled diagram to show the structure of an RNA nucleotide.

b) The diagram shows a molecule of an enzyme called ribonuclease. Each amino acid in the protein is indicated by a 3-letter symbol e.g. Arg = arginine.

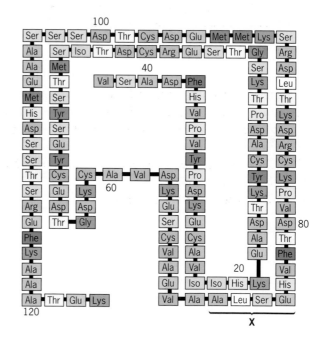

(i) How many nucleotides are there in the mRNA molecule that codes for this enzyme?

(ii) The table gives the mRNA code for the four amino acids in the part of the enzyme labelled **X**.

? EXAM QUESTIONS

Amino acid	Symbol	mRNA code
Alanine	Ala	GCU
Glutamine	Glu	GAG
Leucine	Leu	UUA
Serine	Ser	AGU

Give the DNA code for the part of the enzyme labelled **X**.

c) **(i)** Where does translation occur in a cell?
(ii) Describe what happens during translation.

d) Explain how the structure of DNA is related to its function.

AQA (B) BYB2 Genes and Genetic Engineering January 2001 Q7

5

a) List **three** ways in which transcription differs from translation in protein synthesis.
The diagram represents a polyribosome with several translation sites.

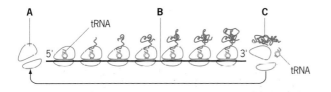

b) Name the structures labelled **A** to **C**.

c) Name **two** molecules, in addition to the molecules shown in the figure, which are required to complete translation.

d) Describe **two** structural features which adapt tRNA to its role in translation.

OCR 9264/3 Biology Linear Paper 3 June 2000 Q4

6

a) Name the organelle where proteins are synthesised from amino acids.

b) The diagram shows a transfer RNA molecule (tRNA).

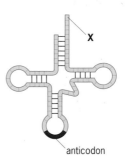

(i) During protein synthesis, which molecule is attached to the tRNA molecule at **X**?
(ii) What is an anticodon?
(iii) Give **one** way in which the structure of a tRNA molecule is different from an mRNA molecule.

AQA (B) BYB2 Genes and Genetic Engineering June 2001 Q5

7 About one in 20 000 humans has the condition albinism. This is caused by the absence of melanin, the dark brown pigment normally present in the human skin, hair, and eyes. Albinism arises from a gene mutation that causes skin cells to produce a different version of the enzyme tyrosinase. This different version of tyrosinase is faulty.
These events are shown in the diagram.

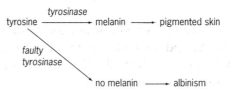

a) Explain how a gene mutation may result in a different version of tyrosinase.

b) The faulty tyrosinase does not produce melanin. Suggest an explanation for this.

AQA (B) BYB2 Genes and Genetic Engineering January 2001 Q6

KEY SKILLS **ASSIGNMENT**

SICKLE CELL ANAEMIA

In addition to the topics covered in this chapter, this Assignment assumes that you know some simple genetics (monohybrid inheritance) and evolution (natural selection).

A small minority of people, particularly those of African origin, inherit sickle cell anaemia (SCA), a condition in which their haemoglobin acts abnormally. SCA is a genetic disease: it results from a gene mutation that is inherited. Carriers of the sickle cell allele are usually perfectly healthy, and in some circumstances even have an advantage over people who possess two normal alleles. However, individuals with two sickle cell alleles have the disease.

The faulty allele makes haemoglobin S instead of the normal haemoglobin A. Problems arise when oxygen tensions are low, as they are in actively respiring tissues. At low oxygen tension, abnormal haemoglobin tends to come out of solution and crystallise, distorting the red blood cells into crescent shapes (Fig 27.A1). This has two main consequences. Firstly, sickle cells tend to form clots which can block blood vessels. Secondly, the spleen destroys abnormal sickle cells at a much greater rate than normal, causing anaemia.

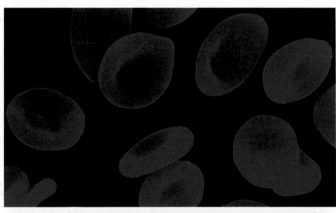

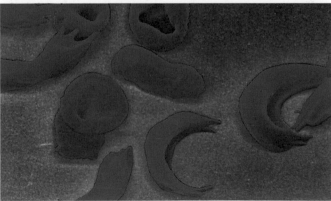

Fig 27.A1 Top: Normal red blood cells.
Bottom: Abnormal red blood cells from someone with sickle cell anaemia. Haemoglobin S tends to polymerise, forming crystal-like structures that distort the red blood cells

1 What is the function of haemoglobin?

Haemoglobin molecules consist of four polypeptide chains: two α-globins and two β-globins. Research shows that sickle cell anaemia results from a fault in just one base in the β-globin gene. Here is the base sequence of the first seven codons of this gene:

Normal base sequence
GUA–CAU–UUA–ACU–CCU–GAA–GAG

Sickle cell sequence
GUA–CAU–UUA–ACU–CCU–GUA–GAG

2 **a)** How can you tell that these codons are from mRNA and not DNA?
b) Which codon has mutated?
c) What is the name given to this type of mutation?
d) Work out the amino acid sequence for both base sequences (you will need to refer to Table 27.2).
e) Explain in general terms how this slight change can produce abnormal haemoglobin.
f) Explain why other types of gene mutation can have a much greater effect on the final protein than the example seen here.

The inheritance of sickle cell anaemia

Generally, people have two copies of each gene, and this is the case with sickle cell anaemia. The two alleles involved are:

Hb^A = coding for normal haemoglobin

Hb^S = coding for sickle cell haemoglobin

So the three possible genotypes are:

Hb^AHb^A Normal

Hb^AHb^S Carriers of the disease but with few, if any, symptoms

Hb^SHb^S Have the disease

3 Sickle cell anaemia is described as an autosomal, recessive disease.
a) What do the terms **(i)** autosomal and **(ii)** recessive mean?
b) Explain why carriers are usually healthy.
c) If two carriers have children, what is the probability that the first child will have the disease?

The sickle cell gene is far commoner than you would expect for one which is selected against. The reason for this is that the carriers usually show a higher resistance to malaria than normal.

4 **a)** What is meant by the phrase 'selected against'?
b) Explain why the frequency of the sickle cell allele is higher than you would expect in some areas.

28

How genes are inherited

Woody Guthrie, the famous American folk singer, suffered from Huntington's disease. His son, Arlo, was born before his father knew he had a disease that could be passed on to his children. Luckily, Arlo is unaffected: he has not inherited his father's faulty gene.
The degeneration caused by Huntington's disease can be revealed by a brain scan. A protein called huntingtin, produced by a faulty gene that was discovered in 1993, damages nerve cells in some areas of the brain, including the cerebral cortex. The gene persists because affected people tend to reproduce before symptoms appear

About eight people in every 100 000 are born with the genetic disorder Huntington's disease. This condition, which involves premature degeneration of the brain, is caused by a single faulty gene on chromosome 4. The slow decline into dementia that develops is fatal.

Diagnosis is difficult because early symptoms are vague: sufferers can have bursts of temper, become more clumsy than usual or have difficulty remembering things. Also, the symptoms do not usually become apparent until the sufferer is in their 30s or 40s, usually after they have passed the gene on to the next generation.

Until recently, people with a family history of the disease knew that they were at risk, but they could never be sure whether they were actually carrying the gene. This led to a dreadful dilemma: should they go ahead and have children and risk passing the gene on, or should they decide to remain childless, only later to experience the anguish of finding out that they were free of Huntington's disease when it was too late to try for a baby?

Today, tests are available that can tell people whether they are carriers of the Huntington's gene. A sample of DNA can even show if an unborn baby carries the disease. This does not make the decisions involved any easier, but it does give affected families accurate information on which to base those decisions.

1 MENDELIAN INHERITANCE

Why don't we look exactly like our parents? Why are brothers and sisters all different (unless identical twins) when born to the same parents? As Fig 28.1 shows, there is often a marked family resemblance between parents and their children and between siblings. However, all new-born babies, except identical twins, are genetically unique. Every new-born baby has around 30 000 genes and because humans – like most animals – are diploid organisms, a baby has two versions of each gene. It inherits one version from its mother and one version from its father.

REMEMBER THIS

Some important terms and concepts relating to inheritance are covered in Chapter 24. If you are new to genetics, read that chapter before you start to study this one.

REMEMBER THIS

A gene is a length of DNA that acts as a template for the production of messenger RNA. mRNA forms a mobile copy of a gene, travelling from the cell's nucleus to the cytoplasm. Ribosomes use this template to synthesise a specific amino acid chain. Complete proteins are made from one or more of these chains.

Fig 28.1 The tendency for certain features to be inherited is shown clearly in this family. However, no two people ever inherit exactly the same set of genes unless they are identical twins

It is difficult to imagine what happens when 30 000 different genes are mixed. Not surprisingly, although a vast amount of genetics research is under way, we are still a long way from completely understanding what happens when those mixed genes are passed on to the next generation. Some genes are activated while others remain hidden, some master genes switch on whole blocks of other genes.

How can we begin to make sense of it all? Amazingly, the man who first worked out the underlying rules did so without any knowledge of genes, chromosomes, DNA or cell division. Known as the father of genetics, **Gregor Mendel** (Fig 28.2) worked with pea plants. These are much simpler organisms than humans and provide a good starting point to understand some of the basic rules of **Mendelian inheritance**.

THE WORK OF GREGOR MENDEL

Gregor Mendel, a Czech monk, was the first to work out the basic laws that govern the inheritance of genes. Before Mendel's work became known, it was widely thought that inheritance was due to 'blending' of different features. So, for example, a tall person married to a short person would produce children who would eventually grow to an intermediate height. Mendel showed it was wrong to assume that characteristics blended and demonstrated how individual characteristics are passed down to the next generation.

For years, Mendel bred and studied the edible pea plant (Fig 28.3). He chose these plants because they had easily observable features, they were easy to cultivate, they had rapid life cycles and their pollination could be controlled. His breeding experiments centred on the inheritance of a few features such as plant height, flower colour and pea shape/appearance.

Mendel was lucky because he chose features that were controlled by single genes. In addition, pea plants tend to fertilise themselves and produce pure breeding homozygous individuals. Had Mendel worked with more complex features or a different species, he might not have been able to work out the underlying laws of inheritance.

Mendel conducted his experiments using a good scientific method. He controlled his experimental conditions carefully, ensuring that plants were pollinated only by the pollen he transferred. He made careful observations and then repeated experiments as many times as was practicable, keeping meticulous records. Mendel was trained in mathematics and he applied statistical methods to his results in a way that was very advanced for the time.

His genius really showed in the way he interpreted his results. Mendel proposed the **particulate theory**, in which he stated that there are individual units of heredity that cannot be diluted, only hidden. Mendel came up with the idea that each pea plant has two factors for each character, one inherited from each parent. Only one unit can be present in the gamete and so be passed on to the next generation. Mendel's 'particles' are, of course, genes.

In 1866, Mendel published a detailed account of his findings, but their significance was not appreciated by the scientific community. In 1884, Mendel died without ever receiving true recognition for his work. In 1900, his work re-emerged thanks to the independent work of three other European scientists – see Table 24.2 in Chapter 24. In the time between 1866 and 1900, the process of meiosis had been discovered and this provided a mechanism that would explain Mendel's observations. All three scientists cited Mendel's original work, which had suggested the basic mechanism of inheritance.

Fig 28.2 Gregor Mendel (1822–1884). Mendel's discoveries, like those of many scientists, were due to a combination of hard work, good scientific method, a touch of genius and a significant amount of luck

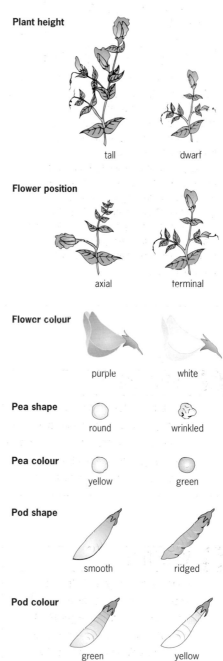

Fig 28.3 The features of pea plants studied by Mendel. Pea plants tend to self-fertilise, producing true-breeding strains. For example, white-flowered plants contain only white-flowered alleles: they are not 'hiding' any purple alleles and so always produce white flowers

MEIOSIS AND MENDEL'S FINDINGS

Mendel's findings can be summarised into two laws, both of which are explained by the process of meiosis (Fig 28.4).

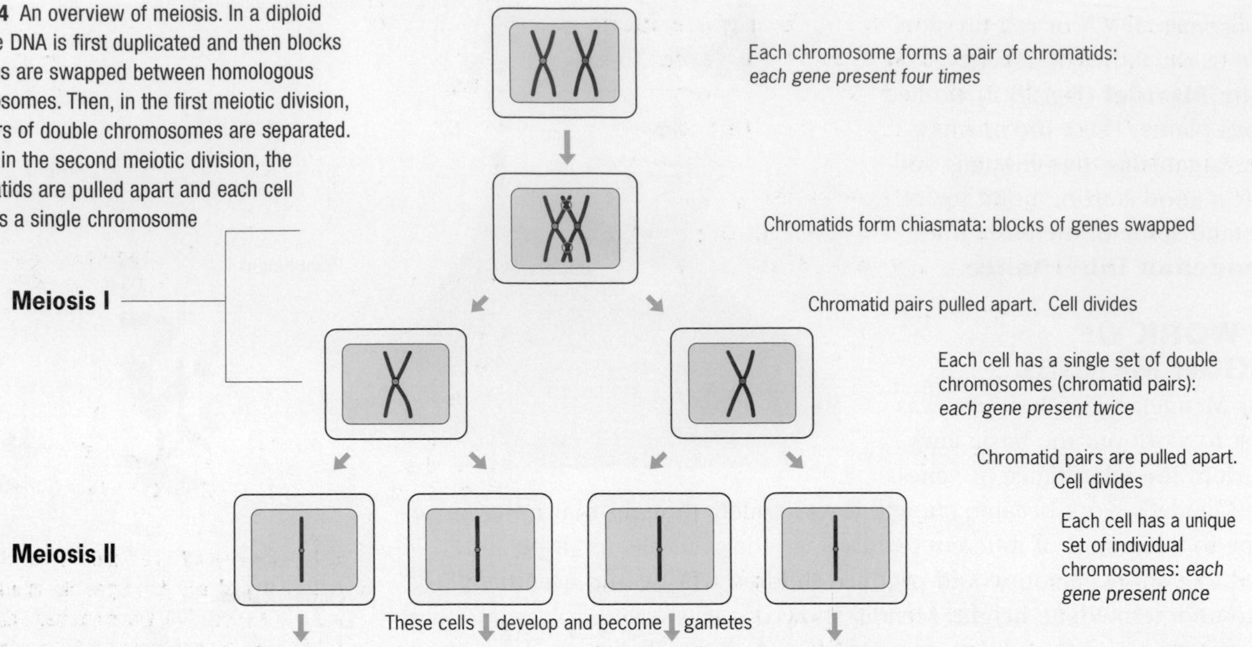

Fig 28.4 An overview of meiosis. In a diploid cell, the DNA is first duplicated and then blocks of genes are swapped between homologous chromosomes. Then, in the first meiotic division, the pairs of double chromosomes are separated. Finally, in the second meiotic division, the chromatids are pulled apart and each cell receives a single chromosome

Each chromosome forms a pair of chromatids: *each gene present four times*

Chromatids form chiasmata: blocks of genes swapped

Chromatid pairs pulled apart. Cell divides

Meiosis I

Each cell has a single set of double chromosomes (chromatid pairs): *each gene present twice*

Chromatid pairs are pulled apart. Cell divides

Meiosis II

Each cell has a unique set of individual chromosomes: *each gene present once*

These cells develop and become gametes

- Mendel's **law of segregation** states:

An organism's characteristics are determined by internal factors that occur in pairs, only one of which can be present in a single gamete.

In other words an organism has two versions of each gene, but only one of these version is passed on to the offspring. This is because meiosis separates homologous pairs of chromosomes, so that only one of each pair passes into the gamete.

- Mendel's **law of independent assortment** states:

Each of a pair of contrasted characteristics can be combined with each of another pair.

Again, this is explained by meiosis; any one of a pair of chromosomes can pass into the gamete with any member of another pair. The significance of this is seen in dihybrid (two-gene) inheritance.

MONOHYBRID INHERITANCE: HOW SINGLE GENES ARE PASSED ON

Monohybrid (single gene) inheritance concerns the inheritance of different alleles (usually two) of a single gene. Like Mendel, we start with the pea plant. Peas have several easily observable features that are controlled by single genes (see Fig 28.3).

For example, pea plants have one gene for height. The height gene has two alleles: tall (T) and dwarf (t). Pea plants are diploid and so have two alleles. There are therefore three possible genotypes:

TT – homozygous for T (homo = same)
Tt – heterozygous (hetero = different)
tt – homozygous for t

Consider what happens when a homozygous tall plant is crossed with a homozygous dwarf plant (Fig 28.5). All the gametes from the tall plant contain a T allele and all those from the dwarf plant have a t allele. These combine at fertilisation to give offspring all with the genotype Tt. The **first generation**, known as the **F₁**, are all tall, because tallness is dominant. However, although they look identical to the tall parent plant, they are different in one very important respect: they are heterozygous and not homozygous.

If two of these heterozygous plants are crossed, half of the gametes from each parent are T and half are t, giving us four possible genotypes in the **second generation**, the **F₂**:

25% are TT

25% are tT

25% are Tt

25% are tt

The first three genotypes give tall plants (the T allele is present) but a quarter of the plants are dwarf (they are homozygous for the t allele).

It is important to realise that chance plays a very important part in genetics. If we took four F₂ pea plants we might expect to get three tall plants and one dwarf, but it is highly likely that we would get something else, such as four and none or two and two. The greater the number of offspring, the higher the chance that the numbers reflect the expected ratio. This is why Mendel had to carry out hundreds of crosses before he became convinced of the underlying ratio – see the Assignment at the end of this chapter.

THE TEST CROSS

It is often important to know whether an individual displaying a dominant trait is homozygous or heterozygous for a particular allele. For example, how can we tell if a tall pea plant is pure breeding, or if it is carrying a hidden dwarf allele? To find out we do a **test cross**. This involves crossing the tall plant with a homozygous dwarf plant (tt). If the tall plant is homozygous (TT), all the offspring produced are tall. However, if it is a heterozygote (Tt), about half the offspring are tall and half are dwarf.

MONOHYBRID INHERITANCE IN HUMANS

Clear-cut examples of monohybrid inheritance in humans are relatively rare, and often involve genetic disease where people inherit one or more faulty alleles. Genetic diseases are often recessive because faulty alleles that fail to make an important protein can be masked by normal ones that function properly. Table 28.1 shows examples of monohybrid inheritance in humans.

In contrast, some genetic disorders such as Huntington's disease (see the Opener of this chapter) are caused by dominant alleles. The alleles concerned code for a product that actively causes damage; the symptoms are not due to an allele not doing its job. Such alleles are dominant because the presence of a normal allele cannot mask the symptoms.

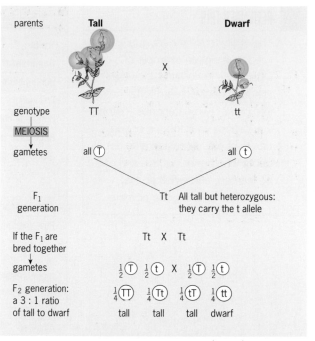

Fig 28.5 Monohybrid inheritance in pea plants. Pure breeding tall and dwarf pea plants are homozygous (TT and tt). When crossed, the plants of the first (F₁) generation are all tall but they are heterozygous (Tt), carrying the allele for dwarfness. If these heterozygous individuals are selfed (bred together) the dwarf form reappears in the next (F₂) generation in a ratio of 1:3 (ie 1 in 4). We say that the heterozygotes do not breed true

SEE QUESTION 1

? TEST YOURSELF

A In pea plants, yellow seeds are dominant to green. If a heterozygous yellow-seeded plant were crossed with a green-seeded plant, what ratio would you expect in the F₁ generation?
B Why is a homozygous dwarf plant used in a test cross? Why not use a homozygous tall plant?

SEE QUESTION 2

REMEMBER THIS

The ability to roll the tongue is often cited as a simple example of inheritance in humans. However, although people seem to fall neatly into one of two categories – either you can roll your tongue or you can't – there is considerable debate about whether or not this skill can be learned, and whether it is controlled by one gene or more.

Table 28.1 Examples of monohybrid inheritance in humans

Traits	Features
Dominant traits	
Huntington's disease	See Opener
Freckles	
Dimple in chin	
Recessive traits	
Sickle cell anaemia	Haemoglobin polymerises, distorting red blood cells into a sickle shape. This leads to circulatory blockages and anaemia. See the Assignment to Chapter 27.
Attached ear lobes	
Phenylketonuria (PKU)	Inability to process the amino acid phenylalanine. Protein intake must be monitored to prevent build-up of the amino acid to harmful concentrations. Can be fatal if not treated.
Haemophilia (also sex linked – see below)	Inability to produce a critical blood-clotting factor, carried on the X chromosome. Haemophilia is now treated by supplying the affected individuals with the clotting factor.
Albinism	Inability to make the pigment melanin.
Rhesus blood group	Presence (Rh^+) or absence (Rh^-) of rhesus protein on red blood cells.
Cystic fibrosis	Excessive mucus production, especially in lungs and pancreas. Breathing and digestion are affected and sufferers are very susceptible to lung infections.
Galactosaemia	Inability to convert galactose to glucose in the liver. All lactose must be avoided to prevent fatal levels of galactose from accumulating.
Tay–Sachs disease	Inability to produce critical lipases, which results in fatty accumulations in the brain.
Lactose intolerance	Inability to breakdown the disaccharide lactose to glucose and galactose. Leads to vomiting, diarrhoea, flatulence. Avoiding lactose prevents symptoms.

SCIENCE FEATURE Cystic fibrosis: an example of monohybrid inheritance in humans

The disease cystic fibrosis is caused by the mutation of a single allele. A normal allele (C) makes a membrane protein essential for the proper functioning of certain epithelial cells. The mutated allele (c) does not code for a functional protein.

Most people have two healthy alleles: their genotype is CC. In the UK population, however, around 1 person in 25 carries a faulty allele. These carriers (Cc) are perfectly healthy because they also possess a normal allele, which makes the working protein.

However, in 1 couple in 625 (1 in 25 x 1 in 25) both partners are carriers. There is a one in four chance that any child they have could inherit both faulty alleles and therefore be a cystic fibrosis sufferer. As 1 cystic fibrosis child in 4 is born to 1 couple in 625, approximately 1 child in every 2500 has this condition.

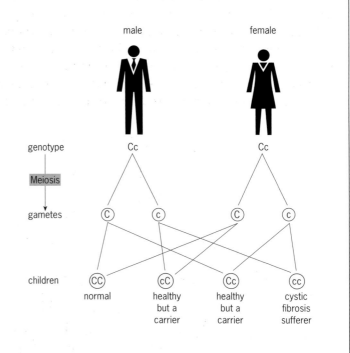

Fig 28.6 The inheritance of cystic fibrosis. If both parents are carriers, there is a 25 per cent chance that any child will be a cystic fibrosis sufferer. There is also an equal chance that a child will inherit no faulty alleles and that his/her descendants will be completely free from the disease. There is a 50 per cent chance of the child being a carrier of cystic fibrosis

INBREEDING AND OUTBREEDING

Everyone agrees that brothers and sisters or other closely related people should not have children. In most human societies such **inbreeding** is illegal for social, moral and biological reasons.

Biologically, inbreeding is undesirable because it reduces genetic variation in the offspring. Inbreeding promotes homozygosity: inbred individuals are homozygous for many more alleles than individuals produced by outbreeding. So, faulty recessive alleles are more likely to be paired up, resulting in a much higher proportion of genetic defects and disease.

In practice, inbreeding tends to happen in small, isolated human and animal populations, or as a consequence of artificial selection. Pedigree dog and cat breeding is a good example of this. To maintain the most desirable features, champion animals are often bred with near relatives. Analysis of the pedigree certificates of the champions in a particular breed will often show that the same individuals crop up again and again. The consequence of this artificial selection is a relatively high proportion of genetic defects.

Outbreeding (the breeding of unrelated individuals) is more desirable as it produces more new genetic combinations and greatly reduces the chance that faulty genes will be expressed. Many organisms have developed mechanisms which prevent inbreeding (Fig 28.7). If pedigree animals are cross-bred, rather than inbred (cross a Siamese cat with a Persian, for example), you often get a fitter, healthier animal. We say that such offspring show **hybrid vigour**, and cross-breeding is a favourite tool of plant breeders who find that hybrids often grow faster and show better disease resistance.

DIHYBRID INHERITANCE: HOW TWO GENES ARE PASSED ON

What happens when a tall, purple-flowered pea plant is crossed with a dwarf, white-flowered individual? Do you always get tall plants with purple flowers, or can the features re-combine, producing tall, white-flowered plants, and dwarf, purple-flowered ones? The simple answer is yes, you can get these recombinants, thanks to the process of meiosis.

The inheritance of two separate genes is called **dihybrid inheritance**. If, as is usual, the two genes are on separate chromosomes, Mendel's second law (page 506) applies: each of a pair of contrasted characteristics can be combined with each of another pair. So, an organism with a genotype of AaBb can produce gametes of AB, Ab, aB or ab. In other words, one allele from each pair passes into the gamete, and all four combinations are possible. This is due to **independent assortment** during meiosis (Fig 28.8).

Fig 28.7 When woodlice lay a batch of eggs, the offspring are either all male or all female. This means that brothers and sisters cannot mate, thus reducing the chances of inbreeding

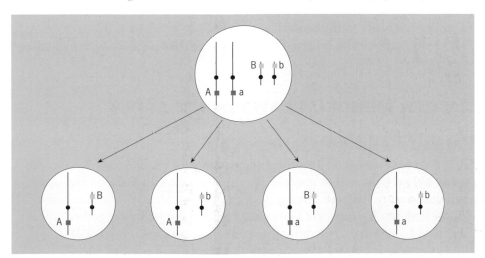

Fig 28.8 Consider a diploid cell with the genotype AaBb. Meiosis produces four different types of gametes by independent assortment

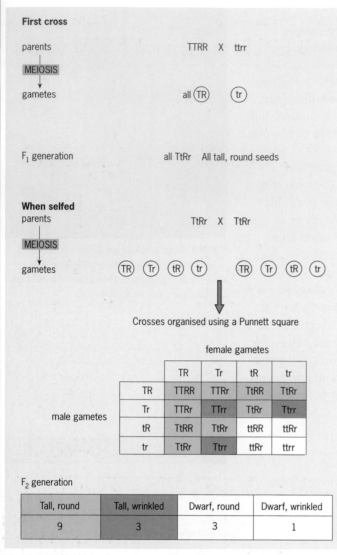

Fig 28.9 Dihybrid inheritance in peas. The grid used to work out the phenotype ratios in the F_2 generation is called a Punnett square

SEE QUESTIONS 3 TO 5

✓ REMEMBER THIS

Humans have over 30 000 genes present on just 23 pairs of chromosomes. So, any single gene is linked with a few thousand others because they are all on the same chromosome.

For an example, let's return to pea plants. In addition to tall and dwarf, we can consider a gene for seed texture. The allele R for round seeds is dominant to r, the allele for wrinkled seeds.

Fig 28.9 shows a cross between a pure breeding, tall, round-seeded plant (TTRR), and a pure breeding dwarf, wrinkled-seeded plant (ttrr). One allele from each pair is passed into the gametes. All of the gametes from the tall, round-seeded plant are TR and all those from the dwarf, wrinkled-seeded plant are tr. All of the F_1 generation have the genotype TtRr and are tall with round seeds.

When we cross two TtRr individuals, things get more complex! Each plant produces four different gametes. Fertilisation is random, producing 16 different combinations in the F_2 generation. To organise our thoughts we can use a **Punnett square** (see Fig 28.9).

DIHYBRID INHERITANCE QUESTIONS IN EXAMINATIONS

Many examination questions require the candidate to work out the ratios of offspring resulting from a particular cross. If you have not practised these questions they can be very time consuming, leading to stress and silly mistakes.

The example given above (TtRr × TtRr) is the most complex dihybrid situation you can get. This is because both male and female are heterozygous for both alleles, and so produce 4 different gametes each, resulting in 16 different combinations. All other genotypes will produce fewer combinations. If you practise these combinations you can learn to work out the gametes and resulting ratios in your head without resorting to Punnett squares and counting the results.

Consider the cross TtRR × Ttrr. The first individual produces the gametes TR and tR in equal amounts. The second one produces Tr and tr in equal amounts. So, there are four possible combinations: the same number as in a monohybrid cross.

You should see that the offspring will be one quarter TTRr, two quarters TtRr and one quarter ttRr, giving three tall plants with round seeds to one dwarf plant with round seeds. You would not get any dwarf plants with wrinkled seeds.

Some exam questions give the ratios of offspring phenotypes and ask you to work backwards to the original genotypes.

If you do not get the expected ratios, consider **linkage** as an explanation (see below).

2 WHEN MENDEL'S LAWS DON'T APPLY

LINKAGE AND CROSSING OVER

So far, we have considered the inheritance of genes on different chromosomes that could be separated at meiosis. We say that genes that are on the same chromosome are **linked**. Such genes cannot obey Mendel's laws because they do not undergo independent assortment. They are inherited together unless separated by crossing over during prophase I of meiosis (Fig 28.10).

Fig 28.10 When two genes are close together on the same chromosome they are almost always inherited together: they are linked. The chance of being separated by crossover depends on the distance between them

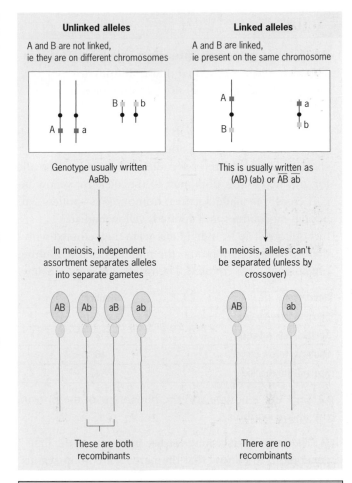

To illustrate linkage, let's consider the sweet pea plant (*Lathyrus odoratus* – a different species from the edible pea we have been looking at). In this species, the allele for purple flowers is dominant to red flowers, and the allele for elongated pollen grains is dominant to round pollen. If pure breeding plants with purple flowers and elongated pollen grains are bred with plants with red flowers and round pollen grains, the F_1 generation all have purple flowers and elongated pollen.

When the F_1 are **selfed** (bred together) you would expect a 9:3:3:1 ratio, but this does not happen. The F_2 generation are mainly like the parents (purple and elongated or red and round) but there are a few recombinants (red and elongated or purple and round). This is because the genes for flower colour and pollen shape are linked (present on the same chromosome) and so are inherited together. Any recombinants we get must therefore be due to crossing over.

Consider the set of results shown in Table 28.2. The total number of offspring is 720, of which 59 (31 + 28) are recombinants. Eight per cent of the offspring result from crossing over and so, by convention, we can say that the **loci** for flower colour and pollen shape are 8 units apart on the chromosome. If the percentage of recombinants were higher, we would know that the loci were further apart on the chromosome.

When genes are not linked, we expect them to be separated 50 per cent of the time because there is a 50 per cent chance that alleles on separate chromosomes will be inherited together. When looking for linkage, geneticists look for a frequency of recombinants significantly less than 50 per cent.

CHROMOSOME MAPS

The frequency with which linked genes are separated by crossover can tell us a lot about the relative positions, or loci, of these genes on the chromosome. Two alleles that are close together are rarely separated by crossing over, while those located on opposite ends of the chromosome are separated almost every time. Analysis of crossover frequency can be used to work out the relative positions of alleles and so make chromosome maps (Fig 28.11).

Genes Y and Z have a crossover frequency of 10 per cent and so are 10 units apart. A third cross between Y and X determines the relative positions of X, Y and Z. If the order of genes is XYZ (as shown), the crossover frequency between X and Y is 13 per cent. If the order is XZY, the crossover frequency is 33 per cent.

Table 28.2 An example of a phenotypic ratio that indicates linkage		
Phenotype	**Approx. expected numbers if no linkage (9:3:3:1)**	**Observed numbers in F_2 generation**
Purple, elongated	405	336
Purple, rounded	135	31
Red, elongated	135	28
Red, rounded	45	325
Total	720	720

✔ **REMEMBER THIS**

The term **locus** is used to describe the position of a gene on a chromosome.

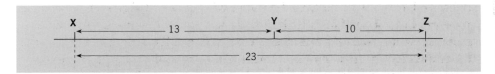

Fig 28.11 The position of genes on chromosomes can be worked out by interpreting crossover frequencies. Consider three genes, X, Y and Z. Genes X and Z are found to have a crossover frequency of 23 per cent, so they are 23 units apart on the chromosome

⬤ **EXAMPLE**

An A-level genetics question involving linkage

This example is a modified question from an UCLES/OCR Modular Biology examination in Applications of Genetics.

The mosquito *Aedes aegypti* has a spotted and a spotless form on a grey or yellow body. The dominant allele is spotless, S, and the recessive allele is spotted, s. The allele for grey body G, is dominant to the allele for yellow body, g. A cross was made between homozygous spotless, grey-bodied mosquitoes and spotted, yellow-bodied mosquitoes. The F_1 individuals were then crossed with the double recessive strain and the numbers of the resulting phenotypes were counted. The results are shown below.

Phenotype	Number
Spotless, grey bodied	442
Spotted, yellow bodied	458
Spotless, yellow bodied	46
Spotted, grey bodied	54

Q State the genotype and the phenotype of the F_1 using the above symbols.

A The question is asking for the genotype of the first generation. You know that the parents are homozygous spotless, grey (genotype: SSGG) and spotted, yellow (genotype: ssgg). The genotype of the F_1 must be SsGg, giving a phenotype of spotless, grey bodied.

Q Using a genetic diagram, explain fully the results shown in the table.

A The table below shows the result of a cross between an F_1 individual, SsGg, and a double recessive which must be ssgg.

So we have:

Parents with genotype	Spotless, grey SsGg	×	Spotted, yellow ssgg	
Gametes	SG Sg sG sg	×	all sg	
Expected ratio	25% SsGg	25% Ssgg	25% ssGg	25% ssgg
Phenotypes	Spotless, grey	Spotless, yellow	Spotted, grey	Spotted, yellow
	1 :	1 :	1 :	1

You should see that the expected results are not the same as those given in the table of results on the left, so there must be an extra complication. The key feature from the table is that there are more of the parental phenotypes than expected, and fewer recombinants. This should scream *LINKAGE* at you!

To produce a really thorough answer you should also work out the percentage of recombinants.

The total number of crosses in the table is 1000, of which 100 (46 + 54) are recombinants. This leaves you with the simple calculation to show that 10 per cent of the mosquitoes are recombinants, and so the two genes are 10 units apart on the chromosome.

3 SEX DETERMINATION

What decides whether a baby will be a boy or a girl?

In humans, sex determination depends on the inheritance of special **sex chromosomes**. Every human cell contains 23 pairs of chromosomes: 22 pairs of **autosomes** and one pair of sex chromosomes (Fig 28.12). Females have two large X chromosomes and so we say they are the **homogametic** sex. Males have one X chromosome and one smaller Y chromosome, and are therefore the **heterogametic** sex.

In the female, all eggs receive an X chromosome during meiosis (Fig 28.13). However, meiosis in the male produces sperm, half of which receive a Y chromosome and half an X. So as the sperm swim for the female's egg it's a real race: if a Y sperm fertilises the female's egg, the offspring is male. If an X sperm wins, the baby will be a female.

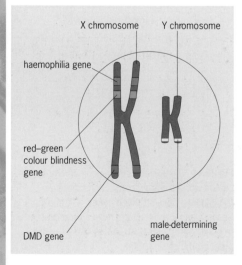

Fig 28.12 The human karyotype of 22 pairs of autosomes and 1 pair of sex chromosomes is shown in Fig 24.2. This diagram shows the sex chromosomes from a male: females have two identical X chromosomes. Note that the Y is much smaller than the X, and carries fewer genes. The Y chromosome does, however, carry the male-determining gene

In humans, about 114 boys are conceived for every 100 girls. We do not yet fully understand the reason for this. Interestingly, however, more male embryos die and so at birth the proportions are down to 106 to 100. By puberty the numbers are equal and in old age the females outnumber males by two to one.

In the 1990s, a group of genetic researchers demonstrated that the Y chromosome carries just 60 functional genes, including a male-determining gene called **SRY**. All embryos are female unless the active SRY imposes maleness upon it. When a male and female embryo share the same blood supply, it is possible that the SRY can produce hormones that can force an XX embryo to develop as a male. This is an extremely rare situation in humans but it is common in cattle.

Androgen Insensitivity Syndrome (AIS) is condition that can cause a genetically male child to develop outwardly as a female. The receptor proteins for the male hormones (androgens) don't work properly, so the embryo fails to read the signals that should turn it into a male. This syndrome may go undetected until the teens, when puberty fails to happen. Genetic analysis will reveal that the 'girl' is in fact a male. Although she will look and act like a girl, she will be unable to have children and will not menstruate.

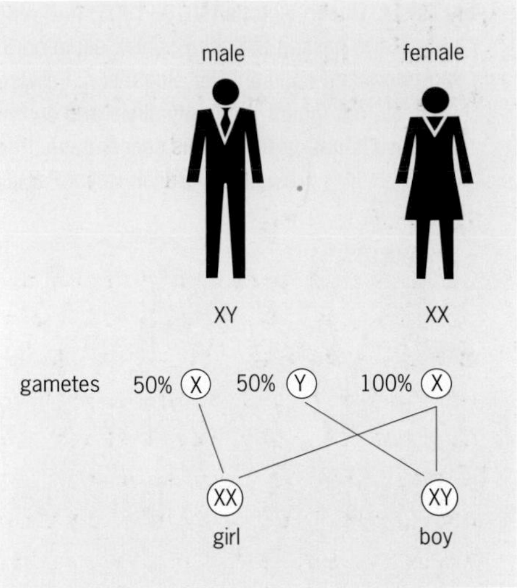

Fig 28.13 Sex determination in humans. Half of the sperm carry an X chromosome, half carry a Y

4 SEX-LINKED INHERITANCE

Genes carried on the sex chromosomes are said to be **sex-linked**. Human females have two X chromosomes, which, like the autosomes, carry two alleles of every gene. Females therefore have two sets of sex-linked alleles. In males, however, the Y chromosome is smaller and cannot mirror all of the genes on the X chromosome, so males have only one set of most sex-linked alleles. This is why males suffer from the effects of X-linked genetic diseases more often than females. There are no known Y-linked diseases or conditions, probably because the Y chromosome carries so few genes.

A classic example of a sex-linked trait transmitted by the X chromosome is **haemophilia**. People suffering from this disease do not make Factor VIII, an essential component in the complex chain reaction of blood clotting (see page 316). In addition to the problems caused if they injure themselves, haemophiliacs suffer from internal bleeding as a result of normal activity. Bleeding at the joints during even light exercise is a particular problem. However, haemophiliacs can usually live a full and active life by having regular injections of Factor VIII.

Haemophilia is caused by a sex-linked recessive allele. Females have a pair of alleles but males possess only one. So, if a male inherits the haemophilia allele, he has the disease since he cannot possess another healthy allele to mask its effect.

Using the following notation:
X^h = the haemophilia allele on the X chromosome
X^H = the healthy allele on the X chromosome
Y = the Y chromosome which does not carry either allele
we can see that the possible genotypes are:
$X^H X^H$ = healthy female
$X^H X^h$ = healthy, carrier female
$X^H Y$ = healthy male
$X^h Y$ = haemophiliac male

SEE QUESTION 6

? TEST YOURSELF

E What would the notation be for a female haemophiliac? Suggest why they are much rarer than male haemophiliacs.

F If a female carrier of the haemophilia allele marries a normal male (as was the case of Queen Victoria and Prince Albert) and they have a son, what is the probability that he will have haemophilia?

Fig 28.14 Queen Victoria (1819–1901) had nine children and passed the haemophilia allele on to two daughters and one son. Her eldest son, Edward VII, did not inherit the faulty allele and so the disease is not carried by his descendants, the present-day British Royal Family

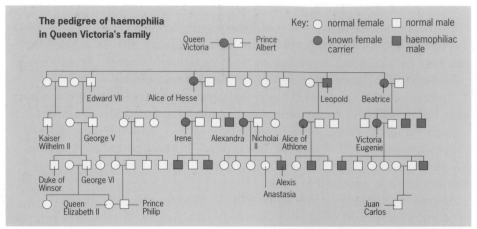

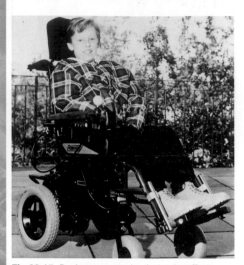

Fig 28.15 Duchenne muscular dystrophy affects mainly boys. The body muscles are gradually replaced by fibrous tissue and become weak, usually confining sufferers to a wheelchair by the age of 10. Few live beyond 20. One male child in 4000 is born with this disease, sometimes caused by a spontaneous mutation, so the mother may not even be a carrier. Gene therapy holds new hope for the future

We can study the inheritance of haemophilia using a **pedigree diagram**. The word pedigree means 'of known ancestry' and applies as much to humans as it does to domestic animals. The pedigree of the Royal Family can be traced back to 1066 and shows that Queen Victoria was a carrier of the haemophilia allele and passed it on to four of her nine children (Fig 28.14). Other sex-linked traits in humans include Duchenne muscular dystrophy (DMD, Figs 28.12 and 28.15) and red–green colour blindness (Fig 28.16).

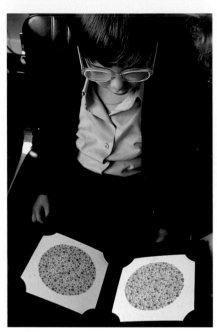

Fig 28.16 A person with red–green colour blindness cannot distinguish between red, green, orange and yellow. On these Ishihara test charts, a person with normal vision would see numbers but a person with red-green colour blindness would see only a random pattern of dots. This condition is caused by a defective allele that codes for the one of the three groups of light-sensitive cone cells in the retina. Eight per cent of males but only 0.4 per cent of females suffer from colour blindness

5 MULTIPLE ALLELES

Sometimes there are more than two alleles for a particular gene. Such genes are said to be **polymorphic.** An estimated 31.7 per cent of human genes are polymorphic – see the Science feature on cheetahs in Chapter 29.

A classic example of a polymorphic gene is the ABO blood group system in humans. The entire human population can be classified into A, B, AB or O depending on the proteins they carry on the membrane of their red blood cells. The ABO system is controlled by one gene (I) with three alleles, I^A, I^B and I^O. The I^A allele codes for A proteins, I^B for B proteins and I^O for no relevant proteins. I^A and I^B are codominant over I^O. Each individual inherits two alleles which combine to produce the blood group as shown in Table 28.3.

Blood group testing can sometimes be used to disprove (but not prove) parentage. Table 28.4 shows, for instance, that a man of blood group AB must pass on either an I^A or an I^B allele, and so cannot possibly be the father of a child with blood group O. DNA fingerprinting, of course, provides a much more accurate (if more expensive) test which can say with a great deal of certainty who *is* the father. The Assignment in Chapter 3 looks at the process of DNA fingerprinting.

Table 28.3 The genetics of the ABO blood group system

Genotype	Phenotype (blood group)
$I^A I^A$ or $I^A I^O$	A
$I^B I^B$ or $I^B I^O$	B
$I^A I^B$	AB
$I^O I^O$	O

Table 28.4 Possible blood groups of children born to parents of particular blood groups

Maternal blood group		Paternal blood group			
		A	**B**	**AB**	**0**
	Genotypes	I^AI^A or I^AI^0	I^BI^B or I^BI^0	I^AI^B	I^0I^0
A	I^AI^A or I^AI^0	A, 0	A, B, AB, 0	A, B, AB	A, 0
B	I^BI^B or I^BI^0	A, B, AB, 0	B, 0	A, B, AB	B, 0
AB	I^AI^B	A, B, AB	A, B, AB	A, B, AB	A, B
0	I^0I^0	A, 0	B, 0	A, B	0

? TEST YOURSELF

G If a child has the blood group B and his mother is blood group O, can his father be a man with blood group A? Say why.

6 POLYGENIC INHERITANCE

So far in this chapter, we have been concerned with **discontinuous inheritance**: all-or-nothing features controlled by single genes. Many features, however, are **continuous** and are controlled by several genes that act together. We describe the inheritance of such characteristics as **polygenic inheritance**.

Skin colour in humans is an example of polygenic inheritance. The depth of skin colour depends on the amount of melanin present. This is determined by at least three genes. As an example of this type of inheritance at work, imagine that the alleles A, B and C contribute cumulatively to the overall amount of melanin but the alleles a, b and c do not. So, someone with the genotype AABBCC would have the darkest skin, while the genotype aabbcc would confer a very pale skin colour.

As Fig 28.17 shows, a cross between two purely homozygous individuals would result in an intermediate (heterozygous) F_1 generation, but the F_2 generation would show a wide variation in skin colour depth. The graph shows a **normal distribution**: most individuals have skin colours in the midrange. This range is one of the characteristic features of continuous variation (see Chapter 24).

✓ REMEMBER THIS

The inheritance of height in humans is polygenic.

? TEST YOURSELF

H Non-identical twins are born to a couple who both have the genotype AaBbCc. Is it possible for one baby to be born with white skin and the other to have black skin? Explain your answer.

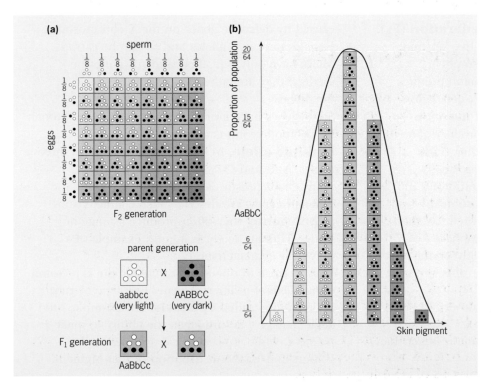

Fig 28.17 Human skin colour is an example of polygenic inheritance which shows a normal distribution in the F_2 generation

(a)
sperm
$\frac{1}{8}$ $\frac{1}{8}$ $\frac{1}{8}$ $\frac{1}{8}$ $\frac{1}{8}$ $\frac{1}{8}$ $\frac{1}{8}$ $\frac{1}{8}$
eggs $\frac{1}{8}$ $\frac{1}{8}$ $\frac{1}{8}$ $\frac{1}{8}$ $\frac{1}{8}$ $\frac{1}{8}$ $\frac{1}{8}$ $\frac{1}{8}$

F_2 generation

parent generation

aabbcc (very light) X AABBCC (very dark)

F_1 generation AaBbCc X AaBbCc

(b)
Proportion of population

$\frac{20}{64}$
$\frac{15}{64}$
$\frac{6}{64}$
$\frac{1}{64}$

AaBbC

Skin pigment

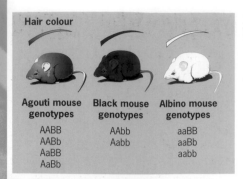

Hair colour

Agouti mouse genotypes	Black mouse genotypes	Albino mouse genotypes
AABB	AABb	aaBB
AABb	Aabb	aaBb
AaBB		aabb
AaBb		

Fig 28.18 The base of the hair of a mouse with agouti coloration is dark brown/black fading to lighter brown towards the tip

? TEST YOURSELF

Two agouti mice, genotypes AaBb, are bred together. What phenotypic ratio would you expect in the next generation?

7 EPISTASIS (WHEN GENES INTERACT)

Up to now, we have discussed the inheritance and effect of genes that act on their own. In practice, however, many genes interact with each other: the presence of one allele often affects the expression of others. We call this interaction **epistasis**. A classic example of epistatic gene interaction is coat colour in mice. The natural coloration of wild mice is called **agouti** and is produced from banded hairs (Fig 28.18).

Two genes are involved, each with a dominant and a recessive allele. We will represent them by the notation Aa and Bb, where A and B are dominant over a and b. The allele A codes for the ability to produce hair pigment: AA and Aa mice have pigmented hairs but all aa individuals are albinos. The B allele codes for the ability to make hair with graduated coloration: BB and Bb mice have graduated hair, bb mice have hair that is all one colour.

Only mice that have both dominant alleles A and B show the agouti coloration. These can have the genotype AABB, AaBB, AABb or AaBb. Mice that are aa cannot make pigment, so, whether they are BB, Bb or bb, no bands are visible. Mice that can make pigments but not banded hair (genotypes AAbb or Aabb) are plain black.

SUMMARY

- Humans, like most organisms, are **diploid**; each cell carries two versions, or **alleles**, of each **gene**. If both alleles are the same, the organism is **homozygous** for that allele. If the pair are different, the organism is **heterozygous**.

- During **meiosis**, only one of the alleles passes into the **gamete**. This is Mendel's first law.

- **Monohybrid inheritance** is the study of one gene. A cross of AA × aa gives all Aa in the **first generation (F_1)**. A cross of Aa × Aa gives a 3:1 phenotypic ratio in the **second generation (F_2)**.

- **Dihybrid inheritance** involves two separate genes, usually on separate chromosomes. Homozygous individuals AABB × aabb produce offspring that are all AaBb in the F_1 generation. However, owing to **independent assortment** during meiosis, AaBb individuals produce gametes of AB, Ab, aB and ab. If bred together, AaBb individuals produce 16 different genotypes, and 4 different phenotypes with a ratio of 9:3:3:1.

- Genes present on the same chromosome are said to be **linked**. Such genes are inherited together unless separated by crossover in prophase I of meiosis. The greater the distance between the genes on the chromosome, the more often they are inherited separately.

- Humans have 23 pairs of chromosomes: 22 pairs of **autosomes** and one pair of **sex chromosomes**. The X chromosome is larger than the Y and carries more genes.

- The sex of a baby is determined at the moment of conception. Because gametes are produced by meiosis, half of the sperm carry a Y chromosome and half carry an X. If a Y sperm fertilises the egg, a male results; X gives a female.

- Genes carried on the sex chromosomes are said to be **sex-linked**. Females have two complete sets of sex-linked genes, but males have only one copy of all but the few genes carried on the Y chromosome.

- Some diseases such as haemophilia are sex-linked; caused by defective genes on the X chromosome. Females have two alleles and so can be carriers. Males have only one allele and so cannot be carriers: they are either healthy or they have the disease.

- Some genes have more than two alleles. The human ABO blood group is an example of such a **multiple allele**. In this case, one gene has three alleles, A, B and O.

- Many characteristics are controlled by many different genes. We describe such characteristics as **polygenic**; they often produce a range of values. Human height is a good example of a polygenic characteristic.

- The presence of one allele can affect the expression of an allele of a different gene. This type of gene interaction is called **epistasis**. For example, one allele may give an organism the ability to make pigment, while another allele determines the colours and/or the distribution of that pigment.

? EXAM QUESTIONS

1 In rabbits, the fat beneath the skin can be either white or yellow. The F_1 generation from a cross between a white-fat rabbit and a yellow-fat rabbit all had white fat. When these offspring were interbred, some of the F_2 generation had white fat and some had yellow fat.

a) What type of inheritance does this character show?

b) Choosing appropriate symbols for the two characters, state the **genotypes** of the parent rabbits.

c) **(i)** Draw a genetic cross diagram to show how the genotypes of the F_1 generation are produced.
(ii) State the likely proportion of each **phenotype** in the F_2 generation.

d) Explain how the genotypes of the white-fat rabbits in the F_2 generation could be determined. Use diagrams where appropriate.

e) What conclusion would you draw if all the F_2 generation yellow-fat rabbits were male?

The yellow colour in the fat is due to xanthophyll, a plant pigment absorbed from the food.

f) Suggest why this pigment is absent from the fat of white-fat rabbits.

OCR 6912 Inheritance Part 1 March 2000 Q1

2 Cystic fibrosis is a common inherited disorder amongst Europeans. It is caused by a single gene. The diagram below is part of a family pedigree showing the inheritance of this disorder.

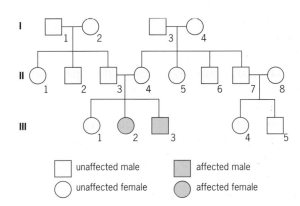

unaffected male   affected male
unaffected female   affected female

a) Describe *one* piece of evidence from the information given on the diagram which suggests that the allele causing this inherited disorder is recessive, and *one* piece of evidence that it is autosomal.

b) If the percentage of children born with cystic fibrosis is known, then it is possible to calculate the frequency of the recessive allele in the population. This information can be used to estimate the percentage of people in the population who are heterozygous for this gene, using the equation below:

$$p^2 + 2pq + q^2 = 1$$

where **p** is the frequency of the normal (dominant) allele
q is the frequency of the recessive allele for cystic fibrosis
p² is the frequency of people homozygous for the normal allele
q² is the frequency of people homozygous for the recessive allele
2pq is the frequency of heterozygous people (carriers)

(i) If **p** is 0.978 and **q** is 0.022, calculate the frequency of the heterozygous genotype in the population.

(ii) Explain the significance of the frequency of the heterozygous genotype of this genetic disorder in the population.

c) A person whose family includes people with cystic fibrosis may wish to know if they or their partner are carriers of this inherited disease, before deciding to have children. Suggest how an individual with the heterozygous genotype for this disorder could be identified.

d) Suggest *two* reasons why chorionic villus sampling is more acceptable than amniocentesis in detecting disorders in a fetus.

Edexcel HB3 June 1999 Q6

3 Eye colour in humans is controlled by a pair of alleles of a gene where the allele giving brown eyes is dominant to the allele giving blue eyes. Both parents of a blue-eyed man, John, were brown-eyed. He married a brown-eyed woman, Sara, whose father had brown eyes and mother blue eyes. Sara had a blue-eyed sister. John and Sara had a brown-eyed child.

a) **(i)** Devise suitable symbols for the alleles for brown eyes and for blue eyes.
(ii) Fill in the boxes and circles on the pedigree (family tree) below to show the genotype of each individual.

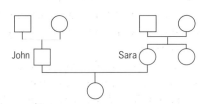

? EXAM QUESTIONS

b) Is it likely that the gene is carried on the X chromosome? Explain your answer.

c) For animals such as mice it is possible to carry out a cross to find out whether an individual showing the dominant allele is heterozygous or homozygous.

(i) Use a genetic cross diagram to show how this might be done.

(ii) In humans such an experiment is not ethically permissible. Explain briefly how it might be possible to determine whether a human was heterozygous or homozygous for a dominant allele.

OCR 6912 Inheritance Part 1 March 2001 Q1

4 Events in meiosis may give rise to genetic variation in daughter cells.

a) Using **diagrams**, show how **(i)** independent assortment and **(ii)** crossing over, may result in such variation for two heterozygous pairs of alleles, P and p, Q and q.

b) Name, and give an example of **one** other source of genetic variation, other than meiosis.

OCR 6912 Inheritance Part 1 November 2001 Q2

5 In *Mirabilis,* the four o'clock plant, one gene with two alleles T and t gives rise to tall or short plants with the tall allele being dominant to short. Another gene with two alleles R and r gives rise to red, white or pink flowers, with neither allele showing dominance to the other.

A pure breeding tall, red-flowered plant was crossed with a short, white-flowered plant and all the F_1 offspring were found to be tall with pink flowers. On intercrossing (selfing) the F_1 an F_2 generation was obtained with six different phenotypes as follows:

tall with red flowers	short with red flowers
tall with pink flowers	short with pink flowers
tall with white flowers	short with white flowers

a) **(i)** Copy and complete the genetic diagram showing clearly the genotypes of the F_1 and F_2 generations together with the appropriate gametes in each case.

parents		TTRR	×	ttrr	
gametes		.....		.....	
F_1			.....		
gametes	TR	.....		tR	.....
	.....	.....	.....	.....	.....
Tr	.....	.....	.....	.....	
	.....	.....	.....	.....	.....
tr	.....	.....	.....	.....	

(ii) Indicate the **proportions of each phenotype** you would expect to find in the F_2.

tall with red flowers	short with red flowers
tall with pink flowers	short with pink flowers
tall with white flowers	short with white flowers

b) **(i)** What name is given to the type of inheritance shown by the **flower colour** gene?

(ii) Suggest how this gene may operate to give red, white or pink flowers.

OCR 6913 Inheritance Part 2 March 2001 Q2

6 In the fruit fly, *Drosophila melanogaster,* the allele **r** for white eyes is sex-linked and is recessive to the allele **R** for the normal red eye colour. Sex-linked genes are carried on the X chromosome. In fruit flies, the sex chromosomes are XX in females and XY in males.

a) When a white-eyed female fly is crossed with a red-eyed male, all of the male offspring have white eyes.

(i) Draw diagrams to represent the chromosomes, and alleles present on them, of the parents in this cross.

(ii) State the genotype and phenotype of the female offspring.

b) A gene which affects the size of the wings in fruit flies is not sex-linked (autosomal). The allele for short (dumpy) wings, **d**, is recessive to the allele, **D**, for normal wings.

A female fly was heterozygous for the genes for eye colour and for wing size. This female was crossed with a white-eyed, short-winged male.

(i) Use a genetic diagram to show the expected results of this cross.

(ii) List the genotypes and full phenotypes of the male offspring produced.

Edexcel B/HB1 June 1999 Q7

KEY SKILLS **ASSIGNMENT**

The chi-squared test

The chi-squared test is a simple statistical method that scientists use to tell if their results are significant or due to chance. For instance, if you were to roll a die, you would have a one in six chance of getting any particular number. If you rolled the die six times, you would probably not get each number once; you may get three twos or some other combination. However, the more times you roll the die, the greater the probability that all six numbers will be represented evenly.

If you are playing a game with someone who gets a higher proportion of sixes than normal, you might suspect them of playing with a loaded die. Without investigating the die itself, how could you decide whether that person was just lucky or that the difference was due to cheating? You could use the chi-squared (χ^2) test.

The nature of science and the null hypothesis

Most people accept that there is a link between smoking and lung cancer but, in science, it is virtually impossible to prove anything absolutely. What you can do is gather support for your hypothesis.

You could do some research into lung cancer victims and divide them into smokers and non-smokers. You could then use the χ^2 test to decide if there was a significant difference in the number of smokers who developed lung cancer compared with the non-smokers.

To use this statistical test you must approach the problem from a specific angle, and this involves developing a null hypothesis.

In the lung cancer example, you would use the hypothesis: 'There is no difference in the incidence of lung cancer between smokers and non-smokers'. You could then perform the χ^2 test (or some more sophisticated statistics) and come up with the conclusion that 'there is a very low probability that these results are due to chance' in which case you can reject the null hypothesis and support the idea that there is a link between smoking and lung cancer.

1 Write a null hypothesis that you might use to investigate the following questions:
a) What is the effect of sunbathing on the incidence of skin cancer?
b) What is the effect of cooking on the vitamin C content of vegetables?

The test itself

The basic idea behind the χ^2 test is that you compare the observed results with those you would expect. The extent to which the results vary is sometimes described as 'goodness of fit'. This means do the actual results fit the results that we had expected?

The formula for the χ^2 test is:

$$\chi^2 = \sum \frac{(O - E)^2}{E}$$

In this formula

O = the observed result,

E = the expected result and

Σ = the sum of.

In practice it is best to use a table like the one below to progressively work out all the steps needed by the formula. In this worked example, a geneticist performed one of Mendel's classic pea experiments and got the following results.

Yellow, round	318
Yellow, wrinkled	103
Green, round	106
Green, wrinkled	33
Total	560

The geneticist wanted to know whether these results did actually reflect the expected 9:3:3:1 ratio, or whether they deviated so significantly that other forces were at work, and that the 9:3:3:1 hypothesis was in doubt. She did the following calculations:

Category	Hypothesis	Observed	Expected	$O - E$	$(O - E)^2$	$\frac{(O - E)^2}{E}$
Yellow, round	9	318	315	3	9	0.028
Yellow, wrinkled	3	103	105	−2	4	0.038
Green, round	3	106	105	1	1	0.009
Green, wrinkled	1	33	35	−2	4	0.114
Total		560				= 0.189

So the value of χ^2 is 0.189. What next? The hard work is now done and we only need one more piece of information; the number of degrees of freedom, which is a measure of the spread of the data. The number of degrees of freedom is always one less than the number of categories of information. So in this case, it is four categories minus one, which equals three degrees of freedom.

KEY SKILLS **ASSIGNMENT**

We now look up the value of 0.189 with three degrees of freedom on a table of χ^2 values. Table 28.A1 is an extract from the χ^2 tables.

Table 28.A1 Extract from χ^2 tables

	Probability (eg 0.5 = 50%)									
	0.99	0.98	0.95	0.9	0.7	0.5	0.3	0.2	0.1	0.02
1 D of F	0.00016	0.0063	0.0039	0.0158	0.148	0.455	1.07	1.64	2.71	5.41
2 D of F	0.0201	0.0404	0.103	0.211	0.713	1.39	2.41	3.22	4.60	7.82
3 D of F	0.115	0.185	0.352	0.584	1.42	2.37	3.66	4.64	6.25	9.84

From Table 28.A1 we can see that our value of 0.189 lies between the 0.98 and 0.95 column. From this we can say that the probability that our results are significant is between 95 and 98 per cent. Put another way, it tells us that the probability that these results are due to chance is between 2 and 5 per cent. By convention, scientists take the cut-off point as 5 per cent. If the test shows that there is a less than 5 per cent probability that your results are due to chance, you can reject your null hypothesis.

2 Use the χ^2 formula to work out the following:
a) A student tosses a coin 100 times. She gets 64 tails and 36 heads. Is the coin biased or are the results due to chance?
b) A geneticist managed to get his hands on a breeding pair of incredibly rare Matabili dung beetles. The male had green eyes and a blue abdomen. The female had pink eyes and an orange abdomen. When he crossed them he got the following results:

Table 28.A2

Phenotype	Numbers of offspring in F_1
Green eyes, blue abdomen	563
Pink eyes, blue abdomen	190
Green eyes, orange abdomen	183
Pink eyes, orange abdomen	64
Total	1000

Do these results conform to the expected 9:3:3:1 ratio?
c) Apart from their diet, these beetles look like being ideal organisms for the study of genetics. What features make an organism suitable for the study of inheritance in the laboratory?
d) Much progress in the field of genetics has been made using the fruit fly *Drosophila*. Find out some of the features that can be studied using *Drosophila*.

Chi-squared in examinations

If you have worked your way through the above examples, you may have concluded that chi-squared involves a lot of arithmetic and is time consuming. The examiners have come to the same conclusion and therefore it is unlikely that you will be set a full investigation to work through from scratch: they want to test your biology, not your maths. However, this test is a very useful tool and examiners will want you to appreciate its uses, and possibly to do some simple calculations.

29

Evolution

The last known surviving kakapo from the mainland is an aged male called Richard Henry who has been transported to an island for his own safety. His 'whuumphing' days may not be over. The New Zealand Department of Conservation is taking no chances with kakapo survival. All individuals are fitted with trackers and if a nesting female goes foraging, a worker will quickly put a thermal blanket over the eggs. 2001 was a record year for kakapo births

On a few remote islands off the coast of New Zealand the world's largest parrot – the kakapo – is just clinging to existence. Looking like a large green football with a beak, the kakapo was once an important part of New Zealand's bird-rich ecosystem. Like many island birds that have evolved in an environment free from predators, the kakapo is flightless. It has also developed a few other odd characteristics.

Locals knew when it was the kakapo's mating season because of its incredibly deep mating call: a sort of giant heartbeat in the night. The male kakapo sits on a hillside and emits a huge 'whuumph' from his vocal sacs. The problem is that, like all bass sounds, it is non-directional. This is a big drawback in a mating call because the females cannot tell where it is coming from.

'Whuuumph; I'm in the mood!'
'Yes, but where are you?'

All is well when numbers are high, but when kakapos are few and far between, the females cannot easily find the males.

Like the dodo, the kakapo's idyllic lifestyle came to an end after human settlers arrived with their animals; their numbers started to plummet. The Polynesian colonists (the Maoris) settled in New Zealand 1000 years ago, bringing dogs and rats. These ate the eggs from the ground-level nests of the kakapo and other flightless birds. More recently, Europeans arrived, adding cats to the predators. Kakapos have not developed any new behaviour to deal with such enemies.

In 1999 there were 62 kakapos left, 26 females and 36 males. These are currently located on five off-shore islands where they have been relocated to protect them from introduced predatory mammals. All the birds have transmitters and their breeding activity is closely monitored; eggs are removed and carefully incubated. The breeding programme is showing signs of success, with some females being able to lay eggs twice a year. The kakapo may yet survive.

1 WHERE DO LIVING THINGS COME FROM?

To most human eyes, the living world seems fairly static. Like produces like: elephants give birth to baby elephants, frogs make frogs. It seems that no new species appear. There are literally millions of different species on Earth. Where did they come from?

THEORIES ON THE ORIGIN OF SPECIES

Ideas about the origins of species – including ourselves – have been central themes in philosophical and religious thought for centuries. Until the nineteenth century, virtually everybody in the Christian world looked to the Bible. The book of Genesis provided the answers: all animals and plants were created at the same time, by the great Creator. Even the eminent taxonomist

Fig 29.1 The discovery in the nineteenth century of fossils such as this *Triceratops* fuelled the idea that the living world was not static. Different layers of rock held different sets of organisms and this led to the acceptance that the living world was not fixed: plants and animals could change with time

Linnaeus, who in 1742 published his system for classifying all living plant species known at the time, said 'there are as many species as God created in the beginning'. Other religions had different versions of the same creationist idea.

The idea that species were not fixed, and that the infinite variety of living things had developed through a slow process of evolution, seems to have arisen first in early Greek philosophy and reappeared from time to time throughout the following centuries. It failed to gain lasting acceptance because no explanation could be found to show how the process might have come about. In addition, in some cultures, any 'non-religious' thought was strongly discouraged, even on penalty of death.

A catalyst in the development of the theory of evolution was the discovery of fossils (Fig 29.1). These were clearly the remains of extinct creatures. Why were they no longer alive?

One Christian explanation said that extinct species were the victims of a great global catastrophe, and that a new set of organisms had been created to replace them. It was even suggested that only the passengers aboard Noah's Ark survived a great global flood. When it was pointed out that different rocks contained different types of plants and animals, it seemed that there would have had to be many such floods, and so the **catastrophe theory** gradually lost credibility.

An early attempt to explain the mechanism of evolution came from the French naturalist Lamarck, who had made extensive studies of different life forms. He was convinced of the process of evolution, but the mechanism he proposed in 1809 to explain it was wrong. Lamarck supposed that organisms adapt to their environment by developing new structures and losing old ones. Such differences, acquired during the animal's life, were then passed on to the next generation. As an example he cited the webbed feet of water birds. Lamarck suggested that, in an effort to swim, the birds extended their toes and so stretched the skin between them. The stretched condition was then inherited and this process was repeated until it produced a fully webbed foot. With no knowledge of genes, chromosomes or DNA, this idea of the **inheritance of acquired characteristics** was perfectly reasonable for the time.

? TEST YOURSELF

A What was wrong with Lamarck's theory of evolution?

2 CHARLES DARWIN AND THE THEORY OF EVOLUTION

REMEMBER THIS

A working definition of evolution is 'a change in the genetic composition of a population over time'. It has also been defined as 'a gradual process in which new species develop from pre-existing ones'. But as we shall see in this chapter, populations can evolve without changing into new species.

REMEMBER THIS

The cornerstone of Darwin's theory is that the mechanism of evolution is natural selection – this is known as **Darwinism**. The application of twentieth-century knowledge about chromosomes, genes and DNA to Darwin's theories is called **neo-Darwinism**.

In a book with the shortened title of *The Origin of Species*, Charles Darwin included these four ideas in his theory of evolution:

- **The living world is changing**, not static.
- **The evolutionary process is usually gradual**, not a series of jumps such as catastrophes and re-creations.
- **The common ancestor theory**. This suggested that closely related species evolved from one basic ancestor. Darwin said that man and apes evolved from a common ancestor, but the popular press took this to mean that humans had recently evolved from apes, leading to cartoons such as the one in Fig 29.2.
- **The theory of natural selection** as a mechanism to explain evolution.

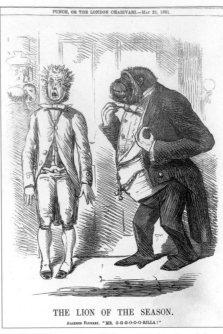

Fig 29.2 This cartoon first appeared in *Punch* in 1861. It shows how poorly people of the time understood Darwin's theories

SCIENCE FEATURE Charles Darwin

Charles Darwin (Fig 29.3) was the son of a doctor. After studying medicine at Edinburgh he changed his mind and prepared for a career in the Church. However, despite going to Cambridge to study theology (religion), Darwin retained his interest in biology and geology.

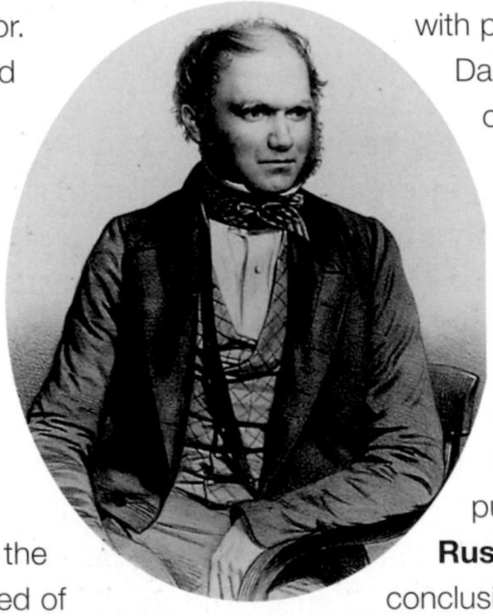

Fig 29.3 Charles Darwin, 1809 to 1892

Darwin was offered the post of naturalist on the HMS *Beagle*, embarking on a scientific survey of South American waters. It was observations he made during this trip that first stimulated the young Darwin into contemplating the origins of species. As well as becoming convinced of the process of evolution, he also developed an idea about how it might happen. This trip was not only the turning point in Darwin's life; it was also the start of the greatest ever revolution in our perception of the living world, and of our place within it.

On returning home, Darwin buried himself in his studies, determined to prove or disprove his ideas about the evolution of species. Like Lamarck, Darwin developed his theories with no knowledge of genes, chromosomes or DNA. Mendel's work with peas (see Chapter 27) was going on during Darwin's lifetime but it remained hidden in an obscure journal.

Even though he managed to build up an overwhelming case for evolution, Darwin was reluctant to publish his theories, knowing the conflict they would cause with the Church. For more than 20 years Darwin carried on gathering evidence to support his theories. He was finally pushed into publication after exchanging letters with **Alfred Russel Wallace**: he too had come to the same conclusion about evolution. They agreed to make a joint announcement, and the Darwin–Wallace paper was presented to the Linnaean Society of London in July 1858.

In November 1859, Darwin published his revolutionary work, *The Origin of Species by Means of Natural Selection*. Few books before or since have caused quite such a controversy. Although Church leaders largely remained quiet, many people bitterly attacked it because it questioned the theory of creation. Darwin was called 'the most dangerous man in England'.

Accurately estimating the age of the Earth is central to the theory of evolution. In 1650 James Usher (then the Archbishop of Armagh) announced that his study of Hebrew literature had led him to the conclusion that the date of creation was the year 4004 BC. Dr Lightfoot, vicar of the University church at Cambridge, took it a step further and proclaimed that the date of creation was 9 a.m. on 23 October, 4004 BC. Baron George Cuvier later stated that the Earth must be much older, more like 70 000 years.

Faced with the large array of different fossils, nineteenth-century geologists decided that the Earth was much older still. In 1830, Charles Lyell suggested that the true age of the Earth would turn out to be millions, rather than thousands of years. He was the first to propose that old rocks could be uncovered or brought to the Earth's surface (by volcanic activity, for example), so revealing the remains of long-dead species.

Ultra-sensitive dating techniques that measure the decay of radioactive isotopes have now put the age of the Earth at about 4 600 000 000 (4.6 billion) years. This is an important breakthrough because, although our brains cannot comprehend such a time scale, it fits in well with the idea of progressive evolution. The Assignment at the end of this chapter should help you to put this geological time scale into perspective.

NATURAL SELECTION: 'SURVIVAL OF THE FITTEST'

Darwin's observations of the living world led to three basic conclusions:
- Generally, organisms produce far more offspring than can possibly survive.
- Living things are locked in a struggle for survival. They compete for food, space, mates, etc. In short, they are trying to eat and not be eaten, so that they can reproduce.
- Individuals of the same species are rarely identical: they show **variation**.

There was nothing particularly original about these observations, or in the idea that organisms evolved. What Darwin did, however, was to put all these factors together and suggest **natural selection** as the mechanism for evolutionary change.

Natural selection is commonly simplified to 'survival of the fittest'. Biologists define fitness as the ability of an organism to survive and reproduce. The fittest organisms survive to produce more offspring than less fit ones. In this way, the characteristic of a population can gradually change from generation to generation.

To illustrate the principle, imagine a population of blackbirds in a woodland. Like most organisms, they reproduce sexually and there is variation in the population. Some have better reflexes than others, forage for food more successfully, others have a better immune system. Normally, there is competition for resources such as food, mates and nesting sites. In this situation, the fittest birds survive and produce more offspring than others. The fitter birds may gather more food and so rear larger families than less fit individuals (Fig 29.4). Or they may rear two clutches of eggs in the time it takes others to produce one. The vital point here is that the *fittest individuals pass their genes on to more of the next generation*.

The severity of an organism's circumstances is called the **selection pressure**. The greater the selection pressure, the faster the evolutionary process. When organisms have a short life cycle, the process can be very swift indeed. In the 1950s farmers in the UK and Australia took the drastic step of introducing a deadly disease, myxomatosis, into the ever-expanding rabbit population (Fig 29.5). This virus killed over 99 per cent of the rabbits: this is selection pressure at a massive level. Only the rabbits whose genes gave them resistance to myxomatosis were able to survive and pass on their genes to the next generation.

Fig 29.4 Blackbirds that can gather the most food tend to raise the greatest number of offspring, and so pass on their genes to more of the next generation

Fig 29.5 When the European rabbit is introduced into countries where it is not endemic, it tends to out-compete the native herbivores. With few predators around, the rabbit population can reach plague proportions. When this has happened in the UK and Australia, farmers faced ruin. The deadly myxoma virus was introduced into the rabbit population, killing 99 per cent. The remaining 1 per cent were naturally resistant to the virus and are the ancestors of today's, largely resistant, rabbit population

B Although we have waged war against species such as locusts, weevils and beetles, we have never succeeded in making any pest species extinct. On the other hand, many larger species such as rhinos, whales and giant pandas are facing extinction, despite our efforts to save them. Give two reasons for this difference.

Fig 29.6 This disused copper mine at Parys Mountain in Anglesey (above) is devoid of life. Copper is a metabolic poison that kills virtually all plants. The grass *Agrostis tenuis* (right) is one of the few that has evolved copper resistance

NATURAL SELECTION IN ACTION

Many people tend to think that evolution is something that happened in the past and that the world of today is the finished product. But this is far from the truth: evolutionary forces are still at work and natural selection is continually changing the characteristics of populations, often with great speed. Consider the following examples:

● The widespread use of antibiotics such as penicillin rapidly led to the development of resistant strains of bacteria. Penicillin works by inhibiting one of the enzymes that the bacteria need in order to manufacture cell walls. However, many resistant strains can make an enzyme, penicillinase, that breaks down the antibiotic.

● The use of warfarin as a rat poison has led to the development of warfarin-resistant rats. Warfarin is an anti-coagulant that kills rats by inducing **haemorrhage** (internal bleeding). Resistant rats have a modified enzyme in their blood-clotting system that allows their blood to clot in the presence of warfarin.

● Copper tolerance in the grass *Agrostis tenuis*. Copper is a metabolic poison that prevents many plants from growing on copper-polluted land (Fig 29.6). However, resistant plants have evolved. These can transport copper out of their cells, so that it accumulates in the cell wall and does not interfere with the plant's metabolism. (In non-polluted areas, the normal type of grass out-competes the copper-resistant strain.)

The organisms in these examples have a short life cycle and high reproductive capacity, and so can respond quickly to the selection pressure placed on them. The effects of selection pressure on organisms that live longer and have a lower reproductive capacity take much longer to become apparent.

DIFFERENT TYPES OF NATURAL SELECTION

From what you have learned about natural selection, it would be tempting to assume that it is just a mechanism for change. However, this is not always the case: it can be a means of keeping things static. Many organisms, crocodiles and sharks for example, have not changed significantly for millions of years.

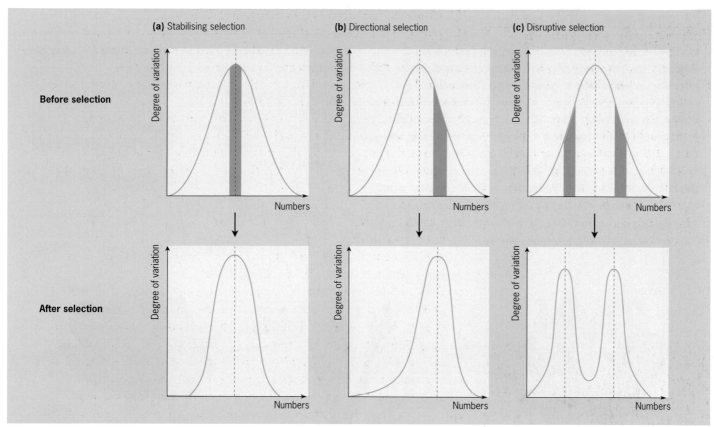

Fig 29.7 The different types of natural selection. Individuals in the shaded areas in the top row of graphs have a selective advantage

There are basically three different types of natural selection:

- **Stabilising selection** (Fig 29.7a). In a favourable and unchanging environment, stabilising selection can lead to a 'standardising' of organisms by selecting against the extremes. An example of this is human birth weight. Studies show that particularly large or small babies have a higher mortality than those in the mid-range.

- **Directional selection** (Fig 29.7b). This mechanism tends to occur following some kind of environmental change that causes selection pressure. The organisms with a particular extreme of phenotype may have an advantage. At some stage, directional selection must have operated to produce the long neck of the giraffe. Probably, in times of food shortage, only the tallest individuals could reach enough food to survive and pass on their genes to the next generation.

- **Disruptive selection** (Fig 29.7c). This type of selection acts at both ends of the distribution, favouring individuals with extreme rather than average characteristics. Rarer than the other two types, disruptive selection is important in the formation of new species, where natural selection acts differently on different populations. An example of this is seen in Darwin's finches (see page 529) where disruptive selection has acted on beak size. In some situations, those with longer, thinner beaks would have succeeded in exploiting insects as a food source, while those with shorter, stronger beaks would have succeeded in cracking seeds. Those with average beaks were adapted for nothing in particular and would have reproduced less successfully.

? TEST YOURSELF

C From your knowledge of the living world, give three examples of directional selection.

⚠ SCIENCE FEATURE Artificial selection

Humans have practised artificial selection, or selective breeding, for thousands of years. We haven't needed a scientific understanding of genetics to do this: animals or plants with the most desirable features are simply crossed or mated. For example, if cows with the highest milk yield are impregnated by the healthiest and fastest-growing bulls, many of the next generation are big, healthy cows that produce large quantities of milk. Those that do not have the desired features are not used for breeding. In this way, artificial selection rather than natural selection is used to bring about great changes in the phenotype in a short time: animals and plants can be 'domesticated' in just a few generations.

Selective breeding is still important in many different areas of food production:

● Increasing the growth rate and the meat, milk or egg production of livestock.
● Increasing the disease resistance of crop plants (increasing the disease resistance of animals by selective breeding is not as reliable).
● Changing the muscle to fat ratio of livestock, so that more lean meat is produced.
● Increasing the yield and nutritive value of crop plants such as wheat.
● Increasing the tolerance of plants and animals to drought, heat or pollutants.

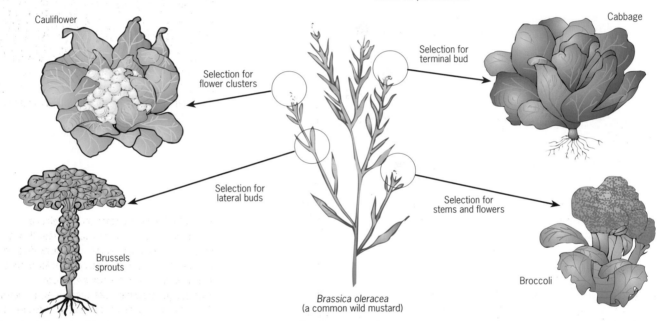

Fig 29.8 Many familiar vegetable types have arisen from the artificial selection of the wild mustard, *Brassica oleracea*. In the original wild population, plants show variation in the size and shape of leaves, stems, terminal buds, lateral buds and flower clusters. Selection for individual features rapidly produces the familiar crops you see here

Fig 29.9 Like all pedigree animals, this champion Abyssinian cat is the product of many generations of artificial selection. Breeders consider the wedge-shaped face a desirable characteristic and so have concentrated on this feature. Breeding animals in this way can lead to problems, and some people strongly disagree with breeding animals for superficial 'perfection'

3 THE EVIDENCE FOR EVOLUTION

Today there are few scientists who seriously doubt the theory of evolution, or that it happens by natural selection. Here we consider some lines of evidence.

FOSSILS

Palaeontology, the study of fossils, supports the theory of evolution. Most fossils are found in sedimentary rocks, formed by layers of silt that preserve the bodies of organisms before they have a chance to decay in the normal way. In many places (see Fig 29.10) it is clear that different layers, or **strata**, represent different time periods. By studying the fossils in these strata we find obvious evidence for the gradual evolution of some species.

GEOGRAPHICAL DISTRIBUTION

A study of the distribution of species presents strong evidence that organisms evolved 'into' their ecosystem rather than being placed there during a single act of creation.

During his time in South America, Darwin encountered a classic piece of evidence for evolution. On the South American mainland, finches have short, straight beaks to crush seeds. Other species have taken advantage of the other food sources. Woodpecker finches, for instance, have long beaks for extracting insects from holes in trees. On the Galapagos Islands, however, things are different. For one reason or another, the finches managed to get to the islands, while many other species of birds did not. In the absence of competition, the finches have evolved to take advantage of the range of food sources on offer (Fig 29.11).

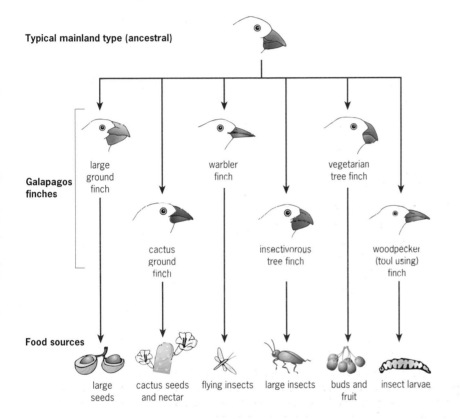

Typical mainland type (ancestral)

Galapagos finches

large ground finch

cactus ground finch

warbler finch

insectivorous tree finch

vegetarian tree finch

woodpecker (tool using) finch

Food sources

large seeds

cactus seeds and nectar

flying insects

large insects

buds and fruit

insect larvae

Fig 29.10 The Grand Canyon in Arizona. The horizontal rock strata can be clearly seen here and span some 500 million years of geological history

Fig 29.11 Darwin's finches on the Galapagos Islands. In a process called adaptive radiation, one common ancestor has evolved into many different species of finch, each exploiting a particular food source, such as seeds, insects and buds. Interestingly, Darwin did not immediately recognise the significance of all of these different species; he simply collected a large number of specimens in a bag and presented them to a bird specialist on his return home

SCIENCE FEATURE Evolution on islands

As Darwin found out on his voyage, islands are particularly interesting places to study evolution. Young, volcanic islands such as the Galapagos (Fig 29.12) are often colonised by plants whose seeds have been blown there by the wind, and by flying animals which have been driven off course by storms. In later years, of course, colonisation by non-flying animals (such as rats and cats) is likely to be a result of human activity.

The evolution of birds on islands can be very spectacular. Free from predators, birds often lose the power of flight and sometimes grow very large. Before the Maori settlers arrived from Polynesia, New Zealand was a bird-dominated ecosystem. Huge flightless birds, **moas**, some as tall as 3.5 metres, evolved to fill the niches normally occupied by mammals. They were easy prey for the Maoris, however, who hunted them to extinction within a few centuries. Ironically, the only survivor of the moa family, the kiwi, has become the country's national emblem.

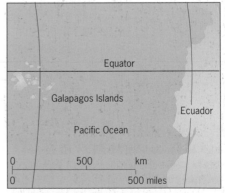

Fig 29.12 The Galapagos Islands are an archipelago situated near the Equator in the Pacific Ocean, about 600 miles west of Ecuador

COMPARATIVE ANATOMY

The anatomy (body plan) of plants and animals can reveal a lot about evolutionary relationships. Fig 29.13 shows one of the classic examples of comparative anatomy, the **pentadactyl limb** (five-fingered limb). This structure appears in various forms throughout the vertebrates: the feet of reptiles and amphibians, the human hand, the bat's wing, the seal's flipper, for example.

Fig 29.13 The vertebrate pentadactyl limb provides strong evidence for evolution. In all these examples, the basic bone layout is the same (in the diagram, the same type of bone is given the same colour in each animal) but the limbs have become modified in different ways according to the way the animal moves

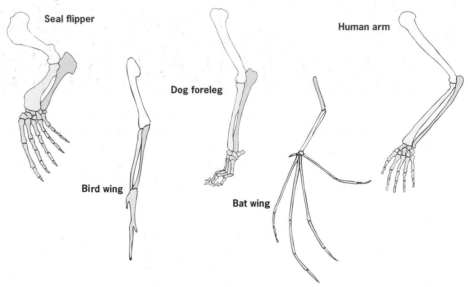

This basic five-fingered plan has been modified in a variety of ways to suit the animal's mode of locomotion. Structures with a common origin but with a different function are said to be **homologous**. So, the wing of a bat is homologous to a human hand. The possession of the pentadactyl limb in so many different organisms is persuasive evidence that these animals have evolved from a common ancestor.

In contrast, totally unrelated structures can become adapted for the same function. The wings of insects and the wings of birds, for example, have a totally different anatomy and origin in the embryo, but are similar in shape because they perform the same job. Such structures are said to be **analogous**.

? TEST YOURSELF

D Are the following pairs of structures analogous or homologous?
a) Trout's fin and dolphin's flipper.
b) Horse's leg and bat's wing.

Fig 29.14 Convergent evolution occurs when unrelated species solve the same evolutionary problem in a similar way
(a) Cacti, which are native only to the Americas, have evolved to cope with desert living. The leaves have become spines which lose no water and protect the plant. The fleshy stem expands and stores water and has taken over the job of photosynthesis

(b) You might guess that this euphorbia from Africa was a cactus, but you would be wrong. The two plants have evolved in different continents and are unrelated; they look very similar because they have developed the same features to survive desert conditions

The process by which unrelated species evolve to resemble each other is called **convergent evolution** (Fig 29.14). A classic example of this is seen in seals (mammals), penguins (birds) and fish. Because of their need to swim efficiently, all have evolved the same streamlined shape. The forelimbs of the seal and the wings of the penguin have both evolved to resemble fins.

Fig 29.14(c) The flipper of the whale and the fin of the shark (right) are a classic example of convergent evolution. The two animals are completely unrelated and the internal structure of forelimbs is totally different. However, they are both the same shape because of the job they have to perform

EMBRYOLOGY

Studying the development of an embryo reveals some interesting evolutionary secrets. The scientist Ernst Haeckl said 'ontogeny recapitulates phylogeny' which means that the evolutionary development of a species (its **phylogeny**) is mirrored by the changes of each individual from birth to maturity (its **ontogeny**).

A number of vertebrate embryos follow a very similar path from fertilisation through several stages of early development – so similar that it is often difficult to tell them apart (Fig 29.15). For example, all vertebrate embryos develop branchial arches; these are relics of the gills possessed by aquatic ancestors. However, although the study of the developing embryo provides further evidence for evolution, it does not give us much information about the actual evolutionary process.

Fig 29.15 Four developmental stages in embryos of fish, turtles, chicken and humans. You can see that the early stages are very similar in all four. For example, in the drawings in the second column, all four embryos have branchial arches, the 'folded' areas just under the head at the front. Only in the later stages do differences become obvious

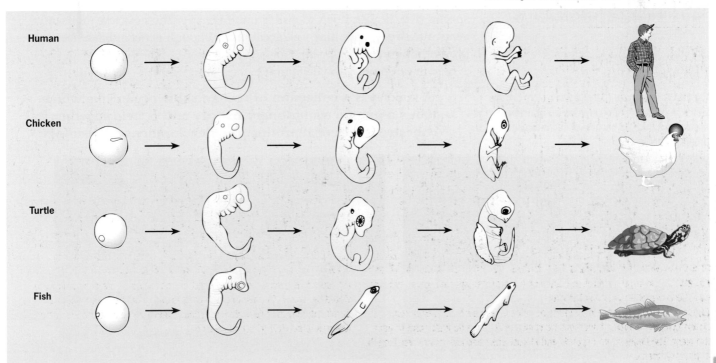

COMPARATIVE BIOCHEMISTRY

The biochemistry of different species shows some striking similarities. Many chemicals such as nucleic acids, ATP and cytochromes, organelles such as ribosomes, and pathways such as respiration, are almost universal. This is strong evidence that all living things have a common ancestry.

4 HOW A NEW SPECIES EVOLVES

We have looked at natural selection and seen that it can lead to stability or change. In this section we examine how natural selection can lead to the formation of a new species.

WHAT IS A SPECIES?

We instinctively think that we know what the word 'species' means, but an accurate definition is difficult, if not impossible. This is a common definition:

> **A species is a population or group of populations of similar organisms that are able to interbreed and produce fertile offspring.**

This would seem to be reasonable; a horse and a donkey can mate to produce a mule (Fig 29.16), which is sterile, so the horse and the donkey can be said to belong to different species. However, this definition runs into trouble. There are many closely related species, wolves, coyotes and domestic dogs, for example, which can interbreed to produce perfectly fertile offspring. The only reason why the wolves and coyotes remain distinct is that the hybrids do not compete as successfully as the pure strains.

The difficulty we have in defining the term species stems from the very nature of the process of evolution. New species evolve from pre-existing ones, and this is usually a slow transition which may take many generations and thousands of years. There are many living species that are still in the process of separating from each other. The wolf and the coyote almost certainly evolved from a common ancestor and for all practical purposes are separate species (Fig 29.17). If left to evolve naturally over the next few thousand years, the two species may well accumulate enough genetic differences to prevent interbreeding.

So, we can modify the definition to:

> **A species is a collection of recognisable organisms which shares a unique evolutionary history and is held together by cohesive forces of reproduction, development and ecology.**

Fig 29.16 'And nothing to look forward to either.' The mule is the product of a mating between a donkey and a horse, two species that have different numbers of chromosomes. It cannot reproduce because it is sterile. Because of its stamina, the mule has been used as a beast of burden for thousands of years

Fig 29.17 The coyote (*Canis latrans* – left) and the wolf (*Canis lupus* – right) are classed as different species, but they are capable of interbreeding and producing fertile offspring

Although more of a mouthful, it is a more satisfactory description because it recognises that all members of a species are usually (but not always) recognisable, and that they share the same 'biology', occupying the same niche in the ecosystem and having particular social structures, courtship, mating habits, etc.

THE DEVELOPMENT OF NEW SPECIES

Speciation, the development of new species, happens when different populations of the same species evolve along different lines. **Populations** are groups of interbreeding individuals of the same species occupying a particular geographical region. Examples of populations may be the water fleas in a pond, all the oaks in a forest or the elephants in a game reserve.

SEE QUESTIONS 1 TO 3

It is thought that speciation usually happens in the following way:

- Part of a population becomes isolated in some way that prevents them from breeding with the rest of the population. Very often, the breakaway population will, by chance, have a different genetic constitution from the original (see genetic drift – page 537).
- The two populations experience different environmental conditions, so that natural selection acts in different ways, causing the genetic make-up of the two populations to differ.
- Eventually, genetic differences accumulate to a point where individuals from separate populations that come together again can no longer interbreed.

It is worth mentioning that new species can also arise from a hybridisation between two different species. This particularly applies to plants, and plant breeders take advantage of this ability to produce new crop varieties.

There are two basic methods of speciation:

- **Allopatric speciation** (*allo* = different, *patris* = country), where the two populations are physically separated.
- **Sympatric speciation**, where the two populations are reproductively isolated in the same environment.

HOW POPULATIONS BECOME ISOLATED

Usually, species evolve when populations become isolated by physical barriers such as stretches of water or mountain ranges. This is known as **geographical isolation** (Fig 29.18).

Fig 29.18 The principle of geographical isolation
(a) The original population of mice has a 50:50 distribution of the alleles A and a, ie A is as common as a. A river changes its course and cuts off a few individuals. In the new, isolated population the allele frequency, by pure chance, is 4:10. This change in allele frequency is called the founder effect. The fewer the individuals in the breakaway founder population, the more dramatic the change in allele frequency is likely to be
(b) The founder population starts to grow, but they are in different conditions from the original population. The alleles A and a were neutral up to now, but then a water-borne disease spreads through the new population, which are confined to marsh land. The aa genotype confers resistance on those who carry it. By natural selection, the A allele is selected against, and so the a allele becomes even more frequent. In this way, the genetic make-up of the two populations diverges

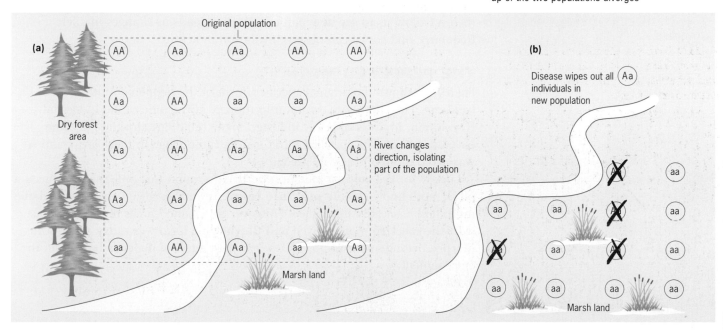

Evidence suggests that geographical isolation is the most common way in which populations become reproductively separated. However, speciation can happen even when two populations overlap. For instance, two populations of frogs may have a different mating call, and the females may respond only to calls of males in 'their' population. Table 29.1 summarises the different isolating mechanisms.

Table 29.1 A summary of isolating mechanisms

Mechanism	How it works
Prezygotic mechanisms, ie gametes don't come into contact	
Geographic	Two populations do not come into contact due to some physical barrier, eg mountain range, water.
Temporal	Different timing, eg flowering times or mating seasons do not coincide.
Behavioural	Individuals of other populations are not acceptable as mates, eg wrong courtship, mating call or pheromones.
Mechanical	Structural differences prevent mating, eg the shapes of the genitals in insects.
Postzygotic mechanisms, ie even if mating takes place	
Hybrid non-viability	Hybrid zygotes fail to develop to sexual maturity.
Hybrid sterility	Hybrids do not produce functional gametes.
Hybrid weakness	Hybrids have reduced survival or reproductive rates.

5 EVOLUTION IN ACTION

The process of evolution is generally slow, and it is no surprise that for centuries people thought that the living world was static. However, there are situations where you can see the process in action. Organisms with short life spans and high reproductive capacities, such as bacteria or insects, can evolve in a very short time indeed – see page 526.

Also of great interest to evolutionary biologists is a **polymorphism**, a situation where two or more different forms, **morphs**, of a species exist in a population. Each morph confers a different survival advantage in different situations. Polymorphisms such as that of wing colour in the peppered moth (below) allow us to see how natural selection leads to changes in allele frequency, and so we can observe evolution in action.

THE PEPPERED MOTH

Fig 29.19 shows the original speckled form of the peppered moth *Biston betularia*. This moth was common in the UK at the time of the Industrial Revolution. In areas of heavy industry, many buildings and trees became covered in soot. The speckled moth lost its camouflage in these areas and so was eaten more often by its predators.

Luckily for the moth, a mutation occurred around this time that produced a black moth known as the **melanic form**. In industrialised areas the melanic moth had a selective advantage. However, in 'cleaner' areas the reverse was true: the speckled moth thrived and the melanic form was rarely seen. Both morphs belong to the same species, but their survival depends largely on the environment.

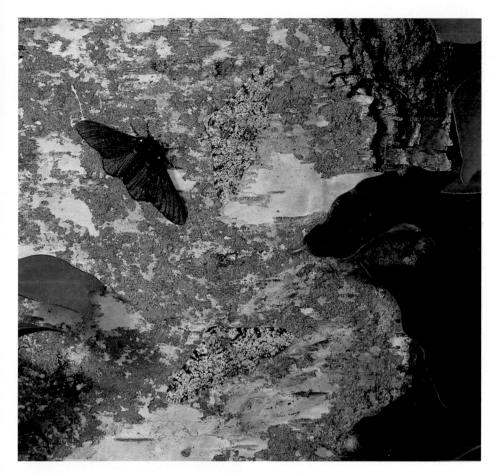

Fig 29.19 The two forms of the peppered moth, normal (speckled) and melanic (black). Against the background of the bark of a tree shown in this photograph, the two speckled moths on the right are almost invisible while the melanic form on the left is far more likely to be seen and caught by a predator. However, it becomes invisible on a black surface such as a sooty wall, and so is more likely to survive in an industrial environment

HOW CAN WE STUDY EVOLUTION?

If evolution is so slow, how can we observe it happening? An important principle here is that evolution involves a *change in allele frequency*. For instance, when myxomatosis was introduced into the rabbit population, the alleles for resistance were selected for, and so they increased in frequency. In any particular situation, if we can show that allele frequencies are changing, we know that evolutionary forces are at work.

Evolutionary studies are usually based on interbreeding populations, which are known as **demes** or **Mendelian populations**. The sum total of all the genes circulating within such a population is called the **gene pool**.

POPULATIONS AND ALLELE FREQUENCY

When the gene pool remains more or less the same, a population is not evolving. Natural selection may be acting, but it is often a force for stability rather than change. However, if the gene pool of a population is changing, natural selection is acting as a force for change and this is evolution in action.

To show how we measure allele frequency, imagine a population of 500 rats. In the population are two alleles for coat colour: black (B) is dominant over brown (b). The rats are diploid – each individual has two alleles – so the population contains 1000 coat colour alleles. If 600 of the alleles are B, then we can say that the frequency of B is 60 per cent, or 0.6 (statisticians prefer decimals). As there are only two alleles, it follows that the frequency of b must be 40 per cent or 0.4. (Remember that the sum of the allele frequencies must be 100%, or 1.)

✔ **REMEMBER THIS**

$p + q = 1$
Frequency of dominant allele + frequency of recessive allele = total number of alleles of that gene.

If one gene is dominant to the other – as in this case – the symbol p is used for the frequency of the dominant allele B, and q is used for the frequency of b. It is clear that $p + q$ must equal 1.

The values of p and q change when a population evolves. To find out if this is happening we use the **Hardy–Weinberg principle**.

THE HARDY–WEINBERG PRINCIPLE

The principle, named after the British mathematician G.H. Hardy and the German biologist W. Weinberg, states that there will be no change in the frequency of alleles in a population so long as the following conditions are met:

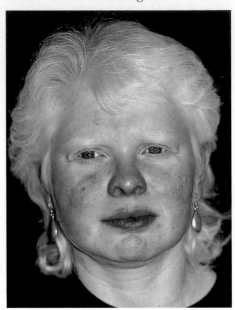

- The population is very large, diploid and reproduces sexually.
- Mating is totally random – there is no tendency for certain genotypes to mate together.
- All genotypes are equally fertile, so there is no natural selection.
- There is no emigration or immigration.
- No mutations occur.

We can use the Hardy–Weinberg rule to estimate allele frequencies using just the information about the frequency of a particular genotype. Consider the example of albinism in humans. One in 10 000 humans are born albinos – they have two copies of a recessive gene which means they

Fig 29.20 This person is an albino. The white hair, pink eyes and pale skin are the result of an inability to make the pigment melanin

cannot make the pigment which gives colour to hair, skin and eyes (Fig 29.20). The frequency of the allele for albinism in the population is difficult to estimate because carriers of the allele are not albinos. So, how can we find the frequency of the albinism allele?

If we call the normal allele A and the albinism allele a, most of the population are AA, some are carriers: Aa or aA, and a small number are albinos: aa. Using the Hardy–Weinberg formula we can estimate the percentage of the population who carry the a allele. Remember that p and q represent the frequency of the A and a alleles respectively. The Hardy–Weinberg equation:

$$p^2 + 2pq + q^2 = 1$$

Put into words, it says:

The frequency of AA individuals, added to the frequencies of Aa, aA and aa, must equal 1.

We also know that $p + q = 1$, and we can use these two equations to work out the proportion of heterozygotes in the population, and any other frequencies involved:

1 in 10 000 is a frequency of 0.01%, or 0.0001 expressed as a decimal.
1 in 10 000 has the genotype aa, so the frequency of aa; qq or q^2, has the value of 0.0001.

So $q = \sqrt{0.0001} = 0.01$.

If $q = 0.01$, then p must be $1 - q$, which is 0.99.

So, in other words, 99 per cent or 0.99 of the alleles in the human population are A.

From this we can calculate that the number of carriers of the albino allele $2pq$ is:

$$2pq = 2 \times 0.99 \times 0.01 = 0.0198 \approx 0.02.$$

So, almost 2 per cent of the population carry the albinism allele.

WHEN ALLELE FREQUENCIES CHANGE

The Hardy–Weinberg principle states that allele frequencies remain constant as long as certain conditions apply. It follows that when these conditions are not met, the allele frequencies can change and the population can evolve. Consider what happens when the conditions listed on page 536 are not met.

- The **population is small**. If the population is small, chance plays a large part in determining which alleles pass to the next generation. The smaller the population, the greater the probability that the genes of one generation will not be accurately represented in the next. The name given to a change in allele frequency caused by chance is **genetic drift**. These are two important causes of genetic drift: **bottlenecks**, when a population declines and then recovers; and the **founder effect**, when a population starts from just a few individuals (see below). As Figs 29.18 and 29.21 show, this can result in a drastic change in the allele frequency.

- **Mutations occur**. Mutations are ultimately the origin of all genetic variation (see Chapter 27). They are rare, and most are harmful or neutral. Occasionally, however, a mutation gives an organism a survival advantage, or the environment changes so that previously harmful/neutral mutations become advantageous. For example, a mutation in a bacterial gene that leads to the production of a modified enzyme could make the bacterium resistant to an antibiotic. That individual would survive, even in the presence of the antibiotic, and pass its resistance gene on to the next generation.

- **Individuals move in or out of populations**, that is, immigration or emigration. If the new individuals breed, this movement may lead to gene flow. When a few individuals move into a new environment and start a population, we see the founder effect. Like the beads in Fig 29.21, the individuals in the founder population are probably not representative of the original population. A group of six human survivors on a desert

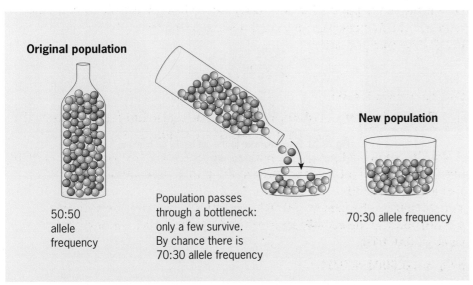

Original population

50:50 allele frequency

Population passes through a bottleneck: only a few survive. By chance there is 70:30 allele frequency

New population

70:30 allele frequency

Fig 29.21 An illustration of genetic drift. In the original 'population' there are 1000 beads; half red, half blue. If ten beads are extracted at random, there is a high probability that there won't be five of each. If these ten 'organisms' then become the founders of a new population, the proportion of colour (the allele frequency) will be significantly changed, in this example to 7:3

island, for example, might have two red-haired people and two blondes. A population founded by these people would have a much higher percentage of red-heads and blondes than the population in their country of origin. In the most extreme case, a new population can be founded by one individual, such as a self-fertilising plant or a pregnant female animal.

● **Mating is non-random.** In many cases, individuals choose mates with a particular genotype. In this case, mating is not random, it is **assortive**. Such assortive mating can change allele frequency.

● **Natural selection occurs.** When there is competition for resources – and this is usually the case in an ecosystem – natural selection acts and only the most successful individuals pass on their genes.

! SCIENCE FEATURE Cheetahs — not fast enough to escape the bottleneck

It is widely known that the cheetah is the fastest land animal and is often hailed as a masterpiece of evolution. It is not so widely known, however, that this magnificent animal is, in genetic terms, on a knife edge. There is so little variation that the world's cheetah population is practically a clone.

About 10 000 years ago there were probably five distinct species of cheetah. Four of them became extinct and the fifth one went through an evolutionary bottleneck in which only a very few individuals survived to produce the population of today. As a result the population is highly inbred – closely related individuals have mated and many alleles have been lost. In addition, any faulty genes are likely to be paired up and expressed rather than masked. It is difficult to breed cheetahs in captivity and a low sperm count is one of the major problems.

Polymorphic genes are those with more than one allele (see page 496 in Chapter 27). In a major study of the world's captive cheetahs, scientists detected no variation whatsoever in any of the 52 gene loci studied. The current estimation is that just 1 per cent of the cheetah's genes are polymorphic, compared with about 23 per cent in the domestic cat and 31.7 per cent in humans (Figs 29.22 and 29.23).

In 1900 there were probably 100 000 cheetahs spread across Africa and Asia. Today the cheetah is virtually extinct in Asia and there are only about 10 000 left in Africa. Of these, the biggest population is about 2500 in Namibia.

Data from Bill Bailey, *Biological Science Review* November 2001.

Fig 29.22 The cheetah, *Acionyx jubatus*

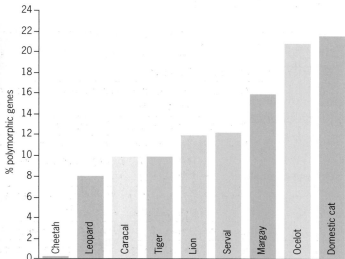

Fig 29.23 Polymorphism in the cat family (in humans, polymorphism is 31.7 per cent)

6 THE DIVERSITY OF LIFE

SEE QUESTION 4

There are millions of different species on Earth, and we have given all the ones we know about a scientific name. The gorilla is simply *Gorilla gorilla*, the chimpanzee is *Pan troglodytes* while we have given ourselves the name *Homo sapiens*, which means 'wise man'. A scientific name tells us both the genus and the species. The generic name has a capital latter while the specific name does not. When written, scientific names should always be in italics or underlined.

We classify organisms according to the similarities between them. Understanding relationships between species allows us to piece together the story of life on Earth. Classification is therefore a bit like a huge family tree over a massive time scale. Just like an address has a house number, a road, a town, etc, we classify organisms into progressively narrower groups. Table 29.2 shows this 'address', which we more correctly call a taxonomic hierarchy. Taxonomy is the science of naming and classifying organisms and the term hierarchy reflects the layered structure of the family tree of living organisms that consists of groups within groups.

This system of classification is called **phylogenetic**, ie based on evolutionary history. It is a bit like a family tree that goes back millions of years. To construct a phylogenetic tree, scientists use anatomical/physical features, fossil records and biochemical analysis of DNA and proteins.

The taxonomic hierarchy is shown in Table 29.2. It has the following key features:

- It consists of a series of groups within groups, from the most general (kingdom) to the most specific (species).
- There is no overlap between the groups. For instance, there is no organism that is part amphibian and part reptile – it's either in one group or the other.
- The groups are based on shared features. The more specific the group, the more shared features there are.

 REMEMBER THIS

It may help to invent a mnemonic to remember the sequence of kingdom, phylum, etc. make up a phrase according to KPCOFGS, such as Keep Putting Cabbages On Fat Greasy Slugs, or something more memorable/rude/personal to you.

REMEMBER THIS

In exam questions on scientific names, remember that the *genus* is more general than the *species*. Two organisms that have the same generic name will have more shared features than two organisms that do not.

Table 29.2 The taxonomic hierarchy

Taxonomic group	Humans	Explanation
Kingdom	Animalia	There are five kingdoms, animals, plants, fungi, protoctista and bacteria.
Phylum	Vertebrata	We are animals with backbones.
Class	Mammalia	There are five classes, fish, amphibians, reptiles, birds and mammals. The defining features of mammals are fur/hair and milk production.
Order	Primates	Primates are adapted for life in the trees. Key primate features are covered in the main text.
Family	Hominidae	Human-like creatures – includes several extinct species.
Genus	*Homo*	No other living species is sufficiently like humans to be included in this genus.
Species	*sapiens*	We are the thinking species, modern man.

SEE QUESTION 5

> ✔ **REMEMBER THIS**
>
> Eukaryotic cells are large with a lot of internal organisation, ie many organelles made from membranes. Prokaryotic cells are much smaller – about a thousandth of the volume, and have little internal organisation

THE FIVE KINGDOMS

Scientists have recently accepted the five kingdom system of classification. All organisms can be placed into one of these five kingdoms. Four contain the **eukaryotic** organisms while a fifth contains the **prokaryotes** – the bacteria.

KINGDOM ANIMALIA (ANIMALS)

- Includes all the vertebrates (mammals, birds, reptiles, amphibians, fish) and invertebrates (insects and other arthropods, molluscs, segmented worms, roundworms, tapeworms/flukes and jellyfish).
- Animals are eukaryotic and multicellular. Their cells have no walls.
- They cannot make food, so move in search of it. They have muscles and nervous systems. The vast majority have a gut.

Fig 29.24(a) The leafy sea dragon is a fish related to the seahorse

Fig 29.24(b) Snow leopard

KINGDOM PLANTAE (PLANTS)

- Includes flowering plants, conifers and ferns along with simpler plants such as mosses, liverworts. NB The algae/seaweeds are classed as protoctistans (see next page), not plants.
- Plants are eukaryotic and multicellular. Cells have walls made from cellulose. There is usually a large vacuole in mature cells.
- Most plants have leaves, stems and roots.
- Most plants possess chlorophyll and make food by photosynthesis. A few are modified to a parasitic way of life.

Fig 29.25(a) Hard shield fern

Fig 29.25(b) Limber pine in the Grand Teton National Park, Wyoming

KINGDOM FUNGI

- Includes filamentous fungi and yeasts.
- Single celled fungi are called yeasts.
- Eukaryotic, but the multicellular fungi do not have separate cells – the nuclei are dotted around in the tissue.
- Walls of fungal cells contain chitin.
- Feed by extracellular digestion – secreting enzymes and absorbing the soluble products. Most feed on dead material, ie are saprophytes. Some are parasitic.

Fig 29.26(a) Group of four poisonous fly agaric mushrooms (*Amanita muscaria*). **(b)** A mature sporangium (fruiting body) of the common bread mould *Mucor* sp.

KINGDOM PROTOCTISTA

The members of this kingdom seem to have very little in common, more like an assembly of organisms that could not be put into any other kingdom – a 'taxonomic dustbin'. The scientific reason for this group is based on their ancestry and evolutionary relationships.

- Includes single-celled protozoans such as amoeba and plasmodium (the malaria parasite), the algae (including seaweeds) and sponges.
- Their cells are eukaryotic, but show a diversity of cell types.
- They feed in a variety of ways.

Fig 29.27(a) Tube sponge. Sponges are multicellular organisms that filter water through their bodies, extracting plankton for nutrition **(b)** Various species of seaweed

KINGDOM PROKARYOTAE

This is the kingdom of bacteria.

- The cells of prokaryotic organisms are much smaller than those of eukaryotes.
- Most feed by extracellular digestion – secreting enzymes and absorbing the soluble products. Most feed on dead material, ie are **saprophytes**. Some are **parasitic**. The cyanobacteria are photosynthetic.
- Prokaryotes reproduce mainly by binary fission (splitting in half), but can reproduce sexually.

✔ REMEMBER THIS

Don't confuse protoctistans with prokaryotes. The words look similar but protoctistans are a kingdom of eukaryotes

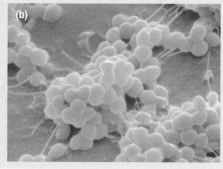

Fig 29.28(a) Compost heap bacteria. Coloured scanning electron micrograph of a dividing spirochaete (green) among other bacteria from a compost heap. **(b)** Colonies of *Staphylococcus epidermidis*. This bacterium is usually non-pathogenic and found all over the human skin, and on many mucous surfaces. It is associated with acne and other minor skin problems

SUMMARY

When you have finished this chapter, you should know and understand the following:

- **Evolution** is defined as a change in the genetic composition of a population over time. New species develop from pre-existing species in this way.
- The basic mechanism of evolution is **natural selection**. This acts on the genetic variation that occurs in a population. Individuals whose genotype gives them an advantage over others are more likely to pass their genes on to the next generation.
- A **population** is a group of interbreeding individuals of the same species in the same geographical area.
- New species are formed when populations become isolated in some way so that they cannot interbreed. Natural selection then acts on the different populations, changing the frequency of **alleles**. With time, the differences between the two populations can become so great that an individual from one cannot interbreed with an individual from the other.
- Humans can speed up the evolutionary process by **artificial selection**. Breeding organisms with desirable features, such as cattle with a high milk yield, changes the allele frequency rapidly. Within only a few generations, organisms are produced which are very different from the original wild stock.

- The evolutionary process is fastest in species which have a large population size, a large reproductive capacity (fecundity) and a rapid life cycle. Many disease-causing and pest organisms fall into this category.
- Many endangered species have a low reproductive capacity and a long life cycle.
- Evolution happens when there is a change in the allele frequency. The Hardy–Weinberg principle states that, in a large population, the allele frequency remains constant unless some outside force acts to change it. These forces are chance (the smaller the population, the greater the effects of chance), migration of organisms, selective mating and natural selection.
- When the conditions for the Hardy–Weinberg equilibrium are not met, the allele frequency changes and the population evolves.
- All organisms can be placed into one of the **five kingdoms**: **Animalia** (animals), **Plantae** (plants), **Fungi**, **Protoctista**, which are all eukaryotes, or **Prokaryotae**.

? EXAM QUESTIONS

1 Read the passage and then answer the questions that follow.

Myxoma virus causes a disease of rabbits called myxomatosis. In the early 1950s, this virus was used to control rabbit populations in Australia. The virus is widespread amongst Brazilian rabbits which suffer only a slight skin disorder. In European rabbits, the virus rapidly leads to a fatal disease. Following infection, death occurs within approximately two weeks. In Australia, the disease caused a 99.5% mortality rate in the rabbit population when it was first introduced in the years 1950–51, but this declined to a 30% mortality rate by 1957–58. By 1959, the rabbit population was increasing again.

A study of the infection was carried out to determine the reason for the decreasing effectiveness of the disease as a means of biological control. A single strain of the virus was used throughout the whole study. Each year, numbers of uninfected rabbits were caught and injected with the virus strain. The symptoms of the disease and mortality rates were monitored and some of the results are presented in the table.

Year of capture	Percentage of rabbits with symptoms		
	severe & fatal	moderate	mild
1953	95	5	0
1954	93	5	2
1955	61	26	13
1956	75	14	11

a) What evidence is there that the geographical isolation of rabbits leads to variation?

b) (i) One hypothesis for the change in mortality rate was that the rabbits were developing an inheritable resistance to the disease. What evidence is there in the passage to support this hypothesis?
(ii) Suggest another hypothesis to account for the decreasing effect of the disease in the wild rabbit population.
In Australia, rabbits are now controlled by the application of a single pesticide chemical.

c) (i) Other than elimination of all rabbits, what might be the long term effect of this on the rabbit population?
(ii) In what way could the use of the pesticide be regarded as an example of artificial selection?

OCR 6912 Inheritance Part 1 March 2000 Q3

2 Read the following passage and answer the questions below, from R. Dawkins, *Unweaving the Rainbow.*

The standard neo-Darwinian view of the evolution of diversity is that a species splits in two when two populations become sufficiently unalike that they can no longer interbreed. Often the populations begin diverging when they chance to be geographically separated. The separation means that they no longer mix their genes sexually and this permits them to evolve in different directions. The divergent evolution might be driven by natural selection (which is likely to push in different directions because of different conditions in the two geographical areas). Or it might consist of random evolutionary drift (since the two populations are not held together genetically by sexual mixing, there is nothing to stop them drifting apart). In either case, when they have evolved sufficiently far apart that they could no longer interbreed even if they were geographically united again, they are defined as belonging to separate species.

a) (i) According to the passage, what commonly starts the divergence of two populations?
(ii) From your knowledge of isolating mechanisms suggest **two** other events which may prevent breeding between two populations.
(iii) When might it be said that populations should be defined as belonging to different species?
Further evolutionary divergence leads to the formation of larger taxonomic groups. The table shows part of the classification of a rat.

b) (i) Complete the names of the principal taxonomic groups in the correct order in the following table.
(ii) Copy and complete the classification of a mammal, **the rat.**

Taxonomic group	Classification of rat
kingdom	
	Rodentia
family	Muridae
	Rattus
	norvegicus

c) Explain briefly why Lamarck's theory of inheritance of acquired characteristics is not acceptable to Darwinist and neo-Darwinist biologists.

OCR 6912 Inheritance Part 1 March 2001 Q4

? EXAM QUESTIONS

3 The purple saxifrage is a flowering plant found in the Arctic. It has two distinct forms, one growing on dry ridges, the other in the valleys:

On dry ridges, where the temperature is higher and the growing season longer, the plants grow in upright tufts. These plants are drought tolerant and store carbohydrate which they can use in times of stress.

In the valleys, where it is colder and the growing season is shorter, the plants grow very close to the ground. These plants have a higher rate of photosynthesis and grow more rapidly.

Both forms usually complete their life cycle in the short Arctic summer. In unfavourable years, plants in the valleys may produce pollen but are unable to form seeds. Some pollen from the valley plants fertilises plants on the ridges. The seeds may then be washed down to the valleys. Those that develop and complete their life cycles are all low-growing plants. When plants from the ridges and the valleys are grown under identical conditions in the laboratory, they have the same growth form as they did when in their original environment.

(Reproduced from *The Biologist*, R.M.M. Crawford by permission of Institute of Biology)

a) What is the evidence in the passage that the two forms of the purple saxifrage: are **(i)** genetically different; **(ii)** belong to the same species?

b) When cross pollination between the low-growing and upright forms occur, both forms of plant develop when the seeds produced are grown in the laboratory. Explain why only low-growing plants develop from the seeds that are washed from the ridges into the valleys.

c) Under the conditions described, the two forms of the purple saxifrage are unlikely to evolve into separate species. Explain why.

d) The effect of global warming on the environment of the Arctic is uncertain. Suggest why species such as the purple saxifrage are likely to survive even if changes in climate do occur.

AQA(NEAB) BY02 Continuity of Life June 2000 Q4

4 The figure shows the way in which four species of monkey are classified.

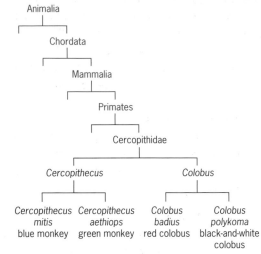

a) This system of classification is described as hierarchical. Explain what is meant by a hierarchical classification.

b) **(i)** To which genus does the greem monkey belong?
(ii) To which family does the red colobus belong?

c) What does the information in the diagram suggest about the similarities and differences in the genes of these four species of monkey?

NEAB Biology Continuity of Life Module Test March 1998 Q3

5

a) Discuss the definition of the term *species*.

b) Outline the main features that are used to classify species of living organism into each of the five kingdoms.

c) Explain briefly the advantages and the disadvantages of the universality of the genetic code to humans.

IBO Biology Higher Level Paper 2 November 1998 Q5

KEY SKILLS **ASSIGNMENT**

THE HISTORY OF THE WORLD

Rocks, and the fossils they contain, can now be dated by radiometric dating techniques, which measure levels of certain isotopes. We can use this technique to construct a geological timescale – a biological history of the world. This Assignment outlines the process of dating using radioactive isotopes and puts the history of life on Earth into perspective.

Dating rocks and fossils

The principle behind radiometric dating is simple: some elements exist as radioactive isotopes that decay at a predictable rate. Measurement of this decay allows us to estimate the age of the rock.

For example, when volcanoes erupt, the lava contains radioactive potassium, ^{40}K. This decays steadily into argon (^{40}Ar). The half-life of ^{40}K is 1.3 billion (1.3×10^9) years. So, in 1.3 billion years, half of the potassium decays. Half of the remaining potassium will decay in the next 1.3 billion years, and so on. By measuring the ratio of radioactive potassium to argon, scientists can estimate the age of the rock strata to the nearest 50 000 years.

In many places, layers of volcanic lava alternate with sedimentary rocks that contain fossils. If we can date the lava, we can also estimate the age of the fossils. Some fossils of primitive bacterium-like organisms are thought to be around 3 500 000 000 years old.

1 a) What does half-life mean?
b) What is the half-life of ^{40}K?
c) What fraction of the original ^{40}K will be left after 3.9 billion years?

Carbon dating

This process can be used to date samples that still contain some organic material. Cosmic radiation is constantly bombarding the atmosphere, turning $^{12}CO_2$ into $^{14}CO_2$. This becomes fixed into organic molecules by the process of photosynthesis and then passes up the food chain. When an organism dies, the ^{14}C decays into ^{14}N, with a half-life of 5730 years. Carbon dating can be used to measure the age of any carbon-containing material up to 50 000 years old.

2 A human skull is found to have only one-sixteenth of the ^{14}C it would have had when 'new'. Estimate its age.

The Earth is currently estimated to be about 4 600 000 000 years old. This timescale is impossible to comprehend, but if we condense this huge span into one year we can begin to put it into perspective.

The Earth calendar

Date	Event
1 Jan	Earth forms
26 Feb	oldest rocks – nothing was permanent before this time
23 March	first life – bacteria-like organisms obtain energy from organic molecules (food) by the process of glycolysis
April–May	organic food runs out; photosynthesis evolves, makes food and releases oxygen as a by-product
30 June	oxygen atmosphere fully developed; aerobic respiration evolves some time after this
17 Sept	first eukaryotes
19 October	first co-ordinated multicellular colonies (like sponges)
5 Nov	first invertebrates (worms and jellyfish)
15 Nov	rapid increase in invertebrate diversity
20 Nov	first fish
21 Nov	first coral reefs
29 Nov	first colonisation of land – vascular plants appear
4 Dec	first amphibians
6 Dec	first trees
7 Dec	first insects
8 Dec	first reptiles
17 Dec	first mammals
19 Dec	first birds and dinosaurs
21 Dec	first flowering plants
25 Dec	last of the dinosaurs
26 Dec	first primates
31 Dec	
14.00	first upright bipedal hominids
21.00	*Homo erectus*
22.45	recent Ice Age starts
23.00	evidence that primitive humans can cook food
23.48	Neanderthal man appears
23.54	*Homo sapiens* appears
23.55	Australian aborigines settle
23.58.49	agriculture invented
23.59.28	beginning of recorded history
23.59.29	pyramids built
23.59.41	Buddha born
23.59.45	Christ born
23.59.53	Battle of Hastings
23.59.57	start of the scientific method
23.59.58	Australia 'discovered' by Cook; the Industrial Revolution
23.59.59	most of the knowledge contained in this book is discovered

KEY SKILLS **ASSIGNMENT**

3 **a)** In our one-year scale, how many seconds represent 1000 years?

b) How long in actual time is represented by one minute, one hour and one day?

c) Use your answer to (a) to estimate how long the dinosaurs lived for.

Fig 29.A1 A satellite map of the world, centred on Europe and Africa. This image is a combination of hundreds of images acquired by NOAA weather satellites from orbits some 820 km above Earth. It didn't always look like this. The shape of the land masses is subject to constant change, shaping patterns of evolution. When land masses separate, two halves of a population can also be separated and proceed to evolve in different directions

4 The rise of humans as the dominant species has been meteoric.

a) What features of humans account for our success?

b) What dangers does this success bring?

5 Write down what you feel to be the four most significant advances made by humans. These could be inventions or social and cultural advances. If you are working in a group, use your own list to provide a discussion.

Fig 29.A2 Fossils of ammonites from Dorset

30

DNA technology

This is 'Rainbow', cc's genetic donor.
Note the differences in coat colour between cc and Rainbow

cc, the first-ever cloned cat shown here at seven weeks old, was born on 22 December 2001, at the College of Veterinary Medicine, Texas A&M University. The announcement of the successful cat cloning was delayed until the animal had completed its shot series and its immune system was fully developed

Allie and cc. Allie is cc's surrogate mother

In February 2002, researchers announced that a cat had been cloned for the first time. The kitten, called cc (the typist's old abbreviation for carbon copy), was born two months earlier, but had to be checked carefully to make sure she was normal and healthy, before the announcement was made.

The cloning of cc highlights some of the technological and ethical issues surrounding cloning technology in this first decade of the twenty-first century. cc was cloned by transplanting DNA from Rainbow, a female three-coloured (tortoiseshell) cat into an egg cell whose nucleus had been removed, and then implanting this embryo into Allie, the surrogate mother. cc is genetically identical to her genetic donor, Rainbow, but she isn't an exact physical copy, underlining the delicate balance between genetics and environment. Coat colour in cats is determined partly genetically, and partly by environmental factors, such as conditions in the womb. In the past couple of years, newspaper reports have hinted that a cult in America, who have developed cloning technology, intend to help a couple whose 10-month-old baby died tragically. The idea is to clone the dead child from frozen cells to 'replace' the child. However, cloning cannot bring the dead back to life; this is a seriously misguided experiment that has been widely condemned by other cloning scientists all over the world.

cc also demonstrates just how difficult and 'wasteful' cloning is. In their first attempt, researchers obtained the cells used to make the clone from the skin cells of a donor cat. Eggs from other cats were used for the next step. Their chromosomes were removed and replaced with the DNA from the skin cells, creating cloned embryos that were then transplanted into surrogate mothers. In total, 188 such nuclear-transfer procedures were performed, which resulted in 82 cloned embryos that were transferred into seven recipient females. Only one cat became pregnant, with a single embryo. But this pregnancy miscarried and the embryo was surgically removed after 44 days. In the next attempt the scientists tried using cells from ovarian tissue to receive the DNA from the cat to be cloned. Five cloned embryos made in this way were implanted into a single surrogate mother. Pregnancy was confirmed by ultrasound after 22 days and a kitten was born on 22 December 2001, 66 days after the embryo was transferred. So, out of 87 implanted cloned embryos, cc is the only one to survive. This is comparable to the success rate in sheep, mice, cows, goats and pigs, but would be unacceptably low for people.

The group in America says that it has dozens of women who have volunteered to be surrogate mothers. But have they been told of the high risk of losing the baby that they are being asked to carry? Are the parents really willing to let so many women go through such heartache, in order to produce a baby that will never replace the one they lost?

Most scientists agree that human cloning for reproductive purposes should not be attempted, but animal cloning looks as though it will expand further in the next few years. The company in Texas that part-funded the research that produced cc hopes to use the technology to provide commercial cloning of companion animals for pet owners. Less controversially, the cloning of prize cattle, and perhaps successful racehorses, might become big business. But perhaps the most acceptable face of animal cloning will be the application of the technology to preserve endangered species.

1 THE NEW GENETICS

Genetics seeks to explain the mechanisms of inheritance and evolution and, like many other 'pure' sciences, it can be applied to solve practical problems. In the 'new genetics' that has developed rapidly in recent years, humans are taking a much more active role in determining how genes pass between individual organisms. Not only can we study DNA, we can manipulate it and use it to our advantage. This branch of genetics is called genetic engineering, or DNA technology. This chapter considers the basic ideas behind this new technology.

Transferring genes between organisms offers many exciting prospects. In agriculture, for example, genes from nitrogen-fixing bacteria could be put into plants which would then fix their own nitrogen and so make some of their own fertiliser. This could greatly improve the efficiency of food production. Crop plants given genes that code for disease resistance or tolerance to pollutants could be grown on previously inhospitable land, such as industrial land or desert margins. DNA technology also has many potential applications in medicine.

The genome of an organism, its entire collection of DNA sequences, can now be explored and mapped. Theoretically, any individual gene can be located, cut out, replaced or inserted into another organism. Of course, this sort of research is regulated very carefully, and the ethical, moral and social consequences need to be taken into account (Fig 30.1).

WHAT IS ALREADY POSSIBLE

We can sequence the genes in whole organisms; several bacteria, some plants, animals and the human genome are now fully sequenced (see the Science feature on page 553).

We can take individual genes and make multiple copies: this is **gene cloning**.

We can extract genes that code for useful products and insert them into other organisms: this is **recombinant DNA technology**. Organisms produced in this way are called **transgenic organisms**. More genes that code for human products could be transferred into microorganisms for mass production. Products already made in this way include insulin, somatostatin, alpha-1-antitrypsin, human growth hormone and blood clotting substances such as Factor VIII. See Chapter 6 for more details.

We are just starting to be able to replace defective genes with healthy ones: this is **gene therapy**. Some successful gene therapy trials have been carried out, but this area is fraught with technical and ethical problems.

We can clone parts of organisms, using **stem cells**. These are undifferentiated cells, found in embryos and in adults, that can be persuaded to turn into particular tissues and organs. Scientists have recently cloned kidneys using stem cells in cows (see the Science feature on page 563).

We can clone whole animals. Many farm animals – goats, sheep, pigs and cows – have been cloned and other animals such as the cat, and endangered species of animals have either just been cloned, or will shortly be cloned.

In this chapter we focus on the techniques involved, and we look briefly at some of the possibilities for the future and at some of the ethical implications of genetic technology.

>
> **REMEMBER THIS**
>
> Genetic engineers manipulate DNA to allow individual genes to be transferred from one organism to another. Often the two organisms are from completely different species.

Fig 30.1 'You were so keen to see what you could do, you didn't stop to think about whether or not you should.' This is a scene and a quote from the film *Jurassic Park*. It is quite unlikely that we will ever be able to recreate dinosaurs, but the quote reflects an important point: all genetic research has the potential for abuse and we should think very carefully about its consequences

2 GENE HUNTING

The first problem that genetic engineers tackle is finding the gene they are interested in. Every human body cell contains 46 chromosomes, each having several thousand genes. So how, for example, do you go about finding the insulin gene that you want to insert into a bacterium? Let's look at three different approaches.

- **Isolate the gene from the rest of the genome.** This has been compared with looking for a needle in a haystack, when the needle is also made from straw. However, there are several sophisticated techniques that can be used to map chromosomes. For instance, if you know part of the base sequence in the required gene you can make a genetic probe. This is a single-stranded piece of matching DNA labelled with a radioactive or fluorescent marker. When mixed with the genomic DNA (DNA from the genome) – in the right conditions – the probe seeks out the complementary base sequence and binds to it, showing you exactly where the gene is (Fig 30.2).

  These techniques can now be combined with the vast database that has been produced by the Human Genome Project (see Section 4 of this chapter). This database of all the genes in the human genome is freely available to all scientists, and they can look up likely genes and gain predictions of function, using computer programs that construct the most likely protein from the DNA sequence. The possibilities are extending so fast, it is impossible to give more than a brief overview in this book.

- **Use messenger RNA.** If you can isolate the mRNA molecule that is a copy of the gene that you are trying to find, you can use it as a genetic probe as described above. You can also use it to make a DNA copy using the enzyme reverse transcriptase. This means essentially that you make an 'artificial' gene. For example, you could find the mRNA that codes for insulin in the cells where the insulin gene is expressed – the cells in the islets of Langerhans. DNA made from mRNA is called complementary DNA, or cDNA. Artificial genes are shorter than those in the genome because they contain no introns (see Chapter 27).

- **Work backwards from the protein.** If you work out the amino acid sequence of the desired protein – a relatively easy process – you can make a piece of DNA that codes for it. It would obviously be a tedious process to make a large gene, but now there is equipment that can synthesise artificial DNA quickly and easily. The end-product of this process is also a short, artificial gene made of cDNA.

✔ REMEMBER THIS

DNA is common to all organisms, and it always has the same basic structure. This allows us to combine genes from organisms as diverse as humans and bacteria.

? TEST YOURSELF

A Suggest what the enzyme reverse transcriptase does.

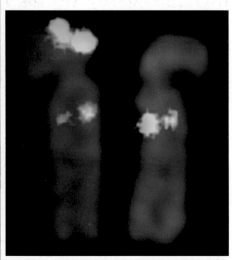

Fig 30.2 The technique called FISH (fluorescent *in situ* hybridisation) was used here – a genetic probe containing a fluorescent marker seeks out the gene. This micrograph shows a deletion on chromosome 17. On the normal chromosome on the left the normal gene shows up as two green dots. These are absent on the chromosome on the right.

3 MANIPULATING DNA USING ENZYMES

Once you have located the gene you want, the next step is to use it. Enzymes are the tools of the trade here. Some of the enzymes used in genetic engineering are shown in Table 30.1. See also Chapter 6.

Table 30.1 Some enzymes commonly used in genetic engineering in addition to those featured in the main text

Enzyme	Function
Exonucleases	Remove terminal base from DNA – useful in analysis of base sequences.
DNase	Makes random cuts in DNA: chopping into varying-sized fragments.
Kinases	Add groups such as phosphates to DNA: this helps with labelling and analysis, particularly when phosphates are made with ^{32}P.
Polymerases	Synthesise nucleic acid molecules from nucleotides. This is essential for the polymerase chain reaction (PCR) – see the Science feature on page 554. Reverse transcriptase is a polymerase that makes DNA from RNA. It occurs naturally in retroviruses such as the human immunodeficiency virus, HIV.

RESTRICTION ENZYMES

Restriction enzymes can be thought of as 'molecular scissors' – they cut DNA strands at specific points (Table 30.2). More properly called **restriction endonucleases**, they are made by bacteria in response to attack by viruses called bacteriophages. Their name reflects their function: they restrict damage by chopping bacteriophage DNA into smaller, non-infectious fragments, and they make their cuts inside the nucleic acid molecules.

There are many different endonuclease enzymes, produced originally by different species of bacteria. Each enzyme cuts DNA at a different base sequence, a point known as the recognition site. The enzymes make staggered cuts in the DNA (Fig 30.3), commonly called sticky ends. These can combine with complementary sticky ends, and this allows lengths of DNA that have been removed from one organism to be spliced – inserted – into the DNA of another. *Eco*RI, for example, is a common endonuclease isolated from the bacterium *Escherichia coli*. This is a very popular enzyme with genetic engineers because it is readily available. It is also reliable: it cuts DNA at its recognition site with precise accuracy, showing very little **star activity** (cutting DNA at sites other than the recognition site).

Table 30.2 The target sites of some restriction enzymes

Enzyme	Bacterial origin	Recognition site
*Eco*RI	*E. coli*	G⌐AATTC CTTAA⌐G
*Hind*III	*H. influenzae*	A⌐AGCTT TTCGA⌐A
*Bam*HI	*B. amyloliquefaciens*	G⌐GATCC CCTAG⌐G

SEE QUESTION 1

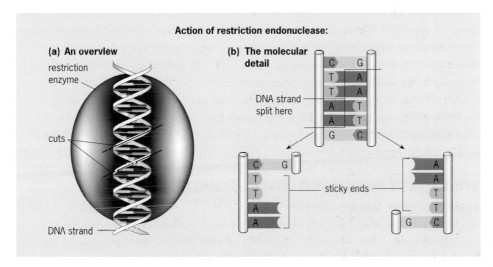

Action of restriction endonuclease:

(a) An overview

restriction enzyme

cuts

DNA strand

(b) The molecular detail

C — G
T — A
T — A
A — T
A — T
G — C

DNA strand split here

sticky ends

C — G
T
T
A
A

A
A
T
T
G — C

Fig 30.3 Endonuclease enzymes are wonderful tools for the genetic engineer because not only do they cut DNA, but they make staggered cuts, leaving 'sticky' ends which make re-joining DNA strands much easier. Endonucleases cut at the sugar–phosphate bonds in the DNA molecule, leaving only the hydrogen bonds intact. At room temperature and above, these bonds are not enough to hold the molecule together, and so it splits, producing a 'sticky end'

You might be asking, 'If restriction enzymes cut DNA, why is the bacterium's own DNA not damaged?' The answer lies in a process called **methylation**: the bacteria add a methyl ($-CH_3$) group to the sequences in their own DNA that could act as recognition sites. This prevents the restriction enzymes doing any damage, presumably because the methylated region no longer fits into the active site of the enzyme.

DNA LIGASES

The sticky ends of the DNA strands produced by the action of restriction enzymes are joined together by enzymes called DNA ligases (ligate means to join/bind). The normal function of a DNA ligase enzyme is to join strands of DNA during replication (see Chapter 27). They are also involved in DNA repair. Genetic engineers use them as 'molecular glue' and they are essential partners for restriction enzymes. An example is T4 DNA ligase, which is made by the bacterium *E. coli*.

✓ **REMEMBER THIS**

The prefix *endo-* means inside; the prefix *exo-* means outside. An endonuclease cuts up a DNA strand by making cuts along the whole length of the molecule. In contrast, an exonuclease cuts DNA by chopping off the nucleotides at the ends of the molecule.

❓ **TEST YOURSELF**

B Different restriction enzymes have differently sized recognition sites. Those in Table 30.2 are described as 'six cutters' because their recognition sites are six bases long, but four or five cutters are also common. Why do you think four cutters produce a larger number of fragments than five cutters?

4 THE HUMAN GENOME PROJECT

The 1990s saw the beginning of the most ambitious project ever to be undertaken in molecular biology: the Human Genome Project (HGP). The aim of this project was simple: to map out the sequence of DNA that makes up all the human genes, but it was a phenomenal task. Although it was estimated that the project would take until 2005 to complete, it was actually published ahead of schedule, in the summer of 2001. Technology advanced very rapidly, enabling much of the sequencing work to be done much faster than had been anticipated.

THE MAIN RESULTS:
WHAT THE HUMAN GENOME SEQUENCE IS

The human genome is divided into 46 chromosomes. Each of these chromosomes has now been completely sequenced; the complete gene sequence from the DNA, pooled from several human individuals, has been read to show the order of the 3.2 million base pairs it contains (Figs 30.4 and 30.5). Obviously, we are all different, but when the overall size of the genome is taken into account those differences are less significant than you might think. Between humans, DNA differs by only 0.2 per cent, or 1 in 500 bases. These differences are being mapped and can reveal some interesting information about disease risk, but the overall sequence is still sufficiently similar in all humans to make the entire sequence very useful.

Surprisingly, there are fewer genes that had been expected – only about 30 000, compared with the 100 000 genes that had been estimated about 10 years ago. A nematode worm has 18 000 genes, a fruit fly has 13 000, so it's quality, not quantity, that obviously counts!

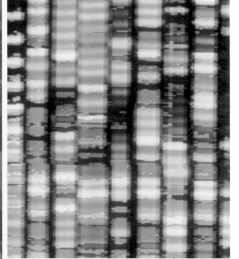

Fig 30.4 Computer analysis of DNA sequencing autoradiograms, used to decode the base sequence of lengths of DNA

WHAT ARE WE USING THE SEQUENCE FOR?

FINDING OUT WHAT GENES DO

The application of the knowledge that has been gained by decoding the human genome is the most exciting part of this project. We are now at the very start of doing this: new technologies are now being developed to monitor the expression of different genes in disease. For example, researchers looking at colon cancer can already monitor how different genes are turned on at each different stage of cancer development, and have plotted out lots of pathways that lead to cancer. Discovering the function of many of the genes in the human genome will produce exciting new information that should help us understand more about how the body works, and how it develops disease.

GENETIC SCREENING

In the short term, the genome is also enabling scientists to set up more and better genetic tests that give advance warning to people of their disease risk. Hopefully, this will allow people to choose a lifestyle to help prevent the disease, and to start preventive drug therapy where necessary. People with a known genetic tendency to develop high cholesterol, for example, could choose to eat a low-fat diet, and perhaps take drugs such as statins, to lower their cholesterol in middle age. This would prevent them succumbing to atherosclerosis, and the forms of heart disease that follow.

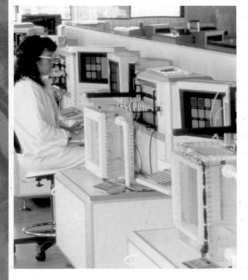

Fig 30.5 Automated electrophoresis equipment used to study the structure of DNA

CHOOSING AND DEVELOPING DRUGS MORE EFFECTIVELY

In the medium term – the next 15 years or so – genome sequence information is likely to become important in prescribing drugs. At the moment, if you have a particular disease, there is a range of drugs that you can be given, but this is often fairly narrow – and they are not tailored to your individual genetic background. The result is adverse reactions that kill hundreds of thousands of people worldwide each year. The new data about the human genome will make it possible to identify which people will react badly to which drugs, so that doctors will be able to avoid them and try something else. It may also become possible to design drugs that are better suited to people with a particular genetic background. Knowing all the genes in the human genome will give researchers new targets to aim for.

GENE THERAPY

Gene therapy, using the genes themselves as medicines, is in many ways the most obvious application of the human genome data. But it is also the most controversial, with a number of deaths linked to experimental treatments (see the Science feature on page 558).

Gene therapy is currently being developed for single gene disorders such as cystic fibrosis, Huntington's disease, sickle cell anaemia and muscular dystrophy. The basic concept is that the abnormal gene can be cut out and replaced by a healthy version, delivered by a tamed virus. Other multifactorial diseases, such as heart disease may also be treatable to some extent when the therapy is developed further.

GERMLINE ENGINEERING

The most long-term application of the human genome sequence may be germline engineering. This is really highly theoretical at the moment, and is likely to require much thought and ethical debate between scientists, governments and the public before it becomes a reality. The idea here is to identify an abnormal gene and then to correct it in the sex cells – the eggs and sperm – that are used to pass genetic information to offspring. No subsequent generation would then be afflicted by their ancestors' gene defect.

Fig 30.6(a) The US geneticist Dr Craig Venter, founder of The Institute of Genomic Research (TIGR) in Maryland, USA. He is standing in the DNA sequencing laboratory at TIGR

Fig 30.6(b) The British molecular biologist Dr John Sulston, director of the Sanger Centre in Cambridge, UK. He is standing in the main sequencing room

---

⚠ SCIENCE FEATURE Human conflict and the human genome map

The image of a scientist as an unfeeling and slightly weird boffin is a common stereotype, but scientists are very human, as the background to this enormously successful scientific project demonstrates. The draft human genome, the sequence of most of the human chromosomes, was announced jointly in June 2000 by the Human Genome Project, a publicly funded programme that included scientists from all over the world, and Celera Genomics, a US private company that used a different technique to produce essentially the same data. Simultaneous publication of the research behind the announcement followed in February 2001. This public show of togetherness masked a bitter dispute that rumbled on during the year before the announcement, and which was still rumbling on when this chapter was written (in March 2002).

The two groups used a slightly different technique to read the 3.2 billion base pairs of DNA that make up the human genome.

The HGP method broke the genome up into manageable segments that were mapped before sequencing took place. Celera took a more sledgehammer approach: it broke the genome into tiny pieces and then used massive computing power to reassemble the read fragments. The arguments originally were about the quality of the work and the time it took. The HGP camp were put on the defensive because Celera managed to produce the sequence in a fraction of the 10 years it took them, and for all the public money that they spent. Celera were criticised for the quality of their work, and for 'cribbing' essential information from the publicly available HGP work, which was updated as work progressed. Many scientists hoped that the joint announcement of the results would be the end of the rows, but a paper in a leading medical journal contained an article from leading HGP scientists, again criticising the quality of the Celera sequence.

! SCIENCE FEATURE The polymerase chain reaction [PCR]

The first recombinant DNA molecules were produced in the early 1970s when bacteria were made to copy foreign DNA along with their own. However, in 1983 an American called Cary Mullis cloned DNA without bacteria using the relevant enzymes. This process, the polymerase chain reaction (PCR), is gene cloning in a test-tube. It allows DNA to be amplified (copied many times).

The idea behind PCR is very simple (Fig 30.7). You mix together your original piece of DNA with the enzyme DNA polymerase in a solution of nucleotides. Next, add some primers, short pieces of DNA which act as signals to the enzymes, effectively saying 'start copying here'. You then heat the DNA so that it denatures into two strands. The thermostable DNA polymerase gets to work and produces two identical strands of DNA (see Chapter 27 for the detailed mechanism of DNA replication).

In a second cycle of reactions the two strands become four, and so on. Typically, the cycle is repeated about 20 times and within an hour you have millions of copies of the original piece of DNA.

This technique has many applications. Forensic scientists use it to amplify tiny DNA samples from spots of blood or hair roots, to obtain enough material for forensic analysis by DNA profiling. PCR also allows archaeologists to study small samples of DNA from historical material such as the preserved bodies found in peat bogs or ice graves.

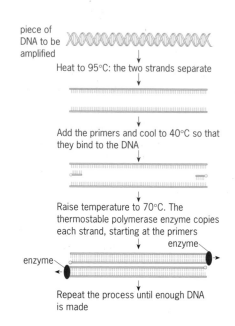

piece of DNA to be amplified

Heat to 95°C: the two strands separate

Add the primers and cool to 40°C so that they bind to the DNA

Raise temperature to 70°C. The thermostable polymerase enzyme copies each strand, starting at the primers

enzyme

enzyme

Repeat the process until enough DNA is made

Fig 30.7 The polymerase chain reaction

SEE QUESTIONS 2 AND 3

5 CLONING GENES

Armed with restriction enzymes and ligases, genetic engineers can create recombinant DNA molecules. However, once a recombinant DNA fragment has been made *in vitro* ('in glass'), it usually needs to be copied repeatedly so that enough molecules are available for analysis or for commercial use.

Although DNA fragments can be multiplied in a test-tube (see the Science feature above on PCR), it is often much easier to put the DNA into organisms such as bacteria which will then adopt the new DNA as their own, and copy it for you. This is called gene cloning. The basic procedure is outlined in Fig 30.8. As an added bonus, the bacteria can also express the gene, making large amounts of the product.

SEE QUESTION 4

✔ REMEMBER THIS

The word *protocol* is often used in genetic engineering. It means 'standard procedure' and so a cloning protocol would be a standard method for cloning a gene.

✔ REMEMBER THIS

To make exact copies of bacteria growing on a Petri dish, the process of **replica plating** can be used. This involves pressing a sheet of velvet or other suitable material on the original colony and then pressing it down on several sterile agar plates. This is commonly used to identify strains of recombinant bacteria – one or more replica plates will contain a particular antibiotic, and only the resistant strains will survive and grow.

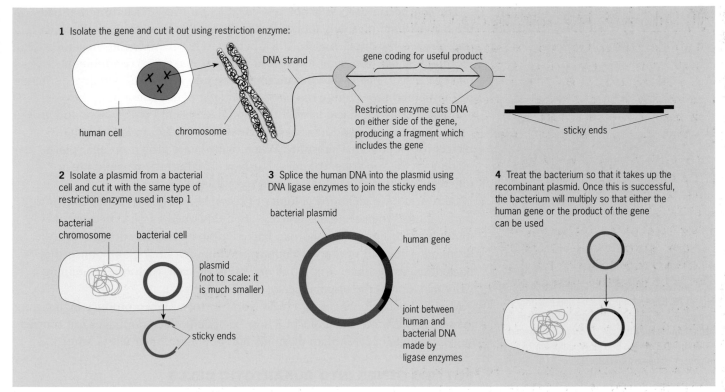

1 Isolate the gene and cut it out using restriction enzyme:

human cell

chromosome

DNA strand

gene coding for useful product

Restriction enzyme cuts DNA on either side of the gene, producing a fragment which includes the gene

sticky ends

2 Isolate a plasmid from a bacterial cell and cut it with the same type of restriction enzyme used in step 1

bacterial chromosome

bacterial cell

plasmid (not to scale: it is much smaller)

sticky ends

3 Splice the human DNA into the plasmid using DNA ligase enzymes to join the sticky ends

bacterial plasmid

human gene

joint between human and bacterial DNA made by ligase enzymes

4 Treat the bacterium so that it takes up the recombinant plasmid. Once this is successful, the bacterium will multiply so that either the human gene or the product of the gene can be used

Fig 30.8 The principles of recombinant DNA technology. A human gene that codes for a useful product, such as insulin, is transferred into a bacterium which then multiplies to form a large population. The bacteria all express the human gene, and large amounts of the gene product can be recovered

USING VECTORS TO TRANSFER GENES

Human genes are often transferred into bacteria or yeast because these microorganisms multiply quickly to form a huge population that expresses the gene and makes large quantities of the gene product. This can then be extracted and purified. The microorganisms are grown in a fermenter, a giant vat with ideal conditions for growth (Fig 30.9).

PUTTING GENES INTO BACTERIA

Putting a human gene into another organism is not an easy task. The first step is to attach the gene to a vector, or carrier. One such vector is a plasmid (Fig 30.10), a tiny circular piece of DNA that occurs in bacteria (see Chapter 1). Like restriction enzymes and ligases, plasmids can be bought commercially.

The plasmid is cut using the same endonuclease enzyme used to cut the human DNA to obtain the gene. Using the same endonuclease is important – the DNA must be cut at the same base sequence to produce complementary sticky ends that will join up. The gene and the plasmid are mixed together and then a DNA ligase is added to join up the sticky ends. The DNA molecule produced is circular, like the original plasmid.

In the second step, the genetic engineer must induce the bacterium to accept the plasmid. One technique involves soaking the bacteria in ice-cold calcium chloride and then incubating them at 42 °C for 2 minutes. Nobody seems to know exactly why this works, but it does. Bacteria which have accepted the plasmid now contain recombinant DNA, and are, by definition, transgenic organisms.

Fig 30.9 Transgenic bacteria or yeasts are cultured in large cylindrical sterile fermenters, although the process is not strictly speaking fermentation. These containers were originally used for making wine and the name has stuck

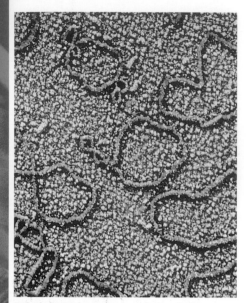

Fig 30.10 False-colour transmission electron micrograph of plasmids of bacterial DNA from the bacterium *Escherichia coli (E. coli)*. This plasmid, called pBR322, is the one most commonly used in genetic engineering work

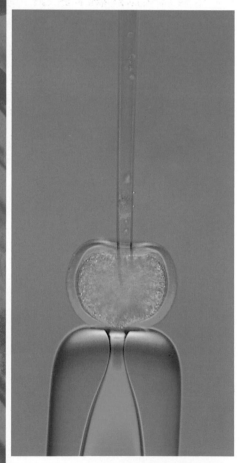

Fig 30.11 Injecting DNA directly into a sheep cell. The cell is held steady by gentle suction from the pipette below, while the even finer pipette above penetrates the cell surface membrane. This technique is used to produce transgenic and cloned animals

The conversion process is not very reliable: for every bacterium that takes up the recombinant plasmid, about 40 000 do not. So how do you tell which ones are recombinant? A clever trick here is to insert two genes into the plasmid: the one you want to use and one that makes the recombinant bacteria easy to detect. For instance, you can add a gene that confers antibiotic resistance and then culture the bacteria on agar plates containing the antibiotic. Only the bacteria with the modified plasmid will survive and grow.

An alternative is to add a second gene that codes for an enzyme that metabolises a coloured substrate. When bacteria are grown on agar plates made with the substrate, colonies that have taken up the plasmid are a different colour from colonies of non-recombinant bacteria.

Once you have identified colonies of recombinant bacteria, you can start pure cultures. Transgenic bacteria are often described as 'sick' because they multiply more slowly than normal bacteria: the population doubles every 30 minutes instead of every 20 minutes. Within ten hours, one transgenic bacterium can produce over a million copies of itself, each one containing a working clone of the original gene.

Modified bacteriophages also make good vectors. They reproduce by inserting their own DNA into the DNA of a host bacterium. Bacteriophages can transfer larger amounts of DNA than plasmids but, at present, their use is limited.

PUTTING GENES INTO EUKARYOTIC CELLS

Usually, a gene is transferred into a different organism so that it can be expressed, making greater volumes of product. This happens only when the gene is able to use the mRNA, ribosomes and Golgi apparatus, etc, for protein synthesis. Some eukaryotic genes are not expressed effectively by prokaryotic cells, and so must be transferred into another eukaryotic cell. Yeast, a single-celled fungus, is often used.

Producing recombinant eukaryotic cells is more difficult than producing recombinant bacteria. The cell walls of fungal cells are a major barrier. But they can be digested away with suitable enzymes to form a protoplast, a cell without a cell wall. In this 'naked' state, the yeast cell will accept plasmids.

Other methods of transferring DNA into eukaryotic cells include the following techniques.

- **Microprojectiles**. It is possible to shoot DNA into host cells. Tiny pellets of metal are coated with DNA and fired at high speed at the target cells. Remarkably, some of the cells recover and accept the foreign DNA.
- **Electroporation**. This involves exposing the host cells to rapid, brief bursts of electricity to create temporary gaps in the cell surface membrane, through which foreign DNA can enter.
- **Liposomes**. In this more subtle method, DNA is inserted into liposomes – spheres formed from a lipid bilayer. The liposomes fuse with the surface membrane of the host cell and introduce the DNA into the cytoplasm.
- **Calcium phosphate precipitation**. Plasmids are mixed with calcium phosphate. As this precipitates, the grains that form contain DNA. These grains can enter cells by endocytosis (see Chapter 4).
- **DNA injection**. DNA can be inserted directly into a cell using a very fine pipette called a micromanipulator that avoids the inevitable hand tremor (see Fig 30.11). Tracey, a transgenic sheep, was created when the gene for alpha-1-antitrypsin was manually injected into the cells of a sheep embryo. This method leaves much to chance: many embryos have to be treated before one takes up the foreign DNA.

6 GENE THERAPY

The new technology of gene therapy promises to revolutionise medicine in the twenty-first century (Fig 30.12). The idea that it can be used to treat genetic diseases such as cystic fibrosis and Tay–Sachs disease has been around for several years but it is taking longer than expected to develop gene therapy into an acceptable and safe treatment.

Recently, scientists have also begun to look beyond just the diseases caused by a defect in a single gene and are considering gene therapy as a potential treatment for all sorts of problems, from cancer and heart disease to AIDS.

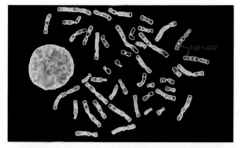
Fig 30.12 Causes and treatment of cystic fibrosis Chromosome pair 7 is highlighted in pink, with the location of the cystic fibrosis gene mutation arrowed

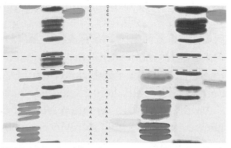
Electrophoresis of the gene sequence shows that base sequence TTC is present in an unaffected person (left) and absent in someone with cystic fibrosis

The healthy gene is isolated before being inserted into a virus. Cystic fibrosis patients use an inhaler with the virus that delivers the healthy gene to lung cells

GENE THERAPY FOR GENETIC DISEASES

The first person to be given gene therapy was Ashanti DeSilva. She became the first patient with severe combined immunodeficiency (SCID) (Fig 30.13) to be treated in September 1990, when she was 4 years old. A team in the USA led by French Anderson gave Ashanti four infusions of cells containing the working gene that she lacked. Over the four months of the treatment, her condition improved.

> ### REMEMBER THIS
> A simple definition of **gene therapy** is the treatment of a genetic disease by giving the individuals copies of healthy genes.

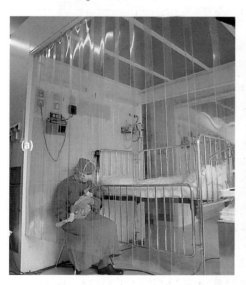

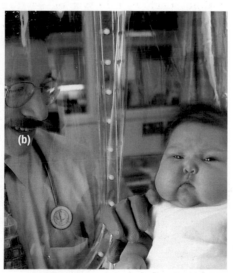

Fig 30.13(a) A SCID baby lives in a sterile compartment and is handled by medical staff wearing full body gowns to avoid passing on any infections
(b) This Apache baby is having gene therapy: some of his stem bone marrow cells are removed, the gene for the enzyme he lacks is inserted into them and they are returned to his body. Then he can produce normal immune cells

With the help of follow-up treatments she has now been transformed from a small girl who was constantly ill and could not leave the house, to a normal, healthy and lively teenager. She continues to do well and, apart from needing regular injections of one of the enzymes she cannot make naturally, she lives a normal life.

Early in 2000, 10 years after the first successful gene therapy trial to treat SCID, researchers in Paris managed to treat two babies, aged 8 and 11 months with gene therapy. The babies both had SCID, but a slightly different form from Ashanti DeSilva. Their treatment involved taking out some bone marrow and selecting out a set of blood stem cells. Stem cells are the cells that differentiate into all sorts of blood cell types, including the white blood cells that make up the immune system. By introducing the gene that the babies were missing into these stem cells, the researchers managed to get the correct gene into all of their immune cells. Both babies developed a fully functional immune system with a few weeks and can now live normal lives. A similar process was used in the summer of 2001 on 18-month-old Rhys Evans at Great Ormond Street Hospital in London.

HOW GENE THERAPY WORKS

One of the main problems that make gene therapy difficult is finding out how to transfer genetic material containing corrected, healthy genes into cells that contain the faulty gene. Scientists have now developed several methods.

Modified viruses are often used as carriers, or vectors. Viruses are useful because they are able to penetrate cells and to introduce their genetic material into the host cell. Before researchers can use viruses in therapy, however, they must remove the genes that code for proteins that viruses use to reproduce themselves. When those genes are replaced with corrective genes, the vector virus becomes a delivery unit that is identical to the original virus on the outside, and can transport useful genes into cells, but cannot cause illness.

Three main approaches to gene therapy are currently possible.

! SCIENCE FEATURE Setbacks plague gene therapy development

Although a handful of young children have received successful gene therapy for severe immunodeficiency syndrome (SCID), few other gene therapy trials have worked very well. A low point for this new technology came in September 1999 when a teenager, Jesse Gelsinger, died during a gene therapy trial. There was a full investigation and there were indications that many research rules had been broken. The investigation was widened to other centres, mainly in the USA, and there were strong suggestions that many poor results in trials were being covered up because of the financial damage they could cause to the companies funding the trials. Many gene therapy programmes in the USA were put on hold. These have now started again, but the rules governing trials have been made stricter.

French Anderson, the pioneer of gene therapy, commented after the recent gene therapy success in April 2000 (see text) that although 340 gene therapy trials were on-going in the USA, only a handful are showing promising results. There are still many problems to overcome and some of them might need unusual solutions. Recent research suggests that lentiviruses may be better gene vectors than other viruses. They can get a gene into cells in the body with higher efficiency and can be modified so that the gene they carry is expressed only in some types of cell. However, a major problem is that lentiviruses are retroviruses and one of the most useful of these is HIV-1 – the virus that causes AIDS. Although the new vectors are extensively modified, they will need to be tested very well to make sure there is no chance of them reverting and becoming able to cause an immunodeficiency in the patients they are used to treat.

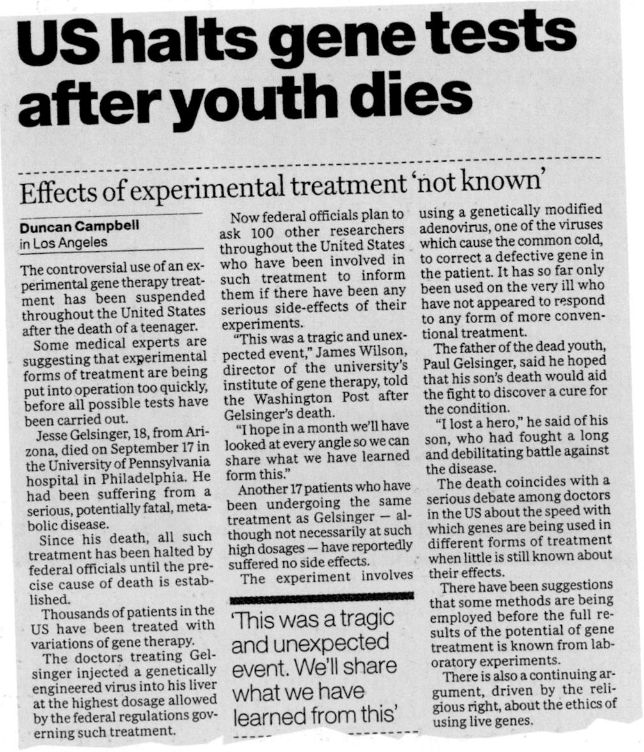

US halts gene tests after youth dies

Effects of experimental treatment 'not known'

Duncan Campbell
in Los Angeles

The controversial use of an experimental gene therapy treatment has been suspended throughout the United States after the death of a teenager.

Some medical experts are suggesting that experimental forms of treatment are being put into operation too quickly, before all possible tests have been carried out.

Jesse Gelsinger, 18, from Arizona, died on September 17 in the University of Pennsylvania hospital in Philadelphia. He had been suffering from a serious, potentially fatal, metabolic disease.

Since his death, all such treatment has been halted by federal officials until the precise cause of death is established.

Thousands of patients in the US have been treated with variations of gene therapy.

The doctors treating Gelsinger injected a genetically engineered virus into his liver at the highest dosage allowed by the federal regulations governing such treatment.

Now federal officials plan to ask 100 other researchers throughout the United States who have been involved in such treatment to inform them if there have been any serious side-effects of their experiments.

"This was a tragic and unexpected event," James Wilson, director of the university's institute of gene therapy, told the Washington Post after Gelsinger's death.

"I hope in a month we'll have looked at every angle so we can share what we have learned form this."

Another 17 patients who have been undergoing the same treatment as Gelsinger — although not necessarily at such high dosages — have reportedly suffered no side effects.

The experiment involves using a genetically modified adenovirus, one of the viruses which cause the common cold, to correct a defective gene in the patient. It has so far only been used on the very ill who have not appeared to respond to any form of more conventional treatment.

The father of the dead youth, Paul Gelsinger, said he hoped that his son's death would aid the fight to discover a cure for the condition.

"I lost a hero," he said of his son, who had fought a long and debilitating battle against the disease.

The death coincides with a serious debate among doctors in the US about the speed with which genes are being used in different forms of treatment when little is still known about their effects.

There have been suggestions that some methods are being employed before the full results of the potential of gene treatment is known from laboratory experiments.

There is also a continuing argument, driven by the religious right, about the ethics of using live genes.

'This was a tragic and unexpected event. We'll share what we have learned from this'

Fig 30.14 Jesse Gelsinger's death was widely reported

EX VIVO GENE THERAPY

The most commonly used technique in gene therapy is known as *ex vivo* because the gene transfer is done outside the body. Cells with defective genes are removed from the patient and given normal copies of the affected DNA before being returned to the body. This therapy has generally targeted blood cells because many genetic defects alter the function of one or other of the blood cells. However, the problem with this approach is that blood cells have limited life spans. Corrected cells can be introduced into the body and they can work well, but after a few weeks they die and the patient's original problems return, making it necessary for many repeat treatments to keep the person healthy.

A better approach is to target the stem cells of the bone marrow. Stem cells survive as long as the patient does, so if these immature cells that give rise to the full array of blood cells in the circulation can be given the correct gene, the treatment should last indefinitely.

Although scientists have been able to obtain stem cells from human bone marrow, it has proved difficult to get genes into these cells. But advances have been made recently. In 1993 a group in Los Angeles treated three new-born babies with SCID (page 557) by inserting genes into their stem cells. Their blood cells began producing the critical enzyme that they lacked and they grew into healthy children.

IN SITU GENE THERAPY

In *in situ* treatment, vectors carrying the corrective genes are introduced directly into the tissue where the genes are needed. This approach makes most sense for problems that are localised, but it cannot correct disorders that affect several systems of the body.

In situ gene therapy is being explored for several diseases. For cystic fibrosis, workers have introduced gene vectors containing healthy copies of the cystic fibrosis gene into the lining of the bronchial tubes. As a first step towards treating muscular dystrophy, other researchers have injected a gene directly into muscle tissue in animals to investigate the possibility of re-engineering the body to make normal muscle proteins. Several teams have also inserted 'suicide' vectors into tumours. These carriers contain a gene that is intended to make cancer cells self-destruct when treated with certain chemotherapy drugs.

IN VIVO GENE THERAPY

This third approach is still in the experimental stages but it is likely to be the gene therapy of the future. Gene carriers will simply be injected into the body, in much the same way that drugs are given now. Once in the body, the carriers will find their target cells, ignore other cell types, and transfer their genetic information efficiently and safely.

PROBLEMS AND ETHICS

The people who have been trying to develop gene therapy techniques have had to overcome many technical problems. *In situ* therapy, for example, is still hampered by a lack of safe ways to implant corrected genes into various organs. *Ex vivo* therapies do not have this problem, but the genes that are transferred do not always yield good quantities of the encoded proteins. And, in both forms of treatment, once the genes enter the cells, they can insert themselves randomly into the DNA of the chromosomes. Sometimes this is harmless, but sometimes it is not. If the gene sent in as therapy manages to disrupt a tumour suppressor gene that normally protects the body against cancer, a tumour could result.

Development of *in vivo* gene therapy has been slow because researchers have found it difficult to create delivery units that will allow the genes to find their way into the correct cell and that can avoid the immune system of the body.

Cost is also a potential hurdle to gene therapy. Because we are still at an early stage of gene therapy research, it is still quite expensive to do and is carried out only in the major medical centres of the developed world. However, French Anderson, the pioneer of the first gene therapy procedure (page 557) thinks that, ultimately, gene therapy will become relatively inexpensive and perhaps in 50 years, may be used regularly and routinely to cure many different medical conditions.

However, the financial and technical challenges of gene therapy are nothing compared to the ethical dilemmas that crop up. So many different questions need to be raised and considered. Here are just a few for you to mull over:

- Should gene therapy be used to prevent disease?
- All the gene therapy carried out so far has affected only the somatic cells, not the germline cells – the eggs or the sperm. Do you think germline gene therapy could ever be justified?
- Where should the line be drawn for genetic therapy? How will it be possible to avoid people trying to use gene therapy to practise eugenics (altering racial characteristics)?
- In 1998, scientists managed to grow human stem cells from embryos in culture for the first time. This raises the possibility of engineering an entire human being. Do you agree that this type of research should continue?

7 STEM CELLS AND CLONING

Cloning is a difficult topic for anyone; it is not only complex biology, but it is also a tricky subject ethically. What you feel about cloning may not just depend on your knowledge of science – it may also depend on your religious, moral or other views. We don't have enough space in this book to do more

---

✔ REMEMBER THIS

Eugenics is the application of genetics to improve human characteristics. It is generally frowned upon, since making genetic alterations to select for cosmetic features, such as blonde hair and blue eyes, is pointless. However, deciding where eugenics end and germline gene therapy and embryo selection for medical reasons begin is a difficult ethical problem.

! SCIENCE FEATURE Gene therapy before birth

The US scientist French Anderson is currently working towards carrying out gene therapy on a fetus that is affected by a genetic disease that is usually fatal before birth.

Because of the enormity of the ethical issues involved, he has raised the possibility of *in utero* gene therapy for public discussion, even before the techniques that would enable it to go ahead have been developed. He argues that if society decides that this sort of treatment should not be allowed, there is little point in spending the time and the money to work on it.

In the next few years, Anderson hopes that it will be possible to take some of the bone marrow stem cells from a fetus with untreatable SCID or who has α-thalassaemia, transform the cells using a corrective gene, and then reintroduce the cells into the fetus at about 15 weeks. Fig 30.15 shows a 15-week-old fetus.

If the therapy works, the child would be born healthy and would never suffer from the symptoms of the disease it has been treated against. However, the problem is that it might not work – if the baby dies before birth, you might say that nothing

has been lost. However, the possibility exists that some sort of partial cure could result and the baby could be born, only to suffer a short life of severe illness and disability.

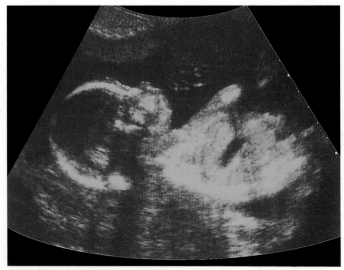

Fig 30.15 The CT scan of a 15-week-old fetus

than a brief introduction. This field is moving so fast, it will also be out of date within months. To keep up, and to find information to help you make your own mind up about cloning, refer to good journals, such as *Scientific American* and *New Scientist* (both are available online), and look at the BBC online Science and Technology news site.

CLONING WHOLE ORGANISMS

In 1997, when scientists succeeded in cloning a sheep, there was a lot of fuss in the media. Was it ethically and morally right to clone animals as advanced as mammals? Would this lead to scientists attempting to clone humans?

The definition of a clone is 'a genetically identical copy' and the term can apply to individual strands of DNA, individual cells or whole organisms. Cloning is, in fact, a common natural process: all organisms made by asexual reproduction are clones of their parent, unless there is a mutation. Even humans make clones – they are more commonly called identical twins.

There is much commercial interest in the cloning process because it can produce exact copies of organisms with desirable characteristics. If, for example, you have a 'perfect' apple tree that produces large amounts of delicious, healthy, blemish-free apples, you can clone it rather than risk losing its valuable genotype in the lottery of sexual reproduction (see Chapter 33).

Cloning plants is easy, and can be done with relatively little equipment in schools and colleges. Cloning mammals, however, is a lot more difficult but technically feasible. There are two approaches to cloning whole organisms, embryo splitting and nuclear transfer.

SEE QUESTIONS 5 AND 6

EMBRYO SPLITTING

Eggs and sperm or fertilised embryos are collected from prize specimens of cattle, for example. Cells can be pulled apart when the embryo is at the 8 or 16 cell stage and each fragment can then develop into a complete embryo. Each embryo is a clone of the original. The process can be repeated many times and then the cloned embryos can be introduced into the uteruses of normal cows. After the normal gestation period, several very average animals give birth to a herd of prize specimens (see Chapter 6 for more details).

NUCLEAR TRANSFER

Cloning is based on nuclear transfer, the same technique scientists have used for some years to copy animals from embryonic cells. Nuclear transfer involves the use of two cells. The recipient cell is normally an unfertilised egg taken from an animal soon after ovulation. Such eggs are poised to begin developing once they are appropriately stimulated. The donor cell is the one to be copied. A researcher working under a high-power microscope holds the recipient egg cell by suction on the end of a fine pipette and uses an extremely fine micropipette to suck out the chromosomes, sausage-shaped bodies that incorporate the cell's DNA (see Fig 30.11). (At this stage, chromosomes are not enclosed in a distinct nucleus.) Then, typically, the donor cell, complete with its nucleus, is fused with the recipient egg. Some fused cells start to develop like a normal embryo and produce offspring if implanted into the uterus of a surrogate mother.

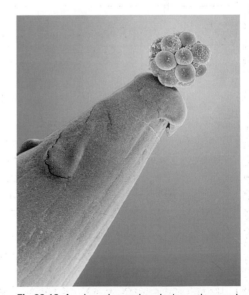

Fig 30.16 A coloured scanning electron micrograph (SEM) of a human embryo at the 10-cell stage on the tip of a pin. The ball of cells (orange) of the embryo is known as a *morula,* a cluster of almost identical, rounded cells, each containing a central nucleus. This 10-cell embryo is about three days old. It is at the early stage of transformation from a single cell to a human composed of millions of cells

TECHNICAL AND ETHICAL PROBLEMS WITH CLONING

Dolly the sheep and cc the cat were created by the process of nuclear transfer, a process that many people feel will give rise to cloned humans at some stage in the future. As we saw in the opener to this chapter, the technique sounds

---

> ### ⚠ SCIENCE FEATURE Will the cloning bubble burst?
>
> Apart from the success rates and embryo wastage that occur in the production of an animal clone, scientists are becoming concerned about the long-term effects of cloning. Dolly the sheep, the first sheep to be cloned, was touted as a perfectly healthy, normal animal. She has had several normal lambs and all seemed to be well. However, in January 2002, it was announced that Dolly has developed premature arthritis. She is only 6, and arthritis is very rare in sheep of this age. This report has been followed by others; in February 2002, a paper reported that cloned mice live significantly shorter lives than mice of the same genetic background produced by normal mating. Researchers are now beginning to think that the trauma to the cells caused during the cloning procedure, where a nucleus is removed from a normal body cell and then transferred into a zygote, could cause problems. The cloned mice all had problems with their immune systems and it may be that we don't understand enough about the delicate process of development at the early embryo stage to find out what has gone wrong.
>
> Steps are now being taken to introduce more gentle cloning methods, which allow two cells to fuse, but it will still be a shock for a cell that is making all the proteins that keep a body cell going one minute to suddenly switch to become a zygote, ready to develop into an embryo. Some leading scientists have called for researchers to be very honest about their cloning experiments – some, particularly those in companies, might not reveal 'failed' cloning experiments because of the damage it could do to the company, but these failures might help to solve some of the puzzles.

✔ REMEMBER THIS

The American cult that says it intends to clone a human baby is based in the USA, but it cannot do its experiments there. Some countries in the world do not have strict regulations on cloning, and scientists from the organisation have set up their operation where they can do so legally, for the moment. Global appreciation of the problem is increasing, and this 'loophole' may be closed during the next few years.

❓ TEST YOURSELF

C There is a common misconception that people can have their cells frozen so that they can be cloned after their death to 'live again'. How would you explain to someone that this is utter rubbish?

---

simple, but we can't do it reliably. It took over two hundred attempts to get Dolly and almost 100 to get cc, the cloned cat. Many laboratories report that only 1 to 2 per cent of cloned embryos survive to become live offspring. Even some clones that survive through birth die shortly afterwards. The recently cloned gaur, an endangered species of wild ox from Asia, lived only a few days before dying from a common dysentery.

This unreliability has two main implications for human cloning that intends to produce living babies. Firstly, each attempt requires an individual egg. Sheep don't tend to complain when you stimulate their ovaries and harvest the eggs, but humans are a different matter. Very few women will give up eggs for any reason other than their own pregnancy, so finding 200 ova for each attempt is always going to be a problem. The women in the USA may have been persuaded to do this by the cult aiming to clone a human (see opener), but whether they will actually do so remains to be seen. Secondly, in humans, failure brings great distress. Miscarriages and malformed offspring always cause great distress to all concerned in the process.

In 2001, the UK government passed a law that allowed scientists to produce human embryo clones solely for the purpose of making replacement tissues that could be used in transplant procedures. This was challenged by the pro-life lobby, but in February 2002, a House of Lords Select Committee upheld this decision, and the Human Fertilisation and Embryology Authority, which regulates such research is now likely to give researchers licences to experiment on early stage human embryos. Any human cloning that intends to produce a full-term pregnancy is banned in the UK, Europe and the USA and many other countries.

STEM CELLS:
THE REASON BEHIND CLONING HUMAN EMBRYOS?

If the objective is not to make a human baby, then why clone? The answer comes down to stem cells. These cells are capable of differentiating into any type of adult body cell. Theoretically, you could say that every cell in the body has the capacity to do this, as they all contain the same genome. However, once a cell has gone through a pathway of differentiation, and has become, say, a brain cell, there is little you can do to shift it back again. Although the adult body does contain some stem cells, they are few and far between and have proved difficult to isolate. A much better source, in terms of numbers of cells and ease of extraction, is to use early stage embryos.

As you can see from the Science feature below, stem cells may be useful for making transplantable organs in culture. Stem cells might be used to treat a variety of serious diseases caused by damage to cells, perhaps including AIDS as well as Parkinson's, muscular dystrophy and diabetes. They may be useful for repairing the damage to spinal cords and nerves that are caused in the many road accidents each year, or for replacing the damaged cells in people with Alzheimer's disease. The possibilities are, apparently, endless.

REMEMBER THIS

Stem cells are the cells in the embryo that have the potential to become all other tissue types. For this reason they are called **totipotent**. There are stem cells in adult bodies, notably in the bone marrow, but these can give rise to only a few tissue types: for this reason they are called **multipotent**. Research on embryonic stem cells therefore has greater potential to repair and replace damaged tissues. Embryonic stem cells have the added advantage that they have not accumulated the mutations that stem cells obtained from adults will have.

! SCIENCE FEATURE The first cloned kidneys

It is only at the beginning of this century that the idea of 'growing' organs for transplant has moved from the realms of science fiction and fantasy into reality. In January 2002, researchers revealed that they had produced a cow kidney using a single skin cell from the ear of an adult cow. This was fused with a donated cow egg that had been stripped of its own genetic material. A small electrical current was then used to stimulate the fused cell to become an embryo. The embryo was rich in stem cells that have the potential to become a wide range of body tissues.

Some of the stem cells were removed from these early embryos and treated with chemicals and growth factors to push them towards development pathways that made them into fully mature kidney cells. They were grown on a biodegradable kidney-shaped scaffold designed by a team from Harvard Medical School. The scientists produced several miniature kidneys each a few centimetres long. These were transplanted back into the adult animal, alongside its existing organs, where they started to produce urine.

Obviously, this study is at a very early stage; it will be some time before fully functioning kidneys are produced that can replace a diseased kidney completely. However, it does provide hope for the future; donation of organs such as kidneys and hearts has always been a problem, and people needing transplants still face long waits to see if a suitable donated organ will become available. One day, in the not too distant future, it may become possible to grow a custom-built kidney, exactly compatible with a particular patient.

SUMMARY

- Genetic engineers isolate, cut out and transfer genes between organisms. Genetic engineering technology has many applications, including making human products such as insulin on a large scale.
- The Human Genome Project has now produced the complete sequence of all 30 000 genes in the human genome. Knowing this sequence will provide a big boost for research in many areas of biology and medicine.
- Strands of DNA are cut using **restriction endonuclease** enzymes, which always cut DNA at a particular base sequence. Strands of DNA can be joined by using DNA **ligase enzymes**.
- In order to transfer DNA into a living cell such as a bacterium, a **vector** (carrier) is needed. Common vectors are **plasmids** (circles of DNA found in bacteria) or viruses.

- Once the foreign DNA is inside the host cell, it is adopted by the host. Genes in the foreign DNA are expressed using the host cell's synthetic machinery (ribosomes, mRNA, etc). In addition, the foreign DNA is **cloned** (copied) every time the host cell divides.
- **Gene therapy** is a technique that replaces faulty genes with healthy ones. It provides great hope for sufferers of genetic diseases such as cystic fibrosis.
- The **polymerase chain reaction** (PCR) allows samples of DNA to be cloned in a test-tube using the necessary enzymes, instead of inside a host cell.
- Many recent developments in this field have made cloning whole animals and organs possible. However, although the technical problems have been overcome, the ethical difficulties still pose a great problem for this area of research.

❓ EXAM QUESTIONS

1 The diagram outlines the way in which the gene for human factor VIII, a protein which is necessary for the clotting of blood, is incorporated into bacterial DNA and inserted into a bacterium.

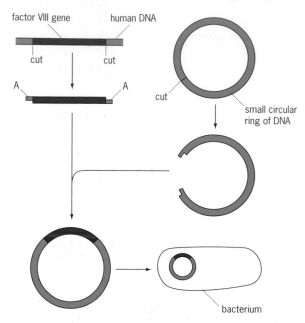

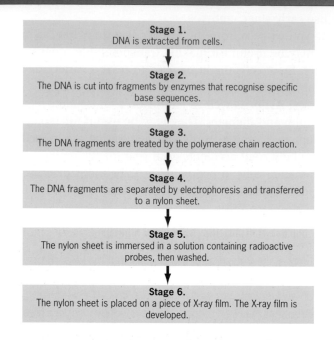

Stage 1.
DNA is extracted from cells.

Stage 2.
The DNA is cut into fragments by enzymes that recognise specific base sequences.

Stage 3.
The DNA fragments are treated by the polymerase chain reaction.

Stage 4.
The DNA fragments are separated by electrophoresis and transferred to a nylon sheet.

Stage 5.
The nylon sheet is immersed in a solution containing radioactive probes, then washed.

Stage 6.
The nylon sheet is placed on a piece of X-ray film. The X-ray film is developed.

AQA (B) BYB2 Genes and Genetic Engineering June 2001 Q6

a) With reference to the diagram, state:
(i) the type of enzyme used to cut the factor VIII gene from human DNA;
(ii) the name of the small circular ring of DNA;
(iii) the term used to describe the regions labelled **A**;
(iv) why these regions have been added to the factor VIII gene;
(v) the word used to describe the bacterial DNA which now contains the human factor VIII gene.

b) Suggest why it is considered better to use genetic engineering as a source of human factor VIII rather than material obtained from human blood.

c) Describe how DNA replicates. (You may use annotated diagrams to assist your explanation if you wish.)

OCR 2801 Biology Foundation January 2001 Q6

2 The flow chart shows one method of finding the sequence of nucleotides in DNA.

a) Name the type of enzyme which is used in **Stage 2.**

b) Why is the polymerase chain reaction used in this procedure (**Stage 3**)?

c) What is meant by electrophoresis (**Stage 4**)?

d) (i) In Stage 5, the nylon sheet is immersed in a solution containing radioactive probes. Explain the function of these radioactive probes (**Stage 5**).
(ii) Explain why only some nucleotide sequences show up when the X-ray film is developed.

3 The diagram illustrates the polymerase chain reaction.

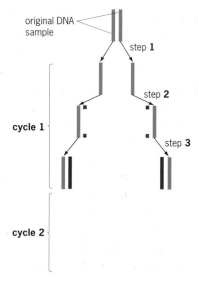

a) (i) What method is used to split the original DNA sample into two strands during Step **1**?
(ii) Describe what happens during Steps **2** and **3** in order to produce a new DNA strand.

b) Draw the DNA molecules that would be present at the end of **Cycle 2**. Use the same method of shading to distinguish original DNA strands and the new DNA strands.

c) Describe **one** example of the use of the polymerase chain reaction.

AQA (B) BYB2 Genes and Genetic Engineering January 2001 Q3

? EXAM QUESTIONS

4

a) **(i)** Describe how a gene may be taken from a mammalian cell and inserted into bacterial cells.
(ii) Describe how bacteria containing the transferred gene can be cultured on a large scale.

b) Mutation of the gene responsible for production of the enzyme galactosidase causes a liver disorder. The liver is unable to metabolise certain carbohydrates, and toxic products accumulate.
(i) Describe **one** way in which the structure of the DNA of a gene may be changed as a result of a mutation.
(ii) The disorder can be treated by introducing the gene for galactosidase into a harmless virus, then injecting the transformed virus into the patient. A liver cell containing DNA from the transformed virus produces galactosidase. Describe how a cell synthesises a protein such as galactosidase.
(iii) The transformed virus enters liver cells but does not usually enter other cells in the body. Suggest **one** explanation for this.

AQA (B) BY82 Genes and Genetic Engineering June 2001 Q7

5 A new variety of tomato has been produced by genetic engineering. This variety contains a synthetic gene that blocks the action of a natural gene that would make the fruit soften rapidly once ripe. It also contains a marker gene.
The marker gene added by the scientists makes this variety of tomato resistant to the antibiotic, kanamycin. It is possible that this gene could be taken up by disease-producing bacteria in the human gut. In humans, kanamycin is used to treat certain types of gut infections.
Using information from the passage, explain the advantages and disadvantages of putting this new variety of tomato on the market.

AQA (B) BYB 2 Genes and Genetic Engineering January 2001 Q4

6 The diagram shows how insulin can be made using genetically modified bacteria.

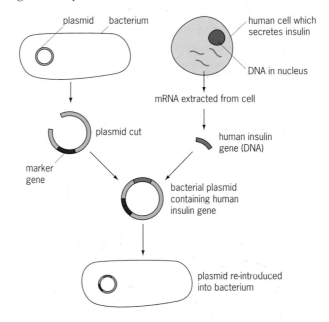

a) **(i)** The human insulin gene is obtained from mRNA, rather than DNA. Suggest why.
(ii) Name the enzyme used to make a single-stranded DNA copy of the mRNA coding for insulin.
(iii) The table shows a sequence of bases from the mRNA coding for insulin. Copy and complete the table to show the sequence of bases you would expect in a single-stranded DNA copy.

mRNA base sequence	U	C	A	A	C	C
DNA base sequence						

b) What is the role of DNA ligase in producing genetically-modified bacteria?

c) The plasmid contains a marker gene coding for antibiotic resistance. Explain the importance of this marker gene.

AQA (A) BYA2 Making Use of Biology January 2001 Q6

KEY SKILLS **ASSIGNMENT**

CYSTIC FIBROSIS AND GENE THERAPY

Cystic fibrosis (CF) is a genetic disease that is due to one defective allele (see Chapter 27). About one person in 25 carries the defective allele, but they also carry the normal allele and do not suffer from the disease.

1 Why don't heterozygotes suffer from the disease?

2 A couple are both carriers for cystic fibrosis.
a) What is the probability that their first child will have cystic fibrosis?
b) What is the probability that their second child will have cystic fibrosis?

The normal allele codes for a protein called cystic fibrosis transmembrane regulator (CFTR). This essential membrane protein in epithelial cells transports chloride ions out of the cells and into mucus. Normally, when chloride ions are secreted, sodium ions follow, and this decreases the water potential of the epithelial mucus. Water follows outwards by osmosis, making normal, watery mucus, which can be moved by the cilia (tiny hairs) lining the airways.

Cystic fibrosis sufferers make a protein that differs in just one of its 1480 amino acids. Although slight, this fault prevents CFTR from functioning normally. Chloride ions and sodium ions cannot be secreted, and so the mucus becomes much thicker than normal. Dead epithelial cells also accumulate in the mucus, adding to the general congestion of the airways.

3 Explain how a fault in just one amino acid can have such a drastic effect.

Sticky mucus is also a real problem in the pancreas. The mucus blocks the pancreatic duct, preventing the secretion of pancreatic juice.

4 Suggest why cystic fibrosis sufferers have to take tablets containing digestive enzymes.

Treatment of CF includes regular physiotherapy in which the chest is patted to dislodge the mucus. Even so, infections are common and most CF sufferers have to take a variety of antibiotics according to the infection they have at the time.

Gene therapy for cystic fibrosis
Now, one exciting possibility is that the faulty genes can be replaced by healthy ones. In 1989 the cystic fibrosis allele was located on chromosome 7. The base sequence of the healthy gene was then compared to the defective allele, and the nature of the fault was narrowed down at the molecular level. This opened up the exciting possibility that if healthy genes could somehow be introduced into the epithelial cells, they might be expressed and so make the correct membrane protein, solving the problem for a time.

The basic steps are as follows:

1. The CFTR gene is isolated, and cut out.
2. The gene is cloned many times.
3. The genes are encapsulated, either by putting them into liposomes (spheres made from lipid) or viruses.
4. The gene particles are inhaled, so that they can pass into the epithelial cells of the lung to be incorporated into the DNA of the cells.

Once in place, if all goes well, the healthy genes are transcribed, making the correct protein.

5 **a)** What is used to cut the CFTR gene out of chromosome 7?
b) Name two different methods of cloning the CFTR gene.
c) Suggest why the pancreatic genes would be more difficult to replace.
d) Suggest why repeat treatments at regular intervals would be necessary.

Germline gene therapy
This area of research is an ethical minefield. So far in this Assignment we have looked at body cell gene therapy – the replacement of the faulty allele in certain cells of the body. However, the eggs or sperm of the sufferer would still carry the defective alleles. In germline gene therapy, the original alleles in the zygote would be changed. The individual would grow and develop with healthy alleles in all cells and all of the individual's children would inherit the healthy allele.

Germline gene therapy sounds attractive but there are problems, so much so that this area of research is banned in the UK, and in many other countries. Ethically, there is the familiar 'where do you stop?' problem. A new treatment for cystic fibrosis or haemophilia would be desirable, but having the technology might allow for all sorts of temptations. Our knowledge of the human genome could reveal, for example, genes that code for intelligence, height or skin colour. Tampering with these characteristics would lead to obvious problems.

Biologically, germline gene therapy is potentially dangerous because we know very little about how genes function in the embryo. Tampering with genes in the zygote could have effects that might become apparent only in later life.

6 In discussion groups, prepare a case which you could present either for, or against, germline gene therapy.

31

How plants work

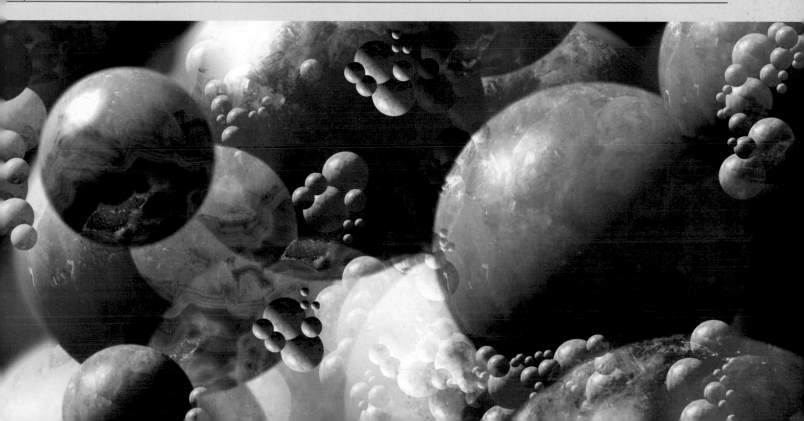

In areas of high salt concentration, normal tomatoes can't absorb salt and so cannot absorb water (the water potential of the soil water is always lower than in the root cells). This new transgenic tomato is able to absorb salt and water together. The salt is transported to the leaves but not to the fruits

One substance that is not in short supply on this planet is salt. Two-thirds of the world is covered in salty water and there are vast deposits on land. This is a big problem because plants can't grow where it's too salty – they can't absorb water efficiently.

The Romans knew this; they would ruin the farmland of their enemies by ploughing salt into the soil. A solution to the salt problem would seem to be to irrigate, because the extra water would dilute the salt, right? Wrong. Often, the extra water raises the salt table, bringing underground reserves to the surface and making matters worse. It has been estimated that about a third of the world's agricultural land is infertile or sub-fertile because of salt.

There are plants that are able to grow in salty conditions, but these are not crop plants. Even if they were, they would taste too salty. However, researchers at the University of California have isolated some genes that give a type of wild mustard, *Arabidopsis*, its salt tolerance. They have been able to put the gene into tomato plants to produce a new variety that absorbs salt but then transports it to the leaves, leaving the all-important fruit tasting normal. Not only will this plant grow in salty areas, but it can be used to gradually reduce the salt in the soils. It is also a possibility that the crop would grow when irrigated with sea water – a big bonus in areas where freshwater is in short supply.

This is just one example of the potential of genetic engineering to help feed the world. There are many 'tolerance genes' that could, in theory, make crop plants easier to grow in poor climates and soils. As with all examples of genetic engineering, there are many tests that need to be done before this plant can be widely grown but, in time, the positive aspects of this technology will outweigh the drawbacks.

This field in Australia shows some of the problems with salt. Originally, this land supported native grass species that pumped up the water and kept the salt table down. In an attempt to make the land more fertile for crops such as wheat, the land was irrigated but this flushed the salt to the surface

1 WHAT IS A PLANT?

There is a tendency to think that anything that photosynthesises is a plant, but this is not true. Three types of organisms can photosynthesise and two of them are definitely not plants. The first are bacteria – the **cyanobacteria** – which are prokaryotes (see Chapter 1). They were probably the first photosynthetic organisms on Earth. The other group consists of the **algae**, including seaweeds. In the modern 5-kingdom system algae are classified as protoctistans; they are eukaryotes but are not 'true' plants (see page 541).

'Proper' plants, or to use the technical term, **higher plants**, are defined as 'multicellular photosynthetic eukaryotes' and include mosses, ferns, conifers and flowering plants (Fig 31.1).

All flowering plants
- develop from embryos protected by the tissues of the parent plant;
- have starch as their main storage compound;
- have cell walls made from cellulose;
- have life cycles that feature alternation of generations (see Chapter 33).

In this chapter we look at the basic structure and function of the flowering plant, focusing on its transport system. Firstly we look at the plant's relationship with water, then at the transport of organic molecules, and finally at the ways in which plants have become adapted to harsh environments.

THE LIFE OF PLANTS

Plants can make their own food, so are fundamentally different from animals that have to find it ready-made. Plants have no need for muscles, complex sense organs or a movable skeleton. They don't need a mouth or intestines, and they have no need to live their lives on our timescale at all.

Yet despite these differences, plants are trying to achieve exactly the same goals as other organisms: to feed and grow, without being eaten, long enough to pass their genes on to the next generation. Make no mistake, there is drama in the plant world; sex, death, deception and chemical warfare to name but a few. It's all happening, but it's usually very, very slow.

Fig 31.1 shows that higher plants evolved from algae, organisms that live their entire life in water. The first land plants relied heavily on water and could only thrive and reproduce in very wet environments. In evolutionary terms, the flowering plants that dominate our planet are relative newcomers and have evolved in close association with insects (see Chapter 33). As you study the biology of the flowering plant, think about the ways in which the 230 000 or so known species have overcome the problems of living on land. Consider the following:
- Compared with water, air offers very little in the way of support. How do plants support themselves without a bony skeleton?
- The structure of the leaf results in plants losing a lot of water. A mature oak tree can lose over 3000 litres (or 600 gallons; about 20 bath tubs full) on a summer's day. How does a plant replace all this water?

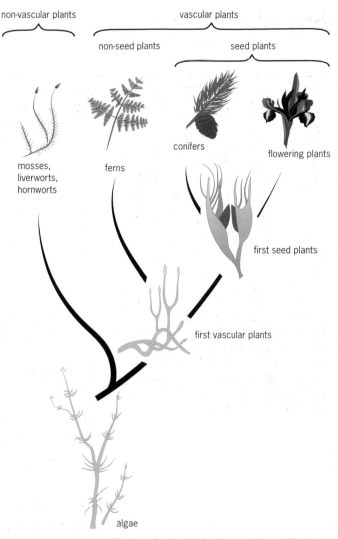

Fig 31.1 Overview of the plant kingdom. The algae gave rise to the plant kingdom. Overall, plant evolution can be seen as a process of overcoming the problems of living on land

Fig 31.2 Tree felling in Australia. Some of the largest trees ever recorded were giant eucalypts in the rainforests of Victoria, Australia. The largest one ever measured was 130 metres (435 feet), ironically measured when it was lying on the ground. These trees were some 30 metres (100 feet) taller than the current record holders, the Californian redwoods

- Some trees are over 100 metres tall (Fig 31.2) – how does water get to the top of tall trees?
- Plants make food in their leaves, but the food is needed in the roots, growing points, flowers, seeds and fruits. How does food get around a plant – which unlike an animal has no blood and no heart to pump it?

Flowering plants are *vascular* plants; they have a transport system. The veins that you can see in leaves and stems are specialised tissues called xylem and phloem that transport substances around the plant.

2 THE STRUCTURE OF A FLOWERING PLANT

Flowering plants, scientifically known as the **angiosperms,** can be further divided into two classes: the **monocotyledons** (**monocots** for short) and the **dicotyledons** (**dicots**). In this chapter we look at the dicots – the largest group of flowering plants. Most dicots are land plants and have specialised structures such as leaves, flowers, fruits and seeds, all supported by a stem. Fig 31.3 on the next page shows the basic structure of a dicot plant.

Generally, the stems of small plants are supported by the turgor of their tissues (see Fig 4.20), and the xylem tissue also provides some support. When water is in short supply the turgor is lost and the plant wilts. These are known as **herbaceous** plants and they tend to have short life cycles – usually only a few years at most. However, turgor cannot support plants over about 1 m tall. Larger plants have developed the supporting tissue we know as wood, which is basically layer upon layer of reinforced xylem that is no longer in use. Tress and shrubs tend to have longer life cycles and survive for many years, sometimes for centuries or even longer. Some trees are hundreds of years old.

For more of an overview of the plant kingdom, see Chapter 29.

3 THE ROOTS

The functions of the root system are to absorb water and minerals, provide the plants with anchorage and support, and often to store food reserves. Roots can also be organs of perennation – survival from one year to the next – and of reproduction. Generally, dicots have a **taproot** system consisting of one large central deep root and several lateral secondary roots (Fig 31.4).

Fig 31.4 Some plants have very deep root systems to gather water. In this dandelion the roots are far more extensive than the aerial parts of the plant

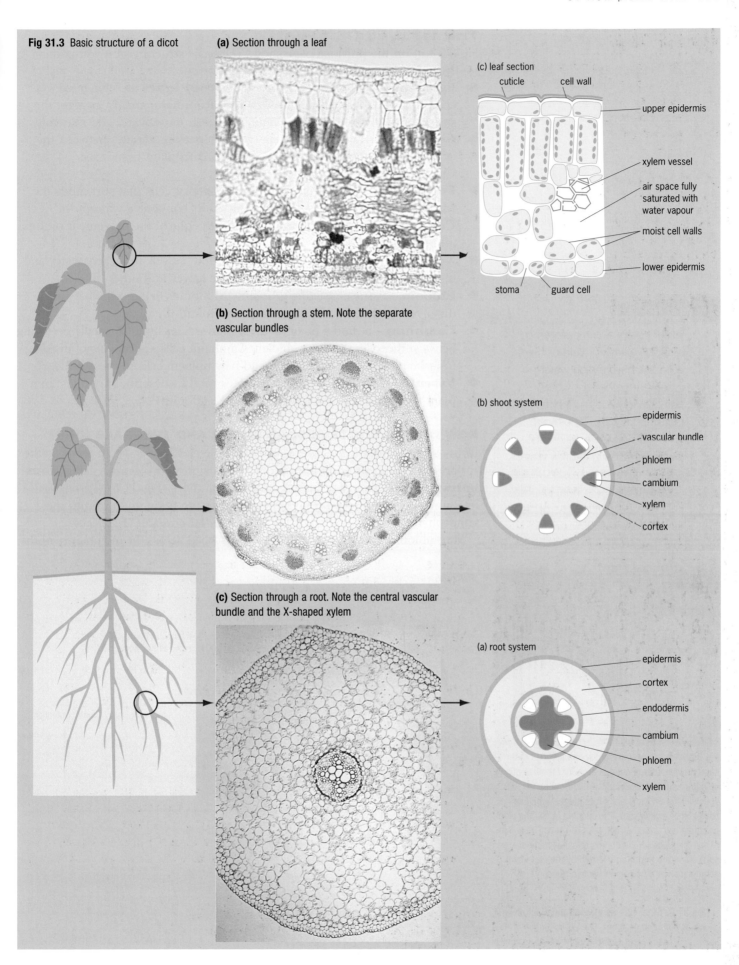

Fig 31.3 Basic structure of a dicot

(a) Section through a leaf

(c) leaf section

cuticle

cell wall

upper epidermis

xylem vessel

air space fully saturated with water vapour

moist cell walls

lower epidermis

stoma

guard cell

(b) Section through a stem. Note the separate vascular bundles

(b) shoot system

epidermis

vascular bundle

phloem

cambium

xylem

cortex

(c) Section through a root. Note the central vascular bundle and the X-shaped xylem

(a) root system

epidermis

cortex

endodermis

cambium

phloem

xylem

THE TISSUES OF A ROOT

Fig 31.3(a) on the previous page shows the basic structure of a young dicot root. The following tissues are of central importance:

- **Epidermis** – literally, the 'outer skin'. **Root hairs** are elongated epidermal cells that project into the soil. Root hairs greatly increase the surface area of the root, allowing it to absorb more water and minerals.
- **Cortex** – a layer of parenchyma (general 'packing tissue') between the epidermis and the vascular tissue. For more on plant tissues see Chapter 2.
- **Endodermis** – the 'inner skin'; a thin layer of cells that surrounds the vascular (conducting) tissue. The endodermis contains a waterproof **Casparian strip** that allows the plant to control the movement of ions into the xylem.
- **Xylem** tissue consists of many dead, hollow vessels that carry water and dissolved mineral ions up the plant to the leaves and other organs.
- **Phloem** – tissue consisting of mainly of tubular living cells that carry dissolved organic materials, eg sucrose, around the plant.
- **Cambium** – a third type of tissue in the vascular bundle, usually occurring between the xylem and the phloem. Cambium cells are capable of dividing to make new xylem to the inside and new phloem cells to the outside.
- **Vascular bundle** – a term sometimes used to describe the 'veins' in a plant that consist of xylem, phloem and cambium

HOW DO THE ROOTS ABSORB WATER AND MINERALS IONS?

Roots absorb water from the soil by osmosis (see Chapter 4). To increase the surface area for absorption, the epidermal cells grow projections called **root hairs** that push their way between the soil particles (Fig 31.5). Throughout its journey from the soil, through the roots, up the plant and out into the atmosphere, water *travels down a water potential gradient* (Fig 31.6).

Mineral ions dissolved in soil water are absorbed across the membranes of the

---

> ### ✔ REMEMBER THIS
>
> Water potential is measured on a negative scale. Dilute or 'weak' solutions have a high water potential, the highest water potential (a value of zero) being that of pure water. The more concentrated the solution, the lower the water potential. Water always moves from a high water potential to a lower one, just like heat moves from regions of high temperature to colder ones.

Fig 31.5 Micrograph of plant root hairs. The root hairs provide a large surface area for the absorption of water and ions. If you pull a plant out of the soil you will strip off most of its root hairs – a bit like having your legs waxed. If re-planted in soil the plant will often wilt until the root hairs re-grow. The action of soil microbes on the humus releases mineral ions such as nitrate that are absorbed by the roots by a combination of diffusion and active transport

Fig 31.6 Water passes through the plant along a water potential gradient. The WP is highest in the soil water and lowest in the atmosphere

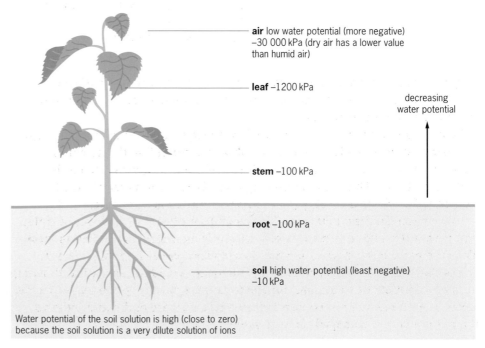

Water passes through the plant along a water potential gradient. The water potential is highest in the soil water and lowest in the atmosphere

air low water potential (more negative) −30 000 kPa (dry air has a lower value than humid air)

leaf −1200 kPa

decreasing water potential

stem −100 kPa

root −100 kPa

soil high water potential (least negative) −10 kPa

Water potential of the soil solution is high (close to zero) because the soil solution is a very dilute solution of ions

root hairs by a combination of facilitated diffusion and active transport. Research has shown that plants absorb some mineral ions by a **proton pump**. The root cells use ATP to secrete H^+ ions into the soil. This results in a slight negative charge inside the root cell that aids the absorption of positive ions such as sodium and potassium. Negative ions such as chloride are absorbed by a **symport** mechanism in which the protons, at higher concentration in the soil, diffuse back into the root cell. The H^+ ions and the chloride ions are absorbed by the same protein at the same time, hence the name 'symport'.

Once absorbed, the ions lower the water potential of the cell cytoplasm, so water passes into the root hairs from the soil by osmosis. Fig 31.7 shows the pathway taken by water from soil, through the root tissue (the cortex) to the xylem in the centre. There are two different pathways that water can take:

● Through the cytoplasm of cells – the **symplast pathway**.
● Between the cells – through the cell walls and spaces around cells – the **apoplast pathway**. There is less resistance to the flow of water through this route.

SEE QUESTIONS 1 AND 2

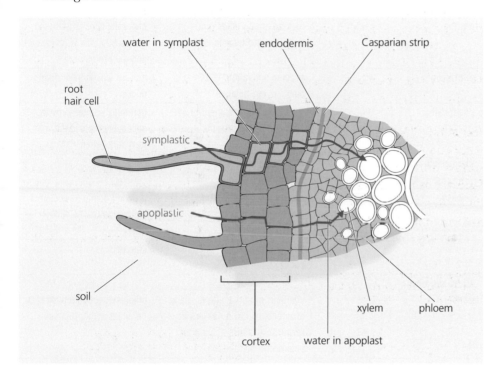

Fig 31.7(a) The pathways taken by water from the soil to the xylem in the centre of the root

Fig 31.7(b) Plants cells are joined by thin threads of cytoplasm called plasmodesmata, which project through holes in the cell wall. These cytoplasmic strands allow materials to pass from cell to cell and are an important part of the symplast pathway

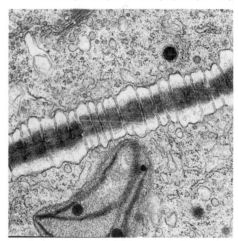

The key difference between the two routes is that the symplast pathway involves living tissue while the apoplast route is non-living; it involves the spaces between the living cells.

Before water and minerals can pass into the xylem they must pass through the **endodermis**, which is a thin membrane surrounding the xylem. A waterproof layer – the **Casparian strip** – fills all the gaps between the endodermal cells. This is made from a waxy, waterproof material called **suberin**. The Casparian strip prevents water and ions from entering the xylem via the apoplast pathway. This is significant because, with the symplast pathway, water and ions must pass through living material and membranes. Thus the plant has a degree of control over what passes into the xylem. It seems that without the Casparian strip the plant would act as a wick, drawing up large amounts of water and mineral ions from the soil to the leaves. As the water evaporated, the ion concentration would increase and could build to the point where it damaged the leaf tissue.

4 WHAT MINERALS DO PLANTS NEED?

A wide variety of dead organic matter accumulates in soil from leaves, faeces, urine, the bodies of animals and so on. The action of decomposers – bacteria and fungi – on this organic matter releases mineral ions into the soil. Many of these are essential to plants, either as components of compounds or as participants in chemical reactions. Some vital minerals are shown in Table 31.1. This 'mush' of dead organic matter and decomposers is called **humus** and is a vital component of soil. There is more about soil in Chapter 35.

Table 31.1 Some vital plant minerals obtained from the soil			
Mineral ion	**Form absorbed by plant**	**Needed for**	**Symptoms of deficiency**
Macronutrients			
nitrate	(NO_3^-)	making amino acids (and proteins) and nucleic acids	generally poor growth; old leaves turn yellow (chlorosis) and die prematurely
potassium	K^+	enzyme activation, water balance, ion balance	mature leaves have dead edges
phosphate	PO_4^{3-}	an essential component of ATP, DNA and some proteins	generally stunted growth, especially roots, plant may turn dark green
magnesium	Mg^{2+}	a central component of chlorophyll	chlorosis
calcium	Ca^{2+}	affects the cytoskeleton, membrane proteins and many enzymes	growing points (meristems) die, young leaves yellow
sulphur	SO_4^{2-}	part of some proteins and coenzymes	chlorosis
Micronutrients			
iron	Fe^{3+}	vital part of active site of many enzymes, electron carriers in respiratory chain, needed for chlorophyll synthesis	chlorosis; young leaves may be white with green veins
chlorine	Cl^-	photosynthesis, ion balance	
manganese	Mn^{2+}	enzyme activator	young leaves pale with dead patches
zinc	Zn^{2+}	enzyme activation, auxin synthesis	mature leaves have many dead spots
copper	Cu^{2+}	active site of many enzymes and electron carriers	shoots fail to develop
molybdenum	MoO_4^{3+}	nitrogen fixation	slow growth

Fig 31.8 Identifying essential nutrients is usually done using hydroponic culture in which soil is replaced by a nutrient solution. In theory the experiment is simple but it must be strictly controlled because some nutrients are needed in only tiny amounts and these may be present as contaminants

It is relatively easy to investigate which minerals are essential to a particular plant species. A batch of seedlings is grown in a solution containing all known nutrients and then transferred to a solution lacking in one particular nutrient. If growth continues as normal, the nutrient can be assumed to be non-essential. However, if growth is abnormal in some way (see Table 31.1) the nutrient is assumed to be essential. A common symptom of deficiency is a yellowing of the leaves called **chlorosis**.

A vital nutrient for all plants is nitrogen. Although this element is abundant as a gas – N_2 forms 80 per cent of the atmosphere – it is in an inert form that cannot generally be used by plants. Nitrogen-fixing bacteria can turn N_2 gas into soluble nitrate (NO_3^-). These bacteria live free in the soil but some plants – the legumes – have developed root nodules that contain nitrogen-fixing bacteria. Find out more about this on page 658.

! SCIENCE FEATURE Mycorrhizas

Over three-quarters of known plant species, and virtually all tree species, have a partner in the soil; a fungus. This association is known as a **mycorrhiza** and is an example of a **mutualistic symbiosis**; an arrangement in which both partners benefit. The fungus receives sugars from photosynthesis and possibly some B-group vitamins while the plant benefits from greatly increased absorption of mineral ions and water.

Young plants become infected with the fungus at an early age and from then on root hair production either stops or is greatly reduced. However, the fungal threads form a network throughout the surrounding soil, providing a much larger surface area than could be achieved with the root hairs alone.

There are two types of mycorrhizae. **Ectotrophic mycorrhizas** surround the root and penetrate the air spaces in between the cells of the cortex, but they do not penetrate the actual cells. This type is common in forest trees such as oak and beech. In contrast, the threads of **endotrophic mycorrhizas** penetrate the walls of the cortical cells. Endotrophic mycorrhizas are the commoner of the two types, being found in the roots of most plant species.

Mycorrhizas have great significance in plant growth and therefore in agriculture, though our knowledge of their biology is still very limited.

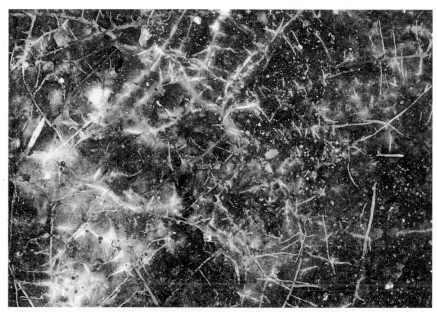

Fig 31.9 Mycorrhizal fungi growing in association with the roots of a small-leaved lime tree

Fig 31.10 Many tree species have mycorrhizas that form a sheath around the roots. These fungi, members of the basidomycete family, reproduce by means of fruiting bodies, commonly known as mushrooms. This is why mushrooms are often seen close to tree trunks

Fig 31.11 Xylem! The stringy bits in celery are xylem vessels – a great source of dietary fibre

SEE QUESTIONS 3 AND 4

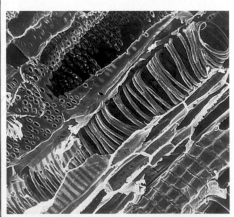

Fig 31.12 Coloured scanning electron micrograph (SEM) of a longitudinal section through xylem vessels in a plant stem. Xylem tissue is responsible for the upward transport of water in the plant, from the roots into the stem and leaves. As seen here (at centre) these elongated vessels may be reinforced and strengthened with spiral bands. Alternatively, they may be perforated with several minute pores (as at lower right) which allow water to diffuse into surrounding tissue. In mature woody plants, xylem makes up the bulk of vascular tissue in the stems and branches, and is a major structural component in the plant

5 XYLEM AND THE TRANSPIRATION STREAM

Transpiration involves vast amounts of water passing through plants and out into the atmosphere. In one investigation in the USA a 15 m maple tree was estimated to have some 177 000 leaves with a total surface area of 675 square metres. On a summer day, the tree lost 220 litres *per hour* to the atmosphere through the leaves. The roots, of course, needed to provide 220 litres per hour to prevent wilting. To understand how all this water gets to the tops of tall trees we need to look at xylem tissue.

XYLEM TISSUE – A RIVER RUNS THROUGH IT

Xylem tissue (Fig 31.11) is a specialised conducting tissue. Ironically, it functions only when it is dead.

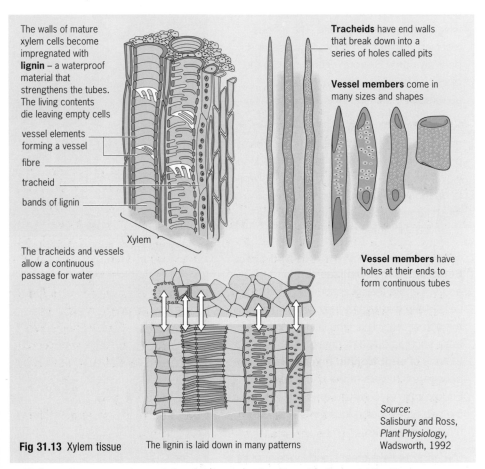

The walls of mature xylem cells become impregnated with **lignin** – a waterproof material that strengthens the tubes. The living contents die leaving empty cells

vessel elements forming a vessel

fibre

tracheid

bands of lignin

Xylem

The tracheids and vessels allow a continuous passage for water

Tracheids have end walls that break down into a series of holes called pits

Vessel members come in many sizes and shapes

Vessel members have holes at their ends to form continuous tubes

The lignin is laid down in many patterns

Source: Salisbury and Ross, *Plant Physiology*, Wadsworth, 1992

Fig 31.13 Xylem tissue

Xylem tissue consists of dead, elongated cells called **vessel elements,** connected end to end to form a network of conducting tubes (Fig 31.12 and Fig 31.13). The development of xylem vessels is shown in Fig 31.14. A second type of xylem vessel – **tracheids** – also form part of the conducting tissue. In evolutionary terns, tracheids are older and less specialised. They differ from vessel elements by being narrower and have having numerous pits (holes) in their side walls that allow sideways movement of fluid.

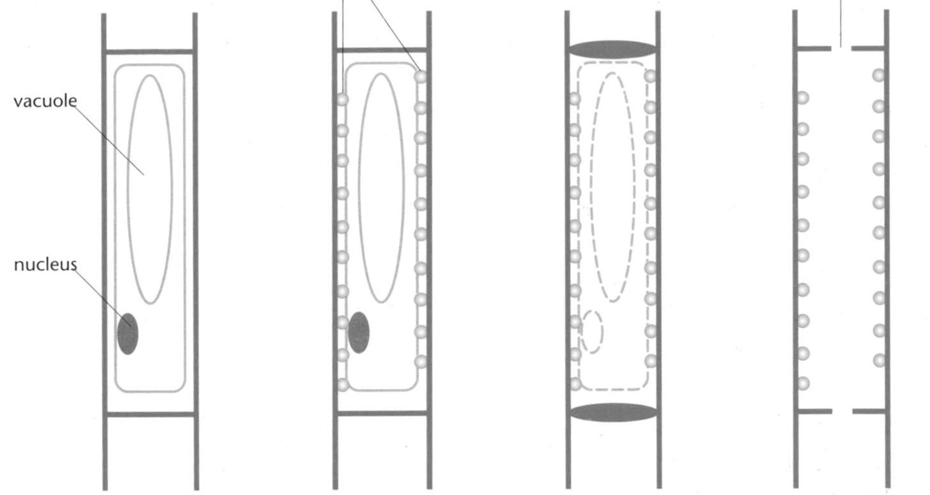

Fig 31.14 Where does xylem come from?

Xylem tissue forms from
unspecialised plant cells

secondary cell wall

perforation plate

vacuole

nucleus

The cells elongate,
becoming cylindrical.
At this stage they only
have a **primary cell wall**
which is relatively thin
and flexible

The cell wall becomes
strengthened by
deposits of cellulose
and lignin around
the side walls – this is
the **secondary cell wall**

The cytoplasm and
nucleus break down,
while the end walls
thicken

A mature cell forms,
with a perforation
plate at each end

So how does water get all the way to the top of trees? Is it pulled or is it pushed? The simple answer is that it's pulled, but experiments have shown that three forces combine to draw water up the plant in the xylem vessels.

- **Capillarity**. Xylem vessels are very thin: 20 μm to 400 μm. When a cut stem is placed in water, fluid is drawn up to about 1 metre – a significant help in smaller plants but nothing like enough to get water to the top of a tall tree.
- **Root pressure**. If the roots created a significant force, fluid would spurt out if the stem was cut. This is not the case, but root pressure can be demonstrated and is a contributory factor (see Fig 31.15d on page 579). Root pressure is created when the cells of the endodermis actively secrete mineral ions into the xylem. This creates a water potential gradient and water follows by osmosis, creating root pressure.
- **Cohesion–tension.** Even when the stem is cut, ruling out the contribution from the two forces above, water still reaches the top. This third force therefore needs looking at in more detail...

THE COHESION–TENSION HYPOTHESIS

When you suck up a drink with a straw, you create a negative pressure with your mouth and a continuous stream of fluid is pulled up through the straw. The fluid in the straw is under negative pressure so if you pierce the straw with a needle air will be sucked in rather then fluid spurting out. Once air is inside the straw, the column is broken and the straw doesn't work.

It's much the same with xylem vessels. There is usually a negative pressure inside the xylem vessels of the stem, because water is *pulled* up the plant from the leaves (Fig 31.6). This mechanism of transpiration is known as the **cohesion–tension hypothesis**.

Water molecules are polar and therefore cohesive ('sticky') – they cling to each other with such force that an enclosed column of water has great tensile strength – it will not break or separate into droplets except under huge tension. The cohesion–tension hypothesis states that as water evaporates from

SEE QUESTION 5

A In the nineteenth century the famous scientist Eduard Strasburger performed a classic experiment. He sawed through a 20 m tree at its base and stuck the cut end into a poisonous solution of picric acid (he must have been a strong man). The passage of the poison up the trunk could be seen by the progressive death of the bark. When the solution reached the leaves they too died, and only then did the upward movement stop – no more poison was drawn from the bucket. What conclusions could you draw from these observations?

SEE QUESTION 6

the leaves, it creates a negative pressure in the xylem that pulls a continuous column of water up from the roots. Water molecules form a continuous column from the surface of the mesophyll cells and down the xylem. Interestingly, as the pull on the column of water increases, the xylem vessels get narrower. The circumference of trees shows a measurable decrease during the day when the plant is actively transpiring.

In plant cells the primary cell wall is a relatively thin layer of cellulose. In xylem cells the secondary cell wall is a complex structure made from two polymers – cellulose and lignin, with some protein. The cellulose forms strong fibres while the lignin forms an interlocking network in between the cellulose – like reinforced concrete. The lignin also makes the xylem vessels impermeable to water – obviously a key feature in a conducting tissue.

Secondary cell walls can be laid down in a variety of patterns; most commonly annular rings (hoops) or spirals. The significance of this pattern is that it prevents collapse inwards, while allowing elongation. In older xylem vessels the pattern is more elaborate, running in all directions, and prevents any further elongation.

THE LEAF, STOMATA AND WATER LOSS

In most plants the leaf is an organ of photosynthesis, and the ways in which it is adapted to maximise this process are covered in Chapter 9. In order to photosynthesise effectively, a leaf must also be an organ of gas exchange. Consequently, it has a large surface area and must be permeable to gases. To aid gas exchange the leaves have many holes – **stomata** – that allow carbon dioxide to diffuse efficiently to the photosynthesising cells, and allow the waste oxygen to leave. The loosely packed mesophyll cells, with their wet cell walls, inevitably lose a lot of water to the environment through the stomata. To minimise water loss the upper epidermis of most plant species secretes a thin layer of wax: the **cuticle**.

Stomata are pores that in most species, but not all, are more numerous on the underside of the leaf than on the top. Each pore, or stoma, is surrounded by two specialised epidermal cells, the **guard cells**, that can alter their shape to open or close the hole. The guard cells are banana shaped, attached at the top and bottom but not along their length. The cell wall of guard cells has radially arranged microfibrils (Fig 31.15b) so when these cells absorb water they cannot swell but can elongate and bend. As they do so, they open the stomata.

The basic mechanism of stomatal opening is in four stages:
- When it gets light, photosynthesis begins, producing ATP.
- ATP powers an active transport mechanism in the guard cell membranes – K^+ ions are pumped into the guard cell.
- This lowers the water potential of the cell, so water moves in by osmosis.
- The turgidity of the guard cell increases, so the cells bend, opening the stoma.

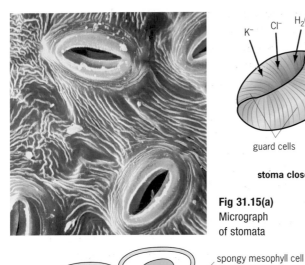

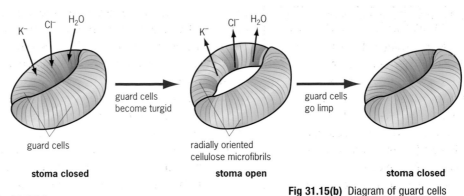

Fig 31.15(a) Micrograph of stomata

Fig 31.15(b) Diagram of guard cells

spongy mesophyll cell

cell wall

cytoplasm

vacuole

air space

epidermal cell

diffusion shell of humid air develops in still conditions

guard cell

cuticle

Fig 31.15(c) in still air, a hemisphere of humid air diffuses out of each stomata. When there are air movements the diffusion shell is blown away and replaced by drier air. This mantains the water potential gradient and increases the rate of transpiration.

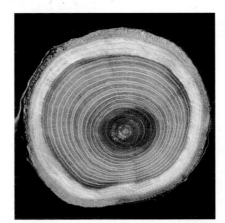

Fig 31.15(d) Guttation. When there is plenty of water in the soil and the air is saturated with water vapour, evaporation is difficult and so the water that arrives at the leaves flows out of the stomata and forms droplets. This phenomenon is evidence for the importance of root pressure because it is seen only in smaller plants

---

! SCIENCE FEATURE Xylem, wood and tree growth

In order for plants to win the race for that all-important light, they had to develop taller and stronger stems. As a result xylem vessels took on a second function – that of support.

Xylem vessels contribute greatly to the familiar substance we call wood – a tissue that develops from a process known as **secondary growth**. Put simply, the cambium cells divide year after year to produce a new layer of xylem vessels outside the old ones (Fig 31.16). Thus every year the trunk and branches of trees and shrubs get thicker as the disused conducting tissue accumulates inside.

It is widely known that you can tell the age of a tree by counting its annual rings. These are a result of growth at different rates during the different seasons. In the spring, when water is plentiful, the new vessel elements and tracheids tend to be large in diameter but with thin walls. As water becomes scarcer during the summer, the vessels are narrower with thicker walls. Thus summer wood is denser and darker than spring wood. The annual rings are much less defined in the wet areas of tropics, ie rainforests, where water supply is relatively constant.

Interestingly, trunks and branches become thicker in response to stress on the tree. Following an observation that trees grown indoors were more slender then their outdoor counterparts, it was found that the stresses and strains caused by wind cause the tree to become more sturdy. These conditions can be reproduced indoors by shaking a tree or even beating the trunk with a mallet.

Fig 31.16 In this section through a *Laburnum* tree trunk, the active xylem (known as the *sapwood*) is the outermost ring, just under the bark, while the phloem forms the inside of the bark. The xylem from previous years is now dead and just has a supporting role. How old do you think this tree is?

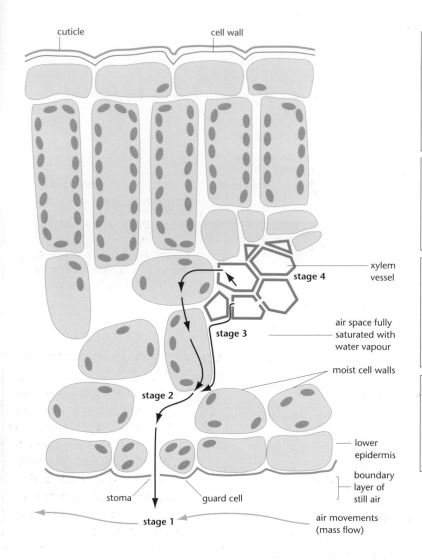

cuticle

cell wall

stage 4

stage 3

stage 2

stage 1

stoma

guard cell

xylem vessel

air space fully saturated with water vapour

moist cell walls

lower epidermis

boundary layer of still air

air movements (mass flow)

Stage 1 Mass flow in air

Wind movements (mass flow) take air, fully saturated with water vapour, away from the leaf surface

The air replacing it contains less water vapour and this maintains a concentration gradient for water vapour leaving the air spaces of a leaf

Stage 2 Diffusion of water in still air

Water vapour diffuses down a concentration gradient through the leaf air spaces, through open stomata, and through the boundary layer of still air on the outside of the leaf

Stage 3 Diffusion of water through cells

Water loss from the surface of cells lowers the water potential inside the cells

Water at a higher potential diffuses from the nearest xylem vessel through leaf cells to replace lost water

Stage 4 Mass flow of water in xylem vessels

Pressure in xylem vessels is lowered as water leaves them

Water moves up the xylem vessel from the roots where the pressure is higher

Fig 31.17 The flow of water through a leaf. Wind movements take away humid air and replace it with drier air, thus maintaining a water potential gradient

WHAT AFFECTS THE RATE OF TRANSPIRATION?

The factors affecting the rate of transpiration are generally those that increase evaporation. The measurement of transpiration rate is covered in the Assignment at the end of the chapter. The commonest factors are:

● **Temperature** – the higher the temperature, the faster water molecules will evaporate.

● **Humidity** (the amount of water vapour in the air) – for water to evaporate, there must be a water potential gradient between the air in the leaf and the atmosphere. Dry air has a very low water potential. The more humid it is, the higher the water potential gradient and the lower the rate of transpiration.

● **Air movement** (ie wind speed) – wind will blow away the small pockets of humid air that develop when the air is still (Fig 31.15c and Fig 31.17). Humid air is replaced by drier air, so the water potential gradient is maintained.

Also, **light** affects transpiration, but more indirectly than the above three factors. In general, light causes stomata to open, greatly increasing the passage of water vapour out of the plant.

✔ **REMEMBER THIS**

The conditions that will increase the speed of drying of clothes on a washing line – warm, dry, windy – will also speed up transpiration.

6 TRANSLOCATION AND PHLOEM

The leaves make most of organic molecules synthesised in a plant, but there are many other areas of the plant that need this 'food' – the roots, the flowers, the seeds and fruits, etc (Fig 31.18); in fact, all the non-green parts. In this section we look at the way in which the products of photosynthesis are moved around the plant.

Fig 31.18 Good food, but how did it get there? The organic molecules contained in this apple were made in the leaves of the apple tree and transported into this fruit by translocation. Fruits like this are 'bribes' to animals to eat the flesh and disperse the seeds (see Chapter 33)

BASIC DEFINITIONS

- **Source** – the parts of a plant that make or store food. The commonest sources are the leaves but other parts that contain chlorophyll, eg the stem, can be a source as can areas of stored food such as tubers.
- **Sink** – the parts of the plant that need food – notably the roots, flowers, fruits and growing points (meristems).
- **Translocation** – the movement of materials, such as sucrose, from sources to sinks via the phloem tissue.

PHLOEM TISSUE – CHAINED TO THE SINK

Phloem tissue is conducting tissue and is similar to xylem in several ways; it is a network of elongated, tubular cells that forms a transport system throughout the plant – xylem and phloem usually occur together. However, phloem tissue is different from xylem in several important ways:

- Its cells are alive – remember that xylem only works when it is dead.
- It moves products of photosynthesis, notably sucrose and amino acids, while xylem just moves water and dissolved minerals.
- Movement in phloem occurs both up and down the stem. Movement in xylem is in one direction only – from the roots to the leaves.
- Fluid in phloem is usually under positive pressure – sap will ooze out if phloem is pierced. Fluid in xylem is usually under negative pressure so air will enter if vessels are pierced.
- Phloem sieve tubes, unlike xylem vessels, do not have lignified walls and are permeable to water and solutes.

As Fig 31.19 shows, phloem tissue is made up from two different types of cells: **sieve elements** and **companion cells**. Sieve elements are joined end to end by a sieve plate (Fig 31.19c) that allows movement of substances through in a continuous stream. Sieve elements have some living cytoplasm around the outside of the cell, but notably there is no nucleus or vacuole. The companion cell in contrast does have a nucleus, is smaller and metabolically very active (also Fig 31.19c). Experiments have shown that the two types of cell are very closely linked. They are products of the same cell division and share many cytoplasmic connections. It seems that the companion cell maintains the sieve element once it has fully differentiated and lost its nucleus. The association is so close that if the sieve element dies, so does the companion cell. The relationship between the two cells is poorly understood, but it seems that the companion cell provides the sieve plate with mRNA and some essential proteins.

REMEMBER THIS

We refer to the organic molecules made by plants in photosynthesis as 'food' because it is simpler than writing 'photosynthetic product' or 'photosynthate'. Strictly speaking (and for exams) the word 'food' should only be used to describe the material eaten by animals and taken into their gut.

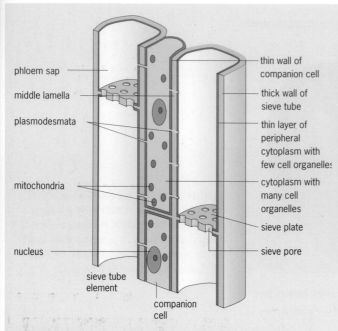

phloem sap

middle lamella

plasmodesmata

mitochondria

nucleus

sieve tube element

companion cell

thin wall of companion cell

thick wall of sieve tube

thin layer of peripheral cytoplasm with few cell organelles

cytoplasm with many cell organelles

sieve plate

sieve pore

Fig 31.19(a) The cells of phloem tissue are living, unlike those of the xylem. Most of the transport of substances happens through the *sieve tubes*, which have cytoplasm but no nucleus. These are elongated cells, 150 to 1000 μm (= 1 mm) long and 10–15 μm in diameter. The companion cells are smaller but metabolically active cells that possess a nucleus, dense cytoplasm and many mitochondria

Fig 31.19 The structure of phloem tissue

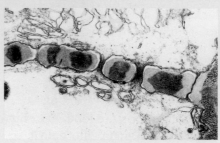

Fig 31.19(b) Two sieve plates can clearly be seen in this false-colour electron micrograph (×1400)

Fig 31.19(c) Transmitting electron micrograph showing the sieve plate between two sieve elements

sieve plate

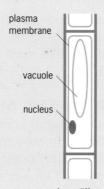

plasma membrane

vacuole

nucleus

An undifferentiated cell elongates

Some of the cytoplasm breaks down, including the nucleus and vacuole, while some endoplasmic reticulum and mitochondria remain

Sieve plates develop and the cytoplasm of adjacent cells join through the pores

Fig 31.19(d) Formation of sieve tubes

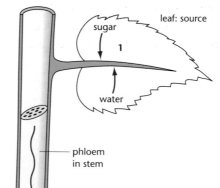

leaf: source

sugar

1

water

phloem in stem

3

sugar

2

water

tuber: example of a sink

Fig 31.20 Mass flow theorem

1 The leaf (the source) makes sucrose and pumps it into the phloem vessels by active transport. The sugar travels via the symplast route through mesophyll and companion cells. This lowers the water potential of the phloem sap in sieve tubes, so water passes into phloem vessels from surrounding tissues. This extra fluid creates hydrostatic pressure in the phloem vessels at the sources

2 The opposite happens in the sinks: sugars are transported out of the phloem, mainly by facilitated diffusion, and into the surrounding cells. This raises the water potential of the phloem sap, so that water follows the sugar into the surrounding cells. This creates lower pressure in the phloem

3 Overall, the higher hydrostatic pressure at the sources and the lower hydrostatic pressure at the sink cause fluid in phloem to move from sources to sink. When the sink is in the roots, water can be recycled back to the sources via the xylem

THE MECHANISM OF TRANSLOCATION

Transport of material in the phloem is rapid; usually 40–250 cm per hour, and can take place up or down the stem. How does fluid move in phloem vessels, in the absence of a heart of other organ to create pressure? One widely accepted theory is the **mass flow hypothesis** or **pressure flow hypothesis**, first proposed by E. Munch in 1930. This is represented by Fig 31.20.

EXPERIMENTAL EVIDENCE FOR TRANSLOCATION

There are several different pieces of evidence to suggest that solutes made in the leaves are distributed to the rest of the plant via the phloem.

RINGING EXPERIMENTS

Phloem tissue of woody plants is situated on the underneath of the bark. If a ring of bark is removed (Fig 31.21), sugar will accumulate and cause a swelling above the ring, suggesting that sugar travels down the stem from the leaves to the roots.

A similar thing happens if steam is applied in a ring around the stem. The living phloem cells are killed by the steam and so translocation stops, just as it does in the ringing experiment. The flow of water in the xylem vessels is unchanged, because the dead xylem vessels are unaffected by the steam.

USING RADIOACTIVE TRACERS

If a particular leaf is given radioactive carbon dioxide (containing ^{14}C), the isotope is incorporated into the sugars. After a while the stem is sectioned and X-rayed to show up the radioactivity, the sugar shows up in exactly the pattern that corresponds to the phloem tissue. This technique will also show that sugars made in the leaves eventually accumulate in the growing points, the new leaves, and other sinks around the plant.

USING APHIDS

It is often difficult to get a pure sample of phloem sap to analyse, but aphids such as greenfly can locate and pierce phloem cells with great reliability and precision (Fig 31.22). If the aphid is then removed, leaving the stylet (needle-like mouthpart) in place, pure phloem sap oozes from the cut end. Analysis of this fluid shows it to contain mainly sucrose with some amino acids and proteins.

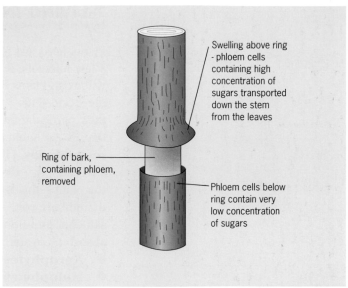

Swelling above ring - phloem cells containing high concentration of sugars transported down the stem from the leaves

Ring of bark, containing phloem, removed

Phloem cells below ring contain very low concentration of sugars

Fig 31.21 Ringing experiment

SEE QUESTION 7

Fig 31.22 The aphids – like this greenfly – can tap into the phloem with great accuracy. Once they pierce the phloem, the sap flows – there is no need to suck, and the aphids get a supply of sugar. A large amount of sap must pass through the aphid in order to extract enough amino acids, so they egest a large amount of sugary water from their anus. This sticky waste is called honeydew

7 LIVING IN EXTREMES

Plants need just four things in order to survive: water, light, a little warmth and a few minerals. If all four are available – even for a few days a year – plant life will exist. Plants have come to colonise many of the habitats of the world (Figs 31.23 and 31.24) and, like animals, have become adapted for their particular environment. So far in this chapter we have discussed plants that live in habitats where water is plentiful – such plants are known as **mesophytes**. However, there are many areas of the world where water is in short supply – **drought** – or where water is abundant but it is not available to the plant – this is **physiological drought**. Two examples of physiological draught are cold – where most of the water exists as ice – or in regions of high salt concentration, where the water potential makes it difficult for the plant to absorb the water. In this section we take a brief look at three plant types:

● **Xerophytes** – plants adapted to arid regions.
● **Halophytes** – plants that can tolerate high salt concentrations.
● **Hydrophytes** – plants that live submerged or partially submerged in water.

SEE QUESTIONS 8 AND 9

XEROPHYTES – CONSERVING WATER IN DRY HABITATS

A xerophyte is a plant that is adapted to reduce water loss. The classic example is the cactus but many other plants show xerophytic adaptations. When water is in short supply the transpiration rate seen in mesophytes cannot be supported. Xerophytes show a range of adaptations to minimise water loss while still being able to photosynthesise. These are summarised in Table 31.2.

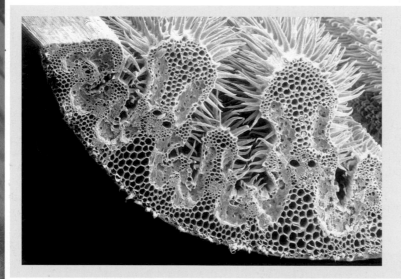

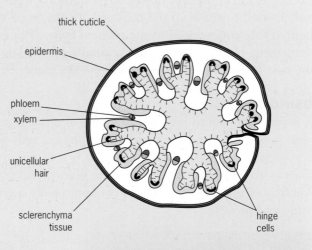

thick cuticle
epidermis
phloem
xylem
unicellular hair
sclerenchyma tissue
hinge cells

Fig 31.23 Marram grass usually grows in exposed sandy areas where it is an important coloniser (see Chapter 33). The tangled roots of marram allow it to bind the shifting sand, and the leaf is a masterpiece of water conservation. The leaf blade is hinged and the more arid the conditions, the more the leaf rolls up

Fig 31.24 Eucalyptus leaves hang vertically so that their flat surfaces get the morning or late afternoon sun, but avoid the scorching rays of the midday sun overhead

REMEMBER THIS

In an exam you may well be given an unfamiliar species of xerophyte. You should be able to work out how it reduces water loss by applying your knowledge of transpiration.

Table 31.2 Adaptations of xerophytic plants	
Adaptation	**How it works**
Swollen stem (eg cacti)	low surface area to volume ratio; acts as a reservoir for water
Leaves reduced to spines (cacti and other succulent plants)	small surface area to reduce water loss; photosynthesis carried out by stem; spines protect stem
Thick cuticle (most xerophytes)	waxy layer is impermeable to water – less evaporation
Low growth form	some plants, even trees, can grow horizontally rather than vertically to avoid the worst of the wind
Leaf rolled inwards, eg marram grass	stomata on inside – humid air trapped inside where it cannot be blown away
Sunken stomata (in pits called **crypts**), eg marram grass	as above, humid air is not blown away
Reduced number of stomata (most xerophytes)	reduced transpiration
Succulence – deep, fleshy leaves	leaves store water
Deep root systems	allow plant to reach water table that may be many metres below; many xerophytes have a high root-to-shoot ratio
Hairs (many cacti and marram grass)	reduce the air movement around the stomata, also protects from cold
CAM photosynthesis	plants open stomata at night, absorb carbon dioxide and store it as an organic acid; then they close the stomata during the day and use the stored carbon for photosynthesis (see Chapter 9)
C4 photosynthesis	makes photosynthesis more efficient by reducing photorespiration (see Chapter 9); makes more efficient use of carbon dioxide and water
Short life cycles	plant can grow, flower and make seeds quickly when water is abundant – dormant seeds can survive the drought; these plants may not show other xerophytic adaptations
Accumulate proline in cell vacuoles	this amino acid lowers the water potential so that more water can be absorbed from soil
Leaves hang vertically, eg eucalyptus (Fig 31.25)	minimises evaporation and overheating by midday sun – photosynthesis mainly morning/later afternoon

HALOPHYTES

Globally, there is no substance that limits plant growth as much as salt – see the chapter opener. Halophytes (literally 'salt plants') have evolved in regions of high salt concentration. There are many different habitats that have problems with salt; ranging from cold, moist **estuaries** and **salt marshes** to hot, dry, salty **deserts**. Human activity has done a lot to increase the problems of salt – irrigation of dry areas can flush salt to the surface – see the opener photo.

Salty environments are areas of physiological drought – water is scarce just as it is in an arid environment. Consequently, many halophytes share adaptations with xerophytes (Table 31.3). The big problem with salt is that it lowers the water potential of the soil so much that water tends to be drawn out of a non-specialised plant. In order to survive, many halophytes species absorb salt (and therefore water), tolerating high concentrations or excreting the salt through the leaves (Fig 31.25).

Fig 31.25 Mangrove plants are important halophytes. This salty mangrove species has glands on its leaves that secrete salt, which appears as white crystals on the leaves

Table 31.3 Adaptations of halophytes	
Adaptation	**How it works**
Cells accumulate salt	lower water potential in roots so that water can be absorbed from salty water
Salt glands in leaves (Fig 31.26)	glands in the leaves secrete excess salt
Accumulate proline in cell vacuoles	this amino acid lowers the water potential so that water can still be absorbed (also seen in xerophytes)
Many xerophytic adaptations	see Table 31.2

HYDROPHYTES

Hydrophytes are flowering plants adapted for life in freshwater. There are very few flowering plants that can survive in the marine environment – the seaweeds are not true plants. Life in water has the advantage of extra support and the obvious fact that water is not in short supply. However, gas exchange is more difficult – there is less oxygen for root respiration – and all but the clearest water can block out a significant amount of available light. Adaptations of hydrophytes are listed in Table 31.4.

Table 31.4 Adaptations of hydrophytes	
Adaptation	**How it works**
Aerenchyma	air spaces in parenchyma tissue conduct oxygen down to submerged and anaerobic areas of plant; also helps flotation
Flexible stems and leaves	plant can bend with the flow of the water; water supports the plant – no need for strengthening tissue that would make the plant more 'brittle'
Breathing roots	aerial roots project up into the air to get oxygen
Reduced roots	no need for the roots to get water and mineral ions
Finely divided underwater leaves	less resistance to the flow of water over the leaf

8 FEED THE WORLD: ADAPTATIONS OF CEREAL CROPS

It may surprise you that over half the world's food supply come from grass. Not the green stuff that cows rely on, but grass *seeds* – rice, wheat and maize. Cereals are grasses that are cultivated for their seeds. There are about a quarter of a million plants species known to science but these three provide us with more food than the rest of them put together (Table 31.5).

The development of agriculture was a crucial step in the cultural evolution of humans. It marked the transition from a hunter–gatherer way of life to more permanent settlements, in which humans grew food plants and kept animals in paddocks. Cereals were among the first cultivated plants, and their selective breeding is covered in Chapter 33.

Table 31.5 The major crops of the world – by area harvested in 1999		
Crop type	**Crop**	**Area harvested (millions of hectares)**
Cereals	wheat	215
	rice	155
	maize	139
	sorghum	43
	millet	36
	oats	13
	rye	10
Non-cereals	soya	72
	sugar cane	19
	potatoes	18

(Source: Food and Agriculture Organisation (FAO) website)

REMEMBER THIS

A hectare is the metric unit of area equal to 2.47 acres; a bit larger than a football pitch. 100 hectares = 1 sq kilometre. A million hectares is a lot of land!

WHAT'S SO SPECIAL ABOUT CEREALS?

Cereal grains have a number of advantages over other plant foods:

● Cereals have a lower water content than most fruit and vegetables and are less water-demanding during growth. Thus they can be grown in more arid environments. Rice is a notable exception.

● The low water content means that cereals are easier to transport and store – they have a far longer shelf life than other plant products.

● They are very productive – yielding more food per area of land than other crops. Throughout the world, the most productive crop a farmer can grow is likely to be a cereal.

● They have a high protein content, 7–15 per cent.

In this section we look at each of the three staples; rice, wheat and maize.

RICE

Most of the world's rice is grown in waterlogged paddy fields (Fig 31.26). Rice-growing regions are hot with a heavy seasonal rainfall. The rice plant grows in heavy, silty soils that drain poorly. Consequently, rice has evolved to cope with being semi-submerged in an oxygen-poor environment.

Rice survives in waterlogged soils because its roots are able to respire anaerobically, and to tolerate the high ethanol levels that result. A specialised tissue, aerenchyma (Fig 31.27), allows oxygen to penetrate down into the roots.

Fig 31.26 Rice seeds are unique in being able to germinate is anaerobic conditions

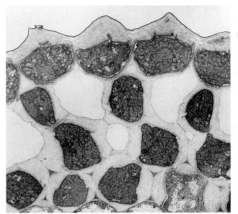

Fig 31.27 Aerenchyma showing air spaces

An advantage of waterlogged anaerobic soil is that **cyanobacteria** fix nitrogen so less fertiliser is needed.

WHEAT

Wheat is a grass that has been cultivated since the earliest times of human settlement. It is grown all over the world in temperate regions. It needs quite a lot of water but is able to tolerate frost. Over the centuries wheat has been selectively bred and has undergone dramatic genetic changes so that the modern varieties are unrecognisable from their ancestors. Find out more in Chapter 33.

Fig 31.28 Modern variety of wheat

Fig 31.29 Maize needs less water than other cereal crops and thrives in bright sunshine

MAIZE

Maize is a tropical crop that thrives in hot, arid conditions. It shows many xerophytic adaptations (Table 31.2) including C4 photosynthesis (see Chapter 9). Maize grows very rapidly, leading to a harvestable crop just 3 months after planting seeds. C4 photosynthesis is much more water efficient than the C3 photosynthesis seen in most temperate crops – C4 plants produce the same amount of food for a smaller input of water.

SUMMARY

- Roots have **root hair** cells to increase the surface area for absorption of water and mineral ions from the soil. Water is absorbed by osmosis, mineral ions largely by active transport. Many plants have a mutualism with a soil fungus, a **mycorrhiza**, that aids absorption.

- Water passes through to the centre of the root via two pathways. The **symplast** ('living') pathway consists of movement through cell cytoplasm while the **apoplast** ('non-living') pathway consists of the spaces between cells and their walls. The endodermis surrounding the vascular bundle contains a Casparian strip that blocks the apoplast pathway.

- **Transpiration** is the loss of water from the upper surfaces of a leaf.

- The **transpiration stream** refers to the passage of water though the plant, from roots to leaves, through the xylem tissue.

- Throughout the plant, from soil to atmosphere, water travels down a water potential gradient. Water potential is highest in the soil water and lowest in the atmosphere.

- Xylem tissues consists of dead, hollow tubular cells that connect to form a continuous conducting pathway. Water and mineral ions pass up the plant in the xylem.

- The **cohesion–tension hypothesis** explains how water passes up the xylem to the leaves. As water evaporates from the leaves it creates a tension in the column of water below. Water is cohesive (sticky) so a continuous column can be pulled up from the roots.

- Fluid in xylem vessels is therefore under negative pressure, so xylem vessels are reinforced by rings of lignin to prevent collapse.

- In most plants the stomata close at night to prevent water loss. The mechanism of opening can be summarised as: light → photosynthesis → ATP → potassium pumped into guard cell → water follows → guard cells get more turgid, bend, pore opens.

- Most photosynthesis takes place in the leaves, these are the **sources** of organic molecules; 'food'. Areas of the plant that need food are referred to as **sinks**.

- Examples of sinks include the growing points (meristems) at the tips of roots and shoots, flowers, fruits and roots.

- Phloem tissue is living (in contrast to xylem). It consists of two types of cells; **sieve tubes** that do most of the transporting, and **companion cells** that control the activities of the sieve tubes.

- The fluid in the phloem is rich in sucrose and amino acids. Movement of fluid in phloem occurs from source to sink and is called **translocation**. In contrast to xylem, this can happen up or down the plant.

- The mechanism of translocation is the **mass flow hypothesis**, or Munch's theorem.

- Plants become adapted for different habitats. **Mesophytes** are plants that grow in areas of plentiful water. **Xerophytes** are adapted to arid environments. **Halophytes** can tolerate salt while **hydrophytes** live in water.

? EXAM QUESTIONS

1 The diagram shows part of a cross-section through a primary root.

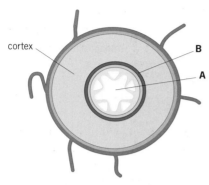

a) Name the tissues labelled **A** and **B**.

b) Water enters root hair cells and moves across the cortex through both apoplast and symplast pathways.
(i) Which part of the cortex cells forms the apoplast pathway?
(ii) Explain in terms of water potential how water enters root hair cells from the soil.

AQA (B) BYB3 Physiology and Transport June 2001 Q1

2 In an investigation, the absorption of water and phosphate ions by the roots of a plant was recorded over a period of 24 hours. The results are shown in the graph below.

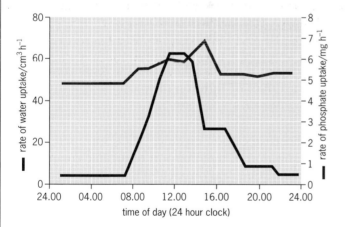

a) **(i)** Describe the changes in the rate of water uptake by the plant during the 24 hour period.
(ii) Suggest an explanation for the changes shown in the rate of water uptake.

b) **(i)** State *one* similarity in the patterns of water uptake and of phosphate uptake as shown by the graph.
(ii) How do the results of this investigation support the view that there is no direct relationship between uptake of phosphate and water absorption?

c) Phosphate ions are used in the production of ATP and of NADP. State *two* other uses of phosphate in a flowering plant.

Edexcel B2 June 1999 Q6

3

a) Name *two* cell types which are found in xylem tissue.

b) One of the functions of xylem tissue is to provide support. Explain how this function is achieved.

c) Name *one* other tissue which provides support in a plant and explain how the cells in this tissue are adapted for this function.

Edexcel B3 June 1999 Q3

4 Diagram (I) below shows some plant cell types drawn in either transverse of longitudinal section. Diagram (II) is a drawing of part of a transverse section of a stem of a flowering plant.

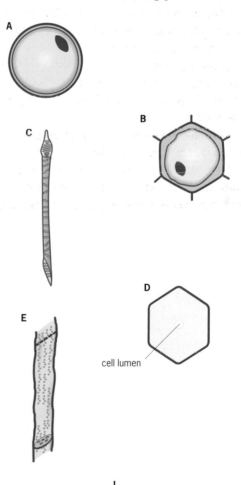

? EXAM QUESTIONS

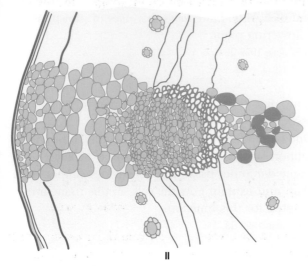

II

a) Identify the cell types **A** to **E** in diagram (I).
b) Draw label lines and use the letters A to E to indicate on diagram (II) **one** location for each cell type.

OCR 6911 Cells and Structures March 2000 Q3

5
a) Describe how the structure of xylem is related to its function.
b) Describe the roles of root pressure and cohesion–tension in moving water through the xylem.
   (i) root pressure
   (ii) cohesion–tension
c) Describe and explain how **three** structural features reduce the rate of transpiration in xerophytic plants.

AQA(B) BYB3 Physiology & Transport June 2001 Q8

6 The graph shows changes in the mean stomatal pore size on leaves of a plant subjected to natural light from the window of a laboratory for a 24 hour period.

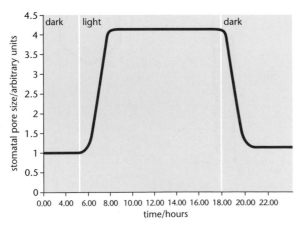

a) What effect does light have on the pore size?

b) A computer simulation was used to predict the effect of different pore sizes on transpiration rate, at two different air speeds. The following results were obtained:

Mean pore size/μm^2	Transpiration rate/μg h^{-1}	
	air speed = 1 m s^{-1}	air speed = 20 m s^{-1}
40	100	480
80	100	720
120	110	880
160	120	980
200	120	1060

(i) On graph paper, plot **two** graphs on the same axes, to show the effect of pore size on transpiration rate, at **two** different air speeds.
(ii) Using your graph, suggest how increasing pore size will affect transpiration rate.
(iii) Explain why the results are different at low and high air speeds.

c) In an experiment, a well watered plant was placed near the window of a laboratory. A fan was arranged to give a constant air speed of 20 m s^{-1}. Measurements of transpiration rate were started at dawn and continued for 24 hours.
(i) Using information from parts (a) and (b), describe how you would expect the transpiration rate to change over the period of the experiment.
(ii) Give **two** other variables which would have to be controlled to ensure a fair test.
(iii) Describe a suitable control for the experiment, which would help to show that the results are due to the **effect of light** on the plant.

WJEC Biology (Modular) B12 June 2001 Q3

7 Radioactive carbon dioxide, $^{14}CO_2$ was used in an investigation of the transport of organic compounds in plants. One leaf on each of two plants was supplied with air containing $^{14}CO_2$. The stem of one plant was ringed below the treated leaf by removing phloem tissue; the other was not ringed. The figures on the diagram show the concentration of radioactive compounds in the treated leaf and in the roots after two hours in sunlight.

? EXAM QUESTIONS

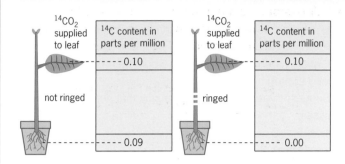

a) Name **one** organic compound transported in plants.

b) Explain what these results show about the transport of organic compounds in plants.

c) Name the process by which organic compounds are transported in plants.

AQA (B) BYB3 Physiology and Transport June 2001 Q6

8 The photomicrograph below shows a transverse section through a leaf of *Erica*, a plant which shows xeromorphic adaptations.

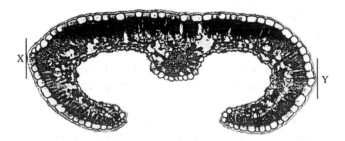

a) Calculate the magnification of this photomicrograph, given that the actual width of the leaf between X and Y is 0.8 mm. Show your working.

b) State *two* xeromorphic features shown by xerophytes such as *Erica*, and, in each case, explain how the feature helps to reduce transpiration.

Edexcel B2 June 1999 Q2

9 The drawing shows a section through a leaf of marram grass.

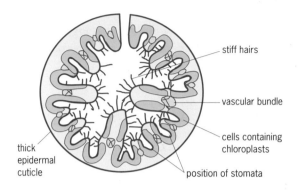

a) Describe **two** features shown on the drawing which help to reduce the rate of transpiration. Explain how each feature achieves this reduction.

b) Marram grass grows on sand dunes. The conditions found in this environment include high salt content in the soil and high air temperatures. Explain how these conditions could affect water uptake by this plant.

   (i) high salt content in the soil

   (ii) high air temperatures

AQA (NEAB) BY03 Physiology June 2000 Q1

KEY SKILLS **ASSIGNMENT**

MEASURING TRANSPIRATION

This assignment features two ways of measuring water loss from the leaves of two plant species, A and B. One species being more adapted to a dry environment than the other.

Experiment 1

One leafy shoot was taken from each plant. They were weighed, hung in the air and re-weighed at 15-minute intervals. The results are shown in Table 31.A1.

Table 31.A1		
Time after start of experiment/ minutes	**Mass of leafy shoot/g**	
	Species A	**Species B**
0	210.0	240.0
15	195.3	233.2
30	184.4	226.1
45	176.4	221.8
60	170.1	216.5
75	166.3	213.1

1 **a)** Plot a graph of mass against time for species A and B.
b) Describe the shape of the curve for species A.
c) Suggest a reason why the curve for species A is not a straight line.

2 Calculate the percentage change in mass for each species after 75 minutes.

3 **a)** What is the name given to plants that are adapted to survive arid environments?
b) Explain which species, A or B, would be more successful in an arid environment.
c) Suggest three structural adaptations that such a plant might possess.

Experiment 2

The apparatus shown in Fig 31.A1 is used to measure transpiration. Known as a **potometer**, it can be thought of as a transparent extension of the xylem. A leafy shoot from species A was taken and the stem cut under water before being placed in the tube of the potometer. The joints were then sealed with Vaseline. The procedure was repeated for species B.

4 Outline how this apparatus works.

5 State two precautions that should be taken when using this apparatus to ensure that the results are comparable for the two different species.

6 What three measurements are required to calculate the rate of transpiration in cm^3 water/min/g in Experiment 2?

7 Explain how the syringe can be used to measure the *volume* of water lost, and to re-set the experiment at the same time.

8 Suggest one reason why the figures obtained for water loss in experiment 2 were likely to have been higher than those from experiment 1 for the same species.

Fig 31.A1 The potometer

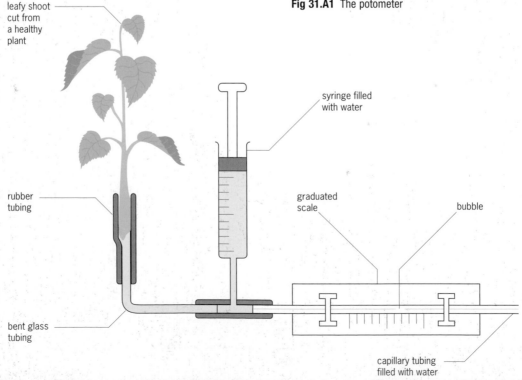

leafy shoot cut from a healthy plant

syringe filled with water

rubber tubing

graduated scale

bubble

bent glass tubing

capillary tubing filled with water

32

How plants grow and respond

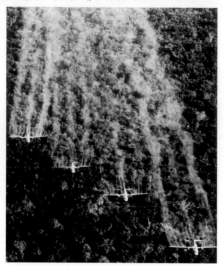

One of the most controversial weapons used by the US army in the Vietnam War wasn't a bomb or an explosive, but a mixture of two synthetic plant hormones. Named 2,4-D and 2,4,5-T (trust us – you don't want the full names), these two artificial auxins were given the name Agent Orange after the orange band on the drums they were stored in.

In high concentration, these hormones act as a defoliant – within a few hours leaves start to fall off the broadleafed (dicot) plants. Agent Orange causes plants to outgrow themselves – the rates of respiration and growth of the leaves are so rapid that the stalks can't bear the weight. This was an attempt to remove the jungle cover from Vietcong guerrillas.

The biggest problem with Agent Orange was that it contained a contaminant – dioxin. This is a particularly nasty chemical that causes birth defects, severe skin problems, leukaemias and other types of cancer. Both American and Vietnamese people were exposed to dioxin, to the extent that the medical, legal and environmental issues that it caused are still being dealt with today.

Spraying Agent Orange. The high concentration of auxin makes the leaves fall off within a few hours, and the active ingredients were so persistent that the land they were used on is still contaminated today

1 INTRODUCTION – AN OVERVIEW OF PLANT GROWTH

Fig 32.1(a) The plant grows from a fertilised cell, just as animals do. All the cells contain all of the genes, so growth and development are brought about by the selective activation of specific genes

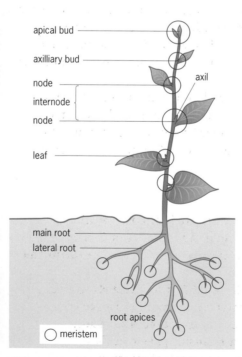

Fig 32.1(b) The basic external anatomy of a plant. The terms apical bud, internodal length and meristem are referred to in the text. The meristems are the growing points of the plant – the areas of cell division

In this chapter and the next we consider the life story of a flowering plant. First we look at the different types of plant growth substances, then we look at their role in the life of a plant. Starting with the seed, we cover germination, growth, the control of growth and flowering. Then in Chapter 33 we cover plant reproduction until we come full circle to seeds again. Later in this chapter we consider the following:

- How do seeds germinate? Why do some stay dormant for years?
- How do seeds know which way is up, so that roots grow down and stems grow up?
- What controls the development of a plant – roots, stem, leaf, flowers and fruits?
- How do plants respond to light?
- How do plants know when to flower?
- What makes the leaves fall in the autumn?

Though plants have no obvious sense organs, muscles or a nervous system, it would be a mistake to think that they are insensitive. Plants respond to their environment in many different ways. They can detect light, gravity, water and other chemicals.

Plants are constantly changing their shape or direction of growth. This movement can result from programmed changes within the plants, eg to form flowers, or as a response to an environmental factor such as light or touch.

Growth in plants is different from growth in animals. As you have grown from a small child all the tissues in your body have grown to some extent, and this has been brought about by cell division (mitosis). In contrast, cell division in plants is restricted to specific areas called **meristems**, and cells then expand and develop into the different tissues of the plant, such as xylem, phloem and mesophyll. Generally, plants grow and develop by a mixture of **cell division**, **expansion** and **differentiation**. Three factors interact to control the development of a plant:

- The genes it has inherited.
- The environment.
- Plant growth substances.

In the next section we look at each of the main types of plant growth substances.

✔ **REMEMBER THIS**

Plant tissues are discussed in detail in Chapter 2.

2 PLANT GROWTH SUBSTANCES

The growth and development of plants are controlled by a range of compounds that fall into five main groups: **auxins**, **gibberellins**, **cytokinins**, **ethylene** and **abscisic acid**. The first three are families of compounds, while the last two are single, specific compounds. Whether they should be called **hormones** or **plant growth substances** is a matter of opinion. The definition of a hormone is 'a substance that is made by one part of an organism and has an effect on another'. In many cases the above compounds act as hormones, but in others they work directly at the site of production. In many cases we aren't clear how they work at all, so don't lose sleep over a name.

Generally, the plant growth substances work in combination. Sometimes they work **antagonistically**, ie having opposing effects, and sometimes they work **synergistically**, ie their combined effects are greater than the sum of their individual effects. 'Two plus two equals fifteen' is one way to describe synergy.

AUXINS

The first plant growth substances to be investigated were the **auxins**, starting with work done by Charles Darwin and his son Francis. The Assignment at the end of this chapter outlines some of the early experiments that led to their discovery.

In 1934 the first auxin to be chemically identified was **indole acetic acid**. Since then a variety of related substances have been discovered including indoles, naphthyls, phenoxyacetic acds and benzoic acids. The defoliants mentioned in the opener are phenoxyacetic acids.

Generally, auxins are substances that promote cell elongation and differentiation. They are made in shoot tips, embryos, young leaves, flowers, fruits and pollen. The actual pathway of auxin synthesis is still unknown. Movement of auxin down the plant is polar, ie in one particular direction, and results in a concentration gradient from tip to root.

SEE QUESTION 1

THE EFFECTS OF AUXINS

- **Elongation of cells**. Auxins stimulate the secretion of H^+ ions, lowering the pH in and around cell walls. This causes a loosening of the bonds between cellulose fibrils, thus weakening cell walls and allowing the cell to expand in a particular direction.
- **Gene activation**.
- **Apical dominance** – the apical bud produces auxin, which inhibits the growth of lateral buds, so that plants tend to grow tall with minimal branching. If the apical bud is removed the apical dominance is lifted and more lateral buds become active.
- **Differentiation** of xylem and phloem.
- Inhibition of **abscission** (leaf and fruit fall).
- Promotion of **fruit development** (auxin is released by seeds).
- **Formation of roots**.

From this list of functions it can be seen that artificial auxins are useful in a variety of situations. Examples include:

- rooting solutions (Fig 32.2);
- to encourage fruit development;
- to inhibit the sprouting of potatoes; and
- as weedkillers (see the opener).

Fig 32.2 Rooting solution contains auxins. This gardener is dipping a cut stem in rooting solution before planting the stem

GIBBERELLINS

Gibberellins were discovered in 1920 by a group of Japanese scientists who were investigating a disease called 'foolish seedling' in which rice plants became tall and pale, and hardly produced any rice. It was found that the seedlings were infected by the fungus *Gibberella* (now called *Fusarium*) which was producing a growth-promoting chemical. The active ingredient turned out to be **gibberellic acid**, and since then about 50 different gibberellins have been isolated.

The principal mode of action of gibberellins is to promote cell elongation. This is seen dramatically in rosette plants such as cabbages and lettuce. These plants have virtually no stem between the leaf node, so they appear to be a mass of leaves. If, however, these plants are treated with just a tiny amount of gibberellin, the cells of the stem elongate greatly so that the **internodal length** (the gap between leaf buds) increases. The overall effect is spectacular (Fig 32.3).

No gibberelin   With gibberelin

Fig 32.3 If you want to cause havoc in your neighbour's cabbage patch, sprinkle the cabbages with a little gibberellin. These cabbages show what happens – they bolt. Two-metre cabbages are always good for a laugh, if not for a meal

Gibberellins also have a variety of other functions. They are involved in seed germination (see below) and in the production of α-amylase – the enzyme that turns starch into maltose in cell germination. This process is the basis of malting in the brewing industry so gibberellins are used to speed up the process. They are also used to promote the growth of large seedless grapes.

Interestingly, when gibberellin action is deliberately blocked by synthetic inhibitors, the result is often a sturdy dwarf plant with deep green leaves, greater disease resistance and other desirable characteristics. Many dwarf plants are mutants that are unable to make gibberellin. The exact mode of action of gibberellins is still unclear, but they are produced in tiny amounts, suggesting that they work at a fundamental cellular level, switching genes on and off.

CYTOKININS

Cytokinins were first discovered by two US researchers, Skoog and Miller, when they found that an old sample of DNA from herring sperm would promote cell division in cultures of carrot root tissue. The same effect was seen when the carrot cells were grown in coconut milk, which is actually liquid endosperm. The active ingredient in both cases turned out to be similar to the base adenine, and was given the name **kinetin**. This turned out to be an artificial cytokinin, because it has never been found in plant tissues. Natural cytokinins identified so far include **zeatin** (from maize) and isopentenyl adenine.

The major effect of cytokinin is to stimulate cell division. When used with auxin, cytokinins can help culture plants. A high cytokinin to auxin ratio will stimulate the formation of shoots, buds and leaves, while a low cytokinin to auxin ratio will lead to root formation. If you use the two together, you can stimulate the development of whole plantlets from samples of tissue (Fig 32.4).

> ✓ **REMEMBER THIS**
> *In vivo* means in a living organism, while *in vitro* means in a test-tube Literally, it means 'in glass'.

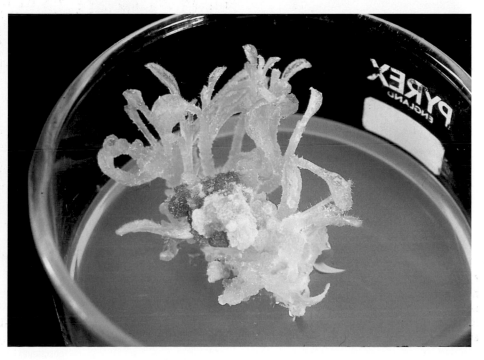

Fig 32.4 Just a small piece of tissue from a plant can be used to clone a new one. This technique allows growers to produce large numbers of new plants without having to wait for flowers to form and seeds to germinate. It also takes a lot less space

Fig 32.5 The phrase 'one bad apple spoils the whole barrel' is perfectly true. The one rotting apple will release large amounts of ethene that will accelerate the ripening and rotting of the others

✔ **REMEMBER THIS**

Trees that shed their leaves every year are called deciduous. In the UK, autumn is the season of leaf abscission.

SEE QUESTION 2

ETHENE (OR ETHYLENE)

Ethene is a very simple gas – formula C_2H_4 – whose most widely known role in plants is the ripening of fruit. One of the first clues that ethene could be a plant growth substance came when it was noticed that plants growing near natural gas street lamps **abscised** (fell off) earlier than those further away. We now know that ethene, a product of natural gas combustion, was responsible. Overall, ethene promotes **ripening** and **senescence** (ageing) of plant tissues.

Ethene is used to control the ripening of fruit – this is the single most important use of a plant hormone in the food business today. The action of ethene is antagonistic to carbon dioxide, so the control of fruit ripening can be reliably inhibited until precisely the right time by using first carbon dioxide and then ethene.

ABSCISIC ACID

The falling of leaves and fruit is called **abscission**. In the 1960s, two US scientists found a substance that appeared to make the leaves fall off cotton plants. They called this substance **abscisin**, and when it was isolated and analysed it was given the name **abscisic acid**. This compound is largely responsible for the fall of leaves in the autumn, and of fruit when ripe.

Abscisic acid is one compound rather than a class of closely related ones. It is more of a growth inhibitor than a growth promoter, and as such it is antagonistic to the first three classes of growth regulators described above. Abscisic acid has a variety of roles in plant development. In addition to leaf fall, it plays an important part in maintaining seed dormancy (see below) and the inhibition of growth during times of physiological stress, eg drought.

Fig 32.6(a) Why do leaves fall? It is an important survival mechanism – when the temperature and/or light intensity is too low for efficient photosynthesis, the leaves would be a liability in terms of water loss. Before the leaves are shed the useful compounds, such as proteins and chlorophyll, are broken down and stored in the stem. The colours produced are due to products of chlorophyll breakdown

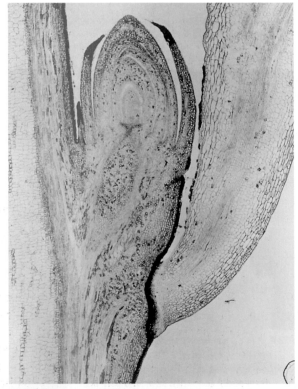

Fig 32.6(b) Abscission. When a leaf is going to be shed an abscission layer (the black band) forms across the petiole (leaf stalk). The new bud can be seen above the 'old' leaf

3 THE LIFE OF PLANTS

In this section we look at the development and growth of a plant from seed to flower.

WHAT IS A SEED?

A seed is similar to an egg in many ways: both contain an embryo organism together with a food store inside a protective coat. Fig 32.7 shows the basic features of a seed:

- The **testa** – the protective seed coat. This varies from a thin membrane to something as tough as a brazil nut.
- The **cotyledons** – or seed leaves. These are usually fleshy swellings rather than true leaves, though they do develop into 'proper' leaves in some species. In some species the seed leaves contain the food reserves. As their names suggest, seeds of monocotyledons have one cotyledon while those of the dicotyledons have two.
- The **endosperm** – a separate tissue that contains the food reserves to nourish the embryo. The **aleurone** layer is the metabolically active outer layer of the endosperm.
- The **embryo**. This has two parts: the **plumule** (young shoot) and **radicle** (young root).
- The **micropyle** – a tiny hole just above the point where the seed was attached to the fruit. This is the main site of water entry.

There are many different types of seed. Some, eg broad bean, are nourished by starch in two large cotyledons, but there is no endosperm (Fig 32.7a). In monocotyledons such as wheat the single cotyledon is a small flat structure and the embryo is nourished by the starchy endosperm. Flour, and all of the products made from it, is basically squashed endosperm. A third seed type is shown by the castor oil seed, which has two small seed leaves (it's a dicot) and the food store consists of an oily endosperm.

Fig 32.7(a) In this broad bean seedling germination has just started, and the stored food reserves are being used to make new root and shoot tissues

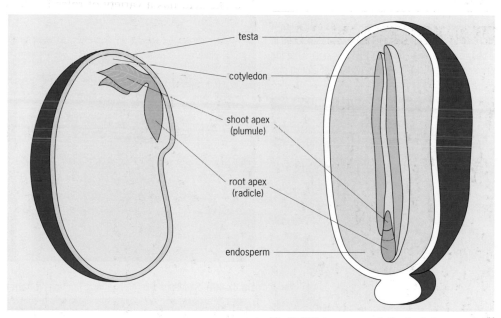

testa

cotyledon

shoot apex
(plumule)

root apex
(radicle)

endosperm

Fig 32.7(b) In some seeds the food stores are held in the cotyledons (left), while in others they are in the endosperm (right)

SEEDS AND DORMANCY

Seeds need water, oxygen and a suitable temperature before they can germinate. There are some seeds, however, than won't germinate even when given these conditions. Such seeds have evolved delaying tactics to ensure that they don't germinate at the wrong time, eg in the autumn, with little chance of surviving the winter, or during a dry season. Seeds can remain dormant for hundreds or even thousands of years. The present record is held by the arctic lupin. Its seeds may land in frozen soil where, without the necessary warmth, they remain dormant almost indefinitely. Vegetation rots very slowly so it accumulates year upon year and the seeds get buried in the permafrost. Lupin seeds as old as ten thousand years have been excavated and, though most are dead, a few have germinated.

Dormancy is defined as a period of metabolic inactivity, in which the normal life processes are suspended. In seeds, dormancy is usually due to the properties of the seed coat. Once the seed coat has been weakened or penetrated in some way, the embryo inside will start to grow. There are three principal mechanisms for maintaining dormancy:

● Keep water or oxygen out of the embryo by having an impermeable seed coat. The breaking of dormancy involves weakening the seed coat in some way. It may be frost, fire (Fig 32.8) or the need to pass through the digestive system of an animal (Fig 32.9). This modification of the seed coat is called **scarification**.

● Mechanical restraint of the embryo. Some seeds maintain dormancy by physically preventing the embryo from expanding. If the seed coat is cut away or dissolved by chemicals, the embryo will germinate. In nature it is often the action of soil microbes that brings about the necessary weakening.

● Growth inhibitors. The fruit or seed may contain an inhibitor – often abscisic acid – that may need be washed out by water before germination can begin. Fire can also destroy the inhibitor. Some seeds accumulate growth promoters that eventually overcome the effect of the inhibitor.

Some seeds, usually tiny ones with few food reserves, will only germinate in the light. In nature such seeds are often weeds that germinate rapidly in freshly ploughed soil or disturbed soil. Other seeds need a specific period of cold before they will germinate. Conifer seeds, for example, need a period of one to two months at 5 °C. This mandatory cold spell is known as **stratification**.

SEE QUESTION 3

? TEST YOURSELF

A Some small seeds with few food reserves will only germinate in the light. Suggest an explanation for this.

Fig 32.8(a) In Australia the *Banksia* is just one of the species that won't release its seeds until there is a fire. The seeds of the *Banksia* are enclosed in tough, oyster-like capsules that only open after the extreme heat of a bush fire.

Fig 32.8(b) Seedlings that grow after a fire benefit from fertile soil and little competition

Fig 32.9 The seeds of the acacia tree are rich in protein and plant-eating animals love them. However, they will seldom germinate unless they pass through the digestive system of a large herbivore such as an elephant. It was thought that the action of the digestive juices weakened the seed wall but research showed that most seed don't germinate because they become infested with beetle larvae. The digestive juices kill the larvae, allowing the seed to germinate. A pile of dung is always a good place for a plant to germinate

HOW DO SEEDS GERMINATE?

The first process in germination (Fig 32.10) is usually water uptake. This is called **imbibition** (imbibe means 'to drink'). This process of hydration makes the tissues more liquid and allows the molecules to move around. The embryo then synthesises **gibberellins** that diffuse into the **aleurone layer**. Here, they bring about the breakdown of stored proteins into amino acids. The amino acids are used to synthesise digestive enzymes that pass into the endosperm (or into the seed leaves) and begin to break down the food reserves. Finally, the embryo synthesises new cells from these food reserves.

WHICH WAY IS UP?

The next step for a germinating seed is to find out which way is up. The young shoot must grow upwards, reach light and start photosynthesising before its food stores run out. The roots need to grow downwards in a search for water and mineral ions.

Fig 32.10(a) Most seeds are instant plants; just add water and stand back. These wheat seedlings show the early stages of germination. The bottom one is the seed and the next one up shows the radicle (young root) beginning to emerge from the testa. In the third one up root hairs can be seen on the radicle and in the top one the radicle is well developed. Inside the seed the starch stores are being mobilised

Fig 32.10(b) The role of gibberellins in breaking dormancy and initiating germination

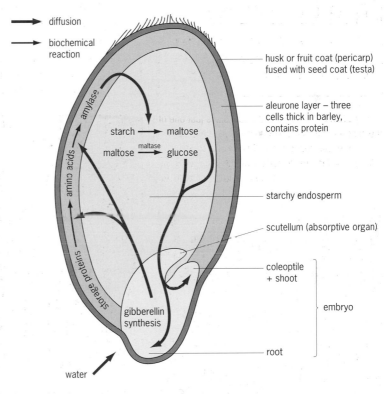

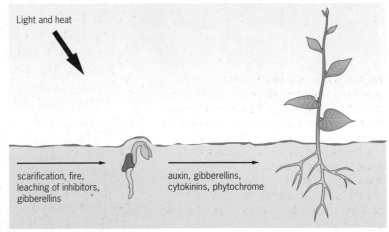

Fig 32.10(c) An overview of the role of hormones in plant development

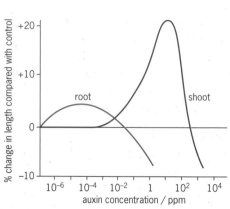

Fig 32.11(a) The negative gravitropism of the shoots is due to an accumulation of auxin on the underside of the young stem

Fig 32.11(b) The effects of auxin concentration on root and shoot growth

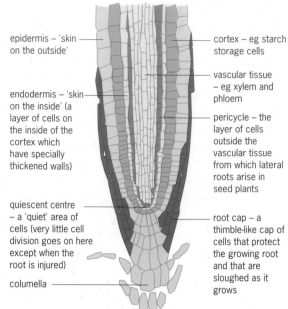

epidermis – 'skin on the outside'

cortex – eg starch storage cells

vascular tissue – eg xylem and phloem

endodermis – 'skin on the inside' (a layer of cells on the inside of the cortex which have specially thickened walls)

pericycle – the layer of cells outside the vascular tissue from which lateral roots arise in seed plants

quiescent centre – a 'quiet' area of cells (very little cell division goes on here except when the root is injured)

root cap – a thimble-like cap of cells that protect the growing root and that are sloughed as it grows

columella

Fig 32.11(c) Diagram of root tip showing columella cells

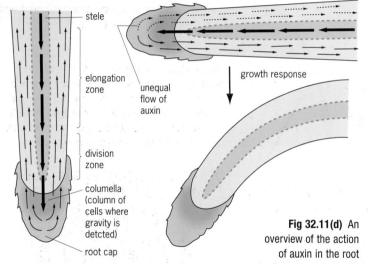

stele

elongation zone

unequal flow of auxin

division zone

columella (column of cells where gravity is detcted)

root cap

growth response

Fig 32.11(d) An overview of the action of auxin in the root

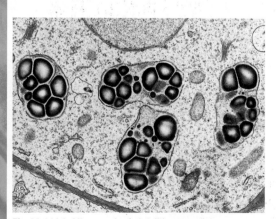

Fig 32.11(e) Micrograph of statolith organelles

The stimulus that most seedlings respond to is gravity. In a response called a **gravitropism** or **geotropism** the roots grow downwards – a **positive gravitropism** – while the shoot grows upwards – a **negative gravitropism**. Fig 32.11 shows a seedling that has germinated with its root and shoot growing horizontally. What mechanism is responsible for the shoot bending up while the root curves down?

The response is brought about largely by auxin. Investigations have shown that the auxin is produced in the tip and passes down the plumule where it accumulates on the lower side of the shoot, stimulating cell elongation along the underside and causing the shoot to bend upwards.

A simple explanation for the roots is more difficult. Experiments have shown that the removal of the root cap prevents the root from responding to gravity, but the exact mechanism is still the subject of research. We know that the root caps of some plant species contain specialised **columella cells** that contain large dense organelles called **statoliths** (Fig 32.11c and e). These are starch-containing plastids (the family of organelles that also includes chloroplasts). Whatever the angle of the root, the statoliths accumulate on the underside of the cells, so it is assumed that they have a role to play in gravity perception. However, while the stimulus is perceived by the columella cells, the response is brought about by the cells of the epidermis, which elongate faster on the upper surface than on the lower one. Exactly how the signal is transduced is not clear, but the tropic response appears to be due to an uneven distribution of auxin. One proposed model states that auxin passes down the root in the phloem (Fig 32.11d) and then accumulates in the tissues on the underside of the root. Auxin inhibits root cell elongation, so the greater concentration of auxin on the underside of the root leaves the cells on the upper side to elongate faster, producing downward growth.

PLANT MOVEMENTS

Most plant movement tends to be slow and results from cell elongation. More rapid and spectacular movements can be brought about by specialised hinge cells, such as those seen in the Venus flytrap (Fig 32.12b) or by mechanisms in which pressure slowly builds up and is rapidly released.

There are two basic plant movement responses:

- **Nastic movements**, which are *non-directional* responses such as a flower closing at night. The response is not towards or away from the stimulus.
- **Tropisms**, which are *directional* responses, such as a plant turning its leaves towards the sun. Tropisms can be positive – towards a stimulus – or negative. Different tropic responses include **hydrotropism** (water), **thigmotropism** (touch), **gravitropism** and **phototropism** (see above).

Fig 32.12(b) The Venus flytrap (*Dionaea muscipula*) is a carnivorous plant that lives in nitrogen-poor swamps. To get some vital nitrogen it catches flies and to do so it must match the speeds seen in the animal kingdom. Amazingly, the fly trap can count. When a fly touches one of the sensitive hairs twice there is a rapid change in the turgor of special hinge cells resulting in closure of the 'jaws'. One touch could be a false alarm, so the trap would shut in vain. Two touches are a surer sign that there really is a fly to catch

Fig 32.12 Nastic movements

Fig 32.12(a) *Mimosa* (the sensitive plant) is a good example of a nastic movement. Its leaves fold up when touched. Quite how it does this is still something of a mystery

Fig 32.12(c) Clematis is a climbing plant. The winding of the tendrils of this clematis plant are a directional response to touch, a process given the wonderful name of positive thigmotropism. Growth is slowed on the side of the tendril that is touched, causing the tendril to wrap around the object

GROWTH IN PLANTS

The way in which plants grow is fundamentally different from animals. From your own experience you will know that growth occurs all over your body. But this does not happen in plants. Plants have specialised regions – **meristems** – that are areas of cell division. Usually there is an apical meristem at the top of the shoot and a root meristem just behind the tip of the root (Fig 32.13). Cambium tissue in vascular bundles is also meristematic – it gives rise to xylem and phloem.

Following cell division, plant growth is a process of cell expansion or elongation, and differentiation. From small, simple cells produced by meristematic tissue, the process of differentiation gives rise to all of the different tissues of a plant, such as xylem, phloem, mesophyll, parenchyma and collenchyma. Look at Chapter 2 for more on plant tissues.

THE CONTROL OF FLOWERING

It had long been a mystery how plants can control the timing of flowering. It's not a matter of the plant reaching a certain stage of maturity – even plants of a different age will flower at the same time. Synchronised flowering is vital because it allows cross-pollination. But how it is achieved?

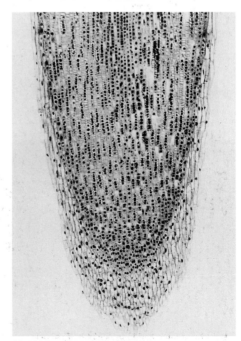

Fig 32.13 Micrograph of the root tip, showing the root cap. Mitosis produces small, square cells with relatively large nuclei, so the areas where the nuclei are close together are likely to be areas of most recent cell division. As the cells elongate, the nuclei get further apart

? **TEST YOURSELF**

B Grasses have **intercalary meristems** at the base of the stem. How does this help them survive being grazed by animals?

Fig 32.14(a) At Christmas the shops are full of strange-looking plants with a crown of red leaves at the tip – the poinsettia. People of a certain age feel obliged to buy them. Poinsettias initiate flowers when days become shorter than nights. In the Northern Hemisphere, this day length condition naturally occurs during mid to late September, and in the Southern Hemisphere during March. Once short days begin, it is important that nights continue to get longer for the leaves to develop their colour

SEE QUESTION 4

Table 32.1 Some examples of short and long day plants

Short day	Long day	Day neutral
chrysanthemum	barley (winter)	antirrhinum
strawberry	wheat (winter)	string bean
rice	oat	cucumber
coffee	rye grass	tomato
tobacco	spinach	
soya bean		

SEE QUESTION 5

Control over the timing of flowering has been the subject of much research, as control over this process obviously of great commercial interest. To understand flowering it is necessary to study **photoperiodism**; the amount of darkness and daylight a plant is exposed to. It used to be thought that the length of day was vital, but it is now clear that *length of darkness* that a plant is exposed to plays a critical role.

There are basically three categories of plants based on their response to light:

- **Short day plants** require need uninterrupted darkness of a certain threshold length before they will flower. Think of them as 'long night' plants (Fig 32.14a).
- **Long day plants** will only flower if the period of darkness is less than a certain critical length each day. They are 'short night' plants (Fig 32.14b).
- **Day neutral plants** that will flower whatever the photoperiod – they use some other internal or external signal to flower (Fig 32.14c).

Fig 32.14(b) Spinach is a nutritious vegetable but it can't flower in the tropics. This is because spinach is a long day plant and the days simply don't get long enough. Spinach needs a day length of 14 hours in order to flower

Fig 32.14(c) Tomatoes are day neutral plants. Most varieties will flower when they reach a certain stage of maturity – the 14th node, ie after the 13th leaf, the 14th one will be a flower that will ultimately bear fruit

Examples of the three types of plant are given in Table 32.1.

At the equator day and night are roughly of equal length (12 hours) and this varies little throughout the year, but the further you go from the equator, the greater the seasonal variation in day length. This variation is reliable, as opposed to rainfall or temperature and so is used by many organisms to synchronise their life cycles. In the UK the length of day varies from about 17 hours on 21 June to about 8 hours on 21 December.

Fig 32.15 Flowering is a big physiological switch for a plant; it has to stop producing lateral buds and leaves and change to making flowers, followed by seeds and fruits. These chrysanthemums are short day plants; they flower in late summer and autumn when the nights exceed a certain critical value

PHYTOCHROME

The light-sensitive substance in plants is called **phytochrome**. This blue-green compound, found in minute amounts of the tips of plants, consists of a light-absorbing pigment attached to a protein. Phytochrome exists in two interconvertible forms. One form absorbs light of 660 nm, ie **red light**, and is called P_R (R for red) or P_{660}. When it absorbs red light it is converted into P_{FR}, or P_{730}. P_{FR} is converted back into P_R by far red light (Fig 32.16a).

Fig 32.16 When plants are given light of different wavelengths to see which stimulate flowering an action spectrum is obtained with peaks at 660 and 730 nm. The absorption spectrum for phytochrome matches the action spectrum very closely, strongly suggesting that phytochrome is indeed the stuff that makes things happen

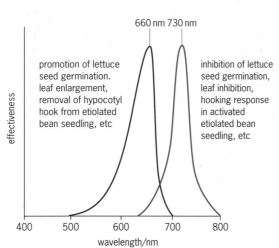

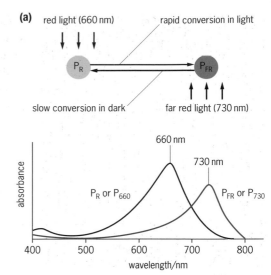

Sunlight contains more red light than far red light, so most of the phytochrome is in the P_{FR} form during the day. This is usually the more physiologically active form, P_R being more stable and inert.

- In short day plants P_{FR} inhibits flowering. As the days shorten and the nights get longer, a threshold is reached when enough P_{FR} is broken down to remove the inhibition
- In long day plants P_{FR} stimulates flowering. As the days get longer and the nights shorter, the level of P_{FR} will increase until it reaches the threshold needed to stimulate flowering.

In the UK, most of the plants that flower in late summer or autumn are short day plants – they will flower when the night lengths exceed a threshold value, eg 14 hours. In contrast, long day plants will not flower unless the day lengths exceed a certain threshold. Such plants often flower in the spring as the days get longer.

? TEST YOURSELF

C Look at Fig 32.16(b). Why do you think that phytochrome appears blue–green?

✔ REMEMBER THIS

Phytochrome simply means 'plant pigment'.

Much research has been carried out on the cocklebur, a short day (long night) plant from North America. Flowering in short day plants is inhibited by PFR, and the longer the night, the more PFR is broken down. The question is: just how long do the nights need to be? An investigation was carried out into the effect of night length, and of flashes of red and far red light. The results are shown in Fig 32.17.

HOW DOES PHYTOCHROME WORK?

The story is beginning to unfold thanks to work done in *Arabidopsis*, the 'lab rat' of the plant world. *Arabidopsis* is a long day plant, so flowering is stimulated by P_{FR}. It has been found that:

- In sunlight, P_R is converted into P_{FR}, the active form. P_{FR} moves from the cytoplasm into the nucleus.
- P_{FR} combines with a protein called **PIF3** (phytochrome interacting factor 3).
- PIF3 is a **transcription factor** that switches on certain genes.
- These genes initiate flowering.

The flowering mechanism in short day plants, where P_{FR} inhibits flowering, is less clear.

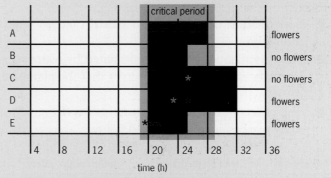

★ = flash of red light (660 nm)
★ = flash of far red (FR) light (730 nm)

Fig 32.17 Five experiments with photoperiodism in the cocklebur
In experiment A plants have 8.5 hours of dark (the threshold) so they flower.
In experiment B plants have less than 8.5 hours – they don't flower.
In experiment C plants have a flash of red (660 nm) light after 8.5 hours which turns all their P_R back to P_{FR} – they don't flower.
In experiment D plants get a flash of red light in the critical dark period, but then a flash of far red light restored all the phytochrome to P_R – so the plant flowers.
Experiment E gave the plants intense exposure to far red light. This reduced the threshold by 2 hours because the P_{FR} was broken down more rapidly.
Many more experiments have been carried out in which different combinations of red and far red flashes were given. The simple over-riding fact is that *the last flash is critical*; if it's far red, the plant flowers. If it's red, it doesn't

? **TEST YOURSELF**

D What would be the benefits of isolating the hormone that stimulates flowering?

THE SEARCH FOR FLORIGEN

Experiments in which different leaves were 'blindfolded' show that each individual leaf is capable of timing the night. However, while the reception is done by the leaves, the flowering response occurs elsewhere in the plant. There must be a signal that travels from leaf to the flower buds. Evidence suggests that this message is chemical in nature (see Fig 32.18) and it has been give the name **florigen.** However, nobody has succeeded in isolating this chemical. Whoever succeeds will be sure of a degree of fame and financial security, courtesy of the florists and other plant growers the world over.

Fig 32.18 The evidence for florigen in the cocklebur (a short day plant). The plant on the left will not flower if kept under long days and short nights. However, if just one leaf is subjected to short days and long nights, the plant will flower. What is the message that passes from leaf to flower buds?

burrs

single leaf masked on long night/ short day routine

plant does not flower

plant flowers

SUMMARY

- Seeds consist of an **embryo plant** together with food stored in a protective coat.
- Seed germination us usually initiated by water. Enzymes digest the food reserves that the plant uses for growth.
- The effects of the various **plant growth substances** are summarised below.

Hormone	Site of production	Some notable effects
Auxin	embryo, young leaves and growing points, particularly apical bud	promotes cell elongation; apical dominance (see assignment) – inhibits formation of lateral buds
Gibberellins	embryo, young leaves and growing points	breaks seed dormancy, promotes germination and seed development by stimulating cell elongation
Cytokinins	roots	promotes cell division
Abscisic acid	root cap, older leaves, stem	imposes seed dormancy and inhibits germination
Ethylene	ripening fruit, older tissues	promotes ripening of fruit

- Flowering and other plant responses are commonly controlled by day length. This is called **photoperiodism.** Usually, it is the length of night/darkness that is critical.
- The pigment **phytochrome** is responsible for the **photoperiodic response**. This exists in two interchangeable forms, P_R and P_{FR}. The latter is the more metabolically active.
- Plants fall into one of three categories according to their photoperiodic response:
  - **Short day plants** need long nights; they won't flower until P_{FR} levels reach a threshold. These plants generally flower in the summer/autumn, as the nights get progressively longer.
  - **Long day plants** need short nights. They won't flower unless the period of darkness is less than a certain critical length each day.
  - **Day neutral plants** use some other environmental or internal factor to trigger flowering.
- There is evidence for a flowering hormone, **florigen,** but it has not been isolated despite much research.

? EXAM QUESTIONS

1 The graph shows the effect of different concentrations of auxin on the growth of roots and shoots.

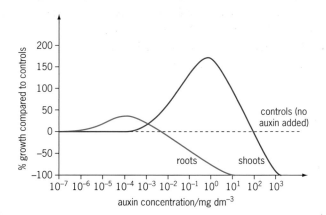

a) With reference to the graph describe:
(i) one similarity between the effect of auxin on roots and shoots;
(ii) two differences between the effects observed in roots and shoots.
b) (i) Which region of the shoot is most sensitive to auxin?
(ii) Explain how auxin exerts its effects on the cells of the shoot.
c) Explain the role of auxin in the phototropic response in shoots.

OCR 6917 Plant Studies March 2000 Q4

2 Read through the following passage, which refers to plant growth substances, and for each dash write the most appropriate word or words to complete the account.

Stem elongation is promoted by auxins. It is also promoted by ——— which can stimulate dwarf varieties of some plants to achieve normal height. These two plant growth substances can act together and enhance each other's activities, an effect known as ———.
Auxins promote apical dominance, whereas ——— have the opposite effect, promoting the growth of lateral buds. These two plant growth substances stimulate ——— in apical meristems and cambium. The growth of a plant is inhibited by abscisic acid which also promotes the closure of ——— during conditions of water stress.

Edexcel B3 June 1999 Q4

3 Dormancy in some seeds is broken by exposure to light. Germination was investigated in Grand Rapids lettuce seeds under different light conditions. Batches of 50 seeds were sown and exposed to different wavelengths of light. In a control batch, kept in the dark, 50% of the seeds germinated. The table shows the percentage germination at each wavelength.

Wavelength of light/nm	% germination
500	34
550	72
600	98
650	98
700	48
750	4
800	30

a) Plot these data on graph paper.
b) (i) On your graph, rule and label a horizontal line at the percentage germination achieved with the control batch.
(ii) Suggest an explanation for the results obtained at 750 nm.
(iii) Which range of wavelengths was most effective in promoting germination?
c) Name the pigment that detects the wavelength of light during germination of these seeds.
d) State **two** ways in which the reliability of this investigation could be improved.
e) State **two** treatments, **other than exposure to light**, which may break dormancy in seeds.

OCR 6917 Plant Studies March 2000 Q5

4 The chrysanthemum plant (White Wonder) has a critical day length of 16 hours. A number of plants were treated to the conditions shown in the table.

Lighting regime	Plant			
	A	B	C	D
top				
light/hr	18	6	6	18
dark/hr	6	18	18	6
bottom				
light/hr	18	6	18	6
dark/hr	6	18	6	18

? EXAM QUESTIONS

In each case the top leaves were removed from the stem, leaving the stem apex intact. In **C and D** the lower part of the stem (with the leaves) was enclosed in a light-proof box, which was subjected to a different light regime to the top of the plant (without the leaves). The plants were kept in these conditions for two weeks. The results are shown in the diagram below.

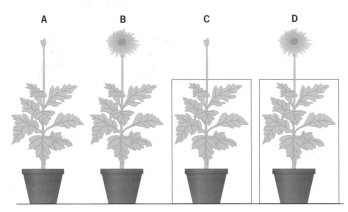

a) What is meant by a *critical day length of 16 hours*?

b) Why was it important to leave the stem apex intact?

c) What conclusion may be drawn from the results of plants **A** and **B**?

d) Use the information in the diagram and table to explain why plant **D** flowered but plant **C** did not.

e) Explain how the phytochrome system is involved in controlling flowering in plants such as chrysanthemums.

OCR 6917 Plant Studies March 2001 Q7

5 The diagram shows the response of a flowering plant to differing periods of illumination.

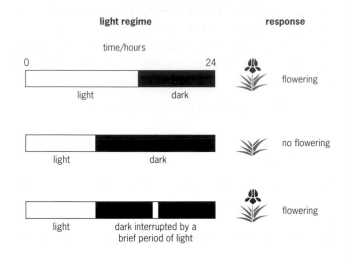

a) Name the receptor substance that is involved in the response.

b) The receptor substance involved in this response has two molecular forms P_R and P_{FR} which are inter-convertible, $P_R \rightleftharpoons P_{FR}$
Explain in terms of these molecular forms:
(i) why the plant flowered when the dark period was short, but did not flower when the dark period was long;
(ii) why the plant flowered when the dark period was interrupted by a brief period of light.

c) A handbook about indoor plants has the following advice on how to ensure that Poinsettia plants can be made to flower at Christmas. *'From the end of September cover with a black polythene bag from early evening and remove next morning so that the plant is kept in total darkness for 14 hours. Continue daily for 8 weeks, then treat normally. Your Poinsettia will then bloom at Christmas time.'*
Suggest how this treatment ensures that the plants will flower at Christmas.

AQA(NEAB) BY04 Biological Basis of Behaviour June 2000 Q2

KEY SKILLS **ASSIGNMENT**

INVESTIGATING AUXINS

The first well-documented investigations into phototropism were performed by Charles Darwin and his son Francis in the 1880s, when the great man was in his 70s. They had observed plants bending towards light and wanted to find out which part of the plant was light-sensitive.

1 What is the plant's response to light called?

2 What happens in the cells of the plant stem to produce this response?

The Darwins' experiment is shown in Fig 32.A1. They worked with grass seedlings, which have a protective covering over the young stem called a **coleoptile**. They illuminated the seedlings from one side, and 'blindfolded' different parts of the coleoptile.

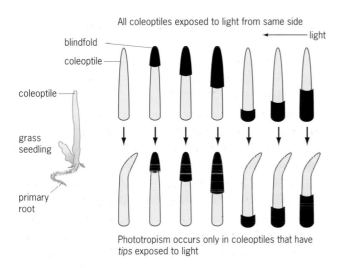

All coleoptiles exposed to light from same side

Phototropism occurs only in coleoptiles that have *tips* exposed to light

Fig 32.A1 The Darwins' experiment

3 What two conclusions can you draw from the Darwin's blindfolding experiment in Fig 32.A1?

Building on Darwin's work, other investigators found that the message is a chemical substance that passes down from the tip to the stem. This substance, it was thought, diffuses down from the tip and has an effect on the stem below.

The Dutch botanist Frits W Went succeeded in isolating the substance responsible. One of his early experiments is shown in Fig 32.A2. This substance was given the name auxin (from the Greek auxene = to increase) and when the chemical nature was investigated it was found to be indole acetic acid.

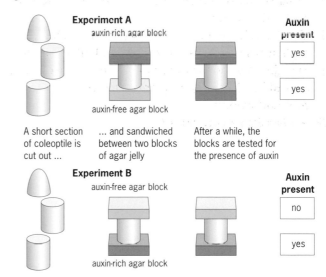

1. Tip removed and placed on gelatin
2. Gelatin placed on one edge of another decapitated coleoptile
3. Substance diffuses down one side from gelatin
4. Coleoptile curves away from gelatin as it grows

diffusion of growth substance

gelatin

oat coleoptile

Fig 32.A2 Went's experiment where he illuminates a coleoptile, collects the auxin in an agar block, then places the block on an non-illuminated coleoptile. That, too, curves

4 What conclusion can you draw from Went's experiment in Fig 32.A2?

It was assumed that auxin diffused down the stem from its site of production. In order to confirm this, the experiment shown in Fig 32.A3 was performed

Experiment A
auxin-rich agar block

auxin-free agar block

Auxin present
yes
yes

A short section of coleoptile is cut out ...

... and sandwiched between two blocks of agar jelly

After a while, the blocks are tested for the presence of auxin

Experiment B
auxin-free agar block

auxin-rich agar block

Auxin present
no
yes

Fig 32.A3 In Experiment A, an auxin-rich block is place at the top end of the coleoptile section and an auxin-free block is placed at the bottom end. In Experiment B the position of the two blocks is reversed

KEY SKILLS **ASSIGNMENT**

5 From the results to this experiment, is the movement of auxin due to diffusion? Explain your answer.

Apical dominance

We know that auxin is made in the apical bud at the tip of the stem, and passed down so that there is a concentration gradient from shoot to root tip. This gradient has important effects on the growth and development of new buds and on the appearance of the whole plant.

6 Look at the experiment in Fig 32.A4, which demonstrates a well-known phenomenon called **apical dominance**.

a) What effect does the apical bud have on the other lateral buds?

b) What happens when the apical bud is removed?

c) Suggest a situation when this may happen in nature.

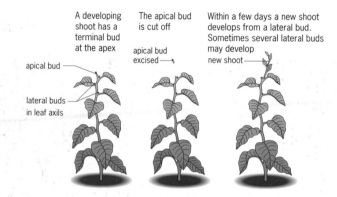

A developing shoot has a terminal bud at the apex

apical bud

lateral buds in leaf axils

The apical bud is cut off

apical bud excised

Within a few days a new shoot develops from a lateral bud. Sometimes several lateral buds may develop

new shoot

Fig 32.A4 An experiment to investigate what happens when the apical bud of a plant is removed

7 **a)** Look at the experiment shown in Fig 32.A5. What does this experiment suggest about the way the terminal bud works in preventing later bud development?

b) What further investigations might you do to confirm your hypothesis?

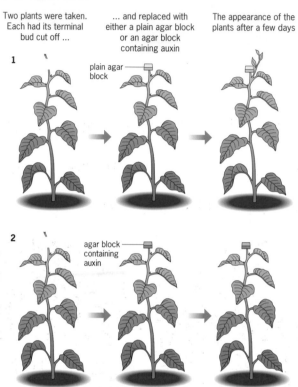

Two plants were taken. Each had its terminal bud cut off ...

... and replaced with either a plain agar block or an agar block containing auxin

The appearance of the plants after a few days

1

plain agar block

2

agar block containing auxin

Fig 32.A5 A further investigation into the effect of the terminal bud

33

Reproduction in flowering plants

Reproduction in flowering plants

The mirror orchid flower is an effective mimic of the rear end of a female bee

Sex is difficult when you live on land and can't move. Sperm can't swim over dry land, so how can genetic information pass from one individual to another? The flowering plants have evolved in close association with the insects, and many plants use animal couriers to transfer their sex cells for them.

But how do plants attract the insect? It pays to advertise: brightly coloured flowers promise freebies such as pollen and nectar, but one of the most bizarre strategies has to be the promise of sex.

The mirror orchid of the western Mediterranean has flowers of metallic blue fringed by a yellow band and surrounded by hairs. On either side are two striped lobes that look like wings. In short, mirror orchid flowers are an almost perfect mimic of the rear end of a female bee. And if that wasn't enough, the flower oozes the pheromone of a sexually receptive female.

Unable to resist, the male homes in and lands on the flower in the correct position for copulation. Before he realises his mistake, another part of the flower, arched like the sting of a scorpion, bows over and deposits two pollen sacs onto the bee's back. If the next orchid he visits has already passed its pollen sacs to another courier, the column will pick up the pollen sacs and the orchid is fertilised.

1 SEX IS DIFFICULT WHEN YOU'RE ROOTED TO THE SPOT

Reproduction has the same function in all organisms – to continue the species. Plants have evolved a wide variety of both sexual and asexual methods of reproduction, and individual species often employ both strategies at some stage in their life cycle.

Any organism that reproduces sexually must transfer its genetic material – not an easy task for an organism that cannot move, such as a plant. Sexual reproduction allows a plant species to increase in number and it also introduces genetic variation. Without this variation, plants could not evolve and adapt to changes in their environment. Sometimes, however, asexual reproduction allows rapid growth when conditions are favourable.

In this chapter we start with the events of sexual reproduction in flowering plants (Fig 33.1 provides a useful overview). We then study individual stages in more detail; we look at reproductive strategies and at the way flowering plants pollinate and fertilise. The chapter ends with a description of asexual reproduction and its commercial applications.

Sexual reproduction in flowering plants has many stages. In the first, flowers are formed that have male and female parts. Unlike the sex organs of animals, which form gametes straight away, plant sex organs first produce male and female **spores**. The male parts, the stamens, make male microspores, more usually called **pollen**. The female parts, the **ovules**, form female spores called **megaspores**. The two commonest methods of pollen transfer are to use insects or to use the wind. You will find more about this later in the chapter.

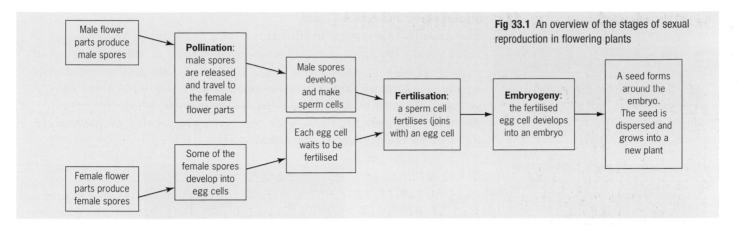

Fig 33.1 An overview of the stages of sexual reproduction in flowering plants

Pollination occurs when pollen from the stamens reaches the female parts of another flower of the same species. The pollen grain produces **sperm** nuclei, which pass through a pollen tube to the **egg** that forms inside the female megaspore. When sperm meets egg, **fertilisation** occurs. A fertilised egg grows into an **embryo** plant, enclosed as a **seed**. The ovary develops into a fruit, whose role is to disperse the seeds away from the parent plant. The seed protects the embryo, supplies it with food and often carries it to a new site, where it can grow into a new plant.

✔ REMEMBER THIS

Flowering plants make up a phylum of plants called the **Angiospermophyta**, commonly called the angiosperms. They have evolved to be the most advanced type of plant. The word angiosperm means 'hidden seed': the seeds of angiosperms are hidden inside the ovary, not exposed as in gymnosperms (eg conifers).

! SCIENCE FEATURE Alternation of generations – a simple explanation

Mammals such as ourselves have very simple life cycles. Adults with diploid cells produce haploid eggs and sperm by meiosis. Fertilisation restores the diploid number, and the resulting baby is unmistakably human.

The situation is very different in simple plants, such as mosses. The 'adult' plant, called the **sporophyte**, produces spores by meiosis. The spores grow on their own – no need for fertilisation. The spores grow into a completely different-looking plant – the **gametophyte** generation. The gametophyte usually grows out of the sporophyte, almost like a parasite. This generation grows up and, being already haploid, makes gametes (eggs and sperm) *by mitosis*. The eggs and sperm fertilise to produce a diploid organism like the original adult, the sporophyte again. This is the basic idea behind the **alternation of generations**.

All plant life cycles show alternation of generations, but as flowering plants evolved the system was modified so that the sporophyte generation dominates and the gametophyte generation is hardly noticeable; it is simply the germinated pollen grain and the female embryo that produce the male and female gametes. At fertilisation, two gametes come together and form an embryo sporophyte plant again.

Fig 33.2 These are two generations of *Polytrichum*. The gametophyte plant is the green plant and the sporophyte plant is the capsules above the gametophyte. Mosses are limited to wet environments, because their tissues aren't resistant to drying out and their male gametes must swim to find the female gametes

FLOWER FORMATION

The structure of a generalised flower is shown in Fig 33.3. All flowers are made up of a few basic parts, often arranged in rings or **whorls** at the end of a flower stalk, or **axis**. The centre of the flower contains the reproductive structures, sometimes either male or female, but usually both. These are surrounded by the protective envelope, the **perianth**, which consists of **petals** and **sepals**. In some plants, the petals and sepals look identical: in this case we call both structures **tepals**.

<table><tr><td>

? **TEST YOURSELF**

A The flowers of protea in Fig 33.4 are large and quite robust. Suggest why.
</td></tr></table>

Fig 33.3 The main parts of a generalised flower

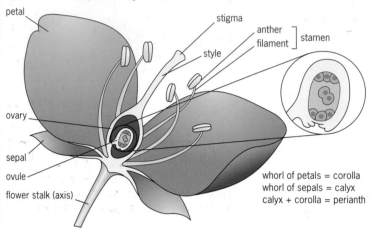

whorl of petals = corolla
whorl of sepals = calyx
calyx + corolla = perianth

Fig 33.4 Protea is a primitive flower. The petals and sepals, which are quite difficult to tell apart, have evolved from the leaves that surround the flower. This large South African flower is normally pollinated by sugarbirds

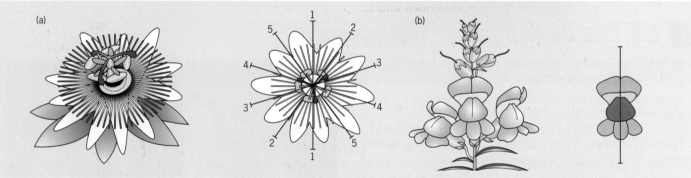

Fig 33.5(a) A flower with more than one plane of symmetry is said to be actinomorphic. This passion flower has five main planes of symmetry

Fig 33.5(b) A flower such as this snapdragon with a single plane of symmetry is said to be zygomorphic

✔ **REMEMBER THIS**

A **diploid** cell has two of each chromosome. In mitosis, one diploid cell divides to give two diploid daughter cells. These are genetically identical to each other and to the parent. In meiosis, one diploid cell divides to give four **haploid** cells, cells with half the number of chromosomes of the parent. Meiosis also involves genetic 'mixing', so that each of the haploid cells produced are genetically different from each other and from the original parent. See Chapter 25 for more detailed information about mitosis and meiosis.

Some flowers have not changed much since flowering plants first evolved. Flowers of primitive species such as the protea (Fig 33.4) usually have many petals and sepals. The numbers of sepals and petals present varies from flower to flower. More primitive flowers are usually **radially symmetrical**, or **actinomorphic**, as shown in Fig 33.5(a).

Other species have evolved quite a lot. Clover flowers, for example, have fewer petals, and the number is constant from flower to flower. Flowers that have evolved for insect or bird pollination (see the Science feature on page 620) are often **bilaterally symmetrical**, or **zygomorphic**, as shown in Fig 33.5(b). Although insect pollination is an important evolutionary step, many highly advanced plants, such as grasses, remain wind pollinated.

REPRODUCTIVE STRUCTURES: THE STAMEN

The size and shape of the reproductive parts of a flower vary between species, but both male and female parts always show the same basic structures. Fig 33.6 shows the structure of the male part, the **stamen**.

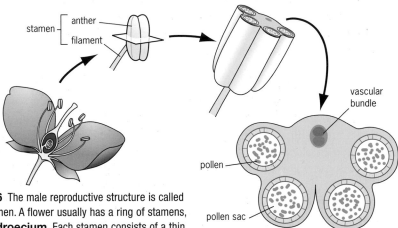

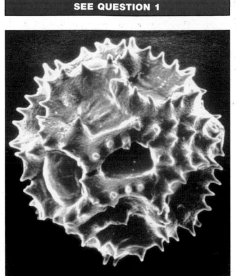

SEE QUESTION 1

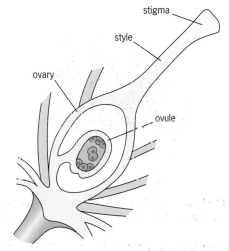

Fig 33.7 Dandelion pollen grain. Most pollen has a highly sculptured surface. It is often possible to identify a plant from a single grain of its pollen

Fig 33.6 The male reproductive structure is called the stamen. A flower usually has a ring of stamens, the **androecium**. Each stamen consists of a thin stalk, the **filament**, which supports the **anther**, a structure that contains four **pollen sacs**

Within the pollen sacs of the stamen, each *diploid* pollen mother cell divides by *meiosis* to form four *haploid* pollen cells. These mature into pollen grains (Fig 33.6) using the nutrients released as the surrounding tissue breaks down. When the pollen is ripe, the sacs split open and the pollen is released. By this point, the *haploid* pollen cell nucleus has usually divided once more, by *mitosis*, to form a haploid **generative nucleus** and a haploid **vegetative nucleus**. The generative nucleus forms the two **male gametes**, the **sperm cells**.

REPRODUCTIVE STRUCTURES: THE OVULE

Fig 33.8 shows the female structure, the **carpel**. Part of the carpel is stretched out into a long style. At the end of the style is a special surface called the **stigma**, which accepts pollen. The swollen part of the carpel is the **ovary**. Each ovary contains one or more **ovules**, depending on the plant.

The way the ovule develops is shown in Fig 33.9. Each ovule is made up of a rugby ball-shaped **nucellus**, surrounded by two layers of cells called **integuments**. Just one of the cells inside the nucellus becomes the **megaspore mother cell**. This divides by meiosis to give four haploid **megaspores**. Three do not develop. The surviving cell becomes the megaspore that develops into the **embryo sac**.

Fig 33.8 The female reproductive part of the flower is the carpel which consists of one or more ovaries. This carpel has just one ovary which contains a mature ovule and is ready for fertilisation

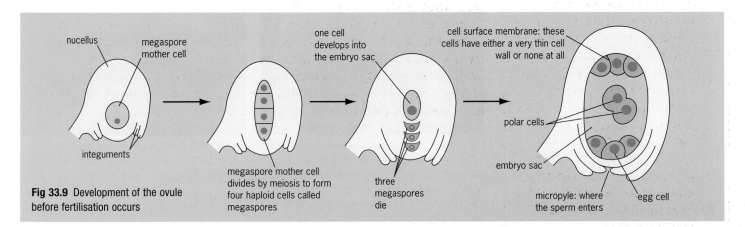

Fig 33.9 Development of the ovule before fertilisation occurs

Inside the embryo sac there are further nuclear divisions but the cytoplasm does not divide to form separate cells (ie there is no cell wall formation). Eventually, the embryo sac contains eight nuclei, three of which play a key role in fertilisation. Two of them, called polar nuclei, fuse together to form a diploid central cell. The third nucleus forms the **egg cell**. This is the female gamete.

POLLINATION

SEE QUESTION 2

The reproductive structures of the flowering plant have now formed. As plants cannot move to find a mate, they must transport their pollen, often over large distances, to make contact with a female plant from the same species. Once the pollen has made it to the stigma, the male gamete must make a smaller but equally crucial journey to meet the female gamete within the ovule.

Most plants are pollinated either by wind or by insects, but there are other **vectors** (pollen carriers), including bats, mice, slugs, birds, small reptiles and water.

WIND POLLINATION

Wind pollination, or **anemophily**, is the most straightforward method of pollination. The anthers of a plant ripen in dry weather and release pollen grains into the air. Pollen often spreads to a radius of 5000 metres around the parent plant, and some pollen grains are known to have travelled thousands of kilometres. The chance of any pollen grain landing on a stigma of the right plant is slim, so large amounts of pollen are produced.

Wind pollination works well for plants that exist in large numbers in a particular area. Grasses and many woodland trees, such as oak, birch and hazel, are successful wind-pollinated plants. Their flowers are not 'showy': they do not need to be. Instead, they have large feathery stigmas that stick out from the flower to catch pollen, as seen in Figs 33.10(a) and (b).

? TEST YOURSELF

B Look at Fig 33.10(b). What advantage does maize gain by having such large stigmas?

Fig 33.10(a) Hazel produces clusters of tiny flowers called catkins early in the year, so pollen grains do not get trapped by the leaves which develop later. Many trees make flowers (ie blossom) early in the year for the same reason. In spring, much of the pollen in the air comes from trees

Fig 33.10(b) Maize is a very large grass: its huge tassels are amongst the largest stigmas in the plant kingdom

INSECT POLLINATION

Insect-pollinated (or **entomophilous**) plants are pollinated by insect vectors (carriers), rather than the wind. The shape of their flowers has evolved to ensure that pollen brushes off on to the visitor and is later transferred to a stigma when the insect visits another flower. Such flowers need to be seen and so they advertise themselves with large white or brightly coloured petals. But, if the plant invests large amounts of energy in making large petals, it has fewer resources available to make gametes. So, many plants make 'cheaper', smaller flowers and crowd them together in clusters, called **inflorescences** (Fig 33.11).

Pollen grains are rich in oils and proteins, and producing large amounts drains resources from the plant. Insect-pollinated flowers tend to produce small numbers of larger and stickier pollen grains. Insect pollination is more efficient than wind pollination because less pollen gets wasted and so less is needed. Since pollen itself is nutritious, a plant may have to sacrifice some of it to the insect as a reward for successful pollination. But some plants have found that a more economical bribe of sugary water, nectar, can be just as effective. If insects get plenty of nectar, they are more likely to visit another flower of the same type. Both insect and plant profit. This mutual benefit is the basis of many plant–insect relationships (see the Science feature on page 620).

Fig 33.11 Some flowers have inflorescences which, at first glance, look like a single flower. Outer 'showy' flowers attract pollinators to the mass of smaller flowers crowded on to the flat disc

FERTILISATION

With pollen grains being produced in astronomical numbers, it is inevitable that much of the pollen that arrives on a particular flower will be of the wrong species. There are very precise biochemical recognition processes that ensure that only pollen of the same species will be able to complete the fertilisation process.

? TEST YOURSELF

C Is pollen from insect-pollinated plants likely to cause hay fever? Give a reason for your answer.

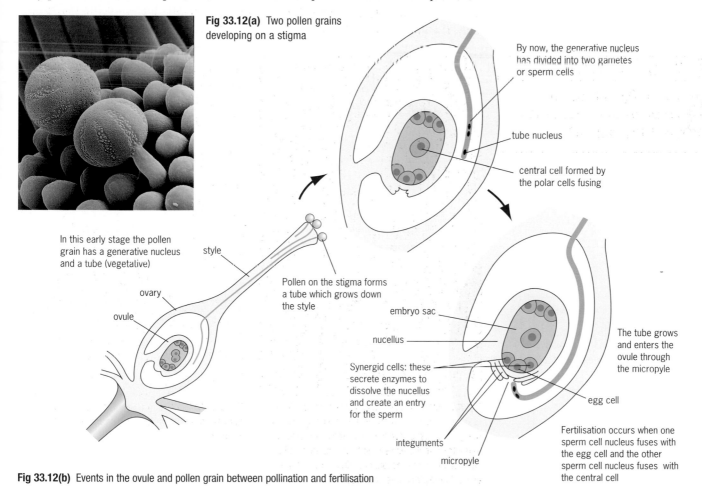

Fig 33.12(a) Two pollen grains developing on a stigma

By now, the generative nucleus has divided into two gametes or sperm cells

tube nucleus

central cell formed by the polar cells fusing

In this early stage the pollen grain has a generative nucleus and a tube (vegetative)

style

ovary

ovule

Pollen on the stigma forms a tube which grows down the style

embryo sac

nucellus

Synergid cells: these secrete enzymes to dissolve the nucellus and create an entry for the sperm

integuments

micropyle

The tube grows and enters the ovule through the micropyle

egg cell

Fertilisation occurs when one sperm cell nucleus fuses with the egg cell and the other sperm cell nucleus fuses with the central cell

Fig 33.12(b) Events in the ovule and pollen grain between pollination and fertilisation

When pollen lands on a stigma of the same species, pollination is complete and **fertilisation** can start. The process is shown in Fig 33.12. The pollen grain draws water out of the stigma by osmosis and swells. This bulge develops into a **pollen tube**, which penetrates the style and grows towards the ovule.

Each pollen grain contains two nuclei, the generative nucleus and the tube (or vegetative) nucleus. The tube nucleus directs the growth of the pollen tube. The generative nucleus forms two sperm cells. The nuclei are not separated from each other by cellulose walls: they simply share the cytoplasm, so they do not really occupy separate cells. The sperm and tube nuclei move down the pollen tube as it grows and they enter the ovule through a gap, the **micropyle**, to reach the embryo sac.

The embryo sac opens, the sperm nuclei enter, and then a **double fertilisation** occurs. In one fertilisation, one of the sperm nuclei fertilises the egg cell. The product of this fusion becomes the zygote, the diploid cell that develops into the embryo. In the second fertilisation, another sperm fuses with the central cell. This fusion produces a **triploid** cell (it has three copies of each chromosome: two from the diploid central cell and one from the haploid sperm). This triploid cell divides by meiosis to form the endosperm tissue. The endosperm contains the food supply that will nourish the growing embryo. Sexual reproduction in flowering plants always involves a double fertilisation.

Fig 33.13 is an overall summary of spore production, gamete production and fertilisation in a flowering plant. After fertilisation in most plants, the flower withers and the ovary swells to form the fruit containing the seeds. Inside each seed, the embryo develops.

? TEST YOURSELF

D Most of the cells of a plant are diploid: they contain two sets of chromosomes. How many sets of chromosomes are in
a) the sperm cell,
b) the central cell and therefore
c) the cell produced when the sperm and central cell fuse to form the endosperm tissue?

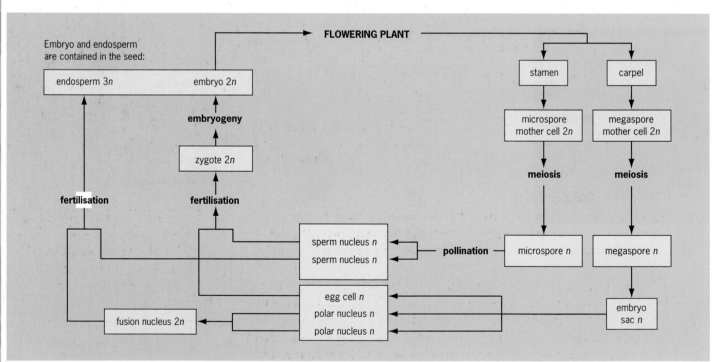

Fig 33.13 Summary of the life cycle of a flowering plant

DEVELOPMENT OF THE EMBRYO

At fertilisation, the egg and sperm nuclei fuse to form a diploid nucleus, the **zygote**, which develops into an embryo plant (Fig 33.14). This consists of a tiny root, or **radicle**, connected to the young shoot, the **plumule**, which later develops leaves. The first leaves, the **cotyledons**, are different from all later leaves. They are usually a different shape and have a simpler network of veins. Sometimes they are swollen with food reserves that nourish the embryo – see page 599 in Chapter 32.

When the central cell is fertilised it divides to form a special triploid tissue called the **endosperm**, which usually nourishes the embryo. In some plants, such as the maize in Fig 33.15(a), endosperm accumulates and is stored inside a **seed**, ready to feed the young plant when the seed **germinates**. In other species, such as peas (Fig 33.15b) and peanuts, the endosperm food reserves are used by the embryo at a much earlier stage and any spare food is transferred and stored within the cotyledons.

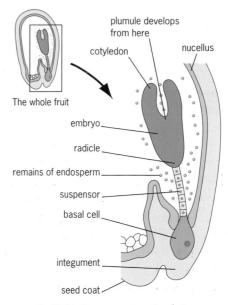

Fig 33.14 The very early embryo often appears round (but never forms a hollow ball of cells like an animal blastula). Soon the first leaves, the cotyledons, and the young root, the radicle, start to take shape, as in this shepherd's purse embryo. The embryo grows as the seed coat develops and the fruit ripens

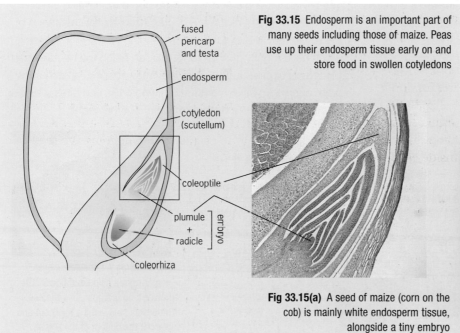

Fig 33.15 Endosperm is an important part of many seeds including those of maize. Peas use up their endosperm tissue early on and store food in swollen cotyledons

Fig 33.15(a) A seed of maize (corn on the cob) is mainly white endosperm tissue, alongside a tiny embryo

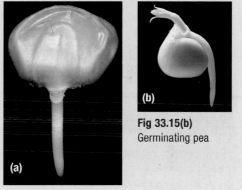

Fig 33.15(b) Germinating pea

FRUITS AND SEEDS

Once the seed is released from the parent it can develop into a new plant. But most seeds don't start growing, or germinating, straight away. If seeds are **dispersed**, the parent is not overcrowded by offspring competing for light, water and nutrients.

Seeds survive dispersal because the **testa**, a tough coating, prevents them from drying out or being digested by animals that eat them. The testa forms when the outer part of the ovule is strengthened by the polysaccharides **cutin** and **lignin**. The testa is sometimes so tough that the seed needs harsh treatment, such as freezing, soaking or a period of wear and tear in the soil, before it can germinate. This ensures that seeds do not germinate until the following year, long after the winter cold, or the summer drought. There's more about germination in Chapter 32.

Some seeds are light enough to be carried on the wind, like pollen. But most seeds need help to disperse. This help comes from the **fruit** that surrounds them. In some plants, the ovary wall develops into a fleshy, soft

? TEST YOURSELF

E Are peas monocots or dicots? (See Fig 33.15 for clues.)

F Why are many fruits brightly coloured?

SEE QUESTION 3

fruit, designed to entice animals to eat it. After digestion, the seeds pass out in the faeces, perhaps many miles away, still protected by the testa. Other fruits have explosive mechanisms to throw the seeds about, and others have hooks to catch on animals' feet or fur to 'hitch a lift'.

! SCIENCE FEATURE The intimate relationship between plants and their pollinators

Insects and flowering plants have evolved together. Some species are now completely dependent on each other: without the plant the insect doesn't get food, and without the insect the plant cannot reproduce. Most plant–insect relationships are not as narrow as this: most plants attract a range of potential pollinators. For each species the range is never very great, because particular colours, shapes, scents, opening times, flowering seasons and so on all tend to be attractive to particular groups of insects.

Red clover, for example, must be **cross-pollinated** (pollinated by pollen from another plant) before it can produce seeds. It is pollinated by a handful of different insects, but bees are the most important.

Red clover is adapted to pollination by bees in several ways. As Fig 33.16(a) shows, the flowers develop as inflorescences.

Each flower in the cluster has a long tube with a tiny hole at the top that gives access to the nectar, as shown in Fig 33.16(b). A visiting bumblebee inserts its proboscis and long tongue into the tube to feed on the nectar. As the bee does this, it brushes its underside against the anthers and picks up sticky pollen. When the bee visits another red clover flower, its underside brushes against the stigma, some of the pollen is transferred, and so pollination is accomplished.

The tiny access hole and the length of the tube make it hard for other insects to reach the nectar. Alternatively, the shape of the flower may prevent them from landing on it. Honeybees, which have slightly shorter tongues, can reach the nectar when the level is high enough. If not, they may adopt a 'smash and grab' approach, as in Fig 33.16(c), where the bee bites a hole in the flower near the base and sucks out the nectar.

Fig 33.16(a) An inflorescence of red clover, *Trifolium pratense*. Individual parts of the cluster are small, but the inflorescence is easily spotted by a flying bee

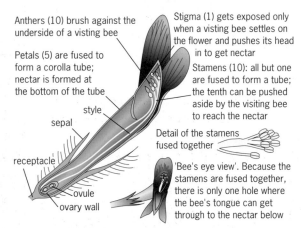

Anthers (10) brush against the underside of a visiting bee

Petals (5) are fused to form a corolla tube; nectar is formed at the bottom of the tube

style

sepal

receptacle

ovule
ovary wall

Stigma (1) gets exposed only when a visiting bee settles on the flower and pushes its head in to get nectar

Stamens (10): all but one are fused to form a tube; the tenth can be pushed aside by the visiting bee to reach the nectar

Detail of the stamens fused together

'Bee's eye view'. Because the stamens are fused together, there is only one hole where the bee's tongue can get through to the nectar below

Fig 33.16(b) A single flower from red clover

Fig 33.16(c) A honeybee steals nectar from a field bean flower. The flower tube is longer than the bee's proboscis so it takes a short cut and pierces the bottom of the flower. Short-tongued bees rob red clover flowers in the same way

---

2 PLANTS AND SEX: STRATEGIES IN REPRODUCTION

Sex in plants and animals is vital for two reasons: it creates new offspring and it shuffles the genes, giving variation. If plants remained genetically the same they would never evolve. That said, sometimes a plant will remain unpollinated – perhaps it grows too far away from others of the same species. It is better to have some offspring instead of none at all, so many plants in this situation pollinate and fertilise themselves.

Self-fertilisation leads to less variation among the offspring, so this creates a dilemma. If self-pollination is too easy, a plant might end up doing this all the time, and so miss out on the chance to shuffle its genes with others. To ensure cross-fertilisation and the advantages that this brings, many plants have developed strategies that make self-fertilisation less likely than 'mating' – but not completely impossible, just in case.

CAREFUL TIMING

Many plants have a definite flowering season. Within that season, individual plants produce the stamens and carpels of their flowers at slightly different times to increase the chance of cross-pollination (Fig 33.17). In the system called **protandry**, the stamens develop first and pollen is shed from the anthers before the stigma of the flowers on the same plant can accept it. When the female parts develop before the male, this is **protogyny**.

DIFFERENT FLOWER SHAPES

Primroses make two types of flower, the **thrum** and **pin** forms, though never together on the same plant: this is known as **heterostyly**. Thrum and pin flowers differ in the heights of the stamens and the stigma within the flower (Fig 33.18). Insect visitors to thrum flowers pick up pollen at a point on their proboscis that is level with the stigmas in pin flowers. Similarly, pollen from pin flowers is more likely to be transferred to stigmas in thrum flowers.

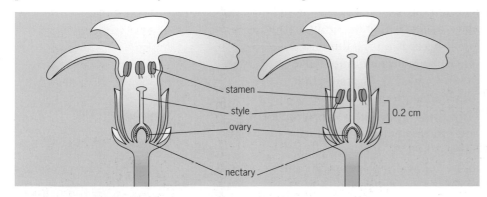

KEEPING THE SEXES SEPARATE

Some species have separate male and female flowers, but both may grow on the same plant (Fig 33.19). Other species are **dioecious**: they have separate sexes, so self-pollination is impossible (Fig 33.20).

PLANTS CAN BE CHOOSY ABOUT THE POLLEN THEY ACCEPT

Even if a plant species does pollinate itself, it can make self-fertilisation more difficult. In some plants, special genes slow up the rate at which the pollen tube can grow. Pollen from a genetically different parent may develop more rapidly and overtake a tube that is growing from its own pollen.

Recently, researchers have also found that some plants can recognise the pollen of *related* plants, as well as their own pollen, and also prevent it from growing properly. This **kin recognition** enables them to avoid self-fertilisation and fertilisation by close relatives.

Fig 33.17 In this plantain flower, the stigmas at the top are ready to receive pollen but the stamens round them are not fully ripened

? TEST YOURSELF

G Why is it important that thrum and pin flowers are produced on different plants?

Fig 33.18 Structure of the thrum and pin primrose flowers

Fig 33.19 Courgettes have separate (larger) male and (smaller) female flowers on the same plant. Insect visitors have to travel from flower to flower, so cross-pollination is made more likely

Fig 33.20 The common holly, *Ilex aquifolium*, has plants that are of separate sexes, so cross-pollination is assured. The photograph on the left shows male flowers; the photograph on the right shows female flowers

ASEXUAL REPRODUCTION

Many flowering plants can reproduce **asexually** as well as sexually. While sexual reproduction adds variety to the species, asexual reproduction allows plants to increase their numbers or to survive between growing seasons, without going through the complex processes of gamete formation, pollination, fertilisation and seed production. It gives them extra flexibility: they can still reproduce and spread, even outside the flowering season.

VEGETATIVE PROPAGATION

Many flowering plants can produce new plants without forming flowers and seeds. A single parent gives rise to offspring that are genetically identical, both to it and to each other. We call this type of reproduction **vegetative propagation**.

Along much of our coastline there are extensive sand dunes. Many of the dunes nearest the sea are held in place by marram grass. Only marram grass succeeds on the dunes where other plants become smothered by the shifting sand. Marram grass forms underground stems, **rhizomes**, which grow very rapidly and produce new roots and leaves at intervals (Fig 33.21). A single individual will quickly spread and colonise a newly formed dune.

Other plants reproduce asexually because they can regenerate and repair themselves: fragments of stems and leaves broken off by passing animals may fall to the ground, root and grow into new plants. Gardeners and farmers take advantage of this ability in the well-known practice of 'taking cuttings'. Many modern varieties of apple are maintained by taking cuttings and then grafting the young plants on to a special rootstock that keeps the tree short. Vines are often grafted on to special roots that are resistant to *Phylloxera*, a small weevil that nearly ruined the wine industry in France in the nineteenth century, and that now threatens to do the same in California.

Plants can reproduce asexually by the way that they overwinter. During the growing season they develop specialised underground stems, or rhizomes, which they pack with stored food. When cold winter temperatures make the plant die back, the rhizomes remain alive and then use the stored food to produce new shoots in the next growing season. **Corms**, **bulbs** and **tubers** (Fig 33.22) are other modified shoots adapted for food storage over winter. All these structures branch from **buds** (the 'eyes' on potatoes are buds). Each bud can form a new individual the following season.

Fig 33.21 Marram grass can grow in shifting sand. The lines of plants grow from a rhizome, a horizontal stem just below the surface

SEE QUESTION 4

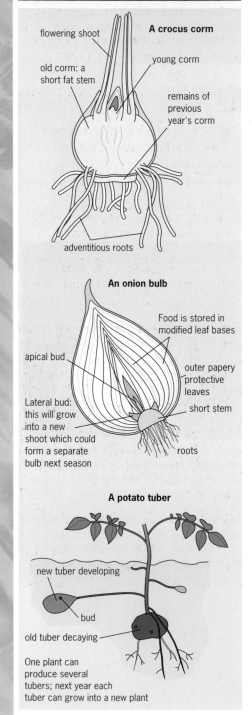

A crocus corm

flowering shoot

old corm: a short fat stem

young corm

remains of previous year's corm

adventitious roots

An onion bulb

Food is stored in modified leaf bases

apical bud

outer papery protective leaves

Lateral bud: this will grow into a new shoot which could form a separate bulb next season

short stem

roots

A potato tuber

new tuber developing

bud

old tuber decaying

One plant can produce several tubers; next year each tuber can grow into a new plant

Fig 33.22 The structure of a typical corm, bulb and tuber

3 COMMERCIAL PLANT BREEDING

Most of our food crops are flowering plant species that have been selectively bred over many generations. When you have got a crop with exactly the characteristics you want, why risk losing it in the lottery of sexual reproduction?

Many of our crops naturally reproduce asexually. Potatoes make tubers, strawberries make runners and onions make bulbs – all are organs of vegetative reproduction. Once selective breeding has produced the desired characteristics, such as large, sweet strawberries that have good frost resistance, the genotype can be maintained generation after generation by **vegetative propagation** in which we take advantage of the plant's natural methods of asexual reproduction.

For crops that don't reproduce asexually, it is necessary to be a little more devious. Some plants will grow from cuttings: you simply take a piece of the parent plant, separate it, put it into soil and it will grow roots. For more reluctant species, rooting powder containing an **auxin** (see page 596 in Chapter 32) can be used.

Fig 33.23 This strawberry plant has grown several runners with a small plant at the end. The plantlets will take root and grow into new plants. Each will have exactly the same genes as the parent

In plants that won't take root so readily, it is often easier to **graft** (Fig 33.24). This is especially common in fruit trees. A shoot from the required crop, called the **scion**, is inserted into a slit in the bark of the **rootstock** plant so that the cambium tissues match. There is not the same problem of rejection that you get in animals, and the cambium is able to differentiate into new xylem and phloem. Once the 'tubes' are connected, the rootstock provides the graft with water and minerals, while the graft provides the stock with sugars and other products of photosynthesis. You can even have two (or more) different grafts on the same rootstock so, for example, you could have Granny Smiths and Cox's Orange Pippins growing on the same apple tree. This even works with apples and pears. The fruit produced is a result of the genes of the graft, not of the rootstock.

A good example of this is seen in the growth of apricots. Apricot branches are often grafted on to a plum tree rootstock – this will produce more apricots, and be more disease resistant, than a 'pure' apricot plant.

Table 33.1 The pros and cons of vegetative reproduction

Advantages	Disadvantages
Growth is faster because the plants often has food reserves, eg in a tuber	Vegetative structures (tubers, bulbs, runners) are bulky – difficult to transport and store compared with seeds
It is easier and cheaper than collecting and growing seeds	Tubers, etc, have a high water content – more likely to rot, ie shorter shelf life than seeds
Genetic stability is maintained – all plants are clones of the parents. This is the key advantage	Tubers, etc, are more disease prone. If original stock is contaminated, the whole crop may be infected
Plants grow at the same rate and can be harvested at the same time	No variation – being clones, if one individual gets a disease they are all likely to be susceptible

Fig 33.24 Grafting basically allows a big, tough tree trunk with its well-established root system to support a more 'delicate' crop. In this example of a pear tree, the scion is bound in place with raffia

MICROPROPAGATION

The techniques of vegetative propagation described above have been practised for decades or centuries, but now valuable crops can be cloned by **micropropagation**. This technique involves **tissue culture** in which new plants are grown from tiny samples taken from virtually anywhere on the parent plant. The samples are placed on a sterile nutrient medium that also contains a carefully balanced mixture of hormones (auxin and cytokinins – see page 597 in Chapter 32).

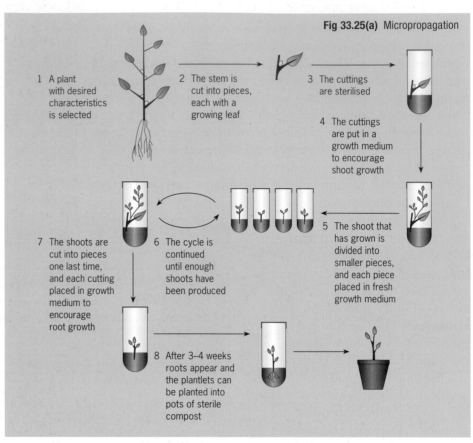

Fig 33.25(a) Micropropagation

1 A plant with desired characteristics is selected

2 The stem is cut into pieces, each with a growing leaf

3 The cuttings are sterilised

4 The cuttings are put in a growth medium to encourage shoot growth

5 The shoot that has grown is divided into smaller pieces, and each piece placed in fresh growth medium

6 The cycle is continued until enough shoots have been produced

7 The shoots are cut into pieces one last time, and each cutting placed in growth medium to encourage root growth

8 After 3–4 weeks roots appear and the plantlets can be planted into pots of sterile compost

Fig 33.25(b) Constant environment growth rooms are used for raising cultured plant material such as oil palm plantlets. The sterile conditions mean the plantlets are kept free from disease. Environmental conditions, eg temperature, are kept at the best levels for healthy growth.

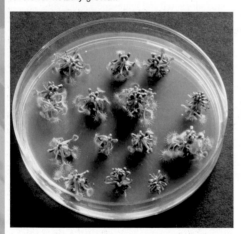

Fig 33.25(c) Round-leaved sundews, *Drosera rotundifolia*, being grown from tissue cultures. The jelly in the Petri dish is a sterile soil substitute containing the nutrients needed by the growing plants. These plants are clones grown from tiny fragments or single cells of a parent plant in the micropropagation process

This technique works well with crops that are difficult to reproduce by more traditional methods, and can be used to produce virus-free crops. But perhaps the biggest advantage is that many generations of plants can be reared quickly and in a minute space (Fig 33.25).

SUMMARY

After reading this chapter, you should know and understand the following:

- Sexual reproduction enables plants to pass their genes on, add to the number of the species, increase variation and evolve.
- Flowering plants produce **spores** before they produce **gametes**. There are two types, **microspores** and **megaspores**.
- Microspores or **pollen grains** are transferred by wind or an animal vector to the stigma of another flower, often on different plant of the same species.
- The pollen grain grows into a tube that penetrates the female part of the flower. **Sperm nuclei** travel along this tube to meet the **egg**.

- **Double fertilisation** occurs in plants: this leads to a **zygote** and a **triploid cell**. The zygote grows into an **embryo**, the triploid cell develops into a food store, the **endosperm**.
- The **ovule** develops into a seed, which protects the embryo for a while. The **ovary** develops into a fruit.
- There are a variety of means to promote cross-pollination.
- Many plants can self-fertilise, but many have strategies to make sure that they cross-pollinate.
- Many plants reproduce vegetatively, by **asexual reproduction**.
- Commercial plant growers can **clone** plants, and keep all desirable characteristics, by taking cuttings, grafting or by **micropropagation**.

? EXAM QUESTIONS

1 The diagram shows the structure of a mature anther seen in cross (transverse) section. In the spaces provided name the structures indicated by the letters **P**, **Q**, **R** and **S**.

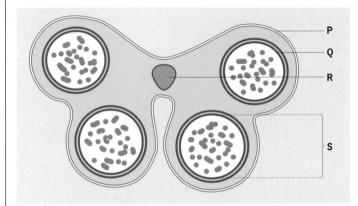

b) Sketch the diagram, and label **one** structure which has diploid cells with the letter **D**, and **one** structure which has haploid cells with the letter **H**.

c) Mature anthers dehisce (split open) when ripe.
   (i) On the diagram use a **labelled arrow** to show clearly where dehiscence occurs.
   (ii) Explain how the process of dehiscence is brought about.

d) Following dehiscence, pollen is dispersed.
   (i) Define what is meant by the term *pollination*.
   (ii) Outline the process of pollination in a **named** herbaceous plant.

OCR 6917 Plant Studies March 2001 Q1

2 The diagram shows a grass spike, *Poa* sp, with three flowers.

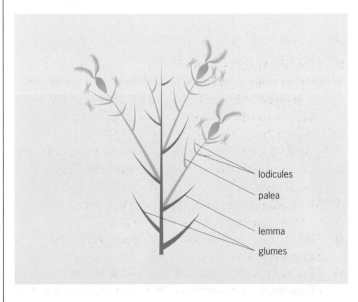

a) Sketch the diagram and label:
   (i) **one** ovary;
   (ii) **one** stigma;
   (iii) with a ruled line and the letter **M**, **one** site where **both** mitosis **and** meiosis occur.

b) State the pollination mechanism of this grass, and give **three** reasons for your answer.

c) Suggest **one** function for the glumes in grass flowers.

OCR 6917 Plant Studies March 2000 Q1

3 Give an account of the events following pollination until the formation of a seed in a flowering plant.
[essay question]

Edexcel B3 June 1999 Q8

4

a) Explain what is meant by *vegetative propagation*.

b) An apricot tree is usually produced by grafting a bud from an apricot stem on to the stem of a young plum tree. The drawing shows the stages in this process.

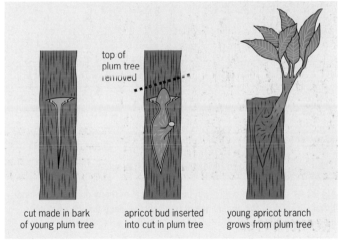

cut made in bark of young plum tree | apricot bud inserted into cut in plum tree | young apricot branch grows from plum tree

(i) Explain why the branch which grows from the bud will produce apricot fruits.
(ii) The apricot bud is grafted on to a plum tree rather than on to another apricot tree. Suggest **one** reason for this.

AQA (B) BYB2 Genes and Genetic Engineering June 2001 Q4

KEY SKILLS **ASSIGNMENT**

POLLEN IN PEAT

Pollen grains have an outer wall that is resistant to decay – so resistant that pollen grains that are hundreds of years old can be found in peat and lake mud. Researchers can tell a plant species just from its pollen: by digging cores of peat or soil and studying the pollen, pollen analysts can build up a picture of the plants that were growing in that place hundreds or thousands of years ago. In many sedimentary rocks pollen provides the only fossils that can be used to study changes in vegetation. The study of pollen is called **palynology**, and has been used to piece together how vegetation in Europe has changed since the last ice age 10 000 years ago.

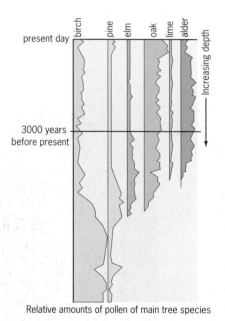

Relative amounts of pollen of main tree species

Fig 33.A1 The pollen data given here is typical of many places in lowland Britain. It reflects the abundance of tree species since the ice retreated about 10 000 years ago

1 **a)** What species of tree was most abundant during the coldest period immediately following the ice age?

b) Today, birch and pine trees can be found in exposed areas in some of the coldest parts of Britain, but trees such as elm, oak and lime cannot live in such conditions. Between 7500 and 5000 years ago the proportions of pollen changed considerably. So some plants must have done better and some fared worse. Suggest a cause.

c) One tree species seems to have suffered a decline at a time when there was no evidence for a change in climate. It has been suggested that the tree might have been eaten by cattle, or perhaps the land where it grew was cleared for crops. What tree species seems to have suffered a decline at the start of the bronze age?

Pollen grains can help archaeologists too

2 **a)** Look at Fig 33.A2. What does this suggest about the abundance of oak trees at this site?

b) Do any other species seem to have the same pattern as oak?

Fig 33.A2 This pollen diagram is from soil at a site where archaeologists were trying to understand the history of the area. From around 300 AD there was a decline in the amount of oak pollen found at this site

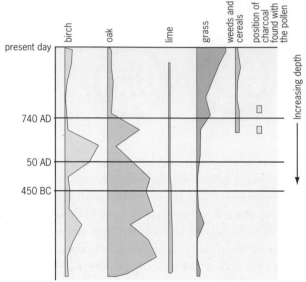

Relative amounts of pollen of selected species

3 Researchers had to consider several theories about the decline of oak, including:

1 It was due to a deterioration in the climate.

2 The land was cleared to make room for human settlement and cultivation.

3 Oak suffered from a disease.

a) What pollen evidence is there in favour of the second theory?

b) Is there any evidence against theories 1 or 3?

4 Is there anything to suggest that the land in the vicinity was cleared by humans? Explain your answer.

Fig 33.A3 Pollen grain of the oak, *Quercus robor* (left), and a tiger lily (right)

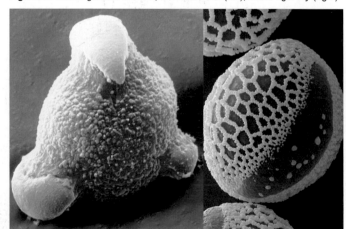

34

Ecology: the basics

Competition for a mate | Rainfall | Parasites | Number of other species in ecosystem | Seasonal changes in climate | Plant producers, eg grass | Competition for food

Number of animals in population | | | | | How much water is available | Hours of sunlight

Temperature | | | | | | Altitude of environment

Catastrophes – drought, hurricanes, fire, etc | General climate | | | | | Interaction with mate

Action of decomposers on body after death | | | Concentration of minerals in the soil | Time needed to rear young | Competition for space and territory

Fig 34.1 Ecology is the study of whole organisms: we learn how an organism interacts with individuals of its own species and other species, and with the physical environment that surrounds it. The jigsaw pieces show just a few of the factors that influence the lives of these zebras in the savannah grasslands of Africa

REMEMBER THIS

The word ecology is loosely used to mean environmental concern or environmental conservation. These topics are important, and they depend heavily on knowledge gained through ecology, but they are not the same thing. We define ecology as: **'the scientific investigation of living organisms in their natural surroundings'**.

1 THE SCIENCE OF ECOLOGY

Ecology is a branch of biology. The word ecology was first used by Ernst Haeckel in the 1860s. It comes from two Greek words: *oikos* meaning home and *logos* meaning understanding. His definition of ecology was 'the knowledge of the sum of the relations of organisms to the surrounding outer world, to organic and inorganic conditions of existence'. Put more simply, ecology is the study of organisms in their natural surroundings. It looks at how they are adapted to their environment and how they interact with both the living and non-living world around them.

Each organism and each physical feature of an environment is separate, but they all interact and interlock, forming a complex system that is a bit like a three-dimensional jigsaw puzzle: ecologists study different parts of the puzzle and then try to work out how it is put together (Fig 34.1).

ECOLOGY AND OTHER BRANCHES OF SCIENCE

Ecology is closely related to two other branches of science: **environmental biology** and **environmental science**. The first of these looks more generally at living things in the environment. The second uses the physical, biological and Earth sciences to help us to explore our environment.

It is also impossible to study ecology without bearing in mind the genetics, evolution and behaviour of the organisms at the centre of a study. Ecological genetics and behavioural ecology are beyond the scope of this book but it is important to remember that each organism you study in the context of its environment has developed to fit that environment through the process of natural selection (see Chapter 29). And, every time something in its environment changes, an organism may respond by showing an altered behaviour pattern. Some aspects of animal behaviour are covered in Chapter 23.

PRACTICAL ECOLOGY

As in other branches of science, progress in ecology depends heavily on scientific investigation. Looking back at our definition of ecology, it should come as no surprise that many of these studies are done outside. Field studies (Fig 34.2) involve a great deal of observation, but some experimental studies are also carried out in natural conditions. Others are carried out in the laboratory, where they can be more strictly controlled. Ecologists use these three basic approaches together to gain a complete picture of the organism or the environment under investigation.

Fig 34.2 Safety and field trips. Ecologists conduct studies outside and must consider safety when working near water or in isolated areas. Here, a student is identifying organisms from a pond, and two students are measuring the rate of water flow in a stream

The exciting thing about ecology is that it is still a relatively young science. In just over 100 years of ecological study, we have barely begun to even scratch the surface. Many environments of the world remain unexplored and there are many organisms that we know of, but whose habits and living conditions are known only sketchily, or are a complete mystery. Unlike biochemistry or physiology, ecology offers an A-level student the chance to do original practical work, and perhaps to make an important scientific discovery.

The Assignment in Chapter 35 is based on the work of real students on an ecology field trip.

! SCIENCE FEATURE Planning a field study

Before you can start taking measurements, you have to decide what kind of ecological investigation you want to do. There are two main types. **Synecology** is the study of an entire community of animals and plants: nettles, for example (Fig 34.3). **Autecology** is the study of a single animal or plant species. It includes an in-depth look at the habits of an organism together with its immediate environment (Fig 34.4).

Every study should be carefully planned. Like other scientists, ecologists follow a basic procedure called the **scientific method**. This involves five steps:

● They first put forward a **hypothesis**. A hypothesis is a statement that explains a particular problem or set of observations. The study sets out to test the hypothesis.

● The second step involves **data collection** (see pages 666 and 680).

● The third step is **data analysis**. Ecologists must summarise their data and put it into a form that allows conclusions to be made. Often, mathematical comparisons are made between different sets of data (see Appendix).

● In the fourth step, the data is used to **draw conclusions** (see Appendix).

● The final step involves presenting the results in a **report**, so that others can examine the study (see Appendix). Eventually, the results of investigations might be presented at conferences or written up in scientific journals. Some studies prove completely that a hypothesis is wrong. But even when a great deal of evidence exists to support a hypothesis and it has become widely accepted, nature is too full of surprises for scientists to say that anything is 100 per cent proven.

Fig 34.3 Caterpillars of the peacock butterfly feeding on a stinging nettle. About 100 species of invertebrate are associated with nettles: approximately 90 per cent of these are insects, but you are also likely to find woodlice, harvestmen, spiders and snails. You can use the plant and its animal partners to study a whole range of topics such as food chains, the effect of biotic and abiotic factors and population ecology

Fig 34.4
Autecological studies have been carried out on species as diverse as ladybirds, vampire bats and puffins

2 TERMS AND CONCEPTS IN ECOLOGY

The relationships between an organism and the physical features of its environment, and between all the other organisms that live with it, are incredibly complex. Any description of them uses terms that you have probably not come across before. So, before getting into detail about any particular aspect of ecology, the next section gives an overview of some basic terms and concepts.

THE BIOSPHERE AND ITS DIVISION INTO BIOMES

All the living organisms that we currently know of live on Earth. But not all parts of the Earth support life. The part that does forms a sort of 'skin', the **biosphere**, which is shown in Fig 34.5.

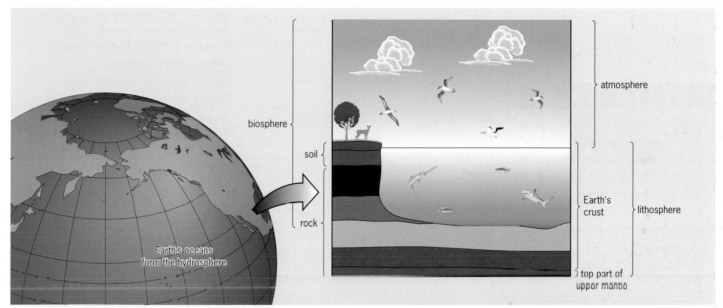

The biosphere is made up of lots of different areas that have very different environmental conditions. We call each of these fairly broad areas **biomes**. The major biomes of the world are listed in Table 34.1 on the next page. Many of the best-studied biomes are the **terrestrial biomes** (those that exist on land) – Fig 34.7 on the next page – but there are also **aquatic biomes** (those that exist in water) (Fig 34.6).

Fig 34.5 The biosphere is the part of the Earth's surface that supports life. It extends from the bottom of the deepest oceans into the part of the Earth's atmosphere that contains breathable air: in total a vertical distance of between 30 and 40 kilometres.
It covers most of the hydrosphere, the surface layers of the lithosphere and some of the atmosphere. The hydrosphere is the part of the Earth's surface that is covered by water. The lithosphere is the layer of soil and rock that forms the Earth's crust. The atmosphere is the blanket of gases that envelop the planet, maintaining a mean temperature of about 7 °C. This temperature is crucial to life because it allows most of the water on Earth to exist as a liquid

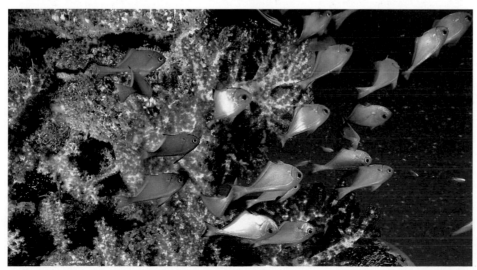

Fig 34.6 A coral reef is one of the major aquatic biomes of the world

THE MAJOR BIOMES

The biomes of the world show great extremes of environmental conditions.

Table 34.1 The major biomes of the world

Biome	Type	Major features
Savannah	terrestrial	Tropical grassland with few trees. Example: the Serengeti in Africa.
Temperate grassland	terrestrial	Grassland with hot summers and cold winters. Some broad-leaved plants. Example: the prairies in the USA.
Desert	terrestrial	Hot and dry. A few highly specialised plants and animals. Example: the Kalahari desert in Africa.
Tundra	terrestrial	Cold for most of the year: very short growing season. No trees: some specialised plants. Example: Northern Canada.
Tropical rainforest	terrestrial	Tropical climate. Lush vegetation. Great diversity of animal and plant life. Example: forests of equatorial Africa.
Temperate rainforest	terrestrial	Cool and wet. Tallest trees in the world. Example: redwood forests of North America.
Temperate deciduous forest	terrestrial	Hot summers and cold winters. Dominated by broad-leaved deciduous trees. Example: oak woodlands of central and western Europe.
Lakes and ponds	aquatic: freshwater	Large bodies of standing freshwater. Example: Great Lakes of North America.
Streams and rivers	aquatic: freshwater	Flowing freshwater. Example: the Amazon.
Marine rocky shore	aquatic: sea water	The border between the land and the sea. Most rocky shores show zonation: there are characteristic bands of different environmental conditions that lead to colonisation by very different plants and animals.
Coral reef	aquatic: sea water	The tropical rainforests of the seas. Reefs form in warm, shallow waters and form a biome that contains an incredible diversity of living organisms.

Fig 34.7 More biomes of the world

(b) Arctic tundra

(c) Desert

(a) Rainforest

THE ECOSYSTEM CONCEPT

The **ecosystem** is the basic functional unit of ecology. It is a single working unit that consists of a group of interrelated organisms and their physical environment (Fig 34.8). We can divide any one of the biomes described above into many different ecosystems.

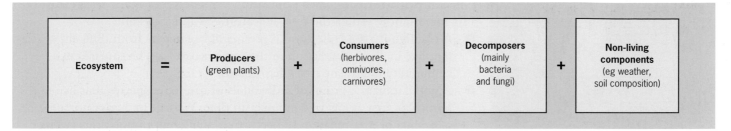

Fig 34.8 A schematic diagram showing the essential features of an ecosystem

When we study a particular organism in its ecosystem, we look at the physical features that might affect it, such as rainfall, soil type, temperature and so on. These physical or **abiotic** features help to determine what range and type and numbers of living organisms live inside the ecosystem. Of course, we also look at the living or **biotic** part of the ecosystem, to see how our chosen organism relates to the other organisms present.

Later in this section, we look at the abiotic and biotic features that affect the distribution of organisms in an ecosystem in much more detail (see Chapter 35).

LEVELS OF ORGANISATION IN AN ECOSYSTEM

Individual living organisms may look randomly arranged in an ecosystem, but they are organised into recognisable units. Fig 34.9 provides an overview of organisation in an ecosystem.

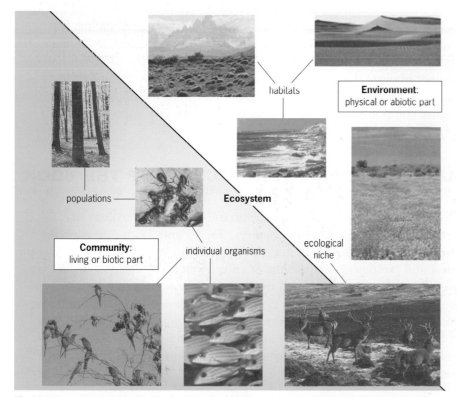

Fig 34.9 An ecosystem has two basic parts: the abiotic part and the biotic part. Within both of these, there are further levels of organisation

Fig 34.10 This individual stoat is part of a larger population. Like other members of the group, she is genetically unique. She has evolved to become well adapted to her woodland environment and, if her particular genes allow her to live a long life, she will pass her genes on to several new members of the next generation

FROM ECOSYSTEM TO INDIVIDUAL

We can study the human body and narrow down our investigation from a body system to an important organ, to the level of the individual cell. In a similar way, we can focus in on the components of an ecosystem. Every ecosystem contains a **community** of organisms, which consist of many different **populations**. Each population is made up of many **individuals**. It is important to define these basic features of an ecosystem:

- A community is a collection of groups of organisms from different species that live in close association in the same ecosystem. Organisms in a community interrelate (see Chapter 36).
- A population is a group of individuals of the same species that live in a particular area at any one moment in time. We can look at a population of rabbits or a population of beech trees. Different populations interact in an ecosystem, forming a community.
- An individual is a single organism within a population. Some populations contain organisms that are genetically identical. Some plant populations, for example, are identical because they have developed as a result of asexual reproduction (see page 622). But most populations contain individuals that are the result of sexual reproduction: these individuals show genetic variation (Fig 34.10 and see Chapters 26 and 33).

ENVIRONMENT, HABITAT AND ECOLOGICAL NICHE

The divisions listed above relate primarily to the *living* components of an ecosystem. We can also sub-divide an ecosystem with reference to its *non-living* components. The term **environment** describes the overall physical surroundings that occur in a biome or an ecosystem (although it is often used more generally: people speak of the environment of the Earth).

Within the environment of a single ecosystem are **habitats**, individual areas in which particular groups or individual organisms live. The habitat of any organism is its normal home. Some organisms, the giant panda, for example, have only one habitat – bamboo forest. Others, such as humans, live in many different ecosystems and biomes and so have many different habitats.

Another important term is **ecological niche**. To describe an organism's niche, we need to show how that organism relates to the physical and biological components of its surroundings. It is not just *where* it lives, it is also *how* it lives and *what* it does.

3 AN OVERVIEW OF THIS SECTION OF THE BOOK

In this section of the book, we provide an introduction to ecology at advanced level. In the next three chapters, 35, 36 and 37, we describe the characteristics of an ecosystem in greater detail. In Chapter 35, we look at the physical and biological components of an ecosystem. In Chapter 36, we see how energy is transferred through the living component, and how nutrients cycle through both the living and non-living components. In Chapter 37 we examine how organisms within communities and populations interact. In Chapter 38 we find out how human activity can influence the local and global environment and look at some of the important issues that concern ecologists today.

35

The biology of ecosystems

The biology of ecosystems

A rabbit suffering from the symptoms of myxomatosis, the deadly viral disease that wiped out most of the rabbit population. Most rabbits today are resistant to the virus

The abiotic and biotic parts of any ecosystem interact in many different ways, some subtle and some not so subtle. Australia provides one of the most dramatic examples of how disrupting the delicate balance of an ecosystem can devastate the environment and the populations of living organisms that depend on it.

Two hundred years ago, Australia was hot, dry and covered predominantly by grasslands, with some areas of forest. When European settlers arrived they started to clear the forests and brought in two alien species of mammal: sheep and rabbits. The rabbits in particular multiplied rapidly in conditions that suited them very well, and were soon eating the grass faster than it could grow.

Within ten years of the introduction of just 24 rabbits, the wild population had grown to millions and started destroying crops and wild vegetation as well as any grass that was left. They out-competed native mammal species, driving many of them to extinction. This plague continued until the late 1950s, when farmers introduced the myxomatosis virus. This killed over 99 per cent of the rabbits, with a few surviving because they were naturally resistant. Today, Australia is in danger of being over-run by another plague of rabbits.

1 THE BASIC FEATURES OF AN ECOSYSTEM

Fig 35.1 A schematic diagram showing a garden pond as an ecosystem. Every living organism in the pond is affected by abiotic factors such as temperature, light levels and mineral availability, and by biotic components – other living organisms

Before looking in detail at the components of an ecosystem, we shall consider an example of a real ecosystem. Fig 35.1 shows the main features of a familiar aquatic ecosystem, the garden pond. This is simplified (a real pond contains many more organisms), but you can see that an ecosystem is made up of physical or abiotic components such as water, mineral nutrients, soil and rock, and also living or biotic components such as fish, snails, insects and water plants. (See Chapter 34 for an introduction to the terms abiotic and biotic.)

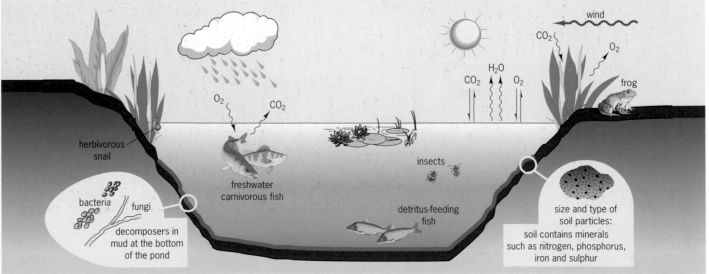
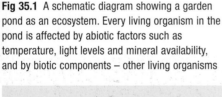

Think about one of the insects in the pond. It can survive and live there only if the physical conditions in the pond suit it. So, the pH and the temperature of the water must be within a suitable range. The insect also needs to interact with other organisms: it needs to eat smaller organisms or to feed on the sap of plants; and it needs to be with other insects of the same species so that it can find a mate and reproduce. And there is a 'grey area': some of its physical needs, for instance dissolved oxygen, are only available if the insect lives with other organisms, such as a large population of phytoplankton which release oxygen as a product of photosynthesis.

2 THE ABIOTIC COMPONENTS OF AN ECOSYSTEM

The abiotic components of an ecosystem give it a physical form, providing places for organisms to live. Physical conditions help to determine where each organism survives best: plants adapted to dry, almost desert conditions will not thrive if a river changes its course and the environment becomes boggy and wet. The nature of the physical environment in any ecosystem depends on:

- the rock and soil types present;
- the type of landscape: mountain range, flood plain, etc;
- the position on Earth: altitude, latitude, etc;
- the climate and weather, including light, temperature, water availability, wind and water movements;
- the potential for catastrophe: fire, flood, etc.

Some abiotic factors, such as the position of an ecosystem on Earth, remain unchanged as time passes. Others, such as the landscape and soil type, change very slowly. Although the basic climate of an ecosystem is usually pretty constant, you can get daily and often hourly changes in weather. Catastrophes such as fires and floods are sudden and short-lived, but they can produce severe, long-lasting effects (Fig 35.2).

Fig 35.2 In some ecosystems, catastrophic events are necessary. Some species of plant, such as *Banksia* from Western Australia (below) reproduce only after a serious fire (below left). *Banksia* seeds are protected by closed capsules that can be broken open only by the heat of the flames: without exposure to fire, the seeds cannot germinate

Fig 35.3 Soil erosion is a serious problem in many parts of Australia. Originally, the soil was held in place by grass on which only kangaroos grazed. With human settlers came huge numbers of domesticated sheep and a rabbit population that grew at a phenomenal rate. Both of these animals ate the grass faster than it could grow. With no plant roots to hold the soil together it blew away or was washed away by rain. The result is the semi-desert conditions that exist there today

SOIL AS AN ABIOTIC FACTOR

The effects of subtle differences in soil type are important to individual species of plants and animals. To gain an overview of how soil type affects plant populations and communities, it is useful to look at two ecosystems with extreme soil compositions. Fig 35.3 shows a semi-desert in central Australia. The 'soil' here is mainly sand, with little organic matter and virtually no microorganisms. This soil cannot easily support plant life: it does not retain any water that falls on it, it warms up very quickly to high temperatures and it contains hardly any mineral nutrients.

The plants that survive here have developed extreme strategies to cope with the extreme environment. Many have very short growing seasons, completing their whole life cycle in the 30 days that follow a decent spell of rain. Others survive long droughts as underground bulbs. Only a few, the succulents, manage to survive above ground for most of the year (Fig 35.4). Because so few plants live in a desert soil, few animals can live there either, and there is little material available for decomposition.

Our second ecosystem example, a peat bog (Fig 35.5), is the opposite of a desert. Peat bog soil is permanently wet and cold, it contains little air and, as a result, the organic matter in it cannot decompose. The peat soil is very deficient in nutrients, which remain locked up in the undecomposed matter.

Fig 35.4 Although these Australian succulents (the plants with the vivid magenta flowers, bottom left) are close to the sea, the sand dunes on which they grow retain little water, and are baked by the Sun all day. We say that these and other succulents are xeromorphic (see Chapter 31): they are adapted to living in a hot, dry environment

Fig 35.5 Areas that contain peat bogs get a lot of rain. Since rainwater contains only small quantities of mineral ions, rain does not add to the mineral content of the peat. In addition, rain leaches any nutrients that are in the peat. Most plants fail to grow in this sort of soil: their roots become waterlogged and they cannot absorb enough minerals. As sphagnum mosses grow in a peat bog, their lower parts decay and add to the mounting pile of organic debris below them

Sphagnum mosses are among the few plants that achieve any success in a bog ecosystem. A few other species manage to survive by obtaining minerals by 'alternative' methods: carnivorous plants, such as Sundew catch insects, and plants such as the bog myrtle have nitrogen-fixing bacteria in root nodules (see Chapter 36).

Soil type affects plants directly: they depend on soil to provide water and nutrients to manufacture food and to anchor them in place. The composition of soil may also directly affect some animals in the ecosystem, for example invertebrates such as worms and snails (Fig 35.6). However, more usually it affects animals indirectly: only those herbivores that can eat the plants that grow in a particular type of soil can live successfully in that area.

Soils at either extreme are not able to sustain many species of living organism, but soils that fall somewhere between these two extremes support more diverse ecosystems. Most of the soils in Britain, for example, are **loams**, soils that retain moisture year-round and contain a reasonable balance of different sized particles and decomposing organic matter.

THE MAIN SOIL TYPES

The soil is the upper, weathered layer of the Earth's crust. The importance of soil cannot be overstated: it underpins the Earth's ecosystems and therefore the agriculture and economy of all countries in the world. All soils contain the following:

- Mineral particles of various sizes. These come from the rock of the area which, when weathered, produces **sand**, **silt** and **clay**. Sand particles are the largest (60–2000 μm in diameter), silt particles are of intermediate size (2–60 μm in diameter) and clay particles are the smallest (less than 2 μm in diameter). We can classify and determine soil types by looking at how much of each type of particle is present in a soil sample (see Fig 35.8).
- Organic matter in various stages of decomposition. Organic matter that has been completely broken down by fungi and bacteria is called humus. Soils with a high humus content retain water much better and usually have a higher mineral content than soils with little organic matter.
- A collection of living organisms. Soil contains an enormous number of bacteria and fungi and some invertebrates such as earthworms, soil arthropods and nematodes.

These components are present in different proportions in different types of soil. As Fig 35.9 shows, there are many soil sub-types that are usually put into three main categories:

- **Sandy soils** are light, heat up quickly and allow water to drain away very rapidly. Sandy soils are made up of relatively large particles with relatively large air spaces. A soil made up of more than 90 per cent sand particles is called just 'sand'.
- **Clay soils** and **silty soils** tend to be cold, heavy and often waterlogged. They contain small particles with small air spaces. A clay soil is any soil with more than 40 per cent clay particles.
- **Loam**, an intermediate soil type, is dark and has a good crumb structure. Loam has a mixture of particles of different sizes and usually a fairly high humus content. Most agricultural and garden soils in Britain are loams.

Fig 35.6 The distribution of snails and slugs is sometimes determined by soil composition. Snails are usually found on chalky soils (they need the calcium to build their shells), slugs are more common on acid, non-chalky soils

! SCIENCE FEATURE Testing for soil type

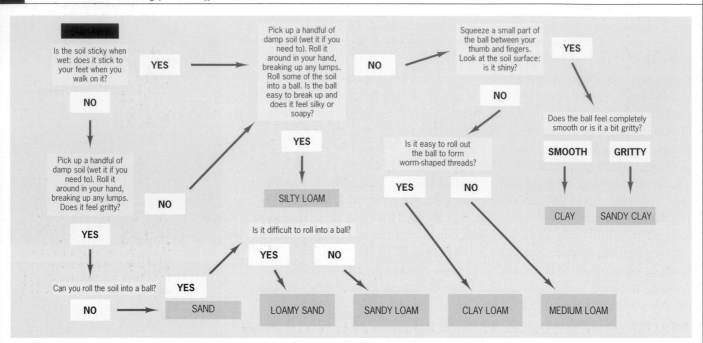

Fig 35.7 A simple key you can use to identify soil types. To carry out the test, take a desertspoonful of moist soil (wet it if necessary to obtain maximum stickiness) and knead it thoroughly between finger and thumb until it 'crumbs'. Then, go through the key

You can find out what type of soil you have at a particular study site using a key devised by the Agricultural Development and Advisory Service of the Department for the Environment, Food and Rural Affairs (Fig 35.7).

Studying soil composition

You can demonstrate the basic composition of a soil sample by the **sedimentation method**. Mix a small handful of soil with water, shake well and pour into a large measuring cylinder. The inorganic contents settle out in order of size, heaviest particles first. You can then work out the relative amounts of sand, silt, and clay in the sample (Fig 35.8).

You can also measure the **moisture content** of soil. Take another small handful and weigh it carefully. Then put it in an oven set at 105 °C and leave it overnight to dry out. Weigh again and then ensure the sample is fully dried by returning it to the oven and leaving it for another two hours, before weighing again. You may need to repeat this last step until the weight of your sample stops decreasing.

The value of knowing how much water is present in a sample is fairly limited because the amount of water in any soil varies according to what time of day it is, how deep you dig for your soil sample and whether it has just rained. A much more useful measure is the **field capacity** of the soil type: the amount of water that it can hold. Field capacity is a fixed value for each soil type.

To determine the field capacity of a soil sample, dry it thoroughly, weigh it, and then pack it into a perforated container such as a kitchen sieve. Gently pour a known volume of water on to the soil and measure the volume of water that comes through the bottom of the sieve. The amount of water left behind is the field capacity of the dry mass of soil you have used (units = $cm^3\ g^{-1}$).

Soils of different types contain varying amounts of **organic matter**. To measure **organic content**, weigh and heat a sample of dried soil (you could use the one that you produced in the moisture content analysis) in an oven at 450 °C. This burns off all the organic material present. Weigh again after burning to work out how much organic matter your soil contains.

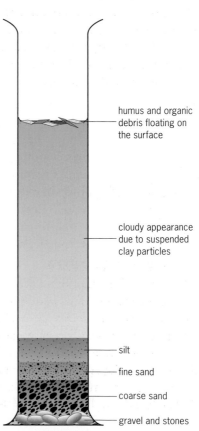

humus and organic debris floating on the surface

cloudy appearance due to suspended clay particles

silt

fine sand

coarse sand

gravel and stones

Fig 35.8 The sedimentation method of soil analysis

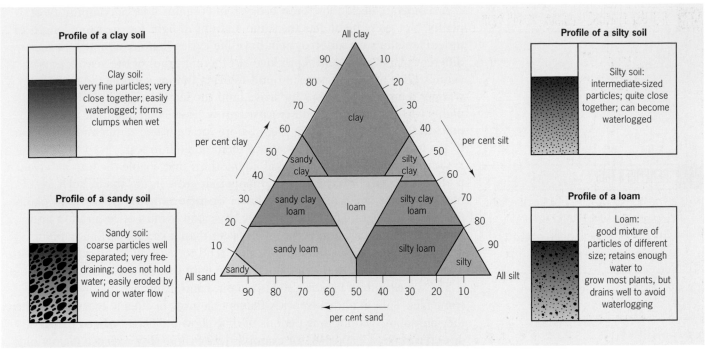

Profile of a clay soil

Clay soil: very fine particles; very close together; easily waterlogged; forms clumps when wet

Profile of a sandy soil

Sandy soil: coarse particles well separated; very free-draining; does not hold water; easily eroded by wind or water flow

Profile of a silty soil

Silty soil: intermediate-sized particles; quite close together; can become waterlogged

Profile of a loam

Loam: good mixture of particles of different size; retains enough water to grow most plants, but drains well to avoid waterlogging

Fig 35.9 A summary of the different soil types that can exist. All three extreme soil types can be improved: you can, for example, make a clay soil lighter and more fertile by digging coarse sand or fine gravel into it, and adding plenty of humus

THE EFFECT OF TEMPERATURE

The temperature range that living organisms can tolerate is narrow: most organisms live in temperatures between 0 °C and 40 °C, although some specialists manage to survive outside this range, between about –40 °C and 250 °C (Fig 35.10). Although large swings in temperature can cause problems, such as when a pond freezes during a sudden period of cold, living organisms have evolved to cope with the usual range of temperatures that occur in their normal environments. The strategies used by ectotherms and endotherms to survive in hot and cold environments are covered in detail in Chapter 15.

? TEST YOURSELF

D Why is it an advantage for plants such as daffodils to survive the winter as underground bulbs rather than as seeds?

Fig 35.10 Left: Bacteria that live in deep sea ecosystems next to hydrothermal vents in the ocean floor can survive temperatures as high as 250 °C. Right: Polar bears and other animals which live in polar regions can survive temperatures of –40 °C. They are well adapted to their environment, having thick fur and a thick layer of body fat

THE EFFECT OF LIGHT

Light often affects the distribution of animals in a particular ecosystem because, like soil type, it determines which plants grow there. As we saw in Chapter 9, light is *the* crucial environmental factor that affects photosynthesis. Light also affects and controls many other processes in plants, such as flowering, by interacting with the chemical **phytochrome** (see Chapter 33). The length of daylight also determines the behavioural patterns of animals, whether they are **diurnal** (active in the day) or **nocturnal** (active at night).

The effects of light depend on its:

- intensity
- wavelength
- duration
- direction

A temperate deciduous forest ecosystem contains a large community of plants. Not every plant gets the same amount of light for the same length of time. Plants in such an ecosystem have life cycles that are adapted so that different plants grow and reproduce at different times of the year. Plants cannot grow at temperatures around freezing, but as soon as the slightly warmer temperatures of spring arrive, and the length of daylight increases, plants such as primroses and crocuses grow. They complete their life cycle quickly, flowering and setting seed, before the light and temperature increase to the levels at which the trees above them come into leaf.

Other plants show physiological adaptations that enable them to survive under the cover of overhead trees. Plants that grow in the shade, such as bluebells, ferns and mosses have a lower **compensation point** than plants that grow in bright light. Shade plants have slower rates of respiration and reach their compensation point much faster than a sun plant. Shade plants often show anatomical adaptations: they tend to have thick, broad leaves with a large surface area, to trap as much of the light that reaches them as possible.

The wavelength of light is also important. Land plants use mainly red, and some blue, light in photosynthesis. Plants that grow in aquatic ecosystems where they are submerged for long periods of time show a different **absorption spectrum** (see Chapter 9): for example, lower-shore seaweeds use mainly blue light as this is the wavelength of light that penetrates the water at this level.

WATER AVAILABILITY

Water is an essential component of all living cells (see Chapter 4). In aquatic ecosystems, a long period of drought can cause a pond or stream to dry up, so killing the vast majority of living organisms that are part of the system. In terrestrial ecosystems, plants and animals are able to survive with smaller quantities of water, and some specialised organisms, such as succulents, cacti, camels and desert rats, can live without water for extended periods. Plants that have become adapted to dry conditions are covered in Chapter 31; some of the animals that live in hot, dry environments are described in Chapter 15.

On a less extreme scale, the amount of water that a particular plant receives in a season can affect its root growth. If freshly planted shrubs are watered every day with relatively small amounts of water (as would happen in a drought when there is a hosepipe ban), the plant develops a shallow root system that spreads sideways in the soil (Fig 35.11a). If they are watered less frequently, but with larger quantities of water, the root system extends deeper into the soil, deep enough to reach the level of groundwater (Fig 35.11b). If the watering stops and the drought continues, as might happen when gardeners go away on holiday, the deep-rooted plants stand a much better chance of survival.

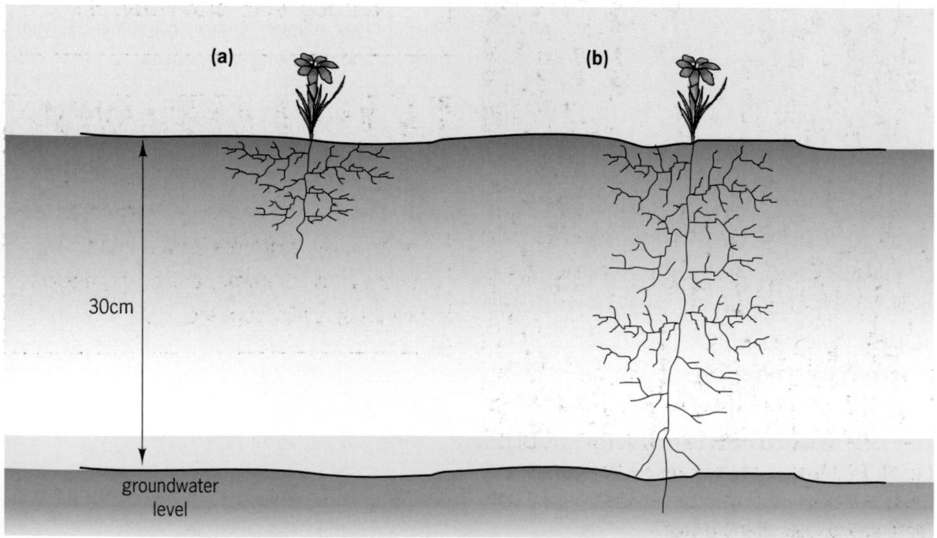

> **REMEMBER THIS**
>
> The compensation point is the light intensity at which the rate of photosynthesis and the rate of respiration are equal.

Fig 35.11(a) An annual garden plant develops a mass of shallow roots when watered 'little and often'
Fig 35.11(b) The same plant develops a much deeper root system if it is given a large amount of water once or twice a week

WIND AND WATER MOVEMENTS

Most organisms find it difficult to survive in an environment that is permanently windswept, since, in windy conditions, plants are likely to lose a lot of water through transpiration. This is why plants that live in such conditions, for example plants adapted to life in the Swiss Alps, have developed fleshy leaves with thick cuticles.

Animals are also affected by air movement (Fig 35.12). Windswept ecosystems tend to support fewer species, and the populations of individual species are small. In general, aquatic ecosystems that have strong water currents are also less diverse. This is because the current sweeps away and dislodges organisms from their habitats. Both wind and water movements can increase suddenly, causing the catastrophe of a gale, hurricane, typhoon, a flood or tidal wave (Fig 35.13).

Fig 35.12 Heat loss is more of a serious problem than water loss for animals in cold, windswept conditions

Fig 35.13 Catastrophic events, such as the hurricane that devastated this mangrove, can wipe out whole populations and even whole communities

3 BIOTIC COMPONENTS OF AN ECOSYSTEM

The biotic environment that surrounds an organism in an ecosystem results from the activities of all the other organisms living there. Complex relationships, called **intraspecific interactions**, occur between organisms of the same species; between organisms from different species, they are called **interspecific interactions**.

The main types of activity that make up intraspecific and interspecific interactions are summarised in Table 35.1, and described in more detail in Chapter 37.

Fig 35.14 The female garden warbler makes regular trips back to her nest with food for the tiny chicks

Fig 35.15 Meercat sentries keep a look out for predators while other members of the group search for food and look after the young

? TEST YOURSELF

E A male stickleback guards his nest by threatening and chasing other males from his territory. Is this an example of intraspecific aggression or interspecific aggression?

Table 35.1 Types of interactions that occur between organisms in an ecosystem

Activity	Type of interaction	What happens
Reproduction	intraspecific	Location of, selection of and competition for a mate.
Caring for young	intraspecific	Both parents, one parent or, more rarely, older siblings feed the young, keep them warm and sheltered and protect them from predators (Fig 35.14).
Social behaviour	intraspecific	Animals co-operate to find food and to defend themselves against competitors or predators (Fig 35.15).
Competition	intraspecific	Organisms compete for resources such as food, space, light, water and mineral nutrients.
Reproduction	interspecific	Animal vectors pollinate some species of plant and help to disperse the fruits and seeds of others (Fig 35.16).
Caring for young	interspecific	Rare, but some species rear the young of another species. The cuckoo, for example, lays its eggs in the nests of smaller birds that then rear the fledglings.
Mutualism	interspecific	Two organisms both benefit from a long-term association.
Parasitism	interspecific	Individuals from one species use another species to provide food and shelter while giving nothing in return. Many hosts actually suffer from the interaction: many parasites cause disease.
Predation	interspecific	All animals obtain food by eating another organism. In its widest sense, predation also applies to herbivores feeding on plants.
Protection	interspecific	Many species make use of other species to hide from predators. Some copy the markings of other species to mimic warning signals that predators avoid.
Competition	interspecific	Organisms in a community compete for resources such as food, space, light, water and mineral nutrients.
Defence	interspecific	Organisms have developed various strategies to defend themselves against predators, parasites and competitors including mimicry (Fig 35.17).

Fig 35.16 Rufous humming birds drink nectar from the scarlet gilia, pollinating it at the same time

Fig 35.17 The hover fly (upper) mimics the markings of the much more dangerous wasp (lower) to fool other organisms into thinking that it, too, has a powerful sting

! SCIENCE FEATURE Measuring abiotic factors related to climate

Temperature

You can use a standard laboratory **mercury thermometer** to measure air temperatures. (Take care not to place the bulb in direct sunlight or you will get unusually high readings.) To measure soil temperatures or temperatures of leaf litter, you should use a more accurate and robust instrument such as an **electronic thermometer** or **thermistor**.

Wind speed

You can measure wind speed using an anemometer: this has cup-shaped propellers attached to a small current generator (Fig 35.18a). When wind rotates the propellers, the generator provides a reading of wind speed. A flow meter (Fig 35.18b) uses the same principle to measure the speed of water flow in a river or stream.

Humidity

The amount of water that animals and plants lose from their surface depends on the level of air humidity (the amount of water held in the air). You can measure relative humidity using a hygrometer. If you don't have one, you can use blotting paper that has soaked in 5 per cent cobalt thiocyanate solution and been left to dry. The paper changes from red to blue when the relative humidity of air is between 50 and 90 per cent. It is not very accurate but it does allow you to find out, for example, which of two rock crevices is damper.

Light levels

To measure light intensity, you can use the kind of electronic light meter that is used by photographers.

? TEST YOURSELF

F List two precautions you must take when measuring air temperature with a mercury thermometer.

Fig 35.18 The anemometer **(a)** and the flow meter **(b)** work on the same principle

SUMMARY

- An ecosystem is made up of **abiotic,** or physical, components and **biotic,** or living, components.
- Abiotic factors that affect an ecosystem include **climate** and **weather** (including humidity, temperature, air and water movements), soil type, rock type, natural catastrophes, water and mineral availability and the position of the ecosystem on Earth.
- Soil type helps to determine the type of plant that grows in an ecosystem: this affects which herbivores can live there and, in turn, which carnivores live there.
- The size of rock particles and the amount of **humus** present determines soil type.
- There are three main soil types: **clay/silt**, **loam** and **sand**. Loams are the most fertile soils because they contain a balance of particles and are rich in humus.

- Temperature, light availability, humidity, water availability and wind and water movements are abiotic factors that you need to study in more detail. You can measure them in the field using simple techniques.
- Biotic factors describe the interactions between living organisms. **Interspecific** interactions occur between organisms of different species; **intraspecific** interactions occur between organisms of the same species.
- Biotic and abiotic factors affect each other: the physical environment determines the distribution of organisms in the ecosystem but living organisms can also change the physical properties of an ecosystem.

KEY SKILLS **ASSIGNMENT**

OFF ON A FIELD TRIP...

James and Yasmin are part of a group of second-year A-level biology students attending a field studies week in May. They visit a hedgerow site to investigate some of the abiotic and biotic components of this ecosystem. Fig 35.A1 shows a rough plan of the site they choose to study.

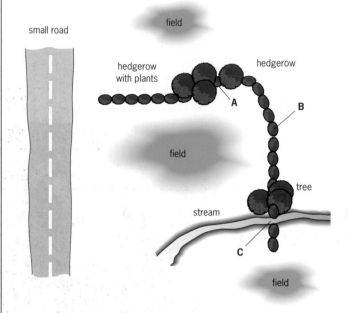

Fig 35.A1 Map showing sites of study along the hedgerow

James wants to investigate whether the distribution of different species of plant corresponds with a particular soil type. So he takes soil samples from the three points marked A, B and C in Fig 35.A1. He looks at the type of soil present, its field capacity and its humus content.

1 **a)** Describe the techniques that he might use to carry out this sort of analysis.
b) What other properties of the soil might James decide to test, if he has access to the right equipment?

2 Table 35.A1 show his results.
a) Which type of soil would you say is present at each point?
b) Give one factor that could explain why the soil at points A and C contain more organic matter than soil taken at point B.
c) James finds bluebells growing near A and C, but not at B. Do you think this distribution could be due only to soil type? What other abiotic factor could be involved?

While James is busy with the soil analysis, Yasmin decides to look at the plants in the hedgerow that are infested with aphids. She finds three elderberry shrubs, one at each of the points A, B and C. Fig 35.A2 shows the distribution of the aphids at the three different sites.

Stems with leaves:

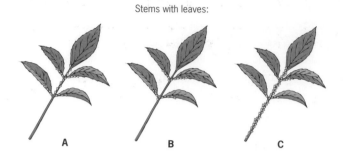

Stems with elderberries starting to form:

Fig 35.A2 Diagram to show where aphids are found on elderberry bushes near points A, B and C

3 **a)** Explain why aphids are distributed differently on the same plant at the three different sites. Which two abiotic factors do you think are responsible for this distribution?
b) What are the three main ways that the aphids can harm the plant?
c) Give an example of how interspecific and intraspecific competition can affect the aphid population.

4 Why are there fewer hedgerows in the British countryside today than, say, 40 years ago? Discuss the long-term environmental effects that result from the destruction of hedgerows.

Table 35.A1 Results of James's soil analysis			
Test	**A**	**B**	**C**
Properties of moist soil when rolled in the hand	Rolls into a ball: feels soapy	Feels gritty: won't roll into a ball	Feels gritty, easy to roll into a ball
Amount of water poured into 100 g dry soil	100 cm^3	100 cm^3	100 cm^3
Amount of water passing through	75 cm^3	90 cm^3	82 cm^3
Starting mass before heating at 450 °C for 4 hours	100 g	100 g	100 g
After heating	81.67 g	94.61 g	74.64 g
After further 30 min	81.59 g	92.71 g	73.98 g

36

Energy transfer and mineral cycling in ecosystems

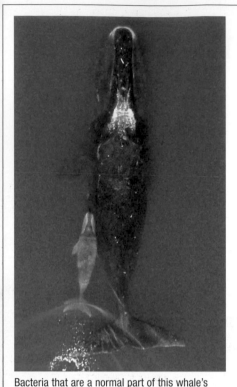

Bacteria that are a normal part of this whale's stomach contents can break down crude oil fractions and PCBs. If these bacteria can be introduced into the stomachs of other species such as dolphins and seals, this could guard them against the worst effects of some pollutants

Ecosystems are sophisticated recycling 'factories' in which elements and simple molecules are continually extracted from 'waste' chemicals and then used to synthesise new chemicals.

Toxic pollutants that are released into the environment by industrial processes cause problems if they cannot be broken down and recycled by living organisms. Pesticides, some oil components and PCBs (polychlorinated biphenyls) are particularly persistent and can build up to dangerous concentrations as they move through the trophic levels of food chains and webs.

One solution to toxic pollutants is to prevent the use or release of such chemicals. However, this is not practicable since many important industrial processes depend on them. Until recently, scientists had drawn a blank in the search for microorganisms that might have enzyme systems weird enough to break such chemicals down into harmless products. But progress is now being made. In the mid-1990s, American scientists discovered bacteria in the stomach of the bowhead whale that can break down naphthalene and anthracene, two persistent and cancer-causing fractions of oil. A couple of the other species found there prefer to feast on PCBs.

Not only could this finding explain why bowhead whales are particularly resistant to the effects of pollutants; it could lead to the identification and mass culture of bacteria that can help clear up the effects of oil spills and industrial contamination.

1 RECYCLING IS NOT NEW

In the last years of the twentieth century, recycling and reusing 'waste' became a normal part of life. However, we humans cannot claim to have thought of it

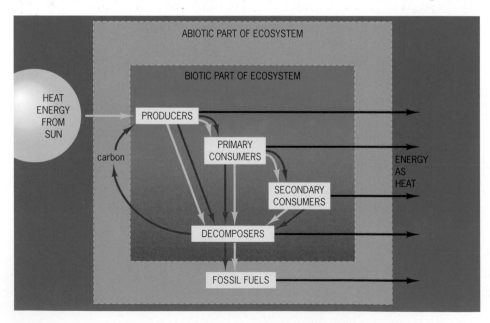

Fig 36.1 The one-way transfer of energy (yellow arrows) and the cyclical movement of mineral nutrients (for example carbon) in an ecosystem

first. Every ecosystem that has ever existed has practised the same principle. Water molecules are constantly cycling through the environment, evaporating from the surface of the Earth, condensing as clouds, falling again as rain and then cycling through various living organisms. When you turn on the tap, many of the water molecules that come out have been through the kidneys of many people, via water treatment systems. Carbon, nitrogen and other minerals also cycle through the environment.

The supply of these vital chemical elements is limited, so they are used again and again. Autotrophs such as plants build them into large, complex molecules, and heterotrophs break the molecules down again, respiring them to gain energy.

The very simple model of an ecosystem in Fig 36.1 illustrates how elements cycle through the compartments of an ecosystem. These processes are all driven by energy from the Sun. As you can see from the diagram, energy is transferred through different organisms in the ecosystem: it is never recycled.

2 ENERGY TRANSFER IN ECOSYSTEMS

Sunlight is the only significant source of energy for most ecosystems. Green plants, and some algae and bacteria, carry out photosynthesis. They use the energy from sunlight to power chemical reactions that combine simple inorganic molecules to form complex organic compounds. We say that the autotrophs in an ecosystem are **producers**: the products that they make form the food that ultimately feeds all other organisms in the system.

Animals, fungi and some protoctists are heterotrophs: they meet their need for energy by feeding on other organisms. We say that they are the **consumers** in an ecosystem. **Primary consumers** are herbivores: they obtain their energy by eating producers. **Secondary consumers** are carnivores or scavengers: they eat primary consumers. **Tertiary consumers** are carnivores that prey on other meat-eaters. This chain of dependence is usually called a **food chain** (Fig 36.2). The distinct levels of each chain are called **trophic levels**.

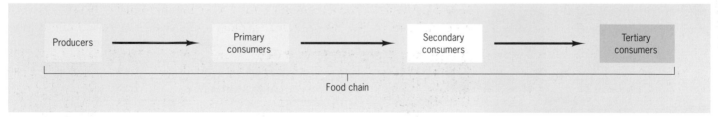

Fig 36.2 A simple flow chart summarising the overall structure of a food chain which has four trophic levels

As one organism eats another, there is a transfer of energy from the bodies of the producers (or consumers) into the bodies of consumers through the trophic levels of a food chain. Energy transfer occurs in one direction only: it is not recycled, and much of the energy escapes from the ecosystem as heat (see Fig 36.1). Energy transfer in ecosystems is relatively inefficient (see Chapter 7).

Fig 36.3 This is what athlete's foot really looks like. Fungi grow into their food (in this case, human skin), secreting enzymes which digest the food externally. The soluble products of digestion are then absorbed through the walls of the hyphae

Decomposers form another major group of heterotrophs in an ecosystem. These are mainly bacteria and fungi. They break down the bodies of dead animals and plants, usually by extracellular digestion (Fig 36.3) and absorb the soluble products. Transfer of energy from producers and consumers to decomposers is also one-way, with a high proportion eventually lost as heat. The process of decomposition allows mineral nutrients to be recycled in the ecosystem (see page 656).

FOOD CHAINS AND FOOD WEBS

Two examples of real food chains are shown in Fig 36.4. Most animals have a varied diet. A food chain is therefore only part of a much larger feeding picture: the food web. Take our top freshwater carnivore, the pike. It does not feed on stickleback alone but also roach and insects such as dragonfly nymphs and pond skaters. Most rivers and lakes also include a wide variety of producers ranging from the microscopic plant plankton to the larger pondweeds, rushes and flowering plants.

Fig 36.5 on the next page shows a simplified example of a food web in the Antarctic.

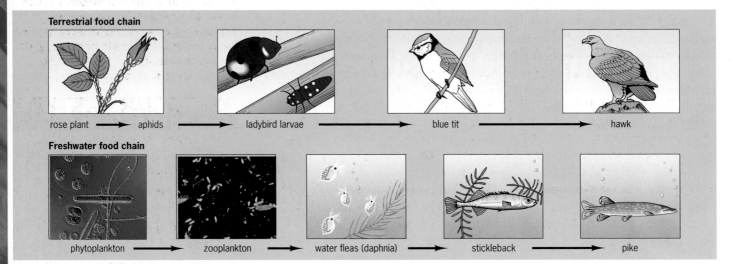

Terrestrial food chain

rose plant → aphids → ladybird larvae → blue tit → hawk

Freshwater food chain

phytoplankton → zooplankton → water fleas (daphnia) → stickleback → pike

Fig 36.4 An example of a terrestrial and a freshwater food chain. Each chain has five trophic levels (most food chains have fewer trophic levels than this)

PYRAMIDS OF NUMBERS

Food chains and food webs tell us a lot about the feeding relationships of organisms in an ecosystem, but they are **qualitative** rather than **quantitative**. That means that we know *which organisms* are part of the chain or web, but we have no idea of the exact *numbers of organisms* involved at each trophic level.

Clearly, numbers are very important: there are many more individual producers than primary consumers, and, as you carry on along a food chain, the numbers of organisms at each trophic level continues to decrease. At the same time, the size of individual organisms usually increases as you go up the trophic levels. It is difficult, if not impossible, to find out exactly how many producers are eaten by a primary consumer and so on, but we use a pyramid of numbers to show the general idea (Fig 36.6).

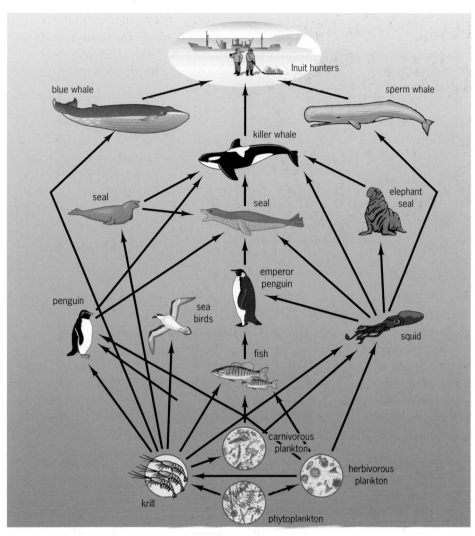

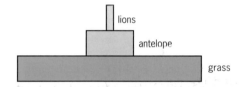

Fig 36.5 A food web of plants and animals in the Antarctic. Humans are usually at the top of food chains and food webs

Fig 36.6 A simple pyramid of numbers. The width of each box represents the relative number of organisms present in each trophic level at any one particular time. In this example, many grass plants are needed to sustain all the antelope that one individual lion must eat to survive

Two odd things can happen when you use pyramids of numbers. Sometimes the base of the pyramid is very narrow, indicating that there are fewer producers than primary consumers. This seems not to make sense, but these examples arise when a few large plants such as trees produce food for thousands of tiny plant-feeders, such as aphids (Fig 36.7a). In pyramids of food chains that have a population of parasites at the top, the upper level can be much larger than the level below it. This is because many parasites can feed on one host (Fig 36.7b). We say that both of these pyramids are **inverted pyramids**.

? TEST YOURSELF

D Draw the pyramids of numbers that you would expect for the following food chains:
a) grass – rabbit – fox
b) hazel trees – squirrel – fleas
How would you describe these two pyramids?

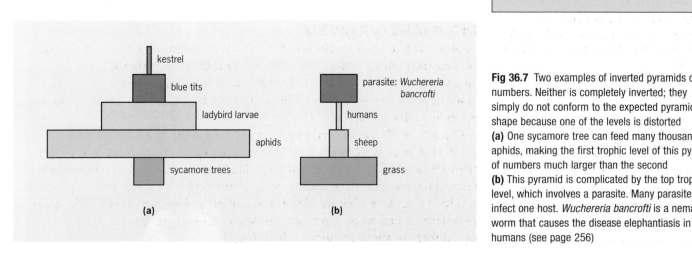

(a)   (b)

Fig 36.7 Two examples of inverted pyramids of numbers. Neither is completely inverted; they simply do not conform to the expected pyramidal shape because one of the levels is distorted
(a) One sycamore tree can feed many thousands of aphids, making the first trophic level of this pyramid of numbers much larger than the second
(b) This pyramid is complicated by the top trophic level, which involves a parasite. Many parasites can infect one host. *Wuchereria bancrofti* is a nematode worm that causes the disease elephantiasis in humans (see page 256)

SCIENCE FEATURE Food webs and pesticides

When developing a new pesticide, scientists look for a chemical that kills only the pest that we want to get rid of, harms no other species, breaks down into something harmless after a short time, does not lead to the development of resistance, and is cheap. This is a pretty tall order: no pesticide yet developed is perfect, but modern pesticides are much safer than some of those developed and used in the 1950s and 1960s.

DDT (dichlorodiphenyltrifluoroethane) was used as a pesticide in many countries after the Second World War. It breaks down slowly, remaining in the environment for between two and five years.

During the 1950s and 1960s it became obvious that DDT was affecting whole food webs. Fig 36.8 shows what happened when DDT was transferred up the trophic levels of a food web in an ecosystem in the USA. DDT is not excreted and concentrates in the fatty tissues of organisms. So, although the concentration of DDT in zooplankton, the primary consumers in the chain, was only 0.04 parts per million, its concentration in the tissues of a top carnivore such as the osprey or bald eagle was 25 parts per million. The bodies of the large birds of prey converted the DDT into a substance that made their eggshells very fragile. As a result, very few birds managed to breed and their numbers fell quickly.

In 1972, DDT was banned and since then, the numbers of the large predatory birds has recovered. However, there are still problems because of the illegal use of DDT.

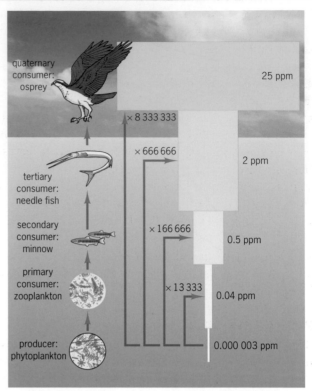

Fig 36.8 Bio-amplification of DDT through an aquatic food chain. DDT accumulates in the fatty tissues and cannot be excreted. The fat of the osprey contains over 8 million times more DDT than the water at the bottom of the food chain in which the producers live

SEE QUESTION 3

Fig 36.9(a) A generalised pyramid of biomass for a terrestrial ecosystem

Fig 36.9(b) In the waters of the Antarctic, the weight of zooplankton (animal plankton) is five times that of the phytoplankton (plant plankton). So, if we draw up a pyramid of biomass for one of the food chains that forms part of the more complex web shown in Fig 36.5, we get an inverted pyramid. This is because the zooplankton eat the phytoplankton so quickly that the plant plankton never gets the chance to attain a large biomass. Instead, phytoplankton has a high turn-over rate, reproducing very quickly. So the biomass of organisms produced each year is large, although the number alive at any one time is often less than the number of consumers

PYRAMIDS OF BIOMASS

Ecologists often use **pyramids of biomass** (Fig 36.9a), to avoid the problem of inverted pyramids of numbers. Pyramids of biomass give a more accurate model of what happens in a food chain. They show the weight or mass of all the organisms at each trophic level – **dry mass** is the most useful measure. This is obviously extremely difficult to do since it means obtaining the dry mass of organisms means killing them, drying them and measuring the remains. Not surprisingly, very few pyramids of biomass have ever been determined in this way.

However, using samples of organisms and their wet mass allows ecologists to make estimates and devise models. These show that the pyramid of

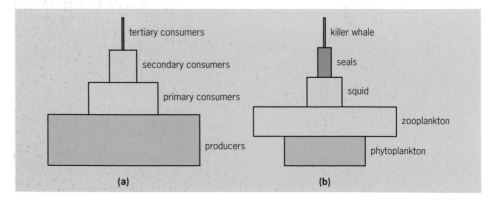

biomass provides a more realistic way of comparing trophic levels. But, even this type of pyramid can be inverted, as Fig 36.9(b) shows.

To obtain an even better model of an ecosystem which avoids inverted pyramids altogether, we must look at the way energy is transferred between the different trophic levels.

EFFICIENCY OF ENERGY TRANSFER BETWEEN DIFFERENT TROPHIC LEVELS

As we saw in Chapter 7, the energy transfer between organisms at different trophic levels of a food chain is never 100 per cent efficient. In fact, less than 30 per cent of the energy available ever reaches the organisms in the next level. There are three main reasons for this:

- Not all the organisms in one trophic level are eaten by organisms from the next level. Many die and provide food for decomposers.
- Not all food that is eaten is digested and not all digested food is absorbed – some is lost as faeces.
- Most of the energy gained from the food that is absorbed is lost through the processes of excretion (production of metabolic waste) and cell respiration.

PYRAMIDS OF ENERGY

Fig 36.10 shows a typical pyramid of energy. It shows the amount of energy that is transferred up the food chain and how much is lost as heat at each level. The more levels in the chain, the less energy there is available at the top level. Consequently, very few food chains have more than four or five levels. It also explains why the top carnivores such as the big cats, tigers and large whales are the first to become endangered species when their ecosystems come under pressure from humans.

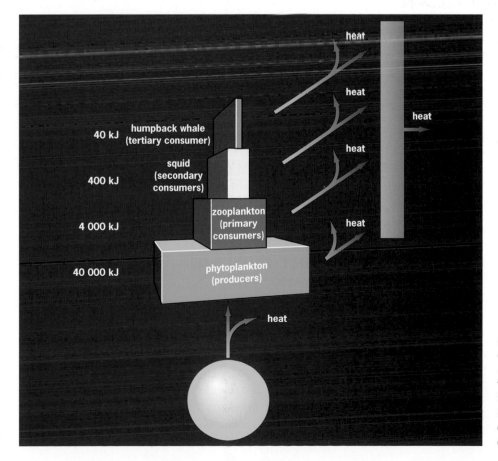

Fig 36.10 A pyramid of energy for an Antarctic food chain similar to the one shown as a pyramid of biomass in Fig 36.9(b). As you go up the pyramid, there is a 90 per cent loss of energy at each trophic level. This energy, of course, does not disappear: it is 'lost' as heat. For a more detailed investigation of energy transfer, see the Assignment at the end of this chapter

! SCIENCE FEATURE Feeding the world: productivity in ecosystems

The rate at which the producers in an ecosystem capture light energy to build biomass is known as the **gross primary productivity** of that ecosystem. This is a theoretical maximum. In reality, all producers use energy for respiration and other life processes. When we subtract the energy that they use to live, taking it away from the energy that they capture, we find the **net primary productivity** of the ecosystem. This is the amount of energy that is available to the consumers of the system.

net primary productivity = gross primary productivity – energy used by producers to live

Different ecosystems of the world have different primary productivities. As Fig 36.11 shows, an ecosystem that is part of tropical rainforest is one of the most productive in the world. A major reason for this is that tropical rainforests grow around the Equator of the Earth, where more light energy is available than further north or south. Ecosystems that have high levels of water and mineral nutrients, such as an estuary or a marsh swamp, are many times more productive than deserts that are dry and severely lacking in nutrients.

The primary productivity of an ecosystem limits the number and variety of consumers that can survive there. This explains why tropical rainforest ecosystems support more diverse animal populations than other parts of the world. As rainforests are cut down to make room for expanding human populations, the potential of the rainforest to produce enough food to support its populations decreases. As less energy is available in food webs and food chains, the numbers of animals at the higher levels are likely to fall. More species are likely to become endangered and, eventually, extinct.

Many scientists think that the conflict between the need to use the land for humans to live on and the need to leave enough land to produce food will reach a crisis in many areas of the world in the next 50 years. The famines that are tragically common in desert areas may well become more of a feature of life in many other parts of the world, as primary productivity cannot keep pace with the growth of human populations.

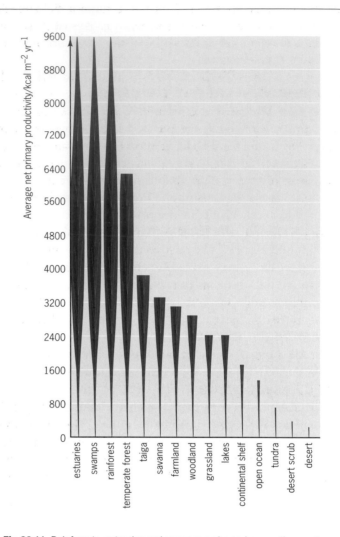

Fig 36.11 Rainforests, estuaries and swamps and marshes are the most productive ecosystems in the world. Open oceans, arctic tundra and deserts are the least productive

The scale of this problem is so large that it is difficult for any individual country to tackle alone. However, many countries worldwide are now working together to try to ensure that the worst outcome does not happen.

3 MINERAL NUTRIENT CYCLING

All organisms need a supply of energy to carry out the processes of life, but they also need to obtain the mineral nutrients (carbon, nitrogen, phosphorus, etc) that make up the complex chemicals in their bodies. Unlike energy, mineral nutrients are recycled over and over again, passing between organisms in the same ecosystem, and also other ecosystems of the world. So the carbon, oxygen and nitrogen atoms in our bodies could have been part of the soil a year ago. They may have made up the proteins of some long extinct dinosaur. And one popular anecdote says that, during the course of our lives, each of us breathes out six carbon atoms that were once part of Napoleon Bonaparte.

Producers accumulate mineral nutrients. These then pass along the food chain and are finally released back into the abiotic part of an ecosystem by decomposers. Decomposers replenish supplies of these elements in the soil and in water, and they also help to return them to the atmosphere.

WHY LIVING ORGANISMS NEED MINERAL NUTRIENTS

As we saw in Chapter 3, the chemicals of life, carbohydrates, proteins, fats and nucleic acids, contain mainly carbon, hydrogen and oxygen. Proteins also always contain nitrogen and often sulphur, and nucleic acids contain phosphorus. In addition to these elements, living things need small amounts of calcium, sodium, potassium, iron, copper, chlorine, magnesium and other trace elements. (Table 10.5 in Chapter 10 gives a list of minerals that the human body needs.)

In this chapter we concentrate on the cycling of carbon and nitrogen.

CYCLING OF INDIVIDUAL MINERAL NUTRIENTS

Mineral nutrients occur in four basic compartments in an ecosystem:

- The organic compartment: living organisms and their debris (faeces, etc).
- As available nutrients that are held on the surfaces of clay particles in soil or in solution, where they can be taken up by plants and microorganisms.
- As nutrients that are temporarily unavailable because they are bound up in soil and rocks.
- In the atmosphere, as gases that occur in the air.

THE CARBON CYCLE

Life on Earth is based on carbon compounds (see Chapter 3). Carbon circulates between the abiotic and biotic parts of the environment, as shown in Fig 36.12. Fig 36.13 gives an exam-friendly version of this diagram. Autotrophs capture carbon during photosynthesis, and all living organisms return it to the atmosphere as they respire and also when they die and decompose.

Fig 36.12 The carbon cycle. Photosynthetic organisms fix about 80 billion tonnes of carbon each year. But this represents less than 1 per cent of the total carbon in the biosphere. Carbon can be locked up for long periods of time in limestone rocks, inorganic solutions (containing soluble carbonates) and fossil fuels such as coal: these reservoirs of carbon are often called carbon sinks. When volcanoes erupt, they return carbon dioxide to the carbon cycle

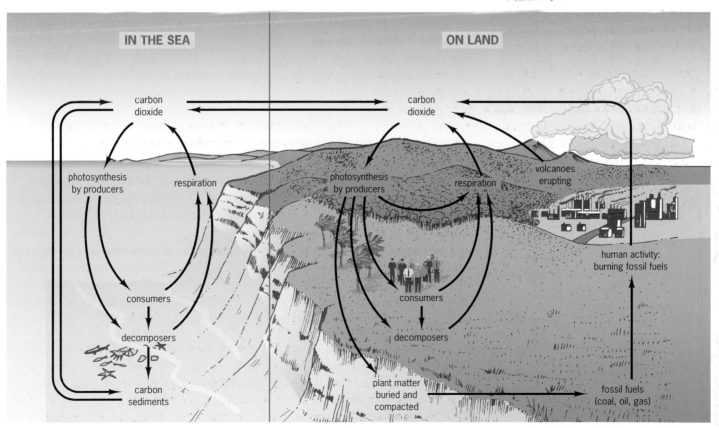

Fig 36.13 An exam/revision version of the carbon cycle. You should find this diagram easy to draw. Learn it so that you can reproduce it in exams

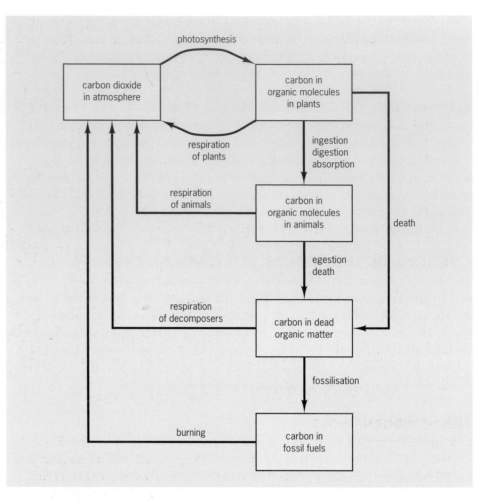

THE ROLE OF MICROORGANISMS IN THE CARBON CYCLE

By looking at Fig 36.12 or Fig 36.13, you should be able to see that carbon would still cycle through the biosphere for some time if decomposers were removed. However, because of their general role in decomposition, decomposers play a key role in the carbon cycle. Bacteria and fungi feed by secreting digestive enzymes on to organic material to digest it. Some of the 'freed' mineral nutrients are absorbed by these microorganisms. Others escape into the soil.

HOW HUMANS INFLUENCE THE CARBON CYCLE

Humans influence the carbon cycle by two activities not practised by other species. Firstly, humans burn fossil fuels such as coal, oil and gas. This has been going on in a big way since the start of the Industrial Revolution in Europe. Burning such fuels is the only way that their 'locked' carbon is released back into the atmosphere.

The second activity is cutting down forests, particularly rainforests in recent times, to make room for farms, fields and homes. This increases the amount of carbon dioxide in the atmosphere because it reduces the total number of photosynthetic organisms, the 'trappers' of CO_2. The possible effects of these activities are discussed in the Science feature on pages 658–659.

SEE QUESTION 4

✔ REMEMBER THIS

Decomposers stop organic material accumulating in an ecosystem. They also ensure that essential nutrients, such as nitrates, sulphates and phosphates, are released back into the soil.

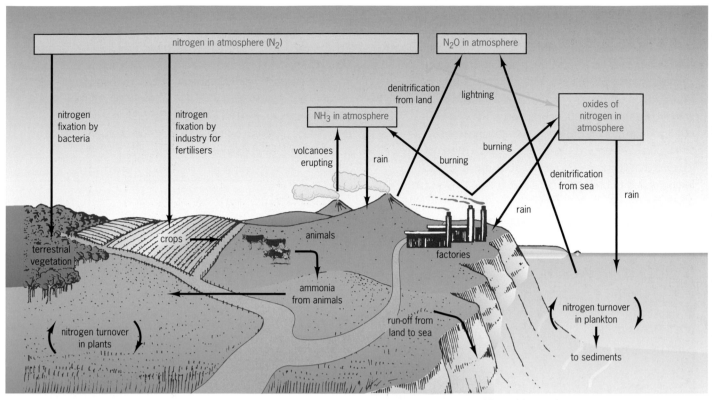

Fig 36.14 The nitrogen cycle. Bacteria play an important role in the movement of nitrogen in the biosphere

THE NITROGEN CYCLE

Like carbon, nitrogen is vital to organisms for the formation of proteins, nucleic acids and their products. Although nitrogen is all around us (it makes up 79 per cent of the air we breathe), most living organisms cannot access this supply. Nitrogen gas is unreactive because the two atoms that make up the N_2 molecule are bound together by a strong triple bond. It can, however, be **fixed** by microorganisms, and then it cycles through ecosystems, as shown in Fig 36.14. Again, an exam-friendly diagram of this cycle is also included (Fig 36.15).

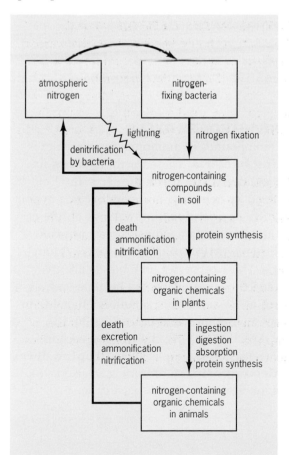

Fig 36.15 An exam/revision version of the nitrogen cycle. You should find this diagram easy to draw. Learn it so that you can reproduce it in exams

SEE QUESTION 5

! SCIENCE FEATURE Carbon dioxide, the greenhouse effect and global warming

Carbon dioxide is one of the main greenhouse gases. Together with other gases, it reduces the heat that can escape from the atmosphere (Fig 36.16). This effect is often called the **greenhouse effect**. Although pollution could be increasing the greenhouse effect, this phenomenon is not caused by pollution: it is a natural effect due to the presence of an atmosphere. In fact, having an atmosphere is one of the reasons that life developed on Earth: it allows surface temperatures to be much more stable than they would otherwise be.

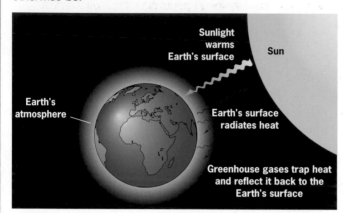

Fig 36.16 The greenhouse effect. Energy from the Sun is either transmitted, reflected or trapped by the atmosphere. Energy radiated back into space is at longer wavelengths (in the infrared region) than the energy arriving from the Sun, and the consequence is that more heat enters than can escape. Greenhouse gases (CO_2, CH_4, nitrous oxides, CFCs), together with water vapour in clouds, contribute to the 'blanketing' effect of the atmosphere

Accurate measurements of atmospheric carbon dioxide concentrations began in the mid-1950s, in Hawaii and at the South Pole. These sites were chosen because they are far away from any major sources of pollution. The curve in Fig 36.17 clearly shows an annual cycle of peaks and troughs corresponding to summer and winter. Superimposed on this is a small but steady rise in overall CO_2 levels. Levels of other greenhouse gases such as methane, nitrogen oxides and chlorofluorocarbons (CFCs) have also increased during the same period.

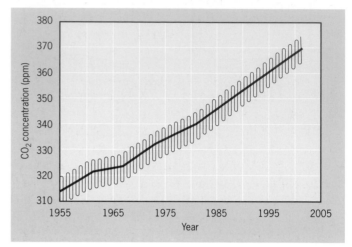

Fig 36.17 The rising level of carbon dioxide in the atmosphere. Data were collected at the Mauna Loa observatory in Hawaii. There is good evidence to suggest that changes in the average temperature at the surface of the Earth are closely linked to changes in the levels of greenhouse gases

What are the likely effects of global warming?

Many scientists think that an increase in carbon dioxide in the atmosphere will cause global warming. They believe that the average temperatures on the Earth's surface will increase, causing widespread disturbances in world climate.

PROCESSES INVOLVED IN THE NITROGEN CYCLE

Nitrogen-rich waste from animals (compounds of ammonia, for example) is converted into nitrates by **nitrifying bacteria**. These microorganisms fix nitrogen by converting atmospheric nitrogen into useful products. In terrestrial ecosystems, nitrogen is fixed by free-living nitrifying bacteria in soil and by **symbiotic bacteria** that live in **nodules**, the swellings on the roots of legumes (plants like peas and beans). In aquatic ecosystems, some blue-green bacteria fix nitrogen. Nitrogen returns to the atmosphere through the action of **denitrifying** bacteria. These processes, in more detail, are:

● **Nitrogen fixation**. This is carried out mainly by soil bacteria such as *Azotobacter*. Some bacteria, such as *Rhizobium* are symbiotic: they live in the root nodules of some plants such as peas and beans (Fig 36.18).

● **Ammonification**. This is the breakdown of organic nitrogen such as proteins and urea into ammonia. It is also known as **saprotrophic decay** and is generally thought of as **decomposition**, or 'rotting'. Bacteria and fungi carry out most of the ammonification in the nitrogen cycle.

! SCIENCE FEATURE Carbon dioxide, the greenhouse effect and global warming continued

Using all the evidence and data available at the moment, a computer model has been developed to predict what might happen in the future. This projection shows that the mean surface temperature might increase.

It is important to remember that any projection is an informed guess: we cannot say exactly what will happen. Nevertheless, the world's scientific community has reached consensus that if greenhouse gas concentrations continue to rise at current levels, global mean temperature will rise by 2 °C above pre-industrial levels by the year 2030.

The effect of any global warming that does occur will not be felt evenly around the globe. There will be greater warming at the high latitudes (at the poles, for example) and at middle latitudes than at the equator. The Northern Hemisphere will warm more than the Southern Hemisphere, because it has more land mass (land heats up more quickly than water). More areas of the world would have extreme heat waves and more forest and grassland fires. The average sea level would rise, perhaps by 2–4 centimetres every 10 years. Many low-lying areas could become flooded and maybe even submerged. This would cause huge problems for developing countries and small islands, which are much more vulnerable to the adverse effects of climate change.

What is being done about the problem of global warming?

Everyone agrees that cutting emissions is a good idea. This means reducing the amount of fossil fuels that we burn: we could make a big difference by improving the efficiency of heating systems and by reusing 'waste' heat from industry. Wherever possible, we should use renewable energy sources such as wind, water and solar power. Although many people object on safety grounds, making more use of nuclear energy would also help to reduce emissions of greenhouse gases.

In addition, we should limit the destruction of forests and use methods of agriculture that do not irreversibly damage the land. Of course, all of this is cancelled out if the human population continues to increase, so efforts should also be made to stabilise the world's population, to prevent further pressure on energy resources.

What are the problems?

The United Nations Framework Convention on climate change agreed that industrialised countries would reduce CO_2 emissions first, providing developing countries space for development. The Kyoto Protocol (1997) set out that industrialised countries would reduce their emissions by 5.2 per cent compared with 1990 levels by 2012. However, not all industrialised countries have put their full backing behind this policy and other strategies are becoming necessary.

Industrialised countries have put forward the idea of creating carbon sinks by planting new forests that will act as a sink for carbon dioxide. Whether this would work or not remains to be seen, but the commitment to reduce emission of greenhouse gases during 2008–2012 is being put in jeopardy by the USA and Japan, who propose to use carbon sinks to meet their reduction commitments without curbing emissions from fossil fuels.

- **Nitrification**. Nitrifying bacteria such as *Nitrosomonas* oxidise ammonia to nitrite (NO_2^-). Other bacteria, such as *Nitrobacter* oxidise nitrite to nitrate (NO_3^-). These oxidation reactions release energy that the bacteria use to make food (they are chemoautotrophs).
- **Denitrification**. Bacteria such as *Pseudomonas denitrificans* obtain energy by using the nitrite or nitrate ion as an electron acceptor for the oxidation of organic compounds. They live in conditions with low oxygen levels such as waterlogged soils, and can be important in reducing nitrate levels further.

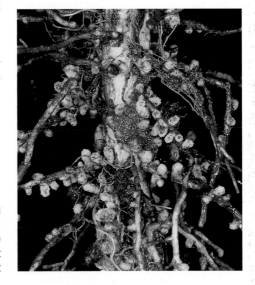

Fig 36.18 Legumes such as peas and beans have nodules on their roots. These house nitrogen-fixing species of the bacterium *Rhizobium*. The bacteria convert nitrogen gas from the atmosphere into ammonia. This passes out of the root nodules, dissolves in soil water and is then taken up by plant roots. The relationship between the bacteria and the plants is therefore symbiotic because both benefit: the plants get a steady supply of nitrogen and the bacteria get a home and a constant supply of sugars from the plant

SUMMARY

- Energy transfer through organisms in an ecosystem is one-way. Elements such as carbon and nitrogen cycle between living organisms and the abiotic environment.

- In a **food chain**, energy is passed from **producers** to **primary consumers** to **secondary** and **tertiary consumers**. Each level of the food chain is called a **trophic level**. In most ecosystems, many food chains are linked together in a **food web**.

- **Decomposers** break down the dead bodies of producers and consumers and play an important role in **mineral cycling**. Most decomposers are bacteria and fungi.

- There are three main types of **ecological pyramid**: **pyramids of numbers**, **pyramids of biomass** and **pyramids of energy**. Pyramids of energy show that over 90 per cent of the energy transferred to the next trophic level is lost as heat.

- The **carbon** and **nitrogen cycles** show how these elements circulate between the biotic and abiotic parts of an ecosystem. You should appreciate the overall cycle and be able to draw a clear summary diagram.

- Carbon dioxide is a **greenhouse gas**: together with other gases in the atmosphere, it acts as a heat insulator. The presence of CO_2 in the atmosphere helps to maintain temperatures on Earth that are warm enough to sustain life.

- Human activities have led to an increased concentration of CO_2 in the atmosphere. Many scientists believe that this is causing a general increase in average temperatures throughout the world, a phenomenon commonly called **global warming**.

? EXAM QUESTIONS

1

a) Explain the meaning of the terms *producer* and *trophic level*.

The table shows the estimated energy content for four trophic levels of a grassland ecosystem.

	Energy content/kJ m^{-2}
Producers	5600
Herbivores	125
Omnivores	15
Carnivores	10

b) (i) Calculate the percentage energy decrease between the producers and herbivores. (Show your working and give your answer to the nearest whole number.)

(ii) Explain why the energy content of the herbivores is less than that of the producers.

c) Suggest **two** factors which might **reduce** the productivity of the producers.

d) Suggest

(i) why it can be a problem placing omnivores into a trophic level;

(ii) a reason for the difference in energy content between the omnivores and the carnivores in this ecosystem.

OCR 2801 Biology Foundation January 2001 Q3

2 Study the food webs in the diagram at the top of the next page, which represent the surface waters of the oceans in two different parts of the world, then answer the questions that follow.

a) Study the diagram and, for each of the trophic levels 1 to 5, write down an appropriate term.

b) From the Antarctic food web, write a food chain containing **four** organisms.

c) From the cold temperate food web write a food chain containing **seven** organisms.

d) Suggest why the cold temperate waters support a more complex food web, with longer food chains, than the Antarctic waters.

e) (i) is a disease caused a sudden decrease in the numbers of *copepods* in the Antarctic, describe the effects this would have on the food web.

(ii) Would a sudden decrease in *copepod* numbers have the same impact on the cold temperate food web? Give a reason for your answer.

OCR 6916 Environmental Biology March 2001 Q4

3 A group of students sampled the animal life in a pond. They plotted their data as a pyramid of biomass as shown below. The biomass is expressed as mg per m^3.

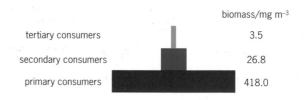

	biomass/mg m^{-3}
tertiary consumers	3.5
secondary consumers	26.8
primary consumers	418.0

a) Describe how the biomass values would have been determined.

? EXAM QUESTIONS

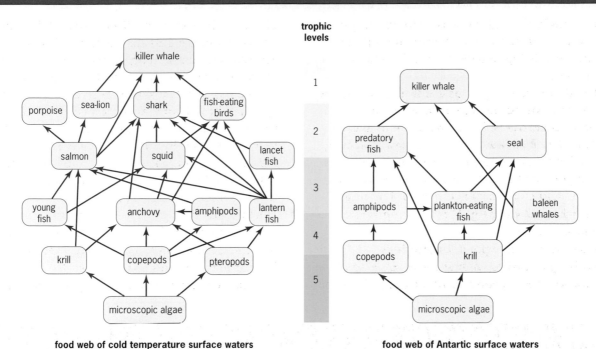

food web of cold temperature surface waters

food web of Antartic surface waters

b) (i) Calculate the percentage decrease in biomass between the primary consumers and the secondary consumers. Show your working.
(ii) Give *two* reasons for the decrease in biomass between each trophic level.
Edexcel B2 June 1999 QA

4 The diagram represents several important aspects of the carbon cycle. Study the diagram and then answer the questions that follow.

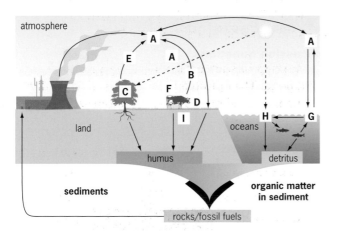

a) Explain briefly (in biological terms) why the chemical compound **A**, around tree **C**, has been shown to vary between 0.03% and 0.04% over a 24 hour period.
b) What biological process is represented by **B**?
c) Name the general class of storage biochemical represented by **D**.
d) When might process **E** be most easily detected?
e) Name a storage compound of carbon found in the liver of animal **F**.
f) Organic compounds of dead organisms are represented by **I**. State the term given to the microorganisms of the carbon cycle which utilise these compounds.
g) Suggest what chemical compound **G** represents.
h) **H** represents phytoplankton. Suggest a reason why this could reduce the concentration of **A** on a long-term basis.
i) **A** and **G** are said to be 'in dynamic equilibrium'. Suggest what this means.
OCR 6910 Molecules and Life June 2000 Q6

5 Give an account of the ways in which nitrogen is cycled in ecosystems. **[essay question]**
Edexcel B2 June 1999 Q8

KEY SKILLS **ASSIGNMENT**

WHY ARE BIG FIERCE ANIMALS SO RARE?

We normally think of large predators such as lions or killer whales as being the most successful organism in their ecosystem because they can take the food that they need without worrying about predators themselves. But if you look at their numbers relative to the numbers of other animals further down the food chain, it is clear that these top carnivores are really quite rare. In this Assignment, we see why.

1 Look at Fig 36.A1. This shows an energy flow diagram for a food chain in the Serengeti.
a) What is process Z?
b) What are organisms X?
c) How many trophic levels are there in the food chain?
d) Calculate the percentage efficiency with which light energy is converted into food energy in the producers.
e) What percentage of the light energy absorbed by plants is eventually available for secondary productivity in the lion?

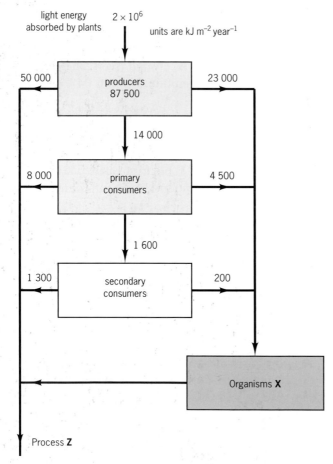

Fig 36.A1 In this schematic diagram, the producers are the grass plants which make up most of the vegetation. The primary consumers are deer and the secondary consumers are lions

2 Look at Fig 36.A2.
a) For every 3000 kJ of food eaten, what amount of energy is available to the deer for growth and repair?
b) What amount is available to the lion for this purpose?
c) Why do you think the deer loses a higher proportion of energy in faeces and urine than the lion?
d) Why should the lion use up more energy in cell respiration than the deer?
e) Suggest why the proportion of energy available for secondary productivity is about three times greater in the lion than in the deer.

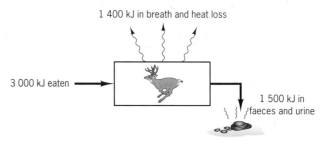

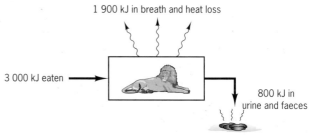

Fig 36.A2 A comparison of the proportion of energy lost as heat and in faeces and urine in a deer and a lion which eat 3000 kJ of food

3 From the information you have gained so far, why are large carnivores so rare?

Discussion questions

4 Do you think that the fact that humans are omnivores has anything to do with their evolutionary success?

5 Some people hold the view that, if humans were all vegetarian, the world food shortage would be largely solved. Do you agree or disagree, and why?

6 Several species of fish, such as trout and turbot, can be farmed. These fish are carnivores. Do you think this is an efficient way of producing food? Discuss the reasons for your answer.

37

Populations

Gatherings of people like this are common today but could not even have been imagined a hundred thousand years ago

For hundreds of thousands of years, the population of human beings on Earth grew at a steady rate. But in the past 200 years, the world's population has increased from 1 billion to over 6 billion people. World population is currently growing by 80 to 90 million people each year. At 20.02.02 GMT on 20.02.2002 (an interesting time point), the world population was estimated at 6 206 852 159 people. It will probably reach 6.5 billion in 2005, and experts estimate that it will pass the 7 billion mark at the end of the year 2012.

This rate of growth is projected to continue, leading to a global population of around 9 billion people by 2030. The rate of increase and the sheer massive numbers of human beings that this population explosion will generate are unprecedented in the history of the planet. Even with our constantly developing technology and ability to adapt, making the Earth's finite resources stretch to accommodate everyone will be a huge challenge.

Trying to predict what will happen to the population further in the future is difficult because it depends on so many factors. If the population carries on increasing at the current rate, a century from now there could be as many as 15 billion people.

But this is the worst case scenario: United Nations projections assume that family planning will become widespread all over the world and that birth rates everywhere will fall. Even taking into account that life expectancy would continue to increase as medicine continues to advance, the UN estimates that population growth will slow down during the early part of this century and will stabilise around the year 2100.

1 HOW POPULATIONS GROW

Populations are changing constantly. Imagine a new population starting in a particular area with a handful of individuals and ideal environmental conditions. A good example would be a pair of healthy rabbits introduced on to a small island that had plenty of edible vegetation and no predators. The population would grow slowly at first: this is phase A in the graph shown in Fig 37.1.

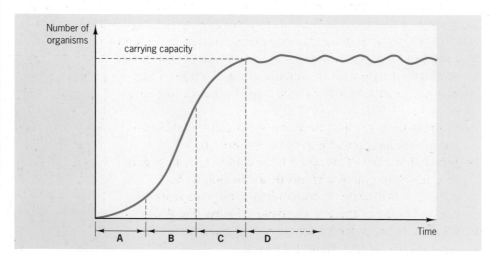

REMEMBER THIS

The **biotic potential** is the maximum rate at which a population can increase in size when there are no limits on its growth.

The **environmental resistance** which affects a population is made up of all the factors that act together to limit its size.

When the biotic potential and the environmental resistance reach a balance, the size of the population levels off: it reaches its **carrying capacity**.

Fig 37.1 This graph shows an idealised curve of population growth. As the curve A to C looks like a flattened S, it is called an S-shaped or sigmoid curve. The text describes phases A to D in detail

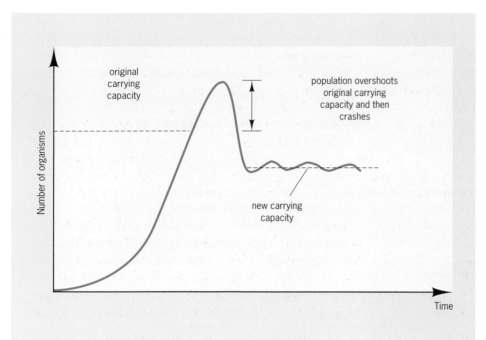

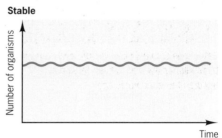

Fig 37.2 A population crash follows a sudden increase in numbers which exceeds the carrying capacity. If the environment is damaged by the temporary overshoot, there may be a new, reduced carrying capacity

Fig 37.3 Stable, irruptive and cyclic population curves

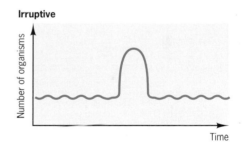

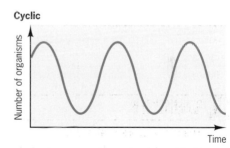

After a few months, the population would start to grow very rapidly and would double at regular intervals. This type of growth is called **exponential growth**, and is illustrated as phase B in the same graph. As more rabbits were born, they would then breed and there would be a huge rabbit population in a couple of years.

At some point, the population would be prevented from increasing further by various **limiting factors** in the environment – phase C. When food, space and water supply are stretched to their limit, more rabbits would die and they might breed less successfully. (If possible, they would leave the environment.) Because there is less food to go around, they would eventually eat the plants faster than the plants could grow.

Most populations stabilise at some point and the growth curve becomes flat. The number of individuals in the population at this point is the **carrying capacity** of the environment – phase D. Sometimes there is a sudden increase in numbers and a population overshoots its carrying capacity (Fig 37.2). The environment cannot provide the resources for this number of organisms and a population crash usually follows. The population dies back below the level of the carrying capacity, and then takes time to stabilise. If the overshoot has damaged the environment, the population might stabilise at a different level.

The two growth curves described in Figs 37.1 and 37.2 occur in laboratory situations but you will not find them in real ecosystems. Fig 37.3 shows the three main types of growth curve that occur in natural conditions:

- **Stable**: the population size remains at roughly the same level over a long time. A study of a population of trees in a well-established woodland would produce this sort of curve.
- **Irruptive**: the population size is basically stable, with the occasional dramatic increase followed by a population crash. This type of population growth occurs in algae. A population is mostly stable but, in hot weather, when water is contaminated by phosphate or nitrate fertilisers, there is an **algal bloom** (Fig 37.4).
- **Cyclic**: the population increases and decreases in a regular cycle of growth and die-back.

Fig 37.4 An algal bloom on a dyke

! SCIENCE FEATURE Studying populations

When we study populations, it is useful to know how many organisms live in a particular area. We can estimate population density (the number of organisms per unit area) using the following techniques:

Direct counting. Useful for large organisms in an open habitat. Ecologists count large mammals, nesting birds or woodland trees in this way.

Sampling. By taking a sample of organisms from a small area, we can estimate the number of organisms that live in a much larger area. The sample must be representative of the whole habitat. Table 37.1 lists a range of techniques.

Table 37.1 Sampling techniques for studying populations		
Technique	**What you do**	**Organisms you can sample**
Sweep netting	wave a net in the air	flying insects
Kick sampling	kick stones and gravel on a river bed and catch the disturbed organisms with a net placed downstream	aquatic invertebrates
Trapping	set sticky traps such as sheets of paper painted with honey	flying insects
	set light traps	moths and other nocturnal insects
	set pitfall traps, Fig 37.5(a)	insects that crawl on the ground
	set Longworth traps, Fig 37.5(b)	small mammals
Quadrat sampling	mark out a small area of ground with a wire grid	animal and plant species in an area of ground
Transect sampling	mark out a line or narrow belt and study along it	changes in landscape and living organism distribution up a slope, for example
Indirect techniques	use data from dead animals (beaver pelts, dead whales), count rabbit droppings, molehills or the exit holes of beetle larvae in wood	specific animal species

Capturing and marking. We can use this technique in field biology to give reliable estimates of large populations. It involves:

- taking a sample and marking the animals as they are caught,
- releasing the marked animals back into the general population,
- counting the marked animals in a second sample captured later.

Generally, the percentage of marked animals that are recaptured in the second sample tells us what percentage of the whole population we managed to mark in the first capture. So, if our second sample contains 5 per cent marked animals, we assume that our original sample represents 5 per cent of the total population of those animals in that area. This can be described by a formula called the Lincoln index:

$$\text{Population size, } P = \frac{\left(\begin{array}{c}\text{number of organisms}\\\text{in 1st sample}\end{array}\right) \times \left(\begin{array}{c}\text{number of organisms}\\\text{in 2nd sample}\end{array}\right)}{\text{number of marked organisms recaptured}}$$

You must remember that this is a rough estimate, not an actual value and, to make sure that your results are as accurate as possible, you need to take the following precautions:

- Make sure that the population you intend to study is fairly stable: choosing a migratory bird at the start of winter is not a good idea.
- Capture animals randomly: do not select a particular group.
- Mix marked individuals evenly with unmarked individuals before releasing your sample back into the general population (this usually means leaving enough time for the animals to mix in with the population).
- Use marks that cannot be washed, rubbed or licked off between capture and recapture.
- Do not make the marks too obvious: the marked animals may be more easily seen and eaten by predators.

Fig 37.5(a) Pitfall traps are set into the ground to catch crawling insects

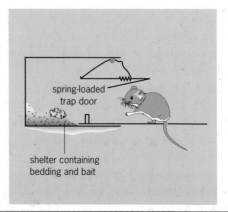

Fig 37.5(b) A Longworth trap

spring-loaded trap door

shelter containing bedding and bait

WHAT LIMITS POPULATION SIZE?

In an ecosystem, **abiotic** (physical) factors in the environment can limit population size. When we look at the whole range of abiotic factors that affect population size, we can classify them into three general categories:

- Limiting factors that are *always present*. These include space and light. A cliff that supports a population of kittiwakes, for example, is a fixed size and can accommodate a fixed number of birds (Fig 37.6).
- Limiting factors that *vary predictably over time*. These include changes in temperature and weather patterns in different seasons which lead to population movement or a change in behaviour among different organisms. As winter approaches, birds that live in an ecosystem in the summer might migrate to a warmer place (Fig 37.7). Small mammals such as dormice hibernate.
- Limiting factors that are *unpredictable*. Some unpredictable limiting factors are **density dependent**: they cause a greater effect if there is a high population than if the population is quite small. Sudden changes in the availability of water and nutrients, or a localised natural disaster, for example, can devastate an overcrowded population but leave a widely dispersed population relatively untouched.
- Other factors are **density independent**: they can devastate a population, whatever its size. Fires, floods, severe frosts and other catastrophes, for example, might destroy part or the whole of an ecosystem without warning (Fig 37.8).

Fig 37.6 The size of the ledges is an unchanging limiting factor. The kittiwakes can get nesting sites only by displacing other birds

Fig 37.7 Arctic terns are the world's champion travellers. In their search for food and the right breeding conditions, they range from deep inside the Arctic Circle, over the Atlantic to the pack ice of Antarctica

Fig 37.8 Catastrophes, including those caused by human activity such as oil spills, have a density-independent effect on populations

2 INTERACTIONS BETWEEN ORGANISMS

Every living organism in an ecosystem affects the others. Some of these interactions are **intraspecific**: they take place between organisms of the same species. Others are **interspecific**: they take place between organisms of different species.

SCIENCE FEATURE Growth and bacterial populations

Bacteria are among the fastest growing organisms on Earth. A population of *Escherichia coli*, for example, can double in size in only 20 minutes, if the conditions are very favourable. This doubling time is called the **growth rate** of the bacterial population. The doubling time is a fixed characteristic of each type of bacteria, and it can be used to identify unknown cultures.

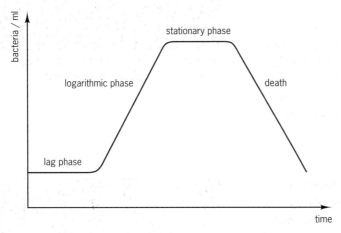

Fig 37.9 A typical bacterial growth curve

A typical growth curve for a population of bacteria is shown in Fig 37.9. There are four basic stages of growth:

The **lag phase** occurs just after some bacteria have been placed in a new container of culture medium. During this period, the bacteria are adjusting to their new surroundings, new food source and new temperature. Their doubling time is relatively slow.

When the **exponential** or **log phase** begins, the bacteria really get going and multiply very quickly. They have plenty of food and waste has not built up to dangerous levels. They make as many new bacterial cells as they can.

After a period of rapid growth, the population then enters the **stationary phase**. The bacteria run out of room or space, or they have produced toxins that limit their growth. The bacteria are still dividing, but the growth rate is equal to the death rate, and no increase in numbers occurs.

In the **death phase**, the balance shifts again, as more bacteria die than are produced by cell division. For the first time since the bacteria were placed in the new culture, their population shrinks.

Interspecific interactions in a bacterial population

The curve shown in Fig 37.10 describes the growth of a single population of bacteria. When more than one type of bacteria is present in the culture fluid, the two species compete. This is an example of an interspecific interaction.

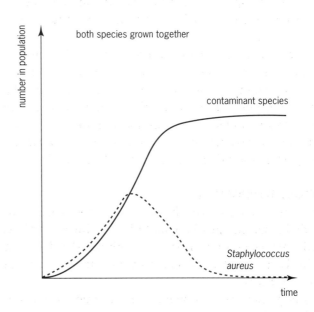

Fig 37.10 Interspecific competition between two bacterial populations

In microbiology laboratories, it is common for cultures to become accidentally contaminated. If a researcher is trying to grow up a culture of *Staphylococcus aureus* to isolate polysaccharides to study their potential as vaccines, it can be disastrous for a second species of bacteria to gain access to the culture. Fig 37.10 shows what can happen when the contaminant is a bacterial species with a faster doubling time. Both species grow equally well at first, but the contaminant species enters the exponential growth phase first, and its numbers increase at a faster rate than the *S. aureus*. The contaminant species may also be more comfortable metabolising different sugars in the medium, and can take more advantage of the food available. After a time, the *S. aureus* population enters a short stationary phase, followed quickly by a decline in population. By the end of the culture time, the researcher returns to the incubator to find a pure culture, not of the *S. aureus* that was expected, but of the rogue contaminant.

INTRASPECIFIC INTERACTIONS

Relationships between organisms of the same species in an ecosystem fall into three main categories:

● Reproduction and care of the young.
● Social behaviour.
● Competition for resources.

REPRODUCTION AND CARE OF THE YOUNG

Organisms of the same species interact at many different stages of sexual reproduction. In some species the relationship is brief and minimal: a wind-pollinated plant simply releases pollen into the air and accepts pollen from another plant of the same species. In others, the association is complex and long term. For example, swans, like humans, find a mate and stay together for many years, sometimes for life (Fig 37.11). Both parents feed and care for the young and, to some extent, provide companionship and help for each other.

Fig 37.11 The female swan carries her young, while the male displays to deter intruders

In both examples, the level of interaction achieves the aim of reproduction: producing enough surviving offspring to maintain the population level in the ecosystem.

In many animal species, reproduction involves interactions with more than just one other member of the same species. Many animals compete with each other for a suitable mate: for example, male red deer fight fiercely for the control of groups of females. Animals that are more social, such as termites and meerkats, provide joint care for each other's young. Some adults in the group, not necessarily just the parents, guard, transport and protect the young, while the others go off to search for food.

SOCIAL BEHAVIOUR

Many animals are social: they live in groups rather than as a collection of individuals that just happen to live in the same place. Group living has its advantages and disadvantages. Animals that live in groups are usually less likely to be eaten by predators than solitary animals. There are several reasons for this:

● A large number of potential prey in one place can confuse a predator – the predator cannot decide which one to attack.
● Different individuals in a large group of animals can perform different tasks: some can look for food, others care for the young and some can act as look-outs for predators. Large groups therefore spot approaching predators much sooner than animals on their own and so can take more successful evading action.
● A large number of animals are more successful than just a few at defending against an attacking predator. If a predator threatens, elephants group together, forming a circle to protect their young, which are herded into the centre of the group.

Animals that live in groups also stand more chance of hunting or finding food successfully (Fig 37.12, and the Science feature on the next page). Hyenas, for example, hunt in packs and can bring down a deer that is much larger than one hyena would be prepared to tackle. Other group associations can help to protect animals from the cold as they huddle together.

Fig 37.12 There is evidence that bees can communicate the position of good pollen sources by elaborate dancing displays to other bees in the hive. Here, a worker bee with full pollen sacs is dancing to indicate the amount, distance and direction of its source of pollen

SCIENCE FEATURE Costs and benefits of social behaviour

Prairie dogs are rabbit-sized rodents with small ears and short legs (Fig 37.13). They live in large groups in the plains of North America digging a complex of tunnels up to 30 metres long. There are two species of prairie dog, the black-tailed and white-tailed forms, and each community or 'town' may contain up to a thousand animals.

Prairie dogs graze above ground during the day when there are many predators around such as coyotes, badgers and hawks. They have 'sentries' sitting upright outside the burrows keeping a watchful eye for predators. The appearance of a predator provokes a series of whistling barks from the sentries.

A cost–benefit analysis has been carried out on the social behaviour of the prairie dog. An anti-predator response was tested by dragging a stuffed badger through a colony. In the larger groups of prairie dog, the predator was detected sooner. Black-tailed prairie dogs live in larger groups than their white-tailed cousins. Consequently the black-tails can afford to spend significantly less time scanning their surroundings for predators (35 per cent of their average daily time budget compared with 45 per cent in the white-tails). Less time being vigilant means more time to feed.

So why don't white-tailed prairie dogs live in large groups as well? One answer is that living in large groups has costs as well as benefits. Within either species, individuals in large groups fight more often than those in small groups. Also disease-spreading fleas are more common in large groups (3.3 fleas per black-tailed burrow, 0.5 fleas per white-tailed burrow).

The evolution of group behaviour therefore depends on balancing advantages against disadvantages. Other factors may also be involved – black-tailed prairie dogs live in open plains, for example, and, since there are many predators around, it is safer for them to live in larger groups.

Fig 37.13 A prairie dog keeping guard as sentry

COMPETITION FOR RESOURCES

Most of the disadvantages of group living arise because animals have to compete with each other for vital resources. Hyenas may make a bigger kill when they hunt in a pack, but the meat has to fill more stomachs. If food becomes scarce, rivalry within the pack becomes intense and some may need to find a new habitat to survive. Animals of the same species also compete for mates, shelter, water and social position.

INTERSPECIFIC INTERACTIONS

A full list of the interactions that can occur between individuals of different species in an ecosystem is given in Chapter 35, Table 35.1. In this chapter we will look in detail at four of these:
- Predator–prey relationships.
- Parasitism and disease.
- Mutualism and commensalism.
- Competing for resources.

PREDATOR–PREY RELATIONSHIPS

A basic definition of a predator, which includes animals that feed on plants, is:

**An organism that eats another organism
to obtain energy and nutrients.**

The relationship between predator and prey is based on a constant battle. The actual 'catching and eating' part of the battle is only a small part of the story. The more serious conflict involves evolution.

All species evolve by the processes of genetic mutation and natural selection (see Chapter 29): prey species are constantly evolving and becoming better able to cope with predators; predator species are constantly evolving to out-manoeuvre their prey (Fig 37.14). Sometimes, either the predator or prey loses the long-term battle. If the prey evolves to completely avoid a predator, and that predator has no alternative food supply, the predator species can be driven to extinction. If a predator out-evolves its prey, it will do well in the short term. But, if the prey species becomes extinct, the predator then needs to find an alternative food supply to survive.

In general, predators thrive when they develop adaptations that allow them to eat well and make the most of their food. Herbivores, animals that eat plants, often have specialised mouthparts and digestive system to gain the most nourishment from the plants available to them. Many herbivores are ruminants: their digestive system has four stomachs and they chew their food twice, once after eating it and once after regurgitating it.

Prey species continue to survive because they have evolved defence mechanisms against predators. In animals, these include **camouflage**, **spines** and **barbs** and **chemical deterrents** (Fig 37.15). Many, particularly insects, produce foul tasting or poisonous chemicals and usually advertise this by the colours and patterns on their bodies and wings.

Although plants cannot run away from grazing animals, they are not totally defenceless. Their defences include thorns, tough tissues and thick waxy cuticles. Many plants have also developed **secondary compounds**, chemicals that have no metabolic role but which put off potential grazers. Examples include tannins in oak leaves and cardiac glycosides in foxglove plants.

Fig 37.14 Bats have evolved sophisticated sonar systems to detect flying insects with great accuracy. Some insects, however, have evolved the ability to detect bat sonar signals and respond by closing their wings and dropping out of the sky. This makes it much harder for the bat to catch them and they gain an important survival advantage over other insect species

Fig 37.15 Animals use many different techniques to avoid capture by predators

The camouflage of this iguana matches the colours of the trunk and lichens in a Peruvian rainforest

The deadly sting of a scorpion deters predators as well as being useful for hunting food

When under threat, a porcupine runs backwards at the enemy with quills raised. They can fall off, and an embedded quill can cause a septic wound that would kill a lion

PARASITISM AND DISEASE

Predators usually kill their prey: parasites have a different strategy because they depend on their prey, the **host** (a different species), staying alive. They live and reproduce using the resources of the host organism that is damaged in the process. This is why many parasites cause disease in humans and other animals (Table 37.2). The more successful parasites cause less damage to their host. The tapeworm, for example, can survive in the human digestive system for many years, causing relatively minor damage in the affected person.

Table 37.2 Some major parasitic diseases and the parasites that cause them		
Disease	Parasite	Symptoms and effects
Malaria	*Plasmodium* (4 different species of this protoctist parasite cause malaria)	Regular cycles of fever: disease transmitted to people by the *Anopheles* mosquito (see Science feature on page 674).
Sleeping sickness	*Trypanosoma* species (also a protoctist parasite)	Transmitted to people by insects such as the tsetse fly which bite infected animals.
Leishmaniasis (also called kala-azar and oriental sore)	*Leishmania* species (another protoctist parasite)	Transmitted to humans via sandflies.
Schistosomiasis (also called bilharzia)	*Schistosoma* (blood fluke)	Flukes live in water (their intermediate host is a water snail) and infect humans who bathe.
Tapeworm infection	*Taenia solium* (tapeworm)	Parasite lives in intestine: passed to humans in infected and undercooked meat.
Toxoplasmosis	*Toxoplasma* species (protoctist parasite)	Causes weakness and fever – passed from mother to baby across placenta. Can be transmitted by domestic cats.

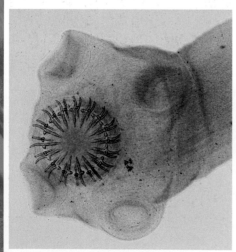

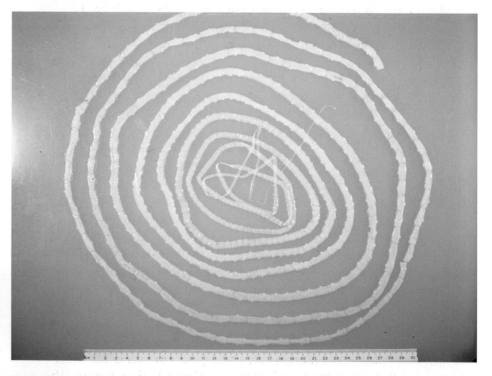

Fig 37.16 Above: The beef tapeworm. The head is at the thin end, attached to many hundreds of segments. They break off and pass out of the host (cattle), to infect another victim. The ruler is 30 cm long. Left: At the head end (the scolex) are four suckers and a ring of hooks that attach the tapeworm to the gut wall of its host

PARASITIC ADAPTATIONS

Parasites show very specialised features that fit them to their lifestyle
(Fig 37.16). All parasites show some general adaptations:

- They have ways of getting into the body of the host. Fungi that parasitise
  plants produce the enzyme **cellulase**, which breaks down plant cell
  walls. The schistosome (blood fluke), which causes bilharzia, can bore
  through human skin.
- They have structures that enable them to attach to their host. Tapeworms
  have a ring of hooks called a **scolex** that they insert into the wall of the
  host's intestine.
- They have lost the organ systems that they no longer need. Tapeworms
  and other gut parasites have, for example, lost their own digestive system
  because they can absorb already digested food across their body surface.
  We call this loss of structure and function **parasitic degeneration**.
- They protect themselves against the internal defences of the host.
  Trypanosomes, which cause sleeping sickness in humans, produce an outer
  layer which varies constantly in its structure. Just as the host's body starts to
  make antibodies (see Chapter 17) to one type of coating, the trypanosome
  produces a different coat which the antibodies cannot recognise.
- They often have complex life cycles that allow them to infect new hosts.
  See the Science feature box on malaria on the next page.
- They show great **fecundity**, or capacity for reproduction. Many
  tapeworm segments break off every day and pass out of the body in the
  faeces. Each segment can contain hundreds or even thousands of eggs.

MUTUALISM AND COMMENSALISM

The term **mutualism** is used to describe a relationship between two
organisms from different species that receive mutual benefit from the
association. Neither is harmed and both do better by being together than they
would alone. The African bird, the oxpecker, for example, cleans the skin of
the rhinoceros, removing all the parasites such as ticks and mites. The bird
eats the insects, gaining the advantage of the nutrients.

Two organisms that are **commensal** live together and, while one of the
partners benefits, the other does not. The relationship between clownfish and
sea anemones is a good example of commensalism (Fig 37.17). The fish hide
from predators in the mass of the anemone's stinging tentacles. Although the
anemone is unharmed, it does not seem to benefit in any way.

COMPETING FOR RESOURCES

Species can compete in different ways. Some organisms stop the organisms
of other species from getting near a particular resource. Wrens, for example,
defend food plants in their territory. In other situations, two species have
access to the resource, but one is better at exploiting it.

Some species have evolved to **coexist**, sharing the resources that are
available in an ecosystem. This is termed **symbiosis**. Some plants, for
example, have a mesh of roots close to the surface of the soil to trap water
from short showers of rain. Other neighbouring plants have long roots that
take advantage of deeper groundwater. Both species are using the same
resource, but they are sharing it in a way that allows both to benefit. This is
called **resource partitioning** and it also occurs between animal species.

Fig 37.17 There are 26 species of clownfish which all associate with sea anemones. This colourful fish hides from predators among the venom-producing tentacles of the sea anemone

⚠ SCIENCE FEATURE The fight against malaria

Malaria is a disease caused by a parasite. Four species of protoctist of the genus *Plasmodium* are responsible for 200 to 300 million cases of malaria every year, worldwide. The disease is common in the tropical and sub-tropical regions of the world – the home of 40 per cent of the world's population. The death rate is appalling, somewhere between one million and five million people each year, many of them children under 5.

Malaria is extremely difficult to control for several reasons. Firstly, it has a complex life cycle (Fig 37.18). Any four of the *Plasmodium* species that cause it can be passed on to human hosts by any one of 60 species of the *Anopheles* mosquito. Draining swamps and marshes and using insecticides to spray mosquito-breeding areas was temporarily successful and reduced the spread of malaria during the 1950s and 1960s. However, not every species of carrier mosquito was wiped out. Many of them have developed resistance to the early insecticides and are now much more difficult to control.

Secondly, the parasite itself is a tricky customer. It produces symptoms that get worse and then better in cycles. Anti-malarial drugs target some but not all stages in the life cycle (Fig 37.19). The sufferer therefore needs to take the drugs over a long period of time, or the parasite 'hides' from their effects. The common anti-malarial drugs have been used so widely that they have put selection pressure on the parasite and it has now developed resistance to many of them.

Although new drugs are being developed all the time (some of the most recent are based on extracts of rainforest plants), workers now concentrate on prevention. Any stagnant water that can be eliminated is removed and mosquito nets in houses are dipped in a safe insecticide such as permethrin. Biological methods of control of the mosquito larvae are being used – fish such as guppies are being bred to eat them. Vegetation around houses is cleared, and trees are being planted in marsh areas to destroy the wet breeding grounds of the mosquito.

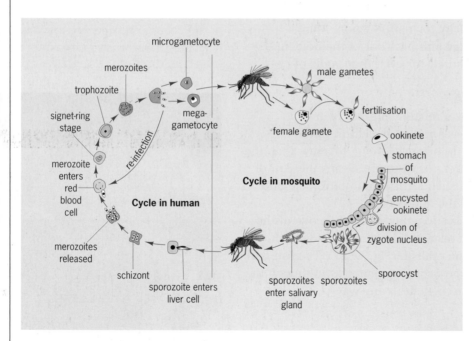

Fig 37.18 The life cycle of the malarial parasite, *Plasmodium vivax*. The *Anopheles* mosquito is the first host; humans are the second host. People with malaria get regular cycles of illness because merozoites are continually released. When the parasite is 'hiding' in the red blood cells, symptoms are not too bad: it is the sudden rush of foreign proteins in the bloodstream that leads to the bouts of fever that are typical of malaria

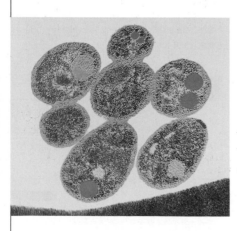

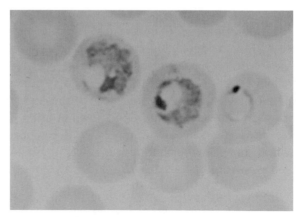

Fig 37.19 The merozoite stage (left) and signet ring stage (right) of *Plasmodium vivax* in human red blood cells

3 A WIDER LOOK AT COMMUNITIES IN AN ECOSYSTEM

We have seen that ecosystems are dynamic systems and that the physical parts of an ecosystem, such as the weather, affect and interact with the organisms that live there. In this section we look at how an ecosystem forms.

Imagine a completely physical environment such as the bare rock and ash left behind after a volcano has erupted and then become quiet. After the ground cools and water re-enters the environment, some plants start to grow there. This process is called **colonisation**. After a time, other species of plant establish themselves and eventually, probably after many years, woodland forms. This process is called **succession**.

Succession also describes the more minor changes in the vegetation of an ecosystem that occur after a less catastrophic environmental change. For example, after a small patch of trees in woodland die after a series of hot, dry summers, the land becomes repopulated by the same sorts of trees very quickly because their seeds are already present in fertile ground.

To distinguish between these two types of situation, we say that the process that results in new vegetation on a bare site is **primary succession**. This takes time, and the final appearance of the community depends on which plant spores and seeds come into the area from neighbouring ecosystems. The process that repopulates a previously well-vegetated area after a minor environmental change is called **secondary succession**.

Strictly, succession is used only to describe the development of and changes in plant communities. In any ecosystem, however, the plant communities help to form the conditions that enable other classes of organism (animals, bacteria, fungi, etc) to colonise the environment. Once a stable community has formed, we say that a **climax community** is established.

SEE QUESTION 2

A DEVELOPING ECOSYSTEM

In order to understand the stages that occur in succession, think of an environment consisting just of bare rock (Fig 37.20).

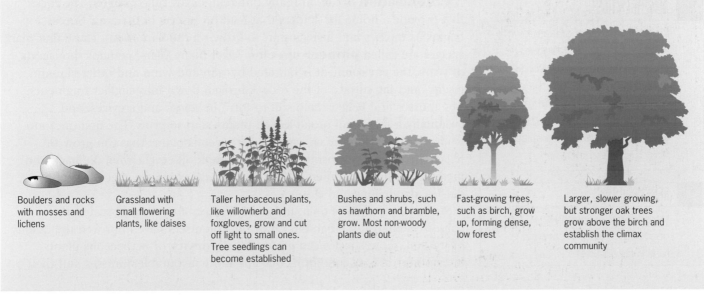

Boulders and rocks with mosses and lichens

Grassland with small flowering plants, like daises

Taller herbaceous plants, like willowherb and foxgloves, grow and cut off light to small ones. Tree seedlings can become established

Bushes and shrubs, such as hawthorn and bramble, grow. Most non-woody plants die out

Fast-growing trees, such as birch, grow up, forming dense, low forest

Larger, slower growing, but stronger oak trees grow above the birch and establish the climax community

Fig 37.20 Succession in a bare site to woodland

! **SCIENCE FEATURE** Studying succession

The process of succession normally takes decades or centuries, but there are two common ways of studying the process:

● Clear a patch of ground and watch what happens to the bare soil.

● Study a sand dune system near a beach and observe the changes that occur as you move inland.

1 Clearing a patch of ground

Remove all the plants, fence it off from grazing animals, and observe the changes. The problem is that it takes at least 50 years, but eventually the forest is re-established. There are two conditions that need to be met if the ecosystem is to develop as normal:

● The soil must initially have relatively low humus content.

● There must be no grazing animals that can get on to the plot.

The climax community that develops depends on the climate. The process outlined in the text on this page will not happen, for example, on exposed hillsides where the soil is too thin or the wind too harsh. Other examples of climax vegetation around the world include rainforest, cloud forest, tundra, grassland and coniferous forest.

2 Studying a sand dune system

Sand dunes are useful areas to study because you can observe colonisation and succession

without having to wait 50 years. You can see some of the changes associated with the development of ecosystems as you simply walk inland (Fig 37.21). Near the sea, the dunes are at their youngest and wind tends to pile up the sand. As a result, the profile changes from year to year. Sand is a very difficult medium for plants; water and nutrients drain straight through, and the constant shifting makes it very difficult for the roots to anchor the plant. However some pioneer species, notably marram grass, have a dense root system that binds the sand together. This holds water and humus particles and makes the whole dune more permanent. Once the marram grass has made the environment less hostile, other plants such as ragwort, willow and grasses can take over. As you move inland the sand gets darker because the humus content increases, as does the species diversity.

Fig 37.21 Sand dune succession

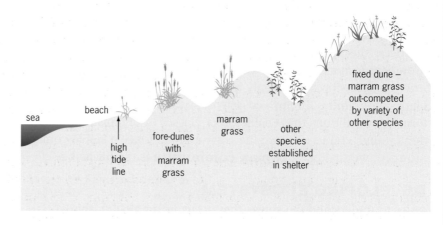

sea — beach — high tide line — fore-dunes with marram grass — marram grass — other species established in shelter — fixed dune – marram grass out-competed by variety of other species

 REMEMBER THIS

Lichens consist of algae growing inside a fungus. The fungus provides anchorage and protection from drying out. In turn the algae can photosynthesise and give the fungus organic compounds that would otherwise be unavailable from the bare rock.

✔ **REMEMBER THIS**

Herbaceous plants have no woody tissue, in contrast to shrubs and trees.

Colonisation occurs in many different stages called **seres**. The rocks first provide a home for spiders that feed on insects in the area. Succession begins as mosses and lichens start to grow on the bare rocks. These first plant species are called **pioneer species**. Over many years, perhaps thousands of years, the environment is battered by rain and wind and other climatic events, and the surface of the rock is broken down into smaller fragments. The mosses and lichens help soil to form in cracks and crevices, and eventually a few small broad-leaved plants start to grow. The first ones are what we might think of as weeds – herbaceous plants that can grow in relatively nutrient-poor soil. They have a rapid life cycle, then die back and increase the humus content of the soil. Typical coloniser species are grasses, daisies, dandelions and clover.

Succession proper is then under way. Succession occurs because the colonisers change the habitat. Once the colonisers have improved the quality of the soil, more species can grow. The **diversity** of herbaceous plants increases greatly as conditions become more favourable. Grasses and then plants with root nodules that have nitrogen-fixing bacteria start to grow. Taller plants cut off the light and so out-compete the shorter ones. The greater diversity of plants attracts more insects that in turn attract birds and small mammals. At this stage grazing can have a marked effect and prevent any further succession. Many plants, including tree saplings, have their growing points at the top of the stem, and herbivores such as rabbits and sheep prevent any further growth. In contrast grasses grow from the base of the stem, so they thrive despite constant grazing.

ESTABLISHMENT OF THE CLIMAX COMMUNITY

In this phase, small woody plants – shrubs such as hawthorn and bramble – begin to dominate. In turn these are out-competed for light by fast growing tree species like birch that form a low, dense forest. Eventually the large but slow growing trees – notably the oak – begin to dominate until the climax community is established and there is *no further succession*.

4 BIODIVERSITY

Biodiversity, a word that is the popular contraction of the term **biological diversity**, is used generally to refer to the variety of life on Earth. More specifically, we use it to describe the numbers of species present in a particular habitat or to show the range of types of organism found there (Fig 37.22). Take the sea bed, for example. It has a large number of species but it is largely made up of a few basic types: annelid worms, molluscs and echinoderms. A rocky shore may have the same number of species but there is often a greater variety of animals and plants: arthropods, annelid worms, molluscs, echinoderms, cnidarians, fish, birds and seaweeds. We say that the shore has a greater **species diversity**: it has a greater **species richness** (number of different species) and a greater **species disparity** (range of different species types), when compared to the sea bed. One of the most diverse ecosystems in the world is the tropical rainforest (Fig 37.23).

> ✓ **REMEMBER THIS**
>
> It is a good idea to learn one example of succession in detail, and make sure that you can name some plant *species* at each stage, rather then talking vaguely about 'trees and shrubs', etc.

> ✓ **REMEMBER THIS**
>
> Details of statistical tests are given in the Appendix to the book.

SEE QUESTIONS 3 AND 4

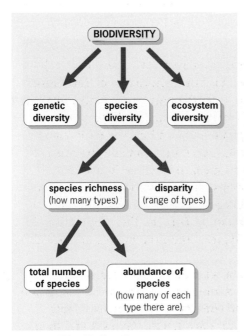

Fig 37.22 What biodiversity means

Fig 37.23 The tropical rainforests of the world encompass a massive diversity of all forms of life

Diversity can be used to indicate the 'biological health' of a particular habitat. A slow increase in plant species diversity is normal for stable habitats such as hedgerows. If a habitat suddenly begins to lose its animal or plant types, ecologists become worried and search for causes (a pollution incident, for example).

You might also have heard of the term **genetic diversity**. Different species are obviously genetically different from each other, but there is also genetic variation within a species (no two people look identical, for example).

A third form is **ecosystem diversity**. The environment is an important factor that determines the number and types of species present in an ecosystem. At its simplest, ecosystem diversity is described as the number of different types of habitat in a given area.

WHY BIODIVERSITY IS IMPORTANT

When a species becomes extinct, many other species can be affected: no type of living organism exists independently. No one can really predict the long-term effects of repeated extinctions. Organisms have, of course, always become extinct, because of evolutionary pressures. What concerns us today is the rate at which species are disappearing because of the impact of human populations.

Although the diverse world of animals and plants provides us with a massive resource of food, we actually use only 30 species or so, to provide 90 per cent of all the food we eat worldwide. Plants have already given us many useful medicines (Fig 37.24 and see Science feature below) and there are undoubtedly more compounds still to be found. In addition, many different animals and plants stabilise soil and so minimise erosion, support fisheries, provide income through tourism and provide beauty and interest.

Fig 37.24 Aspirin, the most commonly used drug in the world, was originally extracted from the bark of the willow

---

⚠ SCIENCE FEATURE Medicines from the rainforest

The weather conditions in the tropical rainforests around the Equator have encouraged the evolution of areas of high plant and animal diversity. The indigenous human populations who live and work in the rainforests need to make a living, and working for developers who sell timber and other products to the developed world is an obvious means of doing this. The consequences for the environment are serious. Some scientists estimate that three species of living things are being made extinct every hour, because the rainforest is being destroyed.

But what difference can a few plants make? Possibly the difference between life and death for many people. Screening plants from the rainforest has shown that they contain many compounds that protect them against predators and parasites. These compounds might also be useful in human medicine. Since 1990, at least four compounds have been identified that might provide drugs for use against the virus that causes AIDS (Acquired Immune Deficiency Syndrome).

The work of developing medicines needs sophisticated chemical and pharmaceutical techniques, but it cannot be done unless plants that are collected for screening can be

identified accurately and consistently. Speed is also important. If destruction continues at its present rate, the rainforests will have been completely wiped out by the end of the century.

Fig 37.25 Scientists who travel to remote areas of rainforest to search for and identify plants which may be useful in medicine are called ethnobotanists. They work with local traditional healers to narrow down the choice of plants to study. Samples are sent back to laboratories for testing. Research programmes are now being set up to share the profits from successful compounds, so that indigenous people can benefit from the discovery, instead of suffering exploitation

MEASURING BIODIVERSITY

Estimates for the total number of species that still exist range from 1 million to 50 million. We base our estimates on the knowledge we already have of numbers and types of species found in particular habitats and on the rate at which our knowledge is increasing. Between 1978 and 1987, 367 new vertebrates, 173 new annelids and 7222 new species of insect were identified.

WHY DO WE NEED TO MEASURE BIODIVERSITY?

There is more to measuring biodiversity than just being able to impress your friends by knowing how many species of living things there are on Earth. We need to measure the diversity:

● To compare the diversity of the same habitat over time, to assess, for example, whether pollution is damaging that habitat.
● To compare two different habitats at the same time, to find out which is the more diverse.
● To check whether a new habitat is being colonised by the number of species that it should. This would be important in the case of a polluted river or lake that had been 'cleaned up' and then repopulated with living organisms.

PUTTING A NUMBER ON BIODIVERSITY – THE SIMPSON DIVERSITY INDEX

Being able to put a number on the diversity of a habitat makes the life of an ecologist much easier. A **diversity index** allows us to estimate the variety of living things in a particular area. However, we must make sure that the value we use accurately reflects all aspects of diversity. We could, for example, simply count the number of different species present:

Stream 1 (100 animals)	16 species
Stream 2 (100 animals)	10 species

The conclusion would be that stream 1 was more diverse. But there is a problem: this method gives an idea of species richness, but it does not reflect how many of each species are present. In stream 1, there may be 85 animals of one species and just 1 of each of the others. In stream 2, there may be 10 animals of each species. Ecologists would say that stream 2 was the more diverse habitat, and so our first conclusion would be wrong.

To avoid this problem, we can use the Simpson index. Its formula takes into account that diversity depends on the number of individuals of each species present and the species richness (the number of different species present). The higher the value of the index, the greater the variety of living organisms found in the area.

$$D = \frac{N(N-1)}{\Sigma n(n-1)}$$

where D is the Simpson diversity index, N is the total number of individuals of all species found, n is the total number of organisms of a particular species found, and Σ means 'the sum of'.

? TEST YOURSELF

D The Simpson index was used to calculate the diversity of an oak wood and a conifer plantation. The values obtained were 19.6 for the oak wood and 12.04 for the conifer plantation. Which habitat is the more diverse?

● EXAMPLE

Q Assume that you have studied a particular habitat and you have recorded the types of 20 animals you have found. The different types are represented by letters of the alphabet, and the record shows five species:

A A A A B B A C C B B A B B A D C C D E

What is the diversity of this habitat?

A Putting the figures into the equation for the Simpson index:

$$D = \frac{20 \times 19}{(7 \times 6) + (6 \times 5) + (4 \times 3) + (2 \times 1) + (1 \times 0)} = \frac{380}{86} = 4.42$$

! SCIENCE FEATURE Quadrats and transects

Two of the sampling techniques described in Table 37.1 are used a lot by plant ecologists. They carry out quadrat sampling to survey a relatively small area, and transect recording to describe differences in vegetation that occur over a larger area of land.

Quadrat sampling

The basic technique of quadrat sampling is shown in Fig 37.26. When you are sampling using a quadrat you may come across problems: sometimes, for example, you may not be able to count how many individual plants there are in a dense clump. So you can use any of these methods:

● Estimate plant *density* by counting the number of individuals per unit area.

● Estimate plant *frequency* by looking at the distribution of individuals per unit area (are plants abundant or rare?).

● Assessing *plant cover* by measuring the proportion of the ground covered by a plant species.

Transect recording

A transect is an imaginary line along which you make careful and systematic observations. There are two types of transect, the **belt transect** and the **line transect** (Fig 37.27). You can use transect recording to study how communities and ecosystems change along an environmental gradient, for example, through a woodland, along a rocky shore, or up the side of a hill.

Fig 37.26 Quadrat sampling

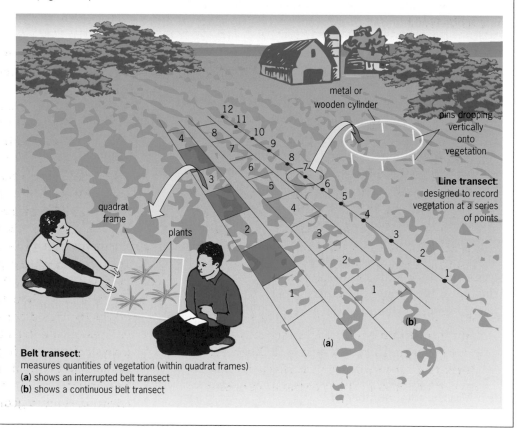

metal or wooden cylinder

pins dropping vertically onto vegetation

Line transect: designed to record vegetation at a series of points

quadrat frame

plants

(a)

(b)

Belt transect:
measures quantities of vegetation (within quadrat frames)
(a) shows an interrupted belt transect
(b) shows a continuous belt transect

Fig 37.27 Types of transect recording

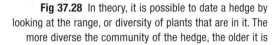

SCIENCE FEATURE How old is my hedge?

As you next walk down a country lane, take a good look at the hedges that you pass. Many of these structures date from the Middle Ages. They were planted by medieval farmers as boundaries to mark off their land and to keep in livestock. They probably planted shrubs such as hawthorn a few feet apart. The gaps at the base of these plants then allowed waves of colonisers.

First of all, fast growing, non-woody annual plants such as typical garden weeds would have thrived: chickweed, groundsel and shepherd's purse for example. Then, non-woody perennials such as bluebells, primroses, stinging nettles, dandelions and thistles may have become established. These plants can produce food reserves to survive the winter. Eventually, woody perennials, shrubs and small trees such as elder, holly, and even some of the larger forest trees could have invaded the hedge.

Fig 37.28 In theory, it is possible to date a hedge by looking at the range, or diversity of plants that are in it. The more diverse the community of the hedge, the older it is

SUMMARY

By the end of this chapter you should know and understand the following:

- A **population** grows **exponentially** in ideal conditions. Growth slows down because of **limiting factors** in the environment. This produces an **S-shaped** or **sigmoid growth curve**.

- When the size of the population becomes stable, we say that this is the **carrying capacity** of the environment. Sometimes, population growth exceeds the carrying capacity and a **population crash** follows. The population may then settle at a new carrying capacity.

- Three main types of growth curve occur in nature: **stable**, **irruptive** and **cyclic**.

- Some factors that limit population growth are **density dependent**: their effects are worse in a large, closely packed population. Others are **density independent**: their effects do not depend on the population distribution or size.

- **Intraspecific interactions** occur between organisms of the same species; **interspecific interactions** occur between organisms of different species. You should know some examples of each type.

- The living part of an ecosystem develops by the process of **succession**. If the area has never been **colonised** before, a newly formed volcanic island, for example, the process is called **primary succession**. If the area was previously covered with vegetation, a forest site that suffers a severe fire, for example, **secondary succession** occurs.

- There are three types of **biological diversity**: **species diversity**, **genetic diversity** and **ecosystem diversity**.

- Measuring biodiversity is important because it allows us to monitor a habitat over time, to compare two different habitats and to find out if a new habitat is being colonised by the correct number and type of species. This is vital if we are to assess the impact of our 'clean-up' efforts after environmental tragedies such as oil spills.

❓ EXAM QUESTIONS

1 A population of brown trout in a river was studied over a period of ten years. Each year the size of the population was measured in early summer, when it included the newly hatched young, and again at the end of the summer.
The results are shown in the graph.

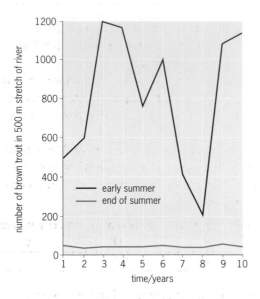

It is suggested that the change in population size between early and late summer is due mainly to one or more density dependent factors.
(i) Explain what is meant by a *density dependent factor*.
(ii) Describe and explain the evidence from the graph which suggests that the change is mainly due to the density dependent factors.
b) Brown trout only grow well in water which is above pH 5.5 Explain **one** way in which the pH of the water might become more acidic than this, other than by pollution or acid rain.

AQA (NEAB) BY05 Ecology June 2000 Q4

2
a) Give the ecological term that best describes each of the following.
(i) all the living organisms in a particular habitat
(ii) the particular set of conditions in a habitat to which a species is adapted
(iii) two species using the same resource in the same habitat
(iv) movement of a species between two habitats to make use of their resources at different times.
b) Explain what is meant by *succession* in a habitat.

AQA (NEAB) BY05 Ecology June 2000 Q1

3 The distribution of two species of plant, greater plantain and white deadnettle, on a roadside verge was investigated. The results are shown in the bar graph.

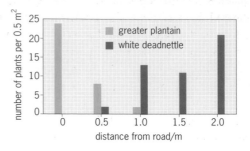

a) **(i)** Describe how the data on distribution could have been obtained.
(ii) Describe what the data show about the distribution of greater plantain.
b) The drawings show the two species of plant.

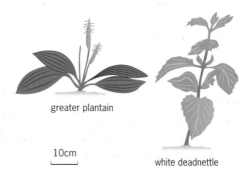

Use evidence from the drawings to suggest a hypothesis that explains the distribution of greater plantain.
c) A diversity index for all of the plants growing on this roadside verge was calculated at three-year intervals. The table shows the results.

Year	Diversity index
1990	6.1
1993	4.5
1996	3.1

Explain what these results indicate about the plant community on this verge.

AQA (NEAB) BY05 Ecology June 2000 Q3

? EXAM QUESTIONS

4 The dog whelk, *Nucella lapillus,* is a marine snail found on rocky shores. It feeds on barnacles and molluscs by drilling holes in their shells or prising them open.

An investigation was carried out into the choice of prey species by the dog whelk. Sixty dog whelks were collected from a rocky shore where they had been feeding on barnacles. The dog whelks were given a mixture of two species of barnacle, *Semibalanus balanoides* and *Elminius modestus.* After 10 days, the number of each species of barnacle which had been eaten and the number remaining uneaten were recorded.

The observed results are shown in the table below.

Species	Number eaten	Number remaining	Total
S. balanoides	137	146	283
E. modestus	166	249	415
Total	303	395	698

Data from Barnett BE (1979)

A chi-squared (χ^2) test was then carried out on these results to test the null hypothesis that dog whelks eat equal percentages of each of the two barnacle species.

The expected results for this test are shown in the table below.

Species	Number eaten	Number remaining	Total
S. balanoides	122.9	160.1	283
E. modestus	180.1	234.9	415
Total	303.0	395.0	698

The formula for calculating chi-squared (χ^2) is given below.

$$\chi^2 = \sum \frac{(O-E)^2}{E}$$

where O = observed values, E = expected values.

a) (i) Use the formula to calculate the value of χ^2. Show your working.
(ii) The table below gives some values for χ^2 with one degree of freedom.

P(%)	20	10	5	2	1
χ^2	1.64	2.70	3.84	5.41	6.63

Does your calculated value of χ^2 enable you to accept or to reject the null hypothesis? Give a reason for your answer.

b) The dog whelks ate 48% of the *S. balanoides* and 40% of the *E. modestus.* Suggest possible long-term effects this could have on the populations of the two species.

c) The investigation was repeated using dog whelks which had not been fed for a period of time. The results for this second experiment are summarised in the table below.

Species	Number available	Number eaten	Percentage eaten
E. balanoides	296	161	54
E. modestus	365	191	

(i) Calculate the percentage of the available *E. modestus* which was eaten. Write your answer in the space in the table.
(ii) Describe the effects of starvation on the feeding behaviour of dog whelks.

d) The barnacles feed by filtering plankton from sea water. The plankton contains microscopic plants (phytoplankton) and microscopic animals (zooplankton) which feed on these plants.
(i) Construct a food web to show these feeding relationships.
(ii) Name the trophic levels occupied by the barnacles in this food web.

Edexcel B5/HB5 June 1999 Q2

KEY SKILLS **ASSIGNMENT**

AN UNCERTAIN FUTURE

In Chapter 29 we looked at how humans evolved from their primate ancestors. In this Assignment, we investigate what has happened to the human population over the past 12 000 years and we raise some questions about what will happen in the next 12 000.

We start our story at the start of the 'Agricultural Revolution' in about 10 000 BC, when there were around 10 million people on Earth. People had changed from being hunter–gatherers to being farmers. With this new settled lifestyle, there was a dramatic increase in the birth rate and a decline in the death rate.

1 **a)** Suggest why nomadic travellers might have had fewer children than farmers living in permanent settlements.
b) Why would life expectancy be lower in hunter–gatherer societies?

By the time of the First Egyptian Dynasty (3000 BC) the global human population had increased from about 10 million people to 100 million. By the height of the Roman Empire and the birth of Jesus Christ, there were 300 million. Civilisations flourished in three main regions of the world: Southwest Asia, China and the American continent.

2 What properties do you think these areas might have had that enabled them to support large numbers of people?

Table 37.A1 shows how population continued to increase until 1800, when it reached the 1 billion milestone.

Table 37.A1 Population trends to the first billion humans						
Year	0	1000	1250	1500	1750	1800
Population (in billions)	0.30	0.31	0.40	0.50	0.79	1.00

3 Plot the figures given in Table 37.A1 as a line graph. Extend the time axis to the year 2100 AD.
a) Estimate how many people would have been alive in the year 2000 if the rate of growth had remained constant after 1800 AD.
b) Current estimates put the world population at 7 billion in the year 2012. Why is your estimate less than this?
c) If population trends continue like this, in which year will the number of people on Earth reach 11.6 billion, the estimated maximum population that the planet's known resources can support?

Staggeringly, adding yet another billion people after 1800 took only 123 years, and another billion was added 33 years after that. Since 1960 we have been in the middle of an incredible population explosion with another billion people being added every 12 years.

4 Experts think that population growth will slow down during this century and stabilise at around 11 billion people. What factors are likely to be responsible?

Most of the Earth's land surface is now inhabited by humans. Table 37.A2 shows the 15 most populated countries today and gives an estimate of how their individual populations will grow in the next 50 years.

Table 37.A2 Human population, in millions, of the 15 most populated countries in the world with estimates of their population in the year 2050 AD			
2000		**2050**	
Country	**Population**	**Country**	**Population**
China	1232	India	1530
India	945	China	1514
USA	269	Pakistan	357
Indonesia	200	USA	348
Brazil	161	Nigeria	339
Russia	148	Indonesia	318
Pakistan	140	Brazil	243
Japan	125	Bangladesh	218
Bangladesh	120	Ethiopia	213
Nigeria	115	Iran	170

5 Explain the different reasons why China, Indonesia and the USA all have a lower place in 2050 than in 2000.

6 Imagine that it is the year 2100. You have been asked to write a book telling the history of the previous century. The population of Earth stands at 3 billion people. Explain in about one thousand words what happened.

Fig 37.A1 The suffering caused by starvation, famine and overcrowding seems likely to continue as the human population continues to increase exponentially

38

Human activity
and the environment

Many problems in the environment happen when people use a resource for today, hoping that everything will be OK tomorrow. In the Philippines sea-horse fishing is very important to the poor fishermen and their families: they can sell sea-horses to traders who then sell them on to the lucrative Chinese medicine market in Hong Kong and elsewhere. Without the sea-horses, many of the people would not be able to feed their families.

As fishing continues, the sea-horses have started to disappear. Fewer sea-horses means that the fishermen catch fewer and earn less and so start to catch smaller and younger sea-horses to supplement their income. The population is further depleted and a vicious circle is set up.

In a ground-breaking environmental project, scientists have studied the life cycle and behaviour of sea-horses and are working with local fishermen to try to change fishing habits, so that the sea-horse population recovers.

Rather than take a standard environmentalist hard-line view and try to persuade the villagers to stop fishing altogether to protect all the sea-horses, the team recognises the conflict between human need and the needs of the environment. They have shown the fishermen that an enforced 'safe area' will allow sea-horses to breed without threat, and have taught them to avoid catching pregnant male sea-horses. In this way, the scientists have proved to the local community that they can help to restore sea-horse numbers. Not only will the sea-horse population be saved for the future, the fishermen will get back their source of income and will be free from the worries of starvation.

Sea-horses are unique in the animal kingdom: the male not the female gets pregnant and gives birth

1 WHY DO HUMANS AFFECT THEIR WORLD SO MUCH?

Fig 38.1 We think of the British countryside as 'natural' but, in fact, it is almost entirely artificially made. Only a few areas in Scotland, Wales and Northern England remain in their 'original' form – all farmland and most forests are the result of human influence on the environment

We humans are a unique species: we live in most of the ecosystems of the world, adapting our habitats to suit us (Fig 38.1), rather than having to adapt to the environment as most other species do. And, unlike other species, our behaviour affects the environment on a global as well as a local level.

The main human activity that affects the world environment is farming. Close behind are industry and mining for fuels such as coal, oil and gas. The more technologically advanced the society, the greater the overall environmental impact of each individual person. So, for example, a child born in Britain into a fairly well-off family may live for 80 years or more. In that time the amount of resources they use is enormous: think of the furniture, electrical appliances, cars, fuel, food, clothing, etc used up by someone like this. At the other extreme, a child born in a poor family in India or South America might live only 20 years. They could live in poorly built housing, or be homeless and forced to live on the streets, probably never get quite enough to eat, and would certainly never use a car or any significant amount of fuel.

In this chapter, we take a brief look at some of the complex issues involved in controlling the effect people have on the ecology of the Earth. There is room only to give an overview of some of the problems and potential solutions.

2 FARMING AND ITS EFFECTS ON THE ENVIRONMENT

It is very common to hear environmentalists complain that farming is destroying hedgerows and wildlife habitats, that woodlands and forests are being cut down to make new farmland and that it should all be stopped. In addition, farmers are often portrayed as hard-line businesspeople who put profit before anything else. There is a clear conflict here, but the issue is not as simple as these points of view suggest.

We must consider five basic facts:
● People need to have food to eat: staple foods need to be cheap.
● There are more people on Earth to feed than ever before.
● Farmers need to make a living.
● Farming needs land.
● Some farming techniques are damaging the environment so badly that the capacity of the land to produce food is being reduced.

The key to solving the problem of food production is to develop farming methods that are efficient enough to provide the amount of food that we need in the short term, but that protect the environment to ensure that food production levels can be maintained, or even increased, in the future.

FOOD PRODUCTION ON A GLOBAL SCALE

On a global scale, food production has increased during the past 50 years (Fig 38.2). Until the early 1980s, this increase matched the increase in world population, so that, overall, more people had enough to eat than ever before. There were still, of course, countries and areas where food supply did not meet demand and many millions of people were malnourished or starving.

Since 1984, population growth has increased faster than food production. Part of the reason for this has been the damaging effects of farming practices which produce food in the short term, but lead to long-term environmental damage, particularly soil erosion and desertification (see page 696). Very poor countries where food production has fallen since the mid-1980s have tried to compensate by importing food from Europe and the USA where more food is produced than is needed by the local population. However, this only increases the debt of the poor countries, leading to more poverty and, in the long term, more malnutrition and starvation.

Many scientists believe that soil erosion, water shortages and pollution from excessive use of pesticides and fertilisers will lead to a crisis in food production during the twenty-first century. Only a change in farming techniques might prevent this.

THE CONCEPT OF SUSTAINABLE AGRICULTURE

The problem of providing food for an increasing world population without reducing the capacity of the land to produce food is a difficult one, but it has some solutions. By using farming techniques that are sustainable, rather than successful in the short term only, many areas of the world should be able to continue food production at the right level for the foreseeable future.

Moves towards sustainable agriculture would include:
● Using new varieties of crop plants, perhaps produced by genetic engineering, which can survive and give good yields in poor soils, dry conditions and without the need for expensive pesticides and fertilisers (see Chapter 30).

? TEST YOURSELF

A What is the difference between a sustainable resource and an unsustainable one? Think of an example of each.

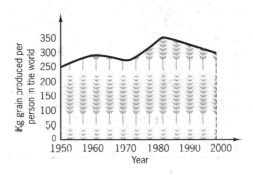

Fig 38.2 The amount of grain produced per person in the world between 1950 and 1999

SEE QUESTION 1

- Making use of non-traditional food sources – more plants that have nitrogen-fixing bacteria in their roots (these need no fertiliser), insects (many cultures of the world eat insects and their larvae as protein-rich delicacies) and unusual animals. In Europe, for example, ostrich farming is becoming more widespread.
- Reducing the amount of fish caught in the seas to levels at which the sizes of the populations do not continue to decrease. At the same time, the number of fish farms could increase.
- Encouraging and rewarding farming methods that do not lead to soil erosion and other environmental problems.
- Using biological pest control (Fig 38.3), rather than artificial pesticides which can persist in the environment.
- Encouraging farmers in poor countries to grow a selection of crops that can feed local people, rather than cash crops such as tobacco and coffee for export.
- Increasing financial and technological aid to poor countries to allow them to set up more sustainable methods of farming and food production.

FARMING IN THE TWENTY-FIRST CENTURY

MONOCULTURE

Until relatively recently, farms were diverse businesses where the farmers would grow several different crops, keep a variety of animals and perhaps have an orchard or two. The produce would be sold in local markets. But in today's supermarket-dominated food industry, farmers are encouraged to grow a single crop such as oilseed rape or wheat, so that the supermarkets have fewer growers to deal with. The word **monoculture** refers to the growing of the same crop in large fields, often year after year. A particular crop will rapidly deplete the soil of particular nutrients at particular depths, and often lead to the accumulation of pests. Without humus to bind it together, soil becomes more powdery and is likely to be eroded by wind and rain, leading to dust storms and the accumulation of silt in rivers, lakes and oceans.

REMOVAL OF HEDGEROWS

Hedgerows can be hundreds of years old, and represent an important habitat for many species. Farmers remove hedgerows for a variety of reasons:
- So that they can use large machinery more efficiently.
- To increase the area for growing crops.
- To avoid the need for maintenance (cutting, etc).
- To prevent them shading the crops.
- To remove what is often seen as a reservoir of crop pests.

The loss of hedgerows has several disadvantages. Hedges, as well as being aesthetically desirable, shelter crops from wind and therefore minimise soil erosion. Without hedgerows, the species diversity of the countryside is lowered as many species of insect, plant, bird and small mammals lose their habitat.

Fig 38.3 Until recently the only biological control for whitefly, a common plague on greenhouse-grown tomatoes (top) was *Encarsia formosa* (centre), a parasitic wasp which eats only adult flies, leaving larvae to develop and so cause a recurring problem. Gardeners and commercial growers can now obtain the beetle *Delphastus pusillus* (bottom) which kills larvae as well as adult flies, so providing much more effective and long-lasting control

USE OF FERTILISERS

One of the key differences between agriculture and a natural ecosystem is that harvesting removes nutrients from the soil. Fertilisers are used to replace these nutrients. There are two types of fertilisers:

- Organic – faeces and urine of animals and humans (sewage sludge or farmyard manure).
- Inorganic – liquid or (usually) pellets containing mineral ions, mainly nitrate, phosphate and potassium.

Both types are added to crops with the same basic aim: to increase crop yield. Both provide mineral ions, but with manure the release of ions is gradual as the manure is decomposed by microbial action. The advantages and disadvantages are given in Table 38.1.

Table 38.1 Organic v inorganic fertiliser

Type of fertiliser	Advantages	Disadvantages
Organic	cheap – farms often generate their own not easily lost by leaching improves soil – better humus levels, water retention, aeration and texture	variable (usually low) nutrient content slow release of nutrients difficult/expensive to store and handle may contain plant or animal pathogens – causing disease may contain metal residues that are passed into food chain requires heavy machinery that can compact soil
Inorganic	exact composition known – soil balance can be controlled easy to store and handle can be applied with light machinery – avoids soil compaction	expensive – it's a commercial product most components soluble – rapid leaching into rivers applied in concentrated form – can cause osmotic damage to plants can cause acidification of soil

BALANCE OF FOOD PRODUCTION AND CONSERVATION

For economic reasons, farmers are under great pressure to:

- Concentrate on a small number of crops and grow these in large areas of monoculture.
- Remove hedgerows to make more growing space and to make the use of heavy machinery easier.
- Use large amounts of chemical fertilisers to increase yield from the depleted soil.
- Drain marshy areas and remove woodland.
- Use a wide variety of pesticides.

The two activities of producing food and conservation would seem to be exclusive, but there is some room for compromise. There are several strategies that farmers can adopt to minimise damage to the environment.

SOME METHODS OF CREATING A MORE SUSTAINABLE AGRICULTURE

- Use more organic manure, which improves soil structure by providing more humus so retaining more water. As the humus decays the nutrients are released slowly so there is less chance of them being leached away.
- Delay the application of chemical fertilisers until the main growing season so that more is absorbed before it is washed away.
- Leave crop stubble over the winter (ie plough later) so that there is less bare soil to blow away.
- Rotate crops – growing a different crop year on year makes better use of minerals available at each soil depth. It also prevents the accumulation of crop-specific pests.
- Set aside – leave areas of wilderness to develop.
- Stop destroying hedgerows.

3 HUMAN ACTIVITIES AND POLLUTION

Pollution occurs when substances are released into the environment in harmful amounts, usually as a direct result of human activity (Fig 38.4). Most pollution results from excessively high concentrations of substances, but there are exceptions.

Some pollutants are harmless substances already found in the natural environment. They cause problems when they are produced in large quantities by human activity; examples include carbon dioxide and organic waste (sewage). Other pollutants produced by human activity are compounds that are known to be toxic, such as pesticides. These often remain in the environment for a long time because there is no living organism to break them down. A third type of pollutant is a chemical that passes all the safety tests, but then has totally unexpected effects, as chlorofluorocarbons have done. Pollutants can affect the air, the sea, waterways and the land.

Fig 38.4 Examples of how we humans pollute our environment.

Fig 38.4(a) The outlet pipe in the background is discharging raw sewage which is being washed up onto this beach

Fig 38.4(b) Effluent from the chemical works is polluting the river which supplies it with vital water

Fig 38.4(c) This landfill site will eventually be levelled and should look more pleasant, but methane production under the surface might still cause problems

Fig 38.4(d) Pollutants in smoke from industrial plants add to the acid content of rain

AIR POLLUTION: ACID RAIN

To many people, walking through a high Swiss meadow in summer during a shower of rain sounds idyllic. Peace and quiet, fresh mountain air and refreshing rain. But what if the rain has the pH of vinegar or lemon juice (Fig 38.5)?

In Switzerland, Norway, Sweden and other parts of Scandinavia, acid rain falls on some of the most unspoilt natural landscape in the world. It starts off, as far as 1000 kilometres away, as emissions of sulphur and nitrogen oxides released from the more highly industrial areas of Europe. Although there are some natural sources of these gases, by far the largest amount is generated by burning fossil fuels: the petrol used to power road vehicles and the oil, coal and gas used in power stations (Fig 38.6).

Fig 38.5 Unpolluted rain has a pH of about 5.6 because of the CO_2 that occurs naturally in the atmosphere: rainwater in most industrialised countries has a pH of between 4 and 4.5. Some cities in the USA are bathed in acid fog with a pH as low as 2.3 which is 1000 times as acidic as normal rain, and the same acidity as lemon juice

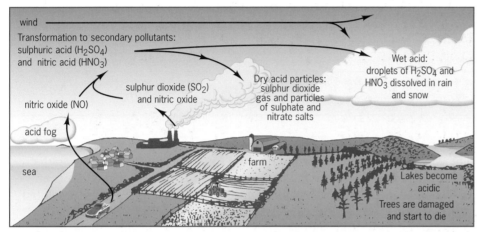

Fig 38.6 As they are carried in the air, oxides of sulphur and nitrogen form secondary pollutants such as nitric acid vapour, droplets of sulphuric acid and particles of nitrates and sulphates. These chemicals reach the surface of the Earth in two forms: wet, as acid rain, snow or cloud vapour; or dry as acidic particles

Normally, rainwater is mildly acidic because of the presence of dissolved carbon dioxide. The pH is normally no lower than 5.6. Any precipitation (rain, hail, sleet or snow) below this value is defined as acid rain. The cause of acid rain is a reaction between certain gases and the water that makes up clouds. The following gases are mainly responsible:

- Sulphur dioxide, which reacts with water to make sulphuric acid (H_2SO_4). Sulphur dioxide is emitted naturally from volcanoes but the vast majority comes from burning fossil fuels, notably coal, that contain a lot of sulphur. Power stations and metal smelting industries are the major sources of sulphur dioxide, not petrol or diesel engines.

- The oxides of nitrogen, nitrous oxide (N_2O), nitric oxide (NO) and nitrogen dioxide (NO_2). Collectively known as NO_x, these gases combine with water to make nitric acid (HNO_3). NO_x gases are made in engines when nitrogen in the air combines with oxygen during combustion.

Perhaps the most extreme example of acid rain was a 'smog' (smoke/fog) that descended on London for a whole week in 1953. Its pH was measured at 1.6 and was responsible for the deaths of 4000 people who suffered from breathing problems (asthma, bronchitis and emphysema).

The effects of acid rain on the environment are still causing arguments. It is quite difficult to get good evidence to demonstrate a definite link between the occurrence of acid rain and the effects that ecologists and environmental scientists have noticed. For example, many European and North American forests have been damaged during since the 1970s, especially at high altitudes and at the edges of some forests, which have large areas of dying trees (Fig 38.7). In Britain the pattern of loss of lichen species seems to follow the

Fig 38.7 These trees in the Czech Republic have been killed by acid rain that has resulted from emissions of sulphur dioxide and oxides of nitrogen given off by heavy industry all over Europe

distribution of sulphur dioxide pollution. In fact, lichens are so sensitive to this sort of pollution that they are used as pollution indicators. Other possible effects of acid rain are listed in Table 38.2.

Many countries, including Britain, have now recognised acid rain as a serious problem and have been reducing their emissions of sulphur dioxide since the mid-1970s.

Table 38.2 Possible effects of acid rain	
Possible effect of acid rain	**How it might happen**
Fish deaths in Scandinavian lakes and elsewhere	Indirect mercury poisoning: more acidic water is thought to convert inorganic mercury compounds in lake sediments into compounds which are soluble in the fatty tissues of fish.
	Indirect aluminium poisoning: in acidic conditions, more aluminium ions are released from the soil and washed into lakes. Fish respond by making excess mucus which clogs their gills with fatal results.
Damage to statues and buildings	Chemical reactions between components of acid rain and rock.
Overstimulation of plant growth	Acid rain falling on the ground means excess nitrogen. The plants then use other nutrients faster, reducing soil fertility.
Thought to make human respiratory problems, such as asthma, worse	Irritation of surface membranes in lungs and bronchi.

AIR POLLUTION: THE OZONE LAYER

The effects of acid rain occur hundreds of kilometres away from the source of the pollutants that cause it. This is bad enough, but other pollutants affect the whole planet. Examples include the effect of greenhouse gases on global warming (see Chapter 36) and ozone depletion.

Ozone, O_3, is a form of oxygen: its molecules contain three oxygen atoms, rather than the two that occur in molecules of atmospheric oxygen. Both gases have been a major feature of the Earth's atmosphere for the past 450 million years. Oxygen is vital to life: it is the fuel for aerobic cell respiration. The role of ozone is less direct: it forms a sort of sunscreen high up in the atmosphere, shielding the biosphere from the harmful effects of ultraviolet light that beams down from the Sun.

Chlorine- and bromine-containing compounds, mainly chlorofluorocarbons (CFCs), that were released into the atmosphere during the twentieth century seem to be causing a measurable thinning of the ozone layer (Fig 38.8). This could cause serious problems in the future.

Fig 38.8 In the mid-1980s, scientists discovered something unexpected: sunlight on the Antarctic in spring causing immense destruction of ozone and a huge hole appearing in that part of the ozone layer. This 'ozone hole' has been noticed every year since, and appears to be growing

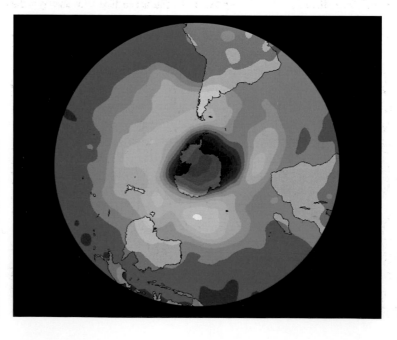

Originally, CFCs were thought to be 'ideal' chemicals: they were non-flammable, non-toxic, non-corrosive, cheap to make and could be used as coolants in fridges, propellants in aerosol spray cans, and in the manufacture of the packing material Styrofoam. However, in the mid-1970s, scientists showed that CFCs rise slowly into the stratosphere and are then converted by the action of ultraviolet light into chlorine atoms. These accelerate the breakdown of ozone into O_2 and O. CFC molecules can stay in the stratosphere for about 100 years, and each one can destroy hundreds of thousands of molecules of ozone.

CFCs have now been phased out in most countries of the world, but their effects continue. Even if no more CFCs were put into the air from tomorrow, the ozone layer would still take about 100 years to recover from the worst of the damage. Less ozone in the stratosphere allows more of the harmful ultraviolet radiation from sunlight to reach the Earth's surface. Scientists predict that this will lead to a huge increase in skin cancers and cataracts (Fig 38.9).

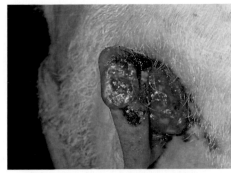

Fig 38.9 In countries such as Australia, New Zealand, South Africa and Chile, which are protected only by a very thin layer of ozone for many months of the year, there have already been many more cases of skin cancer than ever before. Children in Australian schools are required by law to wear hats and blocking creams whenever they go outside

WATER POLLUTION: GETTING INTO HOT WATER

When you think of water pollution, you probably think of chemical effluent (Fig 38.10), outflows of raw sewage or huge oil spills. While these undoubtedly are examples of pollution, probably the most overlooked cause of damage to inland aquatic ecosystems is the warm water released as a by-product of many industrial processes. Fish and water-living invertebrates are killed by warm water, not because they are scalded directly, but because warm water can carry much less dissolved oxygen than cool water and they 'suffocate' (Fig 38.11).

Fig 38.10 The chemical outflow is on its way to the river Mersey

Fig 38.11 As well as causing the death of fish directly, by depriving them of oxygen, warm water can also encourage the growth of parasites which would otherwise not be a problem

SEE QUESTIONS 2 AND 3

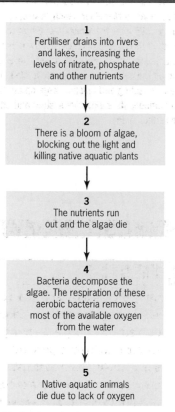

1
Fertiliser drains into rivers and lakes, increasing the levels of nitrate, phosphate and other nutrients

↓

2
There is a bloom of algae, blocking out the light and killing native aquatic plants

↓

3
The nutrients run out and the algae die

↓

4
Bacteria decompose the algae. The respiration of these aerobic bacteria removes most of the available oxygen from the water

↓

5
Native aquatic animals die due to lack of oxygen

Fig 38.12 A summary of eutrophication

? TEST YOURSELF

B Describe how you would measure BOD.

WATER POLLUTION AND SEWAGE

Sewage is good stuff. It's a natural product that is an essential part of most ecosystems. Call it what you will, but faeces and urine from any source are a mixture of undigested food, cells from the gut lining, urea, salt and lots of living bacteria and protoctists. When acted upon by decomposers (or saprophytes), sewage releases vital nutrients such as nitrate that are essential for renewed plant (or algal) growth.

Problems with sewage start when there's too much of it to break down. If the entire sewage production of a village or town is pumped into one stream or river, the natural balance is upset.

Contamination of rivers, lakes and seas by sewage causes two main problems. Firstly, it can introduce potentially dangerous pathogens (organisms that can cause disease) into the environment (see Fig 38.4a). This is a particular problem if the water is drunk by animals or people. The second problem is that sewage and other organic wastes are decomposed by the action of aerobic bacteria. These use large quantities of oxygen, leaving less for other organisms living in the water. Sewage contamination can lead to **eutrophication** (Fig 38.12).

The term **eutrophic** means over-fertile. A eutrophic river or lake contains more nitrates and phosphates than normal, and abnormal growth, particularly of algae, can occur. Note that the algae themselves do not reduce the oxygen content. They photosynthesise, so actually increase the oxygen levels for a short while. The opposite of eutrophic is **oligotrophic**: this means under-fertile. An oligotrophic river cannot sustain much life.

BIOLOGICAL OXYGEN DEMAND

It is easy to find out the oxygen requirements of water from a particular site by taking a sample of the water and measuring its oxygen content, usually with an electronic meter. After keeping it sealed and in the dark for 5 days at 20 °C, the oxygen content is measured again. The rate at which the oxygen has been used up is the **biological oxygen demand** (**BOD**). The BOD of unpolluted river water is about 3 mg O_2 per litre (dm^3) of water, per day. Raw sewage uses over 100 times more oxygen in the same time (Table 38.3).

Although pollution by sewage is a major cause of eutrophication, there are others. If large amounts of nitrate- and phosphate-rich fertilisers are put on to farmland, the excess can run off and contaminate local rivers and lakes. Industrial processes which pollute local waterways with large amounts of concentrated sugars or other organic waste products can also cause the BOD of the water to increase significantly.

Find out more about eutrophication in the Assignment at the end of this chapter.

Table 38.3 Typical BOD values per day of different types of water

Water type	BOD value per day
Clean river water	3 mg dm^{-3}
Water from polluted stream	10 mg dm^{-3}
Domestic sewage, untreated	250–350 mg dm^{-3}

CONTAMINATION OF WATER BY OIL AND DETERGENTS

The names *Exxon Valdez*, *Braer* and *Sea Empress* may sound familiar, even if you can't quite remember that they are oil tankers that have famously run aground causing huge oil spills (Fig 38.13). These pollution disasters make headline news but, in 1993, a study by Friends of the Earth estimated that over 1000 times the amount of oil carried by the *Exxon Valdez* (which polluted the coast of Alaska in 1989) is lost in the USA each year – not through national catastrophes, but through the normal operations of washing tankers and releasing the oily water, from pipeline and storage tank leaks and from accidental loss from offshore oil wells.

The effects of oil spills on ocean ecosystems depend on:

● The amount of oil released.
● How far away from the shore the spill happens.
● The weather conditions, water temperature and speed of ocean currents.
● The type of oil released: an area affected by a spill of refined oil takes over three times longer to recover than a similar area that has a crude oil spill.

In the clean-up operation following a large spill, oil is dispersed by detergents. The aim is to spread the oil molecules to limit damage to the local environment. Unfortunately, many detergents are extremely toxic and lead to many sea-bird deaths.

SPOILING THE LAND

People can pollute areas of land directly by dumping toxic waste, by industrial processes such as mining and by simple neglect: leaving non-biodegradable rubbish behind after a picnic at a local beauty spot, for example. However, human activity, particularly farming, is indirectly causing two main problems: **deforestation** and **desertification**. Both have far-reaching global effects.

DEFORESTATION

Deforestation has been practised by humans for centuries. As human population and technology have increased, so has the need to use wood. (The discovery of fire has had an enormous impact.) Many European forests were cleared by various different settlers from about the fourteenth century onwards. Today, the human population is larger than ever before and the process of deforestation has now spread to all parts of the world, particularly the tropical rainforests.

Deforestation is a process that results from the conflict between immediate human need and the necessity to protect the environment from serious and long-term damage. Tropical hardwoods are prized as building materials throughout the world, and other types of wood are used to make the thousands of tonnes of paper that we use every day. Many people in the poor countries of the world depend on wood or charcoal for fuel, and wood, twigs, crop residues and grass are used for fuel in many of the 'better off' countries such as China, Brazil and Egypt. As well as cutting down forests to use the wood they provide, deforestation is also carried on to clear land for farming. People have cleared large areas of forest for plantations of other trees (rubber and palm oil), and for cattle ranches and enormous wheat fields, in order to fulfil the needs of the human population for more raw materials and more food.

38.13 This cormorant was caught in the oil spill that happened at Manobier beach, Wales, in February 1996 after the tanker *Sea Empress* ran aground. Birds covered in oil cannot swim very easily and sometimes tire and drown. Those which end up on the beach try desperately to clean their feathers, ingesting and inhaling large amounts of oil which damages their digestive system and often causes pneumonia

SEE QUESTION 4

✔ REMEMBER THIS

Many countries worldwide are now working together to try to reduce deforestation: people are being encouraged to manage forests **sustainably**. This means using different techniques to make sure we have the wood we need without destroying large areas of forest. People are also being persuaded to use **paper substitutes**. There is growing investment in **rainforest communities** to find new medicines from plants and **eco-tourism** is being promoted. If people pay to see wildlife in its natural surroundings, this contributes to the income of an area and makes it less likely that habitats are destroyed.

The consequences of deforestation are potentially serious. Locally, clearing the land of trees and plants leads to increased soil erosion (see below). Globally, reducing the biomass of productive trees may contribute to the greenhouse effect (see Chapter 36) as less carbon dioxide is used in photosynthesis. Also, loss of rainforest, one of the most diverse ecosystems of the world, is likely to affect humans and other organisms in ways we cannot predict, although we already know that many of the plants found in rainforests are useful sources of powerful drugs (see Chapter 37). Balancing the needs of people with those of the environment is difficult.

DESERTIFICATION

Soil is an important abiotic factor that determines the distribution of organisms in an ecosystem: few plants can grow without the anchorage of soil or the water and minerals it supplies to them. When the soil is lost, a process called **soil erosion** occurs. This is a worldwide problem.

Soil erosion is a natural process: wind and water movements move surface debris and topsoil from one place to another. In ecosystems that are undisturbed by human activity, healthy plant roots 'bind' the soil, and soil is lost and made at about the same rate. Problems occur when the soil is removed faster than it is formed: this can happen because of farming, deforestation, building, over-grazing by animals, physical damage by off-road vehicles or fire and some kinds of mining.

Farming is probably the worst culprit. Topsoil is eroding on about a third of the world's cropland. Each year, the land must feed 90 million more people with about 25 billion tonnes less topsoil. In Africa, soil erosion is now 20 times faster than it was 30 years ago. The cattle-producing areas are now severely affected by desertification (Fig 38.14). But, there are solutions to soil erosion and desertification, and many of these are starting to be put into practice.

Fig 38.14 Areas of severe desertification occur all over the world. Desertification is defined as the process that reduces the productive potential of hot dry areas by 10 per cent or more. Moderate desertification is a 10–25 per cent drop in productivity: severe desertification is a 25–50 per cent drop. At its most severe, desertification actually creates deserts

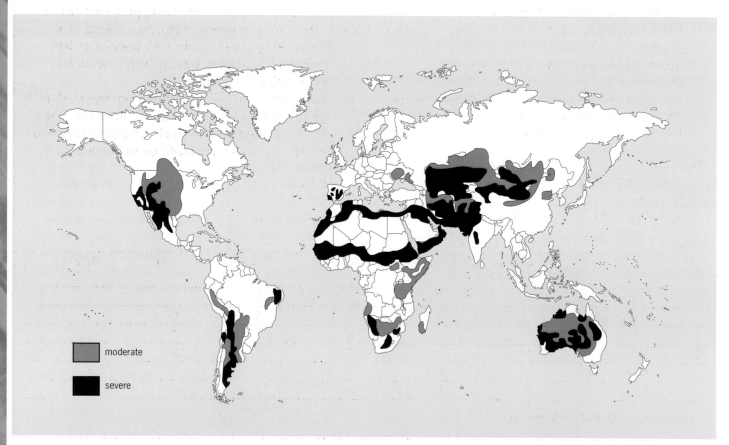

moderate

severe

SUMMARY

After reading this chapter you should know and understand the following:

- Humans influence the environment of the Earth on a *global* as well as a *local* level.
- Farming is the major human activity that causes environmental damage.
- People need to farm to feed the increasing human population but many farming techniques cause long-term damage to the land, lowering its capacity to grow food in the future.
- To avoid this problem it is important that farmers start to use **sustainable** methods of agriculture.
- Many human activities cause **pollution**. Pollutants can affect the air, the sea and rivers and lakes, and the land.

- Acid rain, use of CFCs and other substances that destroy the ozone layer, and increased production of greenhouse gases all contribute to air pollution.
- Water can be contaminated by organic pollution. This can lead to a sequence of events called **eutrophication**. Heat, oil and other substances can also pollute waterways.
- When forests are cleared to make new farmland, fewer trees are available to photosynthesise, we lose species diversity and the land becomes vulnerable to **erosion**.
- Severe soil erosion can lead to **desertification**. Many sustainable methods of agriculture try to minimise the effects of soil erosion.

? EXAM QUESTIONS

1 In 1968, an experiment was started at Nettlecombe Court in Somerset, to determine the long term effects of four different methods of managing grassland. Three of the treatments in the experimental plots are listed below.

Mown frequently (every two weeks) in the growing season
Mown annually in June
Not mown at any time

Quadrats were used to estimate percentage cover. The results for certain species in 1970 and 1990 are shown in the table below.

a) Describe the changes that took place between 1970 and 1990 in the plot that was mown frequently.

b) (i) In 1990, both clover and moss were most abundant in the frequently mown plot. Suggest how these plants might have benefited from the mowing of the plot.
(ii) Which species seems to benefit most from annual mowing compared with other treatments? Use figures from the table to support your answer.

c) In the fourth treatment, a plot was cleared of all vegetation by removing the turf. It was then left unmown. Suggest *two* ways in which plants may colonise this cleared plot.

d) During the period of this experiment, tree seedlings regularly appeared in the unmown plots but none grew into larger trees or shrubs. Suggest reasons for the failure of these tree seedlings to develop into mature plants.

Edexcel B2 June 1999 Q7

Treatment		Percentage cover					
		Yorkshire fog (a grass)	Cock's-foot (a grass)	Moss	Buttercup	Clover	Yarrow
1970	Mown frequently	12	5	1	36	2	0
	Mown annually	1	27	0	2	0	0
	Not mown	29	78	0	19	0	0
1990	Mown frequently	21	2	42	7	30	15
	Mown annually	34	10	26	7	6	24
	Not mown	42	11	2	13	4	14

Data courtesy of J Crothers, Field Studies Council

❓ EXAM QUESTIONS

2 The figure represents a simplified version of the nitrogen cycle

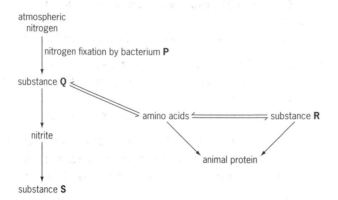

a) Give a suitable name for each of **P**, **Q**, **R** and **S**.

b) State **precisely** why substance **S** is important for the growth of crops.

Large quantities of substance **S** are applied to agricultural land as a component of chemical fertilisers. There is concern about its effect on the environment because much of **S** that is applied to the land eventually leaches out of the soil and into ground water.

c) If substance **S** passes into drinking water, suggest **one** harmful effect it can have on the human body. If **S** and other residues of chemical fertilisers accumulate in slow-flowing water or lakes, rapid growth of algae may occur (an algal 'bloom').

d) **(i)** State the name given to the process by which this happens.
(ii) Explain briefly how this process could destroy most of the community of a lake.

e) Apart from the loss of **S** due to leaching, why is it necessary for farmers to add chemical fertilisers to their fields every year?

OCR 6910 Molecules & Life March 2000 Q3

3 Read the following passage then answer the questions that follow.

Lake Victoria in Africa is known as the lake that smokes. The 'smoke' is in fact huge clouds of insects called non-biting midges, which spend their larval and pupal stages in the lake. The larvae feed by filtering algae from the water. In recent times these swarms of midges have increased and are so dense that they cause distress and breathing difficulties to people and their animals living on the shores. There have also been changes within the lake. About forty years ago a few fish called Nile

Perch were introduced to improve the fishing industry. They are large and tasty to eat, but they are also fierce predators. They are now the dominant species in the lake but more than half of its native species of small fish have become extinct. Some of these small fish feed on algae and some eat midge larvae. The extinction of many of the fish feeding on algae has resulted in algal blooms in the lake. Many more people now live on the shores of the lake, making a living from fishing. During the last forty years the total productivity of Lake Victoria has increased, but its biodiversity has decreased dramatically.

a) On the axes below sketch a curve that represents the growth of the Nile Perch population in Lake Victoria over the last 40 years.

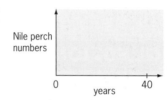

b) Suggest **two** reasons why the native species of fish have declined.

c) Explain **two** effects that the algal blooms might have on the other life on the lake.

d) Suggest a reason for the recent increase in the swarm of midges.

e) **(i)** Explain what is meant by the term *productivity*, used in the last sentence of the passage.
(ii) Explain how the increased human presence on the shores could have contributed to the increase in **primary** productivity of the lake.

OCR 6916 Environmental Biology March 2001 Q3

4 Deforestation of both the tropical rain forests and the cold temperate coniferous forests still continues at a very high rate.

a) Give **four** reasons why these forests may be cut down.

b) Explain how deforestation leads to soil erosion.

c) Name **two** other effects that deforestation can have on the environment.

OCR 6916 Environmental Biology March 2001 Q1

KEY SKILLS **ASSIGNMENT**

JOURNEY ALONG A RIVER

As Fig 38.A1 shows, rivers tend to be relatively unpolluted near to their source. As they flow on towards the sea they are contaminated with organic pollution and industrial effluent, creating increased levels of pollution and eutrophication.

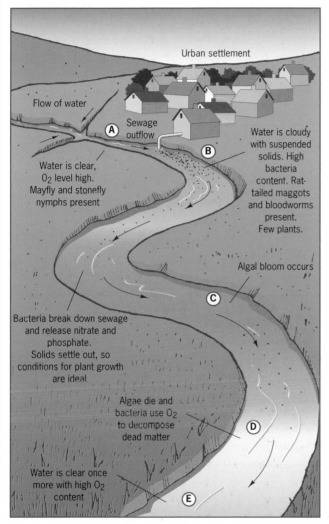

Fig 38.A1 This diagram shows the different conditions found along a river

It is often difficult to tell exactly how polluted the water is at any point, just by looking at it. We need to sample the water and sample the organisms that live there to find out whether the water is safe.

One easy method which allows us to find out which organisms live in a particular part of the river is the kick sample (see page 660). The number and diversity of organisms found are used to give the stretch of waterway a diversity index (see page 679).

1 What sort of diversity index would you expect to find in:
a) a stretch of unpolluted river, such as point A in Fig 38.A1?
b) a stretch of polluted river, such as point B in Fig 38.A1?

2 **a)** What factor is likely to be the most important in determining whether water can support a large or a small number of species?
b) What method would you use to measure this factor?

3 Initially, the water at point B is turbid (cloudy and murky). From your knowledge of the nature of sewage, explain why few animals or plants can grow in this zone of the river.

4 What do you think the BOD value may be for the water in this part of the river?

5 A student who was feeling particularly brave did a kick sample in this area and found very few species. They were the classic indicator species of heavy organic pollution: rat-tailed maggots, bloodworms and tubifex worms – see Fig 38.A2.

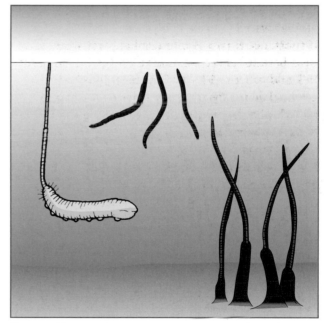

Fig 38.A2 The larva of a rat-tailed maggot, some bloodworms and tubifex worms. The tails of the tubifex worms protrude from tubes that the worms make using mucus and debris

a) What is an indicator species?
b) Explain why each of the three species shown in Fig 38.A2 is able to tolerate low oxygen levels.
c) When the water was tested in the laboratory, large numbers of *E. coli* bacteria were found. Using this and the other evidence, what would you conclude about the sewage outfall?

KEY SKILLS **ASSIGNMENT**

More sophisticated investigations of the water further below the sewage outlet produced the data shown in Fig 38.A3.

Research and discussion questions

Find out about the basic principles behind sewage treatment. What major health problems does effective sewage treatment prevent?

Find out what the current laws are in the UK that control the discharge of sewage into rivers and into the sea. Do you think they are strict enough?

6 Look at Fig 38.A3(b) and (c).
a) Use your knowledge of the nitrogen cycle to explain why there is first a peak in the levels of dissolved NH_4^+, followed by a peak in NO_3^-.
b) The population of sewage fungus falls to 0, just before point C. What does this tell you about the metabolism of the sewage fungus?
c) Why does the level of bacteria not fall to 0?

7 At point C, the water begins to look like pea soup – conditions have become perfect for the growth of algae. Contrary to what you might first think, the problem here is that the water is too fertile – it has become eutrophic.
a) Use the graphs shown in Fig 38.A3(a) and (b) to list three factors that are responsible for the algal bloom.
b) Explain why the algal bloom is harmful to the normal clean-water organisms.
c) Why do the populations of the different invertebrates present in the river water peak at different times?

8 Look at Fig 38.A3(d).
a) Find out the characteristics of tubifex worms and blood worms that allow them to flourish in the river between points B and E.
b) How do the detritivores help to change the water conditions between points C and E? Find out the names of some common detritivores that might be found in this sort of river.

Once the algae have absorbed the available nutrients, they start to die. Once again, bacterial activity becomes a problem, as decomposers rot the dead algae and use up large amounts of oxygen. After this, if there is no more organic discharge, the water further downstream begins to recover. Eventually, the water becomes clean once more

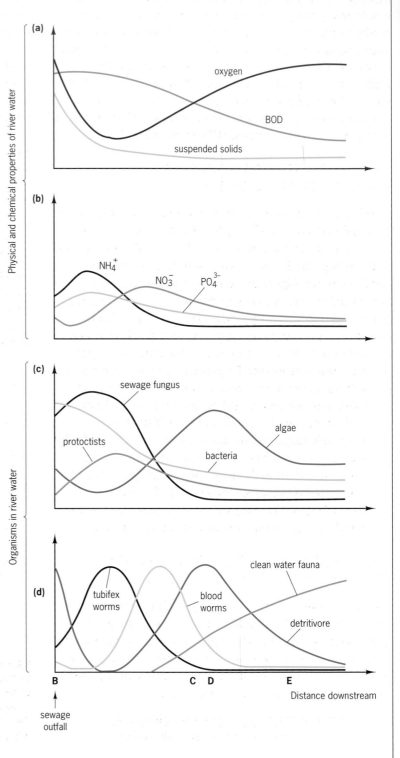

Fig 38.A3 Changes in the river after a sewage outfall

APPENDIX: USING STATISTICS IN BIOLOGY

Statistics is the only useful way of scientifically answering the question 'so what?' Suppose you counted the number of earthworms in a square metre of soil sample. If you did this just once, you might extract six worms. All that this would really tell you is that some soil samples contain earthworms. If you did the experiment many times, you might be able to draw some conclusions about the density of the earthworm population in that particular field. However, if you did the experiment in two separate fields, you might end up with two very different collections of raw data, and this is where the application of statistics will help you answer the question 'Well, so what?'

THE NULL HYPOTHESIS AND EXPERIMENTAL DESIGN

In science, progress is made by a process of **conjecture** and **refutation**, which in plain English means thinking up an explanation for an observation; 'could this possibly be?', then gathering data, analysing it and coming to the conclusion 'it's highly unlikely' or 'possibly, it could'. A vital point here is that you can never prove anything, but you can devise a **hypothesis**, test it and fail to disprove it, so you gather support for a particular idea.

A hypothesis is a testable idea. There are two types:

● The **experimental hypothesis** states that there will be a significant connection between cause and effect. An example could be 'alcohol slows down an individual's ability to do mental arithmetic'.

● The **null hypothesis** takes the opposite standpoint, such as 'alcohol has no effect on an individual's ability to do mental arithmetic'. A key point here is that **statistical tests only test the null hypothesis.** Statistical tests are needed to tell you whether your results are due to chance or not.

THE MEAN, AND THE STANDARD DEVIATION FROM THE MEAN

These are different types of average values. Averaging is very useful in biology, particularly on ecology fieldtrips: it helps us to make the most of our results from sampling.

If we take the mean of several samples, we can get a more accurate estimate of the total number of organisms in the population that we are studying.

$$\text{Mean} = \frac{\text{total number of animals caught}}{\text{total number of samples taken}}$$

Distribution around the mean

We can work out the mean of a sample, but this does not tell us anything about the *range* of values in our sample. For example, a mean value for neck length in 200 swans might be 47 cm. The swan with the longest neck might have a neck 50 cm long, the one with the shortest might have a neck 44 cm long. Or something really unusual might be happening to the swan population and some swans might have necks only 20 cm long, while others had necks over 60 cm long. The mean itself does not tell us which is the real situation.

Normal distribution

Statistical tests usually involve situations in which there is a normal distribution. Fig A.1 gives an example of a normal distribution. The exact shape of the normal distribution depends on the mean and the **standard deviation** (**SD**), the range of values around the mean. Differences in standard deviation affect the shape of the distribution. All the curves in Fig A.2 represent normal distributions. Although the distribution is symmetrical in all the curves shown, the distribution changes as the standard deviation changes. When the SD increases, the curve becomes flatter; if the SD decreases, the curve is taller.

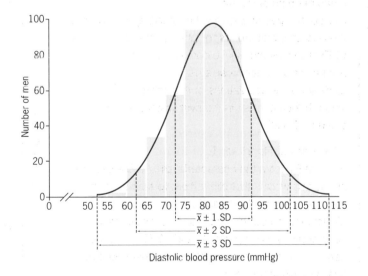

Fig A.1 Normal distribution of blood pressure. The mean blood pressure is 82 mmHg and the standard deviation is 10 mmHg. This means that 68 per cent of the sample have blood pressures between 72 and 92 mmHg and 95 per cent of the sample have blood pressures between 62 and 102 mmHg. It follows that if we sample, there is only a 0.05 per cent (1 in 20 chance) of selecting a male with a blood pressure less than 62 or greater than 102 mmHg

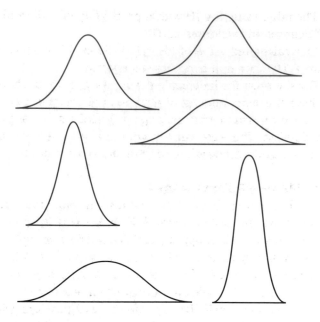

Fig A.2 In a normal distribution:
- Measurements greater than the mean and measurements less than the mean are equally common
- Small deviations from the mean are much more common than large ones
- 68 per cent of all the measurements fall within a range of ± 1 standard deviation from the mean and 95 per cent within ± 2 standard deviations

Standard deviation

We can tell how much individual values deviate away from the mean value by working out the standard deviation:

$$\text{standard deviation} = \sqrt{\frac{\Sigma(x - \bar{x})^2}{n - 1}}$$

where Σ = the sum of, x = measured value, $\bar{x}$ = mean value, and n = number of measurements.

To do this you would:

1 Take each of your measured values and subtract the mean value from each of them.
2 Square each value you obtain in step 1.
3 Add all your squared values together. This gives you the $[\Sigma(x - \bar{x})^2]$ part of the formula.
4 Divide the number you get in step 3 by the number of measurements on which your mean was based minus one. Work out the square root of the number you get. This is your standard deviation from the mean.
5 You can now show your mean value, plus or minus the standard deviation to indicate the range of values in your original sample.

SIGNIFICANCE OF DATA

The *t*-test

The *t*-test is used when we make a range of measurements rather than counting numbers. For example we might wish to compare the thickness of the leaves of a species of plant growing on the north side with those growing on the south

side of a wall. It is only really useful if you have more than about 20 data points – fewer than that, and you might be better off using the Mann–Whitney *U*-test (see later).

Table A.1 shows the results of one such set of measurements.

Table A.1	
Thickness of leaves on south side of wall/μm	**Thickness of leaves on north side of wall/μm**
300	180
270	210
220	220
250	160
210	210
250	190
290	160
190	180
220	200
270	210

The statistical test you can use to find if there is a difference between two means is the *t*-test. The *t*-test compares both the mean and the standard deviation of two populations whose distribution curves overlap (Fig A.3).

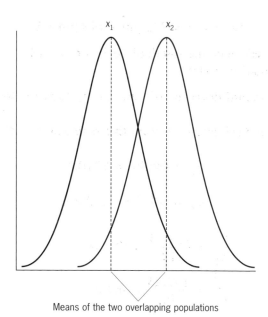

Means of the two overlapping populations

Fig A.3 What the *t*-test measures

(You would not be asked to calculate a *t*-test in an exam so you need not learn the formula. But you may need to use a *t*-test in an investigation.) The *t*-test compares the means and standard deviations of two populations. The *t*-test calculates:

$$\frac{\text{difference in means of two populations}}{\text{variability of the two populations}}$$

The variability of the population is the square of the standard deviation, so the formula for the *t*-test is

$$t = \frac{|x_1 - x_2|}{\sqrt{\dfrac{\sigma_1^{\,2}}{n_1} + \dfrac{\sigma_2^{\,2}}{n_2}}}$$

where t = *t*-test value
$|x_1 - x_2|$ = the difference between the mean value of population 1 (x_1) and the mean value of population 2 (x_2)
$\sigma_1^{\,2}$ = the variance (square of the standard deviation) of population 1
$\sigma_2^{\,2}$ = the variance (square of the standard deviation) of population 2
n_1 = the number of measurements in population 1
n_2 = the number of measurements in population 2

You will see that this formula takes into account both the size of the differences and the size of the sample.

The null hypothesis is that there is no significant difference between the thickness of leaves on the north and south facing walls.

First calculate the mean thickness of each population of leaves

Mean thickness of south facing leaves (x_1) = 247 μm

Mean thickness of north facing leaves (x_2) = 192 μm

$$|x_1 - x_2| = 247\ \mu m - 192\ \mu m = 55$$

Now calculate the standard deviation of each of population leaves.

Standard deviation of south facing leaves = 36.2

Standard deviation of north facing leaves = 21.5

$$\sigma_1^{\,2} = (36.2)^2 = 1310.4$$

$$\sigma_2^{\,2} = (21.5)^2 = 462.25$$

$$\frac{\sigma_1^{\,2}}{n_1} = \frac{1310.4}{10} = 131.04$$

$$\frac{\sigma_2^{\,2}}{n_2} = \frac{462.25}{10} = 46.22$$

$$t = \frac{55}{\sqrt{131.04 + 46.22}} = 4.131$$

In the *t*-test, the degrees of freedom are given by the sum of the number of measurements in both groups minus 2.
In this case, the number of degrees of freedom is

$$(10 + 10 - 2) = 18$$

We now compare our values for *t* with the values of *t* in a table, available in all data books.

The table value for 10 with a probability of 0.05 and 18 degrees of freedom is 2.101.

Our calculated value of *t* is 4.131. Since this is greater than 2.101, the null hypothesis is rejected.

For the *t*-test for independent samples you do not have to have the same number of data points in each group. We have to assume that the population follows a normal distribution. The *t*-test can be performed knowing just the means, standard deviation and number of data points.

The Mann–Whitney *U*-test

If you have fewer than 20 data points, but more than 6, you can use this test to establish if differences between two groups of data might be significant. The way the Mann–Whitney test works is that it assigns an artificial value to each score in the data fields. This is a way of ironing out the 'patchiness' of data when number of samples is only small. You then do the statistical analysis on the artificial values – called rank scores – and look up the result in a table. It has limitations: it can only be used when the number of values in the data set is between 6 and 20. If it is below 6, you really don't have enough data to analyse, if it is above 20, you ought to use the *t*-test instead. This test is more approximate than tests done on larger data sets, but it is important because it still provides an answer to the essential question 'Do my results mean anything?'

The Spearman rank correlation

Basically, this is another test where you rank the values you obtain – not unlike the Mann–Whitney test – and then do your stats on the ranked values. You use this test to prove or disprove that there is a correlation between any two observed variables. A typical application would be to analyse the points on a **scattergram** (Fig A.4). Sometimes the data from these things is fairly clear – the scattergram points lie plainly on a straight line, showing you that there is a correlation, either positive or negative, between two sets of data.

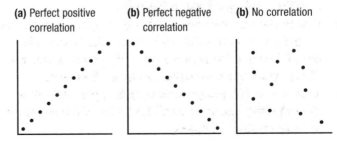

(a) Perfect positive correlation **(b)** Perfect negative correlation **(b)** No correlation

Fig A.4 Correlation

The chi-squared test

When we are comparing two or more sets of data, we need to find out if any differences between them are significant; are they different enough to tell us to look for an explanation?

Chi-squared is one statistical test that is used to decide whether results are **significant** or simply due to chance. Basically, the observed results are compared with those that you would expect of they were due to chance. The chi-squared test gives you a result in terms of probability. By convention, if the probability that your results are due to chance is **less than 5 per cent** (or 0.05 as a decimal), you can say that they are significant and so reject the null hypothesis.

The chi-squared test: a case study

A group of scientists carried out an investigation into the preferred resting places of leopards in a game park in Tanzania, to see if there was a difference between males and females. They tracked 200 leopards by sight, by radio collar and by analysis of their droppings. Their findings are shown in Table A.2.

The null hypothesis for this investigation is 'There is no difference in the distribution of male and female leopards at rest'.

Table A.2 The distribution of leopards

	In trees		In rocks		In open grassland		Row total
Males	a	19	b	56	c	25	100
Females	d	51	e	26	f	23	100
Column total		70		82		48	200 (grand total)

The formula for the chi-squared test is

$$\chi^2 = \sum \frac{(O-E)^2}{E}$$

where $\sum$ = the sum of all, O = observed and E = expected.

Having got our observed results from the field studies, how do we work out the expected results? For each category of box, we can work out the expected value E from the formula

$$E = \frac{\text{Row total} \times \text{column total}}{\text{Grand total}}$$

For each box...

$$\text{Box a} = \frac{100 \times 70}{200} = 35$$

$$\text{Box b} = \frac{100 \times 82}{200} = 41$$

$$\text{Box c} = \frac{100 \times 48}{200} = 24$$

$$\text{Box d} = \frac{100 \times 70}{200} = 35$$

$$\text{Box e} = \frac{100 \times 82}{200} = 41$$

$$\text{Box f} = \frac{100 \times 48}{200} = 24$$

Having both the expected and observed values, one of the easiest ways to work out the value is χ^2 is to fill in a table like Table A.3.

Table A.3 Working out the value of χ^2

Box	Observed	Expected	$O - E$	$(O - E)^2$	$\dfrac{(O - E)^2}{E}$
a	19	35	−16	256	7.3
b	56	41	15	225	5.5
c	25	24	1	1	0.04
d	51	35	16	256	7.3
e	26	41	−15	225	5.5
f	23	24	−1	1	0.04
					Total = 25.68

The next step is to work out the degrees of freedom (D of F). This is a measure of the spread of the data; the more categories, the more degrees of freedom. The formula is

D of F = (no. of rows − 1) × (no. of columns − 1)

which in this case is

(2 − 1) × (3 − 1) = 1 × 2 = 2

So we have a value of 25.68 with 2 degrees of freedom.

What next? To get a probability value (the whole point of the test) we look up these values in a table of chi-squared values.

Table A.4 shows that, with two degrees of freedom, the chances of obtaining a χ^2 value of 25.68 is less than 1 per cent. This means there is a less than 1 per cent probability that our results are due to chance (we needed a value of 9.2 for 1 per cent probability and 5.99 for the vital threshold of 5 per cent probability). Values as high as 25 are only found when there is a marked difference in the observed and expected frequencies. From this was can reject the null hypothesis and accept the experimental hypothesis, ie there is a difference in the habitats preferred by male and female leopards. Note that we have *gained support* for our theory, but we can *never prove* it to be correct.

Chi-squared in examinations

You will not be asked to work through any examples of χ^2 in examinations – it's far too time consuming and the examiners want to test your biology, not your maths. However, you may be expected to know the basic principles and uses of χ^2.

Table A.4 Chi-squared values

Degrees of freedom	Probability					
	0.50	0.25	0.10	0.05	0.02	0.01
1	0.45	1.32	2.71	3.84	5.41	6.64
2	1.39	2.77	4.61	5.99	7.82	9.21
3	2.37	4.11	6.25	7.82	9.84	11.34
4	3.36	5.39	7.78	9.49	11.67	13.28

ANSWERS TO TEST YOURSELF QUESTIONS

CHAPTER 1

A (a) The full stop is about 0.3 mm = 300 µm. (b) 1 million.
B Dispersal of oil pollution, breakdown of waste plastic.
C 1000.
D They are all secretory. You would recognise them by their prominent ER and Golgi bodies.
E Because the ER is a network of thin membranes which would not show up under a light microscope.
F All its tissues would be destroyed.
G We could destroy tumours.
H Because they could digest the organelle/cell that made them.
I To respire and make ATP/energy for swimming; sperm swim incredible distances and use a vast amount of energy.
J (a) 500, (b) 200, (c) 5000.
K The root system, because it receives no light.

CHAPTER 2

A Ciliated epithelium.
B Smooth muscle because its contraction is slowest.
C It fatigues too quickly.
D The reproductive system.

CHAPTER 3

A 3. No, because there is no dot at the right position for C.
B Q=25/50=0.5, R–15/50=0.3. R is the least soluble.
C Sucrose consists of glucose and fructose, both reducing sugars. In the sucrose molecule, the reactive groups of the two component sugars point inwards and so are 'hidden' in the 3D structure. They only become available to react with Benedict's solution when the sucrose has been split apart.
D The glycosidic link, which is between carbon 1 and carbon 4.
E Links: sucrose 1,2; lactose 1,4.
F It would cause an osmotic imbalance – the cytoplasm would become too concentrated.
G Amylose because it has more ends to add or remove glucose units from.
H Because it is broken down/built up even quicker – suiting the metabolic needs of animals.
I Because few organisms have the enzymes that can degrade cellulose.
J Oleic and linoleic acid.
K Oleic and linoleic. They are both unsaturated.
L 46 x 5 cm = 2.3 metres.
M TAGCAATGG.
N Bases (RNA has U instead of T), sugars (RNA has ribose, DNA has deoxyribose), strands (DNA has 2, RNA has 1), life-span (DNA is long, RNA is short).

CHAPTER 4

A 1 mm.
B Water-loving, water-hating. In a phospholipid, the phosphorus end is attracted to water, the lipid end repels water.
C Because the atoms/molecules in liquids and gases are free to move, but those in solids aren't free to move.
D To maintain the diffusion gradient.
E Lung, intestine, leaf, kidney, placenta, root.
F Because water passes out of the bacteria by osmosis, dehydrating and killing them.
G Because there can never be a less concentrated solution, pure water will never absorb water by osmosis.
H You must have a solution to compare it to: hypertonic to what?
I Its cytoplasm is isotonic with sea water, so it doesn't need one.
J Cyanide blocks aerobic respiration, so virtually no ATP is available.

CHAPTER 5

A Protein synthesis.
B Protein, lipid, nucleic acids.
C Because they all contain the same food type: starch.
D Tears, sweat, in the lysosomes of some white cells.
E It stimulates glycogen production.
F Peptide bonds.
G Proteins are denatured. This is irreversible.
H Because that is the optimum temperature for enzymes.

I It would be denatured by the acid.
J There would be no activity at all: the plot would go along the x-axis.

CHAPTER 6

A The castrated ram has no testes and does not produce testosterone or pheromones; he has nothing with which to turn on the ewes, and they would just ignore him.
B Discussion question, no answer provided.
C Fermentation is a vague word, and in biotechnology is taken to mean the metabolics/respiration of microbes. Fermentation can be either aerobic or anaerobic. Glycolysis is the first, anaerobic, process in respiration.

CHAPTER 7

A Small animals have a large surface area:volume ratio and so lose heat faster than larger organisms. They compensate by using most of the energy produced by respiration to maintain their body temperature.
B We still need to maintain body temperature, breathe, keep our hearts beating and keep cellular processes going. All this requires energy.
C False. Some heterotrophs eat other heterotrophs (eg lions eat gazelles).
D (a) It fixed carbon from carbon dioxide by photosynthesis. (b) From respiration. (c) Most lost as heat. The main compounds produced are carbon dioxide and water.
E When the temperature of the surroundings is the same as body temperature.
F (a) Their energy requirement goes down but they probably don't eat less, therefore more of their food energy is stored as fat. (b) About 800 kJ per day. (c) Energy needs of a moderately active woman plus the energy needs of a new-born baby, who gets all his/her energy from milk.
G (a) Carbohydrates: 12 500/17 = 735 g. (b) Lipids: 12 500/38 = 329 g.

CHAPTER 8

A More gas can be exchanged per unit time. Thin-walled, moist, good vascular network.
B Organic 'fuel' is oxidised to release energy and do useful work, and it produces heat as waste.
C It would take too long to break glucose down. ATP releases energy when one phosphate group is removed.
D The first uses a substrate as phosphate source, the second uses free phosphate and is powered by energy gained from redox reactions in the electron transport chain.
E An extra electron.
F It goes into the blood, is taken to the lungs and is breathed out.
G Because each glucose provides two pyruvates and therefore two acetyl-CoAs.
H These more metabolically active cells need more ATP to power their processes and therefore need more mitochondria (the sites of ATP synthesis in the cell).
I Active transport requires energy.
J (a) Each one of the 20 amino acids has a different number of carbon atoms. In respiration, the reactions which break each one down produce different volumes of CO_2 and use different volumes of O_2. (b) CO_2 is produced but no O_2 is used: any value divided by 0 is infinity.
K It prevents them being eaten.
L 50 units.
M The end-products are different (though the processes are similar).

CHAPTER 9

A Provided – water, carbon dioxide; taken away – oxygen, glucose/sugar.
B Oxidation is the loss of electrons, reduction is gain (OILRIG).
C When a molecule is reduced it gains energy.

CHAPTER 10

A Small organisms have a large surface area:volume ratio and so lose heat faster than larger organisms. They compensate by using most of the energy produced by respiration to maintain their body temperature.

B Benedict's test: add blue Benedict's and heat; will give an orange-red precipitate if sugar is present.
C Biuret test: will change from blue to lilac if protein is present.
D The symptoms are those of vitamin A poisoning.
E The meal is reasonably high in energy, complex carbohydrate and fat. It would be a good basis for a lunchtime meal for an adult man. The brown bread provides some protein, fibre and most of the B vitamins; the cheese and butter provide mainly fat, and vitamin A. It lacks vitamin C, iron and other minerals; it could be improved by the addition of a side salad, or just by adding some salad – tomatoes, green salad leaves and cucumber to the sandwich, or by following it up with a few pieces of fruit.
F Exercise increases basal metabolic rate and the muscle to fat ratio; both will use up more energy, so a better diet will prevent more fat being added, while exercise will help take existing fat off.

CHAPTER 11

A They would have to eat smaller, more frequent meals as they would have no stomach to store large meals.
B About 2–3.
C Endopeptidase cuts within the chain, exopeptidase removes the end amino acid.
D (a) You would expect rennin production to decrease with age. It does. Young mammals on milk diets need it but older animals do not. (b) No, since only mammals produce milk.
E So they do not self-digest the organs that produced them.
F The small intestine is a very absorptive region and can take in many of the nutrients produced by bacteria; only a small amount of absorption takes place in the colon, so nutrients produced here will just be lost in the faeces.

CHAPTER 12

A The gills collapse: their surfaces stick together like the bristles of a wet brush.
B 7.5 litres min^{-1}.
C One five thousandth.
D Dead space would be enormous and diver would breathe the same air in and out. Also, the chest muscles are not strong enough to keep expanding the lungs against the water pressure at that depth.

CHAPTER 13

A (a) 6 and (b) 15 litres per minute.
B (a) Increase. (b) Decrease.
C 120 mmHg = systolic pressure, 70 mmHg = diastolic pressure.
D (a) Increase. (b) Decrease.
E Hb 'steals' oxygen from an oxygen-rich environment and delivers it to tissues that are oxygen-poor.

CHAPTER 14

A Because they detect a change in internal conditions and then adjust it back to the normal level.
B The urine of diabetic dogs has high concentrations of sugar. This is very attractive to flies.
C They are metabolically active and are involved in a rapid exchange of large amounts of materials.
D Any excess of the daily need would be deaminated and excreted.

CHAPTER 15

A Humidity changes our ability to lose heat by sweating. Low humidity means a higher upper lethal temperature because we can sweat and lose heat by evaporation more effectively.
B They must be below, otherwise they would disappear as old skin was shed and replaced by new skin.
C Piloerection because we have so few hairs.
D To replace the sodium and chloride ions that they lose constantly by sweating.

CHAPTER 16

A 7.5 litres.
B 14.4 times per hour.
C Cortical, since an otter is not short of water.
D Blood pressure would drop, filtration would become ineffective, kidneys would suffer damage and fail.

E The loop of Henle is shaped like a hairpin, and it behaves in the same way as a countercurrent multiplier.

F Dark yellow urine indicates dehydration. Brown = trouble!

G Much more lost in faeces – dehydration.

H It makes the blood more concentrated, either by adding salt or removing water or both.

I Proximal and distal tubules, ascending limb of loop of Henle.

CHAPTER 17

A Warm, moist, lots of food.

B Mucous secretions containing dust and dead cells, which should be swept by cilia to the exterior, remain trapped in the airways to cause congestion and infection.

C Anticoagulants stop blood clotting so that leech can feed freely. Uses: microsurgery, plastic surgery, dispersal of blood clots, treatment of chronic skin ulcers.

D Two basophils, a lymphocyte and a monocyte.

E Not likely to cause an allergic reaction.

F Some of the donors would have been blood group O, some of the recipients would have been blood group AB, and so no adverse reactions would have occurred.

G AB blood has A and B antigens on the surface of the red blood cells. These react with antibodies present in the blood of people from the other three groups.

H Agglutination occurs when red cells stick together due to antibodies making them sticky. Blood clotting is the complex reaction that turns fibrinogen into a fibrin mesh, in which red cells get stuck.

I Fungi are eukaryotes while antibiotics work by interfering with prokaryote metabolism, and don't affect eukaryotes.

J If the course is not finished, not all the bacteria will be killed. The most susceptible die first, leaving the more resistant ones to breed back. The behaviour therefore speeds up the evolution of superbugs.

CHAPTER 18

A We are using the lactic anaerobic system. This means that respiration can't be completed and so the intermediate compound (lactate) builds up.

CHAPTER 19

A A neurone is a single specialised nerve cell, a nerve is a bundle of neurones in connective tissue.

B **(a)** Active transport mechanism needs specific membrane protein and ATP. **(b)** It moves particles against a diffusion gradient rather than down a diffusion gradient.

C **(a)** To ensure that a neurone responds to a specific minimum level of stimulation. **(b)** Continuous depolarisation and fatigue of neurone.

D Myelinated nerves conduct impulses more quickly.

E Narrow to make diffusion of the neurotransmitter as fast as possible and high resistance to ensure that the action potential can't jump across.

F Because the transmitter is only made on one side.

G No action potential.

H So they don't tear and can be easily stitched up.

CHAPTER 20

A Discussion question, no answer provided.

B Right motor area.

CHAPTER 21

A They have to support more weight.

B The spinal cord.

C Ligaments connect bones together at joints; ligaments need to be flexible to allow the joined bones to move.

D A tendon joins a muscle to a bone. As the muscle contracts it exerts a force on the bone, moving it. If the tendon were elastic, the bone would move less than the change in muscle length.

E **(a)** Smooth, **(b)** smooth, **(c)** skeletal, **(d)** cardiac, **(e)** skeletal.

CHAPTER 22

A The digestive juices undiluted by food would attack the stomach lining. This could cause an ulcer.

CHAPTER 23

A **(a)** Chemoreceptor; exteroreceptor.
(b) Chemoreceptor; exteroreceptor.

B Exteroreceptors since they respond to stimuli from outside the body.

C The light-sensitive film.

D **(a)** The ciliary muscles relax. **(b)** The lens becomes thinner due to tension on the suspensory ligaments. **(c)** The pupil dilates.

E Night sight (ie vision in dim light) needs rods that work properly. Vitamin A is necessary for the formation of retinal in rhodopsin.

F **(a)** Blue, green and red cones. **(b)** Blue cones only.

G We get used to the sound of a ticking clock in a room.

H Instrumental conditioning (using negative reinforcement).

CHAPTER 24

A No: the species certainly goes on but individuals frequently perish.

B Tall.

CHAPTER 25

A DNA contains the genetic code, RNA is built up on it, then RNA is used as the instructions for making proteins.

B $n = 23$.

C G_1, S and G_2.

D The amount of DNA would be halved at each division.

E **(a)** In ovaries and testes respectively. **(b)** Because it reduces the chromosome number (by half).

F In a human cell there are 23 bivalents.

CHAPTER 26

A Heavily pregnant females cannot hunt.

B It would secrete more GnRF, thus stimulating more gonadotrophin release and therefore more oestrogen release.

C Less energy and resources are wasted; the changes of successful fertilisation are increased.

D They both contain the same amount of DNA. Both gametes are haploid – they contain only one copy of each chromosome.

E Day 21.

CHAPTER 27

A **(a)** UAUGCGAUA. **(b)** 3.

B They eat and digest protein, absorb amino acids, transport them in the circulation from the gut to cells.

C Anabolic.

D 61 different tRNAs carry only 20 different amino acids.

E Six: all ^{14}N strands, two hybrids.

F Substitution.

G With no spindle, the chromosomes double but cannot separate, so a cell with twice the normal number of chromosomes results.

CHAPTER 28

A 50:50 (half Yy and half yy).

B If you use a plant that is TT, all the offspring of a cross with a Tt or a TT plant will be tall and you wouldn't gain any information.

C 700 000/2500 = 280.

D In previous centuries, villagers married people from the next village. Because we now have more transport, people move around the world more easily and so people meet and mate with people to whom they are unlikely to be related.

E XhXh. Female haemophiliacs must be a double recessive whose mother is a female carrier and whose father is a male sufferer. The chance of two such people meeting and marrying is unlikely and male haemophiliacs often don't live long enough to reproduce.

F One in two chance.

G No. If the mother is O and the father is A, the child could be only A or O.

H Yes. One could be aabbcc (white) and the other could be AABBCC (very dark).

I 9 agouti, 3 black, 4 albino.

CHAPTER 29

A Acquired characteristics don't affect an organism's DNA and so cannot be passed on to the next generation.

B Insects have short life-spans and high reproductive capacity; whales and rhinos have long life-spans and only produce a handful of offspring during their life time.

C Peppered moth, *Biston betularia*: selection of melanic form following industrial revolution. Hawks: selection for great visual acuity. Cheetah: selection for high speed running.

D **(a)** Analogous. **(b)** Homologous.

CHAPTER 30

A It performs transcription in reverse: it makes DNA from RNA.

B Because the smaller recognition site is more common in any DNA sequence.

C An individual has a brain with memories – these memories are not contained in the cell used for cloning. In cloning you may create a new brain but not the memories.

CHAPTER 31

A Fluid moves up the stem, the pull is created by the leaves.

CHAPTER 32

A They haven't enough food stores to grow out of the soil – they must photosynthesise almost immediately.

B When the tops are chewed off, the growing tip is still intact.

C Because it absorbs the red part of the spectrum.

D Plants could be made to flower and fruit without having to give them specific conditions of light and dark. No need to use artificial light, which is expensive.

CHAPTER 33

A They are pollinated by birds: their flapping would damage more delicate flowers.

B It maximises the chances of catching the pollen.

C No. They produce less pollen than wind pollinated plants; the grains are larger, sticky and are not carried by the air in significant amounts.

D **(a)** One set, **(b)** two sets, **(c)** three sets.

E Dicots.

F Bright colours attract animals, such as birds, which eat the fruit and disperse the seeds.

G Otherwise an insect visitor could move from thrum to pin on the same plant and the plant would then be self-fertilised.

CHAPTER 34

A Temperate deciduous forest.

B Abiotic: **(a)**, **(c)**, **(d)**. Biotic: **(b)**, **(e)**, **(f)**.

C Habitat is where an organism lives, the term ecological niche also describes what it does and how it relates to its physical environment and to other organisms.

CHAPTER 35

A Clay has lots of water and has very little air. Water heats up slowly, so a wet clay soil often feels cold. It is heavy: its particles pack together more tightly than more open sandy soil or loam.

B Loam. Sand lets water and minerals run through too easily. Clay is impervious and does not allow a build-up of nutrients.

C Water content would be different. Values measured depend on conditions, eg, on a hot day, soil would have less water in it at the end of the day because of evaporation.

D Bulbs have a large food supply which gives the young plant a head start so that it can complete its life cycle early, flowering before its light supply is cut off by taller plants coming into leaf.

E Intraspecific (within species) aggression.

F Keep out of direct sunlight, away from wind chill, ensure the thermometer does not get wet – evaporation cools.

CHAPTER 36

A The transfer of energy from one organism to another.

C 7 levels. Unusual: most food chains have only 4 or 5 trophic levels.

D **(a)** Normal pyramid, **(b)** inverted pyramid.

E Mammals are homoiothermic: they use much of their energy to maintain body temperature.

CHAPTER 37

A 305 snails.

B Small mammals have a high metabolic rate and need to feed often. Left without food and water, they die in a few hours. Put a small amount of food and some bedding in the traps.

C Intra: within a species. Inter: between different species.

D The oak wood.

CHAPTER 38

A Sustainable: used but continually replaced, eg solar power, hydroelectric power. Unsustainable: when used it is gone forever, eg coal deposits, oil, gas.

B Take two water samples A and B. Record the oxygen concentration of A. Incubate B in the dark at 20 °C for five days. Then measure the oxygen concentration of B. The BOD is B – A.

ACKNOWLEDGEMENTS

The publisher thanks examination boards (OCR, AQA, IBO, Edexcel, WJEC, AEB, UCLES, NEAB) for their permission to reproduce examination questions.

Every effort has been made to contact the holders of copyright material, but if any have been inadvertently overlooked the publishers will be pleased to make the necessary arrangements at the first opportunity.

Photographs

The publishers would like to thank the following for permission to reproduce photographs (T = Top, B = Bottom, C = Centre, L= Left, R = Right):

A-Z Botanical Collection/M Nimmo, 29.6R; Allsport/Allsport UK, 8.A2, T Duffy, 3.26, T Lewis, 8.7, S Forster, 8.A1, 18.4, M Powell, 10.17, A Bello, 13.A3, M Steele, 18.1, B Radford, 18.7; Heather Angel, 8.19, 9.3, 10.6, 26.1, 29.25a, 29.27b, 33.4, 33.20, 33.23, 34.3, 34.7a, 37.24; Aquarius Library, 30.1; Ardea/S Roberts, 16.17, P Morris, 20.2, 35.2L, J-P Ferraro, 22.1C, 32.8aL&b, 35.3, B Sage, 32.8aR, I Beanes, 33.10a, J Mason, 33.24, S Gooders, 34.2L, C Haagner, 35.15, Weisser, 35.16, Kjaer, 35.17T, D&K Urry, 37.7; BBC Natural History Unit/L M Stone, 22.10, M Barton, 23.30, D Wechsler, 37.25; From: the F G Banting Papers, Thomas Fisher Rare Book Library, University of Toronto, 14.4; Biophoto Associates, 1.15, 1.16, 1.18, 1.20, 1.28, 1.30, 1.33, 1.34, 1.A2, 2.4, 3.16, 4.20, 4.21, 4.22, 4.27, 4.A1, 7.6B, 7.11a, 7.11b, 8.14a, 8.26, 9.4, 10.16, 11.2L, 11.3L, 11.5L, 11.6L, 14.5d, 14.7, 17.5, 19.13b, 20.4, 21.11b, 21.17T, 22.8a, 23.3b, 25.1a, 25.6, 25.9, 25.10L, 25.11T, 25.13, 25.16b, 31.3, 31.7b, 31.15d, 31.19c, 31.27, 32.4L, 32.6b, 32.11e, 32.13, 33.2, 33.10b, 33.11, 33.16c, 33.17, 36.3, 37.16L, 37.19R; www.JohnBirdsall.co.uk, 15.A2, 19.A3, 28.1; © 1992-2002 Warren Bishop, M.D., and The University of Iowa. All rights reserved, 11.2R, 11.5R; Anthony Blake Photo Library/T Hill, 6.A1a, Maximilian, p.212; From: The Discovery of Insulin by Michael Bliss, published by Paul Harris Publishing, Edinburgh, 1983, p.270; Chris Bonington Picture Library/D Scott, 12.16; Mike Boyle, 17.A2; The Three Graces by Reubens, Prado, Madrid/Bridgeman Art Library, London, 3.21; Bruce Coleman Ltd, p.188, 10.3T, 23.16R, 23.17, p.522, H Reinhard, 7.5, 10.10, J Watt, 12.6e, K Taylor, 12.6f, 27.9CL, 27.14b, 29.14b, 29.19, 37.12, 37.14, W Layer, p.398, F Labhardt, 22.1R, 27.9BR, R Williams, 23.16L, P Clement, 26.6, J Cancalosi, 26.7, A Purcell, 27.9CR, J Grayson, 27.9BL, A Bacchella, 29.9, S Krasemann, 29.17L, Dr F Sauer, 36.4L, D Davies, 36.4CL, J Burton, 37.17, J & D Bartlett, p.686; Bubbles Photo Library, 22.1L, 22.A2, I West, 10.20; Cephas, 8.21, M Schilder, 6.11R, TOP/C Adam, 6.11B, Stockfood, 8.21, A Proust, 8.21, M Rock, 8.23, S. Boreham, 8.24; Martyn Chillmaid, 12.3, 20.16, 31.11; Cleveland Museum of Natural History, Ohio, 24.3T; Corbis/S Chenn, 18.5; Corbis/Bettmann, p.594; Ecoscene/Papilio, 33.16a, 36.4; Environmental Images/R Brook, 38.4b&c, 38.10, 38.11R, P Glendell, 38.11L, G Burns, 38.13; Vivien Fifield, 29.3; Format Photographers/J Harrison, 10.19; Geophotos/Tony Waltham, 29.A2, 31.25, 31.26; Gettyimages, 18.10, 22.2, 24.5, p.664; GlucoWatch® Biographer (registered trademark of Cygnus Inc, Redwood City, California, USA), 6.20; The Ronald Grant Archive, 27.14a; Sally & Richard Greenhill Photo Library, 15.6L; Professor D Grierson, Nottingham University, 6.10; GSF Photo Library, 29.6T; Robert Harding Picture Library/J Miller, 32.6a; © David Hoffman, 15.6R; Holt Studios International/P Peacock, 6.4, N Cattlin, 9.14, 9.15, 31.10, 31.22, 31.28, 31.29, 32.14a, 32.15, 33.15a, 33.19, 33.25b, 36.18, 38.3C&B, I Spence, 31.2, I Speed, 32.5, C&T Stuart, 32.9, A Morant, 32.12c, G Roberts, 38.1, R Anthony, 38.3T; Courtesy Kobal Collection/20th Century Fox/Paramount, 15.7; Courtesy Kobal Collection/Michaels, Darren/Columbia, 24.1; Andrew Lambert Photographs, 3.4, 3.10, 3.13, 3.17, p.90, 5.A1, 10.13, 16.2, 18.8; Frank Lane Picture Agency/D Hosking, 7.3, 35.2R, Foto Natura Stock, 9.1, W Wisnievski, p.284, L Batten, 15.4a, M Ranjit, 19.A1, R Wilmshurst, 23.31, 29.4, 37.4, T Davidson, 26.4, L Lewis, 26.8, M J Thomas, 28.7, S Hosking, 29.5, E & D Hosking, 29.16, T Whittaker, 29.17R, D P Wilson, 35.5b, B Borrell, 35.6, A Wharton, 35.12, P Perry, 37.11, M Gore, 35.13, 37.15R; Lawrence Berkeley Laboratory, University of California, 9.13, Lommond Mountain Rescue Team/G Baird, 15.A1; MPL International Ltd/R Adshead, p.168; Meteorological Office/JFP Galvin, 35.18a; Microscopix/A Syred, 12.8c; The Muscular Dystrophy Group, 28.15; NASA, p.384; NHM Photo Library, 20.12, 29.1; NHPA/M Stelzner, 6.3, Agence Nature, 6.7, A Rouse, 8.2, B Jones & M Shumlock, 10.2, 29.27a, A Bannister, 10.9, B Wood, 12.6b, 15.8, J Goodman, 12.6d, P Atkinson, 12.7a, N Wu, 20.1, S Robinson, 23.28, M Harvey, 29.22L, ANT, 29.24a, T Kitchin & V Hurst, 29.24b, J Shaw, 29.25b, L Campbell, 29.26a, D Zupanc, 31.24; Nature and Science AG/J Aribert 4.19, F-L Vaduz, 16.7a; From: Damasio H, Grabowski T, Frank R, Galaburda AM; Damasio AR: The return of Phineas Gage: Clues about the brain from a famous patient. Science, 264: 1102-1105, 1994. Dept of Neurology & Image Analysis Facility, University of Iowa, p.366; Novo Nordisk A/S, 14.A1, 14.A2; Oxford Scientific Films Ltd/O Newman, 3.23, G I Bernard, 7.1, 11.10, R Blythe, 7.4, Kent and Donna Dannen Photo Research Inc., 7.7, T Shepherd, 12.8B, D Broomhall, 12.8, H Pooley, 15.2, P Parks, 21.1a, D J Cox, 21.2a, S Stammers, 23.1, J A L Cooke, 23.2, L E Lauber, 23.11, W Shattil & B Rozinski, 23.29, 37.13, M Hill, 34.1, Tui de Roy, 34.7b, S Osolinski, 34.7c, R Redfern, 34.10, D H Thompson, p.636; Panos Pictures/T Page, 10.14; Dr Colin Paterson, Ninewells Hospital, Dundee, 3.40; Photos Horticultural, 32.14c; Planet Earth/J Lee, 4.A2, A Stevens, 4.A3, 8.1TR, Purdy & Matthews, 7.6C, T Dressler, 7.6TC, P Chippendale, 12.2, P Oxford, 12.5, K Lucas, 12.6a, P Sayers, 15.1, A Kerstitch, 21.1b, P Rowlands, p.412, J Greenfield, 26.3, E Darack Photography, 29.14a, M Snyderman, 29.14cL, J Jackson, 29.14cR, C Weston, 35.5a, W M Smithey, 35.10L, P Osford, 35.10R, H Charles, 35.14, J Lythgoe, 35.17B, M Welby, 37.6, A Bartschi, 37.15C, N Garbutt, 37.15L; Popperfoto, 22.9, Reuter, 10.A2; Professional Sport/T Hindley, 22.A2; Purdue News Service – Photo by David Umbeger, p.28; Purdue Agricultural Communications Service. Photo donated by Dean Fletcher, Savannah River Ecology Laboratory, p.462; Redferns/Michael Och's Archive, p.504; Rex Features Ltd, 3.22, 10.12, 16.1, p.486; Roselin Institute, Edinburgh, 25.1b&c, 30.11; Royal Holloway College, University of London, EM Unit, 1.23, 1.31; Science Photo Library, 1.25, 1.A4, 13.13, p.314, 25.5, 28.2, A B Dowsett, p.2, 8.15, D Scharf, 1.3, 1.4, 1.6, 8.22, 12.A1, 25.3, 29.28b, Moredun Animal Health Ltd, 1.7, 37.19L, Dr G Murti, 1.9, 27.A1T, 30.10, Omikron, 1.11, 27.A1B, D Fawcett, 1.13, CNRI, 1.14, 4.6, 6.24, 11.4R, 13.4c, 13.A2, 19.10b, 20.8b, 24.2, p.446B, 25.7, 26.11, Quest, 1.A1, 3.19, D Phillips, 1.A3, 2.1, S Sinclair, 2.A1T, M Devlin, 2.A1C, 20.6a, John Radcliffe Hospital, 2.A1B, BSIP Leca, 2.A3, Detlev van Ravensway, p.38, C D Winters, 3.6, C Pouderas, Eurelios, 3.20, K Eward, 3.34, B Nelson/Custom Medical Stock, p.68, J C Revy, 5.2, 8.5, 17.11, 18.2, 30.5, 30.12, 31.12, 31.19b, 32.12a, W Baumeister, 5.8, A Bartel, 5.15, 24.3CR, J Mason, p.106, D Leah, 6.2, F Sauze, 6.14, Saturn Stills, 6.19, 14.2, 17.13, H Young, 6.25, 25.16a, P Plailly, 6.26, 30.4, Rosenfeld Images Ltd, 6.A1c, 33.25c, SCIMAT, 6.A1d, Tom van Sant/Geosphere Project, Santa Monica, p.134, Dr T Brain, 7.2, J Howard, 7.6TL, 34.4, St Bartholomew's Hospital, 7.14, M Dohrn, Royal College of Surgeons, 8.1TL, 12.9b, © Health & Safety Laboratory, 8.3, Dr K Porter, 8.14b, S Ford, 8.25, 19.A2, A Syred, 9.2, 10.4, 21.10b, 23.3a, 31.15a, 31.23, 33.A3R, O Burriel, 10.11a&b, 10.21, C Pedrazzini, 10.18, Dr C Liguory/CNRI, 11.3R, J Burbridge, 11.4l, D M Martin MD, 11.6R, M Clarke, 10.15, 11.13, 26.23, B Lehnhausen, p.228, S Stammers, 12.6c, 33.4a, A & H Frieder Michler, 12.13a, BSIP, Edwige, 12.14, J Durham, 12.A2, Biophoto Associates, 13.4a, 19.3R, G Tompkinson, 13.11, 23.9, E Reschke, Peter Arnold Inc, 13.17, A Crump, TDR, WHO, 13.16, E Grave, 13.21a&b, 37.16R, J L Martra; Publiophoto Diffusion, 14.5a, Dr R Clark & M R Goff, 15.5, D Parker, 15.9, 37.23, H Morgan, p.298, 20.14, 20.17a, 26.A1, 30.6a, Prof P Motta, Dept of Anatomy, University, 'La Sapienza', Rome, 16.7b, 21.10c, 23.10, 26.12b, M Chillmaid, 16.15, Dr P Marazzi, 17.1, 29.20, 38.9, NIBSC, 17.4, Dr V Bradbury, 17.6, Dr A Brody, 17.7, Dr K Lounatmaa, 17.18, J King-Holmes, p.332, 24.3BL, 30.6b, 30.9, M Kage, 19.3L, 23.4, D Lovegrove, 19.9, Eye of Science, 19.13a, 29.28a, 33.A3L, Department of Clinical Radiology, Salisbury Hospital, 19.18, GJLP-CNRI, 20.6b, G Watson, 20.8a, D Roberts, 21.1c, Dr J Burgess, 21.2b, 31.9, 32.7a, 32.12b, 33.12a, 33.15b, M Walker, 21.2c, P Plailly, Eurelios, 21.8, p.436, P Menzel, 21.19, 24.3BR, 30.13b, H C Robinson, 22.3, Professors P Motta & T Naguro, 23.6, P Jude, 23.7, A McClenaghan, 23.19b, Petit Format, 24.3CL, S Fraser, Royal Victoria Infirmary, Newcastle upon Tyne, p.446T, P M Motta & J van Blerkom, 25.2, M Sklar, 25.4, C Nuridsany & M Perennou, 26.2, 31.5, P M Motta, G Macchiavelli, S A Nottola, 26.15b, S Terry, 26.18, 31.16, 32.14b, Dr Y Nikas, 26.A2, 30.16, A Tsiaras, 26.A3, A Hart-Davis, 28.16, 32.10a, 32.11a, National Library of Medicine, 29.2, Microfield Scientific Ltd, 29.26b, T Craddock, 29.10, 38.5, © 1995 Worldstat International and J Knighton, 29.A1, Dept of Clinical Cytogenetics, Addenbrookes Hospital, 30.2, K Guldbrandsen, 30.13a, M Kulyuk, 30.15, L Chace, 31.4, J Lepore, 32.1a, V Fleming, p.612, BSIP, Sercomi, 33.7, S Fraser, 37.8, 38.4a, 38.7, Y Hamel, 38.4d, NOAA, 38.8; Science and Society Picture Library, 12.17; www.shoutpictures.com, p.246; Still Pictures/N Cobbing, 6.6, E Parker, 6.17, J Schytte, 6.18, C Pye-Smith, 6.A1b, J Isaac, p.144, F Polking, 8.A3, M Carwardine, 13.1, M Edwards, 16.3, p.346, p.568B, F Bavendam, 34.6, H Schwarzbach, 37.A1; Photo courtesy of the College of Veterinary Medicine, Texas A&M University, p.548; C&S Thompson, 16.4, 33.21; Illustration provided courtesy of UC Davies Health System, Sacramento, CA, USA, 2.3, David Walker, 9.10; John Walmsley Photography, 34.2B, 35.18b, 37.26; The Wellcome Trust, 10.A1, 17.8, 27.12; Robin Westlake, US National Marine Fisheries Service, Southwest Fisheries Science Center, p.648; Woodfall Wild Images/B Gibbons, 37.27; Zeneca Agrochemicals & Seeds, 6.9, p.568T

Front cover: Computer-generated model of DNA molecules
Image supplied by: Doug Struthers/Gettyimages

INDEX